W9-CSD-281

OTHER VOLUMES IN THIS SERIES

I

Tables of the cumulative non-central chi-square distribution
by G. E. HAYNAM, Z. GOVINDARAJULU and F. C. LEONE
Table I (Power of the chi-square test)
Table II (Non-centrality parameter)

Tables of the exact sampling distribution of the two-sample Kolmogorov-Smirnov criterion $D_{mn}(m \leqslant n)$ by P. J. KIM and R. I. JENNRICH
Table I (Upper tail areas)
Table II (Critical values)

Critical values and probability levels for the Wilcoxon Rank Sum Test and the Wilcoxon Signed Rank Test
by FRANK WILCOXON, S. K. KATTI and ROBERTA A. WILCOX
Table I (Critical values and probability levels for the Wilcoxon Rank Sum Test)
Table II (Probability levels for the Wilcoxon Signed Rank Test)

The null distribution of the first three product-moment statistics for exponential, half-gamma, and normal scores by P. A. W. LEWIS and A. S. GOODMAN
Tables 1–3 (Normal with lag 1, 2, 3)
Tables 4–6 (Exponential with lag 1, 2, 3)
Tables 7–9 (Half-gamma with lag 1, 2, 3)

Tables to facilitate the use of orthogonal polynomials for two types of error structures
by KIRKLAND B. STEWART
Table I (Independent error model)
Table II (Cumulative error model)

II

Probability integral of the doubly noncentral *t*-distribution with degrees of freedom n and non-centrality parameters δ and λ by WILLIAM G. BULGREN

Doubly noncentral F distribution–Tables and applications by M. L. TIKU

Tables of expected sample size for curtailed fixed sample size tests of a Bernoulli parameter
by COLIN R. BLYTH and DAVID HUTCHINSON

Zonal polynomials of order 1 through 12 by A. M. PARKHURST and A. T. JAMES

III

Tables of the two factor and three factor generalized incomplete modified Bessel distributions
by BERNARD HARRIS and ANDREW P. SOMS

Sample size requirement: Single and double classification experiments
by KIMIKO O. BOWMAN and MARVIN A. KASTENBAUM

Passage time distributions for Gaussian Markov (Ornstein-Uhlenbeck) statistical processes
by J. KEILSON and H. F. ROSS

Exact probability levels for the Kruskal-Wallis test
by RONALD L. IMAN, DANA QUADE and DOUGLAS A. ALEXANDER

Tables of confidence limits for linear functions of the normal mean and variance
by C. E. LAND

IV

Dirichlet distribution–Type 1
by MILTON SOBEL, V. R. R. UPPULURI and K. FRANKOWSKI

V

Variances and covariances of the normal order statistics for sample sizes 2 to 50
by G. L. TIETJEN, D. K. KAHANER and R. J. BECKMAN

Means, variances and covariances of the normal order statistics in the presence of an outlier
by H. A. DAVID, W. J. KENNEDY and R. D. KNIGHT

Tables for obtaining optimal confidence intervals involving the chi-square distribution
by G. RANDALL MURDOCK and WILLIAM O. WILLIFORD

OTHER VOLUMES IN THIS SERIES

VI

The distribution of the size of the maximum cluster of points on a line
by NORMAN D. NEFF and JOSEPH I. NAUS

VII

The product of two normally distributed random variables
by WILLIAM W. MEEKER, JR., LARRY W. CORWELL and LEO A. AROIAN
Table I (Power of the chi-square test). Table II (Non-centrality parameter).
Table I (Upper tail areas). Table II (Critical values).
Table I (Independent error model). Table II (Cumulative error model).

VIII

Expected sizes of a selected subset in paired comparison experiments
by B. J. TRAWINSKI

Tables of admissible and optimal balanced treatment incomplete block (BTIB) designs for comparing treatments with a control
by ROBERT E. BECHHOFER AND AJIT C. TAMHANE

Expected values and variances and covariances of order statistics for a family of symmetric distributions (student's *t*)
by M. L. TIKU AND S. KUMRA

Selected Tables in Mathematical Statistics

Volume IX

DIRICHLET INTEGRALS OF TYPE 2 AND THEIR APPLICATIONS

by

MILTON SOBEL, V.R.R. UPPULURI, and K. FRANKOWSKI

Edited by the Institute of Mathematical Statistics

Coeditors
R. E. Odeh
University of Victoria
and
J. M. Davenport
Texas Tech University

Managing Editor
N. S. Pearson
Bell Communications Research

AMERICAN MATHEMATICAL SOCIETY
PROVIDENCE, RHODE ISLAND

1980 *Mathematics Subject Classification*
Primary 62Q05; Secondary 62E15, 62H10

International Standard Serial Number 0094-8837
International Standard Book Number 0-8218-1909-7
Library of Congress Card Number 74-6283

This volume was printed directly from author-prepared copy.
The paper used in this book is acid-free and falls within the guidelines established to ensure permanence and durability.

Dedicated to

Marc J. Sobel

Ram Y. Uppuluri

Michael Frankowski

Table of Contents

Preface

This volume of mathematical tables has been prepared under the aegis of the Institute of Mathematical Statistics. The Institute of Mathematical Statistics is a professional society for mathematically oriented statisticians. The purpose of the Institute is to encourage the development, dissemination, and application of mathematical statistics. The Committee on Mathematical Tables of the Institute of Mathematical Statistics is responsible for preparing and editing this series of tables. The Institute of Mathematical Statistics has entered into an agreement with the American Mathematical Society to jointly publish this series of volumes. At the time of this writing, submissions for future volumes are being solicited. No set number has been established for this series. The editors will consider publishing as many volumes as are necessary to disseminate meritorious material.

Potential authors should consider the following rules when submitting material.

1. The manuscript must be prepared by the author in a form acceptable for photo-offset. The author should assume that nothing will be set in type although the editors reserve the right to make editorial changes. This includes both the introductory material and the tables. A computer tape of the tables will be required for the checking process *and* the final printing (assuming it is accepted). The authors should contact the editors prior to submission concerning the requirements for the computer tape.

2. While there are no fixed upper and lower limits on the length of tables, authors should be aware that the purpose of this series is to provide an outlet for tables of high quality and utility which are too long to be accepted by a technical journal but too short for separate publication in book form.

3. The author must, whenever applicable, include in his introduction the following:

   (a) He should give the formula used in the calculation, and the computational procedure (or algorithm) used to generate his tables. Generally speaking FORTRAN or ALGOL programs will not be included but the description of the algorithm used should be complete enough that such programs can be easily prepared.

(b) A recommendation for interpolation in the tables should be given. The author should give the number of figures of accuracy which can be obtained with linear (and higher degree) interpolation.

(c) Adequate references must be given.

(d) The author should give the accuracy of the table and his method of rounding.

(e) In considering possible formats for his tables, the author should attempt to give as much information as possible in as little space as possible. Generally speaking, critical values of a distribution convey more information than the distribution itself, but each case must be judged on its own merits. The text portion of the tables (including column headings, titles, etc.) must be proportional to the size 5–1/4″ by 8–1/4″. Tables may be printed proportional to the size 8–1/4″ by 5–1/4″ (i.e., turned sideways on the page) when absolutely necessary; but this should be avoided and every attempt made to orient the tables in a vertical manner.

(f) The table should adequately cover the entire function. Asymptotic results should be given and tabulated if informative.

(g) Examples of the use of the tables must be included.

4. The author should submit as accurate a tabulation as he/she can. The table will be checked before publication, and any excess of errors will be considered grounds for rejection. The manuscript introduction will be subjected to refereeing. Since an inadequate introduction may lead to rejection, the author should strive for an informative manuscript, which not only establishes a need for the tables, but also explains in detail how to use the tables.

5. Authors having tables they wish to submit should send two copies to:

Dr. Robert E. Odeh, Coeditor
Department of Mathematics
Univerity of Victoria
Victoria, B. C., Canada V8W 2Y2

At the same time, a third copy should be sent to:

James M. Davenport, Coeditor
Department of Mathematics
Texas Tech University
P. O. Box 4319
Lubbock, Texas 79409

Additional copies may be required, as needed for the editorial process. After the editorial process is complete, a camera-ready copy must be prepared for the publisher.

Authors should check several current issues of *The Institute of Mathematical Statistics Bulletin* and *The AMSTAT News* for any up-to-date announcements about submissions to this series.

Acknowledgments

The tables included in the present volume were checked at the Universiy of Victoria. Dr. R. E. Odeh arranged for, and directed this checking with the assistance of Mr. Bruce Wilson. The editors and the Institute of Mathematical Statistics wish to express their great appreciation for this invaluable assistance. So many other people have contributed to the instigation and preparation of this volume that it would be impossible to record their names here. To all these people, who will remain anonymous, the editors and the Institute also wish to express their thanks.

Authors' Acknowledgements

The authors, the editor and the Institute of Mathematical Statistics wish to thank Professor R. E. Odeh for checking the tables and pointing out some errors (that were corrected). The authors want to thank Dr. Mark R. Frolli, Mission Research Inc., Santa Barbara, for the construction of Tables A and B some time ago. In addition, we want to thank Professor Pinyuen Chen, Department of Mathematics, Syracuse University, for assisting Professor Frankowski and Professor Sobel in the final stages of preparing the tables.

DIRICHLET INTEGRALS OF TYPE 2 AND THEIR APPLICATIONS

Milton Sobel*
University of California at Santa Barbara

V.R.R. Uppuluri
Union Carbide Nuclear Division, Oak Ridge

and

K. Frankowski
University of Minnesota

ABSTRACT

This volume deals with incomplete Dirichlet integrals of type 2 and is a companion book to Volume 4 of this series (by the same authors) which deals with incomplete Dirichlet integrals of type 1. As in the previous volume

1) there are several new contributions present, some of which concern the development of new algorithms that made these tables possible,
2) there are many examples given to illustrate the use of the tables,
3) applications of these integrals are given to two types of problems: some that would be classified as being in the area of probability and also to some that are primarily statistical in nature,
4) there is already evidence that these tables and the associatedwrite-up will serve as a catalytic agent for further research,
5) the probabilistic interpretation of the Dirichlet integral plays a major role in the direction we take and in the development of tables.

An important area of application of these integrals is to ranking and selection problems dealing with the multinomial distribution, especially when the statistic of major interest is related to the minimum or maximum frequency among the cells and the stopping rule is of the type used in inverse sampling. In the tables most attention is to the homogeneous multinomial; however much of the analysis attempts to get away from homogeneity.

1980 Mathematics Subject Classification: Primary 62Q05; Secondary 62E15, 62H10.
Supported by a subcontract from the Union Carbide Corporation. The leading author * was also supported by a grant from the National Science Foundation #MCS82-02247 at the University of California, Santa Barbara.
Received by the editors April 1982 and in revised form August 1983 and October 1983.

§1. Introduction and Motivation

The write-up that starts in the next section (§2) is a logical development of type 2 Dirichlet integrals and the applications come up later throughout the book, but mainly in Chapter 5. Since many of the applications are to elementary problems that need little explanation, we decided to motivate the exposition by introducing some elementary applications of the tables here at the outset, even though the Dirichlet integrals are not yet defined. Hence the reader should regard this first section as purely motivational and overlook the fact that it may contain concepts not yet defined.

Consider a multinomial distribution with ten cells marked $0,1,\ldots,9$ each having probability $1/10$. In our previous book (Vol. 4 of this series) on type 1 Dirichlet we included a "favorite" problem (#21 on page 58) on the number of natural numbers $< 10^9$ which have the odd digits (i.e., 1,3,5,7,9) each appearing at least once and related it to the probability of getting frequencies $f_\alpha \geqq 1$ for all of the 5 cells $\alpha = 1,3,5,7,9$ in 9 observations on the above multinomial. The answers, found with the help of Dirichlet type 1, were 49,974,120 and .049974120, respectively, and the relation is obvious. Using this same multinomial background we now, as our first problem, ask for the expected waiting time (or number of observations) needed to get for the first time $f_\alpha \geqq 1$ for all five α-values, $\alpha = 1,3,5,7,9$. Using the result given in (5.8) below with $b = 5$, $p = 1/10$, $\gamma = 1$ and $\underset{\sim}{r} = \underset{\sim}{a} = (1,1,1,1)$, which we write as $r = a = 1$, the answer is simply given by the single term

$$(1.1) \qquad \mu^{[1]} = \mu = \frac{b}{p} C_a^{(b-1)}(1,2) = 50(.456667) = 22.833333 = \frac{137}{6},$$

where the C-value defined (in the text) was obtained from Table A below. (cf. Page A5). Here a denotes the ratio of the probability of cell #α to that of cell #9 ($\alpha = 1,3,5,7$) and hence $a = 1$ in this example. For the special case $a = 1$ it is shown in Section 5.7 (cf. (5.132)) that

C can also be written in terms of sums of powers of reciprocals of the natural numbers; in particular, for $a = 1$ and $m = 2$ we also obtain for μ

$$(1.2)\qquad \mu = 10 \sum_{j=1}^{5} \frac{1}{j} = \frac{137}{6},$$

but it should be noted that this simple result (1.2) does <u>not</u> hold for $a \neq 1$.

As a second illustration of the use of our tables suppose in the above setting we ask for the waiting time until (for the first time) any three of the five conditions (or quotas) above are satisfied. In other words, we are waiting for the 3rd largest frequency to reach the value one (for the first time). Using the result in (5.95) below with $b = 5$, $t = 3$, $r = \gamma = 1$ and $p = 1/10$ we obtain the simple answer

$$(1.3)\qquad \mu_{3,1}^{[1]} = \mu = \frac{t\binom{b}{t}}{p} CD_1^{(t-1,b-t)}(1,2) = 300CD_1^{(2,2)}(1,2) = 7.833333,$$

where the CD-value (defined in the text) was obtained from either table A or B by using (4.26) below which reduces it to a sum of C (or D) values.

<u>Remark</u>: Although the above case with $f_\alpha \geqq 1$ for $\alpha = 1,3,5,7$ and 9 can also be handled as a sum of (independent, non-identical) geometrical chance variables, the methods and formulas derived in the text can be used (similarly) for $f_\alpha \geqq r$ with $r > 1$, in which case the method based on a sum of geometric random variables breaks down.

A related third problem is to consider the analogous finite population (hypergeometric) problem in which the digits $0,1,\ldots,9$ are each repeated α times on cards (or balls in an urn) and we sample the entire deck of size 10α without replacement. We seek the answer (expected waiting time) to the same question as in (1.1) above as a function of α. Then the limit of our answer as $\alpha \to \infty$ must be the specific result in (1.1) above. The solution below does <u>not</u> use our tables directly, but it provides a sequence <u>whose limit</u> (1.1) <u>is obtainable from our tables</u>.

Solution: Using inclusion-exclusion, the probability P_x that we stop sampling in exactly x observations is

$$(1.4)\quad P_x = 5\left[\frac{\binom{9\alpha}{x-1} - \binom{4}{1}\binom{8\alpha}{x-1} + \binom{4}{2}\binom{7\alpha}{x-1} - \binom{4}{3}\binom{6\alpha}{x-1} + \binom{5\alpha}{x-1}}{\binom{10\alpha}{x-1}}\right]\frac{\alpha}{10\alpha+1-x}.$$

If we multiply by x and sum each of these five terms separately from 1 to ∞, it is not difficult to show using simple binomial identities that the (unsigned) j^{th} sum $(j = 1,2,3,4,5)$ is equal to $\binom{5}{j}\frac{10\alpha + 1}{j\alpha + 1}$. Hence our desired answer μ is

$$(1.5)\quad \mu = (10\alpha+1)\sum_{j=1}^{5}\frac{(-1)^{j-1}\binom{5}{j}}{j\alpha + 1} = (10\alpha+1)\int_0^1 [1 - (1-x^{\alpha})^5]dx$$

$$= (10\alpha+1)\left[1 - \frac{1}{\alpha}\int_0^1 (1-y)^5 y^{\frac{1}{\alpha}-1}dy\right] = (10\alpha+1)\left[1 - \frac{120\Gamma(1 + 1/\alpha)}{\Gamma(6 + 1/\alpha)}\right].$$

For $\alpha = 10$ and 20 this gives 20.0578 and 21.3663, respectively, and the limit as $\alpha \to \infty$ is 22.8333 from (1.1) or from (1.5). Using a Γ-table it is now possible to find out how large an α is needed to get within ϵ of the limiting result for any given $\epsilon > 0$; we omit this calculation.

As a fourth problem suppose we want the probability P (in the same multinomial background) that all the odd digits will show up before any even digit shows up. If we combine the 5 even cells to form one enlarged cell with probability 1/2 then we want the probability when the first $(m=1)$ observation falls in this counting cell that each of the $b = 5$ quotas $f_\alpha \geqq 1$ for $\alpha = 1,3,5,7,9$ is satisfied. Hence by our probability interpretation for C given in the text, the answer P is simply given by the single term

$$(1.6)\qquad P = C_{1/5}^{(5)}(1,1) = .00397,$$

where Table A was used again in (1.6)(cf. Page A1).

As a fifth problem we wish to find the probability P that all of the 5 odd digits have frequency less than 3 when two even digits (not necessarily different) are obtained. As in the previous problem we combine the even cells, but the stopping number now is $m = 2$. Since the 5 conditions are all $f_\alpha < 3$ we use the D-integral with common $r = 3$, $b = 5$, and $a = 1/5$. Hence the answer is simply

$$P = D_{1/5}^{(5)}(3,2) = .92925, \tag{1.7}$$

where Table B was used in this case (cf. Page B19).

For the sixth problem we make a minor change in the fifth problem and ask for the probability P that all of the 5 odd digits have frequency less than 3 when two different even digits are obtained (for the first time). This problem is more difficult and requires some analysis before the tables can be used. We first note that the answer should be a little less than that of (1.7), i.e., (1.7) is an upper bound for our answer. Using a multiplier of 5(4) = 20, we can assume that the even cells observed are two specific cells in a specific order, so that the second even cell is our counting cell and $m = 1$. Then we have one condition on each of the $b = 9$ remaining cells and, after relabeling the cells in accordance with the CD integral notation in (4.18), these conditions are $f_1 = f_2 = f_3 = 0$, $f_4 \geqq 1$ and $f_\alpha < 3$ for $\alpha = 5,6,7,8,9$. We use our probability interpretation of the CD integral in (4.18) with $j = 3$, $c = 1$, the common $a = 1$ and $r_1 = r_2 = r_3 = 0$, $r_4 = 1$, $r_5 = r_6 = r_7 = r_8 = r_9 = 3$. Then $R = 16$, $\underline{A}_3 = 3$ and using (4.18), our answer can be written simply as

$$P = 20CD_1^{(1,5;3)}(\underset{\sim}{r},1) \tag{1.8}$$

but this has to be expressed in terms of C and/or D integrals in order to use our tables. We first reduce the 3rd superscript j in (4.18) to zero (and then omit it) by using the straightforward general reduction formula

$$(1.9)\qquad CD_{\underset{\sim}{a}}^{(c,b-j-c;j)}(\underset{\sim}{r},m) =$$

$$\begin{bmatrix} m-1+R_j \\ m-1,r_1,r_2,\ldots,r_j \end{bmatrix} \left(\frac{1}{1+\underline{A}_j}\right)^m \prod_{i=1}^{j} \left(\frac{a_i}{1+\underline{A}_j}\right)^{r_i} CD_{\underset{\sim}{a}'}^{(c,b-j-c)}(\underset{\sim}{r}',m+R_j),$$

where $R_j = r_1 + r_2 + \ldots + r_j$, $\underset{\sim}{r}' = (r_{j+1}, r_{j+2}, \ldots, r_b)$ and

$\underset{\sim}{a}' = \left(\frac{1}{1+\underline{A}_j}\right)(a_{j+1}, a_{j+2}, \ldots, a_b)$.

For our problem, using (1.8), this gives

$$(1.10)\quad P = 20(\tfrac{1}{4})CD_{1/4}^{(1,5)}(1,3,3,3,3,3;1) = 5\left[D_{1/4}^{(5)}(3,1) - D_{1/4}^{(6)}(1,3,3,3,3,3;1)\right],$$

where we replaced $\int_0^{\frac{1}{4}}$ by $\int_0^{\infty} - \int_{\frac{1}{4}}^{\infty}$ to get the last member of (1.10). The variable with $r = 1$ in (1.10) is easily integrated out and we obtain

$$(1.11)\quad P = 5\left[D_{1/4}^{(5)}(3,1) - \frac{4}{5} D_{1/5}^{(5)}(3,1)\right] = 5[.96489157 - .8(.97878737)] = .90930837,$$

where both D-values in (1.11) were taken from Table B (cf. Pages B20, B19). Hence we reduced the answer in (1.7) by approximately .929 - .909 = .02.

The fact that this last problem worked out as it did (with only two D-values, both in our tables) was unforseen and is a tribute to the richness of the Dirichlet integral concept and the general usefulness of our tables, without which the problem would be much more difficult to attack successfully. After reading some of the following sections the reader is invited to make up his/her own multinomial problem and see if he/she can solve it using the Dirichlet integral concept and the tables in this book.

In our last (seventh) problem we generalize the multinomial background to at least two types of categories and find the distribution of the number of cells of type 1 observed if we stop as soon as we observe j different cells of type 2. Suppose the k cells consist of at least $t \geqq 2$ types with k_i of type i having cell probability p_i $(i = 1,2,\ldots,t)$ and such that $\sum_{i=1}^{t} k_i p_i \leqq 1$ and $\sum_{i=1}^{t} k_i \leqq k$; in the previous examples $t = 2$, $k_1 = k_2 = 5$ and $k = 10$. Here we have a common r with $m = 1$ and $k_2 - j$ cells of type 2 and $k_1 - x$ cells of type 1, all having frequency zero. To specify the counting cell and the quota cells we need only use the multiplier $\binom{k_1}{x} k_2 \binom{k_2-1}{j-1}$ and let $p_1/p_2 = a' > 0$. Then, letting $k' = k_1 + k_2$ and letting X denote the number of cells of type 1 observed at stopping time,

$$(1.12) \qquad P\{X = x|j\} = \binom{k_1}{x} k_2 \binom{k_2-1}{j-1} C_{\underset{\sim}{a}}^{(k'-1,k'-j-x)}(1,0;1),$$

where $\underset{\sim}{a} = (a',\ldots,a',1,\ldots,1)$ has k_1 values equal to a' and k_2-1 ones. Using the reduction formula (4.8) with j replaced by $k'-j-x$, $\underline{A}_j = (k_1-x)a' + k_2-j'$ and $\underline{R}_j = 0$ we have for the C-function alone in (1.12)

$$(1.13) \qquad C_{\underset{\sim}{a}}^{(k'-1,k'-j-x)}(1,0;1) = \frac{1}{1+(k_1-x)a'+k_2-j}\, C^{(x+j-1)}_{\underset{\sim}{a}/[1+(k_1-x)a'+k_2-j]}(1,1),$$

where $\underset{\approx}{a} = (a',\ldots,a',1,\ldots,1)$ has x values equal to a' and $j-1$ ones. Suppose now, for simplicity, that $a' = 1$ as in our previous example. Then our result from (1.12) is

$$(1.14) \qquad P\{X = x|j\} = \frac{\binom{k_1}{x} k_2 \binom{k_2-1}{j-1}}{1+k'-x-j}\, C^{(x+j-1)}_{(1+k-x-j)^{-1}}(1,1).$$

By straightforward integration of the right side of (2.2a) for $r = m = 1$ we obtain

$$(1.15) \qquad C_a^{(b)}(1,1) = \Gamma(b+1) \int_0^a \cdots \int_0^a \frac{\prod_1^b dx_i}{\left(1+\sum_1^b x_i\right)^{b+1}} = \sum_{\alpha=0}^{b} (-1)^{\alpha} \frac{\binom{b}{\alpha}}{1+\alpha a}$$

$$= \int_0^1 (1-x^a)^b\, dx = \frac{1}{a} \int_0^1 (1-y)^b y^{(1/a)-1} dy = \frac{b!\,\Gamma(1+\frac{1}{a})}{\Gamma(b+1+\frac{1}{a})}.$$

Using this result in (1.14) with $b = x + j - 1$ and $a = (1+k'-x-j)^{-1}$ we obtain the desired result

$$(1.16)\quad P\{X = x|j\} = \frac{k_2\binom{k_2-1}{j-1}}{k_1 + k_2}\ \frac{\binom{k_1}{x}}{\binom{k_1+k_2-1}{x+j-1}}$$

$$= \binom{k_1+k_2}{k_1}^{-1}\binom{x+j-1}{j-1}\binom{k_1+k_2-x-j}{k_2-j},$$

and it is not difficult to check that the sum on x from 0 to k_1 is 1. For the case $k_1 = k_2 = j = 5$ the results for $x = 0, 1, 2, 3, 4, 5$ resp., are

$$(1.17)\quad \frac{2}{9}, \frac{5}{18}, \frac{5}{21}, \frac{10}{63}, \frac{5}{63}, \frac{1}{42}$$

$$\left(\text{or } \frac{56}{252}, \frac{70}{252}, \frac{60}{252}, \frac{40}{252}, \frac{20}{252}, \frac{6}{252}\right),$$

and their sum is one. The distributions in (1.16) are not all distinct since we can replace k_2-j by $j'-1$ and k_1-x by x' and get the same result, i.e., the same set of probabilities.

In this last example we have an explicit result (1.16) and do not have to use our tables, but the C-integral still plays an important role in the analysis. Moreover for $a' \neq 1$ and j moderate we would integrate $j-1$ times in (1.13) and write the answer as a sum of C-integrals, each having a common a-value; we omit these details.

Throughout this write-up, especially in Chapter 5, many more applications of different types are considered; in Chapter 5 we also use playing cards as a background for elementary problems. The reader is invited to look at some of these applications immediately if he/she wants to be motivated by more problems that can be simply solved with type 2 Dirichlet tables. Corresponding problems for type 1 Dirichlet can be found in Vol. 4 of this series. We do _not_ claim that all these problems can be solved only with the use of Dirichlet integrals but we _do_ claim that the use of these integrals makes many interesting problems easily solvable which would otherwise be much more difficult (if not impossible) to manage. In other words we believe that the Dirichlet integral is a useful tool that should find its way into elementary texts and be more widely used.

§2 Incomplete Type-2 Dirichlet Integrals

This book is a companion volume to Vol. 4 of Selected Tables in Mathematical Statistics by the same authors [18]; the latter deals with Dirichlet, type-1 and the present one deals with Dirichlet, type-2. We feel that an exhaustive study, including tables, of the Dirichlet distributions was long overdue since this distribution has arisen (and keeps on appearing) in many different applications. Dirichlet integrals already appeared in Wilks [20] and in an article by Olkin and Sobel [16]. The type-2 Dirichlet was applied to the area of ranking problems in Cacoullos and Sobel [4], in Rao and Sobel [17], in Gupta and Sobel [11], [12] and also in Gibbons, Olkin and Sobel [9] (cf. Chapter 5, page 152 and Appendix G).

Our treatment of the type-2 Dirichlet is carried out in a manner that is analogous to the treatment of type-1 in [18] with similar notation and with an extension to waiting time problems for some of the illustrative examples Anderson, Sobel and Uppuluri [2] and Sobel and Liu [19]. Thus the type-2 distribution is naturally related to type-1 and our basic type-2 integrals have already been mentioned in some of the problems in [18] (cf. C-integrals and D-integrals in the index of [18]).

It should be noted that both type-1 and type-2 Dirichlet distributions are related to interesting multinomial and combinatorial problems. For example they enable us to generalize the well-known identity between the tail of a binomial sum and the incomplete beta integral. It is our hope that these two volumes will furnish tools and inspire further research in these fascinating areas of probability and statistics.

2.1 Definitions, Preliminaries, Relation of C to D.

The incomplete beta distribution (in 1-dimension) can be written in two equivalent forms. Using the usual notation $I_p(r,s)$ for the incomplete beta function we have

$$I_p(r,s) = \frac{\Gamma(r+s)}{\Gamma(r)\Gamma(s)} \int_0^p x^{r-1}(1-x)^{s-1}dx$$

(2.1)

$$= \frac{\Gamma(r+s)}{\Gamma(r)\Gamma(s)} \int_0^{p/q} \frac{y^{r-1}dy}{(1+y)^{r+s}}$$

where $0 \leq p \leq 1$ and $q = 1-p$.

As soon as we consider the generalization of these two forms to $b \geq 2$ dimensions, the simple relation in (2.1) no longer holds. The generalization of the first integral in (2.1) to a b-fold Dirichlet integral, which is called type-1 was considered in [18].

In the present volume we generalize the form of the second integral in (2.1) to b dimensions and refer to the resulting integral as the incomplete Dirichlet integral of type-2. Our principal interest is in the lower tail (only) form and the upper tail (only) form and we define these respectively by

$$C_a^{(b)}(r,m) = \frac{\Gamma(m+br)}{\Gamma^b(r)\Gamma(m)} \int_0^a \cdots \int_0^a \frac{\prod_{i=1}^{b} x_i^{r-1}dx_i}{(1+x_1+\cdots+x_b)^{m+br}}, \tag{2.2a}$$

$$D_a^{(b)}(r,m) = \frac{\Gamma(m+br)}{\Gamma^b(r)\Gamma(m)} \int_a^\infty \cdots \int_a^\infty \frac{\prod_{i=1}^{b} x_i^{r-1}dx_i}{(1+x_1+\cdots+x_b)^{m+br}}, \tag{2.2b}$$

where we assume that $a \geq 0$, b is an integer and m, b, r are all positive; the only exception to the latter is the trivial case $b = 0$ (with no integrand and in this case we take the value as 1 for any m, any r and the result does not depend on a). In the applications and in our tables the values of m and r will also be positive integers. Half-integer values of m and r are also of interest but these cases will be considered separately in Section 4.7. Mixed integrals (called CD) are considered in section 4.5 and in certain applications.

These integrals (2.2) represent straightforward generalizations of the usual incomplete beta functions and in fact, if we set b = 1 in (2.2a) and (2.2b)

$$C_a^{(1)}(r,m) = I_{\frac{a}{1+a}}(r,m) = 1 - D_a^{(1)}(r,m) \tag{2.3a}$$

$$= I^{(1)}_{\frac{a}{1+a}}(r,m+r-1) = D^{(1)}_{1/a}(m,r) , \tag{2.3b}$$

where the I-notation in (2.3b) was introduced in [18] and defined for general b , any $p \leq 1/b$ and any $n \geq br$ by

$$I_p^{(b)}(r,n) = \frac{n!}{\Gamma^b(r)(n-br)!} \int_0^p \cdots \int_0^p \left(1 - \sum_{i=1}^{b} x_i\right)^{n-br} \prod_{i=1}^{n} x_i^{r-1} dx_i . \tag{2.4}$$

The proof that

$$C_\infty^{(b)}(r,m) = D_0^{(b)}(r,m) = 1 \quad \text{for all} \quad b,\ r,\ m \tag{2.5}$$

will be a consequence of later results and also of the probability interpretation in Section 2.3 and we defer it.

The relation of C and D to each other is one of inclusion-exclusion given by

$$C_a^{(b)}(r,m) = \sum_{\alpha=0}^{b} (-1)^\alpha \binom{b}{\alpha} D_a^{(\alpha)}(r,m) , \tag{2.6a}$$

$$D_a^{(b)}(r,m) = \sum_{\alpha=0}^{b} (-1)^\alpha \binom{b}{\alpha} C_a^{(\alpha)}(r,m) , \tag{2.6b}$$

where $C_a^{(0)}(r,m) = D_a^{(0)}(r,m) = 1$ for all a, r, m.

It is sufficient to consider (2.6b). To prove (2.6b) consider the C integral in (2.2a) with different limits of integration $a_1, a_2, \ldots, a_b$, and denoted by $C_{\underset{\sim}{a}}^{(b)}(r,m)$. We show that for $a_b = \infty$ we again obtain a C integral with the same parameters except that b is reduced by 1. We have

$$(2.7)\quad C^{(b)}_{(a_1,a_2,\ldots,a_{b-1},\infty)}(r,m)$$

$$= \frac{\Gamma(m+br)}{\Gamma^b(r)\Gamma(m)} \int_0^{a_1}\cdots\int_0^{a_{b-1}}\int_0^{\infty} \frac{\left[\prod_{i=1}^{b-1} x_i^{r-1}dx_i\right] x_b^{r-1}dx_b}{(1+x_1+\cdots+x_{b-1}+x_b)^{m+br}}$$

$$= \frac{\Gamma(m+(b-1)r)}{\Gamma^{b-1}(r)\Gamma(m)} \int_0^{a_1}\cdots\int_0^{a_{b-1}} \frac{\prod_{1}^{b-1} x_i^{r-1}dx_i}{(1+x_1+\cdots+x_{b-1})^{m+(b-1)r}} = C^{(b-1)}_{(a_1,\ldots,a_{b-1})}(r,m)$$

which is obtained by making the transformation $y = x_b/(1 + x_1+\cdots+x_b)$, and using the well-known result for the complete beta function. Combining this result (2.7) with (2.5) and the fact that C is a left tail (only) integral and D is a right tail (only) integral with the same integrand, we now obtain by an inclusion-exclusion argument the desired result in (2.6b)

2.2 Relation to negative binomial and negative multinomial sums.

We first illustrate that for $b = 1$ the C and D functions can be used for either tail of the negative binomial distribution and then generalize this by taking $b \geq 2$ and showing that the associated C and D functions represent tails of the negative multinomial distribution. In fact by the relations (2.4) of [18] we have

$$(2.8a)\qquad C_a^{(1)}(r,m) = \left(\frac{1}{1+a}\right)^m \sum_{i=r}^{\infty} \frac{\Gamma(m+i)}{\Gamma(m)i!} \left(\frac{a}{1+a}\right)^i$$

$$= (\frac{a}{1+a})^r \sum_{j=0}^{m-1} \frac{\Gamma(r+j)}{\Gamma(r)j!} (\frac{1}{1+a})^j ,$$

(2.8b) $$D_a^{(1)}(r,m) = (\frac{1}{1+a})^m \sum_{i=0}^{r-1} \frac{\Gamma(m+i)}{\Gamma(m)i!} (\frac{a}{1+a})^i$$

$$= (\frac{a}{1+a})^r \sum_{j=m}^{\infty} \frac{\Gamma(r+j)}{\Gamma(r)j!} (\frac{1}{1+a})^j .$$

The first summation in (2.8b) can be interpreted as the probability of an event associated with the following stopping time problem. We have $b = 2$ cells, one of which (called the counting cell) has probability $1/(1+a)$ and the other (called the blue cell) has probability $a/(1+a)$. We stop, as soon as the counting cell contains exactly m observations and we consider the probability that at the stopping time the blue cell contained at least r observations. We refer to such problems as counting cell problems.

In a similar manner the more general C and D functions in (2.2) can also be interpreted as probabilities associated with a counting cell problem when there are b blue cells present. Our starting point is the probability that each of the b blue cells contains at least r observations at stopping time i.e., when the counting cell has just received its mth observation. We will first show that P equals a type-2 Dirichlet integral and afterwards show that it is also equal to a negative multinomial sum. In the first part of the proof we utilize the probability interpretation of $I = I_p^{(b,j)}(r,s,n)$ defined in (4.75) of [18] that j specified cells contain exactly r observations and $b-j$ cells each contain at least s observations. To implement this we take $j = 1$ (i.e., one cell C_0 is specified as cell zero and called the counting cell) and use (4.75) of [18] to write the probability that C_0 has a frequency of $m-1$ observations in $n-1$ trials. To bring the notation of (4.75) of [18] into line with the present discussion we replace r by m-1, s by r, n by n-1, and since the counting cell will not be counted as a blue cell below, we also change b-1 to b and set j equal to 0. Let the cell probabilities be $p_o = (1+ba)^{-1}$ for the count-cell and $p = a/(1+ba)$ for the blue cells. Then we obtain from (4.75) of [18],

$$I = P[N_0 = m-1,\ \mathrm{Min}(N_1,N_2,\ldots,N_b) \geq r \mid n-1,\ b,\ \frac{a}{1+ba},\ \frac{1}{1+ba}]$$

(2.9)
$$= \frac{p_o^{m-1}(n-1)!}{\Gamma(m)\Gamma^b(r)(n-m-br)!} \int_0^p \cdots \int_0^p (1-p_o-\sum_{\alpha=1}^{b} x_\alpha)^{n-m-br} \prod_{\alpha=1}^{b} x_\alpha^{r-1} dx_\alpha \ .$$

Multiplying by p_o for the event that the last observation falls into the counting cell, and summing on n from $m+br$ to ∞, we obtain for the desired P the result

$$(2.10) \qquad P = \sum_{n=m+br}^{\infty} \frac{p_o^m(n-1)!}{\Gamma(m)\Gamma^b(r)(n-m-br)!} \int_0^p \cdots \int_0^p (1-p_o-\sum_{\alpha=1}^{b} x_\alpha)^{n-m-br} \prod_{\alpha=1}^{b} x_\alpha^{r-1} dx_\alpha .$$

After summing on n under the integral sign, we obtain

$$(2.11) \qquad P = \frac{\Gamma(m+br)p_o^m}{\Gamma(m)\Gamma^b(r)p_o^{m+br}} \int_0^p \cdots \int_0^p \frac{\prod_{\alpha=1}^{b} x_\alpha^{r-1} dx_\alpha}{[1+\sum_{\alpha=1}^{b}(x_\alpha/p_o)]^{m+br}}$$

We now set $y_\alpha = x_\alpha/p_o$ and noting that $a = p/p_o$, we have

$$(2.12) \qquad P = \frac{\Gamma(m+br)}{\Gamma^b(r)\Gamma(m)} \int_0^a \cdots \int_0^a \frac{\prod_{\alpha=1}^{b} y_\alpha^{r-1} dy_\alpha}{(1+y_1+\ldots+y_b)^{m+br}} = C_a^{(b)}(r,m) \ .$$

To show that P is also a negative multinomial sum we again start with $I = I_p^{(b,j)}(r,s,n)$ in (4.48) of [18] with $j = 1$, $r = m-1$, and (s, n and $b-1$) replaced by (r, $n-1$, and b), respectively, and with p as before. Then I can be written as a multinomial probability in the form

(2.13) $$I = \sum_{x_1=r}^{n} \cdots \sum_{x_b=r}^{n} \left[{n-1 \atop m-1, x_1, \ldots, x_b} \right] p_0^{m-1} p^{x_1+x_2+\ldots+x_b} ,$$

where the multinomial coefficient is zero unless $n = m+x_1+\ldots+x_b$. Multiplying by p_0 and summing on n , we obtain

(2.14) $$P = \left(\frac{1}{1+ba}\right)^m \sum_{x_1 \geq r} \cdots \sum_{x_b \geq r} \left[{m-1+\sum_{\alpha=1}^{b} x_\alpha \atop m-1, x_1, \ldots, x_b} \right] \left(\frac{a}{1+ba}\right)^{x_1+\cdots+x_b}$$

which is the desired negative multinomial sum for P .

A similar discussion (which is omitted) gives us the analogous relation for the D-type integral, namely

(2.15) $$D_a^{(b)}(r,m) = \frac{\Gamma(m+br)}{\Gamma^b(r)\Gamma(m)} \int_a^\infty \cdots \int_a^\infty \frac{\prod_{\alpha=1}^{b} x_\alpha^{r-1} dx_\alpha}{(1+x_1+\ldots+x_b)^{m+br}}$$

$$= \frac{1}{(1+ba)^m} \sum_{x_1<r} \cdots \sum_{x_b<r} \left[{m-1+\sum_{\alpha=1}^{b} x_i \atop m-1, x_1, \ldots, x_b} \right] \left(\frac{a}{1+ba}\right)^{x_1+\cdots+x_b} .$$

An alternative derivation of these results can be found in Olkin and Sobel [16].

2.3 The Generalizations $C_a^{(b,j)}(r,m)$ and $D_a^{(b,j)}(r,m)$, and their Probability Interpretation

In (2.7) of [18] we introduced a new parameter j which helped us to obtain recurrence relations and we introduce j here in an analogous manner. For any integers b,j $(0 \leq j \leq b)$, any positive values r, a, m we define the $(b-j)$-fold integrals

(2.16a) $$C_a^{(b,j)}(r,m) = \frac{\Gamma(m+br)}{\Gamma(m)\Gamma^b(r)} \left(\frac{a^r}{r}\right)^j \int_0^a \cdots \int_0^a \frac{\prod_{j+1}^{b} x_i^{r-1} dx_i}{(1+ja+\sum_{j+1}^{b} x_i)^{m+br}} ,$$

$$(2.16b) \qquad D_a^{(b,j)}(r,m) = \frac{\Gamma(m+br)}{\Gamma(m)\Gamma^b(r)} \left(\frac{a^r}{r}\right)^j \int_a^\infty \cdots \int_a^\infty \frac{\prod_{j+1}^{b} x_i^{r-1} dx_i}{(1+ja+\sum_{j+1}^{b} x_i)^{m+br}} .$$

For $j = 0$ we drop the second superscript and note that the result is consistent with (2.2).

In order to develop the probabilistic interpretation of (2.16) we start with the interpretation of (4.48) in [18] with j specified cells having frequency (exactly) r, one specified cell having frequency $m-1$ and the remaining $b-j$ blue cells each having frequency at least r, all based on $n-1$ observations; let p denote the cell probability for the b blue cells and let $p_0 = 1-bp$ denote the cell probability for the sink, which we now call a counting cell. Since the last observation is in the counting cell, we multiply by p_0 and obtain the probability that in exactly n observations j specified blue cells have frequency r and the remaining $b-j$ blue cells each has frequency at least r; this is given by

$$(2.17) \qquad \frac{(n-1)!(p^r/r)^j p_0^m/(m-1)}{\Gamma(m-1)\Gamma^b(r)(n-m-br)!} \int_0^p \cdots \int_0^p (1-jp-p_0 - \sum_{j+1}^{b} x_i)^{n-m-br} \prod_{j+1}^{b} x_i^{r-1} dx_i .$$

Summing (2.17) over n from $m+br$ to ∞ gives us the probability at stopping time defined by "the first time we have m observations in the counting cell" that j specified blue cells have frequency r and the other blue cells have frequency at least r; by a simple sum of a negative binomial we obtain for (2.17) and hence for (2.16a)

$$(2.18) \qquad \frac{\Gamma(m+br)}{\Gamma(m)\Gamma^b(r)} \left(\frac{a^r}{r}\right)^j \int_0^a \cdots \int_0^a \frac{\prod_{j+1}^{b} y_i^{r-1} dy_i}{(1+ja+\sum_{j+1}^{b} y_i)^{m+br}} = C_a^{(b,j)}(r,m),$$

where $a = \frac{p}{p_0}$ and the transformation $y_i = x_i/p_0$ was made in the very last step.

Essentially the same method can be used to show that $D_a^{(b,j)}(r,m)$ can be interpreted as the probability at stopping time (i.e., when the counting cell first gets m observations) that j specified blue cells have frequency r and the remaining blue cells each has frequency less than r ; we omit the details.

2.3.1 An Alternate Generalization of $C_a^{(b)}(r,m)$ and $D_a^{(b)}(r,m)$.

It turns out to be convenient and interesting for some purposes, e.g., for calculation, to consider the alternative generalization

$$(2.19a)\qquad \mathcal{C}_a^{(b,j)}(r,m) = \frac{(1+ja)^m\Gamma(m+(b-j)r)}{\Gamma(m)\Gamma^{b-j}(r)} \int_0^a \cdots \int_0^a \frac{\prod_{j+1}^{b} y_i^{r-1}dy_i}{(1+ja+\sum_{j+1}^{b} y_i)^{m+(b-j)r}} ,$$

$$(2.19b)\qquad \mathcal{D}_a^{(b,j)}(r,m) = \frac{(1+ja)^m\Gamma(m+(b-j)r)}{\Gamma(m)\Gamma^{b-j}(r)} \int_a^\infty \cdots \int_a^\infty \frac{\prod_{j+1}^{b} y_i^{r-1}dy_i}{(1+ja+\sum_{j+1}^{b} y_i)^{m+(b-j)r}} .$$

Note first of all that for $j = 0$ the resulting integral is the same as before. These integrals also have a probability interpretation using a similar idea of stopping when a counting cell first obtains m observations, namely it is the same as before except that at the outset the j specified cells are combined with the counting cell and we wait until the first time this union contains m observations. Then the interpretation is the probability that the $b-j$ remaining cells have frequency at least r for (2.19a) and less than r for (2.19b).

Here we give a few properties of $\mathcal{C}$ and $\mathcal{D}$ which hold for all values of the parameters shown:

(2.20) $$\mathcal{C}_0^{(b,j)}(r,m) = 0 \ , \quad \mathcal{D}_0^{(b,j)}(r,m) = 1 \ ;$$

(2.21) $$\mathcal{C}_a^{(b,j)}(r,m) = \mathcal{C}_{a/(1+ja)}^{(b-j,0)}(r,m) \ , \quad \mathcal{D}_a^{(b,j)}(r,m) = \mathcal{D}_{a/(1+ja)}^{(b-j,0)}(r,m) \ ;$$

(2.22) $$\mathcal{C}_a^{(b,j)}(1,m) = \sum_{\alpha=j}^{b} (-1)^{\alpha-j}\binom{b-j}{\alpha-j}\left(\frac{1+ja}{1+\alpha a}\right)^m \ , \quad \mathcal{D}_a^{(b,j)}(1,m) = \left(\frac{1+ja}{1+ba}\right)^m \ ;$$

(2.23a) $$\mathcal{C}_a^{(b,j)}(r,m) = \sum_{\alpha=j}^{b} (-1)^{\alpha-j}\binom{b-j}{\alpha-j}\mathcal{D}_a^{(\alpha,j)}(r,m) \ ,$$

(2.23b) $$\mathcal{D}_a^{(b,j)}(r,m) = \sum_{\alpha=j}^{b} (-1)^{\alpha-j}\binom{b-j}{\alpha-j}\mathcal{C}_a^{(\alpha,j)}(r,m) \ .$$

Proofs:

The inclusion-exclusion results (2.23) are a direct consequence of (2.21) and (2.6). The first result in (2.20) is obvious from (2.19a); the second result in (2.20) follows from (2.23b) and the first result in (2.20). For (2.21) we prove only the first result since the second is quite similar: we factor out $1+ja$ from the integrand in (2.19a) and set $x_i = y_i/(1+ja)$ $(i = j+1,\ldots,b)$ and the desired result follows.

The probability interpretation given above for $\mathcal{C}$ and $\mathcal{D}$ can now be confirmed by the following argument. From (2.21) $\mathcal{C}_{a/(1+ja)}^{(b-j,0)}(r,m) = C_{a/(1+ja)}^{(b-j)}(r,m)$, which already was shown to have a probability interpretation. We use this to give a probability interpretation for $\mathcal{C}_a^{(b,j)}(r,m)$. We first show that $\mathcal{C}_{a/(1+ja)}^{(b-j,0)}(r,m)$ is the probability that each cell in specified set of $b-j$ blue cells has frequency at least r. To see this, we find the common blue cell probability p' (corresponding to p in (2.9)) when $a' = a/(1+ja)$ and $b' = b-j$. Putting the latter into $p' = a'/(1+b'a')$ we find that $p' = a/(1+ba) = p$. The associated counting cell probability p_0' is $1-b'p' = 1 - (b-j)p = (1+ja)/(1+ba)$ $= [1/(1+ba)] + j[a/(1+ba)]$ which confirms the fact that we have combined the j cells and the old counting cell to form a new counting cell. Since r and m did not change in value, the interpretation of the left hand side of the 1st result in (2.21) is as stated above; a similar argument holds for the 2nd result in (2.21).

To prove (2.22) we start with the 2nd result given there. Under our probability interpretation for $\mathcal{D}$ we form a new counting cell by combining j blue cells and the old counting cell so that the probability for the new counting cell is $(1+ja)/(1+ba)$. Since $r = 1$ indicates that $b-j$ specified cells are to be empty when the new counting cell has frequency m, this means that the first m observations must all fall in the new counting cell and the probability of this is $[(1+ja)/(1+ba)]^m$. If we now use this result in the inclusion-exclusion formula (2.23a) the first result in (2.22) immediately follows.

It is worth noting that b need not represent the total number of blue cells but it does represent the number of blue cells which appear in the probability interpretation. Thus if we actually had 10 blue cells and 1 counting cell and asked for the probability at stopping time (the first time we get m in the counting cell) that five specified cells have frequency at least r, we still use $C_a^{(5)}(r,m)$ where $a = p/p_0$, p is the common blue cell probability and $p_0 = 1-10p$ is the counting cell probability.

2.4 Basic Recurrence Relations.

The main purpose of introducing the additional parameter (as in the case of I and J in [18]) is to be able to develop simple recurrence relations which enable us to calculate all of the multiple integrals described above without a single numerical quadrature. We give recurrence relations for C, D, $\mathcal{C}$ and $\mathcal{D}$ with boundary conditions for each and defer proofs until later.

For any values of r, m, b, j, a with $0 \le j < b$ both integers

$$(2.24a) \quad m(1+ja)C_a^{(b,j)}(r,m+1) = (m+jr)C_a^{(b,j)}(r,m) + r(b-j)C_a^{(b,j+1)}(r,m) ,$$

(2.24b) $$m(1+ja)D_a^{(b,j)}(r,m+1) = (m+jr)D_a^{(b,j)}(r,m) - r(b-j)D_a^{(b,j+1)}(r,m) ,$$

(2.25a) $$\mathcal{C}_a^{(b,j)}(r,m+1) = \mathcal{C}_a^{(b,j)}(r,m) + (b-j)\binom{m+r-1}{m} \frac{a^r(1+ja)^m}{[1+(j+1)a]^{m+r}} \mathcal{C}_a^{(b,j+1)}(r,m+r),$$

(2.25b) $$\mathcal{D}_a^{(b,j)}(r,m+1) = \mathcal{D}_a^{(b,j)}(r,m) - (b-j)\binom{m+r-1}{m} \frac{a^r(1+ja)^m}{[1+(j+1)a]^{m+r}} \mathcal{D}_a^{(b,j+1)}(r,m+r).$$

The boundary conditions for (2.24a) and (2.24b) are for $j = b$

(2.26) $$C_a^{(b,b)}(r,m) = \frac{\Gamma(m+br)}{(r!)^b\Gamma(m)} \left(\frac{a}{1+ba}\right)^{rb} \left(\frac{1}{1+ba}\right)^m = D_a^{(b,b)}(r,m) ,$$

and for $m > r$

(2.27a) $$C_a^{(b,j)}(r,m) = C_a^{(b-1,j)}(r,m) - \frac{1}{\binom{m-1}{r}} \sum_{\alpha=1}^{r} \frac{\binom{m-\alpha-1}{r-\alpha}}{a^\alpha} C_a^{(b,j+1)}(r,m-\alpha),$$

(2.27b) $$D_a^{(b,j)}(r,m) = \frac{1}{\binom{m-1}{r}} \sum_{\alpha=1}^{r} \frac{\binom{m-\alpha-1}{r-\alpha}}{a^\alpha} D_a^{(b,j+1)}(r,m-\alpha) .$$

The boundary conditions for (2.25a) and (2.25b) are much simpler, namely for $j = b$

(2.28) $$\mathcal{C}_a^{(b,b)}(r,m) = 1 = \mathcal{D}_a^{(b,b)}(r,m)$$

and for $m = 1$

(2.29a) $$\mathcal{C}_a^{(b,j)}(r,1) = (b-j)\left[\frac{a}{1+(j+1)a}\right]^r \mathcal{C}_a^{(b,j+1)}(r,r) ,$$

(2.29b) $$\mathcal{D}_a^{(b,j)}(r,1) = 1 - (b-j)\left[\frac{a}{1+(j+1)a}\right]^r \mathcal{D}_a^{(b,j+1)}(r,r) .$$

Proofs:

We prove some of these results, omitting others which are similar. To prove (2.24a) we increase the exponent in the integrand in (2.16a) and obtain

$$C_a^{(b,j)}(r,m) = \frac{\Gamma(m+br)}{\Gamma(m)\Gamma^b(r)} \left(\frac{a^r}{r}\right)^j \int_0^a \cdots \int_0^a \frac{(1+ja+\sum_{j+1}^b x_i) \prod_{j+1}^b x_i^{r-1} dx_i}{(1+ja+\sum_{j+1}^b x_i)^{m+1+br}}$$

$$= \frac{m(1+ja)}{m+br} C_a^{(b,j)}(r,m+1) + \frac{(b-j)\Gamma(m+br)}{\Gamma(m)\Gamma^b(r)} \left(\frac{a^r}{r}\right)^j \int_0^a \cdots \int_0^a \frac{x_{j+1}^r \prod_{j+2}^b x_i^{r-1} dx_i dx_{j+1}}{(1+ja+\sum_{j+1}^b x_i)^{m+1+br}},$$

where we used symmetry with respect to different variables to write the factor $b-j$. Integrating the last expression by parts gives rise to a second term exactly like the one on the left hand side and, substituting the limits in the first term, we obtain

$$\left[1 - \frac{(b-j)r}{m+br}\right] C_a^{(b,j)}(r,m) = \frac{m(1+ja)}{m+br} C_a^{(b,j)}(r,m+1) - \frac{r(b-j)}{m+br} C_a^{(b,j+1)}(r,m) ,$$

which is clearly equivalent to what we have in (2.24a). The proofs of (2.24b), (2.25a) and (2.25b) are quite similar and are omitted. The proofs of (2.26) and (2.28) are trivial from the definitions since there is no integration in these cases. To prove (2.27a) we start with the integral form in (2.16a) and integrate by parts one specific variable repeatedly until it vanishes. This yields r negative terms in addition to one nonzero term in which we substitute limits; each of these terms is identified as a C integral with a simple coefficient. By this straightforward procedure we obtain (2.27a). In (2.27b) the method is quite similar but since the upper limit is ∞ there are no extra nonzero terms to add and the result is given in (2.27b). We defer the proofs of (2.29a) and (2.29b) until after we have proven the so-called single-integral representation of $\mathcal{C}$ and $\mathcal{D}$ in the next section.

2.5 Reduction Formulas, Single-Integral Representation, and Other Results.

A reduction formula is one which reduces the double superscript to a single superscript, or more generally, simply reduces the 2nd superscript. This has already been done for $\mathcal{C}$ and $\mathcal{D}$ in (2.21) but slightly more general results hold, namely for $0 \leq i \leq j$

$$(2.30)\quad \mathcal{C}_a^{(b,j)}(r,m) = \mathcal{C}_{\frac{a}{1+ia}}^{(b-i,j-i)}(r,m) , \quad \mathcal{D}_a^{(b,j)}(r,m) = \mathcal{D}_{\frac{a}{1+ia}}^{(b-i,j-i)}(r,m) .$$

The proofs of these identities in (2.30) are completely analogous to those for (2.21).

For C and D the reduction formulas are as follows: for any i $(0 \leq i \leq j)$

$$(2.31)\quad C_a^{(b,j)}(r,m) = \left[{m+ir-1 \atop r,\ldots,r,m-1}\right]\left(\frac{a}{1+ia}\right)^{ir}\left(\frac{1}{1+ia}\right)^{m} C_{\frac{a}{1+ia}}^{(b-i,j-i)}(r,m+ir),$$

and the same result holds with C replaced by D on both sides of (2.31).

Proof of (2.31):

Using a probabilistic method of proof, the desired event on the left side of (2.31) can be regarded in two steps. In the 1st step we combine i blue cells to form a new counting cell, so that the new a-value is the ratio of the blue cell probability $a/(1+ba)$ to the new counting cell probability $(1+ia)/(1+ba)$, namely $a/(1+ia)$. Then the probability (when the new counting cell gets $m+ir$ observations) that $j-i$ cells have frequency exactly r and $b-j$ cells have frequency at least r is given by the C-function on the right side of (2.31). In the 2nd step (conditioning on the 1st step) we ask that out of $m+ri$ observations in the new counting cell each of the i cells gets exactly r , the old counting cells gets m and also gets the final observation. This is given by the multinomial probability on the right side of (2.31) and the proof is complete. A non-probabilistic proof is also easily obtained, for example by inserting the integral representation in (2.16a) into both sides of (2.31) and simplifying the right side.

Let

(2.32) $$G_r(x) = \frac{1}{\Gamma(r)} \int_0^x t^{r-1}e^{-t}dt$$

denote the usual incomplete gamma function with parameter $r > 0$. Then we can write for any valid values of a , b , j , r , m

(2.33a) $$\mathcal{C}_a^{(b,j)}(r,m) = (1+ja)^m \int_0^\infty e^{-xja}G_r^{b-j}(ax)dG_m(x) ,$$

(2.33b) $$\mathcal{D}_a^{(b,j)}(r,m) = (1+ja)^m \int_0^\infty e^{-xja}[1 - G_r(ax)]^{b-j}dG_m(x) .$$

Proof:

We prove only (2.33a) since (2.33b) is similar. Using the fact that

$$\int_0^\infty e^{-\lambda x}x^{n-1}dx = \frac{\Gamma(n)}{\lambda^n}$$

for $\lambda = 1 + ja + \sum_{j+1}^{b} y_i$ and $n = m + (b-j)r$ and substituting the left side for the right side of the above in the integrand in (2.19a), we can now change the order of integration and replace the common integral with respect to y_i by $G_r(ax)/x^r$. The result is equivalent to (2.33a). Similar results also hold for C and D and we give these without proof:

(2.34a) $$C_a^{(b,j)}(r,m) = \frac{\Gamma(m+jr)}{\Gamma(m)\Gamma^j(r)} \left(\frac{a^r}{r}\right)^j \int_0^\infty e^{-ajx}G_r^{b-j}(ax)dG_{m+jr}(x) ,$$

(2.34b) $$D_a^{(b,j)}(r,m) = \frac{\Gamma(m+jr)}{\Gamma(m)\Gamma^j(r)} \left(\frac{a^r}{r}\right)^j \int_0^\infty e^{-ajx}[1 - G_r(ax)]^{b-j}dG_{m+jr}(x) .$$

We now consider the deferred proof of (2.29a), and (2.29b) is quite similar. Note first of all that $dG_r(x)$ can be written as $e^{(a-1)x}dG_r(ax)/a^r$.

After replacing m by r in (2.33a) and using the above, we integrate by parts once; the first term vanishes and the second term takes the form

$$\frac{(1+ja)^r}{a^r(b-j+1)} \{[1+(j-1)a] \int_0^\infty e^{-x(j-1)a} G_r^{b-j+1}(ax)dG_1(x)\}$$

and the quantity in braces is $\mathcal{C}_a^{(b,j-1)}(r,1)$ by (2.33a). Replacing $j-1$ by j the resulting equality is equivalent to (2.29a).

Some additional results hold for $r = m$ and we consider them here. For $a = 1$ and any r

$$C_1^{(b)}(r,r) = D_1^{(b)}(r,r) = \frac{1}{b+1} \; . \tag{2.35}$$

Proof:

From the result (2.29a) above, using (2.33a), we have for $a < 1$ and $j = 0$

$$\mathcal{C}_a^{(b,0)}(r,r) = C_a^{(b)}(r,r) = \frac{1-a}{a^r(b+1)} \int_0^\infty G_r^{b+1}(ax)e^{-x(1-a)}dx$$

$$= \frac{1}{a^r(b+1)} \int_0^\infty G_r^{b+1}(\frac{ay}{1-a})dG_1(y) \; .$$

Letting $a \to 1$ from below, we obtain (2.35) since $G_r(\infty) = 1$.

Remark:

It is of some interest to note that although $j \geq 0$ in all of the above results, we can also allow negative values of j , for example in (2.21), (2.29a) and (2.29b), provided a is restricted appropriately; thus in (2.21), in (2.29a) and (2.29b) we need $1 + ja > 0$.

An alternative probabilistic proof of (2.35) is quite short and worth including. For $a = 1$ the blue cells and the counting cell all have the same cell probability, namely $1/(b+1)$. Hence for $r = m$ the probability that the counting cell is the last (or first) one to take on r observations is $1/(b+1)$ by a symmetry argument; this proves (2.35).

In addition, it is of interest to note that we can define certain generalized moments and relate them to C and D functions. Consider the integrals

$$\text{(2.36a)}\qquad MC_a^{(b)}(r,m,t) = \int_0^\infty x^t G_r^b(ax)\,dG_m(x)\,,$$

$$\text{(2.36b)}\qquad MD_a^{(b)}(r,m,t) = \int_0^\infty x^t [1-G_r(ax)]^b\,dG_m(x)$$

For $r = m$ and $a = 1$ these represent the tth (raw) moment of the largest (resp., smallest) of $b+1$ independent gamma random variables with common parameter r. We now show that (2.36a) and (2.36b) are easily expressed as C and D functions, respectively, for any positive real r, m and a. Since

$$x^t dG_m(x) = \frac{\Gamma(m+t)}{\Gamma(m)}\,dG_{m+t}(x)\,,$$

we have, using (2.33) for $j = 0$,

$$\text{(2.37a)}\qquad MC_a^{(b)}(r,m,t) = \frac{\Gamma(m+t)}{\Gamma(m)}\,C_a^{(b)}(r,m+t)\,,$$

$$\text{(2.37b)}\qquad MD_a^{(b)}(r,m,t) = \frac{\Gamma(m+t)}{\Gamma(m)}\,D_a^{(b)}(r,m+t)\,,$$

The corresponding moment generating function, using (2.33) with $j = 1$ and (2.21), now takes the form

$$\text{(2.38a)}\qquad \int_0^\infty e^{-ax}G_r^{b-1}(ax)\,dG_m(x) = \frac{\mathcal{C}_a^{(b,1)}(r,m)}{(1+a)^m} = \left(\frac{1}{1+a}\right)^m C_{\frac{a}{1+a}}^{(b-1)}(r,m)\,,$$

$$\text{(2.38b)}\qquad \int_0^\infty e^{-ax}[1-G_r(ax)]^{b-1}\,dG_m(x) = \frac{\mathcal{D}_a^{(b,1)}(r,m)}{(1+a)^m} = \left(\frac{1}{1+a}\right)^m D_{\frac{a}{1+a}}^{(b-1)}(r,m)\,.$$

For the special case $t = 1$, $m = r$ and $a = 1$ we can use (2.35) and (2.37) together with (2.24) and (2.25) to obtain for the first moments $MC_1^{(b)}(r,r,1)$ and $MD_1^{(b)}(r,r,1)$

(2.39a) $$MC_1^{(b)}(r,r,1) = rC_1^{(b)}(r,r+1) = \frac{r}{b+1} + \frac{br}{4^r}\binom{2r-1}{r}C_{\frac{1}{2}}^{(b-1)}(r,2r) ,$$

(2.39b) $$MD_1^{(b)}(r,r,1) = rD_1^{(b)}(r,r+1) = \frac{r}{b+1} - \frac{br}{4^r}\binom{2r-1}{r}D_{\frac{1}{2}}^{(b-1)}(r,2r) .$$

The above results in (2.39a) and (2.39b) will be useful later, especially in dealing with half-integer values of r.

3. Analytic Developments

In this chapter we are concerned with differential properties of the C , D , $\mathcal{C}$ and $\mathcal{D}$ functions which will be useful for the purpose of a Taylor expansion and different approximations. In particular we shall be interested in a normal approximation to both C and D .

3.1 Derivatives of C , D , $\mathcal{C}$ and $\mathcal{D}$ in Terms of Difference Operators.

The partial derivatives with respect to a are related to differences with respect to the second argument m ; we use the symbol $\underset{m}{\Delta}$ as a difference operator and $m^{[\alpha]}$ as an ascending factorial defined by

$$\underset{m}{\Delta}f(m) = f(m+1) - f(m) , \quad m^{[\alpha]} = m(m+1) \cdots (m+ -1).$$

Using these, the main result here is

(3.1) $$\frac{\partial^i}{\partial a^i} C_a^{(b,j)}(r,m) = a^{-i}\, m^{[i]}\, \underset{m}{\Delta}^i\, C_a^{(b,j)}(r,m),$$

and the exact same result holds with D replacing C on both sides of (3.1).

For $\mathcal{C}$ and $\mathcal{D}$ the corresponding formulas are more complicated but they satisfy the following recurrence relation for n = 0,1,...

(3.2) $$a(1+ja)\frac{\partial^{n+1}}{\partial a^{n+1}}\mathcal{C}_a^{(b,j)}(r,m)$$

$$= [m\underset{m}{\Delta} - n\,(1+2ja)]\,\frac{\partial^n}{\partial a^n}\,\mathcal{C}_a^{(b,j)}(r,m) - jn(n-1)\,\frac{\partial^{n-1}}{\partial a^{n-1}}\,\mathcal{C}_a^{(b,j)}(r,m)$$

and no boundary conditions are needed. Thus for $n = 0$ and $n = 1$ we obtain

(3.3) $$a(1+ja)\,\frac{\partial}{\partial a}\,\mathcal{C}_a^{(b,j)}(r,m) = m\underset{m}{\Delta}\,\mathcal{C}_a^{(b,j)}(r,m)\ ,$$

(3.4) $$[a(1+ja)]^2\,\frac{\partial^2}{\partial a^2}\,\mathcal{C}_a^{(b,j)}(r,m) = (m^{[2]}\underset{m}{\Delta}^2 - 2ajm\underset{m}{\Delta})\mathcal{C}_a^{(b,j)}(r,m)\ .$$

Proof:

We prove only the result (3.1) for C and start with the case $i = 1$, namely that

(3.5) $$\frac{\partial}{\partial a}\,C_a^{(b,j)}(r,m) = \frac{m}{a}\,\underset{m}{\Delta}\,C_a^{(b,j)}(r,m)\ .$$

From the integral representation (2.34a) we obtain after differentiating with respect to a

$$\frac{\partial}{\partial a}\,C_a^{(b,j)}(r,m) = \frac{(b-j)r}{a}\,C_a^{(b,j+1)}(r,m) - mjC_a^{(b,j)}(r,m+1) + \frac{rj}{a}\,C_a^{(b,j)}(r,m)\ .$$

If we replace the first term on the right hand side by using the basic recurrence relation (2.24a), we obtain the result (3.5).

To prove the general result (3.1) we first sketch the proof of a

Lemma:

(3.6) $$(m\underset{m}{\Delta} - i+1)\,\cdots\,(m\underset{m}{\Delta} - 1)m\underset{m}{\Delta} = m^{[i]}\underset{m}{\Delta}^i\ .$$

The proof is easily shown by induction using the fact that $\underset{m}{\Delta}\{F(m)G(m)\} = F(m)\underset{m}{\Delta}G(m) + G(m+1)\underset{m}{\Delta}F(m)$; details are omitted. If we iterate differentiation on both sides of (3.5) , we find that

$$\frac{\partial^i}{\partial a^i} C_a^{(b,j)}(r,m) = \frac{1}{a^i} \{(m\underset{m}{\Delta} - i+1) \cdots (m\underset{m}{\Delta}-1)m\underset{m}{\Delta}\} C_a^{(b,j)}(r,m) ,$$

and from (3.6) the desired result (3.1) follows.

To prove (3.2), we actually first prove the special case $n = 0$, which is given by (3.3); this is carried out by using the integral representation (2.33a), just as was done for (3.5). Using the Leibniz rule for differentiating a product, we then obtain the result (3.2) in a straightforward manner; we omit the details.

3.2 Series Expansions; Taylor and Otherwise.

The results in Section 3.1 are quite useful to obtain values of C_a and D_a for a-values not in our table through a Taylor expansion. In fact, from (3.1) we obtain for the expansion of $C_a^{(b)}(r,m)$ about a_o

$$C_a^{(b)}(r,m) = \sum_{\alpha=0}^{\infty} \frac{(a-a_o)^{\alpha}}{\alpha!} \frac{m^{[\alpha]}}{a_o^{\alpha}} \underset{m}{\Delta}^{\alpha} C_{a_o}^{(b)}(r,m) . \tag{3.7}$$

The result for $D_a^{(b)}(r,m)$ is exactly the same in terms of $D_{a_o}^{(b)}(r,m)$.

To illustrate a typical calculation we consider the special case $b = m = r = 2$ and $a = .9$ (taking $a_o = 1$) and work out the exact answer as well as the Taylor expansion to three terms. The exact answer for any a is

$$C_a^{(2)}(2,2) = 1 - \frac{2}{(1+a)^2} - \frac{4a}{(1+a)^3} + \frac{1}{(1+2a)^2} + \frac{4a}{(1+2a)^3} + \frac{6a^2}{(1+2a)^4} \tag{3.8}$$

$$= \frac{a^4(30 + 90a + 80a^2 + 16a^3)}{(1 + 2a)^4(1 + a)^3} ,$$

and this yields .291739 for $a = .9$. The Taylor expansion gives

$$C^{(2)}_{.9}(2,2) \approx \frac{1}{3} + (.9 - 1)2(.2021605) + \frac{(.9-1)^2}{2} 6(-.0399520) = .291703,$$

so that we obtained 4-decimal place accuracy. The latter calculation can be arranged in a standard finite-difference manner so that additional terms can be obtained with a minimal amount of calculation. Thus one extra term gives .291703 + .000029 = .291732, etc. Note that the exact result in (3.8) is a sum of rational functions of a and hence the derivatives (and the differences) will not vanish after a finite number of terms, as did happen in the case of the I and J functions in [18].

An alternative expansion to the above Taylor expansion is more useful when a is close to zero. We derive this by probabilistic methods and note that the same result can also be obtained by a Taylor expansion of $(1 + ab)^{-m} C_a^{(b)}(r,m)$ in powers of $a/(1 + ab)$; the details are omitted.

Clearly the probability that each of the blue cells has frequency at least r at stopping time (when the counting cell first reaches m) is a sum of the disjoint probabilities that the blue cells have a total of $br + \alpha$ and each has at least r $(\alpha = 0,1,\ldots)$. If we write this out for $\alpha = 0$, 1, 2, and 3 and take out common factors, the result is easily seen to be

$$(3.9)\quad C_a^{(b)}(r,m) = \frac{\Gamma(m + br)}{\Gamma(m)(r!)^b} \left(\frac{a}{1+ab}\right)^{rb} \left(\frac{1}{1+ab}\right)^m \left\{1+b \frac{(m+br)}{r+1} \left(\frac{a}{1+ab}\right)\right.$$

$$+ b\left[\frac{b-1}{2(r+1)} + \frac{1}{r+2}\right] \frac{(m+br)^{[2]}}{r+1} \left(\frac{a}{1+ab}\right)^2$$

$$\left. + b\left[\frac{(b-1)(b-2)}{6(r+1)^2} + \frac{b-1}{(r+1)(r+2)} + \frac{1}{(r+2)(r+3)}\right] \frac{(m+br)^{[3]}}{r+1} \left(\frac{a}{1+ab}\right)^3 + \ldots\right\},$$

where $x^{[a]} = x(x+1) \ldots (x+a-1)$.

To illustrate the fact that the Taylor expansion (3.7) is poor, i.e., slowly converging around $a = 0$, we consider the exact formula (3.8) and the new approximation (3.9) for $b = m = r = 2$ at $a = .1$; the latter gives the result

$$(3.10)\quad C^{(2)}_{.1}(2,2) \approx 30\left(\frac{1}{12}\right)^4\left(\frac{10}{12}\right)^2\left[1+4\left(\frac{1}{12}\right)+\frac{35}{3}\left(\frac{1}{12}\right)^2+\frac{7}{15}\left(\frac{1}{3}\right)^3\right] = .00143;$$

the correct answer to 8 decimals obtained from (3.8) is .00144263 . If we tried to compute this value by using the Taylor expansion (3.7) the leading term would be $30a^4 = .003$ at $a = .1$ and the second term is $-240a^5 = -.0024$. Thus the first two terms give .0006, which is quite poor.

It is worth noting that the Taylor expansion given by (3.7) is indeterminate for $a_0 = 0$ but the C-function has the factor a^{br} so that the jth derivative is zero for $j < br$ and hence the indeterminacy vanishes.

The analogue of (3.9) for $D^{(b)}_a(r,m)$ is again obtained by probability considerations; here again we assume that a is close to zero. The probability in this case is a sum of discrete probabilities that the blue cells have a total frequency α and each has frequency less than r $(\alpha = 0,1,\ldots,b(r-1))$. Hence we obtain for $r \geq 4$ the first four terms

$$(3.11)\quad D^{(b)}_a(r,m) \approx$$

$$\left(\frac{1}{1+ab}\right)^m\left[1+\frac{a}{1+ab}\left\{b\binom{m}{1}\right\}+\left(\frac{a}{1+ab}\right)^2\left\{\binom{b}{2}\begin{bmatrix}m+1\\1,1,m-1\end{bmatrix}+b\begin{bmatrix}m+1\\2,m-1\end{bmatrix}\right\}\right.$$

$$\left.+\left(\frac{a}{1+ab}\right)^3\left\{\binom{b}{3}\begin{bmatrix}m+2\\1,1,1,m-1\end{bmatrix}+2\binom{b}{2}\begin{bmatrix}m+2\\1,2,m-1\end{bmatrix}+\binom{b}{1}\begin{bmatrix}m+2\\3,m-1\end{bmatrix}+\ldots\right\}\right.$$

and for $r \leq 3$ we simply omit those terms which have a square bracket containing a subscript $\geq r$.

To illustrate this, we again take $b = m = r = 2$ and $a = .1$. Then (3.11) gives $\left(\frac{5}{6}\right)^2\left(1+\frac{1}{3}+\frac{1}{24}\right) = 275/384 = .95486111\ldots$, which turns out to be exact in this case, since for $b = r = 2$ only the first 3 terms are non-zero.

The result (3.11) shows that $a^m D^{(b)}_a(r,m)$ is a polynomial in the variable $u = a/(1+ab)$; this is also true for $(1+ab)^{-m}D^{(b)}_a(r,m)$ as a function of u.

It follows that $D_a^{(b)}(r,m)$ is a finite polynomial in u and indeed the degree is easily seen to be at most $b(r-1) + m$. Hence the derivatives after this with respect to u must vanish.

This fact could be utilized in writing a Taylor expansion with respect to u, as was done for the I-function in [18].

Another useful expansion which can also be obtained by a probabilistic argument will now be derived. This formula enables one to obtain the C-function from a table of the I_p-function for $p = 1/b$ using negative binomial coefficients. Combining all the blue cells and regarding them as a single blue cell with frequency x, the probability of the event associated with C (and in another equation D) can easily be written in the form

$$(3.12a) \qquad C_a^{(b,j)}(r,m) = \sum_{n=br}^{\infty} C_{ab}^{(1,1)}(x,m) I_{1/b}^{(b,j)}(r,x)$$

$$(3.12b) \qquad D_a^{(b,j)}(r,m) = \sum_{x=0}^{b(r-1)+j} D_{ab}^{(1,1)}(x,m) J_{1/b}^{(b,j)}(r,x),$$

where J is a dual function to I defined in [18] and

$$(3.13) \qquad C_{ab}^{(1,1)}(x,m) = D_{ab}^{(1,1)}(x,m) = \frac{\Gamma(m+x)}{\Gamma(m)x!} \left(\frac{ab}{1+ab}\right)^x \left(\frac{1}{1+ab}\right)^m .$$

3.3 Asymptotic Expansions and approximations.

Regardless of how much detail is put into our table, there will always be a need for results that are outside the limits of the table. Hence there is a definite need for asymptotic expansions in connection with these tables. These expansions enable us to judge how far to calculate exact tables, i.e., where to stop and replace the exact results by asymptotic approximations.

3.3.1 Poisson Type Expansion $(m \to \infty,\ a \to 0,\ ma = \lambda)$.

For small values of a the Taylor expansion is not very useful, although some improvement was achieved in (3.9) and (3.11) by changing the variable a to $a/(1 + ba)$. For small a and large m a more useful approximation is one of Poisson Type in which we assume that $ma = \lambda$ is moderate in size. A definite application of this type of limiting process is developed below in Section 5.2.1. A straightforward generalization for the case of vector arguments will be briefly considered later in Section 4.6.

Starting with the single integral formula in (2.3.3) for $j = 0$, we introduce the standardized variable $y = (x - m)/\sqrt{m}$ to obtain for $\lambda = ma$

$$C_a^{(b)}(r,m) = \int_{-\sqrt{m}}^{\infty} G_r^b\left(\lambda + \frac{y\lambda}{\sqrt{m}}\right) dG_m(m + y\sqrt{m}) . \tag{3.14}$$

Using a Taylor expansion with $h = \dfrac{y\lambda}{\sqrt{m}}$, we have for the integrand

$$\begin{aligned} G_r^b(\lambda + h) = G_r^b(\lambda) &+ hbG_r^{b-1}(\lambda)g_r(\lambda) + \frac{h^2}{2}\Big\{b(b-1)G_r^{b-2}(\lambda)g_r^2(\lambda) \\ &+ bG_r^{b-1}(\lambda)g_{r-1}(\lambda)\left(\frac{r-1-\lambda}{r-1}\right)\Big\} + \mathcal{O}\left(\frac{1}{m^{3/2}}\right) \end{aligned} \tag{3.15}$$

where $g_r(\lambda)$ is the gamma density and $G_r(\lambda)$ is its cdf. Using an Edgeworth expansion, we have for the differential in (3.14)

$$\begin{aligned} &dG_m(m + y\sqrt{m}) \\ &= \varphi(y) - \frac{1}{3\sqrt{m}}\mathcal{H}_3(y)\varphi(y) + \frac{\varphi(y)}{36m}\{2\mathcal{H}_4(y) + 9\mathcal{H}_6(y)\} + \mathcal{O}\left(\frac{1}{m^{3/2}}\right), \end{aligned} \tag{3.16}$$

where $\varphi(y)$ is the standard normal density and $\mathcal{H}_i(y)$ are the Hermite polynomials associated with it.

We note that $\int_{-\infty}^{m} dG_m(m+y\sqrt{m})$ goes to zero exponentially fast. Putting (3.15) and (3.16) into (3.14) we obtain after integration

(3.17a)

$$C_a^{(b)}(r,m) = G_r^b(\lambda) + \frac{\lambda^2 b}{2m} G_r^{b-2}(\lambda) \left\{(b-1)g_r^2(\lambda) + \left(\frac{r-1-\lambda}{r-1}\right) G_r(\lambda)g_{r-1}(\lambda)\right\} + \mathcal{O}\left(\frac{1}{m^{3/2}}\right) .$$

Similarly for the D integral we obtain

(3.17b)

$$D_a^{(b)}(r,m) = [1 - G_r(\lambda)]^b$$

$$+ \frac{\lambda^2 b}{2m} [1-G_r(\lambda)]^{b-2} \left\{(b-1)g_r^2(\lambda) - \left(\frac{r-1-\lambda}{r-1}\right) \left[1-G_r(\lambda)\right] g_{r-1}(\lambda)\right\} + \mathcal{O}\left(\frac{1}{m^{3/2}}\right)$$

Numerical Illustrations

For $r=2$, $m=15$, $a = .1$ and $b = 3$ we obtain from (3.17a)

(3.18) $$C_{.1}^{(3)}(2,15) \approx G_2^3(1.5) + .45e^{-1.5}G_2(1.5) \left\{4.5e^{-1.5} + \frac{1}{2} G_2(1.5)\right\}$$

$$= .086 + .027 = .113$$

and the exact answer is .10185, which is not very close because m and 1/a are not sufficiently large. If we compare this result with the 3-term approximation given in (3.9) we find that the latter gives only .077, which is not as good. Thus the Taylor expansion is not useful for small values of a.

To show that we can get better results we consider one illustration with the D integral with a larger m-value and a smaller a-value. For $b=2$, $r=3$, $m=20$ and $a = 1/18$ we obtain from (3.17b)

(3.19) $$D_{1/18}^{(2)}(3,20) \approx$$

$$\left[1-G_3\left(\frac{10}{9}\right)\right]^2 + \frac{1}{20}\left(\frac{10}{9}\right)^2 \left\{\left(\frac{10}{9}\right)^4 \frac{e^{-20/9}}{4} - \frac{40}{81} e^{-10/9} \left[1-G_3\left(\frac{10}{9}\right)\right]\right\} ,$$

$$= .8067 - .0065 = .8002$$

which is correct to almost three decimal places; the exact answer is .800833... . It is worth noting that the leading term in (3.19) is already correct to almost two decimal places and the result is much better than in (3.18).

3.3.2 Asymptotics for m and r both Large

The results in (3.17) hold for large m and assume that r is moderate; we now assume that m and r are both large, although not necessarily equal, and that we have a common a that is neither very small nor very large. In this situation the asymptotics for the special case $m = r$ was investigated in [4] and a correction term was added in [9] (cf p. 151). We derive the leading term for m not necessarily equal to r and quote the correction term in [9] for m=r.

Our starting point is again the single integral formula but we now use the Edgeworth approximation for both the integrand and the differential in (3.14). Making the substitution $w = \sqrt{\frac{2m-1}{2}} \ln(1 + \frac{y}{\sqrt{m}})$ as in (5.1) of [4] in both parts of (3.14), we obtain for the C integral

$$g_m^*(w) \approx \varphi(w) + \frac{\mu_3}{6\sigma^3}(w^3-3w)\varphi(w) + \ldots \tag{3.20}$$

$$G_r^*(z) \approx \Phi(z) - \frac{\mu_3}{6\sigma^3}(z^2-1)\varphi(z) + \ldots \tag{3.21}$$

where $z = w\sqrt{\frac{2r-1}{2m-1}} + d$ and $d = -\sqrt{\frac{2r-1}{2}}\,\ell n\,\frac{ma}{r}$ and

$$C_a^{(b)}(r,m) \approx \int_{-\infty}^{\infty} \left[G_r^*\left(w\sqrt{\frac{2r-1}{2m-1}} + d\right)\right]^b g_m^*(w)\,dw. \tag{3.22}$$

It should be pointed out we are depending on a limiting operation in which m and r are both tending to infinity and the quantity ma/r is such that d remains moderate. Thus for $m = r$ and both large we require that a is close to 1 in such a manner that $-\sqrt{r}\,\ell n\, a$ approaches a constant.

Substituting (3.20) and (3.21) into (3.22) and expanding with the exponent b, we obtain for the leading term

$$C_a^{(b)}(r,m) = \int_{-\infty}^{\infty} \Phi^b\left(w\sqrt{\frac{2r-1}{2m-1}} - d\right)d\Phi(w) + \mathcal{O}\left(\max\left\{\frac{1}{\sqrt{r}}, \frac{1}{\sqrt{m}}\right\}\right) \tag{3.23a}$$

and similarly

(3.23b)

$$D_a^{(b)}(r,m) = \int_{-\infty}^{\infty}\left[1-\Phi\left(w\sqrt{\frac{2r-1}{2m-1}} - d\right)\right]^b d\Phi(w) + \mathcal{O}\left(\max\left\{\frac{1}{\sqrt{r}}, \frac{1}{\sqrt{m}}\right\}\right).$$

A first order correction for r not necessarily equal to m can be obtained from the above as is indicated in (5.13) of [4] but the results are lengthy and will not be given here. For the case $r = m$ following the method in Section 5.1 of [4] we obtain

(3.24a)

$$C_a^{(b)}(r,r) = \int_{-\infty}^{\infty} \Phi^b(w-d)d\Phi(w) - \frac{db(b-1)}{6\sqrt{6\pi r}}\,\varphi\left(d\sqrt{\frac{2}{3}}\right)\int_{-\infty}^{\infty} \Phi^{b-2}\left(\frac{y}{\sqrt{3}} - \frac{d}{3}\right)d\Phi(y) + \mathcal{O}\left(\frac{1}{r}\right),$$

(3.24b)

$$D_a^{(b)}(r,r) = \int_{-\infty}^{\infty} [1 - \Phi(w-d)]^b d\Phi(w) - \frac{db(b-1)}{6\sqrt{6\pi r}}\,\varphi\left(d\sqrt{\frac{2}{3}}\right)\int_{-\infty}^{\infty} \Phi^{b-2}\left(\frac{y}{\sqrt{3}} - \frac{d}{3}\right)d\Phi(y) + \mathcal{O}\left(\frac{1}{r}\right).$$

It is important to note that the accuracy obtained by (3.24a) and (3.24b) is much better than is indicated by the symbol $\mathcal{O}\left(\frac{1}{r}\right)$ and we give a typical example below to illustrate this. The reason is that, although d will generally be moderate (say, about 2 or 3) for the limiting operation, the value of $\varphi\left(d\sqrt{\frac{2}{3}}\right)$ is quite small and this makes the correction term small.

Numerical Illustration:

Suppose $b = 2$, $a = 4/3$ and $m = r = 100$. Then

$$(3.25a) \qquad C^{(2)}_{4/3}(100,100) \approx \int_{-\infty}^{\infty} \Phi^2\left(w + \sqrt{99.5}\ \ell n\ \frac{4}{3}\right)d\Phi(w) + \frac{2(2.8696)}{60\sqrt{6\pi}}\ \varphi\left(d\sqrt{\frac{2}{3}}\right)$$

$$= .96125 + .00056 = .96181$$

as opposed to the (5 decimal-place) correct answer .96175; thus we have almost 4 decimal accuracy, while $\mathcal{O}(\frac{1}{r})$ would indicate only 2 decimal accuracy. To illustrate (3.24b) we take $b = 2$, $a = 3/4$ and $m = r = 100$. Then

$$(3.25b) \qquad D^{(2)}_{3/4}(100,100) \approx \int_{-\infty}^{\infty} \left[1-\Phi\left(w+\sqrt{99.5}\ \ell n\ \frac{3}{4}\right)\right]^2 d\Phi(w) - \frac{2(2.8696)}{60\sqrt{6\pi}}\ \varphi\left(d\sqrt{\frac{2}{3}}\right)$$

$$= \int_{-\infty}^{\infty} \Phi^2\left(w + \sqrt{99.5}\ \ell n\ \frac{4}{3}\right)d\Phi(w) - \frac{2(2.8696)}{60\sqrt{6\pi}}\ \varphi\left(d\sqrt{\frac{2}{3}}\right)$$

$$= .96125 - .00056 = .96069$$

as opposed to the (5 decimal-place) correct answer .96062.

As the above illustration shows if we compare $C^{(b)}_{a}(r,r)$ with $D^{(b)}_{1/a}(r,r)$, they approach the same value, namely $\int_{-\infty}^{\infty}\Phi^2(w-d)d\Phi(w)$ as $r \to \infty$ in the limiting operation described above. The exact equality (i.e., identity) holds for $b = 1$ and this is a special case of (2.8).

An alternative method of handling the correction term for the D integral which is much simpler for 'hand calculations' is to use only the leading term (3.24a) and (3.24b) with the quantity d replaced in each case by d' defined by

$$(3.26) \qquad d' = -\sqrt{\frac{2r-1-2\ \ell n\ b}{2}}\ .$$

For the last numerical illustration this gives $d' = 2.8596$ and $D^{(2)}_{3/4}(100,100) \approx .96060$, which is actually closer than the previous result with the Edgeworth correction term.

In all of the above asymptotics we have used the normal ranking and selection integral which was calculated independently by S.S. Gupta [10] and by Roy Milton [15].

4. Generalizations of C and D to Vector Arguments; $\underset{\sim}{a}$, $\underset{\sim}{r}$.

Several methods of generalization are of interest; we restrict our attention to those that are more closely related to C and D.

4.1 Definition and Probability Interpretation

The generalization of most interest to us here is the one that allows us to replace the common r for the b blue cells by a vector $\underset{\sim}{r} = (r_1, \ldots, r_b)$ and/or the common p for the b blue cells by a vector $p = (p_1, \ldots, p_b)$, so that r_i and/or p_i correspond to the ith cell. The need for this generalization already arises in several of the applications discussed in Chapter 5 below and in particular in the 'sharing problem' of section 5.3. The interesting feature of this generalization is that the same probability interpretation holds, namely if the b blue cells have quotas $r_1, \ldots r_b$ and cell probabilities $p_1, \ldots, p_b$, then the probability that the frequency $f_i \geq r_i$ $(i = 1,2,\ldots,b)$ when the counting cell with cell probability $p_0 = 1 - \sum_{i=1}^{b} p_i$ reaches $f_0 = m$ is given by $C_{\underset{\sim}{a}}^{(b)}(\underset{\sim}{r},m)$ where $\underset{\sim}{a} = \left(\frac{p_1}{p_0},\ldots,\frac{p_b}{p_0}\right) = \frac{1}{p_0}\underset{\sim}{p}$. Furthermore if we want $f_\alpha = r_\alpha (\alpha = 1,2,\ldots,j)$ for a specified set of j cells and $f_\beta \geq r_\beta (\beta = j+1,\ldots,b)$ at the first instance when $f_0 = m$, then the probability of this event is

(4.1a)
$$C_{\underset{\sim}{a}}^{(b,j)}(\underset{\sim}{r},m) = \frac{\Gamma(m+R)\prod_{1}^{j}\left(\frac{a_i^{r_i}}{r_i}\right)}{\Gamma(m)\prod_{i=1}^{b}\Gamma(r_i)} \int_0^{a_{j+1}} \cdots \int_0^{a_b} \frac{\prod_{j+1}^{b} x_i^{r_i-1}\,dx_i}{\left(1 + \sum_{1}^{j} a_i + \sum_{j+1}^{b} x_i\right)^{m+R}},$$

where $R = r_1 + \ldots + r_b$. Similarly we have the corresponding results for the D integral with $j \geq 0$ if we replace $f_i \geq r_i$ above by $f_i < r_i$, where

(4.1b)
$$D_{\underset{\sim}{a}}^{(b,j)}(\underset{\sim}{r},m) = \frac{\Gamma(m+R)\prod_{1}^{j}\left(\frac{a_i^{r_i}}{r_i}\right)}{\Gamma(m)\prod_{1}^{b}\Gamma(r_i)} \int_{a_{j+1}}^{\infty}\cdots\int_{a_b}^{\infty} \frac{\prod_{j+1}^{b} x_i^{r_i-1}\,dx_i}{\left(1+\sum_{1}^{j}a_i+\sum_{j+1}^{b}x_i\right)^{m+R}} .$$

We shall first prove (4.1a) since the proof of (4.1b) is similar. We utilize the probability interpretation of $I_{\underset{\sim}{p}}^{(b,j)}(\underset{\sim}{r},n)$, introduced in [19], namely that in n observations j specified cells (to which we assign cell probabilities $p_1,\ldots,p_j$) have exactly r_i observations $(i = 1,2,\ldots,j)$ and the remaining $b-j$ blue cells (with cell probabilities $p_{j+1},\ldots,p_b$) have at least r_i observations $(i = j+1,\ldots,b)$. Hence the probability of the same event happening when a counting cell (with cell probability $p_0 = 1 - \sum_{1}^{b} p_i$) has for the first time exactly m observations is given by

(4.2)
$$\sum_{n=m+R}^{\infty} p_0^m \binom{n-1}{m-1}(1-p_0)^{n-m} \quad I^{(b,j)}_{\left(\frac{p_1}{1-p_0},\ \ldots,\ \frac{p_b}{1-p_0}\right)}(\underset{\sim}{r},n-m) .$$

Putting in the integral expression for I (as defined in [19]) and summing the negative binomial under the integral sign we obtain after one or two obvious transformations the result (4.1a); this proves the probability interpretation.

4.2 Recurrence Relations

The method of deriving these relations is quite similar to those used to develop (2.24) and (2.25) above. Let $\underset{\sim}{r}_{(\alpha;j)} = (r_1,\ldots,r_j,r_\alpha,r_{j+1},\ldots, r_{\alpha-1},r_{\alpha+1},\ldots,r_b)$ and let $\underset{\sim}{a}_{(\alpha;j)}$ denote the corresponding $\underset{\sim}{a}$ vector with the same permutation. Then the recurrence relations are

(4.3a)

$$\left(m+\sum_{1}^{j} a_i\right) C_{\underset{\sim}{a}}^{(b,j)}(\underset{\sim}{r},m+1) = \left(m+\sum_{1}^{j} r_i\right) C_{\underset{\sim}{a}}^{(b,j)}(\underset{\sim}{r},m) + \sum_{\alpha=j+1}^{b} r_\alpha\, C_{\underset{\sim}{a}(\alpha;j)}^{(b,j+1)}(\underset{\sim}{r}(\alpha;j),m),$$

(4.3b)

$$\left(m+\sum_{1}^{j} a_i\right) D_{\underset{\sim}{a}}^{(b,j)}(\underset{\sim}{r},m+1) = \left(m+\sum_{1}^{j} r_i\right) D_{\underset{\sim}{a}}^{(b,j)}(\underset{\sim}{r},m) - \sum_{\alpha=j+1}^{b} r_\alpha\, D_{\underset{\sim}{a}(\alpha;j)}^{(b,j+1)}(\underset{\sim}{r}(\alpha;j),m),$$

The first boundary condition for (4.3a) and (4.3b) for $j = b$ is

$$(4.4)\qquad C_{\underset{\sim}{a}}^{(b,b)}(\underset{\sim}{r},m) = \left[\begin{matrix} m-1+R \\ m-1,r_1,\ldots,r_b\end{matrix}\right]\left(\frac{1}{1+A}\right)^m \prod_{i=1}^{b}\left(\frac{a_i}{1+A}\right)^{r_i} = D_{\underset{\sim}{a}}^{(b,b)}(\underset{\sim}{r},m),$$

where $A = a_1 + \ldots + a_b$.

For the second boundary condition we select any one component i with $j+1 \le i \le b$ for which $m > r_i$ (cf. remark below) and reorder the components in both $\underset{\sim}{a}$ and $\underset{\sim}{r}$ so that i is in the $j+1$ position and $j+1,\ldots,i-1$ are all shifted by one to the right. This permutation is utilized for convenience so that our notation will not get too involved. The superscripts (b,j) in our definition, e.g., (4.1), separates the first j components from the last $b-j$ and treats them differently; the superscripts $(b,j+1)$ e.g., in (4.3) and also below, indicate that the $j+1^{st}$ component is taken from the second set and added to the first set. After relabeling the components, the second boundary conditions for (4.3a) and (4.3b) for $m > r_{j+1}$ are

$$(4.5a)\qquad C_{\underset{\sim}{a}}^{(b,j)}(\underset{\sim}{r},m) = C_{\underset{\sim}{a}_{j+1}}^{(b-1,j)}(\underset{\sim}{r}_{j+1},m) - \frac{1}{\binom{m-1}{r_{j+1}}}\sum_{\alpha=1}^{r_j+1}\frac{\binom{m-\alpha-1}{r_{j+1}-\alpha}}{a_{j+1}^{\alpha}} C_{\underset{\sim}{a}}^{(b,j+1)}(\underset{\sim}{r},m-\alpha),$$

$$(4.5b)\qquad D_{\underset{\sim}{a}}^{(b,j)}(\underset{\sim}{r},m) = \frac{1}{\binom{m-1}{r_{j+1}}}\sum_{\alpha=1}^{r_{j+1}}\frac{\binom{m-\alpha-1}{r_{j+1}-\alpha}}{a_{j+1}^{\alpha}} D_{\underset{\sim}{a}}^{(b,j+1)}(\underset{\sim}{r},m-\alpha),$$

where $\underset{\sim}{a}_i = (a_1,\ldots,a_j;\ a_{j+1},\ldots,a_{i-1},\ a_{i+1},\ldots,a_b)$ without a_i, and $\underset{\sim}{r}_i$ is defined similarly for the first term in (4.5a). An alternative second boundary condition in place of (4.5b) which does not require $m > r_{j+1}$ is as follows: For $0 \leqq j < b$

$$D_{\underset{\sim}{a}}^{(b,j)}(\underset{\sim}{r},m) = \sum_{\alpha=0}^{r_{j+1}-1} D_{\underset{\sim}{a}}^{(b,j+1)}(\underset{\sim}{r}^{(\alpha)},m), \tag{4.5c}$$

where $\underset{\sim}{r}^{(\alpha)}$ is the same as $\underset{\sim}{r}$ except that α is in the $j{+}1^{st}$ position instead of r_{j+1}. This result follows directly from the probability interpretation of D; the corresponding result for C entails an infinite series and is also correct but less useful (and hence omitted).

4.3 Recurrence Relations for Unequal a-Values and Unequal r-Values

For the general case with vectors $\underset{\sim}{a}$ and $\underset{\sim}{r}$ we introduce preliminary recurrence relations which will help to equalize the components of $\underset{\sim}{r}$. Once we obtain equal components of $\underset{\sim}{r}$ (or reduce the number of integrals to one), we then use our tables if the a-values are equal (or there is only one) and otherwise we employ a Taylor series expansion to evaluate the integral. If we end up by this method with a C or D having $j > 0$ then we employ a reduction formula to reduce the j-value to zero so that we can again use our tables.

A simple integration by parts on the variable x_{j+1} in both C and D gives the results for any $r_{j+1} \geq 1$

$$C_{\underset{\sim}{a}}^{(b,j)}(\underset{\sim}{r},m) = C_{\underset{\sim}{a}}^{(b,j)}(\underset{\sim}{r}^{(j+1)},m) + C_{\underset{\sim}{a}}^{(b,j+1)}(\underset{\sim}{r},m), \tag{4.6a}$$

$$D_{\underset{\sim}{a}}^{(b,j)}(\underset{\sim}{r},m) = D_{\underset{\sim}{a}}^{(b,j)}(\underset{\sim}{r}^{(j+1)},m) - D_{\underset{\sim}{a}}^{(b,j+1)}(\underset{\sim}{r},m) \tag{4.6b}$$

where $\underset{\sim}{r}^{(j+1)} = (r_1,\ldots,r_j,r_{j+1}+1,\ r_{j+2},\ldots,r_b)$, i.e., we add one to the $(j+1)^{st}$ component of $\underset{\sim}{r}$, but leave $\underset{\sim}{a}$ unchanged.

Another integration by parts (in the reverse direction) gives slightly different results which are also useful but not quite as simple, namely for $r_{j+1} > 1$ and $m > 1$

(4.7a)

$$C_a^{(b,j)}(r,m) = \begin{cases} C_{\underset{\sim}{a}(j+1)}^{(b-1,j)}(\underset{\sim}{r}_{(j+1)},m) - \dfrac{1}{a_{j+1}(m-1)} C_{\underset{\sim}{a}}^{(b,j+1)}(\underset{\sim}{r},m-1) & \text{for } r_{j+1} = 1 \\ C_{\underset{\sim}{a}}^{(b,j)}(\underset{\sim}{r}_{(j+1)},m) - \dfrac{r_{j+1}}{a_{j+1}(m-1)} C_{\underset{\sim}{a}}^{(b,j+1)}(\underset{\sim}{r},m-1) & \text{for } r_{j+1} > 1 \end{cases}$$

(4.7b)

$$D_{\underset{\sim}{a}}^{(b,j)}(\underset{\sim}{r},m) = D_{\underset{\sim}{a}}^{(b,j)}(\underset{\sim}{r}_{(j+1)},m) + \frac{r_{j+1}}{a_{j+1}(m-1)} D_{\underset{\sim}{a}}^{(b,j+1)}(\underset{\sim}{r},m-1) \quad \text{for all } r_{j+1} \geq 1,$$

where $\underset{\sim}{r}_{(j+1)} = (r_1,\ldots,r_j,r_{j+1}-1,r_{j+2},\ldots,r_b)$, $\underset{\sim}{r}_{(j+1)}$ is $\underset{\sim}{r}$ with the $(j+1)^{st}$ component deleted and $\underset{\sim}{a}_{(j+1)}$ indicates the same operation on $\underset{\sim}{a}$. It should be noted that (4.6) is simpler than (4.7) and has no difficulty with $m = 1$.

The use of the above recurrence relations gives rise to C and D integrals with double superscript (b,j) and we now develop a reduction formula which replaces (b,j) by $(b-j,0) = (b-j)$. The straighforward transformation $y_i = x_i/(1 + \underline{A}_j)(i = j+1,\ldots,b)$ where $\underline{A}_j = a_1 + a_2 + \ldots + a_j$ leads to the result

$$(4.8) \qquad C_{\underset{\sim}{a}}^{(b,j)}(\underset{\sim}{r},m) = \begin{bmatrix} m-1 + \underline{R}_j \\ m-1,r_1,\ldots,r_j \end{bmatrix} \left(\frac{1}{1+\underline{A}_j}\right)^m \prod_{i=1}^{j} \left(\frac{a_i}{1+\underline{A}_j}\right)^{r_i} C_{\frac{\underset{\approx}{a}}{1+\underline{A}_j}}^{(b-j)}(\underset{\approx}{r},m+\underline{R}_j),$$

where $\underline{R}_j = r_1 + r_2 + \ldots + r_j$, and $\underset{\approx}{r} = (r_{j+1},\ldots,r_b)$ and $\underset{\approx}{a}$ is defined similarly. Exactly the same formula holds for $D_{\underset{\sim}{a}}^{(b,j)}(\underset{\sim}{r},m)$ with C replaced by D on both sides of (4.8).

To illustrate the plan of using the above formulas we consider a case of unequal r-values, first with equal a-values and later (after developing a Taylor Series expansion) with unequal a-values. Suppose we want to calculate $D_1^{(2)}(4,2;3)$ with common $a = 1$; using (4.6b) twice we have

(4.9)

$$D_1^{(2)}(4,2;3) = D_1^{(2)}(3;2;3) + D_1^{(2,1)}(3;2;3) = D_1^{(2)}(2,2;3) + D_1^{(2,1)}(2,2;3) + D_1^{(2,1)}(3,2;3)$$

$$= D_1^{(2)}(2,3) + \binom{4}{2}\left(\frac{1}{2}\right)^5 D_{1/2}^{(1)}(2,5) + \binom{5}{2}\left(\frac{1}{2}\right)^6 D_{1/2}^{(1)}(2,6)$$

$$= .16049383 + .1875(.35116598) + (.15625)(.26337449)$$

$$= .26748971,$$

which also can be expressed exactly as the fraction 195/729.

4.4 Taylor Series Expansion for Unequal a-Values

Our plan is to first equalize r-values and then use a Taylor expansion in order to cut down the labor involved. However we now write the Taylor expansion with vectors $\underset{\sim}{r}$ since we can do this without any extra difficulty. Our principal interest is in starting with the case $j = 0$. In order to maximize the rate of convergence of the Taylor expansion we choose to expand $C_{\underset{\sim}{a}}^{(b)}(\underset{\sim}{r},m)$ or $D_{\underset{\sim}{a}}^{(b)}(\underset{\sim}{r},m)$ about the weighted average $\bar{\bar{a}} = \sum_{\alpha=1}^{b} a_\alpha r_\alpha / \sum_{\alpha=1}^{b} r_\alpha$, which of course reduces to the usual average $\bar{a}$ when the r-values are all equal. The first derivatives have the combined coefficient $\Sigma(a_\alpha - \bar{\bar{a}}) = 0$ and hence vanish, so that we only need to calculate higher derivatives. The results needed are

$$\text{(4.10)} \qquad \frac{\partial}{\partial a_\alpha} D_{\underset{\sim}{a}}^{(b,j)}(\underset{\sim}{r},m) = -\frac{r_\alpha}{a_\alpha} D_{\underset{\sim}{a}}^{(b,j+1)}(\underset{\sim}{r},m) \qquad (j+1 \le \alpha \le b),$$

$$\text{(4.11)} \qquad \frac{\partial}{\partial a_\alpha} D_{\underset{\sim}{a}}^{(b,j)}(\underset{\sim}{r},m) = \frac{r_\alpha}{a_\alpha} D_{\underset{\sim}{a}}^{(b,j)}(\underset{\sim}{r},m) - m\, D_{\underset{\sim}{a}}^{(b,j)}(\underset{\sim}{r},m+1) \qquad (1 \le \alpha \le j).$$

The second partial derivatives are given only for $j+1 \le \alpha,\beta \le b$ since we will use them only for $j = 0$; these are

(4.12) $$\frac{\partial^2}{\partial a_\alpha^2} D_{\underset{\sim}{a}}^{(b,j)}(\underset{\sim}{r},m) = \frac{mr_\alpha}{a_\alpha} D_{\underset{\sim}{a}}^{(b,j+1)}(\underset{\sim}{r},m+1) - \frac{r_\alpha(r_\alpha-1)}{a_\alpha^2} D_{\underset{\sim}{a}}^{(b,j+1)}(\underset{\sim}{r},m) ,$$

(4.13) $$\frac{\partial^2}{\partial a_\alpha \partial a_\beta} D_{\underset{\sim}{a}}^{(b,j)}(\underset{\sim}{r},m) = \frac{r_\alpha r_\beta}{a_\alpha a_\beta} D_{\underset{\sim}{a}}^{(b,j+2)}(\underset{\sim}{r},m) \quad \text{for } \alpha \neq \beta .$$

<u>Remark:</u> In the differentiation formulas of (4.10) through (4.12) it should be understood that the $j+1^{st}$ component is replaced by a_α in $\underset{\sim}{a}$ and by r_α in $\underset{\sim}{r}$ without altering the order of the remaining components. In addition, the $j+2^d$ component in (4.13) is replaced by a_β in $\underset{\sim}{a}$ and by r_β in $\underset{\sim}{r}$ without altering the order of the remaining components.

Expanding about $\bar{\bar{a}} = \sum_{\alpha=1}^{b} r_\alpha a_\alpha / \sum_{1}^{b} r_\alpha$, we now obtain for $j = 0$ the Taylor expansion which is good to $\mathcal{O}((a_\alpha - \bar{\bar{a}})^3)$,

(4.14a)

$$D_{\underset{\sim}{a}}^{(b)}(\underset{\sim}{r},m) \simeq D_{\bar{\bar{a}}}^{(b)}(\underset{\sim}{r},m) + \frac{mW_1}{2\bar{\bar{a}}} D_{\bar{\bar{a}}}^{(b,1)}(\underset{\sim}{r},m+1) - \frac{W_2}{2(\bar{\bar{a}})^2} D_{\bar{\bar{a}}}^{(b,1)}(\underset{\sim}{r},m) + \frac{W_3}{(\bar{\bar{a}})^2} D_{\bar{\bar{a}}}^{(b,2)}(\underset{\sim}{r},m),$$

where

(4.15)

$$W_1 = \sum_{\alpha=1}^{b} r_\alpha (a_\alpha - \bar{\bar{a}})^2, \quad W_2 = \sum_{\alpha=1}^{b} r_\alpha (r_\alpha - 1)(a_\alpha - \bar{\bar{a}})^2, \quad W_3 = \sum_{\alpha<\beta} r_\alpha r_\beta (a_\alpha - \bar{\bar{a}})(a_\beta - \bar{\bar{a}}).$$

In a similar manner we can also write a Taylor expansion for the C-integral; since they are similar we need only point out that (4.11) and (4.13) are exactly the same (with C replacing D, of course) and on the right side of (4.10) and (4.12) we interchange all + and - signs. Using the same definitions of W_1, W_2 and W_3, the Taylor expansion up to $\mathcal{O}\{(a_\alpha - \bar{\bar{a}})^3\}$ for C is

(4.14b)

$$C_{\underset{\sim}{a}}^{(b)}(\underset{\sim}{r},m) \approx C_{\bar{\bar{a}}}^{(b)}(\underset{\sim}{r},m) - \frac{mW_1}{2\bar{\bar{a}}} C_{\bar{\bar{a}}}^{(b,1)}(\underset{\sim}{r},m+1) + \frac{W_2}{2(\bar{\bar{a}})^2} C_{\bar{\bar{a}}}^{(b,1)}(\underset{\sim}{r},m) + \frac{W_3}{(\bar{\bar{a}})^2} C_{\bar{\bar{a}}}^{(b,2)}(\underset{\sim}{r},m).$$

Numerical Illustration:

Consider the same case as in (4.9) with $\underset{\sim}{r} = (4,2)$, $b = 2$, and $\underset{\sim}{a} = (.8,1.2)$. As in the first line of (4.9) (except that the subscript 1 is everywhere replaced by $\underset{\sim}{a}$), we reduce this to the sum of three D integrals. Since 2 of these are single integrals that are available exactly from our tables we need only consider the term $D_{\underset{\sim}{a}}^{(2)}(2,2;3)$. We evaluate the latter both exactly and by the Taylor expansion (4.14), obtaining for general $\underset{\approx}{a} = (a_1,a_2)$

$$(4.16) \qquad D_{\underset{\sim}{a}}^{(2)}(2,2;3) = \frac{12a_1a_2}{(1+a_1+a_2)^5} + \frac{3(a_1+a_2)}{(1+a_1+a_2)^4} + \frac{1}{(1+a_1+a_2)^3} = \frac{4.28}{27} = .1585185,$$

(where the last two numerical values are results for $a_1 = .8$ and $a_2 = 1.2$), and from (4.14)

(4.17)

$$D_{\underset{\sim}{a}}^{(2)}(2,2;3) \approx D_1^{(2)}(2;3) + \tfrac{3}{2}(.16)D_1^{(2,1)}(2;4) - \frac{.16}{2} D_1^{(2,1)}(2;3) - .16\, D_1^{(2,2)}(2;3)$$

$$= \frac{13}{3^4} + \frac{3}{2}(.16)\frac{10}{3^5} - \frac{.16}{2}\left(\frac{16}{3^5}\right) - .16\left(\frac{10}{3^5}\right) = \frac{4.28}{27} = .1585185 \quad .$$

In this case the numerators in (4.16) have a combined degree of at most 2 and the second derivatives were sufficient to give exact results; this exactness also depended on the fact that we expanded around an a-value where $a_1 + a_2 = 2\bar{\bar{a}}$. The first term 13/81 gave .160 which was good to 2 decimals; in fact the first term in the Taylor expansion actually gives an upper bound.

4.5 Definition and Properties of a CD-Integral.

In some applications we need a mixed integral in the sense that some of the integrals are from 0 to a while others are from a to ∞. In fact there are many interesting problems which require such integrals with a-values and r-values that are not necessarily equal. Nevertheless our treatment will stress the case of equal a-values and equal r-values; a more complete study of the general case can be found in the PhD Thesis of [14] of Albert H. Liu. The corresponding mixture of I and J type integrals does not appear in [18], but has been found to be very useful in a sparate paper [19] by one of the present authors and Albert H. Liu.

We define the CD-integral by

(4.18)

$$CD_{\underset{\sim}{a}}^{(c,b-j-c;j)}(\underset{\sim}{r},m) = \frac{\Gamma(m+R)}{\Gamma(m)\prod_{1}^{b}\Gamma(r_i)} \prod_{i=1}^{j}\left(\frac{a_i^{r_i}}{r_i}\right) \int_0^{a_{j+1}} \cdots \int_0^{a_{j+c}} \int_{a_{j+c+1}}^{\infty} \cdots \int_{a_b}^{\infty} \frac{\prod_{j+1}^{b} x_i^{r_i-1}\,dx_i}{\left(1+\underline{A}_j + \sum_{j+1}^{b} x_i\right)^{m+R}},$$

where $R = r_1+\ldots+r_b$, $\underline{A}_j = a_1+\ldots+a_j$ and c (where $0 \le c \le b-j$) denotes the number of integrals of type C. This CD-integral depends on the particular combination of components put into the 'second' set $j+1,\ldots,j+c$, i.e., we cannot interchange components between this second set and the third set $j+c+1,\ldots,b$ (even though the $\underset{\sim}{a}$ vector is changed accordingly) without changing the value of the integral. On the other hand we <u>can</u> interchange components within each of these two sets provided the corresponding changes are made in the $\underset{\sim}{a}$-vector. For $c = 0$ this CD-integral reduces to the D-integral and for $c = b-j$ it reduces to the C-integral. For $j = 0$ we simply drop the last superscript as in the case of C and D integrals.

The probability interpretation is again given for the multinomial setting with b blue cells having cell probabilities $a_i/(1+A)$ $(i = 1,2,\ldots,b)$ and one counting cell with probability $1/(1+A)$, where $A = a_1+\ldots+a_b$. The b blue cells have quotas $r_i (i = 1,2,\ldots,b)$ and we are interested in the probability of an event occuring when the counting cell reaches m for the first time. This event is that the frequencies $f_i = r_i$ for $i = 1,2,\ldots,j$ <u>and</u> that

$f_i \geq r_i$ for $i = j+1,\ldots,j+c$ and that $f_i < r_i$ for $i = j+c+1,\ldots,b$. Thus, if we regard m as the quota for the counting cell, we are interested in the probability that the counting cell is exactly the $c+1^{st}$ one to reach its quota, that for a specific set of j cells each has exactly its quota, that for a specific set of c cells each reached its quota and that the remaining b-j-c cells have not reached their quota. The proof of this probability interpretation is analogous to that of (4.1) in Section 4.1 and we omit it.

In Section 4.6 below we give some identities and illustrations of the numerical calculation. The CD-integral plays an important role in the multinomial tournament problem with k players.

4.6 Miscellaneous Identities and Other Results

From the probability interpretation it is easy to show an identity that arises in a multinomial tournament with k players. We introduce it here because 1) It represents a basic identity of some interest per se and 2) It is sometimes useful as an aid to numerical computations or as a check on numerical accuracy.

In the sharing model the first problem is to find the probability that each of the $k = b+1$ players will win the tournament if it were continued. The sum of these probabilities must sum to one even if there is a sink present, provided only that the sink (or draw) probability is less than one. Using the probability interpretation of the D function this can be written for $k = 3$ and for any k respectively as the identity in $p_1,p_2,p_3(p_1+p_2+p_3 \leq 1)$

$$D^{(2)}_{\left(\frac{p_1}{p_3},\frac{p_2}{p_3}\right)}(r_1,r_2;r_3) + D^{(2)}_{\left(\frac{p_2}{p_1},\frac{p_3}{p_1}\right)}(r_2,r_3;r_1) + D^{(2)}_{\left(\frac{p_3}{p_2},\frac{p_1}{p_2}\right)}(r_3,r_1;r_2) \equiv 1, \tag{4.19}$$

$$\sum_{\alpha=1}^{b+1} D^{(b)}_{\left(\frac{p_1}{p_\alpha},\ldots,\frac{p_{\alpha-1}}{p_\alpha},\frac{p_{\alpha+1}}{p_\alpha},\ldots,\frac{p_b}{p_\alpha}\right)}(r_1,\ldots,r_{\alpha-1},r_{\alpha+1},\ldots,r_b;r_\alpha) \equiv 1 \tag{4.20}$$

The same results hold for the C-integrals with a proof that is essentially the same. Here the αth term represents the probability that the αth player is the last one to reach his quota and these must therefore also sum to one. Both the C identity and the D identity in (4.20) can be regarded as special cases of a $CD^{(c,b-c)}$ identity for any given c; we identify k with b+1, m with r_k and p_k with $1 - \sum_{i=1}^{b} p_i$ and use $\underset{\sim}{r}_{\alpha,\beta}$ to denote $(r_{i_1},\ldots,r_{i_b})$ where $i_1,\ldots,i_b$ is a specific permutation of subscripts (indexed by β from the set $(1,2,\ldots,\alpha-1,\alpha+1,\ldots,k)$. Let $\underset{\sim}{a}_{\alpha,\beta}$ denote the corresponding vector $(p_{i_1}/p_\alpha,\ldots,p_{i_b}/p_\alpha)$. Then the generalized "row" identity states that for any c

$$\sum_{\alpha=1}^{k}\sum_{\beta=1}^{b!} CD_{\underset{\sim}{a}_{\alpha,\beta}}^{(c,b-c)}(\underset{\sim}{r}_{\alpha,\beta};r_\alpha) \equiv 1, \tag{4.21}$$

e.g., for b = 2 and c = 1 in the CD notation

$$\begin{aligned} CD^{(1,1)}_{\left(\frac{p_1}{p_3};\frac{p_2}{p_3}\right)}(r_1,r_2;r_3) + CD^{(1,1)}_{\left(\frac{p_2}{p_3};\frac{p_1}{p_3}\right)}(r_2,r_1;r_3) + CD^{(1,1)}_{\left(\frac{p_2}{p_1};\frac{p_3}{p_1}\right)}(r_2,r_3;r_1) \\ + CD^{(1,1)}_{\left(\frac{p_3}{p_1},\frac{p_2}{p_1}\right)}(r_3,r_2;r_1) + CD^{(1,1)}_{\left(\frac{p_3}{p_2},\frac{p_1}{p_2}\right)}(r_3,r_1;r_2) + CD^{(1,1)}_{\left(\frac{p_1}{p_2},\frac{p_3}{p_2}\right)}(r_1,r_3;r_2) \equiv 1. \end{aligned} \tag{4.22}$$

The proof is similar to the previous proof; the αth term $(\alpha = 1,2,\ldots,6)$ indicates that the order of the players reaching their quota is (1,3,2), (2,3,1), (2,1,3), (3,1,2), (3,2,1) and (1,2,3), respectively.

In addition to these "row" identities we also have column identities corresponding to the fact that a fixed player is the αth one to reach his quota $(\alpha = 1,2,\ldots k)$ and these probabilities must also sum to one. Let $\underset{\sim}{r}_{\alpha,\beta}$ denote $(r_{i_1},\ldots,r_{i_{\alpha-1}};r_{i_\alpha},\ldots,r_{i_b})$ where $(i_1,\ldots,i_{\alpha-1})$ is a specific combination of $\alpha-1$ subscripts from the set $(1,2,\ldots,c-1,c+1,\ldots,k)$ and $(i_\alpha,\ldots,i_b)$ are the remaining subscripts in that set. Let $\underset{\sim}{a}_{\alpha,\beta}$ denote the corresponding vector $\left(\frac{p_{i_1}}{p_c},\ldots,\frac{p_{i_{\alpha-1}}}{p_c};\ \frac{p_{i_\alpha}}{p_c},\ldots,\frac{p_{i_b}}{p_c}\right)$. then the "column" identity states that for any c

$$\sum_{\alpha=1}^{k}\ \sum_{\beta=1}^{\binom{b}{\alpha-1}} CD_{\underset{\sim}{a}_{\alpha,\beta}}^{(\alpha-1,b-\alpha+1)}(\underset{\sim}{r}_{\alpha,\beta};r_c) \equiv 1, \tag{4.23}$$

For any k this gives 2^b terms and for $k = 3$ and $c = 1$ (say) we obtain

(4.24)

$$D_{\frac{p_2}{p_1},\frac{p_3}{p_1}}^{(2)}(r_2,r_3;r_1) + \left\{CD_{\left(\frac{p_2}{p_1},\frac{p_3}{p_1}\right)}^{(1,1)}(r_2,r_3;r_1) + CD_{\left(\frac{p_3}{p_1},\frac{p_2}{p_1}\right)}^{(1,1)}(r_3,r_2;r_1)\right\} + C_{\left(\frac{p_2}{p_1},\frac{p_3}{p_1}\right)}^{(2)}(r_2,r_3;r_1) \equiv 1.$$

The principle of inclusion-exclusion was used in [18] to relate the I and J-functions and was given in (2.6) for equal a-values and equal r-values. We now generalize it for vectors and show that some versions of it can be useful for numerical calculation. For any vector $\underset{\sim}{r}$ and $\underset{\sim}{a}$ let $\underset{\sim}{r}_{\alpha,\beta}$ denote $(r_{i_1},\ldots,r_{i_\alpha};r_{i_{\alpha+1}},\ldots,r_{i_b})$ where $i_1,\ldots,i_\alpha$ is any specific combination of size α from the set $1,2,\ldots,b$ and $i_{\alpha+1},\ldots,i_b$ is the remaining set of size $b-\alpha$; let $\underset{\sim}{a}_{\alpha,\beta}$ denote the corresponding vector of a-values. Then we have by inclusion-exclusion (letting β index the subset)

$$C_{\underset{\sim}{a}}^{(b)}(\underset{\sim}{r},m) = \sum_{\alpha=0}^{b}(-1)^{\alpha}\sum_{\beta=1}^{\binom{b}{\alpha}} D_{\underset{\sim}{a}_{\alpha,\beta}}^{(\alpha)}(\underset{\sim}{r}_{\alpha,\beta},m) \tag{4.25}$$

The proof is similar to that given above after (2.6) and we omit it here. An even more useful form of this result is to return to equal a-values and equal r-values and apply it to the CD-function. For any $c(0 \le c \le b)$ we obtain

$$CD_a^{(c,b-c)}(r,m) = \sum_{\alpha=0}^{c}(-1)^{\alpha}\binom{c}{\alpha}D_a^{(\alpha+b-c)}(r,m) = \sum_{\alpha=0}^{b-c}(-1)^{\alpha}\binom{b-c}{\alpha}C_a^{(\alpha+c)}(r,m) \tag{4.26}$$

and in fact the same results hold with a replaced by $\underset{\sim}{a}$ throughout (4.26).

The proof of this identity and many other interesting properties of CD-integrals can be found in the PhD thesis of Albert H. Liu [14].

To illustrate the usage of (4.26) for numerical computation consider $CD_1^{(1,1)}(2,4;3)$ where $a = 1$, $c = 1$, $b = 2$, $r = 2$ and $m = 3$. Then (4.26) gives

$$CD_1^{(1,1)}(2;3) = D_1^{(1)}(2,3) - D_1^{(2)}(2,3) = C_1^{(1)}(2,3) - C_1^{(2)}(2,3) = .15200617 \quad . \tag{4.27}$$

It will be quite useful to have the single integral representations (2.34) for vector arguments and we now develop these. The proof of these is quite similar to that for (2.33a) and we now give them without proof. For any vectors $\underset{\sim}{a}$, $\underset{\sim}{r}$ with b components

$$C_{\underset{\sim}{a}}^{(b)}(\underset{\sim}{r},m) = \int_0^\infty \left[\prod_{i=1}^{b} G_{r_i}(a_i x)\right] dG_m(x) \quad , \tag{4.28a}$$

$$D_{\underset{\sim}{a}}^{(b)}(\underset{\sim}{r},m) = \int_0^\infty \left\{\prod_{i=1}^{b} [1 - G_{r_i}(a_i x)]\right\} dG_m(x). \tag{4.28b}$$

$$CD_{\underset{\sim}{a}}^{(c,b-c)}(\underset{\sim}{r},m) = \int_0^\infty \prod_{i=1}^{c} G_{r_i}(a_i x) \prod_{j=c+1}^{b} [1-G_{r_j}(a_j x)]\, dG_m(x) \ . \tag{4.28c}$$

It follows from (4.28c) that for $m = r_1 = \cdots = r_b = r$ say and $a_1 = \cdots = a_b = 1$ that

$$CD_1^{(c,b-c)}(r,r) = \frac{c!(b-c)!}{(b+1)!} \ .$$

The Poisson type expansion for $C_a^{(b)}(r,m)$ was developed in Section 3.3.1 and we now consider a first term extension of that result for vector arguments. Starting with the single integral representation in (4.28), we make the same substitution $y = (x-m)/\sqrt{m}$ as in (3.14) and obtain for $\lambda_i = ma_i (i = 1,2,\ldots,b)$

$$C_{\underset{\sim}{a}}^{(b)}(\underset{\sim}{r},m) = \int_{-\sqrt{m}}^{\infty} \prod_{i=1}^{b} G_{r_i}\left(\lambda_i + \frac{y\lambda_i}{\sqrt{m}}\right) dG_m(m + y\sqrt{m}). \tag{4.29}$$

Using the normal approximation $dG_m(m + y\sqrt{m}) \underset{\sim}{\approx} \phi(y)$ we obtain after integration

(4.30a) $$C_{\underset{\sim}{a}}^{(b)}(\underset{\sim}{r},m) = \prod_{i=1}^{b} G_{r_i}(\lambda_i) + \mathcal{O}\left(\frac{1}{m}\right)$$

Similarly for the D-integral we obtain

(4.30b) $$D_{\underset{\sim}{a}}^{(b)}(\underset{\sim}{r},m) = \prod_{i=1}^{b} [1 - G_{r_i}(\lambda_i)] + \mathcal{O}\left(\frac{1}{m}\right)$$

Higher order terms can easily be obtained in (4.30a) and (4.30b) by similar methods as in (3.17) but they introduce multiple sums and thus are not as useful.

One useful form of the C and D integrals that sometimes arises is a generalization of the last equality in (2.3) where $C_a^{(1)}(r,m)$ is shown to be equal to $D_{1/a}^{(1)}(m,r)$. To show this more general result for $j = 0$ and vector arguments we simply set $x_\alpha = 1/y_\alpha$ and $x_i y_\alpha = y_i (i = 1,\ldots,\alpha-1,\alpha+1,\ldots,b)$ for any α. Taking the integral over y_α to be the last one, we obtain for any α

(4.31a) $$C_{\underset{\sim}{a}}^{(b)}(\underset{\sim}{r},m) = \int_{1/a_\alpha}^{\infty} \frac{\Gamma(m+r)}{\Gamma(m)\Gamma(r)} \frac{y_\alpha^{m-1}}{(1+y_\alpha)^{m+r_\alpha}} C_{\frac{y_\alpha}{1+y_\alpha}\underset{\sim}{a}_\alpha}^{(b-1)}(\underset{\sim}{r}_\alpha;m+r_\alpha)\, dy_\alpha ,$$

where $\underset{\sim}{r}_\alpha = (r_1,\ldots,r_{\alpha-1},r_{\alpha+1},\ldots,r_b)$ and similarly for $\underset{\sim}{a}_\alpha$. Similarly

(4.31b) $$D_{\underset{\sim}{a}}^{(b)}(\underset{\sim}{r},m) = \int_{0}^{1/a_\alpha} \frac{\Gamma(m+r)}{\Gamma(m)\Gamma(r)} \frac{y_\alpha^{m-1}}{(1+y_\alpha)^{m+r_\alpha}} D_{\frac{y_\alpha}{1+y_\alpha}\underset{\sim}{a}_\alpha}^{(b-1)}(\underset{\sim}{r}_\alpha;m+r_\alpha)\, dy_\alpha .$$

The latter results can also be written in terms of Dirichlet random variables with joint density given by the common integrand (together with Γ constants) for C and D, namely if $X_1,\ldots X_b$ are Dirichlet and $Y_1,\ldots,Y_b$ are defined by $Y_\alpha = 1/X_\alpha$, $Y_i = X_i Y_\alpha/(1 + Y_\alpha)(i + 1 ,\ldots,b;i \neq \alpha)$ Then

(4.32a) $$P\{X_i \leq a_i \ (i = 1,\ldots,b)\} = P\left\{Y_i \leq \frac{Y_\alpha a_i}{1+Y_\alpha}\left(\begin{matrix} i = 1,\ldots,b \\ i \neq \alpha \end{matrix}\right), Y_\alpha \geq \frac{1}{a_\alpha}\right\},$$

(4.32b) $$P\{X_i > a_i \ (i = 1,\ldots,b)\} = P\left\{Y_i > \frac{Y_\alpha a_i}{1+Y_\alpha}\left(\begin{matrix} i = 1,\ldots,b \\ i \neq \alpha \end{matrix}\right), Y_\alpha < \frac{1}{a_\alpha}\right\}$$

4.7 Half-Integer Values for r and m

In the previous discussion of C and D integrals we assumed that the m and r-values are both positive integers. In some applications it is useful to consider half-integers. If we replace each factorial by the corresponding gamma function then the definitions of C and D hold unchanged and most of the analytic results in this monograph remain valid, since the fact that r or m is an integer is rarely used in deriving the analytic formulas, e.g. all the recurrence relations in (2.24) and (2.25) hold. The main exception is in boundary conditions, e.g. (2.27) and (2.29) do not have any meaning for noninteger r . Thus a highly accurate calculation cannot be carried out for noninteger values of r and m by the algorithms we have developed thus far. Approximate techniques such as numerical quadrature, Taylor Series or Interpolation have to be used for such values.

In addition we need the half-integer values to establish the relation of C and D integrals to several standard distributions and this is done in section 4.7.2.

In section 4.7.3 we show that for the special values $r = m = \frac{1}{2}$, the integrals in question reduce to elementary arc tangent functions for both C and D when $b = 1$ and $b = 2$ and more generally for any positive integer b the integrals C and D can be expressed as single integrals involving the standard normal cdf and density.

4.7.1 Motivation for Studying Half-Integer Values of r and m

To motivate the usefulness of this generalization to half-integer values, we consider the distribution (and its complement) of the minimum of b studentized chi-squares, with independent chi-square numerators and a common chi-square denominator, the latter being independent of all the numerators. Let $\chi_\alpha^2(\alpha = 0,1,\ldots b)$ denote $b + 1$ independent chi-squares with $\nu_\alpha = 2r_\alpha$ degrees of freedom, where the r_α are half integers so that ν_α can be odd or even. Let

$$(4.33) \qquad Y_1 = \min_{1\le i\le b} \frac{\chi^2_{v_i}/v_i}{\chi^2_{v_0}/v_0}\ , \qquad Y_2 = \max_{1\le i\le b} \frac{\chi^2_{v_i}/v_i}{\chi^2_{v_0}/v_0}$$

and let $a_\alpha = r_\alpha y/r_0$ where y is a 'running' value of Y. Then the probability that $Y_2 < y$ is

$$(4.34a) \qquad \int_0^\infty \left[\prod_{\alpha=1}^{b} G_{r_\alpha}(xa_\alpha)\right] dG_{r_0}(x) = C_{\underset{\sim}{a}}^{(b)}(\underset{\sim}{r}, r_0)$$

by (4.28a) where $r = (r_1,\ldots,r_b)$, $a = (a_1,\ldots,a_b)$ and r_0 takes the place of the usual m value. Similarly by (4.28b), $P\{Y_1 > y\}$ is

$$(4.34b) \qquad \int_0^\infty \left\{\prod_{\alpha=1}^{b} [1 - G_{r_\alpha}(xa_\alpha)]\right\} dG_{r_0}(x) = D_{\underset{\sim}{a}}^{(b)}(\underset{\sim}{r}, r_0)\ ,$$

where $r_\alpha(\alpha = 0,1,\ldots,b)$ are half integers, i.e., are of the form $i/2$ where i is any positive integer.

4.7.2 Relation to Standard Distributions

It is of interest to note that many of the standard distributions can be expressed in terms of the C and D integrals with half-integer arguments. We consider several different cases below and give an explicit relation in each case.

Relation to the univariate- t cdf F_v with v degrees of freedom

Consider for $a > 0$ and any positive integer v

$$(4.35) \qquad 2F_v(\sqrt{av})-1 = \frac{2\Gamma(\frac{v+1}{2})}{\sqrt{\pi v}\,\Gamma(\frac{v}{2})}\int_0^{\sqrt{av}} \frac{dt}{(1+\frac{t^2}{v})^{(v+1)/2}} = \frac{\Gamma(\frac{v+1}{2})}{\sqrt{\pi v}\,\Gamma(\frac{v}{2})}\int_0^{a} \frac{x^{-\frac{1}{2}}dx}{(1+x)^{(v+1)/2}}$$

$$= C_a^{(1)}(\tfrac{1}{2},\tfrac{v}{2})\ ,$$

where $F_v(x)$ is the cdf of Student's t distribution with v degrees of freedom. Setting $t = \sqrt{av}$, it follows that

(4.36) $$F_\nu(t) = \tfrac{1}{2}[1+C^{(1)}_{t^2/\nu}(\tfrac{1}{2},\tfrac{\nu}{2})] .$$

Relation to the cdf $F_{\nu_1,\nu_2}(a)$ of the F-distribution with ν_1 and ν_2 degrees of freedom

Consider for $a > 0$ and any positive integers ν_1 and ν_2

(4.37)

$$F_{\nu_1,\nu_2}(a) = \frac{\nu_1^{\frac{1}{2}\nu_1}\nu_2^{\frac{1}{2}\nu_2}}{B(\frac{\nu_1}{2},\frac{\nu_2}{2})}\int_0^a \frac{x^{\frac{1}{2}\nu_1-1}\,dx}{(\nu_2+\nu_1 x)^{(\nu_1+\nu_2)/2}} = \frac{\Gamma(\frac{\nu_1+\nu_2}{2})}{\Gamma(\frac{\nu_1}{2})\Gamma(\frac{\nu_2}{2})}\int_0^{\nu_1 a/\nu_2} \frac{y^{\frac{1}{2}\nu_1-1}dy}{(1+y)^{(\nu_1+\nu_2)/2}}$$

$$= C^{(1)}_{\nu_1 a/\nu_2}(\tfrac{\nu_1}{2},\tfrac{\nu_2}{2}) ,$$

with some obvious further simplification if $\nu_1 = \nu_2$. Thus for the special case $b = 1$ a table of $C^{(1)}$ values gives rise to percentage points of the F-distribution.

Relation to the One-Sided Multivariate t with common correlation $\rho = 0$.

For any $h_i \geq 0$ and any positive integers $\nu_i (i = 1,2,\dots,b)$ let $T_1 = T_1(m,b,\underset{\sim}{\nu})$ denote the following one-sided b-fold integral

(4.38) $$T_1 = \frac{\Gamma(m+b/2)}{\Gamma(m)\pi^{b/2}\sqrt{\nu_1\nu_2\cdots\nu_b}}\int_{-\infty}^{h_1}\int_{-\infty}^{h_2}\cdots\int_{-\infty}^{h_b}\frac{\prod_1^b dt_i}{(1+\sum_1^b \frac{t_i^2}{\nu_i})^{m+b/2}}$$

$$= \int_{-\infty}^{0}\cdots\int_{-\infty}^{0} + \Sigma_1\int_0^{h_1}\int_{-\infty}^{0}\cdots\int_{-\infty}^{0} + \Sigma_2 \int_0^{h_1}\int_0^{h_2}\int_{-\infty}^{0}\cdots\int_{-\infty}^{0} + \cdots + \int_0^{h_1}\int_0^{h_2}\cdots\int_0^{h_b},$$

where all integrations are over the same integrand, Σ_1 is a sum of $\binom{b}{1}$ terms, each term with a different set of $b-1$ integrals from $-\infty$ to 0, and Σ_2 is likewise a sum of $\binom{b}{2}$ terms, each terms with a different set of b-2 integrals from $-\infty$ to 0, etc. After making the same transformation $t_i^2/\nu_i = x_i$ $(i = 1,2,\ldots b)$ in each term we obtain from (4.38)

$$(4.39)\qquad T_1 = \frac{1}{2^b}\{1 + \sum_{i=1}^{b} C^{(1)}_{\left(\frac{h_i^2}{\nu_i}\right)}(\tfrac{1}{2},m) + \sum_{\substack{i,j=1\\ i<j}}^{b} C^{(2)}_{\left(\frac{h_i^2}{\nu_i},\frac{h_j^2}{\nu_j}\right)}(\tfrac{1}{2},m) + \cdots + C^{(b)}_{\left(\frac{h_1^2}{\nu_1},\ldots,\frac{h_b^2}{\nu_b}\right)}(\tfrac{1}{2},m)\}$$

For the special case of a common $h \geq 0$ and equal ν_i-values it simplifies to

$$(4.40)\qquad T_1 = \frac{1}{2^b}\sum_{i=0}^{b}\binom{b}{i} C^{(i)}_{\frac{h^2}{\nu}}(\tfrac{1}{2},m) = \sum_{\alpha=0}^{b}(-\tfrac{1}{2})^{\alpha}\binom{b}{\alpha} D^{(\alpha)}_{\frac{h^2}{\nu}}(\tfrac{1}{2},m)\ ,$$

where the last result follows by simply using (2.6a) or by a more complicated geometrical argument. Thus the C-(or D-)tables can be used to construct one-sided multivariate t tables or the associated equicoordinate percentage points (h-values) for a common ν-value.

Relation to the Two-Sided Multivariate t with common correlation $\rho = 0$.

Consider for any $h_i \geq 0$ and any positive integers $\nu_i (i = 1,2,\ldots,b)$

$$(4.41)\qquad T_2 = \frac{\Gamma(m+b/2)}{\Gamma(m)\pi^{b/2}\sqrt{\nu_1\nu_2\cdots\nu_b}}\int_{-h_1}^{h_1}\int_{-h_2}^{h_2}\cdots\int_{-h_b}^{h_b}\frac{\prod_1^b dt_i}{\left(1+\sum_1^b\frac{t_i^2}{\nu_i}\right)^{m+b/2}}$$

$$= \frac{2^b\Gamma(m+b/2)}{\Gamma(m)\pi^{b/2}\sqrt{\nu_1\nu_2\cdots\nu_b}}\int_0^{h_1}\int_0^{h_2}\cdots\int_0^{h_b}\frac{\prod_1^b dt_i}{\left(1+\sum_1^b\frac{t_i^2}{\nu_i}\right)^{m+b/2}}$$

The same transformation as in the previous case gives us the simpler result

$$(4.42)\qquad T_2 = C^{(b)}_{\left(\frac{h_1^2}{\nu_1},\frac{h_2^2}{\nu_2},\ldots,\frac{h_b^2}{\nu_b}\right)}(\tfrac{1}{2},m)$$

Thus for the case of a common h-value and a common ν-value, it would be an easy task to use a C-table to obtain equi-coordinate percentage points (h-values) corresponding to specified values of T_2.

4.7.3 Reduction to standard forms for $r = m = \frac{1}{2}$

In view of the fact that we are missing one boundary condition for computation, we use the case $r = m = \frac{1}{2}$ as a starting value and this leads directly to the cases $r = \frac{1}{2}$, $m = \alpha$ $(\alpha = 3/2, 5/2, 7/2, \ldots)$ by the use of the recurrence relations (2.24), (2.25) and the boundary condition (2.26). Similarly we can start with $r = \frac{1}{2}$, $m = 1$ and this leads to $r = \frac{1}{2}$, $m = \alpha$ $(\alpha = 2,3,\ldots)$. These two starting points are easily related by the use of (2.29), so that the case $r = m = \frac{1}{2}$ is the only case that needs a special consideration.

From either (2.33) or (4.34) we obtain

(4.43a) $$C_a^{(b)}(\tfrac{1}{2},\tfrac{1}{2}) = \int_0^\infty G_{\frac{1}{2}}^b(ax)\,dG_{\frac{1}{2}}(x),$$

(4.43b) $$D_a^{(b)}(\tfrac{1}{2},\tfrac{1}{2}) = \int_0^\infty [1-G_{\frac{1}{2}}(ax)]^b\,dG_{\frac{1}{2}}(x)$$

where $G_{\frac{1}{2}}(s)$ is the incomplete gamma function defined in (2.32) for $r = \frac{1}{2}$, i.e.,

(4.44) $$G_{\frac{1}{2}}(x) = \frac{1}{\Gamma(\frac{1}{2})}\int_0^x \frac{e^{-t}}{\sqrt{t}}\,dt .$$

Using the transformation $t = \frac{1}{2}y^2$ in (4.44) we can express (4.43) in terms of the standard normal cdf Φ and density φ as follows:

(4.45a) $$C_a^{(b)}(\tfrac{1}{2},\tfrac{1}{2}) = 2\int_0^\infty [2\Phi(y\sqrt{a})-1]^b \varphi(y)\,dy ,$$

(4.45b) $$D_a^{(b)}(\tfrac{1}{2},\tfrac{1}{2}) = 2^{b+1}\int_0^\infty [1-\Phi(y\sqrt{a})]^b \varphi(y)\,dy .$$

Although we can obtain explicit expressions in terms of arctan for (4.45) with $b = 1$ and 2, we prefer to obtain them through a differential relationship that we refer to as

Lemma: For any integer $b \geq 1$

$$\frac{d}{da} D_a^{(b)}(\tfrac{1}{2},\tfrac{1}{2}) = \frac{-b}{\pi(a+1)\sqrt{a}} + \frac{b(b-1)}{\pi(a+1)\sqrt{1+2a}} D_{\frac{a}{1+2a}}^{(b-2)}(\tfrac{1}{2},\tfrac{1}{2}) \tag{4.46}$$

Proof. Differentiating (4.45b) with respect to a, and then integrating by parts on y yields the result (4.46) after some straightforward manipulation.

We now substitute $b = 1$ and 2, respectively, in (4.46) and obtain after integration

$$\begin{aligned} D_a^{(1)}(\tfrac{1}{2},\tfrac{1}{2}) &= \frac{2}{\pi} \text{ arc tan } (1/\sqrt{a}), \\ D_a^{(2)}(\tfrac{1}{2},\tfrac{1}{2}) &= \frac{4}{\pi} \text{ arc tan } \left(\frac{1}{2\sqrt{a}+\sqrt{1+2a}}\right) \end{aligned} \tag{4.47}$$

Note that the relationship $D_1^{(b)}(r,r) = \frac{1}{b+1}$ is still satisfied as shown in (2.35). The arc tangents can be expressed in many different forms and it is often not trivial to show their equivalence. This same result $b = 2$ (or its equivalent) was also obtained in (24) of [3] (where $\theta = 1/a$); an explicit result for $r = 3/2$, $b = 2$ and $\theta = 1/a$ also appears in (25) of [3]. Some short tables of $D_a^{(b)}(\tfrac{1}{2},\tfrac{1}{2})$ are also given for $b = 1,2,$ and 3 in a separate paper submitted for publication by one of the present authors.

For $b > 2$ the function $D_a^{(b)}(\tfrac{1}{2},\tfrac{1}{2})$ cannot be expressed in terms of arc tangents and other elementary functions; this result is clear for $b = 3$ and appears to hold for all $b \geq 3$.

The corresponding results for $C_a^{(b)}(\tfrac{1}{2},\tfrac{1}{2})$ are most easily obtained by inclusion-exclusion. They are

$$C_a^{(1)}(\tfrac12,\tfrac12) = 1-D_a^{(1)}(\tfrac12,\tfrac12) = \frac{2}{\pi}\arctan(\sqrt{a}),$$

(4.48)

$$C_a^{(2)}(\tfrac12,\tfrac12) = 1-2D_a^{(1)}(\tfrac12,\tfrac12) + D_a^{(2)}(\tfrac12,\tfrac12) = \frac{4}{\pi}\arctan\left(\frac{a}{a+1+\sqrt{1+2a}}\right).$$

Note again that the relation $C_1^{(b)}(\tfrac12,\tfrac12) = \frac{1}{b+1}$ also holds here.

The results in (4.45) can be generalized since we can express C and D in terms of the normal cdf when $r = \tfrac12$ for any m. We now start with the right hand side of (4.43) with x^t in the integrand and general m and use the same notation as in (2.36) and the same transformation as above. This gives

(4.49a)

$$MC_a^{(b)}(\tfrac12,m,t) = \int_0^\infty x^t G_{\frac12}^b(ax)\,dG_m(x) = \frac{\sqrt{2\pi}}{2^{m+t-1}\Gamma(m)} \int_0^\infty y^{2m+2t-1}[2\Phi(y\sqrt{a})-1]^b \varphi(y)\,dy ,$$

(4.49b)

$$MD_a^{(b)}(\tfrac12,m,t) = \int_0^\infty x^t [1-G_{\frac12}(ax)]^b dG_m(x) = \frac{2^b\sqrt{2\pi}}{2^{m+t-1}\Gamma(m)} \int_0^\infty y^{2m+2t-1}[1-\Phi(y\sqrt{a})]^b \varphi(y)\,dy .$$

For the special case $m = \tfrac12$, 1 and $a = 1$ these integrals in (4.49) arise in finding the tth moment of the largest (resp., smallest) of $b + 1$ independent folded normal random variables; these are folded at zero so that they represent $|X|$ where X is standard normal. Thus we obtain using the density $f(x)$ and its cdf $F(x)$

$$f(x) = 2\varphi(x) \quad (0 < x < \infty) , \quad F(x) = 2\Phi(x)-1 \quad (0 < x < \infty)$$

and letting $Z = \text{Max}\,(|X_1|,|X_2|, \ldots , |X_{b+1}|)$, we have for $m = \tfrac12$ and 1 the even and odd moments, respectively,

$$\text{(4.50a)} \quad EZ^{2t} = 2(b+1)\int_0^\infty x^{2t}[2\Phi(x)-1]^b \varphi(x)\,dx = \frac{2^t(b+1)\Gamma(t+\tfrac12)}{\sqrt{\pi}}\, C_1^{(b)}(\tfrac12,t+\tfrac12) .$$

(4.50b) $EZ^{2t+1} = 2(b+1)\int_0^\infty x^{2t+1}[2\Phi(x)-1]^b\varphi(x)dx = \frac{2^{t+1}(b+1)(t!)}{\sqrt{2\pi}} C_1^{(b)}(\tfrac{1}{2},t+1)$.

The corresponding results for $Y = \text{Min}(|X_1|, \ldots, |X_{b+1}|)$ are obtained from (4.50) by simply replacing C by D on the right hand side.

§5. Applications.

5.1 Examples to Illustrate the Probability Interpretation

In Sections 2.2 and 2.3 we gave the probability interpretation of C and D integrals. We now give some specific examples to illustrate this application to certain probability problems.

Example 1. Gum packages have either a baseball card enclosed or a blank card (which is not visible before making the purchase). There are b different baseball cards available and they all appear with equal probability p in a random manner, so that 1-bp is the probability of a blank card. If I stop buying when I have m blank cards what is the probability that I have at least 1 of each of the b baseball cards.

Solution: The probability interpretation of $C_a^{(b)}(r,m)$ in (2.3) with $j = 0$ is such that it exactly fits the above problem if we set $r = 1$ and take $a = p/(1-bp)$. Hence the answer is $C_a^{(b)}(1,m)$ with $a = p/(1-bp)$.
For example with $b = 5$ cards, $p = .10$ and $m = 5$ the numerical answer is $C_{1/5}^{(5)}(1,5) = .129643$. For $m = 10$ the result is $C_{1/5}^{(5)}(1,10) = .460266$; for $m = 15$ it is $C_{1/5}^{(5)}(1, 15) = .731790$ (cf. Page A1).

Example 2: Manufactured items are either good or defective but the defective items fail for b different reasons and the ith type of failure has probability p_i of occurring; the probability of a good item $p_o = 1-\sum_{i=1}^{b} p_i$. Assume we have an (either-or) quota for each type of failure (say r_i for the ith type) as a result of which a machine is stopped for repair. If we continue until we get m good items, what is the probability that the machine will not be stopped in that period.

Solution. Let $\underset{\sim}{r} = (r_1, r_2, \ldots, r_b)$ and let $\underset{\sim}{p} = (p_1, p_2, \ldots, p_b)$. Using the probability interpretation for the D integral with vector arguments in Section 4.1 we are looking for the probability that the frequency in the ith cell is less than r_i when we obtain m good and hence the answer is $D_{\underset{\sim}{a}}^{(b)}(\underset{\sim}{r}, m)$ where $\underset{\sim}{a} = (\frac{p_1}{p_o}, \frac{p_2}{p_o}, \ldots, \frac{p_b}{p_o})$. For example if $b = 3$, $r_i = 2$ (so that we stop production if there are two failures in any category), $p_i = .2$ $(i = 1,2,3)$ and $m = 10$, then the required probability of no stoppage before getting 10 good items is $D_{1/2}^{(3)}(2,10) = .003225$ (cf. Page B13).

Example 3. We use the same background as in Example 1, the only changes being that

i) there are no blank cards so that $p = 1/b$ and

ii) we stop buying as soon as we have m cards all of the same type.

Solution. Since the b cells are equally likely with common probability p we can treat any one cell as the counting cell, so that our resulting probability is again C with superscript b-1 and a = 1 , i.e., the answer is $C_1^{(b-1)}(1,m)$. For example with $b = 10$ and $m = 10$ the numerical value is .991752; for $m = 5$ the corresponding numerical answer is .813025 (cf. Page A5).

5.2.1 General Probability and Waiting-time Applications

In many applications we are interested in the probability that the last (or first) cell to reach its quota does so on or before the nth trial. This is a generalization of the well-known coupon collectors problem where each cell has a quota of one and the b cells have a common probability $p = 1/b$. Hence we are interested in the probability that the last cell reaches its quota on or before the nth trial. This is given by $I_{\underset{\sim}{p}}^{(b)}(\underset{\sim}{r}, n)$ defined in [19] by

(5.1) $$I_{\underset{\sim}{p}}^{(b)}(\underset{\sim}{r},n) = \frac{n!}{(n-R)!\prod_{1}^{b}\Gamma(r_i)} \int_{0}^{p_1} \cdots \int_{0}^{p_b} \left(1 - \sum_{1}^{b} x_i\right)^{n-R} \prod_{1}^{b} x_i^{r_i-1}\, dx_i ,$$

where $R = r_1+\ldots+r_b$. We wish to go through a limiting process in which $p_i \to 0 (i = 1,2,\ldots,b)$, $n \to \infty$ and

(5.2) $$p_i \approx \frac{\lambda_i t}{n} \qquad (i = 1,2,\ldots,b).$$

Since the p_i are small, the x_i are also small and neglecting powers higher than 1 we can write

(5.3) $$\left(1 - \sum_{1}^{b} x_i\right)^{n-R} = e^{(n-R)\ln\left(1 - \sum_{1}^{b} x_i\right)} \approx e^{-(n-R)\sum_{1}^{b} x_i}$$

Using this in (5.1) we obtain after an obvious change of variable

(5.4) $$I_{\underset{\sim}{p}}^{(b)}(\underset{\sim}{r},m) \approx \frac{n!}{(n-R)!(n-R)^R} \prod_{i=1}^{b} \int_{0}^{p_i(n-R)} \frac{x_i^{r_i-1} e^{-x_i} dx_i}{\Gamma(r_i)}$$

From (5.4) and (5.2) we pass to the limit as $n \to \infty$ and obtain

(5.5a) $$L_{\underset{\sim}{\lambda}}^{(b)}(\underset{\sim}{r}) = \lim_{n\to\infty} I_{\underset{\sim}{p}}^{(b)}(\underset{\sim}{r},n) = \prod_{i=1}^{b} G_{r_i}(\lambda_i t)$$

where $G_r(x)$ is the cumulative distribution function (cdf) of the gamma distribution with parameter r,

$$G_r(x) = \int_{0}^{x} \frac{y^{r-1}e^{-y}}{\Gamma(r)}\, dx \quad .$$

The corresponding density $\ell_{\underset{\sim}{\lambda}}$ of L in (5.5) is easily seen to be

(5.6a) $$\ell_{\underset{\sim}{\lambda}}^{(b)}(\underset{\sim}{r}) = \sum_{\alpha=1}^{b} \lambda_\alpha g_{r_\alpha}(\lambda_\alpha t) \prod_{\substack{i=1\\ i\neq\alpha}}^{b} G_{r_i}(\lambda_i t) \quad ,$$

where $g_r(x)$ is the density $x^{r-1}e^{-x}/\Gamma(r)$ of $G_r(x)$. This density in (5.7a) can also be obtained directly as the limit under (5.2) of the probability that the last cell to reach its quota does so on exactly the nth trial, i.e., as the limit of

$$\sum_{\alpha=1}^{b} p_\alpha \, I^{(b,1)}_{(p_\alpha;p_1,\ldots,p_{\alpha-1},p_{\alpha+1},\ldots,p_b)}(r_\alpha;r_1,\ldots,r_{\alpha-1},r_{\alpha+1},\ldots,r_b;n-1).$$

The dual of the above results is to consider the first cell to reach its quota. An application of the dual occurs, for example, if we are dealing with redundancies or elements set up in parallel and we are interested in the distribution of the number of trials (or time) to failure of the system. Thus we can have a multinomial model and assume that at the outset the balls are in the cells, say r_i in the ith cell $(i = 1,2,\ldots,b)$. A trial corresponds to the removal of a ball from some cell and the system breaks down when any one cell is depleted. Here we are interested in the probability that the first cell to reach its quota does so on exactly the nth observation.

Using the notation $M^{(b)}_{\underset{\sim}{\lambda}}(\underset{\sim}{r})$ for the limiting cdf and $m^{(b)}_{\underset{\sim}{\lambda}}(\underset{\sim}{r})$ for the limiting density we have as in (5.5a) and (5.7a).

$$\text{(5.5b)} \qquad M^{(b)}_{\underset{\sim}{\lambda}}(\underset{\sim}{r}) = \lim_{n\to\infty} J^{(b)}_{\underset{\sim}{p}}(\underset{\sim}{r},m) = \prod_{i=1}^{b} [1 - G_{r_i}(\lambda_i t)] ,$$

$$\text{(5.6b)} \qquad m^{(b)}_{\underset{\sim}{\lambda}}(\underset{\sim}{r}) = \sum_{\alpha=1}^{b} \lambda_\alpha \, g_{r_\alpha}(\lambda_\alpha t) \prod_{\substack{i=1 \\ i\neq\alpha}}^{b} [1 - G_{r_i}(\lambda_i t)] .$$

It is of interest to compute <u>ordinary</u> moments of the limiting distribution (5.5a) [resp., (5.5b)] and compare them with the <u>ascending factorial</u> (af) moments of the exact distribution of the waiting time until the last [resp., first] cell reaches its quota. From (5.6a) we find that the γth moment of $L^{(b)}_{\underset{\sim}{\lambda}}(\underset{\sim}{r})$ is

$$\text{(5.7)} \qquad \mu_\gamma = \sum_{\alpha=1}^{b} \lambda_\alpha \int_0^\infty t^\gamma \, g_{r_\alpha}(\lambda_\alpha t) \prod_{\substack{i=1 \\ i\neq\alpha}}^{b} G_{r_i}(\lambda_i t)dt$$

$$= \sum_{\alpha=1}^{b} \frac{\Gamma(r_\alpha+\gamma)}{\Gamma(r_\alpha)\lambda_\alpha^\gamma} \int_0^\infty g_{r_\alpha+\gamma}(y) \prod_{\substack{i=1 \\ i\neq\alpha}}^{b} G_{r_i}\left(\frac{\lambda_i}{\lambda_\alpha} y\right) dy$$

$$= \sum_{\alpha=1}^{b} \frac{\Gamma(r_\alpha+\gamma)}{\Gamma(r_\alpha)\lambda_\alpha^\gamma} C_{\underset{\sim}{a}_\alpha^*}^{(b-1)} (\underset{\sim}{r}_\alpha, r_\alpha+\gamma)$$

where $\underset{\sim}{r}_\alpha = (r_1,\ldots,r_{\alpha-1},r_{\alpha+1},\ldots,r_b)$ and $\underset{\sim}{a}_\alpha^* = \left(\frac{\lambda_1}{\lambda_\alpha},\ldots,\frac{\lambda_{\alpha-1}}{\lambda_\alpha},\frac{\lambda_{\alpha+1}}{\lambda_\alpha},\ldots\frac{\lambda_b}{\lambda_\alpha}\right)$ and the last result in (5.7) is obtained from (4.28a). The corresponding result for the ascending factorial moment of the waiting time until the last cell reaches its quota is

$$\mu^{[\gamma]} = \sum_{\alpha=1}^{b} p_\alpha \sum_{n=R}^{\infty} n^{[\gamma]} I_{\underset{\sim}{p}}^{(b,1)} (r_\alpha-1;r_1,\ldots,r_{\alpha-1},r_{\alpha+1},\ldots,r_b;n-1) \quad .$$

[The symbol $\mu^{[\gamma]}$ should not be confused with $m^{[\alpha]}$ defined in Section 3.1]. Using the integral representation for I and summing the negative binomial under the integral sign (with obvious changes of variables) we obtain

$$\mu^{[\gamma]} = \sum_{\alpha=1}^{b} \frac{\Gamma(r_\alpha+\gamma)}{\Gamma(r_\alpha)p_\alpha^\gamma} C_{\underset{\sim}{a}_\alpha}^{(b-1)} (\underset{\sim}{r}_\alpha, r_\alpha + \gamma) \quad , \tag{5.8}$$

where $\underset{\sim}{a}_\alpha = \left(\frac{p_1}{p_\alpha},\ldots,\frac{p_{\alpha-1}}{p_\alpha},\frac{p_{\alpha+1}}{p_\alpha},\ldots,\frac{p_b}{p_\alpha}\right)$. Thus if we identify λ_i with p_i then by setting $\gamma = 1$ we obtain the exact same result for the expected value in the limiting case as in the exact case. In any case the computations all require C-integrals or D-integrals.

The corresponding formulas for the first cell to reach its quota are obtained by simply replacing C by D in (5.7) and (5.8).

For the more interesting special case of equal r-values and equal a-values (resulting from equal p-values) we give these af moment results separately, namely from (5.9) for the time until all b cells reach their common quota r

$$\mu^{[\gamma]} = \frac{b\Gamma(r+\gamma)}{\Gamma(r)p^\gamma} C_a^{(b-1)}(r,r+\gamma) \quad , \tag{5.9a}$$

and from the dual for the time until any one cell reaches the common quota r

$$\mu^{[\gamma]} = \frac{b\Gamma(r+\gamma)}{\Gamma(r)p^\gamma} D_a^{(b-1)}(r,r+\gamma) \quad . \tag{5.9b}$$

The first af moment $(\gamma = 1)$ is clearly the mean μ and the variance is obtained from the relation

$$\sigma^2 = \mu^{[2]} - \mu\ (1+\mu)\ . \tag{5.10}$$

In the examples we consider below we use the sampling of cards with replacement (and shuffling after each replacement); we could easily have used some other structured multinomial model to illustrate these problems.

<u>Problem 1</u>: What is the mean and variance of the waiting time (i.e., the number of cards required to be drawn) with replacement for r aces $(r = 1,2,3,4)$ from an ordinary deck.

<u>Solution</u>: Here we have $b = 1$, $p = \frac{4}{52} = \frac{1}{13}$, so that from (5.10) and either (5.9a) or (5.9b) we obtain

$$\mu = \frac{r}{p} \quad \text{and} \quad \sigma^2 = \frac{r(1-p)}{p^2}$$

It should be noted that because of replacement the r aces need not be from different suits. For the case of insisting on different suits, see the soluton to problem 3 below.

<u>Problem 2</u>: What is the mean and variance of the waiting time (with replacement) for 4 aces <u>and</u> four deuces.

<u>Solution</u>: From (5.9a) with $b = 2$, $a = 1$, $r = 4$ we have (cf. Page A32)

$$\mu = 2\left(\frac{52}{4}\right)\ 4C_1^{(1)}(4,5) = 66.219, \tag{5.11}$$

$$\mu^{[2]} = 2\left(\frac{52}{4}\right)^2 4(5)C_1^{(1)}(4,6) = 5043.60 \tag{5.12}$$

from which we obtain, using (5.10), $\sigma^2 = 592.45$.

Problem 3: What is the mean and variance of the waiting time (with replacement) until 4 of-a-kind have been drawn.

Solution: From (5.9b) with $b = 13$, $r = 4$ and $a = 1$ we have (cf. Page B32)

(5.13) $$\mu = 13(\tfrac{52}{4})\ 4D_1^{(12)}(4,5) = 18.436,$$

(5.14) $$\mu^{[2]} = 13(\tfrac{52}{4})^2 4(5)D_1^{(12)}(4,6) = 385.172$$

and hence, from (5.10), $\sigma^2 = 26.85$.

If we insist in this problem on 4-of-a-kind from 4 different suits again sampling with replacement, the problem is slightly more difficult and the analysis has not been considered here. However the results are interesting and are given here numerically in terms of C and D integrals for purposes of comparison. Using the same notation μ , $\mu^{[2]}$, we have (cf. Page A5)

(5.15) $$\mu = (52)^2 \sum_{\beta=0}^{12} (-1)^\beta \binom{12}{\beta}\ C_1^{(4\beta+3)}(1,2) = 34.799\ ,$$

(5.16) $$\mu^{[2]} = 2(52)^3 \sum_{\beta=0}^{12} (-1)^\beta \binom{12}{\beta}\ C_1^{(4\beta+3)}(1,3) = 1385.64\ ,$$

and hence, from (5.10), $\sigma^2 = 139.88$.

Problem 4: What is the mean and variance of the waiting time (with replacement)until one of each suit has been drawn?

Solution: From (5.9a) with $b = 4$, $r = 1$, $p = \frac{1}{4}$, $a = 1$ we have (cf. Page A5)

(5.17) $$\mu = 16C_1^{(3)}(1,2) = 8.333\ ,$$

(5.18) $$\mu^{[2]} = 128C_1^{(3)}(1,3) = 77.778 ,$$

so that $\sigma^2 = 14.44$.

Problem 5: What is the mean and variance of the waiting time (with replacement) until a pair is drawn?

Solution: From (5.9b) with $b = 13$, $r = 2$, $p = 1/13$, $a = 1$ we have (cf. Page B14)

(5.19) $$\mu = 338\ D_1^{(12)}(2,3) = 5.212 ,$$

$$\mu^{[2]} = 13182\ D_1^{(12)}(2,4) = 36.425 ,$$

and hence $\sigma^2 = 4.044$.

Problem 6: (Coupon Collector Problem)

What is the expected number of tosses of a fair die (and the corresponding variance) needed to observe each of the six sides at least once?

Solution: Using (5.9a) with $a = r = 1$ and $b = 6$ the first two ascending factorial moments $(\gamma = 1,2)$ and the variance are (cf. Page A5)

$$\mu = 36C_1^{(5)}(1,2) = 36\ (.4083\ldots) = 14.700 ,$$

$$\mu^{[2]} = 432C_1^{(5)}(1,3) = 432\ (.62449074) = 269.780 ,$$

$$\sigma^2 = \mu^{[2]} - \mu\ (1+\mu) = 38.99 .$$

The answer for μ can also be expressed as $6(1 + 1/2 + 1/3 + 1/4 + 1/5 + 1/6)$ as is well known. Hence the C integral is related to sums of reciprocals and we study this relationship in Section 5.7.

5.2.2 Comparison of Problems With and Without Replacement

The corresponding problems without replacement do not involve C or D integrals, but they sometimes involve I and J integrals. Many problems of this type have been considered in [2] and [18] and we shall not dwell on these problems here. However we take the same problems considered in the last subsection and express the answers for sampling without replacement algebraically and also numerically for purposes of comparison. Problem numbers refer to the problems in the previous section except that sampling is now without replacement.

Solution of Problem 1: The mean $(\gamma = 1)$ is obtained by simply dividing the 52-4 = 48 non-ace cards into 5 equal parts, taking r parts plus r aces to obtain

$$\mu = r(1 + \frac{48}{5}) = \frac{53r}{5} \qquad (r = 1,2,3,4) . \tag{5.20}$$

For the γth ascending factorial moment we can easily obtain the more general result (details can be found in [2])

$$\mu^{[\gamma]} = \frac{(N+1)^{[\gamma]}}{(M+1)^{[\gamma]}} r^{[\gamma]} , \tag{5.21}$$

where $N = 52$ equals the total number of cards and $M = 4$ equals the number of aces or objects in each of the N/M classes of objects. Hence the variance from (5.21) with $\gamma = 2$ and (5.10) is

$$\sigma^2 = \frac{53(8)}{25} r(5-r) \qquad (r = 1,2,3,4) \tag{5.22}$$

Solution of Problem 2. For the mean we use the same method as in problem 1 dividing 52-8 = 44 cards into 9 parts and taking $r = 8$; this gives

$$\mu = 8(1 + \frac{44}{9}) = \frac{53}{9}(8) = 47.111 . \tag{5.23}$$

Using (5.21) for $\gamma = 2$, $N = 52$ and $M = 8$ we obtain

(5.24) $$\mu^{[2]} = (\frac{53(54)}{9(10)})\ 8(9) = 2289.60\ ;\ \sigma^2 = 23.03\ .$$

Solution of Problem 3:

For general γ it can be shown (see [2] for explanation) that the γth ascending factorial moment is given by

(5.25) $$\mu^{[\gamma]} = (N+1)^{[\gamma]} b!/(1+\frac{\gamma}{M})^{[b]}$$

which gives the results, setting $b = 13$, $M = 4$, $r = 4$

(5.26) $$\mu = 53\ \frac{13!}{\frac{5}{4}\ \frac{9}{4}\ \cdots\ \frac{53}{4}} = 25.003\ ,$$

(5.27) $$\mu^{[2]} = 53(54)\ \frac{13!}{\frac{6}{4}\ \frac{10}{4}\ \cdots\ \frac{54}{4}} = 683.95\ ;\ \sigma^2 = 33.81\ .$$

Solution of Problem 4:

In this case we give only numerical results (and refer the reader to [2] for details); these results are

(5.28) $$\mu = 7.665\ ;\quad \mu^{[2]} = 60.23\ ;\quad \sigma^2 = 9.14\ .$$

Solution of Problem 5:

We again give only numerical results referring the reader to [2] for details; these are

(5.29) $$\mu = 5.697\ ;\ \mu^{[2]} = 42.929\ ;\ \sigma^2 = 4.78\ .$$

Solution of Problem 6:

Assume a common value M for each of the 6 catagories so that $N = 6M$. For general γ we utilize Corollary 4.1.1 of [2] with $F_2(u)$ equal to the cdf of the maximum of 6 independent random variables each of which is the minimum of M independent uniform (0,1) random variables, i.e., $F_2(u) = [1 - (1-\mu)^M]^6$. Then the γth af moment and its limit as $M \to \infty$ are

$$E\{T^{[\gamma]}\} = (6M+1)^{[\gamma]} 6M \int_0^1 u^\gamma (1 - u^{M-1}[1 - (1-u)^M]^5 du$$

$$= \gamma! \sum_{\beta=1}^{6} (-1)^{\beta-1} \binom{6}{\beta} \frac{(6M+1)^{[\gamma]}}{(\beta M+1)^{[\gamma]}} \to \gamma! \sum_{\beta=1}^{6} (-1)^{\beta-1} \binom{6}{\beta} \left(\frac{6}{\beta}\right)^{\gamma} .$$

For $\gamma = 1$ we also have the alternative expression (obtainable from above)

$$\mu = E\{T\} = (6M+1) \left[1 - \frac{6!}{\left(\frac{M+1}{M}\right)^{[6]}} \right] \to 14.700 .$$

For $\gamma = 2$ in the above and from (5.10) we have

$$\mu^{[2]} = E\{T^{[2]}\} = 2 \sum_{\beta=1}^{6} (-1)^{\beta-1} \binom{6}{\beta} \frac{(6M+1)(6M+2)}{(\beta M+1)(\beta M+2)} \to 269.780$$

and $\sigma^2 \to 38.99$, all of which in the limit as $M \to \infty$ check with the previous results for problem 6 with replacement.

It should be noticed that the numerical results for sampling with and without replacement are fairly close together when the total sample size N is large; this applies to both the mean and the variance. Hence the easier results for sampling with replacement can often be used as approximate answers for sampling without replacement and we use this need for some numerical illustrations as our justification for including here the above solutions without replacement.

5.2.3 Banach Match Box Problems

In this section we generalize the results obtained in [2] where the same problem was discussed. A smoker has b boxes of matches in different pockets, the ith box (or pocket) starting with $r_i \geq 1$ matches. He reaches into his i^{th} pocket with probability p_i $(i = 1,2,\ldots,b)$ where $\sum_1^b p_i \leq 1$ and we associate the remaining probability $1-\sum_1^b p_i$ (if it is positive) with a "sink". The sink corresponds to taking a match from his friends when he forgets that he has matches. Omission of the sink corresponds to the case $\sum_1^b p_i = 1$, and this is one more direction of generalization. The goals of interest are to find the expected waiting time (and/or higher moments) required until (i) any one of the b match boxes is emptied or (ii) all of the b match boxes are emptied at least once or (iii) exactly s $(1 \leq s \leq b)$ of the b match boxes are emptied at least once. In the latter two problems we first assume that the multinomial structure is not changed when a box is emptied (Case 1), e.g., say that the empty box is immediately replenished. Later in section 5.3.5 we consider the more difficult case (Case 2) in which empty boxes are immediately discarded and hence the multinomial structure is changed after each such discard.

Let T_s denote the waiting time until s different boxes are emptied in Case 1, so that (i) and (ii) correspond to $s = 1$ and $s = b$, respectively. Let $\mu_s^{[\gamma]} = E\{T_s^{[\gamma]}\}$ denote the γth ascending factorial moment of T_s.

We identify the starting numbers r_i with the quotas we are waiting for in section 5.2.1 and hence the desired result for (i) from (5.8) and the discussion that follows (5.8) shows that for $s = 1$

$$\mu_1^{[\gamma]} = \sum_{\alpha=1}^{b} \frac{\Gamma(r_\alpha+\gamma)}{\Gamma(r_\alpha)p_\alpha^\gamma} D_{\underset{\sim}{a}_\alpha}^{(b-1)}(\underset{\sim}{r}_\alpha, r_\alpha+\gamma) \quad . \tag{5.30}$$

where $\underset{\sim}{a}_\alpha$ and $\underset{\sim}{r}_\alpha$ are defined in (5.2.1) The corresponding result for problem (ii) is obtained from (5.30) by simply replacing D by C . For the case of general s , problem (iii), we use the CD integral defined in

Section 4.5 by (4.18). Let $\underset{\sim}{a}_{\alpha,\beta}$ and $\underset{\sim}{r}_{\alpha,\beta}$ denote similar permutations of the components of the vectors $\underset{\sim}{a}_{\alpha}$ and $\underset{\sim}{r}_{\alpha}$, respectively, so that β runs from 1 to $(b-1)!$ since each has $b-1$ components. Then we obtain for the γth ascending factorial moment of T_s (the derivation is similar to that given in Section 5.3.3 and is omitted here)

$$(5.31)\qquad \mu_s^{[\gamma]} = \sum_{\alpha=1}^{b} \frac{\Gamma(r_i+\gamma)}{\Gamma(r_i)p_i^{\gamma}} \sum_{\beta=1}^{(b-1)!} CD_{\underset{\sim}{a}_{i,\beta}}^{(s-1,b-s)}(\underset{\sim}{r}_{i,\beta}, r_i+\gamma) .$$

For common r and common p with $s = 1 = \gamma$ we obtain from (5.30)

$$(5.32)\qquad \mu = \frac{br}{p} D_1^{(b-1)}(r,r+1)$$

and for the special case $b = 2$ of two boxes, using (2.25b), (2.28) and (2.35) and setting $p = \frac{1}{2}$, we write this in the form

$$(5.33)\qquad \mu = \frac{2r}{p} D_1^{(1)}(r,r+1) = 4r[\tfrac{1}{2} - \binom{2r-1}{r}(\tfrac{1}{2})^{2r}] .$$

For the absent-minded smoker problem where you have to wait to discover that a box is empty (rather than wait only until the last match is drawn), we merely set $r = N+1$ in any of our formulas. Then considering $\mu = 2N+1 - \mu_F$ where μ_F is the expected number remaining in the other box (in the notation of Feller [8]), we find that the resulting μ_F is identical with the expression obtained by Feller (cf. (3.12) in [8]).

To illustrate (5.31) we take $\gamma = 1$, $b = 3$, a common value of r and a common p-value, so that the common $a = 1$. We obtain for $s = 2$

$$(5.34)\qquad \mu_2 = \frac{6r}{p} CD_1^{(1,1)}(r,r+1) = \frac{6r}{p}\{D_1^{(1)}(r,r+1)-D_1^{(2)}(r,r+1)\} .$$

The corresponding results for $s = 1$ and $s = 3$ are

(5.35) $$\mu_1 = \frac{3r}{p} D_1^{(2)}(r,r+1) \; ; \; \mu_3 = \frac{3r}{p} C_1^{(2)}(r,r+1) \; .$$

Hence the expected waiting times to empty $s = 1,2,$ and 3 match boxes respectively, for say $r = 10$ and $p = 1/3$ are (cf. Pages B86, B86, A86)

(5.36) $$\mu_1 = 22.3447 \; , \; \mu_2 = 29.4529 \text{ and } \mu_3 = 38.2024 \; .$$

The corresponding $\mu_s^{[2]}$-values and numerical σ_s^2-values are

(5.37a) $$\mu_1^{[2]} = \frac{3r(r+1)}{p^2} D_1^{(2)}(r,r+2) = 531.9801 \; ; \; \sigma_1^2 = 10.3503 \; ,$$

(5.37b) $$\mu_2^{[2]} = \frac{6r(r+1)}{p^2} CD_1^{(1,1)}(r,r+2) = 907.0025 \; ; \; \sigma_2^2 = 10.0770 \; ,$$

(5.37c) $$\mu_3^{[2]} = \frac{3r(r+1)}{p^2} C_1^{(2)}(r,r+2) = 1531.0174 \; ; \; \sigma_3^2 = 33.3899 \; .$$

(cf. Pages B86, A86 for the first results in (5.37a) and (5.37c), respectively).

5.3.1 Application to Gamma Populations and Life Testing

Assume that we have b different gamma populations with common known gamma parameter r and different scale parameters θ_α which may be unknown, i.e. with density

(5.38) $$g_r(x,\theta_x) = \frac{x^{r-1}}{\theta_\alpha^r \, \Gamma(r)} e^{-x/\theta_\alpha} \qquad (\alpha = 1,2,\ldots,b) \; .$$

Let $G_r(x,\theta_\alpha)$ denote the cdf of $g_r(x,\theta_\alpha)$. The probability that some particular population (say, the ith) will give rise to the smallest of b observations if we take one from each, is given by

(5.39) $$P_1 = \int_0^\infty g_r(x,\theta_i) \prod_{j\neq i} [1-G_r(x,\theta_j)]dx = \int_0^\infty g_r(y) \prod_{j\neq i} [1-G_r(y\frac{\theta_i}{\theta_j})]dy ,$$

where $g_r(x)$ and $G_r(x)$ with one argument refer to the standard case with $\theta = 1$. If we let $\underset{\sim}{a}_i = (\frac{\theta_i}{\theta_1},\ldots\frac{\theta_i}{\theta_{i-1}}, \frac{\theta_i}{\theta_{i+1}},\ldots\frac{\theta_i}{\theta_b})$ then by (4.28b)

(5.40) $$P_1 = D_{\underset{\sim}{a}_i}^{(b-1)}(r,r)$$

Similarly the probability P_b that the ith population gives rise to the largest of the b observations is $C_{\underset{\sim}{a}_i}^{(b-1)}(r,r)$. More generally the probability P_t that the ith population gives rise to the tth smallest is

(5.41) $$P_t = \Sigma \int_0^\infty g_r(y) \prod_{S_{t-1}} [G_r(y\frac{\theta_i}{\theta_j})] \prod_{\bar{S}_{t-1}} [1-G_r(y\frac{\theta_i}{\theta_k})]dy = \Sigma\, CD_{\underset{\sim}{a}_i}^{(t-1,b-t)}(r,r) ,$$

where S_{t-1} is a combination of $t-1$ subscripts out of $b-1$ (all except the ith), $\bar{S}_{t-1}$ is its complement (not including the ith) and the summation is over all such combinations. The CD integral was defined in (4.18) and the derivation of the single integral representation that is used in (5.41) is similar to that of (4.28) and is omitted.

In ranking and selection problems we generally set up a P*-condition such as the following: If $\theta_i/\theta_j < a$ for all $j \neq i$ where a is specified $(0 < a < 1)$ then we want to have a probability of a correct selection P{CS} which is at least P* in value, where P* is also specified $(\frac{1}{b} < P^* < 1)$. If the ith population gives us the smallest observation then we will select it and get a correct selection if $\theta_i < \theta_j$ for all $j \neq i$; hence we now refer to P_1. It usually turns (as it does in this case) that the first step minimization is to make all the θ_j-values equal to θ_i/a or $\theta_i/\theta_j = a$ for all $j \neq i$. In our problem the P{CS} no longer depends on θ-values and there is no second step minimization. Thus the P*-condition, which can now be used to determine the r-value needed, is

$$D_a^{(b-1)}(r,r) = P^* , \tag{5.42}$$

where a,b and P* are all now specified. In applications r is simply related to the number of observations or the size of the experiment and the determination of the r that satisfies (5.42) is useful in planning before any experimentation is carried out.

The dual problem has as its goal the selection of that one of the b populations with the largest θ-value and the rule is to select the population with the largest observation. Then we want to have a $P\{CS\} \geq P^*$ whenever $\theta_i/\theta_j > a$ where a is again specified but $1 < a < \infty$. We again wish to determine r and corresponding to (5.42) we obtain

$$C_a^{(b-1)}(r,r) = P^* . \tag{5.43}$$

Solutions of both (5.42) and (5.43) can easily be found from the tables in this book using B-tables and A-tables respectively as shown for (5.42) in the following illustration; the method for (5.43) is quite similar.

<u>Numerical Illustration</u>: For $b = 6$, $a = 1/3$ and $P^* = .95$ the required value of r to satisfy (5.42) is (with linear interpolation) 9.30 and if we cannot randomize between 9 and 10 then we need to take $r = 10$ (cf Pages B75, B84); for $P^* = .99$ the corresponding value is 14.52 (which is outside of our tables).

One example of how this gets applied is in the area of life testing where, under the assumption of an exponential distribution for the life of a piece of equipment, the expected life is θ_i for the ith population and we wish to select the one with the longest expected life. Suppose we put n items from each of the b exponential populations on a life test (so that each unit operates independently of the others) and wait for exactly r items to fail from each population $(1 \leq r \leq n)$. It has been shown [6], [7] that for each population the sufficient statistic for θ is 2r (total life observed) and that this statistic has the distribution of gamma with gamma parameter r and the same scale parameter θ. Hence, since we prefer to use the sufficient statistic for selecting the population with the largest θ, the above model is directly applicable and (after a and P^* are specified) we would use (5.43) to determine smallest common r value needed and of course the number of units n needed from each population would have to be greater than (or equal to) r.

Another example is to determine which of b different chemicals will produce tumors as quickly as possible. Suppose that the time to produce a tumor is exponentially distributed with scale parameter θ (which also represents the expected time to produce the tumor). We wish to put n rats on test for each drug and use the same sufficient statistic mentioned above for selecting the chemical with the smallest θ-value. After a $(0 < a < 1)$ and P^* are specified we again use (5.42) to determine the smallest required value of r and of course we need $n \geq r$.

Example: Suppose we have $b = 3$ different chemicals and the time to produce a tumor is exponential with θ-values $\theta_1, \theta_2, \theta_3$. If $\theta_1/\theta_j < a = .65$ for $j = 2,3$, then we want to have $P\{CS\} \geq P^* = .95$. From Table G (cf. Page G8) the smallest common value of r needed is 47.09 (or 48 if randomization is not used) and hence we need at least 48 rats for each of the 3 chemicals. In order to carry out any such experiment it would be useful to know prior to experimentation how many rats are needed in total (in this case, 144 rats is the minimum value of the total number needed).

If we were interested in assessing the relative value of drugs in combating a tumor then, under the same assumptions as above, we may be using as our statistic the post-tumor length of life of the rat and we would then be interested in finding the chemical associated with the largest θ-value. In this case we would specify $a > 1$ and use (5.43).

It is of some interest to ask for the probability that a particular one of b gamma populations will give rise to the smallest s^{th} order statistic if we take n independent observations from each and, as above, there is a common known r and different θ-values $\theta_1, \theta_2, \ldots, \theta_b$, which we now also assume to be known. Let

$$(5.44) \qquad h_s(x,\theta_i) = s\binom{n}{s} g_r(x,\theta_i) G_r^{s-1}(x,\theta_i)[1-G_r(x,\theta_i)]^{n-s} \quad (i=1,2,\ldots,b)$$

denote the density of the sth order statistic from the ith population. Then the corresponding cdf $H_s(x,\theta_i)$ is given by

$$(5.45) \qquad H_s(x,\theta_i) = \sum_{j=s}^{n} \binom{n}{j} G_r^j(x,\theta_i)[1-G_r(x,\theta_i)]^{n-j} .$$

The probability that the ith population gives rise to the smallest s^{th} order statistic is

$$(5.46) \qquad P = b \int_0^\infty h_s(x,\theta_i) \prod_{j\neq i}^{b} [1-H_s(x,\theta_j)]dx$$

$$= b \int_0^\infty h_s(y,1) \prod_{j\neq i}^{b} [1-H_s(\frac{y\theta_i}{\theta_j}, 1)]dy .$$

Substituting from (5.44) and (5.45) into (5.46) gives

(5.47)

$$P = bs\binom{n}{s} \int_0^\infty g_r(x) G_r^{s-1}(x)[1-G_r(x)]^{n-s} \prod_{j\neq i}^{b} \{\sum_{\beta=0}^{s-1} \binom{n}{\beta} G_r^\beta(\frac{x\theta_i}{\theta_j})[1-G_r(\frac{x\theta_i}{\theta_j})]^{n-\beta}\}dx.$$

We shall not evaluate this result, but we wish to point out that after expansion every single term is a CD integral with some vector $\underset{\sim}{a}$ (cf. 4.28c) and for small values of b and s the result is manageable.

5.3.2 Selecting the Population with the Smallest Variance.

In this case we start with $k = b+1$ normal populations and we wish to find the one with the smallest variance; a more detailed discussion can be found in [3] and [9]. Here we assume that the population means μ_i are all known (Case 1) or all unknown (Case 2) and, of course, the variances are not known. In this problem we use the usual sufficient statistic from the ith population $(i = 1,2,\ldots,k)$

$$(5.48) \qquad s_i^2 = \frac{\sum_{j=1}^{n}(x_{ij}-\mu_i)^2}{n}, \quad (s_i^2 = \frac{\sum_{j=1}^{n}(x_{ij}-\bar{x}_i)^2}{n-1}) ,$$

according as the μ_i are all known or all unknown. Then $ns_i^2/2$ [or $(n-1)s_i^2/2$] has the gamma distribution with scale parameter (or θ_i-value) equal to σ_i^2. Hence the same formulation as in section (5.4.1) can now be applied. Let $\sigma_{[1]}^2 \leq \sigma_{[2]}^2 \leq \cdots \leq \sigma_{[k]}^2$ denote the ordered variances and let a (with $0 < a < 1$) and P^* (with $1/k < P^* < 1$) be specified. If $\sigma_{[1]}^2/\sigma_{[j]}^2 < a$ for all $j \neq 1$ then we want to have $P\{CS\} \geq P^*$ and a correct selection is obtained if and only if the best population produces the smallest value of s^2. We now have to solve exactly the same condition as in (5.42) with $r = n/2$ in the first case with all μ_i known and $r = (n-1)/2$ is the second case with all μ_i unknown. Indeed it is now desirable to solve (5.42) both for integer and half-integer values of r. The resulting (integer) value of n would then be the common number of observations required from each of the k populations.

Example: Suppose we wish to (i) select the best of k different marksmen with the same gun or (ii) select the best of k different guns using the same marksman or gun crew. Here variability (i.e, the absence of it) is even more important than hitting the bull's eye and hence we wish to select (i) the marksman or (ii) the gun with the smallest variance. The problem is to determine how many times we have to fire the gun to satisfy a P^*-condition. For simplicity, we assume that we measure only vertical (or only horizontal) displacement and thus have a one-dimensional problem. Assuming that these are independent and normally distributed, the above conditions are satisfied with all means unknown. Then for specified a $(0 < a < 1)$ and P^* $(\frac{1}{k} < P^* < 1)$ we can use (5.42) to determine r and the common number of firings needed (i) by each marksman or (ii) by each gun is $n = 2r+1$; here $n-1 = 2r$ denotes the degrees of freedom of the associated chi-square statistic.

This problem was also considered in Chapter 5 of the book [9] where many detailed examples and some tables can be found. It should be pointed out that the P^*-condition was given there in terms of σ rather than σ^2, so that the a-value above corresponds to $(\Delta^*)^2$ in [9] rather than to Δ^*. In addition the tabulated integers are values of the degrees of freedom $\nu = 2r$ which is n in Case 1 and $n-1$ in case 2. The above analysis also applies if we assume a common unknown mean. We then replace $\bar{x}_i$ in (5.48) by the average $\bar{\bar{x}}$ of all the observations and a conservative solution for r is the same as in Case 2. Actually we only need $n = 2r$ observations from each population in this case and one extra observation from any population to satify the P^*-condition but we prefer not to consider any solutions that involve unequal sample sizes for the k populations or any sequential solutions.

The tables needed for this problem are the same as those needed for the subset selection problem in the following subsection. This explains why the tables used in the book [9] were adapted from tables appearing in [11] [12], which deal with the subset selection problem.

The dual problem of selecting the population with the largest variance is also well defined but does not have as many applications; we would then use (5.43) with $a > 1$ as in the life-testing problem.

5.3.3 Selecting a Subset Containing the Population With the Smallest Variance

In this problem of selecting a random-sized subset of the $k = b+1$ given normal populations that contains the best one (i.e., the one with the smallest variance), we use the same integral equation as before, namely

$$D_a^{(b)}(r,r) = P^* , \tag{5.49}$$

but the parameter a has a different interpretation here. Let $\sigma^2_{[1]} \leq \sigma^2_{[2]} \leq \cdots \leq \sigma^2_{[k]}$ denote the ordered variances let $s^2_{[1]} \leq s^2_{[2]} \leq \cdots \leq s^2_{[k]}$ denote the ordered sample variances and let $s^2_{(1)}$ denote the sample variance from the population with the smallest variance $\sigma^2_{[1]}$. The rule for selecting a subset is to put s^2_i in the subset if and only if $s^2_i < \frac{1}{a} s^2_{[1]}$, where a $(0 < a < 1)$ is determined by (5.49) and r is given because a common number of observations were already taken from each of the $k = b+1$ populations. It was shown in [11] that the minimum $P\{CS\}$ for this problem has to be calculated for the configuration of equal variances and is given by (5.49) As before, the parameter r represents $n/2$ or $(n-1)/2$, where n is the common number of observations per population, according as the means are all known or all unknown; here again the half-integer solutions in r of (5.49) are useful.

A generalization of the above model to r values not necessarily equal (i.e., to vectors $\underset{\sim}{r}$) together with a table of a-values for $P^* = .75$, $.90$, $.95$ and $.99$ as a function of $\nu = 2r$ for common r appears as a companion paper [12] to the above reference [11].

In the subset selection formulation it is natural to ask about the size S of the selected subset. Following the argument in [11], let $Y_i = 1$ if the ith population is included and equal 0 otherwise. Then the expected size $E\{S\}$ is easily seen to be

$$E\{S\} = E\{\sum_{i=1}^{k} Y_i\} = \sum_{i=1}^{k} E\{Y_i\} = \sum_{i=1}^{k} P\ \{i^{\text{th}} \text{ population is included}\} \tag{5.50}$$

$$= \sum_{i=1}^{k} \int_0^{\infty} g_r(x) \prod_{\substack{j=1 \\ j\neq i}}^{k} [1-G_r(\theta_{ij} ax)]dx,$$

where $\theta_{ij} = \sigma_i^2/\sigma_j^2$. Hence if we let $\underset{\sim}{a}_i = (\frac{a\theta_i}{\theta_1}, \ldots \frac{a\theta_i}{\theta_{i-1}}, \frac{a\theta_i}{\theta_{i+1}}, \ldots \frac{a\theta_i}{\theta_k})$

then (5.50) becomes

$$E\{S\} = \sum_{i=1}^{k} D_{\underset{\sim}{a}_i}^{(b)}(r,r) , \tag{5.51}$$

where $b = k-1$, and for the special case $\theta_1 = \cdots = \theta_k$ we obtain $kD_a^{(b)}(r,r)$. For the intermediate case $\delta\sigma_{[1]}^2 = \sigma_{[j]}^2$ for all $j \neq 1$ we have

$$E\{S\} = D_{c/\delta}^{(b)}(r,r) + bD_{(\delta a,a,\ldots,a)}^{(b)}(r,r) . \tag{5.52}$$

5.3.4 Inverse Sampling for Ranking Multinomial Cell Probabilities

This problem was considered in [4]; it represents another important application of C and D integrals. In a multinomial model with $k = b+1$ cells the cell probabilities are $p_1,p_2,\ldots,p_k$ with sum equal to 1 . The ordered p-values are $p_{[1]} \leq p_{[2]} \leq \cdots \leq p_{[k]}$ and the goal is to find the 'best' cell, i.e., the one with the largest p-value, $p_{[k]}$. The rule is simply to wait until any one cell has r observations in it and select that one as being best. We wish to determine r so that a P^*-condition is satisfied. If $p_{[k]}/p_{[j]} \geq \theta^*$ for all $j \neq k$ (where $\theta^* > 1$ is preassigned), then we want to have $P(CS) \geq P^*$ (where $\frac{1}{k} < P^* < 1$ is also preassigned); here the definition of a correct selection is obvious since $p_{[k]} > p_{[k-1]}$. It is shown in [4] that the equation we have to solve is to find the smallest integer r^* greater than or equal to the solution in r of

$$D_{1/\theta^*}^{(b)}(r,r) = P^* . \tag{5.53}$$

This is similar to (5.42) except that the parameters have a different interpretation. A short table of solutions of (5.53) was given in [4]. As noted there several of the answers are based on asymptotic results and are rounded to the nearest integer; hence it is not surprising that our present exact results lead to an r-value that is one larger in some of these cases.

Connected with the above problem, the expected value of the number of observations N_r needed for termination is clearly of considerable interest. More generally we give the βth ascending factorial moment of N_r sketching the derivation without giving all the details.

We start with the probability of reaching frequency r in any cell on exactly the nth observation

$$(5.54)\qquad P\{N_r=n\} = \sum_{\beta=1}^{k} \binom{n-1}{r-1} p_\beta^r (1-p_\beta)^{n-r} J^{(b)}_{\frac{1}{1-p_\beta}(\underset{\sim}{p}_\beta)}(r,n-r),$$

where $\underset{\sim}{p}_\beta = (p_1,\ldots,p_{\beta-1},p_{\beta+1},\ldots,p_k)$ and $b = k-1$. Using the inclusion-exclusion result for J from (4.50) in [18], multiplying by t^{-n} and summing on n from r to ∞ under the integral signs of I, we obtain

$$(5.55)\qquad E\{t^{-N_r}\} = \sum_{\alpha=1}^{k} \frac{(-1)^{\alpha-1}\Gamma(\alpha r)}{\Gamma^\alpha(r)} \sum_{S_\alpha(\underset{\sim}{p})} \sum_{\gamma=1}^{\alpha} p_{i_\gamma}^r \int_0^{p_{i_1}} \cdots \int_0^{p_{i_\alpha}} \frac{\prod_{j\neq i_\gamma} x_j^{r-1} dx_j}{(t-1+p_{i_\gamma} + \sum_{j\neq i_\gamma} x_j)^{\alpha r}},$$

where the middle summation is over combinations of α of the components of $\underset{\sim}{p}$ and the integral, product and sum are all missing the i_γ component. We recognize the right hand side integrals as C integrals after a change of variable, and using inclusion-exclusion again, we obtain

$$(5.56)\qquad E\{t^{-N_r}\} = \sum_{\beta=1}^{k} \left(\frac{p_\beta}{t-1+p_\beta}\right)^r \sum_{\alpha=1}^{k} (-1)^{\alpha-1} \sum_{S_{\alpha-1}(\underset{\sim}{p}_\beta)} C^{(\alpha-1)}\left(\frac{p_{i_1}}{t-1+p_\beta}, \ldots, \frac{p_{i_{\alpha-1}}}{t-1+p_\beta}\right)^{(r,r)}$$

$$= \sum_{\beta=1}^{k} \left(\frac{p_\beta}{t-1+p_\beta}\right)^r D^{(b)}_{\frac{1}{t-1+p_\beta}(\underset{\sim}{p}_\beta)}(r,r)$$

$$= \sum_{\beta=1}^{k} p_\beta^r \frac{\Gamma(kr)}{\Gamma^k(r)} \int_{p_1}^{\infty} \cdots \int_{p_{\beta-1}}^{\infty} \int_{p_{\beta+1}}^{\infty} \cdots \int_{p_k}^{\infty} \frac{\prod_{i=1, \neq\beta}^{k} y_i^{r-1} dy_i}{(t-1+p_\beta + \sum_{i=1, \neq\beta}^{k} y_i)^{kr}}.$$

An immediate corollary obtained by setting $t = 1$ in (5.56) is already given in (4.20). Since (5.56) is the ascending factorial moment generating function, we obtain the γth ascending factorial moment by differentiating γ times and setting $t = 1$, obtaining

$$E\{N_r^{[\gamma]}\} = \frac{\Gamma(r+\gamma)}{\Gamma(r)} \sum_{\beta=1}^{k} \frac{1}{p_\beta^{\gamma}} D_{\frac{1}{p_\beta}(\underset{\sim}{p}_\beta)}^{(b)}(r,r+\gamma) \tag{5.57}$$

For the special case of interest $\gamma = 1$ and $p_k = p > q = p_1 = \cdots = p_{k-1}$, the so-called generalized least favorable (GLF) configuration, we have

$$E\{N_r | \mathrm{GLF}\} = \frac{r}{p} D_{\frac{q}{p}}^{(b)}(r,r+1) + \frac{br}{q} D_{(\frac{p}{q},1,\ldots,1)}^{(b)}(r,r+1) . \tag{5.58}$$

To simplify this result even further (because our tables do not include vector values $\underset{\sim}{a}$) we pause to introduce a

<u>Lemma</u>:

$$D_{(a_1,a,\ldots a)}^{(b,1)}(r,m) = \frac{(a_1/a)^r}{(1-a+a_1)^m} D_{\frac{a}{1-a+a_1}}^{(b,1)}(r,m) , \tag{5.59}$$

where D with vector argument $\underset{\sim}{a}$ is defined in (4.1b).

<u>Proof</u>: From (4.8) with $j = 1$ and $\underset{\sim}{a} = (a_1,a,\ldots a)$ we have

$$D_{(a_1,a,\ldots a)}^{(b,1)}(r,m) = \binom{m+r-1}{r}\left(\frac{1}{1+a_1}\right)^m\left(\frac{a_1}{1+a_1}\right)^r D_{\frac{a}{1+a_1}}^{(b-1)}(r,m+r) . \tag{5.60}$$

Using (2.31) in reverse with $i = j = 1$ and a replaced by $a^* = a/(1-a+a_1)$ so that $a^*/(1+a^*) = a/(1+a_1)$ we obtain the desired lemma (5.59) We note that this lemma can be generalized to hold for any j $(1 \leq j \leq b)$, but it does not hold for $j = 0$.

Returning to (5.58) and using (4.5b) the second D can be written in the form

$$(5.61)\quad D^{(b)}_{(\frac{p}{q},1,\ldots,1)}(r,r+1) = \sum_{\alpha=1}^{r} (\tfrac{q}{p})^{\alpha} D^{(b,1)}_{(\frac{p}{q},1,\ldots,1)}(r,r+1-\alpha) = \frac{q}{p} \sum_{\alpha=1}^{r} D^{(b,1)}_{\frac{q}{p}}(r,\alpha) .$$

where the last equality made use of the above lemma.

Substituting this back in (5.58) and using (2.24b) with $j = 0$ yields

$$(5.62)\quad E\{N_r|GLF\} = \frac{r}{p} \{D^{(b)}_{\frac{q}{p}}(r,r+1) + b \sum_{\alpha=1}^{r} D^{(b,1)}_{\frac{q}{p}}(r,\alpha)\}$$

$$= \frac{r}{p}\{D^{(b)}_{\frac{q}{p}}(r,r+1) + \frac{1}{r} \sum_{\alpha=1}^{r} \alpha[D^{(b)}_{\frac{q}{p}}(r,\alpha) - D^{(b)}_{\frac{q}{p}}(r,\alpha+1)]\}$$

Adding similar terms, the last term vanishes and we obtain

$$(5.63)\quad E\{N_r|GLF\} = \frac{1}{p} \sum_{\alpha=1}^{r} D^{(b)}_{\frac{q}{p}}(r,\alpha) = (1+ba) \sum_{\alpha=1}^{r} D_a^{(b)}(r,\alpha) ,$$

since $bq+p = 1$ and we are setting $q = pa$ so that $a < 1$.

These two results (5.58) and (5.63) should be regarded as corrected versions of Theorem 6.1 of [4] where a permutation error was made in Lemma 6.1. Table N is equivalent to (5.63) and can be used to obtain corrected numerical results for $E(T|LF)$ in Table II of [4].

The same method as above can be applied to obtain a corresponding result for the γth ascending factorial moment for $\gamma \geq 1$. Again considering $p_k = p > q = p_1 = \cdots = p_{k-1}$ we obtain for any $\gamma \geq 1$ (omitting the derivation) the two results corresponding to (5.58) and (5.63)

$$(5.64)\quad E\{N_r^{[\gamma]}|GLF\} = \frac{\Gamma(r+\gamma)}{\Gamma(r)} \{\frac{1}{p^{\gamma}} D^{(b)}_{\frac{q}{p}}(r,r+\gamma) + \frac{b}{q^{\gamma}} D^{(b)}_{(\frac{p}{q},1,\ldots,1)}(r,r+\gamma)\}$$

$$= (1+ba)^{\gamma}\, \gamma! \sum_{\alpha=1}^{r} \binom{\gamma+\alpha-2}{\alpha-1} D_a^{(b)}(r,\gamma+\alpha-1) ;$$

this expresses the desired moments (and hence also the variance) in terms of quantities we have calculated.

An important special configuration is the case of all equal p-values (the so-called W configuration) which involves setting a equal to one above. It should be noted that the above results [e.g., (5.63) and (5.64)] hold by continuity in a , even though we have written $a < 1$. Hence from (5.58) and (5.63) for the W-configuration

$$(5.65)\qquad E\{N_r | W\} = r(b+1)^2\, D_1^{(b)}(r,r+1) = (b+1)\sum_{\alpha=1}^{r} D_1^{(b)}(r,\alpha)$$

and the last equality can now be regarded as an interesting identity in D-values. More generally for any $\gamma \geq 1$, we obtain similarly from the two expressions in (5.64)

$$(5.66)\qquad E\{N_r^{[\gamma]} | W\} = \frac{\Gamma(r+\gamma)}{\Gamma(r)} (b+1)^{\gamma+1} D_1^{(b)}(r,r+\gamma) = (b+1)^{\gamma}\, \gamma! \sum_{\alpha=1}^{r} \binom{\gamma+\alpha-2}{\alpha-1} D_1^{(b)}(r,\gamma+\alpha-1)$$

and again the last equality is an interesting identity for D-values. The same identities could have been obtained from (2.27b) as an afterthought. It should be noted that the dual identities for C are somewhat different, e.g., the one corresponding to (5.65) takes the form

$$(5.67)\qquad r(b+1)\, C_1^{(b)}(r,r+1) = \sum_{\alpha=1}^{r} \sum_{\beta=0}^{b} C_1^{(\beta)}(r,\alpha) .$$

Two other limiting cases of interest here are (i) $q \to 0$ (which implies that $a \to 0$) and (ii) $r \to \infty$ for fixed $a \leq 1$. In Case (i) from (5.89) the limiting value of $E\{N_r\}$ is simply r since each D_a-value approaches one. In Case (ii) for $a < 1$ it follows from an easy probabilistic argument that each D_a-value again approaches one and hence the asymptotic value of $E\{N_r\}$ is $r(1+ba)$. For $a = 1$ another easy probabilistic argument shows that the $D_1(r,r+1)$ value approaches $1/(b+1)$ and hence from (5.65) the asymptotic answer is $r(b+1)$. Hence in both of these subcases and also in Case (i) the asymptotic result for $E\{N_r\}$ for any configuration of the form $p_{[k]} = p \geq q = p_{[k-1]} = \cdots = p_{[1]}$ (the so-called GLF configuration in which $q/p = a \leq 1$) is $r(1+ba)$. It is also clear from (5.63) that $r(1+ba)$ is an upper bound for $E\{N_r\}$ for any GLF configuration with fixed b. For small values of a the asymptotic value may be very close even for moderate values of r and, since the approximation improves with r, there may be no need to calculate any D-values in order to approximate $E\{N_r \mid GLF\}$.

The same inverse-sampling stopping rule (i.e., stop when Max frequency reaches r) was used by Shu-Ping Hu in [13] for the problem of <u>selecting a subset</u> of random size S that includes the best cell, i.e., the one with the largest p-value, In her procedure a cell is retained in the selected subset iff at stopping time it has a frequency equal to or greater than r' where r' (with $0 \leq r' < r$) is determined by the P*-condition. In deriving the $P\{CS\}$, $E\{S\}$ and $E\{N_r\}$ she uses a notation consistent with ours and the natural tool for her problem as well as in [4] where $r' = r$ is the D-integral with equal and unequal values of the components of $\underset{\sim}{a}$. In fact $E\{N_r\}$ is exactly the same in both problems since it depends only on the stopping rule.

5.4 Applications of CD Dirichlet Integrals

The CD integrals were introduced in section 4.5. These integrals were defined, studied and used by Albert H. Liu in his thesis [14] and also in a paper by Sobel and Liu [19]. Here we wish to state the most important properties and applications of these integrals. The notations are quite consistent except that we use CD instead of the single letter L and we break up b into c and $b-c$ instead of b_1 and b_2.

One interesting application which is given in the paper [19] referred to above is to wait for the tth cell to reach the frequency r under a multinomial model with a common cell probability p for b cells with no sink. Let T_{tr} denote the waiting time and $\mu_{t,r}^{[\gamma]}$ denote the γth ascending factorial moment of the required waiting time. Then we get

(5.68) $$P\{T_{t,r} = n\} = t\binom{b}{t}\binom{n-1}{r-1}p^r(1-p)^{n-r}\ IJ_{\frac{p}{1-p}}^{(t-1,b-t)}(r,n-r).$$

Multiplying by $n^{[\gamma]}$ and summing on n under the integral sign in the IJ integral (a direct generalization of the I and J integrals where $t-1$ integrals are of Type I and $b-t$ are of type J), we obtain

(5.69) $$\mu_{t,r}^{[\gamma]} = \frac{t\binom{b}{t}\Gamma(r+\gamma)}{p^{\gamma}\Gamma(r)}\ CD_1^{(t-1,b-t)}(r,r+\gamma)$$

If we were waiting for the tth ordered frequency (from the bottom) to reach r (rather than from the top as in [19] then the only change is to switch the two superscripts of CD in (5.69).

Another application studied in [3] is to the problem of selecting the t-populations with the smallest variances from a set of k normal populations, where $1 \leq t < k$. We assume a common fixed number of observations N from each population and define $n = N$ if all the population means are known and $n = N-1$ if they are all unknown, as in Section (5.3.2) The P*-condition is as follows: If $\sigma_{[t]}/\sigma_{[t+1]} < \Delta^*$ where Δ^* is preassigned $(0 < \Delta^* < 1)$ then we want to have a $P(CS) \geq P^*$, where P^* is also preassigned $(\frac{1}{\binom{k}{t}} < P^* < 1)$; here the definition of a correct selection (CS) is unique since $\Delta^* < 1$. For the least favorable configuration (LFC) in which $\sigma_{[1]} = \cdots = \sigma_{[t]} = \Delta^*\sigma_{[t+1]} = \cdots = \Delta^*\sigma_{[k]}$, we have

(5.70) $$P\{CS|LFC\} = (k-t)\ CD_{\underset{\sim}{a}}^{(t,k-t-1)}(\tfrac{n}{2},\tfrac{n}{2}) = t\ CD_{\underset{\sim}{a}^*}^{(t-1,k-t)}(\tfrac{n}{2},\tfrac{n}{2}),$$

where $\underset{\sim}{a} = \{(\frac{1}{\Delta^*})^2,\dots,(\frac{1}{\Delta^*})^2,1\dots,1)$, which has k-t-1 ones, and $\underset{\sim}{a}^* = \{1,\dots,1, (\Delta^*)^2,\dots,(\Delta^*)^2\}$, which has t-1 ones. The second expression is easily obtained from the first one by an integration-by-parts, using the single integral representation. The value of n needed is obtained by setting either of these two expressions equal to P^* . Tables for selected values of P^* and Δ^* are given in Liu's thesis [14].

Another application given by Liu is to a fixed subset size approach for selecting the populations with the smallest variances. Here we select a subset of the k populations of size s and require that it contain the t 'best' populations, i.e., the t with the t smallest variances. The formulation otherwise is similar to that above. The equation determining the common value of n required is

$$(k-s)\binom{k-t}{k-s} \, CD_{\underset{\sim}{a}}^{(s,k-s-1)}\left(\frac{n}{2},\frac{n}{2}\right) = P^* , \tag{5.71}$$

where $\underset{\sim}{a} = \{(\frac{1}{\Delta^*})^2,\dots,(\frac{1}{\Delta^*})^2,1,\dots,1\}$, which has k-t-1 ones. Tables for this problem also are given in Liu's thesis [14].

5.5 Expressing the C-integral in terms of the I-integral

For any positive integers m , b , r and any b-tuple $\underset{\sim}{a} = (a_1,a_2,\dots,a_b)$ with positive components we define the b-dimensional integrals

$$C_{\underset{\sim}{a}}^{(b)}(r,m) = \frac{\Gamma(m+br)}{\Gamma(m)\Gamma^b(r)} \int_0^{a_1}\cdots\int_0^{a_b} \frac{\prod_{i=1}^{b} x_i^{r-1}\, dx_i}{\left(1 + \sum_{i=1}^{b} x_i\right)^{m+br}} , \tag{5.72}$$

$$I_{\underset{\sim}{p}}^{(b)}(r,n) = \frac{\Gamma(n+1)}{\Gamma(n+1-br)\Gamma^b(r)} \int_0^{p_1}\cdots\int_0^{p_b} \left(1 - \sum_{i=1}^{b} x_i\right)^{n-br} \prod_{i=1}^{b} x_i^{r-1} dx_i , \tag{5.73}$$

where $\underset{\sim}{p} = (p_1,p_2,\dots,p_b)$ is a b-tuple with positive components summing to one; for b = 0 both are defined to be equal to 1. Our goal is to develop an exact expression for the C-integral (5.72) as a sum of products of I-integrals (5.73) This is accomplished on two levels of generality according as i) we use arbitrary a_i or ii) we use a common value a for each a_i ; the case $a = 1$ and $m = r$ is an important special case of the latter as it gives rise to a family of identities among the I-integrals because of the result shown earlier

$$C_1^{(b)}(r,r) = \frac{1}{b+1} . \tag{5.74}$$

The basic relation between the C-integral and the I-integrals is the same on both levels, but on level 1 the number of terms required to write the relation grows very rapidly and becomes unmanageable for hand calculation when $b \geq 5$. On level 2 the result is useful for a wider range of b-values.

If we make the transformation

$$y_i = \frac{x_i}{1 + \sum_{\alpha=1}^{b} x_\alpha} \quad (i = 1,2,\ldots,b), \tag{5.75}$$

then it is easily verified that the Jacobian is given by

$$J\left(\frac{y_1,\ldots,y_b}{x_1,\ldots,x_b}\right) = \left(1 + \sum_{i=1}^{b} x_i\right)^{-(b+1)} = \left(1 - \sum_{i=1}^{b} y_i\right)^{b+1} . \tag{5.76}$$

Since (4) can also be written as

$$x_i = \frac{y_i}{1 + \sum_{i=1}^{b} y_i} \quad (i = 1,2,\ldots,b) \tag{5.77}$$

we obtain from (5.72)

(5.78) $$C_{\underset{\sim}{a}}^{(b)}(r,m) = \frac{\Gamma(m+br)}{\Gamma(m)\Gamma^b(r)} \int \cdots \int_V \left(1 - \sum_{\alpha=1}^{b} y_\alpha\right)^{m-1} \prod_{\beta=1}^{b} y_\beta^{r-1} dy_\beta ,$$

where the region of integration V is bounded below by the b coordinate hyperplanes and above by the b additional hyperplanes

(5.79) $$y_i \le a_i \left(1 - \sum_{\alpha=1}^{b} y_\alpha\right) \quad (i = 1,2,\ldots,b) ,$$

i.e., V consists of those points that are 'under' each of these hyperplanes, 'under' meaning on the side containing the origin.

We now use an inclusion-exclusion method to write (5.78) as a sum of products of I-integrals. For $b = 2$ it is clear that we can write (on level 1)

(5.80)

$$C_{\underset{\sim}{a}}^{(2)}(r,m) = \frac{\Gamma(m+2r)}{\Gamma(m)\Gamma^2(r)} \int_0^{A_{12}} \int_0^{A_2(1-y_1)} (1-y_1-y_2)^{m-1} y_1^{r-1} y_2^{r-1} dy_2 dy_1$$

$$+ \frac{\Gamma(m+2r)}{\Gamma(m)\Gamma^2(r)} \int_0^{A_{21}} \int_0^{A_1(1-y_2)} (1-y_1-y_2)^{m-1} y_1^{r-1} y_2^{r-1} dy_1 dy_2$$

$$- \frac{\Gamma(m+2r)}{\Gamma(m)\Gamma^2(r)} \int_0^{A_{12}} \int_0^{A_{21}} (1-y_1-y_2)^{m-1} y_1^{r-1} y_2^{r-1} dy_1 dy_2$$

$$= I_{A_2}^{(1)}(r,m+r-1)\, I_{A_{12}}^{(1)}(r,m+2r-1) + I_{A_1}^{(1)}(r,m+r-1)\, I_{A_{21}}^{(1)}(r,m+2r-1)$$

$$- I_{A_{12},A_{21}}^{(2)}(r,m+2r-1)$$

where $A_i = a_i/(1 + a_i)$ $(i = 1,2)$, $A_{12} = a_1/(1 + a_1 + a_2)$ and $A_{21} = a_2/(1 + a_1 + a_2)$. For level 2 we simply set $a_i = a$; for the special case $a = 1$ and $m = r$ we use (5.74) and obtain the identity

(5.81) $$C_1^{(2)}(r,r) = \frac{1}{3} = 2I_{1/2}^{(1)}(r,2r-1)\ I_{1/3)}^{(1)}(r,3r-1) - I_{1/3}^{(2)}(r,3r-1)\ .$$

For $b = 3$ we now give the result for level 2 ; a general proof of our result is given later. By elementary methods of advanced calculus we obtain for any $a > 0$

(5.82) $$\begin{aligned} C_a^{(3)}(r,m) = {} & 6I_{a/(1+a)}^{(1)}(r,m+r-1)\ I_{a/(1+2a)}^{(1)}(r,m+2r-1) I_{a/(1+3a)}^{(1)}(r,m+3r-1) \\ & - 3I_{a/(1+a)}^{(1)}(r,m+r-1)\ I_{a/(1+3a)}^{(2)}(r,m+3r-1) \\ & - 3I_{a/(1+2a)}^{(2)}(r,m+2r-1)\ I_{a/(1+3a)}^{(1)}(r,m+3r-1) \\ & + I_{a/(1+3a)}^{(3)}(r,m+3r-1)\ . \end{aligned}$$

For $a = 1$ and $m = r$ this reduces to the identity for all r

(5.83) $$\begin{aligned} C_1^{(3)}(r,r) = \frac{1}{4} = {} & 6I_{1/2}^{(1)}(r,2r-1)\ I_{1/3}^{(1)}(r,3r-1)\ I_{1/4}^{(1)}(r,4r-1) \\ & - 3I_{1/2}^{(1)}(r,2r-1)\ I_{1/4}^{(2)}(r,4r-1) \\ & - 3I_{1/3}^{(2)}(r,3r-1)\ I_{1/4}^{(1)}(r,4r-1) + I_{1/4}^{(3)}(r,4r-1)\ . \end{aligned}$$

For $b = 4$ we give the 8-term result only for $a = 1$ and $m = r$, namely

(5.84) $$\begin{aligned} C_1^{(4)}(r,r) = \frac{1}{5} = {} & 24I_{1/2}^{(1)}(r,2r-1)\ I_{1/3}^{(1)}(r,3r-1)\ I_{1/4}^{(1)}(r,4r-1)\ I_{1/5}^{(1)}(r,5r-1) \\ & - 12I_{1/2}^{(1)}(r,2r-1)\ I_{1/3}^{(1)}(r,3r-1)\ I_{1/5}^{(2)}(r,5r-1) \\ & - 12I_{1/2}^{(1)}(r,2r-1)\ I_{1/4}^{(2)}(r,4r-1)\ I_{1/5}^{(1)}(r,5r-1) \\ & - 12I_{1/3}^{(2)}(r,3r-1)\ I_{1/4}^{(1)}(r,4r-1)\ I_{1/5}^{(1)}(r,5r-1) \\ & + 6I_{1/3}^{(2)}(r,3r-1)\ I_{1/5}^{(2)}(r,5r-1) + 4I_{1/2}^{(1)}(r,2r-1)\ I_{1/5}^{(3)}(r,5r-1) \\ & + 4I_{1/4}^{(3)}(r,4r-1)\ I_{1/5}^{(1)}(r,5r-1) - I_{1/5}^{(4)}(r,5r-1) \end{aligned}$$

In general the number of terms for our result (on level 2) is 2^{b-1} and each coefficient corresponds to a composition of b . Thus the 8 terms in (5.84) correspond respectively to the 8 compositions of $b = 4$, namely

(5.85) $$(1,1,1,1),\ (1,1,2),\ (1,2,1),\ (2,1,1),\ (2,2),\ (1,3),\ (3,1),\ (4)\ .$$

Moreover, for each composition $(b_1,\ldots,b_s)$ of b the corresponding coefficient is simply the multinomial coefficient

(5.86) $$\begin{bmatrix} b \\ b_1,\ldots,b_s \end{bmatrix} = \frac{b!}{b_1!b_2!\cdots b_s!}\ .$$

Finally, as a check the (algebraic) sum of these coefficients is 1 for all b . Thus for $b = 6$, if we omit the I-function arguments which are clear from the subscripts, we obtain for $a = 1$ and $m = r$ the 32-term result

(5.87)

$$C_1^{(6)}(r,r) = \frac{1}{7} = 720\ I_{1/2}^{(1)}I_{1/3}^{(1)}I_{1/4}^{(1)}I_{1/5}^{(1)}I_{1/6}^{(1)}I_{1/7}^{(1)} - 360\ I_{1/2}^{(1)}I_{1/3}^{(1)}I_{1/4}^{(1)}I_{1/5}^{(1)}I_{1/7}^{(2)}$$

$$- 360\ I_{1/2}^{(1)}I_{1/3}^{(1)}I_{1/4}^{(1)}I_{1/6}^{(2)}I_{1/7}^{(1)} - 360\ I_{1/2}^{(1)}I_{1/3}^{(1)}I_{1/5}^{(2)}I_{1/6}^{(1)}I_{1/7}^{(1)} - 360\ I_{1/2}^{(1)}I_{1/4}^{(2)}I_{1/5}^{(1)}I_{1/6}^{(1)}I_{1/7}^{(1)}$$

$$- 360\ I_{1/3}^{(2)}I_{1/4}^{(1)}I_{1/5}^{(1)}I_{1/6}^{(1)}I_{1/7}^{(1)} + 180\ I_{1/2}^{(1)}I_{1/3}^{(1)}I_{1/5}^{(2)}I_{1/7}^{(2)} + 180\ I_{1/2}^{(1)}I_{1/4}^{(2)}I_{1/5}^{(1)}I_{1/7}^{(2)}$$

$$+ 180\ I_{1/3}^{(2)}I_{1/4}^{(1)}I_{1/5}^{(1)}I_{1/7}^{(2)} + 180\ I_{1/2}^{(1)}I_{1/4}^{(2)}I_{1/6}^{(2)}I_{1/7}^{(1)} + 180\ I_{1/3}^{(2)}I_{1/4}^{(1)}I_{1/6}^{(2)}I_{1/7}^{(1)}$$

$$+ 180\ I_{1/3}^{(2)}I_{1/5}^{(2)}I_{1/6}^{(1)}I_{1/7}^{(1)} + 120\ I_{1/2}^{(1)}I_{1/3}^{(1)}I_{1/4}^{(1)}I_{1/7}^{(3)} + 120\ I_{1/2}^{(1)}I_{1/3}^{(1)}I_{1/6}^{(3)}I_{1/7}^{(1)}$$

$$+ 120\ I_{1/2}^{(1)}I_{1/5}^{(3)}I_{1/6}^{(1)}I_{1/7}^{(1)} + 120\ I_{1/4}^{(3)}I_{1/5}^{(1)}I_{1/6}^{(1)}I_{1/7}^{(1)} - 90\ I_{1/3}^{(2)}I_{1/5}^{(2)}I_{1/7}^{(2)}$$

$$- 60\ I_{1/2}^{(1)}I_{1/4}^{(2)}I_{1/7}^{(3)} - 60\ I_{1/2}^{(1)}I_{1/5}^{(3)}I_{1/7}^{(2)} - 60\ I_{1/3}^{(2)}I_{1/4}^{(1)}I_{1/7}^{(3)} - 60\ I_{1/3}^{(2)}I_{1/6}^{(3)}I_{1/7}^{(1)}$$

$$- 60\ I_{1/4}^{(3)}I_{1/5}^{(1)}I_{1/7}^{(2)} - 60\ I_{1/4}^{(3)}I_{1/6}^{(2)}I_{1/7}^{(1)} - 30\ I_{1/2}^{(1)}I_{1/3}^{(1)}I_{1/7}^{(4)} - 30\ I_{1/2}^{(1)}I_{1/6}^{(4)}I_{1/7}^{(1)}$$

$$- 30\ I^{(4)}_{1/5}I^{(1)}_{1/6}I^{(1)}_{1/7} + 20\ I^{(3)}_{1/4}I^{(3)}_{1/7} + 15\ I^{(2)}_{1/3}I^{(4)}_{1/7} + 15\ I^{(4)}_{1/5}I^{(2)}_{1/7} + 6\ I^{(1)}_{1/2}I^{(5)}_{1/7}$$

$$+ 6\ I^{(5)}_{1/6}I^{(1)}_{1/7} - I^{(6)}_{1/7} .$$

The corresponding results for a_i or for equal a can be written down for any b. For level 2 we can write the general result in the form

$$C^{(b)}_a(r,m) = \Sigma(-1)^s \begin{bmatrix} b \\ b_1,\ldots,b_s \end{bmatrix} \prod_{j=1}^{s} I^{(b_j)}_{a/(1+B_j a)}(r,m+B_j r-1), \tag{5.88}$$

where $B_j = \sum_{\alpha=1}^{j} b_\alpha$, the outside sum is over the set of compositions $(b_1,\ldots,b_s)$ of b , and s is the number of parts in any particular composition. For level 1 all the coefficients in (5.88) are $+1$ or -1 as in (5.80) and the number of terms grows rapidly so that it quickly becomes unmanageable. In addition, the I-functions that appear for the level 1 result as in (5.80) have upper limits that are not necessarily equal and tables for these generalized I-functions are not yet available. For $a = 1$ and $m = r$ in (5.88) the common value of both sides is $1/(b + 1)$; this gives us non-trivial identities for the I-functions with common upper limits that are reciprocals of an integer as in available tables [14], [18].

Proof of the Partition Formula (5.88) by Induction

Let d denote the dimension of our hypervolume V . Suppose the partition formula (5.88) is true for any dimension less than d ; we wish to show it for d . We "chop" off a piece of V with one hyperplane $x_i = c$ where $c = a/(1 + da)$, the common coordinate of the point where all the hyperplanes meet and we integrate over the remaining hypervolume. This is done for each i $(i = 1,2,\ldots,d)$ and we add the integrals together, removing the overlap later. By the induction assumption this total is

(5.89) $$T_1 = \binom{d}{1} \sum_{\underset{\sim}{b}} (-1)^{\alpha} \begin{bmatrix} d-1 \\ b_1,b_2,\ldots,b_\alpha \end{bmatrix} F_{\underset{\sim}{b}} ,$$

where the summation is over all compositions $\underset{\sim}{b} = (b_1,b_2,\ldots,b_\alpha)$ of $d-1$, α is the number of parts in the composition, and letting $B_j = b_1 + b_2 + \cdots + b_j$ $(j = 1,2,\ldots,\alpha)$, so that $B_\alpha = d - 1$, we can write the d-dimensional integral $F_{\underset{\sim}{b}}$ as

(5.90) $$F_{\underset{\sim}{b}} = I^{(b_1)}_{a/(1+aB_1)}(r,m+B_1r-1)\, I^{(b_2)}_{a/(1+aB_2)}(r,m+B_2r-1) \cdots$$
$$I^{(b_\alpha)}_{1/(1+aB_\alpha)}(r,m+B_\alpha r-1) I^{(1)}_{a/(1+da)}(r,m+dr-1) .$$

The hypervolume remaining after chopping V with any two such hyperplanes was included twice above and has to be removed; hence we add to (5.89) the negative term

(5.91) $$T_2 = - \binom{d}{2} \sum_{b'} (-1)^{\beta} \begin{bmatrix} d-2 \\ b'_1,b'_2,\ldots,b'_\beta \end{bmatrix} F_{\underset{\sim}{b}'} ,$$

where the summation is over all compositions $\underset{\sim}{b}' = (b'_1,b'_2,\ldots,b'_\beta)$ of $d - 2$, β is the number of parts in the composition, and, letting $B'_j = b'_1 + b'_2 + \cdots + b'_j$ $(j = 1,2,\ldots,\beta)$ so that $B'_\beta = d - 2$, we can write the d-dimensional integral $F_{\underset{\sim}{b}'}$ as

(5.92) $$F_{\underset{\sim}{b}'} = I^{(b'_1)}_{a/(1+aB'_1)}(r,m+B'_1r-1)\, I^{(b'_2)}_{a/(1+aB'_2)}(r,m+B_2r-1) \cdots$$
$$I^{(b'_\beta)}_{a/(1+aB'_\beta)}(r,m+B'_\beta r-1)\, I^{(2)}_{a/(1+da)}(r,m+dr-1) .$$

In the next step we add back hypervolumes obtained by chopping V with 3 hyperplanes $x_i = c$, $x_j = c$, $x_k = c$. In the next-to-last step of this inclusion-exclusion argument we have only one term in the sum over partitions; the term becomes

(5.93) $$T_{d-1} = (-1)^{d-2} d I^{(1)}_{a/(1+a)}(r,m+r-1)\, I^{(d-1)}_{a/(1+da)}(r,m+dr-1).$$

In the last step the coefficient is $(-1)^{d-1}$ and the term becomes

(5.94) $$T_d = (-1)^{d-1} I_{a/(1+da)}(r,m+dr-1)\ .$$

Combining these terms, we note that each multinomial coefficient corresponding to a composition of $d - j$ is multiplied by $\binom{d}{j}$ and after cancelling $(d - j)!$ it becomes a multinomial coefficent with d on top, corresponding to a composition of d. Moreover, if we consistently put the new part j at the end of the composition then the compositions are all different compositions of d. There are a total of 2^{d-1} compositions of d and to see that we have all of them we use the induction hypothesis again. The jth term has 2^{d-1-j} type terms $(j = 1,2,\dots,d-1)$ and the dth term adds on one more type, namely the trivial composition (d) with one part. Adding these over j, we obtain 2^{d-1} and it is easily seen that these are all different. Hence our formula contains all the compositions of d and this proves the formula (5.88).

In the corresponding result on level 1 the multinomial coefficient in (5.88) is replaced by sums which arise because the parts b_i of the composition (and their corresponding a-values) are ordered subsets. Then $B_j a$ is replaced by an appropriate sum of the components of $\underset{\sim}{a}$, etc. It is also quite possible that a corresponding result can be carried out for vector values $\underset{\sim}{r}$ of r, but we shall not prove this result for vectors.

Dual Property for D in Term of J.

Without any formal proof we wish to point out that D integrals can also be expressed in terms of J (or I) integrals and indeed the relation between D and J is exactly the same as the relation between C and I. Thus, for example, for $b = 2$ on level 2 we have

(5.95) $$D_a^{(2)}(r,m) = 2J^{(1)}_{\frac{a}{1+a}}(r,m+r-1)\; J^{(1)}_{\frac{a}{1+2a}}(r,m+2r-1) - J^{(2)}_{\frac{a}{1+2a}}(n,m+2r-1)\ ,$$

which is exactly the result obtained from (5.106) for C in terms of I by setting $a_1 = a_2 = a$.

It should be noted that the results for $b = 1$ have not been written explicitly here since they already appear in (2.3) above. Hence these results can be regarded as a generalization of (2.3).

5.6 Application of the C Integral to Truncated Riemann Zeta Series

From (2.22) with $j = 0$ we have

(5.96) $$C_a^{(b)}(1,m) = \sum_{\alpha=0}^{b} (-1)^{\alpha}\binom{b}{\alpha} \frac{1}{(1+\alpha a)^m}\ ,\ \cdot\ D_a^{(b)}(1,m) = \frac{1}{(1+ba)^m}\ ,$$

since for $j = 0$ the functions $\mathcal{C}$ and C are equal and $\mathcal{D}$ and D are also equal. It follows that

(5.97) $$L_\gamma(j) = \sum_{\alpha=0}^{b} (-1)^{\alpha}\binom{b}{\alpha} \frac{1}{(\alpha+j)^{\gamma+1}} = \left(\frac{1}{j}\right)^{\gamma+1} C^{(b)}_{1/j}(1,\gamma+1).$$

We are going to show that the L (and hence also the C) are simply related to the truncated Riemann Zeta sums

(5.98) $$S_i = \sum_{\alpha=1}^{b+1} \frac{1}{\alpha^i}$$

and hence our tables can be used to evaluate them.

We note that for $\gamma = 0,1,2,\ldots$

(5.99) $$(-1)^\gamma \gamma! L_\gamma(j) = \frac{\partial^\gamma}{\partial j^\gamma} L_o(j) = \frac{\partial^\gamma}{\partial j^\gamma} \int_o^1 x^{j-1}(1-x)^b dx = b!\frac{\partial^\gamma}{\partial j^\gamma} \frac{\Gamma(j)}{\Gamma(j+b+1)}$$

Using the well-known function

(5.100) $$\psi(x) = \frac{d}{dx} \log \Gamma(x)$$

and its derivatives (polygamma functions) and letting $R(j)$ denote $\Gamma(j)/\Gamma(j+b+1)$ we obtain after one differentiation

(5.101) $$\frac{\partial}{\partial j} R(j) = [\psi(j) - \psi(j+b+1)]\, R(j) \quad .$$

Differentiating (5.101) α times we obtain with Leibniz's rule

(5.102) $$R(j)^{(\alpha+1)} = \sum_{\beta=0}^{\alpha} \binom{\alpha}{\beta} D_\beta\, R(j)^{(\alpha-\beta)} \quad ,$$

where

(5.103) $$D_\beta = \frac{\partial^\beta}{\partial j^\beta} [\psi(j) - \psi(j+b+1)] = (-1)^{\beta+1}\beta! \sum_{\alpha=j}^{b+j} \frac{1}{\alpha^{\beta+1}}$$

and the last equality appears e.g. in 6.4.3 (page 260) of [1]. Letting $T_{\alpha+1} = R^{(\alpha)}(j)/R(j)$ we obtain from (5.102) for $\alpha = 0,1,2,3,4,5,\ldots$

$$T_1 = D_0 \ , \ T_2 = D_0^2 + D_1 , \ T_3 = D_0^3 + 3D_0 D_1 + D_2 \ ,$$

$$T_4 = D_0^4 + 6D_0^2 D_1 + 4D_0 D_2 + 3D_1^2 + D_3 \ ,$$

(5.104) $$T_5 = D_0^5 + 10D_0^3 D_1 + 10D_0^2 D_2 + 15D_0 D_1^2 + 5D_0 D_3 + 10D_1 D_2 + D_4 ,$$

$$T_6 = D_0^6 + 15D_0^4 D_1 + 20D_0^3 D_2 + 45D_0^2 D_1^2 + 15D_0^2 D_3 + 60D_0 D_1 D_2$$

$$+ 6D_0 D_4 + 15D_1^3 + 15D_1 D_3 + 10D_2^2 + D_5 \ ,\ldots \ .$$

Since we are interested in obtaining (5.98) we set $j = 1$ after having differentiated so that D_β in (5.103) becomes $(-1)^{\beta+1}\beta! S_{\beta+1}$. It follows from (5.97) and (5.99) with $j = 1$ that

(5.105) $$T_\alpha = \frac{(-1)^\alpha \alpha! L_\alpha(1)(b+1)!}{b!} = (-1)^\alpha \alpha!(b+1) C_1^{(b)}(1,\alpha+1) \ .$$

Hence from (5.104) we obtain for $\alpha = 1,2,3$

$$(b+1)C_1^{(b)}(1,2) = -D_0 = S_1 = \sum_{\alpha=1}^{b+1} \frac{1}{\alpha} ,$$

(5.106)
$$2(b+1)C_1^{(b)}(1,3) = D_0^2 + D_1 = S_1^2 + S_2 = (\sum_1^{b+1} \frac{1}{\alpha})^2 + \sum_1^{b+1} \frac{1}{\alpha^2},$$

$$6(b+1)C_1^{(b)}(1,4) = -[D_0^3 + 3D_0 D_1 + D_2] = S_1^3 + 3S_1S_2 + 2S_3 = \dots$$

$$\dots$$

This expresses the $C_1^{(b)}(1,\alpha+1)$ values in terms of truncated Riemann Zeta series. We now invert this result to obtain S_i as a function of C-values.

In view of the structure in (5.106) it is convenient to let $A_\alpha = (b+1)C_1^{(b)}(1,\alpha+1)$ for $\alpha = 1,2,\dots$ and we obtain from (5.106) (5.104)

(5.107)
$$S_1 = A_1, S_2 = 2A_2 - A_1^2 , \; S_3 = 3A_3 - 3A_1A_2 + A_1^3 ,$$

$$S_4 = 4A_4 - A_1^4 + 4A_1^2A_2 - 2A_2^2 - 4A_1A_3 ,$$

$$S_5 = 5A_5 + A_1^5 - 5A_1^3A_2 + 5A_1^2A_3 + 5A_1A_2^2 - 5A_1A_4 - 5A_2A_3 ,$$

$$S_6 = 6A_6 - A_1^6 + 6A_1^4A_2 - 6A_1^3A_3 - 9A_1^2A_2^2 + 6A_1^2A_4$$

$$- 6A_1A_5 + 12A_1A_2A_3 + 2A_2^3 - 3A_3^2 - 6A_2A_4 ,$$

$$\dots$$

These are the desired expressions for S_i given in (5.98) in terms of the C functions. After computing these coefficients of A's in (5.107) it appears that these constants are related to monomial symmetric functions (or ms numbers) and are given in [5] as the last column in each table on pages 81-96 for $i = 2(1)12$, except that for even i all the signs have to be changed so that the new numbers always add to $+1$.

Numerical Illustration

We wish to evaluate S_2 for $b = 15$ i.e., $\sum_{\alpha=1}^{16} \frac{1}{\alpha^2}$. From (5.107) we have

(5.108) $$S_2 = 2A_2 - A_1^2 = 32\{C_1^{(15)}(1,3) - 8[C_1^{(15)}(1,2)]^2\}$$

$$= 32\{.40667735 - 8\ (.21129556)^2\} = 1.584347\ .$$

This is still far from the sum to ∞ which is well known to be $\pi^2/6 = 1.644934$.

It should be noted in (5.107) that for each S_i the number of terms corresponds exactly to the number of partitions of i and the number of parts in the partition determines the sign of that term. Furthermore the sum of the coefficients of the A-values for each S_i in (5.107) must add to one; this is a 'weak' check on the algebra.

References

[1] Abramovitz, A. and Stegun, I.A. (1964), Handbook of Mathematical Functions, National Bureau of Standards, Washington, D.C.

[2] Anderson, K., Sobel, M. and Uppuluri, V.R.R. (1982), Distribution and moments of quota fulfillment times. Canadian Journal of Statistics 10 (to appear).

[3] Bechhofer, R.E. and Sobel, M. (1954), A single-sample multiple-decision procedure for ranking variances of normal populations, Ann. Math. Statist. 25, 273-289.

[4] Cacoullos, T. and Sobel, M. (1966), An inverse-sampling procedure for selecting the most probable event in a multinomial distribution. In Multivariate Analysis I (P.R. Krishaiah, ed.), pp. 423-455, Academic Press, New York.

[5] David, F.N., Kendall, M.G. and Barton, D.E. (1966), Symmetric Function and Allied Tables, Cambridge University Press.

[6] Epstein, B. and Sobel, M. (1953), Life Testing, JASA 48, 486-502.

[7] Epstein, B. and Sobel, M. (1955), Sequential procedures in life testing from an exponential distribution. Ann. Math. Statist. 26, 82-93.

[8] Feller, W. (1957), An Introduction to Probability Theory and its Applications, Vol. I. (Second edition). John Wiley, New York, N.Y.

[9] Gibbons, J. D., Olkin, I. and Sobel, M. (1977), Selecting and Ordering Populations : A New Statistical Methodology. John Wiley, New York, N.Y.

[10] Gupta, S.S. (1963), Probability integrals of the multivariate normal and multivariate t. Ann. Math. Statist. 34, 792-828.

[11] Gupta, S.S. and Sobel, M. (1962a), On selecting a subset containing the population with the smallest variance. Biometrika 49, 495-507.

[12] Gupta, S.S. and Sobel, M., (1962b), On the smallest of several correlated F-statistics. Biometrika 49, 509-523.

[13] Hu, Shu-Ping (1982), Subset selection with inverse sampling procedures and Dirichlet distributions. Ph.D. Dissertation, Department of Mathematics, University of California at Santa Barbara.

[14] Liu, Albert H. (1980), Dirichlet distributions - Type 2 for ranking and selection problems. Ph.D. Dissertation, Department of Mathematics, University of California at Santa Barbara, June 1980.

[15] Milton, R.C. (1963), Tables of the equally correlated multivariate normal probability integral, Tech. Report # 27, Department of Statistics, University of Minnesota, Minneapolis, Minnesota.

[16] Olkin, I. and Sobel, M. (1965), Integral expressions for tail probabilities of the multinomial and negative multinomial distributions. Biometrika 52, 167-179.

[17] Rao, J.S. and Sobel, M. (1980), Incomplete Dirichlet integrals with applications to ordered uniform spacings. Journal of Multivariate Analysis 10, 603-610.

[18] Sobel, M., Uppuluri, V.R.R., and Frankowski, K. (1977), Dirichlet Distribution-Type I. Selected Tables in Mathematical Statistics, Vol. IV, American Mathematical Society, Providence, Rhode Island.

[19] Sobel, M. and Liu, Albert H. (1980), Dirichlet distributions used for different stopping rules. (Submitted).

[20] Wilks, S.S (1962), Mathematical Statistics, John Wiley and Sons, Inc., New York, NY.

INDEX

TABLE A

This table gives the value of the incomplete Dirichlet integral of type 2

$$C_A^{(B)}(R,M) = \frac{\Gamma(M+BR)}{\Gamma^B(R)\Gamma(M)} \int_0^A \cdots \int_0^A \frac{\prod_{i=1}^{B} x_i^{R-1}\, dx_i}{(1 + \sum_{i=1}^{B} x_i)^{M+BR}}$$

as defined in (   ) for $R = 1(1)10$,
$B = 1(1)15.$
$M = 1(1)15$, and
$A = \frac{1}{5}, \frac{1}{4}, \frac{1}{3}, \frac{1}{2}, 1, 2, 3, 4, 5.$

Note: In the table entries less than 10^{-4} are given in exponential notation to avoid too many zeros. Thus .85999\04 represents .000085999.

Note 2: The symbols M, B, R and A which are all capital here in the table correspond to the same lower case symbols in the text.

DIRICHLET C INTEGRAL
1/A = 5 R = 1

M	B= 1	B= 2	B= 3	B= 4	B= 5
1	.16666667	.04761905	.01785714	.00793651	.00396825
2	.30555556	.12131519	.05665391	.02958869	.01677847
3	.42129630	.20702408	.11304272	.06667937	.04172892
4	.51774691	.29580203	.18157746	.11774519	.07973706
5	.59812243	.38217929	.25680315	.17954873	.12964287
6	.66510202	.46301436	.33413235	.24825256	.18894773
7	.72091835	.53670121	.41009567	.32018283	.25456528
8	.76743196	.60262428	.48229390	.39223220	.32339874
9	.80619330	.66078686	.54922876	.46200845	.39270359
10	.83849442	.71156045	.61010314	.52782831	.46026595
11	.86541201	.75551804	.66463373	.58863072	.52444834
12	.88784335	.79332527	.71289306	.64385842	.58415338
13	.90653612	.82567123	.75518487	.69333684	.63874511
14	.92211343	.85322614	.79195035	.73716507	.68795507
15	.93509453	.87661711	.82370038	.77562521	.73179015

M	B= 6	B= 7	B= 8	B= 9	B=10
1	.00216450	.00126263	.00077700	.00049950	.00033300
2	.01013576	.00643862	.00426107	.00291765	.00205610
3	.02736839	.01864765	.01311435	.00947268	.00700048
4	.05593312	.04039751	.02990399	.02260709	.01740489
5	.09613845	.07291306	.05637111	.04431253	.03534332
6	.14676169	.11599143	.09306053	.07565053	.06221479
7	.20556365	.16824189	.13932598	.11658475	.09846143
8	.26983733	.22750590	.19359055	.16608848	.14354613
9	.33685529	.29129305	.25371516	.22241992	.19612865
10	.40417020	.35713805	.31736002	.28345284	.25434478
11	.46977646	.42284379	.38227311	.34698015	.31610169
12	.53216387	.48661383	.44648279	.41094613	.37933132
13	.59029909	.54709690	.50839916	.47359451	.44217345
14	.64356600	.60337054	.56684309	.53354003	.50308450
15	.69168826	.65488921	.62102532	.58978057	.56088188

M	B=11	B=12	B=13	B=14	B=15
1	.00022894	.00016160	.00011671	.85999\04	.64499\04
2	.00148511	.00109585	.00082386	.00062969	.00048837
3	.00527693	.00404720	.00315183	.00248811	.00198818
4	.01361490	.01080087	.00867614	.00704771	.00578283
5	.02855319	.02333192	.01926087	.01604688	.01348087
6	.05169554	.04335330	.03666096	.03123620	.02679737
7	.08384709	.07193715	.06213821	.05400610	.04720392
8	.12489018	.10931576	.09621088	.08510436	.07562925
9	.17386663	.15488108	.13858380	.12451027	.11229001
10	.22919536	.20733822	.18823977	.17146885	.15667414
11	.28894346	.26494192	.24363577	.22464447	.20765189
12	.35108511	.32574888	.30293968	.28233568	.26366473
13	.41370834	.38783791	.36425507	.34269734	.32293917
14	.47515445	.44947312	.42580144	.40393194	.38368375
15	.53409206	.50920413	.48603672	.46443020	.44424358

DIRICHLET C INTEGRAL
1/A = 4 R = 1

M	B= 1	B= 2	B= 3	B= 4	B= 5
1	.20000000	.06666667	.02857143	.01428571	.00793651
2	.36000000	.16444444	.08680272	.05054422	.03160746
3	.48800000	.27229630	.16629997	.10842209	.07428225
4	.59040000	.37833086	.25717035	.18279622	.13456779
5	.67232000	.47632724	.35109473	.26694548	.20811095
6	.73785600	.56350350	.44212706	.35453627	.28945835
7	.79028480	.63909726	.52654286	.44053957	.37339236
8	.83222784	.70347412	.60237054	.52145505	.45564941
9	.86578227	.75757684	.66888753	.59517129	.53316157
10	.89262582	.80259317	.72618994	.66068062	.60400548
11	.91410065	.83976233	.77486382	.71777222	.66720923
12	.93128052	.87026839	.81575150	.76676186	.72251624
13	94502442	.89518707	.84979531	.80827859	.77016199
14	.95601953	.91546456	.87793927	.84310893	.81068807
15	.96481563	.93191491	.90107169	.87209031	.84480042

M	B= 6	B= 7	B= 8	B= 9	B=10
1	.00476190	.00303030	.00202020	.00139860	.00099900
2	.02086924	.01438235	.01026164	.00753455	.00566725
3	.05291705	.03890443	.02935683	.02264228	.01779227
4	.10190749	.07899729	.06245047	.05020180	.04094193
5	.16562957	.13412692	.11023477	.09176309	.07724276
6	.23992684	.20145414	.17104768	.14665242	.12682109
7	.32000615	.27689633	.24161345	.21239467	.18794508
8	.40139211	.35612091	.31795176	.28547265	.25760763
9	.48045378	.43524183	.39614514	.36209207	.33223937
10	.55458480	.51118736	.47283995	.43876368	.40832816
11	.62215946	.58180597	.54548396	.51264695	.48284158
12	.68237353	.64580351	.61236366	.58168160	.55344159
13	.73504661	.70259457	.67251760	.64456806	.61853193
14	.78043148	.75212715	.72559063	.70066061	.67719527
15	.81905285	.79471623	.77167436	.74982398	.72907292

M	B=11	B=12	B=13	B=14	B=15
1	.00073260	.00054945	.00042017	.00032680	.00025800
2	.00435134	.00340087	.00269953	.00217225	.00176925
3	.01420803	.01150624	.00943407	.00782033	.00654642
4	.03381289	.02823623	.02381219	.02025844	.01737170
5	.06566146	.05630515	.04865975	.04234835	.03709011
6	.11051186	.09696018	.08559537	.07598492	.06779654
7	.16729622	.14971221	.13462589	.12159457	.11026867
8	.23352459	.21257149	.19423135	.17808984	.16381170
9	.30591543	.28257945	.26179166	.24319126	.22647977
10	.38101810	.35640844	.33414567	.31393357	.29552225
11	.45568866	.43086860	.40811026	.38718211	.36788530
12	.52737414	.50324776	.48086246	.46004461	.44064265
13	.59422318	.57147933	.55015771	.53013258	.51129259
14	.65506938	.63417187	.61440383	.59567689	.57791177
15	.70933864	.69054695	.67263092	.65553002	.63918934

DIRICHLET C INTEGRAL
1/A = 3 R = 1

M	B= 1	B= 2	B= 3	B= 4	B= 5
1	.25000000	.10000000	.05000000	.02857143	.01785714
2	.43750000	.23500000	.14250000	.09367347	.06524235
3	.57812500	.37225000	.25737500	.18721720	.14147663
4	.68359375	.49678750	.37708125	.29571094	.23787308
5	.76269531	.60315063	.49011594	.40679951	.34345210
6	.82202148	.69069897	.59040745	.51171834	.44861850
7	.86651611	.76102583	.67571664	.60543165	.54662672
8	.89988708	.81657033	.74614348	.68583841	.63363403
9	.92491531	.85990832	.80302590	.75280269	.70811444
10	.94368649	.89341959	.84822275	.80732844	.77012319
11	.95776486	.91915770	.88369022	.85096374	.82064854
12	.96832365	.93882408	.91125715	.88541712	.86112890
13	.97624274	.95379154	.93252435	.91233553	.89313305
14	.98218205	.96514775	.94883605	.93319297	.91817050
15	.98663654	.97374326	.96128965	.94924822	.93759407

M	B= 6	B= 7	B= 8	B= 9	B=10
1	.01190476	.00833333	.00606061	.00454545	.00349650
2	.04746315	.03572421	.02763413	.02186196	.01762378
3	.11013880	.08781443	.07140162	.05901670	.04946449
4	.19529499	.16305082	.13805558	.11829586	.10241170
5	.29406639	.25476172	.22293277	.19677355	.17499774
6	.39710113	.35439931	.31854480	.28810198	.26200100
7	.49678486	.45406919	.41710799	.38485649	.35650522
8	.58801764	.54783310	.51218080	.48034972	.45177022
9	.66808218	.63200745	.59932746	.56958302	.54239545
10	.73610952	.70487890	.67609214	.64946486	.62475654
11	.79246886	.76619187	.74161922	.71858063	.69692892
12	.83824222	.81662712	.79617042	.77677297	.75834742
13	.87483610	.85737341	.84068168	.82470451	.80939133
14	.90372570	.88982001	.87641865	.86349011	.85100578
15	.92630462	.91535923	.90473908	.89442684	.88440659

M	B=11	B=12	B=13	B=14	B=15
1	.00274725	.00219780	.00178571	.00147059	.00122549
2	.01443595	.01198832	.01007533	.00855685	.00733496
3	.04195838	.03596437	.03111017	.02713017	.02383097
4	.08945742	.07875881	.06982469	.06229036	.05588046
5	.15666767	.14108589	.12772442	.11617723	.10612777
6	.23942957	.21976084	.20250401	.18726987	.17374619
7	.33141759	.30908624	.28910207	.27113168	.25490077
8	.42598037	.40260155	.38132039	.36187533	.34404623
9	.51744937	.49447980	.47326241	.45360587	.43534593
10	.60176214	.58030568	.56023506	.54141815	.52373944
11	.67653604	.65728996	.63909217	.62185558	.60550289
12	.74081641	.72411112	.70817007	.69293810	.67836556
13	.79469671	.78057959	.76700280	.75393256	.74133806
14	.83893955	.82726756	.81596792	.80502050	.79440676
15	.87466365	.86518443	.85595634	.84696766	.83820751

DIRICHLET C INTEGRAL

1/A = 2 R = 1

M	B= 1	B= 2	B= 3	B= 4	B= 5
1	.33333333	.16666667	.10000000	.06666667	.04761905
2	.55555556	.36111111	.25666667	.19333333	.15170068
3	.70370370	.53240741	.42211111	.34585185	.29038009
4	.80246914	.66743827	.56930741	.49482222	.43641018
5	.86831276	.76787551	.68844827	.62390626	.57033595
6	.91220850	.84004201	.77940451	.72757176	.68264724
7	.94147234	.89075717	.84621611	.80666799	.77123349
8	.96098156	.92586937	.89400806	.86489471	.83813436
9	.97398771	.94992854	.92756035	.90667180	.88708967
10	.98265847	.96629350	.95080024	.93609076	.92209045
11	.98843898	.97736624	.96673984	.95652348	.94668547
12	.99229265	.98482945	.97759360	.97057023	.96374601
13	.99486177	.98984561	.98494481	.98015328	.97546549
14	.99657451	.99321006	.98990396	.98665373	.98345709
15	.99771634	.99546320	.99323950	.99104425	.98887649

M	B= 6	B= 7	B= 8	B= 9	B=10
1	.03571429	.02777778	.02222222	.01818182	.01515152
2	.12270408	.10160935	.08573192	.07345009	.06373366
3	.24846109	.21582737	.18980828	.16865224	.15116581
4	.38942291	.35084612	.31863855	.29136832	.26800123
5	.52510769	.48638290	.45283403	.42347663	.39756406
6	.64326235	.60840025	.57728701	.54932148	.52402858
7	.73924071	.71016505	.68358944	.65917709	.63665234
8	.81341095	.79046741	.76909182	.74910732	.73036483
9	.86866999	.85129164	.83485168	.81926180	.80444563
10	.90873534	.89597007	.88374639	.87202192	.86075921
11	.93719794	.92803619	.91917823	.91060436	.90229683
12	.95710899	.95064837	.94435434	.93821798	.93223112
13	.97087637	.96638126	.96197587	.95765626	.95341873
14	.98031191	.97721621	.97416814	.97116598	.96820811
15	.98673534	.98461998	.98252961	.98046350	.97842093

M	B=11	B=12	B=13	B=14	B=15
1	.01282051	.01098901	.00952381	.00833333	.00735294
2	.05590087	.04948489	.04415674	.03967882	.03587577
3	.13650967	.12407755	.11342145	.10420362	.09616505
4	.24777176	.23010116	.21454386	.20075133	.18844706
5	.37451909	.35388796	.33530875	.31848907	.30319001
6	.50102712	.48000724	.46071411	.44293598	.42649528
7	.61578692	.59638982	.57829973	.56137926	.54551055
8	.71273746	.69611636	.68040748	.66552895	.65140914
9	.79033668	.77687664	.76401408	.75170344	.73990411
10	.84992497	.83948949	.82942611	.81971077	.81032175
11	.89423962	.88641817	.87881923	.87143067	.86424139
12	.92638628	.92067655	.91509557	.90963746	.90429675
13	.94925989	.94517656	.94116576	.93722472	.93335084
14	.96529300	.96241922	.95958543	.95679034	.95403275
15	.97640125	.97440382	.97242803	.97047332	.96853914

DIRICHLET C INTEGRAL
A = 1 R = 1

M	B= 1	B= 2	B= 3	B= 4	B= 5
1	.50000000	.33333333	.25000000	.20000000	.16666667
2	.75000000	.61111111	.52083333	.45666667	.40833333
3	.87500000	.78703704	.72048611	.66772222	.62449074
4	.93750000	.88734568	.84563079	.81004907	.77912269
5	.96875000	.94161523	.91761912	.89610511	.87660804
6	.98437500	.97012174	.95699609	.94481789	.93344958
7	.99218750	.98483225	.97787321	.97126214	.96496005
8	.99609375	.99233992	.98872324	.98523102	.98185252
9	.99804688	.99614456	.99428923	.99247758	.99070674
10	.99902344	.99806381	.99712016	.99619165	.99527750
11	.99951172	.99902908	.99855185	.99807981	.99761276
12	.99975586	.99951360	.99927316	.99903449	.99879754
13	.99987793	.99975649	.99963566	.99951542	.99939578
14	.99993896	.99987814	.99981752	.99975710	.99969688
15	.99996948	.99993903	.99990866	.99987834	.99984810

M	B= 6	B= 7	B= 8	B= 9	B=10
1	.14285714	.12500000	.11111111	.10000000	.09090909
2	.37040816	.33973214	.31432981	.29289683	.27453430
3	.58819323	.55713559	.53015717	.50643114	.48534961
4	.75184705	.72750812	.70558023	.68566532	.66745480
5	.85878504	.84237542	.82717596	.81302489	.79979125
6	.92278322	.91273224	.90322599	.89420588	.88562273
7	.95893479	.95315947	.94761131	.94227076	.93712094
8	.97857856	.97540118	.97231341	.96930915	.96638295
9	.98897414	.98727752	.98561484	.98398427	.98238415
10	.99437702	.99348958	.99261461	.99175158	.99089999
11	.99715051	.99669289	.99623975	.99579093	.99534630
12	.99856225	.99832858	.99809649	.99786593	.99763687
13	.99927670	.99915819	.99904022	.99892279	.99880589
14	.99963685	.99957702	.99951738	.99945792	.99939864
15	.99981792	.99978781	.99975776	.99972778	.99969785

M	B=11	B=12	B=13	B=14	B=15
1	.08333333	.07692308	.07142857	.06666667	.06250000
2	.25860089	.24462567	.23225445	.22121527	.21129556
3	.46645388	.44939017	.43388048	.41970280	.40667735
4	.65070473	.63521899	.62083767	.60742868	.59488172
5	.78736737	.77566365	.76460465	.75412625	.74417347
6	.87743479	.86960624	.86210612	.85490747	.84798659
7	.93214710	.92733626	.92267697	.91815900	.91377322
8	.96352996	.96074583	.95802662	.95536878	.95276906
9	.98081297	.97926934	.97775201	.97625979	.97479162
10	.99005941	.98922940	.98840959	.98759960	.98679910
11	.99490573	.99446909	.99403627	.99360716	.99318165
12	.99740928	.99718311	.99695834	.99673492	.99651284
13	.99868950	.99857363	.99845825	.99834336	.99822895
14	.99933955	.99928063	.99922189	.99916332	.99910492
15	.99966800	.99963820	.99960846	.99957879	.99954917

DIRICHLET C INTEGRAL
A = 2 R = 1

M	B= 1	B= 2	B= 3	B= 4	B= 5
1	.66666667	.53333333	.45714286	.40634921	.36940837
2	.88888889	.81777778	.76625850	.72626858	.69382674
3	.96296296	.93392593	.90997344	.88956179	.87176769
4	.98765432	.97690864	.96734647	.95870373	.95080045
5	.99588477	.99208955	.98855482	.98523803	.98210734
6	.99862826	.99732052	.99606827	.99486491	.99370513
7	.99954275	.99909831	.99866544	.99824316	.99783061
8	.99984758	.99969773	.99955026	.99940503	.99926190
9	.99994919	.99989890	.99984910	.99979975	.99975086
10	.99998306	.99996623	.99994950	.99993286	.99991631
11	.99999435	.99998873	.99998313	.99997754	.99997197
12	.99999812	.99999624	.99999437	.99999250	.99999063
13	.99999937	.99999875	.99999812	.99999750	.99999687
14	.99999979	.99999958	.99999937	.99999916	.99999896
15	.99999993	.99999986	.99999979	.99999972	.99999965

M	B= 6	B= 7	B= 8	B= 9	B=10
1	.34099234	.31825952	.29953837	.28377319	.27026018
2	.66668564	.64345723	.62322671	.60536073	.58940356
3	.85599215	.84182315	.82896454	.81719592	.80634866
4	.94350750	.93672855	.93038949	.92443193	.91880892
5	.97913813	.97631082	.97360957	.97102127	.96853497
6	.99258460	.99149968	.99044732	.98942489	.98843014
7	.99742707	.99703191	.99664459	.99626460	.99589153
8	.99912076	.99898150	.99884404	.99870828	.99857415
9	.99970239	.99965433	.99960667	.99955938	.99951247
10	.99989986	.99988349	.99986721	.99985101	.99983488
11	.99996643	.99996090	.99995539	.99994989	.99994442
12	.99998877	.99998691	.99998506	.99998321	.99998136
13	.99999625	.99999563	.99999500	.99999438	.99999376
14	.99999875	.99999854	.99999833	.99999812	.99999792
15	.99999958	.99999951	.99999944	.99999937	.99999930

M	B=11	B=12	B=13	B=14	B=15
1	.25850974	.24816935	.23897789	.23073728	.22329414
2	.57501688	.56194297	.54998131	.53897289	.52878970
3	.79629076	.78691685	.77814146	.76989427	.76211670
4	.91348204	.90841944	.90359433	.89898398	.89456890
5	.96614136	.96383248	.96160144	.95944222	.95734953
6	.98746106	.98651592	.98559316	.98469140	.98380941
7	.99552499	.99516463	.99481013	.99446121	.99411760
8	.99844157	.99831050	.99818085	.99805259	.99792565
9	.99946591	.99941969	.99937381	.99932825	.99928300
10	.99981884	.99980288	.99978698	.99977117	.99975542
11	.99993896	.99993351	.99992809	.99992268	.99991728
12	.99997952	.99997768	.99997584	.99997401	.99997218
13	.99999314	.99999253	.99999191	.99999129	.99999067
14	.99999771	.99999750	.99999729	.99999709	.99999688
15	.99999924	.99999917	.99999910	.99999903	.99999896

DIRICHLET C INTEGRAL
A = 3 R = 1

M	B= 1	B= 2	B= 3	B= 4	B= 5
1	.75000000	.64285714	.57857143	.53406593	.50068681
2	.93750000	.89540816	.86372449	.83836614	.81726118
3	.98437500	.97166545	.96087136	.95144788	.94306121
4	.99609375	.99260399	.98943073	.98650897	.98379349
5	.99902344	.99810637	.99723881	.99641344	.99562469
6	.99975586	.99952022	.99929208	.99907064	.99885527
7	.99993896	.99987914	.99982044	.99976276	.99970604
8	.99998474	.99996966	.99995473	.99993997	.99992535
9	.99999619	.99999240	.99998863	.99998489	.99998116
10	.99999905	.99999810	.99999715	.99999621	.99999527
11	.99999976	.99999952	.99999929	.99999905	.99999881
12	.99999994	.99999988	.99999982	.99999976	.99999970
13	.99999999	.99999997	.99999996	.99999994	.99999993
14	1	1	.99999999	.99999999	.99999998
15	1	1	1	1	1

M	B= 6	B= 7	B= 8	B= 9	B=10
1	.47433488	.45277420	.43466323	.41913954	.40561891
2	.79921243	.78346524	.76951316	.75699981	.74566495
3	.93549022	.92857999	.92221732	.91631670	.91081180
4	.98125121	.97885706	.97659147	.97443880	.97238632
5	.99486819	.99414041	.99343846	.99275990	.99210268
6	.99864543	.99844065	.99824056	.99804483	.99785314
7	.99965022	.99959524	.99954105	.99948762	.99943489
8	.99991087	.99989652	.99988230	.99986821	.99985423
9	.99997746	.99997379	.99997013	.99996649	.99996286
10	.99999433	.99999340	.99999246	.99999154	.99999061
11	.99999858	.99999834	.99999811	.99999787	.99999764
12	.99999964	.99999958	.99999953	.99999947	.99999941
13	.99999991	.99999990	.99999988	.99999987	.99999985
14	.99999998	.99999997	.99999997	.99999997	.99999996
15	1	1	1	1	1

M	B=11	B=12	B=13	B=14	B=15
1	.39368895	.38304870	.37347249	.36478708	.35685693
2	.73531271	.72579206	.71698407	.70879344	.70114265
3	.90565006	.90078904	.89619391	.89183576	.88769026
4	.97042349	.96854148	.96673279	.96499100	.96331054
5	.99146506	.99084550	.99024269	.98965544	.98908272
6	.99766526	.99748094	.99729999	.99712221	.99694743
7	.99938284	.99933144	.99928065	.99923046	.99918083
8	.99984036	.99982661	.99981296	.99979941	.99978597
9	.99995926	.99995568	.99995211	.99994856	.99994502
10	.99998969	.99998877	.99998785	.99998694	.99998603
11	.99999740	.99999717	.99999694	.99999670	.99999647
12	.99999935	.99999929	.99999923	.99999917	.99999911
13	.99999984	.99999982	.99999981	.99999979	.99999978
14	.99999996	.99999996	.99999995	.99999995	.99999994
15	.99999999	.99999999	.99999999	.99999999	.99999999

DIRICHLET C INTEGRAL

A = 4 R = 1

M	B= 1	B= 2	B= 3	B= 4	B= 5
1	.80000000	.71111111	.65641026	.61779789	.58837894
2	.96000000	.93234568	.91111988	.89386564	.87931866
3	.99200000	.98537174	.97966006	.97461333	.97007549
4	.99840000	.99695242	.99562223	.99438642	.99322875
5	.99968000	.99937694	.99908811	.99881154	.99854569
6	.99993600	.99987388	.99981344	.99975450	.99969694
7	.99998720	.99997461	.99996221	.99994999	.99993794
8	.99999744	.99999490	.99999239	.99998989	.99998742
9	.99999949	.99999898	.99999847	.99999797	.99999746
10	.99999990	.99999980	.99999969	.99999959	.99999949
11	.99999998	.99999996	.99999994	.99999992	.99999990
12	1	1	.99999999	.99999998	.99999998
13	1	1	1	1	1
14	1	1	1	1	1
15	1	1	1	1	1

M	B= 6	B= 7	B= 8	B= 9	B=10
1	.56484378	.54536641	.52884016	.51454718	.50199725
2	.86673966	.85565783	.84575426	.83680272	.82863673
3	.96594206	.96213915	.95861234	.95532018	.95223034
4	.99213729	.99110287	.99011831	.98917782	.98827666
5	.99828936	.99804155	.99780145	.99756838	.99734175
6	.99964064	.99958550	.99953143	.99947838	.99942627
7	.99992605	.99991431	.99990271	.99989124	.99987990
8	.99998497	.99998253	.99998011	.99997771	.99997532
9	.99999696	.99999647	.99999597	.99999548	.99999499
10	.99999939	.99999929	.99999919	.99999909	.99999899
11	.99999988	.99999986	.99999984	.99999982	.99999980
12	.99999998	.99999997	.99999997	.99999996	.99999996
13	1	1	1	1	1
14	1	1	1	1	1
15	1	1	1	1	1

M	B=11	B=12	B=13	B=14	B=15
1	.49084175	.48082458	.47175241	.46347606	.45587809
2	.82113018	.81418516	.80772417	.80168473	.79601577
3	.94931701	.94655921	.94393968	.94144398	.93905991
4	.98741089	.98657718	.98577270	.98499500	.98424197
5	.99712107	.99690588	.99669582	.99649055	.99628975
6	.99937504	.99932465	.99927505	.99922620	.99917806
7	.99986868	.99985757	.99984658	.99983570	.99982492
8	.99997295	.99997060	.99996826	.99996593	.99996362
9	.99999450	.99999401	.99999352	.99999304	.99999256
10	.99999889	.99999879	.99999869	.99999859	.99999849
11	.99999978	.99999976	.99999974	.99999972	.99999970
12	.99999996	.99999995	.99999995	.99999994	.99999994
13	1	1	.99999999	.99999999	.99999999
14	1	1	1	1	1
15	1	1	1	1	1

DIRICHLET C INTEGRAL
A = 5 R = 1

M	B= 1	B= 2	B= 3	B= 4	B= 5
1	.83333333	.75757576	.71022727	.67640693	.65039128
2	.97222222	.95270891	.93755381	.92511824	.91455182
3	.99537037	.99149206	.98812091	.98512079	.98240660
4	.99922840	.99852509	.99787483	.99726750	.99669592
5	.99987140	.99974901	.99963187	.99951928	.99941069
6	.99997857	.99995770	.99993733	.99991743	.99989794
7	.99999643	.99999291	.99998943	.99998600	.99998262
8	.99999940	.99999881	.99999823	.99999765	.99999707
9	.99999990	.99999980	.99999970	.99999961	.99999951
10	.99999998	.99999997	.99999995	.99999993	.99999992
11	1	1	1	.99999999	.99999999
12	1	1	1	1	1
13	1	1	1	1	1
14	1	1	1	1	1
15	1	1	1	1	1

M	B= 6	B= 7	B= 8	B= 9	B=10
1	.62941091	.61192728	.59700222	.58402391	.57257246
2	.90535372	.89720299	.88988102	.88323195	.87714059
3	.97992102	.97762330	.97548324	.97347778	.97158881
4	.99615480	.99564003	.99514840	.99467730	.99422459
5	.99930566	.99920384	.99910493	.99900867	.99891487
6	.99987883	.99986008	.99984166	.99982355	.99980574
7	.99997927	.99997596	.99997268	.99996944	.99996623
8	.99999649	.99999592	.99999536	.99999479	.99999423
9	.99999941	.99999931	.99999922	.99999912	.99999903
10	.99999990	.99999989	.99999987	.99999985	.99999984
11	.99999998	.99999998	.99999998	.99999998	.99999997
12	1	1	1	1	1
13	1	1	1	1	1
14	1	1	1	1	1
15	1	1	1	1	1

M	B=11	B=12	B=13	B=14	B=15
1	.56234795	.55312913	.54474839	.53707588	.53000909
2	.87151929	.86629978	.86142779	.85685945	.85255879
3	.96980186	.96810510	.96648878	.96494470	.96346594
4	.99378847	.99336743	.99296018	.99256559	.99218270
5	.99882333	.99873388	.99864640	.99856076	.99847684
6	.99978819	.99977091	.99975387	.99973707	.99972049
7	.99996305	.99995990	.99995678	.99995369	.99995062
8	.99999368	.99999312	.99999257	.99999202	.99999148
9	.99999893	.99999883	.99999874	.99999865	.99999855
10	.99999982	.99999980	.99999979	.99999977	.99999976
11	.99999997	.99999997	.99999996	.99999996	.99999996
12	1	1	1	1	1
13	1	1	1	1	1
14	1	1	1	1	1
15	1	1	1	1	1

DIRICHLET C INTEGRAL
1/A = 5 R = 2

M	B= 1	B= 2	B= 3	B= 4	B= 5
1	.02777778	.00307742	.00060596	.00016620	.56700\04
2	.07407407	.01237139	.00312437	.00101927	.00039624
3	.13194444	.02993875	.00924475	.00349022	.00151881
4	.19624486	.05654126	.02060554	.00880912	.00422849
5	.26322445	.09184136	.03844336	.01828621	.00956380
6	.33020405	.13473062	.06340244	.03307282	.01865065
7	.39532310	.18365629	.09549811	.05397174	.03252501
8	.45734124	.23688992	.13419162	.08133180	.05197327
9	.51548325	.29272300	.17852624	.11502931	.07742371
10	.56931845	.34959429	.22728279	.15451928	.10890193
11	.61866737	.40616114	.27912575	.19893049	.14604666
12	.66353004	.46132925	.33272395	.24717930	.18817099
13	.70403105	.51425354	.38683968	.29808187	.23435053
14	.74037811	.56432092	.44038655	.35045178	.28352047
15	.77283085	.61112289	.49246025	.40317580	.33456779

M	B= 6	B= 7	B= 8	B= 9	B=10
1	.22592\04	.10112\04	.49544\05	.26096\05	.14583\05
2	.00017505	.85261\04	.44846\04	.25102\04	.14790\04
3	.00073476	.00038585	.00021634	.00012796	.79125\04
4	.00221641	.00124493	.00073939	.00045980	.00029716
5	.00538223	.00321092	.00200886	.00130748	.00087986
6	.01118054	.00704071	.00461712	.00313236	.00218730
7	.02062665	.01363605	.00933125	.00657424	.00474875
8	.03465789	.02393589	.01702363	.01241368	.00924883
9	.05399855	.03879019	.02856931	.02149599	.01647609
10	.07906184	.05884349	.04473255	.03463173	.02724147
11	.10990281	.08445032	.06606594	.05249377	.04228211
12	.14622245	.11563428	.09283992	.07553090	.06216909
13	.18741359	.15209166	.12501183	.10391193	.08723557
14	.23263428	.19323233	.16223452	.13750653	.11753589
15	.28089336	.23824580	.20389769	.17590153	.15283914

M	B=11	B=12	B=13	B=14	B=15
1	.85620\06	.52411\06	.33251\06	.21759\06	.14630\06
2	.90980\05	.58055\05	.38234\05	.25882\05	.17948\05
3	.50803\04	.33685\04	.22966\04	.16044\04	.11451\04
4	.00019846	.00013633	.95976\04	.69039\04	.50619\04
5	.00060924	.00043242	.00031361	.00023182	.00017428
6	.00156578	.00114536	.00085386	.00064733	.00049815
7	.00350498	.00263623	.00201607	.00156475	.00123063
8	.00702102	.00541817	.00424254	.00336541	.00270092
9	.01283465	.01014195	.00811668	.00657019	.00537314
10	.02173003	.01754991	.01433171	.01182067	.00983764
11	.03446968	.02840387	.02363174	.01983271	.01677588
12	.05170341	.04339954	.03673424	.03132809	.02690160
13	.07388779	.06308295	.05424737	.04695553	.04088728
14	.10122947	.08778327	.07659668	.06721466	.05928804
15	.13365959	.11757198	.10397276	.09239552	.08247573

DIRICHLET C INTEGRAL
1/A = 4 R = 2

M	B= 1	B= 2	B= 3	B= 4	B= 5
1	.04000000	.00592593	.00148457	.00049990	.00020385
2	.10400000	.02281481	.00723281	.00286656	.00132072
3	.18080000	.05295802	.02025991	.00919658	.00470342
4	.26272000	.09608472	.04283084	.02179370	.01219346
5	.34464000	.15018535	.07594224	.04256941	.02574008
6	.42328320	.21236430	.11927220	.07261052	.04696264
7	.49668352	.27950228	.17143558	.11200935	.07681153
8	.56379238	.34868927	.23036851	.15993212	.11540896
9	.62419036	.41746064	.29371287	.21484180	.16207222
10	.67787745	.48388954	.35912970	.27478360	.21547222
11	.72512209	.54658602	.42451571	.33766129	.27386171
12	.76635378	.60464279	.48812633	.40146150	.33531380
13	.80208791	.65755633	.54862287	.46440882	.39792924
14	.83287423	.70514157	.60506556	.52505299	.45999075
15	.85926251	.74745136	.65687291	.58229950	.52005923

M	B= 6	B= 7	B= 8	B= 9	B=10
1	.95097\04	.49015\04	.27284\04	.16145\04	.10041\04
2	.00067832	.00037819	.00022484	.00014076	.91934\04
3	.00262681	.00156973	.00098973	.00065182	.00044506
4	.00732760	.00465611	.00309403	.00213292	.00151615
5	.01649477	.01106748	.00770823	.00553703	.00408218
6	.03184200	.02242250	.01628729	.01214158	.00925238
7	.05473113	.04022980	.03034251	.02338744	.01836405
8	.08590603	.06559627	.05116783	.04064273	.03278977
9	.12537077	.09902250	.07960006	.06495853	.05370664
10	.17241569	.14032504	.11587120	.09688052	.08188902
11	.22575611	.18868262	.15957600	.13635622	.11757332
12	.28373212	.24277427	.20974396	.18274626	.16041796
13	.34452026	.30096515	.26498387	.23492204	.20955472
14	.40632093	.36149887	.32366157	.29141738	.26370833
15	.46750098	.42266554	.38407587	.35059894	.32135235

M	B=11	B=12	B=13	B=14	B=15
1	.65077\05	.43670\05	.30189\05	.21411\05	.15529\05
2	.62212\04	.43384\04	.31045\04	.22719\04	.16955\04
3	.00031326	.00022628	.00016715	.00012591	.96488\04
4	.00110612	.00082519	.00062766	.00048558	.00038134
5	.00307713	.00236442	.00184742	.00146486	.00117678
6	.00718516	.00567213	.00454260	.00368456	.00302264
7	.01465270	.01185627	.00971263	.00804422	.00672798
8	.02681531	.02219210	.01856062	.01567004	.01334174
9	.04491394	.03794150	.03234035	.02778852	.02405091
10	.06988397	.06014842	.05216433	.04555060	.04002232
11	.10219269	.08946093	.07881947	.06984754	.06222358
12	.14175825	.12601915	.11263269	.10116149	.09126442
13	.18795872	.16942731	.15341185	.13948111	.12729227
14	.23971763	.21880596	.20046677	.18429426	.16996005
15	.29563955	.27290411	.25269667	.23465086	.21846526

DIRICHLET C INTEGRAL
1/A = 3 R = 2

M	B= 1	B= 2	B= 3	B= 4	B= 5
1	.06250000	.01300000	.00427778	.00180820	.00089580
2	.15625000	.04690000	.01917222	.00940567	.00520669
3	.26171875	.10231750	.04957403	.02747493	.01669946
4	.36718750	.17501500	.09709361	.05951306	.03914818
5	.46606445	.25872091	.16008741	.10668615	.07504571
6	.55505371	.34711990	.23470552	.16770985	.12488385
7	.63291931	.43491997	.31615822	.23946098	.18714991
8	.69966125	.51823932	.39974462	.31787285	.25884470
9	.75597477	.59458665	.48149318	.39876828	.33621486
10	.80290270	.66264360	.55844363	.47845485	.41544358
11	.84161824	.72198321	.62867085	.55404715	.49315758
12	.87329459	.77279764	.69115781	.62356063	.56671147
13	.89903163	.81566907	.74560257	.68585101	.63427800
14	.91981923	.85139309	.79221574	.74047027	.69480107
15	.93652356	.88085219	.83153970	.78749389	.74787062

M	B= 6	B= 7	B= 8	B= 9	B=10
1	.00049527	.00029676	.00018909	.00012647	.87973\04
2	.00314095	.00202061	.00136639	.00096156	.00069906
3	.01085712	.00743044	.00529498	.00389847	.00294862
4	.02714577	.01960822	.01463317	.01121527	.00878845
5	.05500690	.04164729	.03236972	.02570921	.02079368
6	.09601931	.07574067	.06100989	.05001063	.04160599
7	.14993764	.12256454	.10187143	.08586855	.07325270
8	.21483973	.18113696	.15474307	.13368252	.11660728
9	.28764372	.24910229	.21796063	.19240855	.17116411
10	.36475430	.32326216	.28879651	.25980568	.23515326
11	.44264543	.40015606	.36398961	.33288884	.30590446
12	.51827511	.47654807	.44025744	.40843235	.38031919
13	.58930790	.54975055	.51468871	.48340315	.45532139
14	.65417262	.61777813	.58497835	.55525931	.52820251
15	.71200476	.67936357	.64951431	.62210112	.59682820

M	B=11	B=12	B=13	B=14	B=15
1	.63204\04	.46659\04	.35249\04	.27164\04	.21299\04
2	.00052217	.00039907	.00031103	.00024655	.00019836
3	.00228114	.00179900	.00144244	.00117333	.00096657
4	.00701710	.00569350	.00468429	.00390116	.00328402
5	.01708042	.01421886	.01197524	.01018937	.00874880
6	.03505666	.02986605	.02569129	.02228985	.01948653
7	.06314180	.05492193	.04815531	.04252313	.03778888
8	.10257114	.09089365	.08107507	.07274154	.06560879
9	.15329719	.13811860	.12510827	.11386733	.10408550
10	.21398938	.19566690	.17968508	.16565102	.15325250
11	.28230673	.26152587	.24311077	.22669991	.21200033
12	.35532395	.33297202	.31287968	.29473330	.27827412
13	.42998173	.40700748	.38608810	.36696519	.34942194
14	.50346333	.48075514	.45983740	.44050662	.42258938
15	.57344729	.55174825	.53155182	.51270404	.49507181

DIRICHLET C INTEGRAL
1/A = 2 R = 2

M	B= 1	B= 2	B= 3	B= 4	B= 5
1	.11111111	.03472222	.01563333	.00852099	.00522537
2	.25925926	.11226852	.06110778	.03789819	.02557717
3	.40740741	.22106481	.13889222	.09543514	.06966092
4	.53909465	.34381430	.24110295	.17976103	.13990695
5	.64883402	.46563679	.35538111	.28278677	.23192881
6	.73662551	.57676665	.46992793	.39381078	.33704200
7	.80490779	.67231558	.57608761	.50299266	.44556741
8	.85693238	.75097413	.66888699	.60319060	.54929619
9	.89595082	.81363016	.74645295	.69034901	.64263900
10	.92485337	.86227998	.80905442	.76302977	.72270588
11	.94604857	.89929930	.85819323	.82162669	.78878835
12	.96146327	.92701589	.89591309	.86759963	.84164927
13	.97259610	.94749628	.92434835	.90287806	.88286693
14	.98058890	.96246718	.94546978	.92946339	.91433835
15	.98629805	.97331326	.96096893	.94920112	.93795578

M	B= 6	B= 7	B= 8	B= 9	B=10
1	.00347113	.00244296	.00179609	.00136639	.00106840
2	.01831510	.01370220	.01060254	.00842637	.00684391
3	.05311347	.04185336	.03384191	.02793727	.02345912
4	.11240885	.09255931	.07771725	.06630138	.05731487
5	.19463690	.16631900	.14421111	.12655806	.11219622
6	.29321792	.25846174	.23029211	.20704907	.18758150
7	.39926495	.36114996	.32923966	.30214494	.27886297
8	.50421359	.46590179	.43291522	.40419876	.37896272
9	.60147336	.56552680	.53382070	.50561427	.48033485
10	.68699443	.65508187	.62634495	.60029632	.57654822
11	.75906382	.73197649	.70714921	.68427843	.66311628
12	.81772702	.79556450	.77494310	.75568235	.73763140
13	.86413765	.84654406	.82996407	.81429455	.79944749
14	.90000315	.88638046	.87340423	.86101752	.84917082
15	.92718671	.91685401	.90692291	.89736289	.88814699

M	B=11	B=12	B=13	B=14	B=15
1	.00085440	.00069623	.00057644	.00048382	.00041092
2	.00565955	.00475158	.00404118	.00347556	.00301832
3	.01998147	.01722644	.01500644	.01319111	.01168756
4	.05010236	.04421772	.03934819	.03526893	.03181471
5	.10032632	.09038286	.08195545	.07473977	.06850582
6	.17106692	.15690262	.14463728	.13392655	.12450335
7	.25865124	.24094811	.22532081	.21143041	.19900772
8	.35660340	.33665079	.31873314	.30255230	.28786624
9	.45753186	.43684518	.41798316	.40070675	.38481791
10	.55478746	.53475779	.51624710	.49907804	.48310097
11	.64345783	.62513175	.60799343	.59191972	.57680497
12	.72066287	.70466819	.68955407	.67523973	.66165481
13	.78534717	.77192787	.75913224	.74690984	.73521612
14	.83782080	.82692929	.81646244	.80639010	.79668526
15	.87925123	.87065419	.86233662	.85428116	.84647207

DIRICHLET C INTEGRAL
A = 1 R = 2

M	B= 1	B= 2	B= 3	B= 4	B= 5
1	.25000000	.12962963	.08420139	.06110778	.04737274
2	.50000000	.33333333	.25000000	.20000000	.16666667
3	.68750000	.53549383	.44339554	.38082721	.33519263
4	.81250000	.69770233	.61800934	.55851777	.51193436
5	.89062500	.81280007	.75318902	.70537024	.66576669
6	.93750000	.88826017	.84773340	.81338667	.78365560
7	.96484375	.93512366	.90933675	.88654619	.86612306
8	.98046875	.96312213	.94747566	.93319950	.92005658
9	.98925781	.97937931	.97021156	.96164281	.95358816
10	.99414063	.98861807	.98338489	.97840440	.97364741
11	.99682617	.99378218	.99085352	.98802847	.98529730
12	.99829102	.99663158	.99501733	.99344457	.99191018
13	.99908447	.99818769	.99730835	.99644533	.99559763
14	.99951172	.99903048	.99855589	.99808761	.99762534
15	.99974060	.99948383	.99922956	.99897771	.99872818

M	B= 6	B= 7	B= 8	B= 9	B=10
1	.03836576	.03205142	.02740439	.02385571	.02106592
2	.14285714	.12500000	.11111111	.10000000	.09090909
3	.30025246	.27253662	.24995008	.23114785	.21522419
4	.47419907	.44284390	.41626907	.39338536	.37342176
5	.63218751	.60319746	.57780766	.55530950	.53517885
6	.75750334	.73420627	.71323890	.69420748	.67680929
7	.84762225	.83071590	.81515492	.80074521	.78733239
8	.90786966	.89650205	.88584577	.87581374	.86633462
9	.94598110	.93876831	.93190621	.92535861	.91909504
10	.96909013	.96471287	.96049904	.95643450	.95250707
11	.98265186	.98008518	.97759125	.97516481	.97280127
12	.99041144	.98894597	.98751169	.98610672	.98472942
13	.99476437	.99394477	.99313812	.99234380	.99156122
14	.99716879	.99671772	.99627190	.99583111	.99539517
15	.99848087	.99823572	.99799265	.99775158	.99751247

M	B=11	B=12	B=13	B=14	B=15
1	.01882075	.01697866	.01544265	.01414413	.01303334
2	.08333333	.07692308	.07142857	.06666667	.06250000
3	.20154526	.18965338	.17920925	.16995566	.16169385
4	.35581506	.34014287	.32608140	.31337771	.30183093
5	.51701791	.50051811	.48543563	.47157473	.45877610
6	.66080663	.64600943	.63226338	.61944151	.60743811
7	.77479151	.76301992	.75193222	.74145654	.73153183
8	.85734906	.84880714	.84066640	.83289036	.82544747
9	.91308956	.90731989	.90176665	.89641295	.89124389
10	.94870616	.94502249	.94144787	.93797500	.93459739
11	.97049654	.96824698	.96604932	.96390062	.96179821
12	.98337826	.98205191	.98074912	.97946876	.97820980
13	.99078985	.99002921	.98927885	.98853835	.98780734
14	.99496388	.99453709	.99411463	.99369637	.99328216
15	.99727524	.99703985	.99680623	.99657434	.99634414

DIRICHLET C INTEGRAL
A = 2 R = 2

M	B= 1	B= 2	B= 3	B= 4	B= 5
1	.44444444	.31288889	.25006470	.21219030	.18643837
2	.74074074	.62388148	.55261732	.50300374	.46575060
3	.88888889	.82033778	.77128655	.73339920	.70271950
4	.95473251	.92130502	.89466740	.87248213	.85345872
5	.98216735	.96747071	.95486700	.94378142	.93385732
6	.99314129	.98708386	.98162146	.97662479	.97200595
7	.99740893	.99501703	.99278487	.99068520	.98869808
8	.99903470	.99811784	.99724207	.99640194	.99559320
9	.99964436	.99930030	.99896647	.99864182	.99832550
10	.99987016	.99974305	.99961843	.99949609	.99937587
11	.99995296	.99990655	.99986073	.99981547	.99977073
12	.99998306	.99996628	.99994962	.99993311	.99991671
13	.99999394	.99998791	.99998191	.99997594	.99997000
14	.99999784	.99999569	.99999354	.99999140	.99998927
15	.99999923	.99999847	.99999771	.99999694	.99999618

M	B= 6	B= 7	B= 8	B= 9	B=10
1	.16758530	.15307311	.14148967	.13198642	.12402018
2	.43636922	.41238038	.39228417	.37511096	.36020140
3	.67706210	.65509486	.63594725	.61902002	.60388364
4	.83680395	.82199318	.80866045	.79653982	.78543173
5	.92485570	.91660738	.90898761	.90190141	.89527459
6	.96770168	.96366463	.95985818	.95625329	.95282648
7	.98680832	.98500398	.98327545	.98161485	.98001559
8	.99481246	.99405694	.99332435	.99261271	.99192037
9	.99801678	.99771507	.99741985	.99713066	.99684712
10	.99925763	.99914125	.99902660	.99891361	.99880217
11	.99972649	.99968272	.99963940	.99959651	.99955403
12	.99990044	.99988429	.99986824	.99985231	.99983647
13	.99996410	.99995822	.99995236	.99994653	.99994073
14	.99998715	.99998503	.99998292	.99998081	.99997871
15	.99999542	.99999467	.99999391	.99999316	.99999241

M	B=11	B=12	B=13	B=14	B=15
1	.11722570	.11134752	.10620114	.10164968	.09758924
2	.34708885	.33543225	.32497552	.31552205	.30691794
3	.59022014	.57778778	.56639867	.55590403	.54618408
4	.77518245	.76567096	.75680024	.74849128	.74067886
5	.88904796	.88317347	.87761149	.87232895	.86729792
6	.94955843	.94643302	.94343663	.94055761	.93778595
7	.97847213	.97697970	.97553420	.97413205	.97277011
8	.99124587	.99058795	.98994548	.98931750	.98870310
9	.99656888	.99629562	.99602707	.99576298	.99550312
10	.99869221	.99858367	.99847646	.99837055	.99826586
11	.99951194	.99947023	.99942889	.99938790	.99934725
12	.99982074	.99980511	.99978957	.99977413	.99975877
13	.99993496	.99992920	.99992348	.99991777	.99991209
14	.99997662	.99997453	.99997245	.99997037	.99996830
15	.99999165	.99999090	.99999016	.99998941	.99998866

DIRICHLET C INTEGRAL

A = 3 R = 2

M	B= 1	B= 2	B= 3	B= 4	B= 5
1	.56250000	.44278426	.38065277	.34091651	.31263135
2	.84375000	.76538421	.71422264	.67676534	.64749381
3	.94921875	.91527572	.88961092	.86893243	.85160436
4	.98437500	.97212444	.96192433	.95312866	.94536595
5	.99536133	.99136500	.98781832	.98461017	.98166906
6	.99865723	.99743224	.99629822	.99523769	.99423839
7	.99961853	.99925805	.99891496	.99858667	.99827126
8	.99989319	.99979003	.99969007	.99959292	.99949832
9	.99997044	.99994150	.99991312	.99988527	.99985789
10	.99999189	.99998389	.99997599	.99996818	.99996045
11	.99999779	.99999561	.99999344	.99999128	.99998914
12	.99999940	.99999881	.99999822	.99999763	.99999705
13	.99999984	.99999968	.99999952	.99999936	.99999920
14	.99999996	.99999991	.99999987	.99999983	.99999979
15	.99999999	.99999998	.99999997	.99999995	.99999994

M	B= 6	B= 7	B= 8	B= 9	B=10
1	.29113377	.27405434	.26004368	.24826879	.23818359
2	.62363041	.60358825	.58637922	.57134827	.55803990
3	.83668910	.82359713	.81193285	.80141745	.79184690
4	.93840016	.93207069	.92626271	.92089098	.91589025
5	.97894565	.97640409	.97401733	.97176439	.96962855
6	.99329125	.99238930	.99152708	.99070014	.98990485
7	.99796721	.99767330	.99738854	.99711210	.99684327
8	.99940601	.99931582	.99922756	.99914110	.99905632
9	.99983095	.99980442	.99977828	.99975251	.99972708
10	.99995281	.99994524	.99993775	.99993032	.99992297
11	.99998701	.99998490	.99998280	.99998072	.99997864
12	.99999646	.99999588	.99999531	.99999473	.99999416
13	.99999905	.99999889	.99999873	.99999857	.99999842
14	.99999974	.99999970	.99999966	.99999962	.99999957
15	.99999993	.99999992	.99999991	.99999990	.99999989

M	B=11	B=12	B=13	B=14	B=15
1	.22941299	.22168956	.21481671	.20864621	.20306386
2	.54612526	.53535952	.52555581	.51656841	.50828162
3	.78306728	.77495955	.76742971	.76040223	.75381549
4	.91120937	.90680742	.90265111	.89871293	.89496991
5	.96759626	.96565636	.96379952	.96201784	.96030456
6	.98913815	.98839747	.98768061	.98698565	.98631094
7	.99658145	.99632612	.99607681	.99583313	.99559472
8	.99897311	.99889136	.99881100	.99873195	.99865414
9	.99970197	.99967717	.99965267	.99962845	.99960450
10	.99991567	.99990844	.99990127	.99989415	.99988709
11	.99997658	.99997453	.99997249	.99997046	.99996845
12	.99999359	.99999302	.99999245	.99999189	.99999132
13	.99999826	.99999811	.99999795	.99999780	.99999764
14	.99999953	.99999949	.99999945	.99999941	.99999936
15	.99999988	.99999986	.99999985	.99999984	.99999983

DIRICHLET C INTEGRAL
A = 4 R = 2

M	B= 1	B= 2	B= 3	B= 4	B= 5
1	.64000000	.53377229	.47624702	.43833897	.41072405
2	.89600000	.84092547	.80371468	.77577583	.75350284
3	.97280000	.95388126	.93918126	.92708802	.91678171
4	.99328000	.98785648	.98324385	.97919884	.97557866
5	.99840000	.99699256	.99572313	.99455953	.99348068
6	.99962880	.99928513	.99896313	.99865892	.99836971
7	.99991552	.99983486	.99975742	.99968276	.99961053
8	.99998106	.99996263	.99994466	.99992711	.99990992
9	.99999580	.99999167	.99998761	.99998360	.99997964
10	.99999908	.99999817	.99999726	.99999637	.99999548
11	.99999980	.99999960	.99999940	.99999920	.99999901
12	.99999996	.99999991	.99999987	.99999983	.99999978
13	1	.99999998	.99999997	.99999996	.99999995
14	1	1	1	1	1
15	1	1	1	1	1

M	B= 6	B= 7	B= 8	B= 9	B=10
1	.38933696	.37207346	.35771627	.34550367	.33493034
2	.73504284	.71931912	.70565149	.69358328	.68279339
3	.90778362	.89978810	.89258724	.88603281	.88001519
4	.97229106	.96927237	.96647652	.96386892	.96142285
5	.99247183	.99152210	.99062320	.98976863	.98895317
6	.99809343	.99782847	.99757355	.99732760	.99708978
7	.99954048	.99947238	.99940607	.99934138	.99927819
8	.99989308	.99987654	.99986030	.99984434	.99982862
9	.99997574	.99997188	.99996807	.99996430	.99996058
10	.99999460	.99999372	.99999285	.99999199	.99999114
11	.99999881	.99999862	.99999842	.99999823	.99999804
12	.99999974	.99999970	.99999966	.99999961	.99999957
13	.99999994	.99999994	.99999993	.99999992	.99999991
14	.99999999	.99999999	.99999998	.99999998	.99999998
15	1	1	1	1	1

M	B=11	B=12	B=13	B=14	B=15
1	.32564543	.31739637	.30999585	.30330150	.29720278
2	.67304763	.66417007	.65602529	.64850688	.64152976
3	.87445090	.86927483	.86443515	.85988990	.85560464
4	.95911717	.95693483	.95486187	.95288665	.95099942
5	.98817256	.98742324	.98670224	.98600702	.98533538
6	.99685933	.99663565	.99641819	.99620648	.99600012
7	.99921640	.99915589	.99909660	.99903844	.99898135
8	.99981315	.99979790	.99978286	.99976802	.99975338
9	.99995688	.99995323	.99994961	.99994603	.99994247
10	.99999029	.99998944	.99998860	.99998777	.99998694
11	.99999785	.99999766	.99999747	.99999728	.99999709
12	.99999953	.99999949	.99999945	.99999940	.99999936
13	.99999990	.99999989	.99999988	.99999987	.99999986
14	.99999998	.99999998	.99999997	.99999997	.99999997
15	1	1	1	1	1

DIRICHLET C INTEGRAL
A = 5 R = 2

M	B= 1	B= 2	B= 3	B= 4	B= 5
1	.69444444	.60000835	.54753277	.51232167	.48631163
2	.92592593	.88538781	.85742771	.83611484	.81892001
3	.98379630	.97225571	.96314212	.95555133	.94901701
4	.99665638	.99391166	.99154809	.98945480	.98756591
5	.99933556	.99874404	.99820553	.99770810	.99724389
6	.99987140	.99975134	.99963808	.99953046	.99942762
7	.99997559	.99995215	.99992954	.99990764	.99988638
8	.99999544	.99999098	.99998662	.99998234	.99997815
9	.99999916	.99999832	.99999750	.99999669	.99999589
10	.99999985	.99999969	.99999954	.99999939	.99999924
11	.99999997	.99999994	.99999992	.99999989	.99999986
12	1	.99999999	.99999998	.99999998	.99999997
13	1	1	1	1	1
14	1	1	1	1	1
15	1	1	1	1	1

M	B= 6	B= 7	B= 8	B= 9	B=10
1	.46593801	.44933495	.43541284	.42348426	.41308988
2	.80452744	.79216468	.78133945	.77171875	.76306685
3	.94326384	.93811433	.93344683	.92917397	.92523074
4	.98583846	.98424251	.98275626	.98136323	.98005062
5	.99680730	.99639421	.99600145	.99562650	.99526734
6	.99932894	.99923392	.99914216	.99905333	.99896716
7	.99986569	.99984551	.99982580	.99980652	.99978765
8	.99997403	.99996997	.99996598	.99996204	.99995817
9	.99999510	.99999431	.99999354	.99999277	.99999201
10	.99999909	.99999894	.99999880	.99999865	.99999851
11	.99999983	.99999981	.99999978	.99999975	.99999973
12	.99999997	.99999996	.99999996	.99999995	.99999995
13	1	1	1	1	1
14	1	1	1	1	1
15	1	1	1	1	1

M	B=11	B=12	B=13	B=14	B=15
1	.40390876	.39570848	.38831578	.38159832	.37545288
2	.75521079	.74801990	.74139312	.73525067	.72952845
3	.92156735	.91814475	.91493172	.91190290	.90903735
4	.97880823	.97762782	.97650259	.97542686	.97439585
5	.99492232	.99459005	.99426935	.99395923	.99365882
6	.99888342	.99880191	.99872247	.99864494	.99856920
7	.99976914	.99975098	.99973315	.99971563	.99969839
8	.99995434	.99995056	.99994683	.99994314	.99993950
9	.99999125	.99999050	.99998976	.99998902	.99998829
10	.99999836	.99999822	.99999808	.99999793	.99999779
11	.99999970	.99999967	.99999964	.99999962	.99999959
12	.99999995	.99999994	.99999994	.99999993	.99999993
13	1	.99999999	.99999999	.99999999	.99999999
14	1	1	1	1	1
15	1	1	1	1	1

DIRICHLET C INTEGRAL

1/A = 5 R = 3

M	B= 1	B= 2	B= 3	B= 4	B= 5
1	.00462963	.00021541	.23900\04	.42974\05	10539\05
2	.01620370	.00117892	.00016952	.36360\04	.10162\04
3	.03549383	.00369548	.00066326	.00016623	.52254\04
4	.06228567	.00870853	.00189318	.00054482	.00019043
5	.09577546	.01714317	.00440438	.00143373	.00055167
6	.13484689	.02977144	.00885628	.00321915	.00135145
7	.17825959	.04712173	.01595259	.00640222	.00290932
8	.22477320	.06943659	.02636068	.01156557	.00564862
9	.27322488	.09667260	.04063634	.01932135	.01007745
10	.32257381	.12853104	.05916580	.03025044	.01675234
11	.37192273	.16450756	.08213122	.04484309	.02623004
12	.42052395	.20395007	.10950070	.06344991	.03901491
13	.46777513	.24611707	.14104001	.08624929	.05551005
14	.51320896	.29023048	.17634064	.11323338	.07597857
15	.55647928	.33552019	.21485836	.14421231	.10051944

M	B= 6	B= 7	B= 8	B= 9	B=10
1	.32108\06	.11481\06	.46427\07	.20696\07	.99850\08
2	.34285\05	.13317\05	.57702\06	.27279\06	.13848\06
3	.19350\04	.81111\05	.37466\05	.18710\05	.99612\06
4	.76774\04	.34523\04	.16922\04	.88937\05	.49513\05
5	.00024040	.00011533	.59736\04	.32929\04	.19116\04
6	.00063242	.00032211	.00017560	.00010121	.61103\04
7	.00145342	.00078233	.00044725	.00026872	.00016832
8	.00299646	.00169740	.00101426	.00063350	.00041072
9	.00564889	.00335463	.00208860	.00135265	.00090577
10	.00987864	.00612829	.00396400	.00265553	.00183281
11	.01620516	.01046716	.00701507	.00485009	.00344364
12	.02515866	.01686838	.01168397	.00831927	.00606547
13	.03723321	.02584042	.01845469	.01350552	.01009383
14	.05284076	.03786010	.02781775	.02088434	.01597448
15	.07227211	.05332947	.04022958	.03092936	.02417535

M	B=11	B=12	B=13	B=14	B=15
1	.51427\08	.27983\08	.15956\08	.94754\09	.58331\09
2	.74577\07	.42219\07	.24942\07	.15289\07	.96788\08
3	.55947\06	.32879\06	.20088\06	.12694\06	.82617\07
4	.28929\05	.17612\05	.11108\05	.72252\06	.48281\06
5	.11592\04	.72962\05	.47425\05	.31703\05	.21724\05
6	.38371\04	.24922\04	.16667\04	.11435\04	.80246\05
7	.00010923	.73081\04	.50208\04	.35306\04	.25344\04
8	.00027491	.00018914	.00013330	.95946\04	.70369\04
9	.00062413	.00044088	.00031827	.00023421	.00017531
10	.00129789	.00093985	.00069407	.00052155	.00039802
11	.00250192	.00185460	.00139929	.00107248	.00083361
12	.00451410	.00342060	.00263359	.00205660	.00162655
13	.00768363	.00594407	.00466457	.00370753	.00298084
14	.01242014	.00979687	.00782737	.00632600	.00516573
15	.01917248	.01540163	.01251514	.01027487	.00851441

DIRICHLET C INTEGRAL

1/A = 4 R = 3

M	B= 1	B= 2	B= 3	B= 4	B= 5
1	.00800000	.00056790	.88691\04	.21224\04	.66494\05
2	.02720000	.00295967	.00059030	.00016662	.58952\04
3	.05792000	.00884412	.00217022	.00070783	.00027917
4	.09888000	.01989077	.00582853	.00215884	.00093846
5	.14803200	.03741440	.01277691	.00529525	.00251187
6	.20308224	.06216204	.02424442	.01109990	.00569530
7	.26180250	.09424971	.04127434	.02064460	.01136808
8	.32220047	.13321532	.06456338	.03493858	.02050340
9	.38259845	.17814208	.09437115	.05478082	.03404556
10	.44165425	.22781084	.13050446	.08064743	.05278099
11	.49834782	.28085052	.17236422	.11263167	.07722989
12	.55194901	.33586775	.21903432	.15043995	.10757711
13	.60197679	.39154659	.26939031	.19343549	.14365022
14	.64815628	.44671683	.32220930	.24071523	.18494237
15	.69037753	.50039337	.37626757	.29120307	.23067030

M	B= 6	B= 7	B= 8	B= 9	B=10
1	.25082\05	.10836\05	.51911\06	.26975\06	.14968\06
2	.24446\04	.11402\04	.58219\05	.31938\05	.18573\05
3	.00012613	.63097\04	.34167\04	.19710\04	.11973\04
4	.00045826	.00024443	.00013972	.84442\04	.53427\04
5	.00131626	.00074450	.00044734	.00028229	.00018551
6	.00318200	.00189919	.00119488	.00078481	.00053425
7	.00673250	.00422112	.00277075	.00188860	.00132853
8	.01280335	.00839766	.00573184	.00404336	.00293242
9	.02230880	.01524898	.01078940	.00785662	.00586178
10	.03613314	.02564933	.01875881	.01406668	.01077479
11	.05501551	.04042552	.03047869	.02348303	.01843155
12	.07945077	.06025181	.04671346	.03690262	.02962580
13	.10962335	.08556111	.06805710	.05501599	.04509907
14	.14538272	.11648691	.09485539	.07832036	.06545192
15	.18625967	.15284344	.12715874	.10705415	.09106758

M	B=11	B=12	B=13	B=14	B=15
1	.87658\07	.53709\07	.34194\07	.22498\07	.15231\07
2	.11334\05	.72024\06	.47379\06	.32109\06	.22332\06
3	.75931\05	.49935\05	.33876\05	.23607\05	.16842\05
4	.35124\04	.23854\04	.16658\04	.11918\04	.87100\05
5	.00012613	.88281\04	.63355\04	.46468\04	.34741\04
6	.00037484	.00026989	.00019872	.00014920	.00011397
7	.00095990	.00070968	.00053527	.00041085	.00032026
8	.00217758	.00165036	.00127321	.00099773	.00079277
9	.00446546	.00346369	.00272940	.00218094	.00176440
10	.00840567	.00666269	.00535548	.00435834	.00358621
11	.01470048	.01189004	.00973635	.00806066	.00673911
12	.02411924	.01987886	.01656296	.01393445	.01182532
13	.03742306	.03138873	.02657959	.02269991	.01953576
14	.05527915	.04712611	.04051178	.03508714	.03059459
15	.07817930	.06766188	.05898599	.05175967	.04568807

DIRICHLET C INTEGRAL
1/A = 3 R = 3

M	B= 1	B= 2	B= 3	B= 4	B= 5
1	.01562500	.00181000	.00041457	.00013564	.55299\04
2	.05078125	.00876250	.00251242	.00095528	.00043478
3	.10351563	.02437525	.00843150	.00365041	.00183099
4	.16943359	.05114879	.02072428	.01004298	.00548941
5	.24359131	.08997814	.04169231	.02228622	.01314419
6	.32145691	.14015480	.07281099	.04239529	.02674676
7	.39932251	.19973732	.11442441	.07178835	.04807446
8	.47440720	.26606026	.16573972	.11098390	.07835140
9	.54479909	.33621956	.22504853	.15951607	.11799318
10	.60932499	.40744957	.29006503	.21604962	.16652955
11	.66739830	.47736714	.35828026	.27862717	.22269675
12	.71887238	.54409363	.42726326	.34496887	.28464971
13	.76391219	.60628258	.49487613	.41275695	.35022837
14	.80288895	.66308405	.55939733	.47986219	.41721870
15	.83629760	.71407304	.61956572	.54449405	.48356731

M	B= 6	B= 7	B= 8	B= 9	B=10
1	.26173\04	.13795\04	.78872\05	.48047\05	.30796\05
2	.00022413	.00012657	.76632\04	.49012\04	.32766\04
3	.00101890	.00061245	.00039072	.00026132	.00018163
4	.00327143	.00208071	.00139224	.00097020	.00069896
5	.00832962	.00557558	.00389647	.00281946	.00209950
6	.01790868	.01255417	.00912758	.00683628	.00524789
7	.03381455	.02471378	.01862655	.01439739	.01136549
8	.05759281	.04370510	.03403353	.02707388	.02192741
9	.09021096	.07081362	.05679856	.04638295	.03845756
10	.13186002	.10670004	.08790098	.07351101	.06226931
11	.18191348	.15126338	.12765151	.10908172	.09421892
12	.23903190	.20364747	.17561995	.15302506	.13453360
13	.30136799	.26237117	.23069227	.20456991	.18275059
14	.36681476	.32553742	.29122888	.26234906	.23777118
15	.43324522	.39106915	.35527881	.32458129	.29800612

M	B=11	B=12	B=13	B=14	B=15
1	.20579\05	.14237\05	.10143\05	.74104\06	.55330\06
2	.22719\04	.16242\04	.11916\04	.89400\05	.68384\05
3	.00013033	.96057\04	.72437\04	.55714\04	.43596\04
4	.00051771	.00039256	.00030371	.00023910	.00019112
5	.00160144	.00124677	.00098788	.00079482	.00064813
6	.00411312	.00328147	.00265846	.00218277	.00181347
7	.00913386	.00745394	.00616458	.00515814	.00436077
8	.01803310	.01502758	.01266804	.01078774	.00926944
9	.03230484	.02744511	.02354856	.02038278	.01778051
10	.05333279	.04612090	.04022367	.03534514	.03126750
11	.08214236	.07220028	.06392069	.05695505	.05104139
12	.11920221	.10634559	.09545574	.08614934	.07813256
13	.16432054	.14859976	.13507296	.12334339	.11310125
14	.21665319	.19835415	.18237826	.16833651	.15591975
15	.27481109	.25441925	.23637558	.22031669	.20594900

DIRICHLET C INTEGRAL
1/A = 2 R = 3

M	B= 1	B= 2	B= 3	B= 4	B= 5
1	.03703704	.00766782	.00270836	.00124911	.00067600
2	.11111111	.03276910	.01407476	.00738964	.00439562
3	.20987654	.08088590	.04073495	.02386068	.01540116
4	.31961591	.15143886	.08687645	.05582160	.03866332
5	.42935528	.23908166	.15263359	.10604905	.07805336
6	.53177869	.33626246	.23438599	.17395157	.13489011
7	.62282172	.43546278	.32621733	.25590797	.20750424
8	.70085861	.53056243	.42160562	.34645105	.29180344
9	.76588935	.61736524	.51473623	.43964797	.38240009
10	.81887735	.69354847	.60122754	.53020729	.47379616
11	.86126776	.75831357	.67832924	.61411929	.56128083
12	.89466626	.81194414	.74476728	.68885142	.64141422
13	.92064288	.85539250	.80041487	.75322829	.71213478
14	.94062489	.88994919	.84592370	.80714500	.77260791
15	.95584927	.91700765	.88239464	.85123325	.82294412

M	B= 6	B= 7	B= 8	B= 9	B=10
1	.00040689	.00026416	.00018147	.00013024	.96787\04
2	.00284662	.00196046	.00141448	.00105840	.00081545
3	.01063577	.00771921	.00581988	.00452193	.00360010
4	.02824681	.02147621	.01684137	.01353682	.01110219
5	.05990077	.04745172	.03853797	.03193341	.02690179
6	.10805088	.08874369	.07434937	.06330588	.05463178
7	.17250562	.14623524	.12592606	.10984589	.09686023
8	.25047830	.21826485	.19253837	.17158156	.15422593
9	.33737110	.30107463	.27123253	.24629300	.22516258
10	.42786923	.38973540	.35755952	.33004477	.30624751
11	.51694717	.47916118	.44653568	.41805702	.39296553
12	.60055404	.56491863	.53351538	.50559650	.48058618
13	.67592269	.64369833	.61478459	.58865696	.56490141
14	.74156714	.71345457	.68782730	.66433337	.64268862
15	.79708276	.77330023	.75131732	.73090699	.71188208

M	B=11	B=12	B=13	B=14	B=15
1	.73978\04	.57923\04	.46250\04	.37560\04	.30952\04
2	.00064352	.00051811	.00042427	.00035252	.00029662
3	.00292447	.00241612	.00202507	.00171848	.00147413
4	.00925936	.00783258	.00670645	.00580280	.00506717
5	.02297903	.01986071	.01734033	.01527377	.01355793
6	.04768371	.04202484	.03734962	.03343882	.03013168
7	.08619794	.07731841	.06983254	.06345387	.05796734
8	.13965027	.12726151	.11662125	.10739889	.09934071
9	.20704865	.19136285	.17765933	.16559423	.15489814
10	.28546399	.26715830	.25091487	.23640636	.22337137
11	.37067958	.35074567	.33280452	.31656753	.30180001
12	.45803279	.43757638	.41892603	.40184378	.38613282
13	.54318588	.52324031	.50484238	.48780710	.47197905
14	.62266042	.60405597	.58671386	.57049759	.55529082
15	.69408644	.67738843	.66167603	.64685317	.63283680

DIRICHLET C INTEGRAL
A = 1 R = 3

M	B= 1	B= 2	B= 3	B= 4	B= 5
1	.12500000	.05246914	.03033221	.02036747	.01491292
2	.31250000	.17232510	.11551657	.08552481	.06724568
3	.50000000	.33333333	.25000000	.20000000	.16666667
4	.65625000	.49951417	.40762303	.34649124	.30256633
5	.77343750	.64548802	.56071541	.49940562	.45251240
6	.85546875	.76047263	.69128351	.63774496	.59461147
7	.91015625	.84424178	.79257000	.75034351	.71482771
8	.94531250	.90181909	.86571769	.83489113	.80802821
9	.96728516	.93966777	.91570418	.89450991	.87549737
10	.98071289	.96369446	.94841189	.93451206	.92174594
11	.98876953	.97852903	.96908666	.96030668	.95208839
12	.99353027	.98748518	.98179711	.97641558	.97130143
13	.99630737	.99279416	.98943698	.98621757	.98312123
14	.99790955	.99589389	.99394508	.99205667	.99022331
15	.99882507	.99768096	.99656500	.99547496	.99440894

M	B= 6	B= 7	B= 8	B= 9	B=10
1	.01155087	.00930630	.00771921	.00654737	.00565241
2	.05504711	.04638023	.03993353	.03496718	.03103394
3	.14285714	.12500000	.11111111	.10000000	.09090909
4	.26931662	.24318065	.22204072	.20455380	.18982453
5	.41521804	.38468916	.35913626	.33736644	.31855018
6	.55884005	.52851850	.50237257	.47951391	.45930068
7	.68431508	.65766838	.63409272	.61301117	.59399217
8	.78425645	.76296432	.74370580	.72614524	.71002353
9	.85825344	.84247524	.82793335	.81444941	.80188176
10	.90992988	.89892380	.88861797	.87892454	.86977192
11	.94435436	.93704346	.93010634	.92350242	.91719789
12	.96642348	.96175633	.95727892	.95297354	.94882509
13	.98013587	.97725138	.97445918	.97175189	.96912311
14	.98844045	.98670423	.98501129	.98335866	.98174377
15	.99336529	.99234257	.99133951	.99035498	.98938796

M	B=11	B=12	B=13	B=14	B=15
1	.00495014	.00438670	.00392619	.00354387	.00322215
2	.02784840	.02522027	.02301810	.02114831	.01954253
3	.08333333	.07692308	.07142857	.06666667	.06250000
4	.17723170	.16633003	.15679153	.14836896	.14087224
5	.30209059	.28754588	.27458129	.26293797	.25241225
6	.44125539	.42501412	.41029376	.39687018	.38456324
7	.57670475	.56088980	.54634078	.53289061	.52040225
8	.69513641	.68131992	.66844047	.65638777	.64506967
9	.79011587	.77905770	.76862906	.75876416	.74940712
10	.86110089	.85286181	.84501263	.83751735	.83034489
11	.91116417	.90537685	.89981490	.89446002	.88929619
12	.94482059	.94094876	.93719971	.93356472	.93003607
13	.96656724	.96407933	.96165499	.95929029	.95698170
14	.98016430	.97861818	.97710355	.97561872	.97416217
15	.98843754	.98750288	.98658324	.98567792	.98478628

DIRICHLET C INTEGRAL
A = 2 R = 3

M	B= 1	B= 2	B= 3	B= 4	B= 5
1	.29629630	.18811259	.14233890	.11652395	.09973774
2	.59259259	.45814519	.38589161	.33924522	.30600561
3	.79012346	.68731891	.62177772	.57477504	.53869887
4	.89986283	.83757525	.79261492	.75762493	.72910862
5	.95473251	.92187846	.89590683	.87437259	.85595787
6	.98033836	.96452297	.95115934	.93952517	.92918904
7	.99171874	.98457709	.97824166	.97251710	.96727642
8	.99659605	.99351775	.99068878	.98805984	.98559655
9	.99862826	.99734683	.99613887	.99499259	.99389931
10	.99945620	.99893681	.99843825	.99795782	.99749343
11	.99978737	.99958120	.99938075	.99918542	.99899474
12	.99991783	.99983735	.99975838	.99968082	.99960456
13	.99996857	.99993757	.99990696	.99987673	.99984685
14	.99998808	.99997627	.99996457	.99995296	.99994144
15	.99999552	.99999106	.99998663	.99998222	.99997784

M	B= 6	B= 7	B= 8	B= 9	B=10
1	.08784646	.07892589	.07195340	.06633275	.06169158
2	.28080196	.26085648	.24457062	.23095145	.21934568
3	.50974993	.48577725	.46545351	.44790642	.43253449
4	.70512752	.68449520	.66643319	.65040355	.63601927
5	.83986447	.82556977	.81271183	.80102908	.79032603
6	.91986932	.91137049	.90355056	.89630271	.88954437
7	.96243128	.95791714	.95368512	.94969713	.94592280
8	.98327362	.98107179	.97897590	.97697376	.97505537
9	.99285231	.99184628	.99087691	.98994062	.98903442
10	.99704342	.99660641	.99618127	.99576701	.99536281
11	.99880830	.99862579	.99844690	.99827140	.99809906
12	.99952950	.99945557	.99938271	.99931084	.99923992
13	.99981730	.99978807	.99975913	.99973048	.99970210
14	.99993001	.99991866	.99990740	.99989621	.99988510
15	.99997348	.99996914	.99996483	.99996053	.99995626

M	B=11	B=12	B=13	B=14	B=15
1	.05778469	.05444363	.05154865	.04901212	.04676837
2	.20930368	.20050466	.19271278	.18575031	.17948047
3	.41890681	.40670507	.39568798	.38566867	.37649976
4	.62299289	.61110517	.60018519	.59009723	.58073175
5	.78045264	.77129107	.76274709	.75474409	.74721895
6	.88321021	.87724755	.87161328	.86627158	.86119236
7	.94233744	.93892068	.93565545	.93252732	.92952392
8	.97321241	.97143787	.96972575	.96807088	.96646876
9	.98815575	.98730243	.98647253	.98566438	.98487652
10	.99496795	.99458179	.99420377	.99383339	.99347020
11	.99792970	.99776314	.99759923	.99743783	.99727882
12	.99916990	.99910074	.99903239	.99896483	.99889801
13	.99967398	.99964611	.99961848	.99959108	.99956391
14	.99987407	.99986310	.99985220	.99984136	.99983059
15	.99995200	.99994777	.99994355	.99993935	.99993517

DIRICHLET C INTEGRAL
A = 3 R = 3

M	B= 1	B= 2	B= 3	B= 4	B= 5
1	.42187500	.31040080	.25760640	.22546961	.20333022
2	.73828125	.63614737	.57579365	.53413818	.50288495
3	.89648438	.83885464	.79908000	.76885819	.74458213
4	.96240234	.93659399	.91670220	.90043788	.88664658
5	.98712158	.97703996	.96863233	.96136532	.95493582
6	.99577332	.99217158	.98899456	.98613188	.98351439
7	.99865723	.99744772	.99633769	.99530623	.99433916
8	.99958420	.99919577	.99882925	.99848096	.99814821
9	.99987388	.99975320	.99963713	.99952504	.99941647
10	.99996239	.99992584	.99989021	.99985542	.99982138
11	.99998894	.99997809	.99996741	.99995691	.99994655
12	.99999679	.99999361	.99999048	.99998737	.99998429
13	.99999908	.99999816	.99999725	.99999635	.99999545
14	.99999974	.99999948	.99999922	.99999896	.99999870
15	.99999993	.99999985	.99999978	.99999970	.99999963

M	B= 6	B= 7	B= 8	B= 9	B=10
1	.18690211	.17409165	.16374090	.15515134	.14787356
2	.47817266	.45791308	.44085880	.42620973	.41342444
3	.72435757	.70706688	.69199592	.67866099	.66671957
4	.87465792	.86404562	.85452076	.84587793	.83796563
5	.94915230	.94388485	.93904076	.93455121	.93036356
6	.98109520	.97884066	.97672566	.97473082	.97284083
7	.99342627	.99255986	.99173396	.99094381	.99018551
8	.99782898	.99752168	.99722502	.99693795	.99665960
9	.99931103	.99920843	.99910840	.99901075	.99891528
10	.99978802	.99975530	.99972317	.99969158	.99966051
11	.99993634	.99992627	.99991632	.99990649	.99989677
12	.99998125	.99997822	.99997523	.99997226	.99996932
13	.99999456	.99999367	.99999279	.99999191	.99999104
14	.99999844	.99999818	.99999793	.99999767	.99999742
15	.99999956	.99999948	.99999941	.99999934	.99999927

M	B=11	B=12	B=13	B=14	B=15
1	.14160372	.13612813	.13129152	.12697804	.12309928
2	.40212087	.39202017	.38291311	.37463901	.36707205
3	.65592041	.64607384	.63703328	.62868313	.62093067
4	.83066875	.82389770	.81758133	.81166214	.80609300
5	.92643652	.92273703	.91923821	.91591781	.91275724
6	.97104333	.96932818	.96768692	.96611243	.96459864
7	.98945587	.98875218	.98807214	.98741379	.98677541
8	.99638920	.99612614	.99586984	.99561984	.99537569
9	.99882184	.99873030	.99864052	.99855240	.99846585
10	.99962991	.99959977	.99957005	.99954075	.99951183
11	.99988716	.99987765	.99986824	.99985893	.99984970
12	.99996639	.99996349	.99996061	.99995775	.99995491
13	.99999017	.99998931	.99998845	.99998760	.99998675
14	.99999717	.99999692	.99999667	.99999642	.99999617
15	.99999919	.99999912	.99999905	.99999898	.99999891

DIRICHLET C INTEGRAL
A = 4 R = 3

M	B= 1	B= 2	B= 3	B= 4	B= 5
1	.51200000	.40620800	.35317115	.31962732	.29584230
2	.81920000	.74242095	.69485288	.66091572	.63479525
3	.94208000	.90773552	.88309120	.86382800	.84800604
4	.98304000	.97083318	.96112292	.95299211	.94596436
5	.99532800	.99153866	.98829785	.98544109	.98287240
6	.99876864	.99769143	.99672229	.99583504	.99501286
7	.99968614	.99939795	.99912942	.99887673	.99863724
8	.99992207	.99984826	.99977781	.99971020	.99964506
9	.99998106	.99996275	.99994499	.99992772	.99991087
10	.99999547	.99999104	.99998670	.99998244	.99997824
11	.99999893	.99999788	.99999684	.99999582	.99999480
12	.99999975	.99999951	.99999926	.99999902	.99999878
13	.99999994	.99999989	.99999983	.99999977	.99999972
14	.99999999	.99999997	.99999996	.99999995	.99999994
15	1	1	1	.99999999	.99999999

M	B= 6	B= 7	B= 8	B= 9	B=10
1	.27778024	.26342160	.25162689	.24169688	.23317502
2	.61370875	.59611733	.58108444	.56799968	.55644415
3	.83458042	.82292176	.81262090	.80339650	.79504667
4	.93975659	.93418532	.92912418	.92448206	.92019091
5	.98052986	.97837072	.97636401	.97448646	.97272002
6	.99424415	.99352045	.99283534	.99218383	.99156190
7	.99840900	.99819055	.99798071	.99777855	.99758329
8	.99958209	.99952105	.99946175	.99940403	.99934776
9	.99989442	.99987832	.99986254	.99984706	.99983187
10	.99997411	.99997005	.99996604	.99996208	.99995817
11	.99999380	.99999280	.99999182	.99999084	.99998988
12	.99999854	.99999830	.99999807	.99999783	.99999760
13	.99999966	.99999961	.99999955	.99999950	.99999944
14	.99999992	.99999991	.99999990	.99999988	.99999987
15	.99999998	.99999998	.99999998	.99999997	.99999997

M	B=11	B=12	B=13	B=14	B=15
1	.22574854	.21919482	.21335040	.20809211	.20332508
2	.54611853	.53680190	.52832679	.52056337	.51340910
3	.78742156	.78040684	.77391328	.76786989	.76221924
4	.91619844	.91246356	.90895331	.90564084	.90250395
5	.97105041	.96946609	.96795758	.96651698	.96513763
6	.99096631	.99039436	.98984377	.98931262	.98879926
7	.99739429	.99721099	.99703293	.99685971	.99669097
8	.99929283	.99923914	.99918659	.99913512	.99908466
9	.99981693	.99980225	.99978779	.99977355	.99975951
10	.99995431	.99995050	.99994673	.99994300	.99993931
11	.99998892	.99998797	.99998702	.99998609	.99998516
12	.99999737	.99999714	.99999691	.99999668	.99999646
13	.99999939	.99999933	.99999928	.99999922	.99999917
14	.99999986	.99999985	.99999983	.99999982	.99999981
15	.99999997	.99999997	.99999996	.99999996	.99999996

DIRICHLET C INTEGRAL
A = 5 R = 3

M	B= 1	B= 2	B= 3	B= 4	B= 5
1	.57870370	.48060317	.42970762	.39677026	.37300817
2	.86805556	.80910380	.77153385	.74419486	.72283048
3	.96450617	.94265830	.92661592	.91386465	.90325238
4	.99129801	.98485919	.97964178	.97521168	.97133938
5	.99799597	.99633713	.99489764	.99361422	.99244936
6	.99955883	.99916712	.99881075	.99848150	.99817407
7	.99990613	.99981901	.99973715	.99965957	.99958559
8	.99998055	.99996199	.99994416	.99992696	.99991031
9	.99999606	.99999222	.99998849	.99998484	.99998128
10	.99999921	.99999844	.99999768	.99999693	.99999620
11	.99999985	.99999969	.99999954	.99999939	.99999924
12	.99999997	.99999994	.99999991	.99999988	.99999985
13	1	.99999999	.99999998	.99999998	.99999997
14	1	1	1	1	1
15	1	1	1	1	1

M	B= 6	B= 7	B= 8	B= 9	B=10
1	.35471212	.33999889	.32779309	.31742805	.30846459
2	.70536919	.69064998	.67795834	.66682407	.65692180
3	.89414905	.88617046	.87906420	.87265514	.86681665
4	.96788659	.96476268	.96190460	.95926650	.95681392
5	.99137856	.99038471	.98945532	.98858093	.98775414
6	.99788472	.99761072	.99734996	.99710082	.99686196
7	.99951471	.99944654	.99938078	.99931717	.99925550
8	.99989415	.99987842	.99986308	.99984811	.99983347
9	.99997778	.99997434	.99997097	.99996766	.99996439
10	.99999547	.99999475	.99999404	.99999334	.99999264
11	.99999910	.99999895	.99999881	.99999866	.99999852
12	.99999982	.99999979	.99999977	.99999974	.99999971
13	.99999997	.99999996	.99999995	.99999995	.99999994
14	1	1	1	1	.99999999
15	1	1	1	1	1

M	B=11	B=12	B=13	B=14	B=15
1	.30059945	.29361525	.28735135	.28168600	.27652489
2	.64801726	.63993639	.63254644	.62574389	.61944658
3	.86145402	.85649457	.85188122	.84756832	.84351875
4	.95452016	.95236413	.95032883	.94840031	.94656702
5	.98696908	.98622095	.98550581	.98482036	.98416178
6	.99663230	.99641093	.99619710	.99599014	.99578951
7	.99919561	.99913734	.99908057	.99902519	.99897110
8	.99981913	.99980508	.99979130	.99977777	.99976447
9	.99996117	.99995800	.99995488	.99995179	.99994875
10	.99999196	.99999128	.99999060	.99998994	.99998927
11	.99999838	.99999824	.99999810	.99999796	.99999782
12	.99999968	.99999965	.99999962	.99999960	.99999957
13	.99999994	.99999993	.99999993	.99999992	.99999992
14	.99999999	.99999999	.99999999	.99999998	.99999998
15	1	1	1	1	1

DIRICHLET C INTEGRAL
1/A = 5 R = 4

M	B= 1	B= 2	B= 3	B= 4	B= 5
1	.00077160	.15664\04	.10120\05	.12270\06	.22165\07
2	.00334362	.00010835	.91174\05	.13197\05	.27147\06
3	.00870199	.00041710	.44325\04	.75335\05	.17467\05
4	.01763260	.00117948	.00015427	.30344\04	.78558\05
5	.03065641	.00273380	.00043054	.96755\04	.27736\04
6	.04802149	.00550162	.00102376	.00025987	.81856\04
7	.06972784	.00995479	.00215318	.00061111	.00021004
8	.09556874	.01657397	.00410754	.00129153	.00048130
9	.12517809	.02580567	.00723571	.00249967	.00100410
10	.15807738	.03802411	.01192767	.00449335	.00193512
11	.19371827	.05350187	.01858840	.00758400	.00348435
12	.23151922	.07239164	.02760789	.01212326	.00591470
13	.27089520	.09471918	.03933134	.01848299	.00953514
14	.31128083	.12038641	.05403274	.02703086	.01468729
15	.35214724	.14918287	.07189464	.03810433	.02172649

M	B= 6	B= 7	B= 8	B= 9	B=10
1	.52742\08	.15352\08	.52114\09	.19963\09	.84304\10
2	.71447\07	.22558\07	.81917\08	.33228\08	.14742\08
3	.50487\06	.17202\06	.66564\07	.28502\07	.13252\07
4	.24778\05	.90672\06	.37252\06	.16790\06	.81622\07
5	.94892\05	.37130\05	.16141\05	.76366\06	.38730\06
6	.30213\04	.12589\04	.57720\05	.28592\05	.15096\05
7	.83211\04	.36780\04	.17733\04	.91747\05	.50325\05
8	.00020370	.95169\04	.48111\04	.25939\04	.14754\04
9	.00045204	.00022247	.00011761	.65932\04	.38817\04
10	.00092291	.00047697	.00026302	.00015299	.93069\04
11	.00175383	.00094899	.00054455	.00032802	.00020584
12	.00313098	.00176884	.00105379	.00065613	.00042403
13	.00529088	.00311261	.00192111	.00123415	.00082021
14	.00851640	.00520436	.00332101	.00219751	.00149967
15	.01312691	.00831327	.00547409	.00372487	.00260666

M	B=11	B=12	B=13	B=14	B=15
1	.38576\10	.18894\10	.98589\11	.55777\11	.37308\11
2	.70435\09	.35811\09	.19202\09	.10792\09	.63525\10
3	.65977\08	.34791\08	.19269\08	.11135\08	.66802\09
4	.42260\07	.23076\07	.13187\07	.78384\08	.48223\08
5	.20816\06	.11752\06	.69202\07	.42267\07	.26655\07
6	.84075\06	.49000\06	.29695\06	.18617\06	.12024\06
7	.28995\05	.17420\05	.10851\05	.69752\06	.46093\06
8	.87796\05	.54302\05	.34726\05	.22863\05	.15444\05
9	.23820\04	.15146\04	.99326\05	.66915\05	.46165\05
10	.58809\04	.38397\04	.25792\04	.17762\04	.12504\04
11	.00013374	.89548\04	.61549\04	.43288\04	.31072\04
12	.00028292	.00019404	.00013632	.97830\04	.71540\04
13	.00056121	.00039382	.00028254	.00020670	.00015387
14	.00105100	.00075378	.00055171	.00041112	.00031130
15	.00186884	.00136846	.00102087	.00077423	.00059589

DIRICHLET C INTEGRAL
1/A = 4 R = 4

M	B= 1	B= 2	B= 3	B= 4	B= 5
1	.00160000	.56424\04	.56615\05	.78733\06	.24275\06
2	.00672000	.00037035	.47653\04	.98048\05	.27197\05
3	.01696000	.00135383	.00021666	.51738\04	.16025\04
4	.03334400	.00363867	.00070596	.00019285	.66087\04
5	.05628160	.00802290	.00184646	.00056972	.00021421
6	.08564173	.01537339	.00411944	.00141948	.00058116
7	.12087388	.02651221	.00813845	.00310043	.00137271
8	.16113920	.04211246	.01460126	.00609430	.00289958
9	.20543105	.06262129	.02422034	.01098552	.00558438
10	.25267569	.08821798	.03764526	.01841830	.00995050
11	.30181012	.11880719	.05539061	.02903777	.01659122
12	.35183790	.15404148	.07778047	.04342469	.02612260
13	.40186567	.19336468	.10491599	.06203412	.03912627
14	.45112380	.23606698	.13666782	.08514707	.05609121
15	.49897454	.28134355	.17269126	.11284085	.07736265

M	B= 6	B= 7	B= 8	B= 9	B=10
1	.75415\07	.27751\07	.11606\07	.53642\08	.26863\08
2	.92745\06	.36785\06	.16369\06	.79743\07	.41786\07
3	.59567\05	.25335\05	.11950\05	.61169\06	.33457\06
4	.26605\04	.12077\04	.60158\05	.32265\05	.18378\05
5	.92844\04	.44784\04	.23479\04	.13158\04	.77876\05
6	.00026973	.00013769	.75734\04	.44236\04	.27145\04
7	.00067880	.00036530	.00021017	.00012764	.81049\04
8	.00152059	.00085963	.00051586	.00032497	.00021313
9	.00309243	.00183037	.00114263	.00074504	.00050374
10	.00579533	.00358013	.00231906	.00156190	.00108681
11	.01012524	.00650927	.00436476	.00303048	.00216654
12	.01664681	.01110624	.00769192	.00549519	.00403002
13	.02595173	.01792200	.01279291	.00938734	.00705143
14	.03860687	.02753025	.02021318	.01520888	.01168465
15	.05509954	.04047833	.03051133	.02350250	.01844235

M	B=11	B=12	B=13	B=14	B=15
1	.14364\08	.81117\09	.47947\09	.29425\09	.18564\09
2	.23249\07	.13600\07	.83014\08	.52547\08	.34319\08
3	.19329\06	.11691\06	.73530\07	.47825\07	.32027\07
4	.11003\05	.68698\06	.44457\06	.29673\06	.20345\06
5	.48229\05	.31036\05	.20638\05	.14119\05	.99009\06
6	.17359\04	.11496\04	.78449\05	.54946\05	.39370\05
7	.53428\04	.36360\04	.25431\04	.18216\04	.13324\04
8	.00014460	.00010098	.72305\04	.52912\04	.39469\04
9	.00035119	.00025135	.00018403	.00013744	.00010447
10	.00077745	.00056952	.00042591	.00032432	.00025095
11	.00158799	.00118920	.00090739	.00070383	.00055395
12	.00302244	.00231113	.00179738	.00141884	.00113495
13	.00540425	.00421468	.00333754	.00267886	.00217615
14	.00913982	.00726198	.00584985	.00477007	.00393210
15	.01470555	.01189133	.00973515	.00805784	.00673527

DIRICHLET C INTEGRAL
1/A = 3 R = 4

M	B= 1	B= 2	B= 3	B= 4	B= 5
1	.00390625	.00026050	.42641\04	.11031\04	.37666\05
2	.01562500	.00157960	.00032470	.97584\04	.37148\04
3	.03759766	.00534315	.00133808	.00045964	.00019310
4	.07055664	.01331091	.00395952	.00153258	.00070403
5	.11381531	.02725225	.00942471	.00405911	.00202219
6	.16572571	.04858026	.01917680	.00908826	.00487350
7	.22412491	.07809238	.03463238	.01788207	.01025158
8	.28669548	.11586124	.05693388	.03174536	.01933613
9	.35122138	.16126656	.08675288	.05182078	.03334621
10	.41574728	.21312805	.12418397	.07889997	.05336025
11	.47866004	.26989093	.16873884	.11329257	.08014443
12	.53871312	.32982085	.21942549	.15477514	.11402664
13	.59501289	.39117760	.27488365	.20262072	.15483853
14	.64698190	.45235045	.33354419	.25569267	.20193059
15	.69431083	.51194930	.39378440	.31257768	.25424912

M	B= 6	B= 7	B= 8	B= 9	B=10
1	.15489\05	.72747\06	.37740\06	.21149\06	.12603\06
2	.16611\04	.83435\05	.45759\05	.26878\05	.16680\05
3	.93229\04	.49830\04	.28779\04	.17664\04	.11388\04
4	.00036469	.00020644	.00012510	.80008\04	.53455\04
5	.00111728	.00066689	.00042259	.00028085	.00019403
6	.00285661	.00179059	.00118271	.00081468	.00058079
7	.00634321	.00415969	.00285532	.00203356	.00149293
8	.01257195	.00859448	.00611363	.00449151	.00338912
9	.02268545	.01611354	.01184675	.00895846	.00693494
10	.03783340	.02783545	.02109822	.01638773	.01299220
11	.05900651	.04483750	.03495425	.02783273	.02256053
12	.08688171	.06799978	.05440104	.04432343	.03667451
13	.12171053	.09786760	.08017975	.06672521	.05627277
14	.16327049	.13455947	.11266869	.09560839	.08206395
15	.21088337	.17773592	.15181481	.13115379	.11441571

M	B=11	B=12	B=13	B=14	B=15
1	.78966\07	.51575\07	.34885\07	.24312\07	.17385\07
2	.10828\05	.72979\06	.50770\06	.36290\06	.26556\06
3	.76439\05	.53069\05	.37914\05	.27761\05	.20765\05
4	.37028\04	.26438\04	.19370\04	.14511\04	.11082\04
5	.00013844	.00010150	.76158\04	.58300\04	.45416\04
6	.00042609	.00032026	.00024578	.00019204	.00015244
7	.00112424	.00086509	.00067815	.00054026	.00043654
8	.00261542	.00205742	.00164547	.00133512	.00109713
9	.00547587	.00439774	.00358413	.00295881	.00247041
10	.01048087	.00858230	.00711956	.00597382	.00506325
11	.01856711	.01548244	.01305889	.01112625	.00956472
12	.03074983	.02607980	.02234233	.01931101	.01682319
13	.04800515	.04136321	.03595438	.03149677	.02778391
14	.07113854	.06220336	.05480707	.04861891	.04339213
15	.10066457	.08922874	.07961587	.07145822	.06447648

DIRICHLET C INTEGRAL
1/A = 2 R = 4

M	B= 1	B= 2	B= 3	B= 4	B= 5
1	.01234568	.00174214	.00049300	.00019553	.94604\04
2	.04526749	.00923615	.00318458	.00143665	.00076277
3	.10013717	.02742279	.01118133	.00564555	.00325612
4	.17329675	.06021655	.02832094	.01578031	.00979734
5	.25864960	.10915710	.05798820	.03521818	.02334512
6	.34969263	.17310962	.10203479	.06681116	.04693473
7	.44073566	.24881141	.16024626	.11203550	.08285119
8	.52744331	.33181838	.23045109	.17055930	.13196785
9	.60692532	.41745439	.30909556	.24032081	.19347705
10	.67757600	.50153052	.39200645	.31800368	.26505351
11	.73880659	.58076068	.47509950	.39969610	.34333457
12	.79075981	.65290405	.55489175	.48152958	.42453166
13	.83405417	.71671187	.62877626	.56016246	.50500102
14	.86957774	.77176232	.69508768	.63305695	.58166347
15	.89833492	.81825220	.75302004	.69856294	.65223417

M	B= 6	B= 7	B= 8	B= 9	B=10
1	.52151\04	.31516\04	.20389\04	.13900\04	.98775\05
2	.00045160	.00028900	.00019603	.00013911	.00010234
3	.00205604	.00138618	.00098202	.00072317	.00054935
4	.00655605	.00463465	.00341659	.00260321	.00203709
5	.01645841	.01214601	.00928510	.00729956	.00587072
6	.03467405	.02660347	.02102130	.01700664	.01402678
7	.06382252	.05071475	.04129509	.03429425	.02894666
8	.10551733	.08652896	.07239712	.06157146	.05307977
9	.15989573	.13486769	.11563520	.10048659	.08830892
10	.22553242	.19506021	.17095050	.15146937	.13545073
11	.29971839	.26504025	.23686374	.21355830	.19399231
12	.37897276	.34173275	.31073537	.28454372	.26213093
13	.45961837	.42158629	.38922981	.36135315	.33707820
14	.53827990	.50110092	.46883934	.44054947	.41551902
15	.61222347	.57724215	.54634302	.51881145	.49409634

M	B=11	B=12	B=13	B=14	B=15
1	.72594\05	.54856\05	.42432\05	.33480\05	.26872\05
2	.77527\04	.60180\04	.47683\04	.38450\04	.31477\04
3	.00042807	.00034073	.00027613	.00022725	.00018954
4	.00162951	.00132772	.00109889	.00092184	.00078242
5	.00481149	.00400653	.00338179	.00288808	.00249177
6	.01175665	.00998900	.00858677	.00745643	.00653245
7	.02476800	.02143952	.01874443	.01653104	.01469058
8	.04628587	.04075845	.03619623	.03238331	.02916154
9	.07835047	.07008715	.06314360	.05724470	.05218470
10	.12208327	.11078631	.10113400	.09280752	.08556373
11	.17735634	.16305656	.15064773	.13978968	.13021820
12	.24274326	.22581475	.21091210	.19769790	.18590547
13	.31574461	.29684567	.27998571	.26485104	.25118963
14	.39320010	.37316365	.35506838	.33863910	.32365122
15	.47176479	.45147124	.43293601	.41592995	.40026331

DIRICHLET C INTEGRAL
A = 1 R = 4

M	B= 1	B= 2	B= 3	B= 4	B= 5
1	.06250000	.02166209	.01129060	.00708024	.00493135
2	.18750000	.08632449	.05222452	.03607434	.02694271
3	.34375000	.19560344	.13336965	.09984608	.07913893
4	.50000000	.33333333	.25000000	.20000000	.16666667
5	.63671875	.47755611	.38609738	.32602794	.28325994
6	.74609375	.61032103	.52307460	.46125725	.41468538
7	.82812500	.72148584	.64649813	.58988437	.54511657
8	.88671875	.80806474	.74858497	.70123581	.66221504
9	.92700195	.87177341	.82748817	.79064708	.75920281
10	.95385742	.91657124	.88523107	.85819141	.83441994
11	.97131348	.94693004	.92564684	.90672536	.88967332
12	.98242188	.96688816	.95291526	.94018475	.92847290
13	.98936462	.97968144	.97076073	.96246994	.95471155
14	.99363708	.98770985	.98214550	.97689052	.97190393
15	.99623108	.99265835	.98925443	.98599830	.98287329

M	B= 6	B= 7	B= 8	B= 9	B=10
1	.00367400	.00286827	.00231732	.00192183	.00162701
2	.02118285	.01726938	.01446333	.01236757	.01075146
3	.06518415	.05519412	.04771768	.04192878	.03732433
4	.14285714	.12500000	.11111111	.10000000	.09090909
5	.25111202	.22598339	.20575216	.18908240	.17508882
6	.37808133	.34840356	.32376139	.30291080	.28499602
7	.50853736	.47790850	.45176965	.42912035	.40924824
8	.62924092	.60083987	.57600727	.55402896	.53438006
9	.73184989	.70770353	.68613599	.66668570	.64900339
10	.81322212	.79410639	.77671117	.76076224	.74604647
11	.87414342	.85988010	.84668913	.83441901	.82294903
12	.91761508	.90748585	.89798702	.88903997	.88058067
13	.94741106	.94050992	.93396105	.92772585	.92177223
14	.96715338	.96261271	.95826037	.95407833	.95005130
15	.97986585	.97696473	.97416048	.97144503	.96881141

M	B=11	B=12	B=13	B=14	B=15
1	.00140050	.00122213	.00107877	.00096153	.00086423
2	.00947274	.00843937	.00758938	.00687966	.00627938
3	.03358126	.03048309	.02787954	.02566322	.02375538
4	.08333333	.07692308	.07142857	.06666667	.06250000
5	.16316033	.15286083	.14387035	.13594851	.12891109
6	.26940699	.25569583	.24352552	.23263715	.22282817
7	.39163011	.37587155	.36166873	.34878324	.33702492
8	.51666422	.50057522	.48587197	.47236145	.45988694
9	.63281843	.61791694	.60412701	.59130846	.57934551
10	.73239475	.71967057	.70776210	.69657653	.68603600
11	.81218130	.80203524	.79244364	.78334983	.77470554
12	.87255611	.86492185	.85764023	.85067899	.84401027
13	.91607314	.91060550	.90534946	.90028776	.89540532
14	.94616617	.94241157	.93877760	.93525553	.93183764
15	.96625356	.96376614	.96134445	.95898429	.95668189

DIRICHLET C INTEGRAL
A = 2 R = 4

M	B= 1	B= 2	B= 3	B= 4	B= 5
1	.19753086	.11448573	.08261546	.06557478	.05488214
2	.46090535	.32830030	.26407347	.22503515	.19834330
3	.68038409	.55541266	.48326916	.43460549	.39883724
4	.82670325	.73618102	.67643800	.63258471	.59833744
5	.91205609	.85630930	.81554078	.78349881	.75717892
6	.95757761	.92683694	.90251485	.88231948	.86502296
7	.98033836	.96468445	.95153080	.94011974	.93000640
8	.99117682	.98366674	.97706068	.97112791	.96572196
9	.99614445	.99270283	.98956910	.98667754	.98398358
10	.99835228	.99683083	.99540922	.99406974	.99279969
11	.99930901	.99865546	.99803298	.99743703	.99686419
12	.99971489	.99944061	.99917568	.99891897	.99866960
13	.99988401	.99977109	.99966089	.99955315	.99944766
14	.99995339	.99990763	.99986265	.99981839	.99977478
15	.99998147	.99996318	.99994511	.99992723	.99990955

M	B= 6	B= 7	B= 8	B= 9	B=10
1	.04750425	.04208246	.03791533	.03460319	.03190113
2	.17872083	.16356706	.15143928	.14146723	.13309202
3	.37106751	.34867103	.33009442	.31435083	.30077862
4	.57047702	.54714439	.52717392	.50978938	.49444941
5	.73490012	.71562561	.69867055	.68355894	.66994660
6	.84988375	.83641696	.82428699	.81325131	.80312868
7	.92090363	.91261369	.90499385	.89793724	.89136155
8	.96074268	.95611784	.95179331	.94772723	.94388646
9	.98145514	.97906815	.97680392	.97464759	.97258710
10	.99158952	.99043177	.98932051	.98825089	.98721885
11	.99631176	.99577760	.99525995	.99475731	.99426844
12	.99842685	.99819015	.99795899	.99773294	.99751165
13	.99934423	.99924270	.99914296	.99904487	.99894835
14	.99973180	.99968939	.99964753	.99960619	.99956534
15	.99989204	.99987471	.99985754	.99984052	.99982365

M	B=11	B=12	B=13	B=14	B=15
1	.02965043	.02774358	.02610505	.02468017	.02342836
2	.12593681	.11973746	.11430276	.10949063	.10519304
3	.28891507	.27842534	.26906027	.26062996	.25298682
4	.48076239	.46843650	.45724881	.44702549	.43762855
5	.65757668	.64625248	.63582006	.62615676	.61716327
6	.79377989	.78509563	.77698839	.76938701	.76223283
7	.88520189	.87940607	.87393146	.86874272	.86381020
8	.94024425	.93677860	.93347119	.93030656	.92727156
9	.97061249	.96871540	.96688874	.96512645	.96342327
10	.98622100	.98525444	.98431666	.98340549	.98251902
11	.99379226	.99332783	.99287433	.99243105	.99199734
12	.99729480	.99708209	.99687329	.99666817	.99646652
13	.99885329	.99875963	.99866728	.99857619	.99848629
14	.99952496	.99948502	.99944550	.99940639	.99936767
15	.99980692	.99979033	.99977388	.99975754	.99974134

DIRICHLET C INTEGRAL
A = 3 R= 4

M	B= 1	B= 2	B= 3	B= 4	B= 5
1	.31640625	.21952260	.17676734	.15171248	.13488212
2	.63281250	.51871318	.45641097	.41532492	.38544332
3	.83056641	.75123185	.70049721	.66372034	.63514629
4	.92944336	.88700170	.85645320	.83257315	.81297960
5	.97270203	.95336906	.93816361	.92554482	.91471882
6	.99000549	.98210596	.97546726	.96969301	.96455710
7	.99649429	.99350800	.99087174	.98849362	.98631655
8	.99881172	.99774616	.99677079	.99586601	.99501872
9	.99960834	.99924452	.99890261	.99857867	.99826994
10	.99987388	.99975386	.99963891	.99952828	.99942142
11	.99996018	.99992165	.99988426	.99984785	.99981233
12	.99998764	.99997554	.99996369	.99995205	.99994062
13	.99999622	.99999249	.99998881	.99998518	.99998159
14	.99999886	.99999772	.99999660	.99999549	.99999439
15	.99999966	.99999932	.99999898	.99999865	.99999831

M	B= 6	B= 7	B= 8	B= 9	B=10
1	.12262300	.11320099	.10567732	.09949517	.09430127
2	.36235828	.34377659	.32836796	.31529883	.30401587
3	.61193816	.59249622	.57583328	.56129917	.54844404
4	.79637888	.78198791	.76929589	.75795089	.74770002
5	.90521696	.89673720	.88907252	.88207433	.87563209
6	.95991604	.95567212	.95175531	.94811346	.94470652
7	.98430196	.98242227	.98065691	.97899004	.97740914
8	.99421957	.99346158	.99273938	.99204867	.99138599
9	.99797433	.99769024	.99741637	.99715168	.99689530
10	.99931790	.99921737	.99911956	.99902423	.99893117
11	.99977761	.99974362	.99971031	.99967763	.99964552
12	.99992937	.99991828	.99990736	.99989659	.99988596
13	.99997804	.99997453	.99997106	.99996762	.99996421
14	.99999329	.99999221	.99999113	.99999006	.99998899
15	.99999798	.99999765	.99999732	.99999700	.99999667

M	B=11	B=12	B=13	B=14	B=15
1	.08985959	.08600574	.08262139	.07961890	.07693186
2	.29413489	.28537943	.27754455	.27047460	.26404883
3	.53694436	.52656008	.51710839	.50844708	.50046342
4	.73835538	.72977357	.72184262	.71447344	.70759399
5	.86966119	.86409536	.85888160	.85397689	.84934575
6	.94150301	.93847763	.93560974	.93288224	.93028079
7	.97590412	.97446669	.97308996	.97176809	.97049613
8	.99074847	.99013371	.98953969	.98896465	.98840708
9	.99664649	.99640463	.99616918	.99593966	.99571567
10	.99884021	.99875120	.99866401	.99857853	.99849465
11	.99961397	.99958292	.99955236	.99952225	.99949257
12	.99987546	.99986509	.99985483	.99984470	.99983467
13	.99996083	.99995748	.99995416	.99995087	.99994761
14	.99998794	.99998688	.99998584	.99998480	.99998377
15	.99999635	.99999602	.99999570	.99999538	.99999507

DIRICHLET C INTEGRAL
A = 4 R = 4

M	B= 1	B= 2	B= 3	B= 4	B= 5
1	.40960000	.31126445	.26473022	.23619000	.21635949
2	.73728000	.64553258	.59237933	.55589046	.52853399
3	.90112000	.85055846	.81657118	.79105552	.77068792
4	.96665600	.94519941	.92910013	.91612909	.90523047
5	.98959360	.98183225	.97551554	.97013699	.96542651
6	.99693363	.99441239	.99223398	.99029805	.98854557
7	.99913564	.99837702	.99769233	.99706367	.99647960
8	.99976479	.99954910	.99934820	.99915917	.99897998
9	.99993780	.99987906	.99982312	.99976951	.99971791
10	.99998394	.99996847	.99995350	.99993897	.99992483
11	.99999594	.99999197	.99998809	.99998429	.99998055
12	.99999899	.99999799	.99999701	.99999604	.99999509
13	.99999975	.99999951	.99999926	.99999902	.99999878
14	.99999994	.99999988	.99999982	.99999976	.99999970
15	.99999999	.99999997	.99999996	.99999994	.99999993

M	B= 6	B= 7	B= 8	B= 9	B=10
1	.20152264	.18986374	.18037638	.17245156	.16569618
2	.50687797	.48908818	.47407667	.46114874	.44983558
3	.75377759	.73934691	.72677985	.71566340	.70570730
4	.89581402	.88751402	.88008721	.87336314	.86721749
5	.96122040	.95741084	.95392257	.95070069	.94770381
6	.98693817	.98544917	.98405913	.98275336	.98152042
7	.99593221	.99541575	.99492586	.99445913	.99401283
8	.99880917	.99864563	.99848847	.99833701	.99819065
9	.99966808	.99961981	.99957295	.99952736	.99948293
10	.99991103	.99989754	.99988434	.99987141	.99985872
11	.99997689	.99997328	.99996972	.99996622	.99996277
12	.99999415	.99999321	.99999229	.99999137	.99999047
13	.99999855	.99999831	.99999808	.99999785	.99999762
14	.99999965	.99999959	.99999953	.99999947	.99999942
15	.99999991	.99999990	.99999989	.99999987	.99999986

M	B=11	B=12	B=13	B=14	B=15
1	.15984353	.15470535	.15014440	.14605792	.14236722
2	.43980693	.43082197	.42270011	.41530268	.40852102
3	.69669986	.68848205	.68093140	.67395162	.66746567
4	.86155669	.85630844	.85141570	.84683260	.84252175
5	.94489983	.94226331	.93977373	.93741424	.93517084
6	.98035121	.97923834	.97817573	.97715828	.97618170
7	.99358475	.99317304	.99277615	.99239276	.99202174
8	.99804893	.99791143	.99777781	.99764777	.99752103
9	.99943957	.99939721	.99935576	.99931517	.99927538
10	.99984626	.99983401	.99982197	.99981011	.99979843
11	.99995936	.99995599	.99995267	.99994938	.99994613
12	.99998957	.99998868	.99998780	.99998693	.99998607
13	.99999739	.99999716	.99999694	.99999671	.99999649
14	.99999936	.99999930	.99999925	.99999919	.99999913
15	.99999985	.99999983	.99999982	.99999980	.99999979

DIRICHLET C INTEGRAL
A = 5 R = 4

M	B= 1	B= 2	B= 3	B= 4	B= 5
1	.48225309	.38713915	.34020793	.31063061	.28966114
2	.80375514	.73027232	.68617564	.65517369	.63150821
3	.93771433	.90416826	.88094649	.86314912	.84871440
4	.98236740	.97056733	.96149637	.95405784	.94772031
5	.99539121	.99185049	.98891149	.98637156	.98412041
6	.99886422	.99790935	.99707122	.99631719	.99562771
7	.99973248	.99949377	.99927564	.99907336	.99888385
8	.99993921	.99988278	.99982971	.99977939	.99973136
9	.99998658	.99997380	.99996153	.99994970	.99993826
10	.99999711	.99999431	.99999158	.99998892	.99998632
11	.99999939	.99999879	.99999820	.99999763	.99999706
12	.99999987	.99999975	.99999962	.99999950	.99999938
13	.99999997	.99999995	.99999992	.99999990	.99999987
14	1	.99999999	.99999998	.99999998	.99999997
15	1	1	1	1	1

M	B= 6	B= 7	B= 8	B= 9	B=10
1	.27371947	.26102581	.25057955	.24176798	.23419147
2	.61250104	.59669754	.58322343	.57151405	.56118459
3	.83657353	.82609920	.81689117	.80867822	.80126796
4	.94218152	.93725159	.93280270	.92874441	.92501017
5	.98209011	.98023537	.97852420	.97693306	.97544404
6	.99498986	.99439459	.99383520	.99330661	.99280481
7	.99870496	.99853511	.99837308	.99821792	.99806886
8	.99968531	.99964098	.99959818	.99955675	.99951656
9	.99992716	.99991636	.99990583	.99989555	.99988550
10	.99998378	.99998128	.99997883	.99997642	.99997405
11	.99999650	.99999594	.99999540	.99999486	.99999432
12	.99999926	.99999914	.99999903	.99999891	.99999879
13	.99999985	.99999982	.99999980	.99999977	.99999975
14	.99999997	.99999996	.99999996	.99999995	.99999995
15	1	1	1	1	.99999999

M	B=11	B=12	B=13	B=14	B=15
1	.22757638	.22172800	.21650333	.21179457	.20751867
2	.55196162	.54364418	.53608050	.52915329	.52277012
3	.79451892	.78832401	.78260020	.77728182	.77231602
4	.92154948	.91832305	.91529969	.91245416	.90976577
5	.97404316	.97271927	.97146330	.97026778	.96912645
6	.99232661	.99186938	.99143097	.99100954	.99060355
7	.99792527	.99778661	.99765245	.99752239	.99739613
8	.99947749	.99943946	.99940238	.99936618	.99933081
9	.99987566	.99986601	.99985655	.99984726	.99983813
10	.99997172	.99996942	.99996715	.99996492	.99996271
11	.99999380	.99999327	.99999276	.99999225	.99999174
12	.99999868	.99999856	.99999845	.99999834	.99999823
13	.99999972	.99999970	.99999968	.99999965	.99999963
14	.99999994	.99999994	.99999993	.99999993	.99999992
15	.99999999	.99999999	.99999999	.99999999	.99999998

DIRICHLET C INTEGRAL
1/A = 5 R = 5

M	B= 1	B= 2	B= 3	B= 4	B= 5
1	.00012860	.11651\05	.44700\07	.37160\08	.50171\09
2	.00066444	.97366\05	.48784\06	.48421\07	.74387\08
3	.00200403	.44441\04	.28332\05	.33115\06	.57395\07
4	.00460879	.00014669	.11635\04	.15817\05	.30686\06
5	.00895006	.00039161	.37894\04	.59250\05	.12774\05
6	.01546197	.00089736	.00010412	.18535\04	.44116\05
7	.02450628	.00183040	.00025079	.50363\04	.13152\04
8	.03635002	.00340522	.00054346	.00012208	.34777\04
9	.05115470	.00587800	.00107944	.00026915	.83195\04
10	.06897515	.00953483	.00199267	.00054758	.00018275
11	.08976567	.01467586	.00345571	.00103973	.00037293
12	.11339126	.02159714	.00567800	.00185915	.00071355
13	.13964192	.03057222	.00890010	.00315362	.00128992
14	.16824840	.04183519	.01338404	.00510542	.00221704
15	.19889821	.05556660	.01940071	.00792838	.00364214

M	B= 6	B= 7	B= 8	B= 9	B=10
1	.94444\10	.22634\10	.65102\11	.21550\11	.79593\12
2	.15471\08	.40171\09	.12355\09	.43406\10	.16974\10
3	.13111\07	.36728\08	.12031\08	.44597\09	.18270\09
4	.76586\07	.23053\07	.80184\08	.31281\08	.13395\08
5	.34656\06	.11168\06	.41128\07	.16848\07	.75272\08
6	.12948\05	.44514\06	.17310\06	.74299\07	.34571\07
7	.41579\05	.15198\05	.62248\06	.27937\06	.13515\06
8	.11794\04	.45695\05	.19663\05	.92089\06	.46245\06
9	.30147\04	.12344\04	.55675\05	.27159\05	.14135\05
10	.70501\04	.30421\04	.14350\04	.72779\05	.39198\05
11	.00015263	.69222\04	.34077\04	.17938\04	.99830\05
12	.00030883	.00014684	.75284\04	.41061\04	.23581\04
13	.00058857	.00029266	.00015597	.88004\04	.52083\04
14	.00106336	.00055172	.00030507	.00017779	.00010829
15	.00183123	.00098923	.00056651	.00034051	.00021319

M	B=11	B=12	B=13	B=14	B=15
1	.32884\12	.18262\12	.21143\12	.43689\12	.10223\11
2	.72558\11	.33745\11	.17762\11	.12449\11	.14566\11
3	.81292\10	.38794\10	.19736\10	.10817\10	.67782\11
4	.61958\09	.30579\09	.15957\09	.87568\10	.50648\10
5	.36139\08	.18436\08	.99049\09	.55669\09	.32582\09
6	.17204\07	.90607\08	.50093\08	.28883\08	.17279\08
7	.69610\07	.37805\07	.21487\07	.12703\07	.77733\08
8	.24619\06	.13772\06	.80390\07	.48691\07	.30461\07
9	.77676\06	.44706\06	.26775\06	.16601\06	.10610\06
10	.22207\05	.13135\05	.80643\06	.51142\06	.33370\06
11	.58235\05	.35365\05	.22236\05	.14411\05	.95932\06
12	.14147\04	.88113\05	.56689\05	.37521\05	.25463\05
13	.32099\04	.20484\04	.13473\04	.90998\05	.62915\05
14	.68486\04	.44737\04	.30058\04	.20701\04	.14572\04
15	.00013820	.92325\04	.63313\04	.44430\04	.31823\04

DIRICHLET C INTEGRAL
1/A = 4 R = 5

M	B= 1	B= 2	B= 3	B= 4	B= 5
1	.00032000	.57280\05	.37595\06	.48504\07	.94799\08
2	.00160000	.45311\04	.38218\05	.58159\06	.12810\06
3	.00467200	.00019588	.20691\04	.36632\05	.90168\06
4	.01040640	.00061277	.79272\04	.16129\04	.44020\05
5	.01958144	.00155146	.00024106	.55745\04	.16749\04
6	.03279350	.00337407	.00061899	.00016105	.52922\04
7	.05040957	.00653671	.00139458	.00040456	.00014450
8	.07255550	.01155901	.00282945	.00090754	.00035037
9	.09913061	.01898101	.00526683	.00185362	.00076940
10	.12983963	.02931450	.00912098	.00349766	.00155332
11	.16423372	.04299635	.01485447	.00616674	.00291676
12	.20175456	.06034979	.02294583	.01025131	.00514203
13	.24177678	.08155768	.03385178	.01618617	.00857583
14	.28364618	.10664944	.04796888	.02442285	.01361702
15	.32671186	.13550149	.06559946	.03539637	.02069553

M	B= 6	B= 7	B= 8	B= 9	B=10
1	.24527\08	.77630\09	.28589\09	.11843\09	.53867\10
2	.36334\07	.12378\07	.48441\08	.21126\08	.10049\08
3	.27874\06	.10179\06	.42195\07	.19323\07	.95870\08
4	.14753\05	.57520\06	.25182\06	.12080\06	.62395\07
5	.60552\05	.25114\05	.11579\05	.58054\06	.31158\06
6	.20542\04	.90316\05	.43733\05	.22869\05	.12731\05
7	.59959\04	.27853\04	.14129\04	.76895\05	.44330\05
8	.00015477	.75729\04	.40143\04	.22695\04	.13526\04
9	.00036045	.00018522	.00010236	.59999\04	.36914\04
10	.00076895	.00041380	.00023791	.00014431	.91512\04
11	.00152056	.00085469	.00051010	.00031967	.00020863
12	.00281387	.00164788	.00101888	.00065859	.00044174
13	.00491130	.00298751	.00191124	.00127216	.00087581
14	.00813783	.00513700	.00338953	.00231974	.00163704
15	.01287168	.00840808	.00571579	.00401617	.00290162

M	B=11	B=12	B=13	B=14	B=15
1	.26379\10	.13615\10	.71124\11	.32931\11	.35387\12
2	.51315\09	.27787\09	.15789\09	.93068\10	.55822\10
3	.50802\08	.28440\08	.16677\08	.10172\08	.64121\09
4	.34251\07	.19785\07	.11934\07	.74710\08	.48295\08
5	.17691\06	.10531\06	.65257\07	.41864\07	.27677\07
6	.74658\06	.45738\06	.29087\06	.19103\06	.12903\06
7	.26811\05	.16884\05	.11007\05	.73942\06	.50984\06
8	.84256\05	.54481\05	.36375\05	.24970\05	.17562\05
9	.23651\04	.15685\04	.10714\04	.75097\05	.53837\05
10	.60232\04	.40922\04	.28574\04	.20432\04	.14919\04
11	.00014089	.97963\04	.69855\04	.50919\04	.37841\04
12	.00030572	.00021732	.00015812	.00011740	.88736\04
13	.00062044	.00045047	.00033411	.00025248	.00019397
14	.00118581	.00087849	.00066367	.00051008	.00039804
15	.00214682	.00162132	.00124659	.00097371	.00077128

DIRICHLET C INTEGRAL
1/A = 3 R = 5

M	B= 1	B= 2	B= 3	B= 4	B= 5
1	.00097656	.38245\04	.45453\05	.94166\06	.27219\06
2	.00463867	.00027843	.41614\04	.10009\04	.32212\05
3	.01287842	.00110916	.00020320	.55973\04	.19888\04
4	.02729797	.00320159	.00070322	.00021917	.85320\04
5	.04892731	.00748985	.00193486	.00067484	.00028578
6	.07812691	.01507238	.00450278	.00174016	.00079643
7	.11462641	.02706128	.00921081	.00390888	.00192185
8	.15764368	.04441914	.01699906	.00785715	.00412677
9	.20603810	.06782044	.02883988	.01441010	.00804337
10	.25846540	.09756204	.04561390	.02446961	.01444549
11	.31351406	.13353214	.06799143	.03891426	.02418810
12	.36981382	.17523357	.09634105	.05848982	.03811962
13	.42611359	.22184884	.13067963	.08371064	.05698157
14	.48133067	.27233029	.17066779	.11478874	.08131346
15	.53457571	.32549931	.21564630	.15160097	.11137918

M	B= 6	B= 7	B= 8	B= 9	B=10
1	.98140\07	.41393\07	.19620\07	.10176\07	.56693\08
2	.12610\05	.56799\06	.28422\06	.15431\06	.89417\07
3	.84049\05	.40263\05	.21202\05	.12021\05	.72302\06
4	.38718\04	.19649\04	.10856\04	.64117\05	.39955\05
5	.00013857	.74223\04	.42905\04	.26338\04	.16973\04
6	.00041071	.00023139	.00013957	.88859\04	.59112\04
7	.00104947	.00061989	.00038913	.00025642	.00017579
8	.00237665	.00146721	.00095623	.00065089	.00045910
9	.00486678	.00313096	.00211364	.00148333	.00107475
10	.00915020	.00611747	.00426819	.00308266	.00229093
11	.01598577	.01107745	.00797102	.00591448	.00450178
12	.02620227	.01877280	.01390370	.01058109	.00823700
13	.04061531	.03001499	.02283664	.01779653	.01415005
14	.05993280	.04558257	.03556221	.02833537	.02298105
15	.08466537	.06613070	.05281156	.04296033	.03549617

M	B=11	B=12	B=13	B=14	B=15
1	.33470\08	.20730\08	.13365\08	.89149\09	.61208\09
2	.54633\07	.34882\07	.23111\07	.15804\07	.11106\07
3	.45643\06	.29998\06	.20397\06	.14278\06	.10249\06
4	.26019\05	.17579\05	.12252\05	.87710\06	.64258\06
5	.11384\04	.78965\05	.56354\05	.41215\05	.30790\05
6	.40778\04	.29001\04	.21169\04	.15802\04	.12028\04
7	.00012454	.90710\04	.67652\04	.51498\04	.39906\04
8	.00033357	.00024852	.00018918	.00014672	.00011566
9	.00079980	.00060884	.00047259	.00037309	.00029894
10	.00174388	.00135486	.00107133	.00086021	.00070007
11	.00350091	.00277301	.00223160	.00182094	.00150409
12	.00653638	.00527288	.00431472	.00357501	.00299494
13	.01144438	.00939301	.00780844	.00656437	.00557353
14	.01892268	.01578596	.01332022	.01135300	.00976285
15	.02972359	.02517985	.02154813	.01860606	.01619412

DIRICHLET C INTEGRAL
1/A = 2 R = 5

M	B= 1	B= 2	B= 3	B= 4	B= 5
1	.00411523	.00040269	.92467\04	.31850\04	.13886\04
2	.01783265	.00254609	.00071273	.00027892	.00013324
3	.04526749	.00883561	.00294000	.00128970	.00066944
4	.08794391	.02229094	.00862695	.00419154	.00234601
5	.14484581	.04573886	.02020428	.01075572	.00644582
6	.21312808	.08103607	.04018955	.02321420	.01480002
7	.28899727	.12860819	.07058248	.04384802	.02956199
8	.36847929	.18739528	.11236623	.07447754	.05280644
9	.44796130	.25513201	.16526521	.11602426	.08607389
10	.52449953	.32881241	.22779624	.16827471	.13000252
11	.59593522	.40518798	.29755304	.22989407	.18415112
12	.66087675	.48119015	.37161388	.29865498	.24704573
13	.71860255	.55422252	.44696095	.37179541	.31641405
14	.76892762	.62231377	.52082879	.44640753	.38952882
15	.81206338	.68415214	.59093969	.51977726	.46357213

M	B= 6	B= 7	B= 8	B= 9	B=10
1	.70564\05	.39917\05	.24436\05	.15891\05	.10839\05
2	.72607\04	.43428\04	.27837\04	.18820\04	.13273\04
3	.00038896	.00024497	.00016389	.00011490	.83622\04
4	.00144565	.00095496	.00066481	.00048218	.00036140
5	.00419175	.00289346	.00209007	.00156461	.00120544
6	.01010998	.00726704	.00543179	.00418759	.00331035
7	.02112077	.01575683	.01215529	.00963038	.00779762
8	.03929971	.03033509	.02409209	.01957622	.01620769
9	.06647377	.05293240	.04317731	.03591214	.03035273
10	.10381530	.08504245	.07108779	.06040896	.05204012
11	.15155399	.12737936	.10888053	.09436263	.08272913
12	.20888153	.17968294	.15673348	.13829633	.12321281
13	.27405899	.24071648	.21385560	.19180353	.17341166
14	.34468000	.30844187	.27857878	.25356642	.23232943
15	.41801226	.38031399	.34859479	.32153329	.29817330

M	B=11	B=12	B=13	B=14	B=15
1	.76832\06	.56220\06	.42245\06	.32467\06	.25441\06
2	.96879\05	.72746\05	.55942\05	.43901\05	.35059\05
3	.62739\04	.48276\04	.37946\04	.30374\04	.24698\04
4	.00027826	.00021910	.00017583	.00014341	.00011862
5	.00095097	.00076522	.00062620	.00051988	.00043706
6	.00267170	.00219413	.00182887	.00154402	.00131812
7	.00642870	.00538147	.00456389	.00391435	.00339042
8	.01363033	.01161576	.01001218	.00871554	.00765266
9	.02600175	.02253123	.01971765	.01740437	.01547890
10	.04534997	.03991093	.03542464	.03167745	.02851298
11	.07324259	.06539083	.05880826	.05322773	.04845002
12	.11068249	.10013598	.09115806	.08343938	.07674514
13	.15786611	.14457472	.13309676	.12309777	.11431957
14	.21408740	.19826023	.18440776	.17219005	.16134043
15	.27780505	.25989017	.24401264	.22984575	.21712925

DIRICHLET C INTEGRAL
A = 1 R = 5

M	B= 1	B= 2	B= 3	B= 4	B= 5
1	.03125000	.00905286	.00428802	.00252553	.00168058
2	.10937500	.04235413	.02328085	.01508477	.01074863
3	.22656250	.11008786	.06849466	.04817031	.03644083
4	.36328125	.21084392	.14535601	.10962059	.08735321
5	.50000000	.33333333	.25000000	.20000000	.16666667
6	.62304688	.46240939	.37137802	.31211846	.27019505
7	.72558594	.58462267	.49602081	.43416170	.38806066
8	.80615234	.69111160	.61232409	.55392008	.50836846
9	.86657715	.77795443	.71279045	.66195020	.62069678
10	.91021729	.84506665	.79426403	.75286636	.71810321
11	.94076538	.89468063	.85696266	.82507468	.79749424
12	.96159363	.93002571	.90314107	.87969629	.85889873
13	.97547913	.95443208	.93591508	.91934275	.90432150
14	.98455811	.97084401	.95845494	.94712497	.93666669
15	.99039459	.98163230	.97354521	.96601632	.95895964

M	B= 6	B= 7	B= 8	B= 9	B=10
1	.00120859	.00091695	.00072337	.00058783	.00048894
2	.00814610	.00644584	.00526526	.00440706	.00376049
3	.02893188	.02377147	.02003729	.01722710	.01504609
4	.07224966	.06138252	.05321630	.04687167	.04181046
5	.14285714	.12500000	.11111111	.10000000	.09090909
6	.23883557	.21441904	.19482478	.17872417	.16524043
7	.35213167	.32320047	.29931566	.27920474	.26199899
8	.47155318	.44100231	.41512693	.39285182	.37341937
9	.58626873	.55692204	.53148890	.50915147	.48931628
10	.68826328	.66221521	.63917246	.61856641	.59997301
11	.77323075	.75160156	.73211541	.71440657	.69819531
12	.84020805	.82323751	.80769988	.79337546	.78009216
13	.89057154	.87788563	.86610531	.85510632	.84478918
14	.92694173	.91784438	.90929164	.90121690	.89356574
15	.95230950	.94601430	.94003253	.93433015	.92887884

M	B=11	B=12	B=13	B=14	B=15
1	.00041437	.00035661	.00031087	.00027396	.00024369
2	.00325920	.00286133	.00253929	.00227428	.00205309
3	.01331083	.01190169	.01073762	.00976191	.00893378
4	.03768568	.03426396	.03138283	.02892584	.02680745
5	.08333333	.07692308	.07142857	.06666667	.06250000
6	.15377045	.14388514	.13527047	.12769114	.12096717
7	.24708309	.23400785	.22243696	.21211315	.20283599
8	.35627804	.34101534	.32731570	.31493303	.30367216
9	.47154006	.45548368	.44088239	.42752587	.41524455
10	.58306783	.56759733	.55335974	.54019199	.52796051
11	.68326272	.66943431	.65656868	.64454969	.63328076
12	.76771252	.75612485	.74523714	.73497263	.72526658
13	.83507286	.82589037	.81718562	.80891115	.80102641
14	.88629302	.87936082	.87273692	.86639367	.86030714
15	.92365471	.91863736	.91380925	.90915511	.90466158

DIRICHLET C INTEGRAL
A = 2 R = 5

M	B= 1	B= 2	B= 3	B= 4	B= 5
1	.13168724	.07020713	.04853250	.03746431	.03072922
2	.35116598	.23125919	.17821430	.14759246	.12737688
3	.57064472	.43635145	.36539570	.32000865	.28786217
4	.74135040	.62816380	.55936569	.51142157	.47534711
5	.85515419	.77541638	.72126287	.68073405	.64862021
6	.92343647	.87384432	.83707062	.80787385	.78369949
7	.96137106	.93322883	.91086030	.89221571	.87619518
8	.98124157	.96635028	.95384972	.94300820	.93339920
9	.99117682	.98371712	.97718193	.97132898	.96600656
10	.99596046	.99238346	.98914426	.98616757	.98340319
11	.99819282	.99653738	.99499961	.99355733	.99219494
12	.99920754	.99846340	.99775862	.99708695	.99644380
13	.99965852	.99933204	.99901825	.99871550	.99842247
14	.99985510	.99971475	.99957837	.99944554	.99931590
15	.99993935	.99988005	.99982196	.99976498	.99970900

M	B= 6	B= 7	B= 8	B= 9	B=10
1	.02618610	.02290620	.02042152	.01847048	.01689530
2	.11289900	.10194641	.09332770	.08634106	.08054452
3	.26359436	.24445445	.22886884	.21586459	.20480396
4	.44682854	.42348942	.40389329	.38711156	.37251246
5	.62219265	.59984869	.58056972	.56366968	.54866547
6	.76310285	.74518508	.72934814	.71517376	.70235784
7	.86213306	.84959313	.83827279	.82795294	.81846953
8	.92474859	.91686815	.90962226	.90290954	.89665198
9	.96111181	.95657109	.95232947	.94834467	.94458333
10	.98081547	.97837787	.97606999	.97387569	.97178194
11	.99090089	.98966628	.98848406	.98734854	.98625500
12	.99582565	.99522967	.99465361	.99409556	.99355395
13	.99813815	.99786169	.99759241	.99732971	.99707308
14	.99918916	.99906509	.99894349	.99882418	.99870701
15	.99965396	.99959980	.99954644	.99949385	.99944199

M	B=11	B=12	B=13	B=14	B=15
1	.01559512	.01450240	.01357017	.01276474	.01206131
2	.07564503	.07144006	.06778488	.06457313	.06172475
3	.19524942	.18688957	.17949605	.17289709	.16696073
4	.35964861	.34819266	.33789864	.32857748	.32008104
5	.53520466	.52302268	.51191608	.50172527	.49232284
6	.69067234	.67994189	.67002869	.66082254	.65223392
7	.80969633	.80153403	.79390303	.78673852	.77998694
8	.89078811	.88526856	.88005298	.87510794	.87040542
9	.94101858	.93762839	.93439446	.93130137	.92833602
10	.96977802	.96785500	.96600530	.96422248	.96250096
11	.98519950	.98417869	.98318970	.98223003	.98129752
12	.99302742	.99251479	.99201507	.99152734	.99105083
13	.99682211	.99657639	.99633559	.99609942	.99586760
14	.99859185	.99847858	.99836709	.99825729	.99814909
15	.99939080	.99934027	.99929035	.99924102	.99919225

DIRICHLET C INTEGRAL
A = 3 R = 5

M	B= 1	B= 2	B= 3	B= 4	B= 5
1	.23730469	.15610610	.12234224	.10317804	.09057482
2	.53393555	.41708674	.35740861	.31951589	.29265671
3	.75640869	.65985652	.60207401	.56182524	.53141271
4	.88618469	.82610924	.78545365	.75489349	.73051571
5	.95107269	.91979432	.89650744	.87787823	.86232440
6	.98027229	.96587441	.95433324	.94462200	.93619974
7	.99243879	.98638172	.98124231	.97673863	.97270831
8	.99721849	.99483835	.99272914	.99082026	.98906780
9	.99901088	.99812392	.99731167	.99655773	.99585120
10	.99965813	.99934123	.99904380	.99876226	.99849410
11	.99988466	.99977523	.99967062	.99957012	.99947317
12	.99996189	.99992514	.99988953	.99985494	.99982124
13	.99998764	.99997557	.99996378	.99995222	.99994087
14	.99999605	.99999217	.99998835	.99998458	.99998086
15	.99999876	.99999753	.99999631	.99999511	.99999391

M	B= 6	B= 7	B= 8	B= 9	B=10
1	.08153687	.07467483	.06924955	.06482866	.06114095
2	.27230038	.25616060	.24294132	.23184519	.22235090
3	.50722711	.48730430	.47046476	.45594846	.44323900
4	.71030605	.69309172	.67813055	.66492355	.65311940
5	.84896197	.83724413	.82680798	.81739979	.80883498
6	.92874170	.92203580	.91593504	.91033291	.90514953
7	.96904749	.96568496	.96256939	.95966234	.95693417
8	.98744207	.98592181	.98449112	.98313776	.98185201
9	.99518435	.99455143	.99394798	.99337048	.99281607
10	.99823746	.99799089	.99775325	.99752362	.99730123
11	.99937935	.99928833	.99919983	.99911364	.99902955
12	.99978835	.99975619	.99972470	.99969384	.99966355
13	.99992973	.99991877	.99990797	.99989734	.99988686
14	.99997719	.99997356	.99996997	.99996642	.99996291
15	.99999273	.99999156	.99999040	.99998924	.99998810

M	B=11	B=12	B=13	B=14	B=15
1	.05800692	.05530261	.05293944	.05085222	.04899185
2	.21410111	.20684147	.20038532	.19459204	.18935343
3	.43197067	.42187570	.41275257	.40444614	.39683458
4	.64246163	.63275740	.62385818	.61564717	.60803094
5	.80097491	.79371277	.78696442	.78066241	.77475182
6	.90032334	.89580575	.89155774	.88754743	.88374842
7	.95436147	.95192537	.94961041	.94740375	.94529458
8	.98062605	.97945343	.97832879	.97724757	.97620590
9	.99228240	.99176749	.99126967	.99078750	.99031975
10	.99708544	.99687568	.99667149	.99647245	.99627821
11	.99894741	.99886707	.99878841	.99871132	.99863570
12	.99963380	.99960456	.99957578	.99954746	.99951955
13	.99987652	.99986631	.99985622	.99984626	.99983641
14	.99995943	.99995599	.99995258	.99994920	.99994585
15	.99998696	.99998583	.99998471	.99998360	.99998249

DIRICHLET C INTEGRAL
A = 4 R = 5

M	B= 1	B= 2	B= 3	B= 4	B= 5
1	.32768000	.23953861	.19976335	.17597792	.15972557
2	.65536000	.55486083	.49978797	.46315167	.43626424
3	.85196800	.78615618	.74432273	.71395570	.69027919
4	.94371840	.91124668	.88811691	.87008963	.85530094
5	.98041856	.96700044	.95657092	.94796125	.94059284
6	.99363062	.98872353	.98464673	.98112281	.97799993
7	.99803464	.99639279	.99495820	.99367218	.99249975
8	.99941876	.99890509	.99843897	.99800907	.99760815
9	.99983399	.99968144	.99953907	.99940486	.99927743
10	.99995395	.99991047	.99986905	.99982936	.99979114
11	.99998754	.99997555	.99996396	.99995271	.99994177
12	.99999670	.99999348	.99999034	.99998726	.99998424
13	.99999914	.99999830	.99999747	.99999665	.99999584
14	.99999978	.99999956	.99999935	.99999914	.99999892
15	.99999994	.99999989	.99999983	.99999978	.99999973

M	B= 6	B= 7	B= 8	B= 9	B=10
1	.14771440	.13836612	.13081818	.12455450	.11924500
2	.41531457	.39831871	.38412306	.37200249	.36147399
3	.67096908	.65472323	.64074067	.62849467	.61762091
4	.84275792	.83186645	.82224196	.81362084	.80581431
5	.93413209	.92836730	.92315493	.91839278	.91400531
6	.97518427	.97261304	.97024187	.96803802	.96597660
7	.99141794	.99041063	.98946598	.98857499	.98773060
8	.99723116	.99687443	.99653515	.99621114	.99590064
9	.99915578	.99903914	.99892692	.99881864	.99871389
10	.99975421	.99971844	.99968369	.99964988	.99961691
11	.99993110	.99992068	.99991048	.99990048	.99989068
12	.99998128	.99997837	.99997550	.99997267	.99996989
13	.99999504	.99999426	.99999348	.99999271	.99999194
14	.99999872	.99999851	.99999830	.99999810	.99999790
15	.99999967	.99999962	.99999957	.99999951	.99999946

M	B=11	B=12	B=13	B=14	B=15
1	.11466744	.11066602	.10712781	.10396871	.10112461
2	.35220103	.34394052	.33651168	.32977670	.32362833
3	.60785731	.59900926	.59092844	.58349930	.57663015
4	.79868248	.79211876	.78604002	.78038016	.77508576
5	.90993493	.90613671	.90257486	.89922035	.89604932
6	.96403812	.96220704	.96047072	.95881872	.95724234
7	.98692718	.98616013	.98542564	.98472051	.98404203
8	.99560221	.99531465	.99503696	.99476827	.99450785
9	.99861235	.99851374	.99841783	.99832440	.99823327
10	.99958473	.99955328	.99952249	.99949233	.99946276
11	.99988105	.99987159	.99986228	.99985312	.99984410
12	.99996714	.99996443	.99996175	.99995910	.99995648
13	.99999119	.99999044	.99998970	.99998897	.99998824
14	.99999770	.99999750	.99999730	.99999710	.99999691
15	.99999941	.99999936	.99999931	.99999925	.99999920

DIRICHLET C INTEGRAL

A = 5 R = 5

M	B= 1	B= 2	B= 3	B= 4	B= 5
1	.40187757	.31295227	.27081923	.24482997	.22666351
2	.73677555	.65281233	.60484359	.57204182	.54746844
3	.90422454	.85868167	.82868954	.80638393	.78866921
4	.96934359	.95072174	.93705644	.92617867	.91710804
5	.99104994	.98466898	.97958490	.97531117	.97160086
6	.99756184	.99562540	.99398358	.99254247	.99124949
7	.99937071	.99883266	.99835474	.99792081	.99752103
8	.99984446	.99970456	.99957595	.99945609	.99934332
9	.99996289	.99992834	.99989577	.99986480	.99983519
10	.99999141	.99998321	.99997535	.99996776	.99996041
11	.99999806	.99999618	.99999435	.99999257	.99999082
12	.99999957	.99999915	.99999874	.99999833	.99999793
13	.99999991	.99999982	.99999972	.99999963	.99999955
14	.99999998	.99999996	.99999994	.99999992	.99999990
15	1	1	.99999999	.99999998	.99999998

M	B= 6	B= 7	B= 8	B= 9	B=10
1	.21299538	.20219975	.19337373	.18596974	.17963346
2	.52800579	.51200046	.49847639	.48681214	.47658913
3	.77400528	.76151413	.75064791	.74104193	.73244128
4	.90931151	.90246495	.89635568	.89083623	.88579993
5	.96830888	.96534175	.96263529	.96014331	.95783130
6	.99007157	.98898628	.98797756	.98703347	.98614481
7	.99714884	.99679958	.99646979	.99615681	.99585852
8	.99923647	.99913470	.99903732	.99894383	.99885379
9	.99980674	.99977931	.99975279	.99972707	.99970210
10	.99995327	.99994632	.99993955	.99993293	.99992645
11	.99998911	.99998744	.99998580	.99998418	.99998259
12	.99999754	.99999715	.99999677	.99999639	.99999602
13	.99999946	.99999937	.99999929	.99999920	.99999912
14	.99999988	.99999986	.99999985	.99999983	.99999981
15	.99999998	.99999997	.99999997	.99999996	.99999996

M	B=11	B=12	B=13	B=14	B=15
1	.17412391	.16927057	.16494886	.16106527	.15754805
2	.46751286	.45936875	.45199593	.44527090	.43909690
3	.72466086	.71756204	.71103840	.70500642	.69939942
4	.88116703	.87687626	.87287950	.86913822	.86562110
5	.95567278	.95364690	.95173694	.94992925	.94821255
6	.98530435	.98450626	.98374579	.98301896	.98232247
7	.99557326	.99529961	.99503643	.99478274	.99453769
8	.99876686	.99868275	.99860120	.99852201	.99844500
9	.99967779	.99965409	.99963097	.99960837	.99958626
10	.99992011	.99991389	.99990778	.99990179	.99989589
11	.99998103	.99997948	.99997796	.99997646	.99997498
12	.99999565	.99999528	.99999492	.99999457	.99999421
13	.99999903	.99999895	.99999887	.99999878	.99999870
14	.99999979	.99999977	.99999975	.99999973	.99999972
15	.99999996	.99999995	.99999995	.99999994	.99999994

DIRICHLET C INTEGRAL
1/A = 5 R = 6

M	B= 1	B= 2	B= 3	B= 4	B= 5
1	.21433\04	.87972\07	.20306\08	.11707\09	.11930\10
2	.00012860	.86161\06	.26015\07	.17904\08	.20745\09
3	.00044117	.45486\05	.17565\06	.14261\07	.18652\08
4	.00113578	.17170\04	.83133\06	.78771\07	.11550\07
5	.00243816	.51903\04	.30957\05	.33897\06	.55374\07
6	.00460879	.00013351	.96556\05	.12107\05	.21904\06
7	.00792504	.00030338	.26226\04	.37341\05	.74414\06
8	.01266254	.00062441	.63698\04	.10219\04	.22315\05
9	.01907790	.00118511	.00014100	.25307\04	.60260\05
10	.02739411	.00210188	.00028858	.57556\04	.14876\04
11	.03778937	.00351918	.00055213	.00012161	.33972\04
12	.05038968	.00560714	.00099628	.00024094	.72450\04
13	.06526505	.00855663	.00170766	.00045098	.00014541
14	.08242895	.01257223	.00279676	.00080248	.00027645
15	.10184049	.01786347	.00439832	.00136463	.00050054

M	B= 6	B= 7	B= 8	B= 9	B=10
1	.17911\11	.35433\12	.84806\13	.24002\13	.14439\13
2	.34399\10	.74031\11	.19404\11	.59404\12	.21267\12
3	.33969\09	.78855\10	.22024\10	.71145\11	.25906\11
4	.22997\08	.57375\09	.17016\09	.57838\10	.21956\10
5	.12000\07	.32075\08	.10075\08	.35963\09	.14239\09
6	.51457\07	.14689\07	.48755\08	.18242\08	.75231\09
7	.18875\06	.57383\07	.20080\07	.78606\08	.33720\08
8	.60892\06	.19660\06	.72371\07	.29592\07	.13185\07
9	.17628\05	.60280\06	.23296\06	.99331\07	.45907\07
10	.46496\05	.16799\05	.68022\06	.30195\06	.14455\06
11	.11309\04	.43064\05	.18236\05	.84143\06	.41675\06
12	.25611\04	.10255\04	.45328\05	.21709\05	.11110\05
13	.54427\04	.22866\04	.10531\04	.52275\05	.27610\05
14	.00010927	.48061\04	.23026\04	.11829\04	.64405\05
15	.00020837	.95763\04	.47648\04	.25301\04	.14184\04

M	B=11	B=12	B=13	B=14	B=15
1	.29851\13	.74831\13	.17171\12	.37451\12	.81627\12
2	.10614\12	.10675\12	.18597\12	.38110\12	.81905\12
3	.10602\11	.52125\12	.37841\12	.47526\12	.86630\12
4	.91483\11	.41628\11	.21255\11	.13596\11	.13332\11
5	.61434\10	.28527\10	.14182\10	.76411\11	.47520\11
6	.33633\09	.16095\09	.81697\10	.43785\10	.24937\10
7	.15606\08	.76997\09	.40130\09	.21942\09	.12541\09
8	.63098\08	.32074\08	.17167\08	.96086\09	.55951\09
9	.22691\07	.11872\07	.65213\08	.37365\08	.22215\08
10	.73717\07	.39661\07	.22341\07	.13097\07	.79500\08
11	.21903\06	.12107\06	.69883\07	.41885\07	.25946\07
12	.60113\06	.34106\06	.20158\06	.12344\06	.77989\07
13	.15365\05	.89403\06	.54062\06	.33805\06	.21770\06
14	.36826\05	.21957\05	.13574\05	.86613\06	.56822\06
15	.83249\05	.50820\05	.32098\05	.20885\05	.13951\05

DIRICHLET C INTEGRAL
1/A = 4 R = 6

M	B= 1	B= 2	B= 3	B= 4	B= 5
1	.64000\04	.58993\06	.25634\07	.24718\08	.38741\09
2	.00037120	.54591\05	.30523\06	.34698\07	.61230\08
3	.00123136	.27243\04	.19165\05	.25386\06	.50074\07
4	.00306637	.97259\04	.84406\05	.12888\05	.28227\06
5	.00636938	.00027823	.29268\04	.51016\05	.12328\05
6	.01165421	.00067765	.85063\04	.16773\04	.44463\05
7	.01940528	.00145879	.00021545	.47662\04	.13784\04
8	.03003532	.00284633	.00048832	.00012027	.37754\04
9	.04385438	.00512452	.00100953	.00027487	.93201\04
10	.06105143	.00862723	.00193112	.00057742	.00021054
11	.08168789	.01372067	.00345628	.00112798	.00044039
12	.10570122	.02078056	.00583921	.00206814	.00086114
13	.13291633	.03016650	.00937920	.00358596	.00158640
14	.16306230	.04219652	.01440852	.00591709	.00277127
15	.19579221	.05712430	.02127511	.00934076	.00461585

M	B= 6	B= 7	B= 8	B= 9	B=10
1	.84114\10	.23054\10	.75184\11	.27923\11	.11163\11
2	.14557\08	.42895\09	.14854\09	.58308\10	.25264\10
3	.12973\07	.40954\08	.15014\08	.61857\09	.27980\09
4	.79319\07	.26738\07	.10351\07	.44660\08	.21025\08
5	.37414\06	.13424\06	.54739\07	.24687\07	.12076\07
6	.14514\05	.55261\06	.23681\06	.11142\06	.56544\07
7	.48208\05	.19422\05	.87270\06	.42763\06	.22481\06
8	.14095\04	.59921\05	.28172\05	.14352\05	.78050\06
9	.37017\04	.16562\04	.81305\05	.42990\05	.24152\05
10	.88663\04	.41645\04	.21305\04	.11674\04	.67663\05
11	.00019603	.96431\04	.51313\04	.29090\04	.17374\04
12	.00040397	.00020764	.00011472	.67189\04	.41295\04
13	.00078209	.00041912	.00023999	.00014501	.91609\04
14	.00143191	.00079839	.00047301	.00029444	.00019098
15	.00249329	.00144348	.00088341	.00056578	.00037633

M	B=11	B=12	B=13	B=14	B=15
1	.39860\12	*	*	*	*
2	.11793\10	.57132\11	.25749\11	.50234\12	*
3	.13655\09	.70844\10	.38443\10	.21181\10	.10865\10
4	.10632\08	.57050\09	.32165\09	.18872\09	.11374\09
5	.63153\08	.34924\08	.20247\08	.12219\08	.76265\09
6	.30542\07	.17383\07	.10342\07	.63923\08	.40830\08
7	.12527\06	.73296\07	.44709\07	.28266\07	.18435\07
8	.44811\06	.26929\06	.16826\06	.10872\06	.72337\07
9	.14271\05	.87997\06	.56271\06	.37134\06	.25187\06
10	.41102\05	.25979\05	.16989\05	.11441\05	.79062\06
11	.10838\04	.70157\05	.46877\05	.32196\05	.22653\05
12	.26427\04	.17504\04	.11941\04	.83585\05	.59840\05
13	.60081\04	.40682\04	.28315\04	.20185\04	.14696\04
14	.00012823	.88691\04	.62931\04	.45660\04	.33786\04
15	.00025846	.00018244	.00013188	.97323\04	.73149\04

DIRICHLET C INTEGRAL
1/A = 3 R = 6

M	B= 1	B= 2	B= 3	B= 4	B= 5
1	.00024414	.56905\05	.49632\06	.83069\07	.20475\07
2	.00134277	.48329\04	.53044\05	.10299\05	.28229\06
3	.00422668	.00022157	.29930\04	.66629\05	.20166\05
4	.00999451	.00072746	.00011860	.29954\04	.99435\05
5	.01972771	.00191591	.00037051	.00010514	.38044\04
6	.03432751	.00430116	.00097147	.00030698	.00012039
7	.05440223	.00854501	.00222296	.00077586	.00032797
8	.08021259	.01540624	.00455879	.00174419	.00079075
9	.11166897	.02566494	.00854084	.00355715	.00172140
10	.14836808	.04003574	.01483007	.00668026	.00343539
11	.18965457	.05908567	.02413455	.01168776	.00636064
12	.23469439	.08316951	.03714181	.01922980	.01103123
13	.28254919	.11239088	.05444619	.02998089	.01806195
14	.33224456	.14659199	.07648249	.04457630	.02810479
15	.38282735	.18536983	.10347580	.06354587	.04179183

M	B= 6	B= 7	B= 8	B= 9	B=10
1	.65131\08	.24804\08	.10793\08	.52030\09	.27196\09
2	.97374\07	.39554\07	.18149\07	.91493\08	.49700\08
3	.75064\06	.32410\06	.15640\06	.82271\07	.46359\07
4	.39760\05	.18186\05	.92060\06	.50433\06	.29432\06
5	.16271\04	.78595\05	.41633\05	.23706\05	.14305\05
6	.54849\04	.27894\04	.15426\04	.91131\05	.56777\05
7	.00015857	.84661\04	.48773\04	.29839\04	.19166\04
8	.00040425	.00022596	.00013531	.85587\04	.56593\04
9	.00092729	.00054126	.00033623	.00021950	.00014921
10	.00194368	.00118177	.00076003	.00051128	.00035685
11	.00376810	.00238077	.00158218	.00109511	.00078374
12	.00682252	.00446930	.00306360	.00217847	.00159665
13	.01162995	.00788183	.00556306	.00405814	.00304233
14	.01879044	.01314737	.00953870	.00712841	.00545989
15	.02894013	.02086369	.01553472	.01187719	.00928375

M	B=11	B=12	B=13	B=14	B=15
1	.15172\09	.89224\10	.54685\10	.34484\10	.21990\10
2	.28685\08	.17407\08	.11015\08	.72216\09	.48783\09
3	.27628\07	.17248\07	.11197\07	.75155\08	.51910\08
4	.18086\06	.11603\06	.77179\07	.52954\07	.37318\07
5	.90530\06	.59609\06	.40590\06	.28445\06	.20436\06
6	.36955\05	.24949\05	.17375\05	.12426\05	.90945\06
7	.12814\04	.88612\05	.63055\05	.45986\05	.34264\05
8	.38823\04	.27470\04	.19956\04	.14830\04	.11241\04
9	.00010491	.75875\04	.56224\04	.42544\04	.32785\04
10	.00025683	.00018970	.00014327	.00011030	.86361\04
11	.00057679	.00043467	.00033430	.00026168	.00020803
12	.00120030	.00092204	.00072157	.00057388	.00046295
13	.00233385	.00182584	.00145280	.00117317	.00095974
14	.00426974	.00339895	.00274772	.00225134	.00186664
15	.00739382	.00598409	.00491120	.00408023	.00342666

DIRICHLET C INTEGRAL
1/A = 2 R = 6

M	B= 1	B= 2	B= 3	B= 4	B= 5
1	.00137174	.94177\04	.17700\04	.53304\05	.21055\05
2	.00685871	.00069118	.00015836	.54120\04	.23391\04
3	.01966164	.00274510	.00075009	.00028758	.00013504
4	.04242239	.00782844	.00250255	.00106517	.00053986
5	.07656353	.01795935	.00660315	.00309043	.00168022
6	.12208505	.03522607	.01467216	.00748499	.00434044
7	.17772246	.06134165	.02855417	.01575089	.00968992
8	.24130807	.09727377	.04999350	.02959887	.01922169
9	.31019248	.14304671	.08028720	.05067267	.03457475
10	.38162816	.19774178	.12001327	.08024013	.05727196
11	.45306385	.25966368	.16889963	.11893358	.08843752
12	.52233481	.32660750	.22585210	.16660370	.12856635
13	.58775739	.39615569	.28911565	.22231404	.17740074
14	.64814747	.46594787	.35651768	.28446450	.23393693
15	.70278611	.53388791	.42573523	.35100392	.29655000

M	B= 6	B= 7	B= 8	B= 9	B=10
1	.99083\06	.52667\06	.30610\06	.19045\06	.12501\06
2	.11788\04	.66172\05	.40224\05	.25990\05	.17623\05
3	.72532\04	.42846\04	.27166\04	.18191\04	.12720\04
4	.00030761	.00019059	.00012572	.87067\04	.62679\04
5	.00101134	.00065510	.00044850	.00032060	.00023724
6	.00274862	.00185581	.00131551	.00096881	.00073575
7	.00643111	.00451292	.00330479	.00250287	.00194779
8	.01332199	.00968940	.00731416	.00568644	.00452818
9	.02493788	.01874998	.01455935	.01160005	.00943854
10	.04285045	.03322149	.02648264	.02158765	.01792310
11	.06842695	.05457217	.04457341	.03711541	.03140092
12	.10257244	.08395763	.07013274	.05956139	.05128241
13	.14553745	.12199448	.10403569	.08998000	.07874383
14	.19683511	.16861602	.14654773	.12889584	.11451030
15	.25528083	.22304052	.19723757	.17617513	.15869754

M	B=11	B=12	B=13	B=14	B=15
1	.85671\07	.60831\07	.44491\07	.33366\07	.25567\07
2	.12425\05	.90458\06	.67657\06	.51774\06	.40406\06
3	.92126\05	.68692\05	.52486\05	.40946\05	.32521\05
4	.46571\04	.35522\04	.27699\04	.22010\04	.17777\04
5	.00018059	.00014074	.00011189	.90488\04	.74264\04
6	.00057303	.00045582	.00036912	.00030354	.00025295
7	.00155020	.00125720	.00103602	.00086558	.00073189
8	.00367813	.00303799	.00254528	.00215886	.00185084
9	.00781525	.00656742	.00558902	.00480871	.00417707
10	.01511035	.01290561	.01114623	.00972039	.00854919
11	.02692341	.02334830	.02044727	.01806011	.01607165
12	.04466809	.03929363	.03486279	.03116359	.02804094
13	.06960079	.06204779	.05572664	.05037619	.04580210
14	.10260079	.09260776	.08412483	.07685039	.07055627
15	.14399165	.13146956	.12069616	.11134292	.10315739

DIRICHLET C INTEGRAL
A = 1 R = 6

M	B= 1	B= 2	B= 3	B= 4	B= 5
1	.01562500	.00381516	.00165122	.00091701	.00058477
2	.06250000	.02047664	.01027739	.00627092	.00427638
3	.14453125	.06006624	.03421586	.02267503	.01641654
4	.25390625	.12792085	.08115306	.05780068	.04412816
5	.37695313	.22183715	.15414759	.11686971	.09349464
6	.50000000	.33333333	.25000000	.20000000	.16666667
7	.61279297	.45116472	.36051709	.30189812	.26062530
8	.70947266	.56481312	.47541771	.41370456	.36809156
9	.78802490	.66664000	.58522147	.52572562	.47981818
10	.84912109	.75251511	.68307702	.62976882	.58704320
11	.89494324	.82137239	.76531970	.72041185	.68319646
12	.92826843	.87428284	.83108959	.79518963	.76455669
13	.95187378	.91349200	.88149957	.85405895	.83004076
14	.96821594	.94165760	.91875166	.89857054	.88051297
15	.97930527	.96135329	.94542528	.93106929	.91797809

M	B= 6	B= 7	B= 8	B= 9	B=10
1	.00040692	.00030063	.00023198	.00018499	.00015138
2	.00313221	.00241084	.00192432	.00157920	.00132460
3	.01259288	.01006058	.00828299	.00697911	.00598925
4	.03528256	.02915364	.02468923	.02131097	.01867681
5	.07756538	.06606194	.05739163	.05063843	.04523974
6	.14285714	.12500000	.11111111	.10000000	.09090909
7	.22986542	.20598615	.18686975	.17119432	.15809015
8	.33277171	.30447976	.28122486	.26171756	.24508254
9	.44302934	.41271264	.38718619	.36532244	.34633312
10	.55173779	.52188944	.49620160	.47377662	.45396942
11	.65159349	.62425422	.60025565	.57893925	.55981941
12	.73790710	.71437462	.69334654	.67437332	.65711535
13	.80869616	.78950094	.77207286	.75612406	.74143209
14	.86416347	.84922123	.83546063	.82270760	.81082476
15	.90593127	.89476427	.88435038	.87458963	.86540154

M	B=11	B=12	B=13	B=14	B=15
1	.00012646	.00010744	.92588\04	.80745\04	.71141\04
2	.00113078	.00097943	.00085869	.00076063	.00067975
3	.00521676	.00460009	.00409843	.00368372	.00333616
4	.01657251	.01485760	.01343647	.01224191	.01122544
5	.04083170	.03716898	.03408041	.03144304	.02916641
6	.08333333	.07692308	.07142857	.06666667	.06250000
7	.14696058	.13738203	.12904516	.12171856	.11522551
8	.23070280	.21812982	.20702905	.19714544	.18828083
9	.32964826	.31484410	.30159817	.28966016	.27883239
10	.43630222	.42041220	.40601793	.39289722	.38087182
11	.54252823	.52678066	.51235161	.49906042	.48676002
12	.64130981	.62674916	.61326672	.60072661	.58901666
13	.72782132	.71515058	.70330465	.69218821	.68172151
14	.79970157	.78924768	.77938821	.77006038	.76121096
15	.85672021	.84849093	.84066769	.83321140	.82608852

DIRICHLET C INTEGRAL
A = 2 R = 6

M	B= 1	B= 2	B= 3	B= 4	B= 5
1	.08779150	.04328154	.02874992	.02162966	.01741445
2	.26337449	.16084890	.11905461	.09600829	.08125830
3	.46822131	.33521598	.27040309	.23086878	.20377194
4	.65030737	.52245341	.45047613	.40262536	.36780099
5	.78687192	.68545261	.62126425	.57535106	.54014323
6	.87791495	.80766141	.75876562	.72155466	.69169661
7	.93355236	.88956733	.85650496	.82999520	.80787645
8	.96534517	.93987892	.91950996	.90244866	.88772966
9	.98256627	.96870971	.95706004	.94694163	.93796135
10	.99149573	.98432557	.97805281	.97243937	.96733718
11	.99596046	.99240012	.98918578	.98623834	.98350568
12	.99812518	.99641686	.99483596	.99335755	.99196429
13	.99914740	.99835098	.99759965	.99688589	.99620426
14	.99961920	.99925686	.99890992	.99857623	.99825412
15	.99983263	.99967119	.99951485	.99936301	.99921519

M	B= 6	B= 7	B= 8	B= 9	B=10
1	.01462815	.01264825	.01116767	.01001762	.00909774
2	.07093539	.06326645	.05732060	.05256050	.04865341
3	.18381541	.16838144	.15601482	.14583627	.13728029
4	.34096208	.31944080	.30167384	.28667559	.27378942
5	.51189577	.48849923	.46865603	.45151521	.43649079
6	.66687810	.64572002	.62733530	.61112051	.59664705
7	.78891261	.77232778	.75760194	.74436909	.73236171
8	.87476605	.86317171	.85267789	.84308938	.83425962
9	.92986668	.92248426	.91568912	.90938785	.90350851
10	.96264633	.95829551	.95423166	.95041402	.94681046
11	.98095103	.97854718	.97627332	.97411303	.97205310
12	.99064349	.98938545	.98818252	.98702856	.98591852
13	.99555058	.99492160	.99431468	.99372766	.99315872
14	.99794231	.99763975	.99734559	.99705910	.99677967
15	.99907102	.99893016	.99879236	.99865738	.99852503

M	B=11	B=12	B=13	B=14	B=15
1	.00834460	.00771617	.00718347	.00672589	.00632837
2	.04538180	.04259719	.04019460	.03809762	.03624924
3	.12996519	.12362310	.11805998	.11313150	.10872785
4	.26255852	.25265384	.24383145	.23590613	.22873436
5	.42316339	.41122339	.40043619	.39062016	.38163205
6	.58359996	.57174123	.56088674	.55089131	.54163849
7	.72137823	.71126296	.70189310	.69317009	.68501364
8	.82607558	.81844810	.81130560	.80458965	.79825191
9	.89799440	.89279993	.88788784	.88322725	.87879226
10	.94339515	.94014693	.93704822	.93408424	.93124242
11	.97008270	.96819282	.96637583	.96462525	.96293549
12	.98484816	.98381393	.98281277	.98184204	.98089944
13	.99260633	.99206915	.99154606	.99103603	.99053820
14	.99650677	.99623993	.99597876	.99572289	.99547200
15	.99839513	.99826753	.99814210	.99801872	.99789727

DIRICHLET C INTEGRAL
A = 3 R = 6

M	B= 1	B= 2	B= 3	B= 4	B= 5
1	.17797852	.11143546	.08518160	.07069271	.06134023
2	.44494629	.33180460	.27732099	.24384693	.22064215
3	.67854309	.57024126	.50915886	.46807272	.43776985
4	.83427429	.75743607	.70828712	.67261099	.64485000
5	.92187309	.87685167	.84501671	.82038761	.80032032
6	.96567249	.94261026	.92494979	.91054358	.89833630
7	.98574722	.97505468	.96633368	.95890221	.95239305
8	.99435067	.98975915	.98582399	.98234904	.97921994
9	.99784582	.99598989	.99433664	.99283411	.99144981
10	.99920505	.99849048	.99783469	.99722483	.99665238
11	.99971476	.99945035	.99920210	.99896702	.99874302
12	.99990011	.99980543	.99971500	.99962816	.99954441
13	.99996575	.99993277	.99990087	.99986989	.99983974
14	.99998848	.99997725	.99996629	.99995556	.99994504
15	.99999619	.99999244	.99998876	.99998513	.99998156

M	B= 6	B= 7	B= 8	B= 9	B=10
1	.05472487	.04975585	.04586147	.04271130	.04010006
2	.20334404	.18980647	.17883606	.16970986	.16196109
3	.41410719	.39489577	.37885041	.36515785	.35327377
4	.62226913	.60332678	.58707153	.57287651	.56030778
5	.78340518	.76879966	.75595936	.74451194	.73419143
6	.88772376	.87832510	.86988367	.86221775	.85519367
7	.94658217	.94132135	.93650680	.93206263	.92793147
8	.97636285	.97372674	.97127458	.96897849	.96681687
9	.99016169	.98895389	.98781453	.98673440	.98570621
10	.99611127	.99559699	.99510604	.99463567	.99418363
11	.99852855	.99832241	.99812366	.99793153	.99774538
12	.99946339	.99938480	.99930841	.99923400	.99916143
13	.99981033	.99978159	.99975347	.99972592	.99969888
14	.99993472	.99992457	.99991458	.99990475	.99989506
15	.99997803	.99997455	.99997111	.99996771	.99996435

M	B=11	B=12	B=13	B=14	B=15
1	.03789305	.03599790	.03434901	.03289837	.03161000
2	.15527323	.14942310	.14424818	.13962697	.13546656
3	.34281756	.33351361	.32515650	.31758945	.31069028
4	.54905309	.53888057	.52961358	.52111460	.51327456
5	.72480122	.71619183	.70824699	.70087448	.69399992
6	.84871010	.84268828	.83706574	.83179204	.82682584
7	.92406881	.92043932	.91701446	.91377077	.91068877
8	.96477255	.96283166	.96098279	.95921642	.95752453
9	.98472404	.98378303	.98287909	.98200879	.98116918
10	.99374806	.99332742	.99292037	.99252580	.99214273
11	.99756468	.99738898	.99721788	.99705103	.99688816
12	.99909053	.99902120	.99895331	.99888678	.99882153
13	.99967234	.99964626	.99962060	.99959535	.99957049
14	.99988551	.99987608	.99986678	.99985758	.99984850
15	.99996102	.99995773	.99995447	.99995124	.99994804

DIRICHLET C INTEGRAL
A = 4 R = 6

M	B= 1	B= 2	B= 3	B= 4	B= 5
1	.26214400	.18489456	.15144129	.13187000	.11868989
2	.57671680	.47263430	.41827791	.38307812	.35771033
3	.79691776	.71796295	.67018717	.63650103	.61076130
4	.91435827	.86994715	.83976486	.81691932	.79856489
5	.96720650	.94656204	.93118799	.91884715	.90850011
6	.98834579	.97994865	.97323008	.96757062	.96265179
7	.99609687	.99300246	.99038425	.98809062	.98603649
8	.99875438	.99769717	.99676318	.99591880	.99514370
9	.99961807	.99927774	.99896697	.99867892	.99840915
10	.99988677	.99978232	.99968452	.99959209	.99950412
11	.99996739	.99993655	.99990713	.99987891	.99985171
12	.99999084	.99998202	.99997349	.99996522	.99995716
13	.99999748	.99999503	.99999263	.99999028	.99998798
14	.99999932	.99999865	.99999800	.99999735	.99999671
15	.99999982	.99999964	.99999947	.99999929	.99999912

M	B= 6	B= 7	B= 8	B= 9	B=10
1	.10905263	.10161427	.09564911	.09072705	.08657512
2	.33820988	.32255698	.30959621	.29861069	.28912799
3	.59008565	.57290030	.55825605	.54553843	.53432800
4	.78324490	.77011253	.75863164	.74844124	.73928669
5	.89957207	.89170936	.88467786	.87831413	.87249928
6	.95828494	.95434789	.95075639	.94744961	.94438204
7	.98416830	.98244971	.98085469	.97936387	.97796237
8	.99442435	.99375119	.99311714	.99251675	.99194576
9	.99815456	.99791288	.99768238	.99746168	.99724969
10	.99941997	.99933913	.99926123	.99918593	.99911300
11	.99982540	.99979988	.99977508	.99975093	.99972736
12	.99994931	.99994163	.99993411	.99992675	.99991952
13	.99998572	.99998350	.99998132	.99997916	.99997704
14	.99999608	.99999546	.99999484	.99999424	.99999364
15	.99999895	.99999878	.99999861	.99999844	.99999828

M	B=11	B=12	B=13	B=14	B=15
1	.08301080	.07990680	.07717138	.07473647	.07255041
2	.28082189	.27345861	.26686547	.26091164	.25549585
3	.52432632	.51531392	.50712491	.49963105	.49273123
4	.73098170	.72338583	.71639066	.70991072	.70387747
5	.86714403	.86217951	.85755147	.85321634	.84913860
6	.94151864	.93883180	.93629935	.93390317	.93162827
7	.97663847	.97538274	.97418746	.97304625	.97195373
8	.99140072	.99087881	.99037770	.98989541	.98943024
9	.99704550	.99684835	.99665762	.99647275	.99629328
10	.99904221	.99897337	.99890634	.99884097	.99877714
11	.99970434	.99968182	.99965977	.99963816	.99961695
12	.99991242	.99990545	.99989858	.99989182	.99988517
13	.99997495	.99997288	.99997084	.99996882	.99996683
14	.99999304	.99999245	.99999187	.99999129	.99999071
15	.99999811	.99999795	.99999779	.99999763	.99999747

DIRICHLET C INTEGRAL
A = 5 R = 6

M	B= 1	B= 2	B= 3	B= 4	B= 5
1	.33489798	.25360570	.21640252	.19387198	.17831329
2	.66979595	.57909002	.52940057	.49620972	.47173549
3	.86515311	.80816241	.77222916	.74618407	.72586575
4	.95197851	.92541735	.90672062	.89221998	.88035099
5	.98453803	.97430227	.96644967	.96001099	.95452250
6	.99539121	.99193752	.98910464	.98667422	.98453095
7	.99870746	.99765097	.99673853	.99592665	.99519034
8	.99965496	.99935513	.99908577	.99883906	.99861015
9	.99991157	.99983134	.99975709	.99968752	.99962178
10	.99997810	.99995762	.99993824	.99991976	.99990205
11	.99999473	.99998970	.99998486	.99998019	.99997565
12	.99999876	.99999757	.99999640	.99999526	.99999415
13	.99999972	.99999944	.99999917	.99999890	.99999863
14	.99999994	.99999987	.99999981	.99999975	.99999969
15	.99999999	.99999997	.99999996	.99999994	.99999993

M	B= 6	B= 7	B= 8	B= 9	B=10
1	.16671109	.15761080	.15021285	.14403625	.13877185
2	.45257793	.43696855	.42387844	.41266010	.40288154
3	.70927189	.69528762	.68322968	.67264974	.66323815
4	.87029391	.86156336	.85384731	.84693300	.84066862
5	.94972214	.94544588	.94158356	.93805732	.93480994
6	.98260516	.98085097	.97923633	.97773783	.97633777
7	.99451356	.99388528	.99329749	.99274416	.99222061
8	.99839575	.99819351	.99800167	.99781886	.99764400
9	.99955925	.99949948	.99944213	.99938691	.99933360
10	.99988499	.99986850	.99985254	.99983703	.99982194
11	.99997124	.99996695	.99996275	.99995865	.99995463
12	.99999306	.99999199	.99999094	.99998991	.99998889
13	.99999838	.99999812	.99999787	.99999762	.99999737
14	.99999963	.99999957	.99999951	.99999945	.99999939
15	.99999992	.99999990	.99999989	.99999988	.99999986

M	B=11	B=12	B=13	B=14	B=15
1	.13421055	.13020510	.12664841	.12346036	.12057972
2	.39424124	.38652107	.37955848	.37322933	.36743687
3	.65477231	.64708699	.64005627	.63358206	.62758648
4	.83494201	.82966788	.82477983	.82022515	.81596128
5	.93179801	.92898779	.92635247	.92387038	.92152373
6	.97502239	.97378078	.97260410	.97148508	.97041767
7	.99172314	.99124872	.99079489	.99035956	.98994099
8	.99747621	.99731475	.99715902	.99700851	.99686277
9	.99928201	.99923199	.99918340	.99913612	.99909006
10	.99980724	.99979289	.99977886	.99976514	.99975170
11	.99995070	.99994683	.99994304	.99993931	.99993564
12	.99998789	.99998690	.99998593	.99998497	.99998402
13	.99999713	.99999688	.99999664	.99999641	.99999617
14	.99999934	.99999928	.99999922	.99999917	.99999911
15	.99999985	.99999984	.99999982	.99999981	.99999980

DIRICHLET C INTEGRAL
1/A = 5 R = 7

M	B= 1	B= 2	B= 3	B= 4	B= 5
1	.35722\05	.67147\08	.94130\10	.37943\11	.29383\12
2	.24410\04	.75402\07	.13841\08	.66587\10	.58619\11
3	.93871\04	.45195\06	.10651\07	.60520\09	.60153\10
4	.00026752	.19204\05	.57076\07	.37943\08	.42326\09
5	.00062929	.64858\05	.23923\06	.18442\07	.22960\08
6	.00129254	.18513\04	.83522\06	.74053\07	.10235\07
7	.00239796	.46396\04	.25265\05	.25566\06	.39035\07
8	.00410872	.00010474	.68015\05	.77996\06	.13093\06
9	.00660358	.00021697	.16614\04	.21447\05	.39408\06
10	.01006867	.00041806	.37368\04	.53958\05	.10807\05
11	.01468879	.00075725	.78262\04	.12568\04	.27326\05
12	.02063894	.00130022	.00015402	.27354\04	.64324\05
13	.02807662	.00213056	.00028694	.56066\04	.14207\04
14	.03713534	.00335015	.00050911	.00010891	.29637\04
15	.04791953	.00507832	.00086469	.00020158	.58715\04

M	B= 6	B= 7	B= 8	B= 9	B=10
1	.35223\13	.60002\14	.33654\14	.90473\14	.26296\13
2	.77986\12	.14016\12	.33645\13	.17195\13	.28817\13
3	.87902\11	.17012\11	.40949\12	.12399\12	.63473\13
4	.67646\10	.14070\10	.35788\11	.10733\11	.38585\12
5	.39977\09	.89113\10	.23999\10	.75088\11	.26694\11
6	.19344\08	.46089\09	.13120\09	.42998\10	.15809\10
7	.79799\08	.20270\08	.60876\09	.20880\09	.79777\10
8	.28856\07	.77951\08	.24651\08	.88361\09	.35072\09
9	.93335\07	.26750\07	.88907\08	.33258\08	.13699\08
10	.27422\06	.83191\07	.29009\07	.11308\07	.48283\08
11	.74074\06	.23734\06	.86679\07	.35161\07	.15545\07
12	.18576\05	.62729\06	.23954\06	.10098\06	.46173\07
13	.43592\05	.15483\05	.61721\06	.27005\06	.12757\06
14	.96372\05	.35933\05	.14929\05	.67709\06	.33011\06
15	.20185\04	.78856\05	.34097\05	.16009\05	.80471\06

M	B=11	B=12	B=13	B=14	B=15
1	.66810\13	.15329\12	.32727\12	.66408\12	.13014\11
2	.67678\13	.15361\12	.32736\12	.66405\12	.13012\11
3	.80198\13	.15851\12	.32934\12	.66470\12	.13011\11
4	.20153\12	.20817\12	.35096\12	.67424\12	.13046\11
5	.10949\11	.58721\12	.52283\12	.75599\12	.13437\11
6	.64308\11	.29282\11	.16189\11	.12973\11	.16212\11
7	.33362\10	.15130\10	.75021\11	.42880\11	.32090\11
8	.15151\09	.70351\10	.34886\10	.18576\10	.10995\10
9	.61130\09	.29186\09	.14777\09	.78950\10	.44671\10
10	.22238\08	.10919\08	.56638\09	.30829\09	.17545\09
11	.73833\08	.37259\08	.19805\08	.11013\08	.63744\09
12	.22594\07	.11710\07	.63755\08	.36226\08	.21371\08
13	.64257\07	.34177\07	.19047\07	.11054\07	.66478\08
14	.17100\06	.93266\07	.53173\07	.31505\07	.19307\07
15	.42833\06	.23938\06	.13952\06	.84347\07	.52652\07

DIRICHLET C INTEGRAL
1/A = 4 R = 7

M	B= 1	B= 2	B= 3	B= 4	B= 5
1	.12800\04	.61388\07	.17816\08	.12936\09	.16364\10
2	.84480\04	.65041\06	.24306\07	.20795\08	.29590\09
3	.00031386	.36798\05	.17361\06	.17323\07	.27556\08
4	.00086436	.14765\04	.86404\06	.99607\07	.17608\07
5	.00196536	.47110\04	.33652\05	.44428\06	.86792\07
6	.00390313	.00012710	.10924\04	.16381\05	.35182\06
7	.00700356	.00030120	.30741\04	.51967\05	.12210\05
8	.01160991	.00064331	.77037\04	.14577\04	.37293\05
9	.01805881	.00126137	.00017529	.36883\04	.10229\04
10	.02665733	.00230181	.00036746	.85445\04	.25583\04
11	.03766344	.00395091	.00071781	.00018339	.59044\04
12	.05127100	.00643214	.00131857	.00036813	.00012697
13	.06760007	.00999942	.00229448	.00069642	.00025640
14	.08669251	.01492654	.00380547	.00124967	.00048947
15	.10851245	.02149372	.00604646	.00213853	.00088823

M	B= 6	B= 7	B= 8	B= 9	B=10
1	.30003\11	.71758\12	.20795\12	.62310\13	*
2	.59312\10	.15196\10	.46780\11	.16526\11	.63103\12
3	.60167\09	.16506\09	.53764\10	.20006\10	.82596\11
4	.41712\08	.12218\08	.42014\09	.16373\09	.70510\10
5	.22223\07	.69307\08	.25106\08	.10226\08	.45764\09
6	.97009\07	.32127\07	.12235\07	.52001\08	.24144\08
7	.36130\06	.12674\06	.50639\07	.22423\07	.10787\07
8	.11804\05	.43748\06	.18306\06	.84316\07	.41976\07
9	.34523\05	.13487\05	.58988\06	.28221\06	.14522\06
10	.91788\05	.37712\05	.17211\05	.85403\06	.45369\06
11	.22456\04	.96818\05	.46025\05	.23656\05	.12959\05
12	.51047\04	.23048\04	.11394\04	.60574\05	.34181\05
13	.00010869	.51285\04	.26323\04	.14457\04	.83939\05
14	.00021821	.00010740	.57144\04	.32380\04	.19324\04
15	.00041546	.00021288	.00011725	.68461\04	.41953\04

M	B=11	B=12	B=13	B=14	B=15
1	*	*	*	*	*
2	.21102\12	*	*	*	*
3	.36676\11	.16410\11	.55428\12	*	*
4	.32906\10	.16332\10	.83689\11	.40628\11	.11228\11
5	.22117\09	.11388\09	.61699\10	.34592\10	.19315\10
6	.12054\08	.63927\09	.35669\09	.20755\09	.12462\09
7	.55561\08	.30300\08	.17342\08	.10343\08	.63873\09
8	.22283\07	.12482\07	.73192\08	.44634\08	.28156\08
9	.79367\07	.45629\07	.27388\07	.17060\07	.10975\07
10	.25504\06	.15035\06	.92312\07	.58696\07	.38476\07
11	.74855\06	.45216\06	.28376\06	.18405\06	.12287\06
12	.20269\05	.12535\05	.80350\06	.53133\06	.36101\06
13	.51053\05	.32297\05	.21133\05	.14238\05	.98417\06
14	.12044\04	.77887\05	.51987\05	.35667\05	.25067\05
15	.26772\04	.17684\04	.12033\04	.84015\05	.60004\05

DIRICHLET C INTEGRAL
1/A = 3 R = 7

M	B= 1	B= 2	B= 3	B= 4	B= 5
1	.61035\04	.85493\06	.55151\07	.75051\08	.15859\08
2	.00038147	.82954\05	.67357\06	.10625\06	.24941\07
3	.00134277	.43015\04	.43112\05	.78026\06	.20222\06
4	.00350571	.00015833	.19247\04	.39594\05	.11263\05
5	.00756121	.00046382	.67316\04	.15604\04	.48450\05
6	.01425278	.00114999	.00019644	.50897\04	.17162\04
7	.02429014	.00250704	.00049754	.00014302	.52113\04
8	.03827076	.00493107	.00112359	.00035583	.00013947
9	.05662031	.00891338	.00230679	.00079963	.00033567
10	.07955725	.01501235	.00436922	.00164774	.00073778
11	.10708158	.02381094	.00772247	.00315047	.00149875
12	.13898478	.03586581	.01285378	.00564245	.00284143
13	.17487588	.05165536	.02029833	.00953985	.00506762
14	.21421805	.07153369	.03060035	.01532536	.00855905
15	.25637038	.09569623	.04426763	.02352095	.01376739

M	B= 6	B= 7	B= 8	B= 9	B=10
1	.44713\09	.15436\09	.61872\10	.27795\10	.13643\10
2	.76172\08	.28021\08	.11833\08	.55571\09	.28382\09
3	.66617\07	.26033\07	.11555\07	.56592\08	.29968\08
4	.39862\06	.16501\06	.76809\07	.39162\07	.21469\07
5	.18354\05	.80249\06	.39093\06	.20715\06	.11741\06
6	.69330\05	.31936\05	.16247\05	.89332\06	.52274\06
7	.22375\04	.10830\04	.57428\05	.32712\05	.19738\05
8	.63433\04	.32185\04	.17755\04	.10461\04	.65008\05
9	.00016123	.85549\04	.49005\04	.29824\04	.19063\04
10	.00037310	.00020657	.00012266	.76993\04	.50561\04
11	.00079576	.00045872	.00028186	.00018223	.00012281
12	.00157965	.00094613	.00060058	.00039939	.00027590
13	.00294222	.00182732	.00119642	.00081731	.00057813
14	.00517688	.00332752	.00224371	.00157252	.00113777
15	.00865439	.00574643	.00398449	.00286151	.00211557

M	B=11	B=12	B=13	B=14	B=15
1	.71632\11	.39272\11	.21534\11	.10563\11	.27711\12
2	.15512\09	.89610\10	.54151\10	.33893\10	.21735\10
3	.16907\08	.10052\08	.62438\09	.40250\09	.26775\09
4	.12485\07	.76247\08	.48521\08	.31978\08	.21721\08
5	.70296\07	.44054\07	.28691\07	.19309\07	.13369\07
6	.32189\06	.20681\06	.13773\06	.94583\07	.66700\07
7	.12487\05	.82174\06	.55916\06	.39155\06	.28107\06
8	.42205\05	.28424\05	.19747\05	.14091\05	.10290\05
9	.12689\04	.87375\05	.61931\05	.45002\05	.33415\05
10	.34468\04	.24248\04	.17522\04	.12958\04	.97769\05
11	.85662\04	.61515\04	.45284\04	.34059\04	.26099\04
12	.00019673	.00014409	.00010799	.82552\04	.64212\04
13	.00042102	.00031427	.00023960	.00018606	.00014682
14	.00084547	.00064267	.00049813	.00039269	.00031421
15	.00160270	.00123966	.00097619	.00078081	.00063315

DIRICHLET C INTEGRAL
1/A = 2 R = 7

M	B= 1	B= 2	B= 3	B= 4	B= 5
1	.00045725	.22213\04	.34390\05	.91000\06	.32698\06
2	.00259107	.00018554	.35005\04	.10500\04	.41234\05
3	.00828126	.00082990	.00018716	.63004\04	.26874\04
4	.01966164	.00264068	.00069970	.00026190	.00012063
5	.03862894	.00670278	.00205459	.00084780	.00041941
6	.06644764	.01443553	.00504781	.00227810	.00120424
7	.10353925	.02740839	.01079575	.00528969	.00297348
8	.14946219	.04708343	.02065086	.01091072	.00649241
9	.20303895	.07455442	.03603214	.02039758	.01279329
10	.26256869	.11034731	.05820585	.03509555	.02310700
11	.32606707	.15432291	.08806828	.05624685	.03872674
12	.39148965	.20569111	.12598092	.08478776	.06082628
13	.45691223	.26311930	.17169136	.12118024	.09027032
14	.52065731	.32490190	.22435001	.16531375	.12745841
15	.58136691	.38915319	.28261039	.21649410	.17223486

M	B= 6	B= 7	B= 8	B= 9	B=10
1	.14297\06	.71615\07	.39616\07	.23633\07	.14958\07
2	.19286\05	.10191\05	.58904\06	.36460\06	.23819\06
3	.13389\04	.74409\05	.44834\05	.28742\05	.19352\05
4	.63766\04	.37163\04	.23290\04	.15437\04	.10697\04
5	.00023433	.00014282	.92896\04	.63552\04	.45261\04
6	.00070860	.00045042	.00030346	.00021392	.00015637
7	.00183640	.00121435	.00084568	.00061331	.00045952
8	.00419474	.00287851	.00206816	.00154065	.00118167
9	.00862041	.00612423	.00453116	.00346193	.00271477
10	.01619003	.01188076	.00903570	.00707004	.00566148
11	.02813478	.02127967	.01660673	.01328850	.01085321
12	.04569602	.03554773	.02841847	.02322317	.01932292
13	.06994652	.05585143	.04566511	.03805823	.03222370
14	.10161415	.08312155	.06939696	.05890862	.05069900
15	.14094558	.11789405	.10035369	.08665497	.07572481

M	B=11	B=12	B=13	B=14	B=15
1	.99278\08	.68514\08	.48845\08	.35791\08	.26845\08
2	.16252\06	.11492\06	.83731\07	.62572\07	.47786\07
3	.13557\05	.98133\06	.73006\06	.55593\06	.43189\06
4	.76852\05	.56885\05	.43175\05	.33476\05	.26438\05
5	.33307\04	.25186\04	.19485\04	.15373\04	.12334\04
6	.00011773	.90860\04	.71596\04	.57429\04	.46781\04
7	.00035361	.00027826	.00022313	.00018185	.00015030
8	.00092836	.00074417	.00060680	.00050211	.00042080
9	.00217516	.00177458	.00147023	.00123432	.00104830
10	.00462138	.00383388	.00322480	.00274504	.00236110
11	.00901664	.00759961	.00648487	.00559313	.00486934
12	.01632193	.01396451	.01207963	.01054937	.00929038
13	.02764800	.02399158	.02102243	.01857763	.01653995
14	.04414320	.03881856	.03443048	.03076828	.02767785
15	.06684570	.05952180	.05340076	.04822615	.04380744

DIRICHLET C INTEGRAL
A = 1 R = 7

M	B= 1	B= 2	B= 3	B= 4	B= 5
1	.00781250	.00161780	.00064238	.00033737	.00020663
2	.03515625	.00979151	.00450470	.00259606	.00169833
3	.08984375	.03199890	.01673628	.01047809	.00727579
4	.17187500	.07507657	.04389207	.02957688	.02167045
5	.27441406	.14194580	.09136445	.06569864	.05050795
6	.38720703	.23026036	.16096598	.12253769	.09832527
7	.50000000	.33333333	.25000000	.20000000	.16666667
8	.60473633	.44239680	.35208710	.29399038	.25323844
9	.69638062	.54895472	.45907810	.39758890	.35244068
10	.77275085	.64640760	.56309223	.50291442	.45688291
11	.83384705	.73071663	.65797817	.60287510	.55915662
12	.88105774	.80028800	.73995071	.69229402	.65323436
13	.91646576	.85541551	.80751956	.76828684	.73519249
14	.94234085	.89759137	.86099208	.83004959	.80327762
15	.96082306	.92888721	.90183150	.87832367	.85752655

M	B= 6	B= 7	B= 8	B= 9	B=10
1	.00013938	.00010043	.75897\04	.59462\04	.47915\04
2	.00120454	.00090333	.00070557	.00056842	.00046919
3	.00540234	.00420333	.00338492	.00279860	.00236243
4	.01677532	.01349994	.01118167	.00946949	.00816208
5	.04060134	.03369434	.02863765	.02479503	.02178798
6	.08176600	.06977367	.06071392	.05364367	.04798208
7	.14285714	.12500000	.11111111	.10000000	.09090909
8	.22295430	.19949885	.18075788	.16541544	.15260782
9	.31766034	.28991685	.26719241	.24818703	.23202214
10	.42024723	.39022638	.36506877	.34360877	.32503656
11	.52332277	.49323054	.46748014	.44511129	.42543922
12	.62036069	.59213410	.56751467	.54576833	.52635866
13	.70667165	.68168634	.65951262	.63962559	.62163272
14	.77971314	.75869427	.73974558	.72251360	.70672817
15	.83887522	.82196844	.80650999	.79227439	.77908576

M	B=11	B=12	B=13	B=14	B=15
1	.39490\04	.33151\04	.28260\04	.24405\04	.21311\04
2	.00039491	.00033777	.00029279	.00025670	.00022725
3	.00202802	.00176522	.00155441	.00138234	.00123979
4	.00713662	.00631439	.00564288	.00508582	.00461747
5	.01937828	.01740909	.01577323	.01439515	.01322019
6	.04335266	.03950107	.03624951	.03347011	.03106862
7	.08333333	.07692308	.07142857	.06666667	.06250000
8	.14174363	.13240383	.12428284	.11715238	.11083834
9	.21808084	.20591616	.19519567	.18566662	.17713311
10	.30876974	.29437734	.28153275	.26998341	.25953052
11	.40796018	.39229385	.37814681	.36528845	.35353457
12	.50888242	.49302938	.47855604	.46526806	.45300805
13	.60523287	.59019015	.57631639	.56345918	.55149341
14	.69217802	.67869481	.66614217	.65440809	.64339943
15	.76680404	.75531580	.74452781	.73436246	.72475443

DIRICHLET C INTEGRAL
A = 2 R = 7

M	B= 1	B= 2	B= 3	B= 4	B= 5
1	.05852766	.02678740	.01713742	.01258550	.00995794
2	.19509221	.11078766	.07891966	.06206597	.05157852
3	.37717828	.25288283	.19670363	.16389154	.14206183
4	.55926434	.42507284	.35478997	.31005201	.27846806
5	.71100273	.59264038	.52261633	.47460157	.43887759
6	.82227754	.73189484	.67275594	.62952234	.59583443
7	.89646075	.83455410	.79053334	.75653491	.72894914
8	.94238370	.90343753	.87378003	.84977651	.82960225
9	.96917208	.94628576	.92785113	.91232703	.89887611
10	.98405451	.97133457	.96060883	.95126999	.94296422
11	.99199181	.98524141	.97933425	.97404572	.96923657
12	.99608072	.99263447	.98952743	.98668100	.98404368
13	.99812518	.99642258	.99485062	.99338309	.99200198
14	.99912119	.99830329	.99753380	.99680437	.99610898
15	.99959549	.99921195	.99884576	.99849433	.99815577

M	B= 6	B= 7	B= 8	B= 9	B=10
1	.00825305	.00705907	.00617672	.00549812	.00495995
2	.04439027	.03913690	.03511809	.03193678	.02935071
3	.12633924	.11439260	.10495814	.09728747	.09090730
4	.25468160	.23595600	.22073056	.20804265	.19726253
5	.41087388	.38810912	.36910009	.35289645	.33885638
6	.56846537	.54556292	.52597050	.50892021	.49387770
7	.70581449	.68594593	.66857296	.65316670	.63934862
8	.81220228	.79690861	.78327034	.77096819	.75976755
9	.88698635	.87631929	.86663835	.85777114	.84958777
10	.93546387	.92861240	.92229683	.91643256	.91095437
11	.96481300	.96070813	.95687224	.95326715	.94986278
12	.98157932	.97926127	.97706917	.97498700	.97300191
13	.99069415	.98944961	.98826054	.98712067	.98602482
14	.99544311	.99480322	.99418647	.99359054	.99301348
15	.99782858	.99751157	.99720375	.99690432	.99661257

M	B=11	B=12	B=13	B=14	B=15
1	.00452259	.00416003	.00385452	.00359347	.00336778
2	.02720348	.02538952	.02383487	.02248616	.02130387
3	.08550262	.08085510	.07680828	.07324688	.07008397
4	.18795885	.17982508	.17263668	.16622493	.16046035
5	.32652816	.31558307	.30577521	.29691640	.28885970
6	.48045708	.46837117	.45740086	.44737541	.43815940
7	.62683869	.61542400	.60493901	.59525255	.58625895
8	.74949084	.74000029	.73118688	.72296278	.71525621
9	.84198779	.83489181	.82823588	.82196773	.81604403
10	.90581079	.90096039	.89636928	.89200932	.88785691
11	.94663493	.94356380	.94063293	.93782848	.93513868
12	.97110335	.96928255	.96753214	.96584582	.96421817
13	.98496873	.98394877	.98296185	.98200530	.98107681
14	.99245366	.99190966	.99138028	.99086445	.99036125
15	.99632792	.99604985	.99577791	.99551170	.99525086

DIRICHLET C INTEGRAL
A = 3 R = 7

M	B= 1	B= 2	B= 3	B= 4	B= 5
1	.13348389	.07977546	.05957336	.04870415	.04180549
2	.36708069	.26173805	.21363562	.18493963	.16543784
3	.60067749	.48605620	.42481204	.38488939	.35607332
4	.77587509	.68454878	.62910388	.59011383	.56044374
5	.88537359	.82589462	.78583282	.75578095	.73183367
6	.94559777	.91203185	.88743583	.86794847	.85178683
7	.97570986	.95869918	.94535573	.93428644	.92478542
8	.98969047	.98176159	.97519213	.96952941	.96452454
9	.99580699	.99234975	.98935742	.98669513	.98428325
10	.99835553	.99692761	.99564826	.99448018	.99339990
11	.99937495	.99881096	.99829179	.99780775	.99735240
12	.99976882	.99955422	.99935247	.99916121	.99897878
13	.99991652	.99983739	.99976180	.99968918	.99961914
14	.99997049	.99994208	.99991461	.99988796	.99986201
15	.99998976	.99997980	.99997007	.99996056	.99995124

M	B= 6	B= 7	B= 8	B= 9	B=10
1	.03698600	.03340090	.03061331	.02837335	.02652716
2	.15111258	.14003061	.13113478	.12379312	.11760208
3	.33393349	.31618799	.30152288	.28911952	.27843702
4	.53671678	.51708208	.50042163	.48601154	.47335789
5	.71199205	.69509629	.68041411	.66745402	.65587040
6	.83797173	.82590467	.81519200	.80556047	.79681244
7	.91643964	.90898432	.90223859	.89607292	.89039110
8	.96002336	.95592263	.95214931	.94864957	.94538240
9	.98206983	.98001868	.97810334	.97630380	.97460443
10	.99239139	.99144305	.99054616	.98969395	.98888103
11	.99692116	.99651060	.99611806	.99574143	.99537902
12	.99880394	.99863574	.99847345	.99831644	.99816422
13	.99955135	.99948559	.99942164	.99935934	.99929856
14	.99983671	.99981199	.99978780	.99976410	.99974085
15	.99994209	.99993311	.99992427	.99991557	.99990700

M	B=11	B=12	B=13	B=14	B=15
1	.02497451	.02364711	.02249673	.02148826	.02059550
2	.11229057	.10766894	.10360008	.09998214	.09673766
3	.26910131	.26084404	.25346692	.24681966	.24078599
4	.46210959	.45200877	.44286077	.43451534	.42685423
5	.64541115	.63588689	.62715179	.61909128	.61161370
6	.78880019	.78141036	.77455401	.76816006	.76217077
7	.88511957	.88020072	.87558854	.87124567	.86714131
8	.94231579	.93942417	.93668678	.93408649	.93160899
9	.97299280	.97145880	.96999408	.96859167	.96724564
10	.98810302	.98735629	.98663781	.98594500	.98527567
11	.99502938	.99469135	.99436391	.99404618	.99373742
12	.99801635	.99787248	.99773230	.99759553	.99746194
13	.99923917	.99918107	.99912418	.99906841	.99901369
14	.99971802	.99969559	.99967353	.99965181	.99963043
15	.99989854	.99989021	.99988198	.99987385	.99986582

DIRICHLET C INTEGRAL

A = 4 R = 7

M	B= 1	B= 2	B= 3	B= 4	B= 5
1	.20971520	.14303667	.11520716	.09924183	.08862881
2	.50331648	.39966735	.34780724	.31501190	.29175203
3	.73819750	.64882250	.59708613	.56153666	.53485194
4	.87912612	.82267704	.78592916	.75883576	.73746896
5	.94959043	.92043162	.89956876	.88324544	.86981076
6	.98059472	.96753396	.95745390	.94916493	.94208837
7	.99299644	.98773833	.98342772	.97973366	.97648025
8	.99760279	.99565400	.99397826	.99249265	.99114958
9	.99921501	.99853871	.99793499	.99738482	.99687650
10	.99975242	.99952989	.99932539	.99913487	.99895567
11	.99992439	.99985431	.99978846	.99972602	.99966644
12	.99997755	.99995627	.99993593	.99991638	.99989751
13	.99999349	.99998723	.99998116	.99997527	.99996954
14	.99999815	.99999636	.99999460	.99999288	.99999120
15	.99999949	.99999898	.99999849	.99999800	.99999752

M	B= 6	B= 7	B= 8	B= 9	B=10
1	.08094234	.07505381	.07036019	.06650705	.06327100
2	.27408292	.26003182	.24848587	.23876223	.23041500
3	.51370196	.49630794	.48161483	.46894870	.45785462
4	.71988568	.70498360	.69207788	.68071439	.67057681
5	.85838411	.84843770	.83962926	.83172376	.82455266
6	.93589394	.93037348	.92538659	.92083364	.91664121
7	.97356082	.97090497	.96846345	.96620027	.96408818
8	.98991855	.98877857	.98771447	.98671483	.98577080
9	.99640216	.99595619	.99553438	.99513350	.99475096
10	.99878594	.99862430	.99846968	.99832127	.99817837
11	.99960931	.99955432	.99950123	.99944982	.99939995
12	.99987924	.99986150	.99984423	.99982740	.99981096
13	.99996393	.99995846	.99995309	.99994783	.99994267
14	.99998954	.99998791	.99998631	.99998473	.99998317
15	.99999704	.99999657	.99999611	.99999565	.99999520

M	B=11	B=12	B=13	B=14	B=15
1	.06050354	.05810164	.05599133	.05411800	.05244030
2	.22313869	.21671584	.21098675	.20583105	.20115604
3	.44801192	.43918670	.43120342	.42392725	.41725250
4	.66143618	.65312156	.64550202	.63847513	.63195928
5	.81799082	.81194278	.80633397	.80110504	.79620794
6	.91275350	.90912704	.90572723	.90252610	.89950065
7	.96210603	.96023701	.95846750	.95678633	.95518418
8	.98487537	.98402286	.98320860	.98242870	.98167986
9	.99438471	.99403302	.99369448	.99336787	.99305215
10	.99804044	.99790700	.99777766	.99765208	.99752997
11	.99935148	.99930428	.99925826	.99921332	.99916940
12	.99979489	.99977915	.99976373	.99974860	.99973375
13	.99993760	.99993261	.99992769	.99992286	.99991809
14	.99998163	.99998012	.99997862	.99997714	.99997567
15	.99999475	.99999430	.99999386	.99999342	.99999299

DIRICHLET C INTEGRAL
A = 5 R = 7

M	B= 1	B= 2	B= 3	B= 4	B= 5
1	.27908165	.20589273	.17341123	.15405751	.14083566
2	.60467690	.51045506	.46071314	.42816094	.40448657
3	.82174041	.75444925	.71363543	.68471463	.66250235
4	.93027216	.89506119	.87119553	.85311051	.83854878
5	.97549372	.96041852	.94925271	.94030216	.93279685
6	.99207496	.98647513	.98202442	.97828617	.97504132
7	.99760204	.99573127	.99415929	.99278725	.99156113
8	.99931280	.99873707	.99823176	.99777673	.99736011
9	.99981178	.99964576	.99949506	.99935590	.99922593
10	.99995038	.99990496	.99986266	.99982281	.99978499
11	.99998734	.99997544	.99996414	.99995333	.99994293
12	.99999686	.99999385	.99999096	.99998815	.99998542
13	.99999924	.99999850	.99999779	.99999708	.99999640
14	.99999982	.99999964	.99999947	.99999930	.99999913
15	.99999996	.99999992	.99999988	.99999984	.99999980

M	B= 6	B= 7	B= 8	B= 9	B=10
1	.13105373	.12342845	.11726074	.11213299	.10777833
2	.38614384	.37131820	.35896693	.34844024	.33930814
3	.64457356	.62960456	.61679631	.60563082	.59575389
4	.82636403	.81589256	.80671508	.79854962	.79119732
5	.92631591	.92060220	.91548631	.91085033	.90660868
6	.97216199	.96956612	.96719751	.96501577	.96299079
7	.99044744	.98942374	.98847408	.98758666	.98675248
8	.99697415	.99661343	.99627398	.99595277	.99564744
9	.99910352	.99898750	.99887697	.99877125	.99866977
10	.99974888	.99971425	.99968092	.99964875	.99961761
11	.99993290	.99992318	.99991375	.99990458	.99989563
12	.99998277	.99998018	.99997765	.99997517	.99997274
13	.99999572	.99999506	.99999441	.99999377	.99999314
14	.99999897	.99999881	.99999864	.99999849	.99999833
15	.99999976	.99999972	.99999968	.99999964	.99999960

M	B=11	B=12	B=13	B=14	B=15
1	.10401718	.10072361	.09780633	.09519733	.09284475
2	.33127256	.32411912	.31768890	.31186109	.30654194
3	.58691306	.57892231	.57164080	.56495941	.55879201
4	.78451271	.77838611	.77273278	.76748590	.76259185
5	.90269721	.89906649	.89567758	.89249926	.88950608
6	.96109948	.95932367	.95764879	.95606297	.95455639
7	.98596444	.98521689	.98450519	.98382553	.98317469
8	.99535609	.99507718	.99480943	.99455176	.99430325
9	.99857209	.99847784	.99838668	.99829836	.99821264
10	.99958740	.99955806	.99952950	.99950166	.99947449
11	.99988691	.99987837	.99987003	.99986185	.99985383
12	.99997036	.99996802	.99996571	.99996345	.99996122
13	.99999252	.99999191	.99999130	.99999071	.99999012
14	.99999817	.99999802	.99999787	.99999772	.99999757
15	.99999957	.99999953	.99999949	.99999945	.99999942

DIRICHLET C INTEGRAL
1/A = 5 R = 8

M	B= 1	B= 2	B= 3	B= 4	B= 5
1	.59537\06	.51671\09	.44308\11	.12570\12	.71378\14
2	.45645\05	.65432\08	.73523\10	.24879\11	.16704\12
3	.19449\04	.43895\07	.63501\09	.25402\10	.19289\11
4	.60794\04	.20737\06	.38004\08	.17819\09	.15200\10
5	.00015554	.77393\06	.17707\07	.96534\09	.92023\10
6	.00034504	.24280\05	.68425\07	.43049\08	.45642\09
7	.00068720	.66550\05	.22816\06	.16450\07	.19310\08
8	.00125745	.16358\04	.67455\06	.55364\07	.71643\08
9	.00214847	.36740\04	.18031\05	.16742\06	.23786\07
10	.00346851	.76467\04	.44228\05	.46188\06	.71762\07
11	.00533855	.00014908	.10070\04	.11762\05	.19911\06
12	.00788862	.00027463	.21482\04	.27917\05	.51305\06
13	.01125328	.00048133	.43254\04	.62232\05	.12374\05
14	.01556696	.00080723	.82723\04	.13114\04	.28122\05
15	.02095906	.00130159	.00015105	.26269\04	.60561\05

M	B= 6	B= 7	B= 8	B= 9	B=10
1	*	*	*	*	*
2	.16768\13	*	*	*	*
3	.22616\12	.33535\13	*	*	*
4	.19583\11	.33797\12	.67025\13	.41797\14	*
5	.12931\10	.24113\11	.55208\12	.13840\12	.20975\13
6	.69691\10	.13913\10	.34034\11	.96707\12	.29480\12
7	.31935\09	.68035\10	.17594\10	.52926\11	.17840\11
8	.12794\08	.29018\09	.79095\10	.24925\10	.88186\11
9	.45736\08	.11019\08	.31596\09	.10400\09	.38277\10
10	.14815\07	.37837\08	.11394\08	.39114\09	.14936\09
11	.44019\07	.11893\07	.37552\08	.13427\08	.53122\09
12	.12114\06	.34558\07	.11424\07	.42488\08	.17399\08
13	.31129\06	.93586\07	.32341\07	.12497\07	.52915\08
14	.75195\06	.23781\06	.85788\07	.34402\07	.15047\07
15	.17172\05	.57028\06	.21446\06	.89147\07	.40239\07

M	B=11	B=12	B=13	B=14	B=15
1	*	*	*	*	*
2	*	*	*	*	*
3	*	*	*	*	*
4	*	*	*	*	*
5	*	*	*	*	*
6	.72912\13	*	*	*	*
7	.63983\12	.21453\12	.33368\13	*	.22226\14
8	.34131\11	.13991\11	.57502\12	.22228\12	.13552\12
9	.15429\10	.66388\11	.30600\11	.14566\11	.77830\12
10	.62268\10	.27926\10	.13304\10	.66682\11	.35545\11
11	.22858\09	.10553\09	.51714\10	.26672\10	.14443\10
12	.77181\09	.36622\09	.18404\09	.97176\10	.53630\10
13	.24177\08	.11779\08	.60631\09	.32724\09	.18413\09
14	.70751\08	.35369\08	.18633\08	.10271\08	.58912\09
15	.19457\07	.99728\08	.53738\08	.30238\08	.17672\08

DIRICHLET C INTEGRAL
1/A = 4 R = 8

M	B= 1	B= 2	B= 3	B= 4	B= 5
1	.25600\05	.64381\08	.12563\09	.69079\11	.70964\12
2	.18944\04	.76839\07	.19315\08	.12508\09	.14425\10
3	.77926\04	.48600\06	.15462\07	.11685\08	.15064\09
4	.00023521	.21654\05	.85809\07	.75043\08	.10757\08
5	.00058124	.76248\05	.37090\06	.37238\07	.59057\08
6	.00124562	.22578\04	.13303\05	.15219\06	.26578\07
7	.00239721	.58431\04	.41191\05	.53324\06	.10209\06
8	.00423975	.00013567	.11314\04	.16465\05	.34410\06
9	.00700356	.00028795	.28112\04	.45710\05	.10385\05
10	.01093432	.00056659	.64132\04	.11583\04	.28500\05
11	.01628014	.00104485	.00013589	.27114\04	.71981\05
12	.02327831	.00182139	.00026991	.59187\04	.16894\04
13	.03214266	.00302234	.00050635	.00012143	.37142\04
14	.04305263	.00480135	.00090279	.00023569	.77001\04
15	.05614460	.00733738	.00153787	.00043513	.00015138

M	B= 6	B= 7	B= 8	B= 9	B=10
1	.11123\12	.23128\13	.17458\14	*	*
2	.24495\11	.54767\12	.14618\12	.31837\13	*
3	.27846\10	.66645\11	.19316\11	.63668\12	.20305\12
4	.21575\09	.55129\10	.16896\10	.59525\11	.23149\11
5	.12809\08	.34856\09	.11256\09	.41530\10	.17017\10
6	.62136\08	.17963\08	.60997\09	.23491\09	.10005\09
7	.25645\07	.78579\08	.28008\08	.11240\08	.49622\09
8	.92596\07	.30006\07	.11207\07	.46803\08	.21388\08
9	.29852\06	.10209\06	.39885\07	.17311\07	.81792\08
10	.87266\06	.31428\06	.12824\06	.57769\07	.28193\07
11	.23416\05	.88632\06	.37714\06	.17610\06	.88675\07
12	.58238\05	.23124\05	.10245\05	.49528\06	.25707\06
13	.13535\04	.56270\05	.25919\05	.12957\05	.69254\06
14	.29593\04	.12858\04	.61490\05	.31752\05	.17458\05
15	.61218\04	.27753\04	.13759\04	.73304\05	.41426\05

M	B=11	B=12	B=13	B=14	B=15
1	*	*	*	*	*
2	*	*	*	*	*
3	*	*	*	*	*
4	.91805\12	.25659\12	*	*	*
5	.75411\11	.34621\11	.14426\11	.14973\12	*
6	.46213\10	.22738\10	.11619\10	.57937\11	.21533\11
7	.23682\09	.12051\09	.64572\10	.35841\10	.19910\10
8	.10524\08	.55071\09	.30346\09	.17447\09	.10350\09
9	.41447\08	.22268\08	.12571\08	.74017\09	.45148\09
10	.14699\07	.81003\08	.46790\08	.28141\08	.17523\08
11	.47527\07	.26845\07	.15854\07	.97297\08	.61732\08
12	.14152\06	.81866\07	.49402\07	.30917\07	.19971\07
13	.39127\06	.23166\06	.14275\06	.91048\07	.59846\07
14	.10115\05	.61248\06	.38515\06	.25024\06	.16729\06
15	.24592\05	.15220\05	.97615\06	.64572\06	.43883\06

DIRICHLET C INTEGRAL
1/A = 3 R = 8

M	B= 1	B= 2	B= 3	B= 4	B= 5
1	.15259\04	.12939\06	.62103\08	.69054\09	.12562\09
2	.00010681	.14118\05	.85298\07	.10985\07	.22181\08
3	.00041580	.81690\05	.61051\06	.90250\07	.20114\07
4	.00118828	.33321\04	.30319\05	.51013\06	.12484\06
5	.00278151	.00010750	.11738\04	.22302\05	.59633\06
6	.00564933	.00029186	.37740\04	.80387\05	.23376\05
7	.01030953	.00069312	.00010487	.24867\04	.78294\05
8	.01729984	.00147806	.00025874	.67868\04	.23038\04
9	.02712996	.00288388	.00057812	.00016672	.60774\04
10	.04023678	.00522138	.00118732	.00037431	.00014596
11	.05694798	.00886853	.00226748	.00077725	.00032305
12	.07745718	.01425388	.00406419	.00150709	.00066535
13	.10181186	.02183110	.00688897	.00275031	.00128546
14	.12991341	.03204712	.01111268	.00475488	.00234539
15	.16152765	.04530784	.01715039	.00783114	.00406437

M	B= 6	B= 7	B= 8	B= 9	B=10
1	.31503\10	.98867\11	.36527\11	.15163\11	.66735\12
2	.60199\09	.20120\09	.78423\10	.34381\10	.16516\10
3	.58858\08	.20887\08	.85530\09	.39101\09	.19502\09
4	.39248\07	.14750\07	.63326\08	.30130\08	.15554\08
5	.20075\06	.79701\07	.35806\07	.17703\07	.94463\08
6	.83985\06	.35141\06	.16490\06	.84601\07	.46606\07
7	.29929\05	.13168\05	.64429\06	.34250\06	.19457\06
8	.93421\05	.43123\05	.21963\05	.12081\05	.70699\06
9	.26068\04	.12598\04	.66676\05	.37903\05	.22823\05
10	.66047\04	.33348\04	.18311\04	.10743\04	.66492\05
11	.00015380	.80972\04	.46059\04	.27854\04	.17702\04
12	.00033244	.00018215	.00010717	.66722\04	.43496\04
13	.00067245	.00038274	.00023258	.00014890	.99468\04
14	.00128157	.00075637	.00047403	.00031170	.00021318
15	.00231459	.00141408	.00091275	.00061576	.00043073

M	B=11	B=12	B=13	B=14	B=15
1	.26559\12	.95477\14	*	*	*
2	.85028\11	.45632\11	.24089\11	.10289\11	*
3	.10433\09	.59066\10	.34948\10	.21273\10	.12974\10
4	.85782\09	.49963\09	.30453\09	.19274\09	.12573\09
5	.53630\08	.32053\08	.20001\08	.12947\08	.86462\09
6	.27210\07	.16671\07	.10637\07	.70259\08	.47813\08
7	.11671\06	.73235\07	.47742\07	.32154\07	.22274\07
8	.43525\06	.27952\06	.18605\06	.12768\06	.89981\07
9	.14408\05	.94626\06	.64263\06	.44914\06	.32182\06
10	.43007\05	.28863\05	.19987\05	.14218\05	.10353\05
11	.11720\04	.80317\05	.56674\05	.41011\05	.30334\05
12	.29455\04	.20596\04	.14800\04	.10888\04	.81763\05
13	.68837\04	.49078\04	.35892\04	.26831\04	.20446\04
14	.00015064	.00010943	.81401\04	.61800\04	.47765\04
15	.00031055	.00022971	.00017368	.00013385	.00010488

DIRICHLET C INTEGRAL
1/A = 2 R = 8

M	B= 1	B= 2	B= 3	B= 4	B= 5
1	.00015242	.52735\05	.67584\06	.15774\06	.51721\07
2	.00096530	.49381\04	.77086\05	.20375\05	.72936\06
3	.00340395	.00024563	.45908\04	.13619\04	.52935\05
4	.00882318	.00086286	.00019009	.62771\04	.26354\04
5	.01875843	.00240197	.00061494	.00022429	.00010122
6	.03465483	.00563885	.00165605	.00066233	.00031982
7	.05761630	.01160486	.00386359	.00168298	.00086572
8	.08823160	.02149529	.00802493	.00378324	.00206448
9	.12650072	.03652000	.01513669	.00767723	.00442671
10	.17185671	.05772808	.02632000	.01428185	.00866874
11	.22326016	.08584514	.04268911	.02465246	.01569531
12	.27933666	.12115604	.06519774	.03987262	.02653637
13	.33852852	.16345289	.09449410	.06091537	.04224156
14	.39923812	.21205175	.13081306	.08850185	.06374835
15	.45994771	.26586809	.17392463	.12298494	.09174730

M	B= 6	B= 7	B= 8	B= 9	B=10
1	.21068\07	.99672\08	.52581\08	.30127\08	.18414\08
2	.31748\06	.15830\06	.87184\07	.51793\07	.32652\07
3	.24532\05	.12858\05	.73777\06	.45369\06	.29464\06
4	.12957\04	.71201\05	.42482\05	.27001\05	.18041\05
5	.52620\04	.30240\04	.18726\04	.12283\04	.84331\05
6	.00017522	.00010507	.67399\04	.45557\04	.32102\04
7	.00049835	.00031105	.00020634	.00014352	.00010367
8	.00124494	.00080709	.00055268	.00039501	.00029219
9	.00278847	.00187360	.00132223	.00096979	.00073372
10	.00568862	.00395327	.00287048	.00215765	.00166786
11	.01070165	.00767665	.00572599	.00440529	.00347545
12	.01875267	.01385848	.01060250	.00833838	.00670688
13	.03086425	.02345470	.01837751	.01475629	.01208848
14	.04804786	.03747881	.03003144	.02459030	.02049633
15	.07117427	.05688725	.04655071	.03882428	.03289314

M	B=11	B=12	B=13	B=14	B=15
1	.11853\08	.79599\09	.55357\09	.39623\09	.29000\07
2	.21593\07	.14851\07	.10553\07	.77096\08	.57666\08
3	.19996\06	.14071\06	.10206\06	.75947\07	.57776\07
4	.12552\05	.90289\06	.66789\06	.50593\06	.39115\06
5	.60087\05	.44146\05	.33280\05	.25645\05	.20138\05
6	.23402\04	.17546\04	.13470\04	.10552\04	.84117\05
7	.77246\04	.59053\04	.46136\04	.36719\04	.29697\04
8	.00022230	.00017314	.00013756	.00011116	.91163\04
9	.00056947	.00045152	.00036454	.00029892	.00024845
10	.00131936	.00106406	.00087241	.00072547	.00061075
11	.00279950	.00229481	.00190936	.00160921	.00137152
12	.00549630	.00457574	.00386101	.00329606	.00284252
13	.01006980	.00850764	.00727543	.00628736	.00548362
14	.01734008	.01485635	.01286734	.01125026	.00991807
15	.02823830	.02451621	.02149199	.01900052	.01692295

DIRICHLET C INTEGRAL
A = 1 R = 8

M	B= 1	B= 2	B= 3	B= 4	B= 5
1	.00390625	.00068931	.00025190	.00012539	.73891\04
2	.01953125	.00464246	.00196377	.00107143	.00067367
3	.05468750	.01672613	.00805228	.00477268	.00318416
4	.11328125	.04288943	.02314225	.01477794	.01040656
5	.19384766	.08789973	.05242557	.03578894	.02646853
6	.29052734	.15335740	.09983329	.07233157	.05591492
7	.39526367	.23698686	.16646186	.12713509	.10226171
8	.50000000	.33333333	.25000000	.20000000	.16666667
9	.59819031	.43531540	.34530314	.28764259	.24731987
10	.68547058	.53589637	.44572478	.38448936	.33977125
11	.75965881	.62933440	.54460887	.48400342	.43797925
12	.82035828	.71180204	.63645860	.58002109	.53562475
13	.86841202	.78143548	.71755713	.66771560	.62724799
14	.90537643	.83800884	.78605871	.74404022	.70894576
15	.93309975	.88244497	.84170968	.80770521	.77857996

M	B= 6	B= 7	B= 8	B= 9	B=10
1	.48386\04	.34041\04	.25221\04	.19430\04	.15430\04
2	.00046339	.00033903	.00025942	.00020537	.00016696
3	.00229195	.00173896	.00137128	.00111361	.00092551
4	.00780984	.00612831	.00496971	.00413317	.00350673
5	.02063747	.01670416	.01390193	.01182102	.01022466
6	.04514088	.03759178	.03204267	.02781160	.02449100
7	.08520149	.07281828	.06344582	.05612005	.05024597
8	.14285714	.12500000	.11111111	.10000000	.09090909
9	.21742508	.19431498	.17587897	.16080638	.14823859
10	.30546798	.27819909	.25592764	.23734694	.22157712
11	.40155808	.37185241	.34705691	.32597741	.30778851
12	.49948264	.46930219	.44359951	.42136427	.40188060
13	.59344733	.56460824	.53959044	.51759609	.49804646
14	.67894527	.65284229	.62981210	.60926259	.59075469
15	.75315745	.73064153	.71046755	.69222018	.67558489

M	B=11	B=12	B=13	B=14	B=15
1	.12555\04	.10420\04	.87925\05	.75228\05	.65134\05
2	.00013866	.00011720	.00010051	.87266\04	.76574\04
3	.00078366	.00067380	.00058682	.00051665	.00045915
4	.00302369	.00264220	.00233482	.00208292	.00187349
5	.00896754	.00795600	.00712728	.00643787	.00585677
6	.02182335	.01963854	.01781998	.01628529	.01497468
7	.04543716	.04143217	.03804798	.03515278	.03264933
8	.08333333	.07692308	.07142857	.06666667	.06250000
9	.13758870	.12844143	.12049432	.11352167	.10735154
10	.20800225	.19617727	.18577201	.17653592	.16827521
11	.29189920	.27787383	.26538338	.25417418	.24404712
12	.38462479	.36920337	.35531407	.34272024	.33123356
13	.48050943	.46465437	.45022316	.43701070	.42485159
14	.57395406	.55860046	.54448754	.53144902	.51934905
15	.66031773	.64622568	.63315339	.62097400	.60958257

DIRICHLET C INTEGRAL
A = 2 R = 8

M	B= 1	B= 2	B= 3	B= 4	B= 5
1	.03901844	.01662991	.01026514	.00736780	.00573425
2	.14306762	.07571666	.05199683	.03992919	.03261248
3	.29914139	.18793131	.14110385	.11483370	.09782964
4	.47255669	.33934338	.27419768	.23437722	.20704483
5	.63152071	.50219400	.43048668	.38323574	.34905570
6	.75869193	.65075721	.58425638	.53746607	.50199906
7	.85053781	.77024135	.71619470	.67593550	.64412686
8	.91176840	.85721053	.81766785	.78672223	.76136085
9	.95003752	.91558847	.88903141	.86734611	.84899415
10	.97271551	.95223859	.93562346	.92155410	.90930956
11	.98556638	.97399377	.96419839	.95564376	.94801626
12	.99257594	.98630877	.98081703	.97589445	.97141322
13	.99627543	.99300272	.99005287	.98735059	.98484671
14	.99817261	.99651625	.99498888	.99356423	.99222437
15	.99912119	.99830531	.99753909	.99681375	.99612304

M	B= 6	B= 7	B= 8	B= 9	B=10
1	.00469251	.00397275	.00344668	.00304581	.00273040
2	.02769294	.02415143	.02147535	.01937877	.01768955
3	.08583122	.07686078	.06987015	.06424973	.05961973
4	.18689295	.17129449	.15878723	.14848676	.13982385
5	.32283219	.30187913	.28463153	.27010727	.25765450
6	.47379309	.45059447	.43103294	.41421825	.39954207
7	.61800057	.59594117	.57692655	.56027065	.54549164
8	.73992040	.72138373	.70508295	.69055585	.67746929
9	.83307658	.81901939	.80643240	.79503773	.78463012
10	.89844616	.88866906	.87977131	.87160146	.86404509
11	.94111354	.93479610	.92896303	.92353862	.91846438
12	.96728725	.96345516	.95987126	.95650048	.95331515
13	.98250682	.98030556	.97822359	.97624573	.97435978
14	.99095626	.98975003	.98859796	.98749386	.98643268
15	.99546226	.99482776	.99421663	.99362649	.99305533

M	B=11	B=12	B=13	B=14	B=15
1	.00247583	.00226610	.00209032	.00194087	.00181224
2	.01629783	.01513015	.01413553	.01327745	.01252904
3	.05573058	.05241120	.04954016	.04702878	.04481070
4	.13241398	.12598716	.12034770	.11534997	.11088311
5	.24682115	.23728251	.22879842	.22118693	.21430729
6	.38657192	.37499028	.36455779	.35508997	.34644186
7	.53223877	.52024917	.50932099	.49929609	.49004836
8	.66557525	.65468419	.64464799	.63534877	.62669118
9	.77505354	.76618643	.75793228	.75021306	.74296483
10	.85701341	.85043605	.84425622	.83842737	.83291085
11	.91369410	.90919051	.90492308	.90086645	.89699929
12	.95029304	.94741590	.94466861	.94203839	.93951442
13	.97255574	.97082531	.96916150	.96755835	.96601073
14	.98541021	.98442289	.98346771	.98254204	.98164363
15	.99250151	.99196357	.99144029	.99093060	.99043354

DIRICHLET C INTEGRAL
A = 3 R = 8

M	B= 1	B= 2	B= 3	B= 4	B= 5
1	.10011292	.05723839	.04180940	.03370067	.02863344
2	.30033875	.20505629	.16362427	.13956281	.12350310
3	.52559280	.40945554	.35042914	.31303909	.28657506
4	.71330452	.61063443	.55127271	.51072853	.48049806
5	.84235632	.76877462	.72144759	.68695050	.66001825
6	.91978741	.87442343	.84256421	.81800679	.79804442
7	.96172924	.93677590	.91793715	.90270610	.88988106
8	.98270016	.97014902	.96009593	.95163235	.94428503
9	.99253028	.98665573	.98171850	.97741799	.97358563
10	.99689922	.99430712	.99204237	.99001292	.98816367
11	.99875602	.99766700	.99668547	.99578509	.99494914
12	.99951562	.99907657	.99867097	.99829174	.99793411
13	.99981630	.99964538	.99948440	.99933154	.99918553
14	.99993194	.99986737	.99980563	.99974629	.99968900
15	.99997531	.99995154	.99992855	.99990623	.99988451

M	B= 6	B= 7	B= 8	B= 9	B=10
1	.02513365	.02255341	.02056172	.01897106	.01766690
2	.11186291	.10295221	.09586049	.09004988	.08518025
3	.26653869	.25066470	.23767088	.22676930	.21744524
4	.45669357	.43723613	.42089358	.40688021	.39466669
5	.63805185	.61958248	.60370145	.58980884	.57748859
6	.78124484	.76675627	.75403032	.74269295	.73247754
7	.87878481	.86899522	.86023026	.85229160	.84503417
8	.93777159	.93190822	.92656782	.92165840	.91711104
9	.97011550	.96693585	.96399543	.96125617	.95868887
10	.98645835	.98487143	.98338416	.98198225	.98065450
11	.99416613	.99342770	.99272754	.99206072	.99142331
12	.99759468	.99727091	.99696083	.99666286	.99637573
13	.99904542	.99891049	.99878015	.99865393	.99853146
14	.99963354	.99957969	.99952730	.99947624	.99942640
15	.99986333	.99984262	.99982236	.99980250	.99978302

M	B=11	B=12	B=13	B=14	B=15
1	.01657510	.01564547	.01484273	.01414132	.01352224
2	.08102502	.07742681	.07427257	.07147877	.06898222
3	.20934615	.20222125	.19588655	.19020357	.18506576
4	.38388082	.37425178	.36557683	.35770063	.35050173
5	.56644077	.55644238	.54732326	.53895077	.53121962
6	.72318735	.71467303	.70681854	.69953193	.69273914
7	.83834851	.83214990	.82637137	.82095901	.81586877
8	.91287262	.90890127	.90516333	.90163131	.89828244
9	.95627059	.95398292	.95181081	.94974183	.94776552
10	.97939194	.97818726	.97703439	.97592826	.97486452
11	.99081209	.99022441	.98965803	.98911106	.98858186
12	.99609836	.99582986	.99556948	.99531656	.99507053
13	.99841239	.99829645	.99818341	.99807303	.99796516
14	.99937767	.99932999	.99928327	.99923746	.99919250
15	.99976389	.99974509	.99972659	.99970838	.99969045

DIRICHLET C INTEGRAL
A = 4 R = 8

M	B= 1	B= 2	B= 3	B= 4	B= 5
1	.16777216	.11084969	.08788098	.07493783	.06643519
2	.43620762	.33593775	.28768310	.25781415	.23693248
3	.67779953	.58094238	.52710150	.49095852	.46425861
4	.83886080	.77099108	.72853343	.69797144	.67427064
5	.92744450	.88887400	.86229617	.84198709	.82555290
6	.96996468	.95109706	.93702905	.92571898	.91622184
7	.98839009	.98014900	.97359808	.96810075	.96333487
8	.99576025	.99246598	.98970861	.98731024	.98517346
9	.99852406	.99729757	.99622771	.99526908	.99439507
10	.99950675	.99907588	.99868753	.99833099	.99799958
11	.99984086	.99969661	.99956321	.99943830	.99932033
12	.99995021	.99990383	.99986007	.99981845	.99977863
13	.99998484	.99997043	.99995662	.99994333	.99993047
14	.99999549	.99999114	.99998693	.99998283	.99997884
15	.99999869	.99999741	.99999616	.99999493	.99999373

M	B= 6	B= 7	B= 8	B= 9	B=10
1	.06033061	.05568575	.05200395	.04899551	.04647900
2	.22123826	.20886187	.19876155	.19030431	.18308013
3	.44334884	.42631497	.41203806	.39981171	.38916362
4	.65501410	.63885972	.62498726	.61285998	.60210811
5	.81175658	.79987364	.78944257	.78015102	.77177770
6	.90801568	.90077945	.89430062	.88843074	.88306180
7	.95911147	.95530895	.95184389	.94865627	.94570132
8	.98323794	.98146321	.97982056	.97828877	.97685162
9	.99358834	.99283680	.99213162	.99146611	.99083502
10	.99768875	.99739522	.99711651	.99685071	.99659629
11	.99920819	.99910107	.99899831	.99889942	.99880400
12	.99974035	.99970342	.99966770	.99963306	.99959939
13	.99991800	.99990588	.99989406	.99988253	.99987126
14	.99997493	.99997111	.99996737	.99996369	.99996008
15	.99999255	.99999139	.99999024	.99998912	.99998800

M	B=11	B=12	B=13	B=14	B=15
1	.04433438	.04247880	.04085300	.03941337	.03812702
2	.17681002	.17129655	.16639550	.16199866	.15802302
3	.37976362	.37137259	.36381221	.35694623	.35066841
4	.59246651	.58373882	.57577573	.56846116	.56170316
5	.76416003	.75717504	.75072751	.74474212	.73915835
6	.87811265	.87352065	.86923641	.86522028	.86143991
7	.94294468	.94035930	.93792352	.93561967	.93343313
8	.97549643	.97421301	.97299309	.97182982	.97071746
9	.99023418	.98966019	.98911023	.98858194	.98807332
10	.99635201	.99611682	.99588988	.99567045	.99545789
11	.99871169	.99862222	.99853534	.99845085	.99836856
12	.99956662	.99953467	.99950347	.99947298	.99944313
13	.99986023	.99984942	.99983881	.99982840	.99981817
14	.99995653	.99995304	.99994960	.99994621	.99994287
15	.99998690	.99998582	.99998475	.99998369	.99998264

DIRICHLET C INTEGRAL
A = 5 R = 8

M	B= 1	B= 2	B= 3	B= 4	B= 5
1	.23256804	.16739843	.13927009	.12275517	.11158218
2	.54265876	.44758054	.39902829	.36783105	.34541965
3	.77522680	.69916016	.65461747	.62368800	.60026152
4	.90443126	.86027607	.83137012	.80992082	.79290323
5	.96364998	.94285473	.92794990	.91624756	.90658028
6	.98733746	.97890334	.97239822	.96704149	.96245913
7	.99589128	.99283519	.99033482	.98819210	.98630354
8	.99874255	.99772849	.99685883	.99608853	.99539220
9	.99963357	.99931998	.99904085	.99878682	.99855227
10	.99989758	.99980602	.99972211	.99964406	.99957071
11	.99997238	.99994689	.99992299	.99990038	.99987882
12	.99999278	.99998596	.99997946	.99997322	.99996720
13	.99999816	.99999640	.99999470	.99999304	.99999143
14	.99999954	.99999910	.99999867	.99999825	.99999783
15	.99999989	.99999978	.99999967	.99999957	.99999947

M	B= 6	B= 7	B= 8	B= 9	B=10
1	.10337509	.09701322	.09189088	.08764854	.08405765
2	.32821318	.31440532	.30296924	.29327053	.28489234
3	.58154932	.56605518	.55288769	.54147501	.53142948
4	.77882261	.76683024	.75639791	.74717462	.73891537
5	.89832831	.89112091	.88471783	.87895406	.87371118
6	.95843929	.95484903	.95159889	.94862549	.94588218
7	.98460700	.98306174	.98163936	.98031915	.97908549
8	.99475376	.99416223	.99360972	.99309030	.99259941
9	.99833344	.99812770	.99793307	.99774804	.99757143
10	.99950128	.99943518	.99937197	.99931129	.99925286
11	.99985816	.99983829	.99981910	.99980054	.99978253
12	.99996137	.99995572	.99995023	.99994487	.99993964
13	.99998987	.99998833	.99998683	.99998536	.99998392
14	.99999742	.99999702	.99999663	.99999624	.99999586
15	.99999936	.99999926	.99999916	.99999906	.99999897

M	B=11	B=12	B=13	B=14	B=15
1	.08096506	.07826380	.07587661	.07374607	.07182850
2	.27754743	.27103027	.26518921	.25990943	.25510214
3	.52247686	.51441633	.50709670	.50040144	.49423895
4	.73144238	.72462266	.71835411	.71255668	.70716646
5	.86890122	.86445700	.86032598	.85646629	.85284401
6	.94333350	.94095183	.93871519	.93660581	.93460907
7	.97792626	.97683185	.97579446	.97480772	.97386628
8	.99213343	.99168943	.99126503	.99085821	.99046730
9	.99740226	.99723975	.99708324	.99693218	.99678609
10	.99919646	.99914188	.99908898	.99903760	.99898762
11	.99976502	.99974798	.99973136	.99971513	.99969927
12	.99993453	.99992952	.99992462	.99991981	.99991508
13	.99998250	.99998110	.99997973	.99997838	.99997705
14	.99999549	.99999511	.99999475	.99999438	.99999403
15	.99999887	.99999878	.99999868	.99999859	.99999850

DIRICHLET C INTEGRAL
1/A = 5 R = 9

M	B= 1	B= 2	B= 3	B= 4	B= 5
1	.99229\07	.40014\10	.21113\12	.41242\14	*
2	.84345\06	.56405\09	.39007\11	.93193\13	.43140\14
3	.39444\05	.41871\08	.37361\10	.10570\11	.61117\13
4	.13419\04	.21770\07	.24698\09	.82025\11	.53680\12
5	.37107\04	.88984\07	.12664\08	.49006\10	.35904\11
6	.88430\04	.30439\06	.53664\08	.24032\09	.19612\10
7	.00018822	.90598\06	.19559\07	.10070\08	.91154\10
8	.00036643	.24091\05	.63008\07	.37067\08	.37068\09
9	.00066344	.58334\05	.18299\06	.12229\07	.13459\08
10	.00113095	.13047\04	.48630\06	.36717\07	.44312\08
11	.00183222	.27254\04	.11965\05	.10153\06	.13389\07
12	.00284162	.53640\04	.27512\05	.26106\06	.37495\07
13	.00424356	.00010018	.59570\05	.62910\06	.98091\07
14	.00613079	.00017861	.12223\04	.14302\05	.24135\06
15	.00860217	.00030544	.23895\04	.30844\05	.56164\06

M	B= 6	B= 7	B= 8	B= 9	B=10
1	*	*	*	*	*
2	*	*	*	*	*
3	.46056\14	*	*	*	*
4	.54816\13	.55956\14	*	*	*
5	.40756\12	.61552\13	.82994\14	*	*
6	.24267\11	.40508\12	.81774\13	.15343\13	*
7	.12234\10	.21907\11	.48576\12	.12380\12	.32682\13
8	.53794\10	.10272\10	.24167\11	.66676\12	.20682\12
9	.21062\09	.42770\10	.10605\10	.30755\11	.10101\11
10	.74582\09	.16073\09	.41898\10	.12694\10	.43417\11
11	.24178\08	.55196\09	.15101\09	.47694\10	.16922\10
12	.72467\08	.17492\08	.50156\09	.16491\09	.60606\10
13	.20244\07	.51580\08	.15478\08	.52921\09	.20123\09
14	.53070\07	.14248\07	.44688\08	.15871\08	.62380\09
15	.13130\06	.37084\07	.12141\07	.44741\08	.18162\08

M	B=11	B=12	B=13	B=14	B=15
1	*	.11490\13	.46915\13	.12251\12	.28098\12
2	*	.11490\13	.46914\13	.12251\12	.28097\12
3	*	.11491\13	.46913\13	.12251\12	.28097\12
4	*	.11498\13	.46912\13	.12250\12	.28095\12
5	*	.11583\13	.46936\13	.12249\12	.28090\12
6	*	.12255\13	.47185\13	.12256\12	.28083\12
7	.10538\13	.16614\13	.48938\13	.12324\12	.28097\12
8	.72769\13	.40936\13	.59147\13	.12770\12	.28277\12
9	.36981\12	.16068\12	.11101\12	.15144\12	.29384\12
10	.16436\11	.68974\12	.34665\12	.26265\12	.34852\12
11	.66128\11	.28142\11	.13181\11	.73300\12	.58711\12
12	.24423\10	.10647\10	.49923\11	.25544\11	.15335\11
13	.83561\10	.37382\10	.17848\10	.90736\11	.49948\11
14	.26673\09	.12245\09	.59757\10	.30799\10	.16767\10
15	.79906\09	.37627\09	.18780\09	.98617\10	.54248\10

DIRICHLET C INTEGRAL
1/A = 4 R = 9

M	B= 1	B= 2	B= 3	B= 4	B= 5
1	.51200\06	.67936\09	.89603\11	.37545\12	.32845\13
2	.41984\05	.90176\08	.15325\09	.75462\11	.70961\12
3	.18944\04	.63052\07	.13589\08	.78119\10	.81951\11
4	.62198\04	.30887\06	.83193\08	.55412\09	.64653\10
5	.00016601	.11899\05	.39519\07	.30277\08	.39119\09
6	.00038193	.38375\05	.15521\06	.13585\07	.19353\08
7	.00078499	.10772\04	.52454\06	.52108\07	.81513\08
8	.00147594	.27023\04	.15674\05	.17567\06	.30055\07
9	.00258146	.61753\04	.42244\05	.53104\06	.99001\07
10	.00425203	.00013040	.10423\04	.14617\05	.29586\06
11	.00665765	.00025726	.23823\04	.37074\05	.81193\06
12	.00998179	.00047840	.50906\04	.87487\05	.20662\05
13	.01441396	.00084457	.00010249	.19360\04	.49153\05
14	.02014170	.00142382	.00019564	.40441\04	.11004\04
15	.02734228	.00230352	.00035598	.80184\04	.23315\04

M	B= 6	B= 7	B= 8	B= 9	B=10
1	.68481\14	.45068\14	.36162\14	*	*
2	.10486\12	.23752\13	.83564\14	*	*
3	.12895\11	.27332\12	.73468\13	.20774\13	*
4	.11020\10	.24675\11	.67876\12	.21603\12	.65690\13
5	.72116\10	.17178\10	.49623\11	.16621\11	.61508\12
6	.38479\09	.97391\10	.29574\10	.10346\10	.40420\11
7	.17430\08	.46784\09	.14915\09	.54388\10	.22075\10
8	.68930\08	.19580\08	.65436\09	.24838\09	.10441\09
9	.24287\07	.72871\08	.25491\08	.10059\08	.43734\09
10	.77441\07	.24495\07	.89555\08	.36696\08	.16485\08
11	.22619\06	.75290\07	.28728\07	.12210\07	.56617\08
12	.61120\06	.21370\06	.84981\07	.37420\07	.17895\07
13	.15403\05	.56473\06	.23374\06	.10651\06	.52484\07
14	.36450\05	.13990\05	.60186\06	.28354\06	.14383\06
15	.81463\05	.32678\05	.14593\05	.71002\06	.37045\06

M	B=11	B=12	B=13	B=14	B=15
1	*	*	*	*	*
2	*	*	*	*	*
3	*	*	*	*	*
4	*	*	*	*	*
5	.22008\12	.19126\13	*	*	*
6	.17018\11	.71040\12	.18961\12	*	*
7	.97543\11	.45755\11	.21617\11	.83504\12	*
8	.47694\10	.23299\10	.11954\10	.62179\11	.29047\11
9	.20587\09	.10351\09	.54929\10	.30361\10	.17038\10
10	.79871\09	.41216\09	.22427\09	.12754\09	.75032\10
11	.28208\08	.14924\08	.83079\09	.48296\09	.29118\09
12	.91612\08	.49657\08	.28253\08	.16758\08	.10302\08
13	.27587\07	.15309\07	.88972\08	.53801\08	.33673\08
14	.77564\07	.44041\07	.26128\07	.16099\07	.10251\07
15	.20482\06	.11892\06	.71983\07	.45169\07	.29246\07

DIRICHLET C INTEGRAL
1/A = 3 R = 9

M	B= 1	B= 2	B= 3	B= 4	B= 5
1	.38147\05	.19695\07	.70671\09	.64460\10	.10129\10
2	.29564\04	.23867\06	.10779\07	.11380\08	.19831\09
3	.00012612	.15244\05	.85301\07	.10339\07	.19882\08
4	.00039166	.68255\05	.46640\06	.64409\07	.13604\07
5	.00098912	.24049\04	.19802\05	.30934\06	.71439\07
6	.00215418	.70981\04	.69566\05	.12210\05	.30702\06
7	.00419301	.00018247	.21046\04	.41242\05	.11244\05
8	.00746972	.00041954	.56350\04	.12254\04	.36087\05
9	.01238478	.00087934	.00013620	.32682\04	.10357\04
10	.01934778	.00170437	.00030165	.79446\04	.26997\04
11	.02874783	.00308906	.00061947	.00017815	.64695\04
12	.04092517	.00528183	.00119062	.00037209	.00014393
13	.05614684	.00858133	.00215830	.00072963	.00029970
14	.07458848	.01332639	.00371373	.00135214	.00058800
15	.09632327	.01987997	.00609835	.00238149	.00109330

M	B= 6	B= 7	B= 8	B= 9	B=10
1	.22647\11	.64315\12	.20896\12	.59285\13	*
2	.47973\10	.14601\10	.52549\11	.21347\11	.92733\12
3	.51837\09	.16749\09	.63414\10	.27100\10	.12717\10
4	.38103\08	.13036\08	.51738\09	.23019\09	.11208\09
5	.21431\07	.77456\08	.32166\08	.14870\08	.74847\09
6	.98351\07	.37472\07	.16256\07	.77978\08	.40519\08
7	.38356\06	.15374\06	.69556\07	.34576\07	.18528\07
8	.13073\05	.55007\06	.25916\06	.13334\06	.73610\07
9	.39739\05	.17521\05	.85827\06	.45650\06	.25937\06
10	.10944\04	.50463\05	.25664\05	.14095\05	.82342\06
11	.27640\04	.13305\04	.70150\05	.39735\05	.23846\05
12	.64657\04	.32433\04	.17704\04	.10331\04	.63630\05
13	.00014124	.73702\04	.41596\04	.24980\04	.15775\04
14	.00029008	.00015721	.91613\04	.56556\04	.36590\04
15	.00056343	.00031661	.00019026	.00012062	.79876\04

M	B=11	B=12	B=13	B=14	B=15
1	*	*	*	*	*
2	.37603\12	.44604\13	*	*	*
3	.63897\11	.33133\11	.16171\11	.49100\12	*
4	.58660\10	.32506\10	.18760\10	.10996\10	.62543\11
5	.40359\09	.23030\09	.13767\09	.85421\10	.54478\10
6	.22468\08	.13147\08	.80470\09	.51160\09	.33583\09
7	.10555\07	.63259\08	.39566\08	.25663\08	.17172\08
8	.43043\07	.26402\07	.16861\07	.11144\07	.75879\08
9	.15555\06	.97580\07	.63585\07	.42802\07	.29633\07
10	.50606\06	.32445\06	.21559\06	.14772\06	.10394\06
11	.15007\05	.98264\06	.66547\06	.46388\06	.33156\06
12	.40971\05	.27383\05	.18889\05	.13389\05	.97168\06
13	.10385\04	.70795\05	.49715\05	.35815\05	.26381\05
14	.24609\04	.17101\04	.12218\04	.89417\05	.66818\05
15	.54845\04	.38824\04	.28208\04	.20961\04	.15884\04

DIRICHLET C INTEGRAL
1/A = 2 R = 9

M	B= 1	B= 2	B= 3	B= 4	B= 5
1	.50805\04	.12584\05	.13403\06	.27678\07	.83021\08
2	.00035564	.13054\04	.16927\05	.39549\06	.12939\06
3	.00137174	.71476\04	.11109\04	.29133\05	.10345\05
4	.00385555	.00027478	.50466\04	.14742\04	.56555\05
5	.00882318	.00083261	.00017835	.57629\04	.23778\04
6	.01743373	.00211711	.00052260	.00018554	.81992\04
7	.03082792	.00469745	.00132142	.00051227	.00024148
8	.04996248	.00934019	.00296365	.00124712	.00062471
9	.07547523	.01696532	.00601434	.00273196	.00144889
10	.10760239	.02856013	.01121238	.00546896	.00306007
11	.14615498	.04506451	.01943176	.01012695	.00595821
12	.19054887	.06724895	.03160652	.01751690	.01080223
13	.23987542	.09560763	.04862922	.02853351	.01838641
14	.29299632	.13028505	.07123949	.04406777	.02958511
15	.34864678	.17104658	.09992235	.06490333	.04526973

M	B= 6	B= 7	B= 8	B= 9	B=10
1	.31572\08	.14133\08	.71214\09	.39247\09	.23197\09
2	.52534\07	.24765\07	.13018\07	.74334\08	.45285\08
3	.44696\06	.22136\06	.12117\06	.71586\07	.44904\07
4	.25920\05	.13455\05	.76563\06	.46735\06	.30151\06
5	.11525\04	.62569\05	.36946\05	.23271\05	.15424\05
6	.41903\04	.23741\04	.14524\04	.94266\05	.64122\05
7	.00012977	.76568\04	.48450\04	.32362\04	.22568\04
8	.00035203	.00021588	.00014107	.96855\04	.69173\04
9	.00085394	.00054319	.00036602	.00025798	.00018850
10	.00188158	.00123915	.00085971	.00062131	.00046401
11	.00381290	.00259497	.00185099	.00137004	.00104477
12	.00717774	.00503922	.00369038	.00279428	.00217374
13	.01265691	.00915067	.00687064	.00531593	.00421463
14	.02105316	.01564813	.01202987	.00950054	.00766961
15	.03323191	.02535205	.01992960	.01604810	.01317955

M	B=11	B=12	B=13	B=14	B=15
1	.14501\09	.94894\10	.64442\10	.44989\10	.31784\10
2	.29061\08	.19460\08	.13502\08	.96507\09	.70724\09
3	.29559\07	.20244\07	.14330\07	.10432\07	.77776\08
4	.20341\06	.14236\06	.10275\06	.76115\07	.57660\07
5	.10654\05	.76143\06	.55994\06	.42187\06	.32453\06
6	.45311\05	.33042\05	.24741\05	.18947\05	.14794\05
7	.16300\04	.12119\04	.92343\05	.71842\05	.56907\05
8	.51019\04	.38650\04	.29948\04	.23656\04	.19000\04
9	.00014186	.00010942	.86161\04	.69066\04	.56221\04
10	.00035601	.00027936	.00022343	.00018165	.00014979
11	.00081654	.00065142	.00052886	.00043584	.00036390
12	.00172917	.00140155	.00115427	.00096377	.00081438
13	.00340975	.00280594	.00234284	.00198084	.00169318
14	.00630561	.00526471	.00445398	.00381130	.00329401
15	.01100298	.00931451	.00797974	.00690729	.00603333

DIRICHLET C INTEGRAL
A = 1 R = 9

M	B= 1	B= 2	B= 3	B= 4	B= 5
1	.00195313	.00029483	.99406\04	.46987\04	.26679\04
2	.01074219	.00218627	.00085246	.00044117	.00026701
3	.03271484	.00860910	.00382302	.00214904	.00137959
4	.07299805	.02396094	.01195003	.00724189	.00490778
5	.13342285	.05294979	.02928124	.01899553	.01352844
6	.21197510	.09898083	.05998379	.04137962	.03083655
7	.30361938	.16288380	.10701012	.07800887	.06057665
8	.40180969	.24252280	.17101876	.13096634	.10555445
9	.50000000	.33333333	.25000000	.20000000	.16666667
10	.59273529	.42944285	.33969377	.28240456	.24244350
11	.67619705	.52490622	.43455649	.37358210	.32925809
12	.74827766	.61468671	.52888740	.46801880	.42207847
13	.80834484	.69520866	.61777126	.56032447	.51546427
14	.85686064	.76448770	.69764859	.64604768	.60449198
15	.89498019	.82195131	.76648826	.72212503	.68539248

M	B= 6	B= 7	B= 8	B= 9	B=10
1	.16979\04	.11675\04	.84874\05	.64339\05	.50387\05
2	.00017835	.00012744	.95615\04	.74438\04	.59646\04
3	.00096386	.00071389	.00055176	.00044048	.00036068
4	.00357475	.00273789	.00217573	.00177842	.00148631
5	.01024072	.00809143	.00659915	.00551483	.00469844
6	.02418458	.01966784	.01643271	.01401959	.01216133
7	.04907685	.04098644	.03501968	.03045750	.02686850
8	.08808364	.07537862	.06574778	.05821030	.05215971
9	.14285714	.12500000	.11111111	.10000000	.09090909
10	.21287501	.19005335	.17187140	.15702317	.14465448
11	.29537835	.26852417	.24664462	.22842877	.21299652
12	.38589966	.35650872	.33205791	.31133119	.29349164
13	.47915731	.44898514	.42339438	.40133405	.38206354
14	.57000985	.54074986	.51548511	.49336352	.47377097
15	.65421286	.62724443	.60357156	.58254236	.56367690

M	B=11	B=12	B=13	B=14	B=15
1	.40499\05	.33249\05	.27781\05	.23560\05	.20234\05
2	.48911\04	.40873\04	.34697\04	.29849\04	.25972\04
3	.00030143	.00025617	.00022077	.00019253	.00016962
4	.00126465	.00109205	.00095474	.00084350	.00075198
5	.00406600	.00356445	.00315887	.00282542	.00254736
6	.01069311	.00950829	.00853508	.00772360	.00703815
7	.02397925	.02160861	.01963216	.01796176	.01653336
8	.04720158	.04306866	.03957368	.03658162	.03399278
9	.08333333	.07692308	.07142857	.06666667	.06250000
10	.13418228	.12519446	.11739121	.11054904	.10449779
11	.19973342	.18819648	.17805770	.16906864	.16103745
12	.27794221	.26424421	.25206729	.24115738	.23131555
13	.36504353	.34987047	.33623550	.32389759	.31266550
14	.45625141	.44045761	.42611965	.41302391	.40099870
15	.54661201	.53106660	.51681884	.50369081	.49153774

DIRICHLET C INTEGRAL
A = 2 R = 9

M	B= 1	B= 2	B= 3	B= 4	B= 5
1	.02601229	.01034959	.00617289	.00433454	.00332083
2	.10404918	.05142178	.03409051	.02558823	.02055700
3	.23411065	.13791759	.10005127	.07959669	.06669197
4	.39307468	.26648810	.20851453	.17440593	.15160225
5	.55203870	.41791201	.34805375	.30372551	.27249680
6	.68980752	.56834293	.49776904	.44989708	.41454566
7	.79696105	.69942245	.63723302	.59249871	.55804028
8	.87349928	.80230174	.75318670	.71599313	.68624384
9	.92452477	.87643856	.84098247	.81290130	.78967704
10	.95665193	.92620099	.90245307	.88289900	.86624266
11	.97592823	.95766197	.94273106	.93001784	.91890460
12	.98702670	.97656304	.96766832	.95987419	.95290557
13	.99319252	.98743099	.98237185	.97782919	.97368770
14	.99651258	.99344686	.99068206	.98814791	.98579859
15	.99825165	.99666827	.99520884	.99384793	.99256823

M	B= 6	B= 7	B= 8	B= 9	B=10
1	.00268483	.00225097	.00193712	.00170004	.00151488
2	.01723430	.01487573	.01311385	.01174674	.01065432
3	.05775825	.05117880	.04611426	.04208444	.03879425
4	.13511878	.12255860	.11261683	.10451817	.09777093
5	.24901447	.23055073	.21555302	.20306520	.19246255
6	.38699031	.36469038	.34613811	.33037304	.31674960
7	.53029077	.50723355	.48762406	.47064375	.45572751
8	.66157075	.64057125	.62234742	.60629061	.59196978
9	.76989922	.75269556	.73748811	.72387382	.71156031
10	.85171880	.83883484	.82725370	.81673394	.80709638
11	.90900865	.90007427	.89192116	.88441687	.87746095
12	.94658433	.94078728	.93542506	.93043046	.92575146
13	.96986960	.96631935	.96299559	.95986652	.95690705
14	.98360205	.98153468	.97957850	.97771938	.97594595
15	.99135717	.99020526	.98910510	.98805076	.98703739

M	B=11	B=12	B=13	B=14	B=15
1	.00136641	.00124478	.00114337	.00105754	.00098399
2	.00976077	.00901584	.00838493	.00784343	.00737336
3	.03605202	.03372766	.03172968	.02999176	.02846462
4	.09204713	.08711899	.08282298	.07903846	.07567429
5	.18331764	.17532699	.16826845	.16197527	.15631950
6	.30481527	.29424191	.28478483	.27625722	.26851364
7	.44246995	.43057113	.41980338	.40999009	.40099175
8	.57906890	.56734957	.55662774	.54675844	.53762553
9	.70032867	.69001074	.68047455	.67161461	.66334530
10	.79820419	.78995041	.78224977	.77503315	.76824374
11	.87097526	.86489767	.85917792	.85377473	.84865375
12	.92134686	.91718340	.91323378	.90947528	.90588877
13	.95409695	.95141968	.94886143	.94641061	.94405734
14	.97424893	.97262058	.97105440	.96954484	.96808713
15	.98606095	.98511806	.98420582	.98332173	.98246363

DIRICHLET C INTEGRAL
A = 3 R = 9

M	B= 1	B= 2	B= 3	B= 4	B= 5
1	.07508469	.04114267	.02942541	.02340098	.01969049
2	.24402523	.15974014	.12472463	.10488910	.09186873
3	.45520091	.34144958	.28627957	.25224698	.22859051
4	.64877862	.53829227	.47728562	.43673322	.40706051
5	.79396190	.70754978	.65435994	.61661892	.58770983
6	.88833103	.83054247	.79158692	.76233155	.73899313
7	.94337969	.90908478	.88414561	.86447389	.84820793
8	.97287004	.95435125	.94000920	.92820865	.91813954
9	.98761522	.97835173	.97079450	.96434837	.95869552
10	.99457822	.99022785	.98652400	.98326687	.98034232
11	.99771157	.99577362	.99406530	.99252406	.99111197
12	.99906461	.99823905	.99749058	.99680080	.99615790
13	.99962837	.99928987	.99897599	.99868161	.99840329
14	.99985605	.99972174	.99959497	.99947436	.99935897
15	.99994549	.99989371	.99984414	.99979643	.99975034

M	B= 6	B= 7	B= 8	B= 9	B=10
1	.01715495	.01530114	.01387987	.01275124	.01183040
2	.08254708	.07547998	.06989981	.06535799	.06157336
3	.21091982	.19706854	.18582921	.17646897	.16851397
4	.38402514	.36540837	.34991737	.33673896	.32533151
5	.56447193	.54516058	.52871606	.51444899	.50188713
6	.71963626	.70313798	.68878908	.67611337	.66477586
7	.83433386	.82223553	.81150943	.80187638	.79313499
8	.90933477	.90149806	.89442864	.88798368	.88205765
9	.95364285	.94906293	.94486662	.94098882	.93738032
10	.97767763	.97522304	.97294265	.97080961	.96880320
11	.98980377	.98858158	.98743215	.98634534	.98531315
12	.99555375	.99498236	.99443921	.99392074	.99342412
13	.99813854	.99788551	.99764274	.99740908	.99718358
14	.99924808	.99914115	.99903775	.99893750	.99884013
15	.99970568	.99966229	.99962004	.99957883	.99953859

M	B=11	B=12	B=13	B=14	B=15
1	.01106279	.01041167	.00985132	.00936319	.00893355
2	.05836003	.05558973	.05317084	.05103600	.04913451
3	.16164244	.15562724	.15030281	.14554526	.14125981
4	.31531786	.30642587	.29845333	.29124630	.28468516
5	.49069366	.48062040	.47147954	.46312571	.45544427
6	.65453210	.64519835	.63663310	.62872509	.62138529
7	.78513511	.77776169	.77092464	.76455202	.75858552
8	.87657023	.87145878	.86667343	.86217381	.85792672
9	.93400295	.93082641	.92782622	.92498227	.92227783
10	.96690703	.96510787	.96339490	.96175909	.96019285
11	.98432915	.98338808	.98248555	.98161788	.98078192
12	.99294700	.99248746	.99204385	.99161480	.99119910
13	.99696545	.99675404	.99654878	.99634918	.99615481
14	.99874538	.99865305	.99856293	.99847489	.99838878
15	.99949922	.99946068	.99942289	.99938582	.99934942

DIRICHLET C INTEGRAL

A = 4 R = 9

M	B= 1	B= 2	B= 3	B= 4	B= 5
1	.13421773	.08602817	.06718412	.05673991	.04995380
2	.37580964	.28094728	.23690388	.21017124	.19172651
3	.61740155	.51593481	.46160805	.42590982	.39992050
4	.79456895	.71645529	.66937342	.63622308	.61090760
5	.90086939	.85243218	.82022147	.79614339	.77696173
6	.95614562	.93045031	.91191443	.89732575	.88526327
7	.98194119	.96983698	.96049978	.95281971	.94625982
8	.99299644	.98780427	.98357284	.97995969	.97678543
9	.99741854	.99535361	.99359386	.99204319	.99064754
10	.99908911	.99831744	.99763568	.99701895	.99645236
11	.99969051	.99941674	.99916779	.99893766	.99872256
12	.99989827	.99980531	.99971882	.99963746	.99956030
13	.99996752	.99993712	.99990832	.99988084	.99985447
14	.99998990	.99998027	.99997102	.99996209	.99995345
15	.99999693	.99999396	.99999108	.99998828	.99998555

M	B= 6	B= 7	B= 8	B= 9	B=10
1	.04512081	.04146663	.03858499	.03624053	.03428669
2	.17799825	.16725479	.15854179	.15128450	.14511331
3	.37978826	.36352888	.34999750	.33847886	.32849882
4	.59057753	.57368045	.55928134	.54677552	.53575045
5	.76105114	.74747901	.73566064	.72520539	.71583942
6	.87496503	.86597231	.85798658	.85080216	.84427112
7	.94051448	.93539135	.93076080	.92653091	.92263394
8	.97394223	.97135940	.96898771	.96679130	.96474314
9	.98937279	.98819569	.98709955	.98607192	.98510317
10	.99592605	.99543309	.99496835	.99452790	.99410866
11	.99851986	.99832765	.99814450	.99796926	.99780104
12	.99948671	.99941619	.99934838	.99928296	.99921968
13	.99982906	.99980450	.99978070	.99975757	.99973507
14	.99994505	.99993687	.99992889	.99992109	.99991345
15	.99998287	.99998025	.99997768	.99997515	.99997267

M	B=11	B=12	B=13	B=14	B=15
1	.03262698	.03119506	.02994368	.02883816	.02785243
2	.13977827	.13510344	.13096089	.12725504	.12391285
3	.31972836	.31193070	.30493022	.29859343	.29281670
4	.52591279	.51704685	.50898968	.50161529	.49482442
5	.70736342	.69962783	.69251763	.68594245	.67983005
6	.83828338	.83275476	.82761936	.82282463	.81832797
7	.91901843	.91564429	.91247960	.90949858	.90667999
8	.96282227	.96101206	.95929911	.95767239	.95612274
9	.98418574	.98331350	.98248144	.98168538	.98092183
10	.99370815	.99332434	.99295554	.99260032	.99225749
11	.99763908	.99748278	.99733162	.99718515	.99704299
12	.99915836	.99909881	.99904088	.99898445	.99892941
13	.99971312	.99969169	.99967075	.99965025	.99963016
14	.99990598	.99989864	.99989144	.99988436	.99987740
15	.99997022	.99996782	.99996545	.99996311	.99996080

DIRICHLET C INTEGRAL
A = 5 R = 9

M	B= 1	B= 2	B= 3	B= 4	B= 5
1	.19380670	.13626096	.11205249	.09803041	.08862867
2	.48451675	.39069016	.34421145	.31483947	.29397368
3	.72677512	.64365450	.59652728	.56440135	.54037382
4	.87482191	.82181532	.78821529	.76375770	.74461236
5	.94884530	.92161006	.90268259	.88810460	.87622579
6	.98092210	.96894537	.95996859	.95271241	.94658870
7	.99339642	.98871719	.98498664	.98184493	.97911167
8	.99785153	.99618587	.99478969	.99357263	.99248577
9	.99933656	.99878652	.99830650	.99787587	.99748266
10	.99980408	.99963330	.99947939	.99933801	.99920649
11	.99994433	.99989396	.99984740	.99980380	.99976261
12	.99998471	.99997048	.99995707	.99994431	.99993210
13	.99999592	.99999205	.99998834	.99998477	.99998132
14	.99999894	.99999792	.99999693	.99999597	.99999503
15	.99999973	.99999947	.99999921	.99999896	.99999872

M	B= 6	B= 7	B= 8	B= 9	B=10
1	.08176809	.07647738	.07223540	.06873457	.06578030
2	.27808585	.26541881	.25498324	.24617254	.23859066
3	.52136212	.50573739	.49254040	.48116146	.47119045
4	.72893192	.71568533	.70423937	.69417768	.68521221
5	.86619248	.85750354	.84983885	.84298112	.83677597
6	.94127322	.93656652	.93233645	.92849060	.92496160
7	.97668151	.97448676	.97248103	.97063094	.96891159
8	.99149899	.99059217	.98975106	.98896514	.98822637
9	.99711916	.99678001	.99646130	.99616007	.99587403
10	.99908302	.99896629	.99885535	.99874944	.99864796
11	.99972343	.99968597	.99965003	.99961541	.99958198
12	.99992037	.99990904	.99989808	.99988745	.99987711
13	.99997797	.99997472	.99997155	.99996845	.99996542
14	.99999412	.99999322	.99999234	.99999148	.99999063
15	.99999848	.99999824	.99999800	.99999777	.99999755

M	B=11	B=12	B=13	B=14	B=15
1	.06324272	.06103144	.05908134	.05734420	.05578341
2	.23196616	.22610570	.22086721	.21614350	.21185195
3	.46233907	.45439746	.44720836	.44065111	.43463123
4	.67713552	.66979342	.66306823	.65686815	.65112024
5	.83110964	.82589589	.82106772	.81657204	.81236606
6	.92169876	.91866289	.91582304	.91315432	.91063638
7	.96730383	.96579262	.96436586	.96301368	.96172794
8	.98752846	.98686637	.98623601	.98563398	.98505742
9	.99560132	.99534046	.99509020	.99484952	.99461753
10	.99855042	.99845643	.99836565	.99827779	.99819261
11	.99954962	.99951823	.99948773	.99945805	.99942913
12	.99986704	.99985722	.99984763	.99983825	.99982906
13	.99996246	.99995955	.99995670	.99995390	.99995114
14	.99998980	.99998898	.99998817	.99998738	.99998659
15	.99999732	.99999710	.99999688	.99999666	.99999645

DIRICHLET C INTEGRAL
1/A = 5 R = 10

M	B= 1	B= 2	B= 3	B= 4	B= 5
1	.16538\07	.31148\11	.10236\13	*	*
2	.15436\06	.48365\10	.20684\12	.33657\14	*
3	.78602\06	.39359\09	.21749\11	.43550\13	.14364\14
4	.28916\05	.22335\08	.15738\10	.37141\12	.18229\13
5	.85941\05	.99237\08	.88063\10	.24266\11	.13660\12
6	.21900\04	.36763\07	.40609\09	.12978\10	.81724\12
7	.49621\04	.11809\06	.16062\08	.59179\10	.41413\11
8	.00010242	.33784\06	.56008\08	.23653\09	.18320\10
9	.00019592	.87749\06	.17563\07	.84552\09	.72219\10
10	.00035176	.20994\05	.50282\07	.27451\08	.25769\09
11	.00059850	.46792\05	.13298\06	.81925\08	.84244\09
12	.00097236	.98026\05	.32797\06	.22692\07	.25481\08
13	.00151756	.19443\04	.76017\06	.58804\07	.71887\08
14	.00228643	.36734\04	.16664\05	.14350\06	.19043\07
15	.00333905	.66438\04	.34740\05	.33165\06	.47641\07

M	B= 6	B= 7	B= 8	B= 9	B=10
1	*	*	*	*	.46743\14
2	*	*	*	*	.46743\14
3	*	*	*	*	.46743\14
4	*	*	*	*	.46757\14
5	.11613\13	*	*	*	.46895\14
6	.81247\13	.98697\14	*	*	.47995\14
7	.45135\12	.66639\13	.10899\13	.16580\14	.55265\14
8	.21648\11	.34743\12	.68968\13	.16079\13	.96737\14
9	.92172\11	.15798\11	.33780\12	.85936\13	.30572\13
10	.35426\10	.64546\11	.14579\11	.39015\12	.12514\12
11	.12445\09	.24049\10	.57107\11	.15960\11	.51421\12
12	.40360\09	.82566\10	.20569\10	.59897\11	.19845\11
13	.12181\08	.26338\09	.68738\10	.20829\10	.71301\11
14	.34451\08	.78600\09	.21462\09	.67605\10	.23923\10
15	.91823\08	.22072\08	.62980\09	.20602\09	.75328\10

M	B=11	B=12	B=13	B=14	B=15
1	.25194\13	.80666\13	.21589\12	.52799\12	.12275\11
2	.25194\13	.80666\13	.21589\12	.52799\12	.12275\11
3	.25194\13	.80666\13	.21589\12	.52799\12	.12275\11
4	.25194\13	.80665\13	.21589\12	.52798\12	.12274\11
5	.25197\13	.80662\13	.21588\12	.52796\12	.12274\11
6	.25227\13	.80665\13	.21586\12	.52790\12	.12272\11
7	.25450\13	.80726\13	.21584\12	.52779\12	.12270\11
8	.26786\13	.81176\13	.21594\12	.52765\12	.12264\11
9	.33790\13	.83716\13	.21682\12	.52770\12	.12257\11
10	.66618\13	.96147\13	.22173\12	.52932\12	.12252\11
11	.20625\12	.15083\12	.24454\12	.53878\12	.12277\11
12	.75117\12	.37069\12	.33952\12	.58144\12	.12456\11
13	.27189\11	.11874\11	.70250\12	.75132\12	.13260\11
14	.93398\11	.40119\11	.19904\11	.13716\11	.16355\11
15	.30223\10	.13161\10	.62650\11	.34804\11	.27214\11

DIRICHLET C INTEGRAL

1/A = 4 R = 10

M	B= 1	B= 2	B= 3	B= 4	B= 5
1	.10240\06	.72041\10	.64512\12	.21050\13	.22903\14
2	.92160\06	.10526\08	.12145\10	.45681\12	.35847\13
3	.45261\05	.80623\08	.11815\09	.51855\11	.44482\12
4	.16060\04	.43070\07	.79095\09	.40258\10	.38361\11
5	.46050\04	.18021\06	.40958\08	.24016\09	.25354\10
6	.00011323	.62882\06	.17485\07	.11736\08	.13675\09
7	.00024758	.19032\05	.64046\07	.48918\08	.62671\09
8	.00049325	.51313\05	.20689\06	.17880\07	.25094\08
9	.00091089	.12565\04	.60128\06	.58475\07	.89595\08
10	.00157912	.28349\04	.15960\05	.17377\06	.28968\07
11	.00259483	.59604\04	.39149\05	.47487\06	.85851\07
12	.00407222	.00011783	.89591\05	.12050\05	.23553\06
13	.00614057	.00022063	.19276\04	.28622\05	.60300\06
14	.00894079	.00039362	.39244\04	.64054\05	.14504\05
15	.01262109	.00067252	.76013\04	.13582\04	.32964\05

M	B= 6	B= 7	B= 8	B= 9	B=10
1	.14900\14	.12504\14	*	*	*
2	.56307\14	.19641\14	*	*	*
3	.60739\13	.12148\13	.14601\14	*	*
4	.55824\12	.11050\12	.25764\13	*	*
5	.39848\11	.83371\12	.21434\12	.56542\13	*
6	.23187\10	.51515\11	.14005\11	.43732\12	.13125\12
7	.11436\09	.26952\10	.77008\11	.25482\11	.92727\12
8	.49158\09	.12267\09	.36761\10	.12699\10	.48948\11
9	.18795\08	.49576\09	.15557\09	.55917\10	.22388\10
10	.64922\08	.18069\08	.59295\09	.22141\09	.91695\10
11	.20509\07	.60124\08	.20606\08	.79845\09	.34151\09
12	.59842\07	.18448\07	.65945\08	.26489\08	.11689\08
13	.16259\06	.52625\07	.19596\07	.81511\08	.37080\08
14	.41419\06	.14052\06	.54441\07	.23428\07	.10977\07
15	.99494\06	.35330\06	.14224\06	.63262\07	.30506\07

M	B=11	B=12	B=13	B=14	B=15
1	*	*	*	*	*
2	*	*	*	*	*
3	*	*	*	*	*
4	*	*	*	*	*
5	*	*	*	*	*
6	*	*	*	*	*
7	.31953\12	.92163\14	*	*	*
8	.20197\11	.79586\12	.12488\12	*	*
9	.97485\11	.44692\11	.19823\11	.53261\12	*
10	.41296\10	.19864\10	.99507\11	.48720\11	.17458\11
11	.15836\09	.78483\10	.40999\10	.22131\10	.11753\10
12	.55727\09	.28333\09	.15197\09	.85082\10	.48922\10
13	.18158\08	.94568\09	.51881\09	.29736\09	.17654\09
14	.55173\08	.29412\08	.16481\08	.96375\09	.58433\09
15	.15727\07	.85762\08	.49051\08	.29227\08	.18041\08

DIRICHLET C INTEGRAL
1/A = 3 R = 10

M	B= 1	B= 2	B= 3	B= 4	B= 5
1	.95367\06	.30119\08	.81107\10	.60889\11	.82913\12
2	.81062\05	.40131\07	.13601\08	.11809\09	.17812\10
3	.37611\04	.28043\06	.11790\07	.11754\08	.19558\09
4	.00012612	.13675\05	.70375\07	.80001\08	.14623\08
5	.00034187	.52254\05	.32514\06	.41868\07	.83721\08
6	.00079495	.16662\04	.12393\05	.17963\06	.39142\07
7	.00164447	.46109\04	.40557\05	.65782\06	.15562\06
8	.00310078	.00011374	.11714\04	.21142\05	.54101\06
9	.00542178	.00025497	.30462\04	.60853\05	.16786\05
10	.00890328	.00052702	.72408\04	.15929\04	.47208\05
11	.01386442	.00101583	.00015919	.38378\04	.12182\04
12	.02062960	.00184236	.00032680	.85946\04	.29127\04
13	.02950891	.00316715	.00063131	.00018033	.65058\04
14	.04077881	.00519196	.00115509	.00035686	.00013667
15	.05466492	.00815767	.00201270	.00066986	.00027160

M	B= 6	B= 7	B= 8	B= 9	B=10
1	.16475\12	.38469\13	.10402\14	*	*
2	.38491\11	.10653\11	.34507\12	.10979\12	.16824\14
3	.45544\10	.13424\10	.46995\11	.18609\11	.78322\12
4	.36579\09	.11417\09	.41964\10	.17471\10	.79980\11
5	.22433\08	.73969\09	.28451\09	.12318\09	.58543\10
6	.11205\07	.38952\08	.15651\08	.70318\09	.34517\09
7	.47465\07	.17364\07	.72777\08	.33887\08	.17155\08
8	.17539\06	.67394\07	.29422\07	.14182\07	.73969\08
9	.57697\06	.23246\06	.10556\06	.52612\07	.28247\07
10	.17164\05	.72379\06	.34141\06	.17577\06	.97053\07
11	.46744\05	.20596\05	.10078\05	.53538\06	.30377\06
12	.11770\04	.54100\05	.27428\05	.15018\05	.87485\06
13	.27627\04	.13225\04	.69382\05	.39119\05	.23377\05
14	.60865\04	.30299\04	.16428\04	.95282\05	.58362\05
15	.00012659	.65432\04	.36621\04	.21829\04	.13694\04

M	B=11	B=12	B=13	B=14	B=15
1	*	*	*	*	*
2	*	*	*	*	*
3	.29126\12	*	*	*	*
4	.39016\11	.19098\11	.75740\12	*	*
5	.29953\10	.16195\10	.90078\11	.48537\11	.20713\11
6	.18199\09	.10170\09	.59523\10	.35971\10	.21943\10
7	.92951\09	.53265\09	.31978\09	.19950\09	.12827\09
8	.41141\08	.24133\08	.14803\08	.94297\09	.62022\09
9	.16115\07	.96683\08	.60520\08	.39278\08	.26295\08
10	.56749\07	.34801\07	.22216\07	.14678\07	.99888\08
11	.18192\06	.11396\06	.74150\07	.49844\07	.34461\07
12	.53621\06	.34290\06	.22731\06	.15539\06	.10909\06
13	.14654\05	.95611\06	.64534\06	.44844\06	.31958\06
14	.37392\05	.24876\05	.17088\05	.12065\05	.87239\06
15	.89612\05	.60754\05	.42450\05	.30440\05	.22325\05

DIRICHLET C INTEGRAL
1/A = 2 R = 10

M	B= 1	B= 2	B= 3	B= 4	B= 5
1	.16935\04	.30156\06	.26776\07	.49049\08	.13488\08
2	.00012984	.34322\05	.37084\06	.76796\07	.23013\07
3	.00054380	.20512\04	.26589\05	.61793\06	.20091\06
4	.00164772	.85663\04	.13148\04	.34056\05	.11962\05
5	.00403954	.00028073	.50402\04	.14457\04	.54640\05
6	.00850427	.00076886	.00015967	.50401\04	.20418\04
7	.01594549	.00183040	.00043510	.00015028	.65009\04
8	.02728449	.00389082	.00104840	.00039402	.00018137
9	.04334807	.00752937	.00227904	.00092717	.00045257
10	.06476617	.01346020	.00453803	.00198853	.00102594
11	.09189577	.02248377	.00837653	.00393495	.00213907
12	.12478014	.03541347	.01447166	.00725526	.00414310
13	.16314523	.05298815	.02358629	.01256609	.00751620
14	.20642893	.07578489	.03650542	.02058444	.01286035
15	.25383488	.10414577	.05395679	.03207660	.02087630

M	B= 6	B= 7	B= 8	B= 9	B=10
1	.47968\09	.20348\09	.98072\10	.52058\10	.29801\10
2	.87315\08	.38977\08	.19583\08	.10761\08	.63424\09
3	.81077\07	.38005\07	.19875\07	.11295\07	.68513\08
4	.51199\06	.25149\06	.13667\06	.80228\07	.50036\07
5	.24734\05	.12705\05	.71643\06	.43385\06	.27793\06
6	.97490\05	.52269\05	.30535\05	.19054\05	.12526\05
7	.32657\04	.18240\04	.11023\04	.70793\05	.47710\05
8	.95617\04	.55530\04	.34668\04	.22889\04	.15799\04
9	.00024979	.00015057	.96971\04	.65743\04	.46437\04
10	.00059145	.00036940	.00024508	.00017043	.00012308
11	.00128513	.00083024	.00056668	.00040380	.00029785
12	.00258840	.00172679	.00121100	.00088326	.00066491
13	.00487271	.00335148	.00241186	.00179873	.00138072
14	.00863385	.00611290	.00450854	.00343461	.00268606
15	.01448527	.01054107	.00795826	.00618665	.00492528

M	B=11	B=12	B=13	B=14	B=15
1	.18125\10	.11565\10	.76106\11	.49483\11	.27032\11
2	.39538\09	.25803\09	.17486\09	.12215\09	.87117\10
3	.43792\08	.29217\08	.20202\08	.14395\08	.10520\08
4	.32766\07	.22334\07	.15741\07	.11413\07	.84777\08
5	.18632\06	.12966\06	.93090\07	.68631\07	.51762\07
6	.85887\06	.60977\06	.44572\06	.33397\06	.25562\06
7	.33436\05	.24202\05	.18000\05	.13700\05	.10637\05
8	.11308\04	.83392\05	.63071\05	.48739\05	.38368\05
9	.33915\04	.25467\04	.19576\04	.15351\04	.12248\04
10	.91658\04	.70033\04	.54682\04	.43494\04	.35153\04
11	.00022602	.00017561	.00013920	.00011225	.91866\04
12	.00051372	.00040562	.00032623	.00026657	.00022082
13	.00108536	.00087036	.00070988	.00058749	.00049239
14	.00214674	.00174727	.00144441	.00121014	.00102575
15	.00399928	.00330185	.00276504	.00234408	.00200859

DIRICHLET C INTEGRAL
A = 1 R = 10

M	B= 1	B= 2	B= 3	B= 4	B= 5
1	.00097656	.00012650	.39434\04	.17727\04	.97096\05
2	.00585938	.00102392	.00036879	.00018133	.00010578
3	.01928711	.00437495	.00179535	.00095862	.00059289
4	.04614258	.01313763	.00606442	.00349230	.00228023
5	.08978271	.03115480	.01598621	.00986471	.00677177
6	.15087891	.06217149	.03507494	.02305093	.01657056
7	.22724915	.10866588	.06672193	.04642867	.03481853
8	.31452942	.17099700	.11319742	.08294346	.06465283
9	.40726471	.24718472	.17487978	.13422610	.10836475
10	.50000000	.33333333	.25000000	.20000000	.16666667
11	.58809853	.42447136	.33495673	.27798866	.23833778
12	.66818810	.51549406	.42504232	.36432535	.32036158
13	.73826647	.60194721	.51531455	.45429263	.40848102
14	.79756355	.68051160	.60136495	.54314466	.49796991
15	.84627187	.74916985	.67982908	.62679520	.58439037

M	B= 6	B= 7	B= 8	B= 9	B=10
1	.60114\05	.40431\05	.28859\05	.21540\05	.16644\05
2	.68679\04	.47969\04	.35317\04	.27057\04	.21381\04
3	.00040247	.00029126	.00022081	.00017341	.00013998
4	.00161358	.00120727	.00094086	.00075635	.00062304
5	.00498095	.00384475	.00307500	.00252713	.00212191
6	.01263372	.01003972	.00822715	.00690311	.00590172
7	.02744161	.02240498	.01878132	.01606825	.01397232
8	.05253447	.04397977	.03765299	.03280419	.02898205
9	.09054951	.07757353	.06772449	.06000777	.05380745
10	.14285714	.12500000	.11111111	.10000000	.09090909
11	.20904792	.18647185	.16850578	.15384789	.14164791
12	.28686002	.26037166	.23883511	.22093678	.20579706
13	.37255452	.34346860	.31934136	.29893970	.28141803
14	.46159375	.43149061	.40604891	.38418429	.36513610
15	.54940573	.51986114	.49445450	.47228732	.45271556

M	B=11	B=12	B=13	B=14	B=15
1	.13221\05	.10741\05	.88910\06	.74760\06	.63712\06
2	.17320\04	.14317\04	.12036\04	.10263\04	.88583\05
3	.00011553	.97094\04	.82842\04	.71591\04	.62549\04
4	.00052341	.00044687	.00038670	.00033849	.00029921
5	.00181283	.00157107	.00137797	.00122097	.00109137
6	.00512291	.00450316	.00400046	.00358604	.00323959
7	.01231171	.01096834	.00986250	.00893861	.00815684
8	.02589971	.02336675	.02125203	.01946255	.01793059
9	.04872243	.04448063	.04089123	.03781652	.03515473
10	.08333333	.07692308	.07142857	.06666667	.06250000
11	.13132609	.12247310	.11479141	.10805940	.10210847
12	.19280321	.18151430	.17160444	.16282715	.15499235
13	.26617503	.25277009	.24087223	.23022744	.22063726
14	.34835265	.33342262	.32003214	.30793698	.29694384
15	.43526322	.41956941	.40535462	.39239840	.38052408

DIRICHLET C INTEGRAL

A = 2 R = 10

M	B= 1	B= 2	B= 3	B= 4	B= 5
1	.01734153	.00645424	.00372413	.00256048	.00193224
2	.07514663	.03473976	.02226044	.01634605	.01292563
3	.18112265	.10013929	.07025170	.05467508	.04508243
4	.32242400	.20629853	.15636980	.12804030	.10956481
5	.47550047	.34216022	.27679544	.23678433	.20929406
6	.61837184	.48815185	.41670259	.36990963	.33620317
7	.73743131	.62516144	.55730284	.51011293	.47463633
8	.82814329	.74051392	.68296739	.64076597	.60779499
9	.89239761	.82934866	.78489980	.75074150	.72311773
10	.93523383	.89284787	.86110314	.83568522	.81448735
11	.96236343	.93545862	.91424672	.89664279	.88155519
12	.97880561	.96254592	.94915831	.93769701	.92763459
13	.98839689	.97897814	.97093519	.96386253	.95752069
14	.99380735	.98854868	.98391933	.97975400	.97594991
15	.99677022	.99392733	.99136063	.98900576	.98682077

M	B= 6	B= 7	B= 8	B= 9	B=10
1	.00154414	.00128257	.00109521	.00095483	.00084597
2	.01070463	.00914881	.00799908	.00711500	.00641403
3	.03855965	.03382313	.03021903	.02737901	.02507953
4	.09645608	.08661341	.07891635	.07270997	.06758458
5	.18901076	.17330133	.16069898	.15031593	.14158048
6	.31042441	.28987791	.27299941	.25881035	.24666289
7	.44659868	.42365337	.40438512	.38788032	.37351807
8	.58094724	.55843622	.53914466	.52232962	.50747341
9	.70000643	.68019382	.66289445	.64757115	.63384061
10	.79631342	.78041595	.76629551	.75360131	.74207733
11	.86833299	.85655390	.84592708	.83624310	.82734578
12	.91864199	.91049817	.90304650	.89617151	.88978530
13	.95175395	.94645423	.94154289	.93696062	.93266143
14	.97243757	.96916730	.96610202	.96321319	.96047826
15	.98477628	.98285070	.98102750	.97929374	.97763899

M	B=11	B=12	B=13	B=14	B=15
1	.00075922	.00068854	.00062988	.00058047	.00053828
2	.00584454	.00537262	.00497506	.00463549	.00434201
3	.02317696	.02157471	.02020539	.01902052	.01798431
4	.06327004	.05958059	.05638407	.05358373	.05110701
5	.13410625	.12762182	.12193040	.11688549	.11237556
6	.23610875	.22682653	.21857892	.21118638	.20451024
7	.36085894	.34958162	.33944470	.33026269	.32189027
8	.49420167	.48223547	.47136174	.46141428	.45226113
9	.62141991	.61009442	.59969755	.59009747	.58118813
10	.73153090	.72181328	.71280727	.70441881	.69657131
11	.81911539	.81145795	.80429833	.79757543	.79123890
12	.88381928	.87821879	.87293949	.86794487	.86320449
13	.92860880	.92477320	.92113034	.91765999	.91434509
14	.95787903	.95540058	.95303048	.95075825	.94857497
15	.97605472	.97453384	.97307037	.97165920	.97029596

DIRICHLET C INTEGRAL
A = 3 R = 10

M	B= 1	B= 2	B= 3	B= 4	B= 5
1	.05631351	.02961764	.02075819	.01629650	.01358587
2	.19709730	.12384555	.09469462	.07856132	.06813463
3	.39067501	.28224291	.23192607	.20164748	.18095157
4	.58425272	.46947427	.40880817	.36949542	.34123053
5	.74153460	.64425502	.58684235	.54712651	.51724275
6	.85163192	.78148931	.73604233	.70274712	.67665276
7	.92044275	.87576803	.84444928	.82032417	.80071772
8	.95976322	.93398204	.91467011	.89913067	.88608909
9	.98065222	.96692344	.95605304	.94697015	.93912845
10	.99109672	.98425780	.97858693	.97369282	.96936172
11	.99605786	.99283709	.99006250	.98760114	.98537573
12	.99831292	.99686688	.99558152	.99441447	.99333968
13	.99929951	.99867638	.99810818	.99758217	.99709008
14	.99971691	.99945776	.99921655	.99898962	.99877450
15	.99988835	.99978386	.99968499	.99959074	.99950041

M	B= 6	B= 7	B= 8	B= 9	B=10
1	.01175210	.01042186	.00940851	.00860812	.00795809
2	.06075518	.05521075	.05086501	.04734974	.04443605
3	.16568495	.15383592	.14429897	.13641063	.12974591
4	.31957631	.30225867	.28797293	.27590859	.26553139
5	.49354552	.47406553	.45762632	.44347278	.43109342
6	.65530376	.63730714	.62179845	.60820546	.59613037
7	.78421832	.76998781	.75748716	.74634895	.73631144
8	.87483330	.86492221	.85606244	.84804846	.84073038
9	.93220636	.92599665	.92035710	.91518543	.91040540
10	.96546157	.96190413	.95862699	.95558418	.95274070
11	.98333638	.98144860	.97968732	.97803362	.97647284
12	.99233963	.99140184	.99051699	.98967788	.98887884
13	.99662616	.99618618	.99576692	.99536582	.99498085
14	.99856941	.99837301	.99818424	.99800225	.99782636
15	.99941347	.99932951	.99924822	.99916933	.99909262

M	B=11	B=12	B=13	B=14	B=15
1	.00741840	.00696223	.00657091	.00623100	.00593260
2	.04197365	.03985946	.03802023	.03640239	.03496573
3	.12401853	.11902776	.11462818	.11071157	.10719551
4	.25647258	.24846807	.24132300	.23488989	.22905474
5	.42012688	.41030913	.40144181	.39337240	.38598132
6	.58528588	.57545795	.56648301	.55823336	.55060747
7	.72718154	.71881273	.71109118	.70392671	.69724667
8	.83399538	.82775633	.82194448	.81650459	.81139156
9	.90595850	.90179880	.89788949	.89420059	.89070736
10	.95006921	.94754787	.94515895	.94288782	.94072226
11	.97499330	.97358555	.97224178	.97095550	.96972121
12	.98811527	.98738340	.98668009	.98600266	.98534883
13	.99461032	.99425283	.99390719	.99357241	.99324759
14	.99765599	.99749065	.99732993	.99717347	.99702096
15	.99901790	.99894502	.99887383	.99880422	.99873609

DIRICHLET C INTEGRAL
A = 4 R = 10

M	B= 1	B= 2	B= 3	B= 4	B= 5
1	.10737418	.06684385	.05145568	.04305842	.03765814
2	.32212255	.23394937	.19435899	.17076246	.15467867
3	.55834575	.45489636	.40141637	.36696351	.34221657
4	.74732431	.66053355	.61002074	.57517799	.54894639
5	.87016037	.81184637	.77435525	.74690135	.72534646
6	.93894857	.90560197	.88229824	.86432191	.84967214
7	.97334267	.95648885	.94386185	.93367290	.92509179
8	.98906568	.98134232	.97521167	.97006962	.96561238
9	.99574797	.99247954	.98975843	.98739958	.98530302
10	.99842088	.99712593	.99600488	.99500562	.99409809
11	.99943659	.99895113	.99851732	.99812151	.99775536
12	.99980593	.99963229	.99947311	.99932504	.99918591
13	.99993521	.99987555	.99981973	.99976698	.99971677
14	.99997896	.99995916	.99994034	.99992232	.99990499
15	.99999334	.99998696	.99998083	.99997489	.99996913

M	B= 6	B= 7	B= 8	B= 9	B=10
1	.03384109	.03097208	.02872049	.02689604	.02538084
2	.14281504	.13359628	.12616294	.12000156	.11478413
3	.32323867	.30803299	.29546089	.28481765	.27563975
4	.52810599	.51093249	.49640096	.48385555	.47285275
5	.70766519	.69271682	.67979633	.66843814	.65831900
6	.83730447	.82660238	.81717065	.80874038	.80112017
7	.91765879	.91109042	.90519864	.89985183	.89495409
8	.96166239	.95810563	.95486399	.95188133	.94911584
9	.98340735	.98167161	.98006690	.97857193	.97717047
10	.99326300	.99248701	.99176043	.99107597	.99042795
11	.99741329	.99709130	.99678642	.99649638	.99621934
12	.99905424	.99892891	.99880908	.99869408	.99858339
13	.99966872	.99962256	.99957804	.99953500	.99949329
14	.99988825	.99987203	.99985628	.99984096	.99982602
15	.99996353	.99995806	.99995272	.99994750	.99994238

M	B=11	B=12	B=13	B=14	B=15
1	.02409763	.02299351	.02203092	.02118237	.02042725
2	.11029007	.10636487	.10289667	.09980221	.09701806
3	.26760770	.26049284	.25412647	.24838091	.24315746
4	.46307955	.45430729	.44636422	.43911821	.43246564
5	.64920558	.64092407	.63334162	.62635445	.61988002
6	.79416884	.78777928	.78186820	.77636961	.77123030
7	.89043322	.88623343	.88231068	.87862956	.87516114
8	.94653538	.94411474	.94183368	.93967571	.93762719
9	.97584986	.97459996	.97341255	.97228082	.97119909
10	.98981186	.98922404	.98866147	.98812163	.98760238
11	.99595386	.99569871	.99545289	.99521554	.99498595
12	.99847656	.99837323	.99827307	.99817584	.99808130
13	.99945278	.99941337	.99937498	.99933752	.99930093
14	.99981143	.99979717	.99978322	.99976954	.99975612
15	.99993736	.99993243	.99992759	.99992282	.99991813

DIRICHLET C INTEGRAL

A = 5 R = 10

M	B= 1	B= 2	B= 3	B= 4	B= 5
1	.16150558	.11102364	.09028933	.07843027	.07054503
2	.43068155	.33971398	.29590128	.26863518	.24946332
3	.67742619	.58902948	.54038972	.50779099	.48369055
4	.84192262	.78049242	.74270831	.71568990	.69479986
5	.93102485	.89682782	.87374468	.85628134	.84223096
6	.97260589	.95639777	.94457737	.93518802	.92736365
7	.98993133	.98313862	.97785716	.97348256	.96972316
8	.99653149	.99394899	.99183242	.99001562	.98841198
9	.99886905	.99796185	.99718563	.99649898	.99587876
10	.99964824	.99934970	.99908521	.99884527	.99862424
11	.99989498	.99980197	.99971723	.99963875	.99956523
12	.99996975	.99994209	.99991632	.99989204	.99986899
13	.99999156	.99998365	.99997616	.99996900	.99996212
14	.99999771	.99999553	.99999343	.99999140	.99998944
15	.99999939	.99999881	.99999824	.99999769	.99999715

M	B= 6	B= 7	B= 8	B= 9	B=10
1	.06482631	.06043735	.05693215	.05404889	.05162264
2	.23497588	.22349414	.21408113	.20616638	.19937950
3	.46478555	.44935459	.43639438	.42527257	.41556671
4	.67784979	.66363709	.65143182	.64075864	.63129124
5	.83047841	.82037998	.81152978	.80365525	.79656436
6	.92063837	.91473068	.90945664	.90468905	.90033612
7	.96641281	.96344677	.96075428	.95828504	.95600191
8	.98696951	.98565411	.98444201	.98331588	.98226263
9	.99531040	.99478400	.99429244	.99383040	.99339379
10	.99841842	.99822518	.99804260	.99786921	.99770385
11	.99949583	.99942990	.99936697	.99930667	.99924868
12	.99984697	.99982584	.99980549	.99978584	.99976681
13	.99995549	.99994907	.99994285	.99993679	.99993089
14	.99998753	.99998567	.99998385	.99998207	.99998033
15	.99999662	.99999610	.99999559	.99999509	.99999460

M	B=11	B=12	B=13	B=14	B=15
1	.04954375	.04773613	.04614514	.04473038	.04346130
2	.19346789	.18825238	.18360180	.17941751	.17562370
3	.40678158	.39930335	.39237252	.38606718	.38029221
4	.62279624	.61510116	.60807506	.60161624	.59564424
5	.79011673	.78420658	.77875224	.77368931	.76896612
6	.89632930	.89261598	.88915488	.88591296	.88286339
7	.95387658	.95188695	.95001541	.94824767	.94657195
8	.98127210	.98033623	.97944848	.97860349	.97779679
9	.99297937	.99258452	.99220711	.99184534	.99149772
10	.99754559	.99739367	.99724745	.99710639	.99697005
11	.99919278	.99913876	.99908644	.99903568	.99898635
12	.99974834	.99973039	.99971291	.99969587	.99967923
13	.99992514	.99991951	.99991401	.99990862	.99990333
14	.99997863	.99997695	.99997530	.99997369	.99997209
15	.99999411	.99999364	.99999317	.99999270	.99999224

TABLE B

This table gives the value of the incomplete Dirichlet integral of type 2

$$D_A^{(B)}(R,M) = \frac{\Gamma(M+BR)}{\Gamma^B(R)\Gamma(M)} \int_A^\infty \cdots \int_A^\infty \frac{\prod_{i=1}^{B} x_i^{R-1} dx_i}{\left(1 + \sum_{i=1}^{B} x_i\right)^{M+BR}}$$

as defined in ( ) for $R = 1(1)10$,

$B = 1(1)15$,

$M = 1(1)15$, and

$A = \frac{1}{5}, \frac{1}{4}, \frac{1}{3}, \frac{1}{2}, 1, 2, 3, 4, 5.$

Note 1: In this tables entries less than 10^{-4} are given in exponential notation to avoid too many zeros. Thus .85999\04 represents .000085999.

Note 2: The symbols M, B, R and A which are all capital here in the table correspond to the same lower case symbols in the text.

DIRICHLET D INTEGRAL
1/A = 5 R = 1

M	B= 1	B= 2	B= 3	B= 4	B= 5
1	.83333333	.71428571	.62500000	.55555556	.50000000
2	.69444444	.51020408	.39062500	.30864198	.25000000
3	.57870370	.36443149	.24414063	.17146776	.12500000
4	.48225309	.26030820	.15258789	.09525987	.06250000
5	.40187757	.18593443	.09536743	.05292215	.03125000
6	.33489798	.13281031	.05960464	.02940119	.01562500
7	.27908165	.09486451	.03725290	.01633400	.00781250
8	.23256804	.06776036	.02328306	.00907444	.00390625
9	.19380670	.04840026	.01455192	.00504136	.00195313
10	.16150558	.03457161	.00909495	.00280075	.00097656
11	.13458799	.02469401	.00568434	.00155597	.00048828
12	.11215665	.01763858	.00355271	.00086443	.00024414
13	.09346388	.01259898	.00222045	.00048024	.00012207
14	.07788657	.00899927	.00138778	.00026680	.61035\04
15	.06490547	.00642805	.00086736	.00014822	.30518\04

M	B= 6	B= 7	B= 8	B= 9	B=10
1	.45454545	.41666667	.38461538	.35714286	.33333333
2	.20661157	.17361111	.14792899	.12755102	.11111111
3	.09391435	.07233796	.05689577	.04555394	.03703704
4	.04268834	.03014082	.02188299	.01626926	.01234568
5	.01940379	.01255867	.00841653	.00581045	.00411523
6	.00881991	.00523278	.00323713	.00207516	.00137174
7	.00400905	.00218033	.00124505	.00074113	.00045725
8	.00182229	.00090847	.00047887	.00026469	.00015242
9	.00082832	.00037853	.00018418	.94532\04	.50805\04
10	.00037651	.00015772	.70838\04	.33761\04	.16935\04
11	.00017114	.65717\04	.27245\04	.12058\04	.56450\05
12	.77791\04	.27382\04	.10479\04	.43063\05	.18817\05
13	.35359\04	.11409\04	.40304\05	.15380\05	.62723\06
14	.16072\04	.47538\05	.15501\05	.54927\06	.20908\06
15	.73057\05	.19808\05	.59621\06	.19617\06	.69692\07

M	B=11	B=12	B=13	B=14	B=15
1	.31250000	.29411765	.27777778	.26315789	.25000000
2	.09765625	.08650519	.07716049	.06925208	.06250000
3	.03051758	.02544270	.02143347	.01822423	.01562500
4	.00953674	.00748315	.00595374	.00479585	.00390625
5	.00298023	.00220093	.00165382	.00126207	.00097656
6	.00093132	.00064733	.00045939	.00033212	.00024414
7	.00029104	.00019039	.00012761	.87401\04	.61035\04
8	.90949\04	.55998\04	.35447\04	.23000\04	.15259\04
9	.28422\04	.16470\04	.98464\05	.60527\05	.38147\05
10	.88818\05	.48441\05	.27351\05	.15928\05	.95367\06
11	.27756\05	.14247\05	.75975\06	.41916\06	.23842\06
12	.86736\06	.41904\06	.21104\06	.11031\06	.59605\07
13	.27105\06	.12325\06	.58623\07	.29028\07	.14901\07
14	.84703\07	.36249\07	.16284\07	.76389\08	.37253\08
15	.26470\07	.10661\07	.45234\08	.20102\08	.93132\09

DIRICHLET D INTEGRAL
1/A = 4 R = 1

M	B= 1	B= 2	B= 3	B= 4	B= 5
1	.80000000	.66666667	.57142857	.50000000	.44444444
2	.64000000	.44444444	.32653061	.25000000	.19753086
3	.51200000	.29629630	.18658892	.12500000	.08779150
4	.40960000	.19753086	.10662224	.06250000	.03901844
5	.32768000	.13168724	.06092699	.03125000	.01734153
6	.26214400	.08779150	.03481543	.01562500	.00770735
7	.20971520	.05852766	.01989453	.00781250	.00342549
8	.16777216	.03901844	.01136830	.00390625	.00152244
9	.13421773	.02601229	.00649617	.00195313	.00067664
10	.10737418	.01734153	.00371210	.00097656	.00030073
11	.08589935	.01156102	.00212120	.00048828	.00013366
12	.06871948	.00770735	.00121211	.00024414	.59403\04
13	.05497558	.00513823	.00069264	.00012207	.26401\04
14	.04398047	.00342549	.00039579	.61035\04	.11734\04
15	.03518437	.00228366	.00022617	.30518\04	.52151\05

M	B= 6	B= 7	B= 8	B= 9	B=10
1	.40000000	.36363636	.33333333	.30769231	.28571429
2	.16000000	.13223140	.11111111	.09467456	.08163265
3	.06400000	.04808415	.03703704	.02913063	.02332362
4	.02560000	.01748514	.01234568	.00896327	.00666389
5	.01024000	.00635823	.00411523	.00275793	.00190397
6	.00409600	.00231209	.00137174	.00084859	.00054399
7	.00163840	.00084076	.00045725	.00026111	.00015543
8	.00065536	.00030573	.00015242	.80340\04	.44407\04
9	.00026214	.00011117	.50805\04	.24720\04	.12688\04
10	.00010486	.40427\04	.16935\04	.76062\05	.36251\05
11	.41943\04	.14701\04	.56450\05	.23404\05	.10357\05
12	.16777\04	.53457\05	.18817\05	.72011\06	.29593\06
13	.67109\05	.19439\05	.62723\06	.22157\06	.84550\07
14	.26844\05	.70687\06	.20908\06	.68176\07	.24157\07
15	.10737\05	.25705\06	.69692\07	.20977\07	.69021\08

M	B=11	B=12	B=13	B=14	B=15
1	.26666667	.25000000	.23529412	.22222222	.21052632
2	.07111111	.06250000	.05536332	.04938272	.04432133
3	.01896296	.01562500	.01302666	.01097394	.00933081
4	.00505679	.00390625	.00306510	.00243865	.00196438
5	.00134848	.00097656	.00072120	.00054192	.00041355
6	.00035959	.00024414	.00016969	.00012043	.87064\04
7	.95892\04	.61035\04	.39928\04	.26762\04	.18329\04
8	.25571\04	.15259\04	.93948\05	.59470\05	.38588\05
9	.68190\05	.38147\05	.22105\05	.13216\05	.81238\06
10	.18184\05	.95367\06	.52013\06	.29368\06	.17103\06
11	.48490\06	.23842\06	.12238\06	.65262\07	.36006\07
12	.12931\06	.59605\07	.28796\07	.14503\07	.75801\08
13	.34482\07	.14901\07	.67755\08	.32228\08	.15958\08
14	.91952\08	.37253\08	.15942\08	.71618\09	.33596\09
15	.24521\08	.93132\09	.37512\09	.15915\09	.70729\10

DIRICHLET D INTEGRAL
1/A = 3 R = 1

M	B= 1	B= 2	B= 3	B= 4	B= 5
1	.75000000	.60000000	.50000000	.42857143	.37500000
2	.56250000	.36000000	.25000000	.18367347	.14062500
3	.42187500	.21600000	.12500000	.07871720	.05273438
4	.31640625	.12960000	.06250000	.03373594	.01977539
5	.23730469	.07776000	.03125000	.01445826	.00741577
6	.17797852	.04665600	.01562500	.00619640	.00278091
7	.13348389	.02799360	.00781250	.00265560	.00104284
8	.10011292	.01679616	.00390625	.00113811	.00039107
9	.07508469	.01007770	.00195313	.00048776	.00014665
10	.05631351	.00604662	.00097656	.00020904	.54994\04
11	.04223514	.00362797	.00048828	.89589\04	.20623\04
12	.03167635	.00217678	.00024414	.38395\04	.77335\05
13	.02375726	.00130607	.00012207	.16455\04	.29001\05
14	.01781795	.00078364	.61035\04	.70522\05	.10875\05
15	.01336346	.00047018	.30518\04	.30224\05	.40782\06

M	B= 6	B= 7	B= 8	B= 9	B=10
1	.33333333	.30000000	.27272727	.25000000	.23076923
2	.11111111	.09000000	.07438017	.06250000	.05325444
3	.03703704	.02700000	.02028550	.01562500	.01228949
4	.01234568	.00810000	.00553241	.00390625	.00283604
5	.00411523	.00243000	.00150884	.00097656	.00065447
6	.00137174	.00072900	.00041150	.00024414	.00015103
7	.00045725	.00021870	.00011223	.61035\04	.34853\04
8	.00015242	.65610\04	.30608\04	.15259\04	.80431\05
9	.50805\04	.19683\04	.83475\05	.38147\05	.18561\05
10	.16935\04	.59049\05	.22766\05	.95367\06	.42833\06
11	.56450\05	.17715\05	.62089\06	.23842\06	.98846\07
12	.18817\05	.53144\06	.16933\06	.59605\07	.22811\07
13	.62723\06	.15943\06	.46182\07	.14901\07	.52640\08
14	.20908\06	.47830\07	.12595\07	.37253\08	.12148\08
15	.69692\07	.14349\07	.34350\08	.93132\09	.28033\09

M	B=11	B=12	B=13	B=14	B=15
1	.21428571	.20000000	.18750000	.17647059	.16666667
2	.04591837	.04000000	.03515625	.03114187	.02777778
3	.00983965	.00800000	.00659180	.00549562	.00462963
4	.00210850	.00160000	.00123596	.00096982	.00077160
5	.00045182	.00032000	.00023174	.00017114	.00012860
6	.96819\04	.64000\04	.43452\04	.30202\04	.21433\04
7	.20747\04	.12800\04	.81472\05	.53297\05	.35722\05
8	.44458\05	.25600\05	.15276\05	.94054\06	.59537\06
9	.95266\06	.51200\06	.28643\06	.16598\06	.99229\07
10	.20414\06	.10240\06	.53705\07	.29290\07	.16538\07
11	.43745\07	.20480\07	.10070\07	.51689\08	.27564\08
12	.93739\08	.40960\08	.18881\08	.91215\09	.45939\09
13	.20087\08	.81920\09	.35401\09	.16097\09	.76566\10
14	.43043\09	.16384\09	.66377\10	.28406\10	.12761\10
15	.92236\10	.32768\10	.12446\10	.50129\11	.21268\11

DIRICHLET D INTEGRAL
1/A = 2 R = 1

M	B= 1	B= 2	B= 3	B= 4	B= 5
1	.66666667	.50000000	.40000000	.33333333	.28571429
2	.44444444	.25000000	.16000000	.11111111	.08163265
3	.29629630	.12500000	.06400000	.03703704	.02332362
4	.19753086	.06250000	.02560000	.01234568	.00666389
5	.13168724	.03125000	.01024000	.00411523	.00190397
6	.08779150	.01562500	.00409600	.00137174	.00054399
7	.05852766	.00781250	.00163840	.00045725	.00015543
8	.03901844	.00390625	.00065536	.00015242	.44407\04
9	.02601229	.00195313	.00026214	.50805\04	.12688\04
10	.01734153	.00097656	.00010486	.16935\04	.36251\05
11	.01156102	.00048828	.41943\04	.56450\05	.10357\05
12	.00770735	.00024414	.16777\04	.18817\05	.29593\06
13	.00513823	.00012207	.67109\05	.62723\06	.84550\07
14	.00342549	.61035\04	.26844\05	.20908\06	.24157\07
15	.00228366	.30518\04	.10737\05	.69692\07	.69021\08

M	B= 6	B= 7	B= 8	B= 9	B=10
1	.25000000	.22222222	.20000000	.18181818	.16666667
2	.06250000	.04938272	.04000000	.03305785	.02777778
3	.01562500	.01097394	.00800000	.00601052	.00462963
4	.00390625	.00243865	.00160000	.00109282	.00077160
5	.00097656	.00054192	.00032000	.00019869	.00012860
6	.00024414	.00012043	.64000\04	.36126\04	.21433\04
7	.61035\04	.26762\04	.12800\04	.65684\05	.35722\05
8	.15259\04	.59470\05	.25600\05	.11943\05	.59537\06
9	.38147\05	.13216\05	.51200\06	.21714\06	.99229\07
10	.95367\06	.29368\06	.10240\06	.39480\07	.16538\07
11	.23842\06	.65262\07	.20480\07	.71781\08	.27564\08
12	.59605\07	.14503\07	.40960\08	.13051\08	.45939\09
13	.14901\07	.32228\08	.81920\09	.23729\09	.76566\10
14	.37253\08	.71618\09	.16384\09	.43144\10	.12761\10
15	.93132\09	.15915\09	.32768\10	.78444\11	.21268\11

M	B=11	B=12	B=13	B=14	B=15
1	.15384615	.14285714	.13333333	.12500000	.11764706
2	.02366864	.02040816	.01777778	.01562500	.01384083
3	.00364133	.00291545	.00237037	.00195313	.00162833
4	.00056020	.00041649	.00031605	.00024414	.00019157
5	.86185\04	.59499\04	.42140\04	.30518\04	.22537\04
6	.13259\04	.84999\05	.56187\05	.38147\05	.26515\05
7	.20399\05	.12143\05	.74915\06	.47684\06	.31194\06
8	.31383\06	.17347\06	.99887\07	.59605\07	.36699\07
9	.48281\07	.24781\07	.13318\07	.74506\08	.43175\08
10	.74279\08	.35401\08	.17758\08	.93132\09	.50794\09
11	.11428\08	.50573\09	.23677\09	.11642\09	.59757\10
12	.17581\09	.72248\10	.31569\10	.14552\10	.70303\11
13	.27047\10	.10321\10	.42092\11	.18190\11	.82709\12
14	.41611\11	.14744\11	.56123\12	.22737\12	.97305\13
15	.64018\12	.21063\12	.74831\13	.28422\13	.11448\13

DIRICHLET D INTEGRAL
A = 1 R = 1

M	B= 1	B= 2	B= 3	B= 4	B= 5
1	.50000000	.33333333	.25000000	.20000000	.16666667
2	.25000000	.11111111	.06250000	.04000000	.02777778
3	.12500000	.03703704	.01562500	.00800000	.00462963
4	.06250000	.01234568	.00390625	.00160000	.00077160
5	.03125000	.00411523	.00097656	.00032000	.00012860
6	.01562500	.00137174	.00024414	.64000\04	.21433\04
7	.00781250	.00045725	.61035\04	.12800\04	.35722\05
8	.00390625	.00015242	.15259\04	.25600\05	.59537\06
9	.00195313	.50805\04	.38147\05	.51200\06	.99229\07
10	.00097656	.16935\04	.95367\06	.10240\06	.16538\07
11	.00048828	.56450\05	.23842\06	.20480\07	.27564\08
12	.00024414	.18817\05	.59605\07	.40960\08	.45939\09
13	.00012207	.62723\06	.14901\07	.81920\09	.76566\10
14	.61035\04	.20908\06	.37253\08	.16384\09	.12761\10
15	.30518\04	.69692\07	.93132\09	.32768\10	.21268\11

M	B= 6	B= 7	B= 8	B= 9	B=10
1	.14285714	.12500000	.11111111	.10000000	.09090909
2	.02040816	.01562500	.01234568	.01000000	.00826446
3	.00291545	.00195313	.00137174	.00100000	.00075131
4	.00041649	.00024414	.00015242	.00010000	.68301\04
5	.59499\04	.30518\04	.16935\04	.10000\04	.62092\05
6	.84999\05	.38147\05	.18817\05	.10000\05	.56447\06
7	.12143\05	.47684\06	.20908\06	.10000\06	.51316\07
8	.17347\06	.59605\07	.23231\07	.10000\07	.46651\08
9	.24781\07	.74506\08	.25812\08	.10000\08	.42410\09
10	.35401\08	.93132\09	.28680\09	.10000\09	.38554\10
11	.50573\09	.11642\09	.31866\10	.10000\10	.35049\11
12	.72248\10	.14552\10	.35407\11	.10000\11	.31863\12
13	.10321\10	.18190\11	.39341\12	.10000\12	.28966\13
14	.14744\11	.22737\12	.43712\13	.10000\13	.26333\14
15	.21063\12	.28422\13	.48569\14	.10000\14	*

M	B=11	B=12	B=13	B=14	B=15
1	.08333333	.07692308	.07142857	.06666667	.06250000
2	.00694444	.00591716	.00510204	.00444444	.00390625
3	.00057870	.00045517	.00036443	.00029630	.00024414
4	.48225\04	.35013\04	.26031\04	.19753\04	.15259\04
5	.40188\05	.26933\05	.18593\05	.13169\05	.95367\06
6	.33490\06	.20718\06	.13281\06	.87791\07	.59605\07
7	.27908\07	.15937\07	.94865\08	.58528\08	.37253\08
8	.23257\08	.12259\08	.67760\09	.39018\09	.23283\09
9	.19381\09	.94300\10	.48400\10	.26012\10	.14552\10
10	.16151\10	.72538\11	.34572\11	.17342\11	.90949\12
11	.13459\11	.55799\12	.24694\12	.11561\12	.56843\13
12	.11216\12	.42922\13	.17639\13	.77073\14	.35527\14
13	.93464\14	.33017\14	.12599\14	*	*
14	*	*	*	*	*
15	*	*	*	*	*

DIRICHLET D INTEGRAL
A = 2 R = 1

M	B= 1	B= 2	B= 3	B= 4	B= 5
1	.33333333	.20000000	.14285714	.11111111	.09090909
2	.11111111	.04000000	.02040816	.01234568	.00826446
3	.03703704	.00800000	.00291545	.00137174	.00075131
4	.01234568	.00160000	.00041649	.00015242	.68301\04
5	.00411523	.00032000	.59499\04	.16935\04	.62092\05
6	.00137174	.64000\04	.84999\05	.18817\05	.56447\06
7	.00045725	.12800\04	.12143\05	.20908\06	.51316\07
8	.00015242	.25600\05	.17347\06	.23231\07	.46651\08
9	.50805\04	.51200\06	.24781\07	.25812\08	.42410\09
10	.16935\04	.10240\06	.35401\08	.28680\09	.38554\10
11	.56450\05	.20480\07	.50573\09	.31866\10	.35049\11
12	.18817\05	.40960\08	.72248\10	.35407\11	.31863\12
13	.62723\06	.81920\09	.10321\10	.39341\12	.28966\13
14	.20908\06	.16384\09	.14744\11	.43712\13	.26333\14
15	.69692\07	.32768\10	.21063\12	.48569\14	*

M	B= 6	B= 7	B= 8	B= 9	B=10
1	.07692308	.06666667	.05882353	.05263158	.04761905
2	.00591716	.00444444	.00346021	.00277008	.00226757
3	.00045517	.00029630	.00020354	.00014579	.00010798
4	.35013\04	.19753\04	.11973\04	.76734\05	.51419\05
5	.26933\05	.13169\05	.70430\06	.40386\06	.24485\06
6	.20718\06	.87791\07	.41429\07	.21256\07	.11660\07
7	.15937\07	.58528\08	.24370\08	.11187\08	.55522\09
8	.12259\08	.39018\09	.14335\09	.58880\10	.26439\10
9	.94300\10	.26012\10	.84326\11	.30990\11	.12590\11
10	.72538\11	.17342\11	.49603\12	.16310\12	.59952\13
11	.55799\12	.11561\12	.29178\13	.85844\14	.28549\14
12	.42922\13	.77073\14	.17164\14	*	*
13	.33017\14	*	*	*	*
14	*	*	*	*	*
15	*	*	*	*	*

M	B=11	B=12	B=13	B=14	B=15
1	.04347826	.04000000	.03703704	.03448276	.03225806
2	.00189036	.00160000	.00137174	.00118906	.00104058
3	.82190\04	.64000\04	.50805\04	.41002\04	.33567\04
4	.35735\05	.25600\05	.18817\05	.14139\05	.10828\05
5	.15537\06	.10240\06	.69692\07	.48754\07	.34929\07
6	.67551\08	.40960\08	.25812\08	.16812\08	.11268\08
7	.29370\09	.16384\09	.95599\10	.57971\10	.36347\10
8	.12770\10	.65536\11	.35407\11	.19990\11	.11725\11
9	.55520\12	.26214\12	.13114\12	.68932\13	.37822\13
10	.24139\13	.10486\13	.48569\14	.23769\14	.12201\14
11	.10495\14	*	*	*	*
12	*	*	*	*	*
13	*	*	*	*	*
14	*	*	*	*	*
15	*	*	*	*	*

DIRICHLET D INTEGRAL
A = 3 R = 1

M	B= 1	B= 2	B= 3	B= 4	B= 5
1	.25000000	.14285714	.10000000	.07692308	.06250000
2	.06250000	.02040816	.01000000	.00591716	.00390625
3	.01562500	.00291545	.00100000	.00045517	.00024414
4	.00390625	.00041649	.00010000	.35013\04	.15259\04
5	.00097656	.59499\04	.10000\04	.26933\05	.95367\06
6	.00024414	.84999\05	.10000\05	.20718\06	.59605\07
7	.61035\04	.12143\05	.10000\06	.15937\07	.37253\08
8	.15259\04	.17347\06	.10000\07	.12259\08	.23283\09
9	.38147\05	.24781\07	.10000\08	.94300\10	.14552\10
10	.95367\06	.35401\08	.10000\09	.72538\11	.90949\12
11	.23842\06	.50573\09	.10000\10	.55799\12	.56843\13
12	.59605\07	.72248\10	.10000\11	.42922\13	.35527\14
13	.14901\07	.10321\10	.10000\12	.33017\14	*
14	.37253\08	.14744\11	.10000\13	*	*
15	.93132\09	.21063\12	.10000\14	*	*

M	B= 6	B= 7	B= 8	B= 9	B=10
1	.05263158	.04545455	.04000000	.03571429	.03225806
2	.00277008	.00206612	.00160000	.00127551	.00104058
3	.00014579	.93914\04	.64000\04	.45554\04	.33567\04
4	.76734\05	.42688\05	.25600\05	.16269\05	.10828\05
5	.40386\06	.19404\06	.10240\06	.58105\07	.34929\07
6	.21256\07	.88199\08	.40960\08	.20752\08	.11268\08
7	.11187\08	.40090\09	.16384\09	.74113\10	.36347\10
8	.58880\10	.18223\10	.65536\11	.26469\11	.11725\11
9	.30990\11	.82832\12	.26214\12	.94532\13	.37822\13
10	.16310\12	.37651\13	.10486\13	.33761\14	.12201\14
11	.85844\14	.17114\14	*	*	*
12	*	*	*	*	*
13	*	*	*	*	*
14	*	*	*	*	*
15	*	*	*	*	*

M	B=11	B=12	B=13	B=14	B=15
1	.02941176	.02702703	.02500000	.02325581	.02173913
2	.00086505	.00073046	.00062500	.00054083	.00047259
3	.25443\04	.19742\04	.15625\04	.12578\04	.10274\04
4	.74831\06	.53357\06	.39063\06	.29250\06	.22334\06
5	.22009\07	.14421\07	.97656\08	.68023\08	.48552\08
6	.64733\09	.38975\09	.24414\09	.15819\09	.10555\09
7	.19039\10	.10534\10	.61035\11	.36789\11	.22945\11
8	.55998\12	.28470\12	.15259\12	.85556\13	.49881\13
9	.16470\13	.76946\14	.38147\14	.19897\14	.10844\14
10	*	*	*	*	*
11	*	*	*	*	*
12	*	*	*	*	*
13	*	*	*	*	*
14	*	*	*	*	*
15	*	*	*	*	*

DIRICHLET D INTEGRAL
A = 4 R = 1

M	B= 1	B= 2	B= 3	B= 4	B= 5
1	.20000000	.11111111	.07692308	.05882353	.04761905
2	.04000000	.01234568	.00591716	.00346021	.00226757
3	.00800000	.00137174	.00045517	.00020354	.00010798
4	.00160000	.00015242	.35013\04	.11973\04	.51419\05
5	.00032000	.16935\04	.26933\05	.70430\06	.24485\06
6	.64000\04	.18817\05	.20718\06	.41429\07	.11660\07
7	.12800\04	.20908\06	.15937\07	.24370\08	.55522\09
8	.25600\05	.23231\07	.12259\08	.14335\09	.26439\10
9	.51200\06	.25812\08	.94300\10	.84326\11	.12590\11
10	.10240\06	.28680\09	.72538\11	.49603\12	.59952\13
11	.20480\07	.31866\10	.55799\12	.29178\13	.28549\14
12	.40960\08	.35407\11	.42922\13	.17164\14	*
13	.81920\09	.39341\12	.33017\14	*	*
14	.16384\09	.43712\13	*	*	*
15	.32768\10	.48569\14	*	*	*

M	B= 6	B= 7	B= 8	B= 9	B=10
1	.04000000	.03448276	.03030303	.02702703	.02439024
2	.00160000	.00118906	.00091827	.00073046	.00059488
3	.64000\04	.41002\04	.27826\04	.19742\04	.14509\04
4	.25600\05	.14139\05	.84323\06	.53357\06	.35389\06
5	.10240\06	.48754\07	.25552\07	.14421\07	.86314\08
6	.40960\08	.16812\08	.77431\09	.38975\09	.21052\09
7	.16384\09	.57971\10	.23464\10	.10534\10	.51347\11
8	.65536\11	.19990\11	.71103\12	.28470\12	.12524\12
9	.26214\12	.68932\13	.21546\13	.76946\14	.30545\14
10	.10486\13	.23769\14	*	*	*
11	*	*	*	*	*
12	*	*	*	*	*
13	*	*	*	*	*
14	*	*	*	*	*
15	*	*	*	*	*

M	B=11	B=12	B=13	B=14	B=15
1	.02222222	.02040816	.01886792	.01754386	.01639344
2	.00049383	.00041649	.00035600	.00030779	.00026874
3	.10974\04	.84999\05	.67170\05	.53998\05	.44057\05
4	.24387\06	.17347\06	.12673\06	.94733\07	.72224\07
5	.54192\08	.35401\08	.23912\08	.16620\08	.11840\08
6	.12043\09	.72248\10	.45117\10	.29158\10	.19410\10
7	.26762\11	.14744\11	.85127\12	.51154\12	.31819\12
8	.59470\13	.30091\13	.16062\13	.89743\14	.52163\14
9	.13216\14	*	*	*	*
10	*	*	*	*	*
11	*	*	*	*	*
12	*	*	*	*	*
13	*	*	*	*	*
14	*	*	*	*	*
15	*	*	*	*	*

DIRICHLET D INTEGRAL
A = 5 R = 1

M	B= 1	B= 2	B= 3	B= 4	B= 5
1	.16666667	.09090909	.06250000	.04761905	.03846154
2	.02777778	.00826446	.00390625	.00226757	.00147929
3	.00462963	.00075131	.00024414	.00010798	.56896\04
4	.00077160	.68301\04	.15259\04	.51419\05	.21883\05
5	.00012860	.62092\05	.95367\06	.24485\06	.84165\07
6	.21433\04	.56447\06	.59605\07	.11660\07	.32371\08
7	.35722\05	.51316\07	.37253\08	.55522\09	.12450\09
8	.59537\06	.46651\08	.23283\09	.26439\10	.47887\11
9	.99229\07	.42410\09	.14552\10	.12590\11	.18418\12
10	.16538\07	.38554\10	.90949\12	.59952\13	.70838\14
11	.27564\08	.35049\11	.56843\13	.28549\14	*
12	.45939\09	.31863\12	.35527\14	*	*
13	.76566\10	.28966\13	*	*	*
14	.12761\10	.26333\14	*	*	*
15	.21268\11	*	*	*	*

M	B= 6	B= 7	B= 8	B= 9	B=10
1	.03225806	.02777778	.02439024	.02173913	.01960784
2	.00104058	.00077160	.00059488	.00047259	.00038447
3	.33567\04	.21433\04	.14509\04	.10274\04	.75386\05
4	.10828\05	.59537\06	.35389\06	.22334\06	.14782\06
5	.34929\07	.16538\07	.86314\08	.48552\08	.28983\08
6	.11268\08	.45939\09	.21052\09	.10555\09	.56830\10
7	.36347\10	.12761\10	.51347\11	.22945\11	.11143\11
8	.11725\11	.35447\12	.12524\12	.49881\13	.21849\13
9	.37822\13	.98464\14	.30545\14	.10844\14	*
10	.12201\14	*	*	*	*
11	*	*	*	*	*
12	*	*	*	*	*
13	*	*	*	*	*
14	*	*	*	*	*
15	*	*	*	*	*

M	B=11	B=12	B=13	B=14	B=15
1	.01785714	.01639344	.01515152	.01408451	.01315789
2	.00031888	.00026874	.00022957	.00019837	.00017313
3	.56942\05	.44057\05	.34783\05	.27940\05	.22780\05
4	.10168\06	.72224\07	.52702\07	.39352\07	.29974\07
5	.18158\08	.11840\08	.79851\09	.55425\09	.39440\09
6	.32424\10	.19410\10	.12099\10	.78064\11	.51894\11
7	.57901\12	.31819\12	.18331\12	.10995\12	.68282\13
8	.10339\13	.52163\14	.27775\14	.15486\14	*
9	*	*	*	*	*
10	*	*	*	*	*
11	*	*	*	*	*
12	*	*	*	*	*
13	*	*	*	*	*
14	*	*	*	*	*
15	*	*	*	*	*

DIRICHLET D INTEGRAL
1/A = 5 R = 2

M	B= 1	B= 2	B= 3	B= 4	B= 5
1	.97222222	.94752187	.92529297	.90509577	.88660000
2	.92592593	.86422324	.81176758	.76645385	.72680000
3	.86805556	.76604986	.68473816	.61836591	.56315000
4	.80375514	.66405154	.56028366	.48065508	.41795000
5	.73677555	.56539246	.44740736	.36266312	.29972500
6	.66979595	.47432253	.35017729	.26703060	.20897500
7	.60467690	.39301010	.26950147	.19262465	.14230000
8	.54265876	.32220743	.20445441	.13653986	.09496250
9	.48451675	.26175650	.15319301	.09532936	.06227422
10	.43068155	.21095739	.11354473	.06568005	.04021719
11	.38133263	.16882639	.08335554	.04472484	.02562285
12	.33646996	.13426918	.06067369	.03013885	.01612832
13	.29596895	.10619144	.04382779	.02012019	.01004219
14	.25962189	.08356469	.03144186	.01331862	.00619150
15	.22716915	.06546120	.02241588	.00874875	.00378337

M	B= 6	B= 7	B= 8	B= 9	B=10
1	.86955079	.85374704	.83902709	.82525896	.81233341
2	.69172583	.66042160	.63226602	.60677292	.58355545
3	.51649465	.47655633	.44198796	.41178143	.38516679
4	.36752119	.32624935	.29196848	.26313116	.23860485
5	.25169910	.21422212	.18442032	.16033714	.14060242
6	.16705509	.13593723	.11229557	.09397541	.07953495
7	.10799626	.08382276	.06630751	.05331316	.04347484
8	.06826415	.05043641	.03813697	.02941344	.02307735
9	.04231669	.02970953	.02143983	.01583956	.01194193
10	.02578725	.01717686	.01181385	.00835022	.00604276
11	.01547817	.00976800	.00639494	.00431960	.00299745
12	.00916551	.00547319	.00340695	.00219704	.00146056
13	.00536172	.00302612	.00178921	.00110051	.00070031
14	.00310214	.00165304	.00092747	.00054366	.00033090
15	.00177688	.00089310	.00047509	.00026519	.00015428

M	B=11	B=12	B=13	B=14	B=15
1	.80015900	.78865835	.77776541	.76742332	.75758281
2	.56230125	.54275488	.52470510	.50797545	.49241719
3	.36154551	.34044516	.32148774	.30436714	.28883281
4	.21754325	.19930236	.18338426	.16939851	.15703503
5	.12423315	.11050908	.09889267	.08897613	.08044556
6	.06798227	.05861834	.05094007	.04457855	.03925876
7	.03589313	.02995901	.02525016	.02146751	.01839527
8	.01837936	.01483234	.01211114	.00999353	.00832444
9	.00916421	.00714309	.00564528	.00451704	.00365468
10	.00446367	.00335732	.00256595	.00198935	.00156225
11	.00212936	.00154418	.00114045	.00085610	.00065212
12	.00099700	.00069657	.00049678	.00036085	.00026645
13	.00045900	.00030875	.00021249	.00014926	.00010679
14	.00020810	.00013468	.89398\04	.60695\04	.42049\04
15	.93034\04	.57896\04	.37045\04	.24297\04	.16292\04

DIRICHLET D INTEGRAL
1/A = 4 R = 2

M	B= 1	B= 2	B= 3	B= 4	B= 5
1	.96000000	.92592593	.89629321	.87011719	.84670923
2	.89600000	.81481481	.74921164	.69482422	.64883214
3	.81920000	.69135802	.59621416	.52270508	.46426059
4	.73728000	.57064472	.45726331	.37609863	.31571377
5	.65536000	.46090535	.34069381	.26135254	.20633804
6	.57671680	.36579790	.24797109	.17657471	.13059494
7	.50331648	.28613524	.17702071	.11654663	.08048460
8	.43620762	.22110451	.12432216	.07542419	.04849737
9	.37580964	.16907992	.08609796	.04799271	.02866266
10	.32212255	.12813464	.05890657	.03009224	.01665693
11	.27487791	.09634183	.03987607	.01862621	.00953743
12	.23364622	.07193524	.02674072	.01139784	.00538948
13	.19791209	.05338051	.01778238	.00690365	.00300984
14	.16712577	.03939310	.01173645	.00414324	.00166313
15	.14073749	.02892634	.00769363	.00246596	.00091018

M	B= 6	B= 7	B= 8	B= 9	B=10
1	.82556800	.80631677	.78866509	.77238423	.75729083
2	.60931840	.57492716	.54466825	.51779917	.49375071
3	.41672704	.37734777	.34422064	.31599055	.29166599
4	.26940621	.23301344	.20382804	.18002114	.16031780
5	.16669082	.13721737	.11474021	.09722786	.08333314
6	.09954525	.07776517	.06200349	.05029615	.04140596
7	.05772358	.04269090	.03238710	.02510276	.01981640
8	.03264867	.02281212	.01643692	.01215353	.00918666
9	.01807445	.01190964	.00813734	.00573174	.00414329
10	.00982075	.00609280	.00394201	.00264181	.00182423
11	.00524889	.00306165	.00187338	.00119316	.00078624
12	.00276456	.00151414	.00087520	.00052921	.00033248
13	.00143707	.00073817	.00040264	.00023093	.00013821
14	.00073821	.00035525	.00018268	.99296\04	.56568\04
15	.00037515	.00016897	.81843\04	.42125\04	.22827\04

M	B=11	B=12	B=13	B=14	B=15
1	.74323560	.73009540	.71776746	.70616514	.69521473
2	.47207818	.45242845	.43451709	.41811213	.40302228
3	.27050481	.25194059	.23553361	.22093737	.20787520
4	.14380515	.12981443	.11784586	.10751934	.09854133
5	.07213461	.06298515	.05541967	.04909719	.04376320
6	.03452561	.02911214	.02479074	.02129665	.01843910
7	.01589287	.01292354	.01063765	.00885119	.00743617
8	.00707737	.00554354	.00440591	.00354731	.00288917
9	.00306270	.00230838	.00176983	.00137761	.00108688
10	.00129253	.00093656	.00069210	.00052043	.00039745
11	.00053349	.00037132	.00026428	.00019184	.00014173
12	.00021587	.00014422	.98788\04	.69184\04	.49415\04
13	.85798\04	.54979\04	.36224\04	.24460\04	.16881\04
14	.33552\04	.20609\04	.13053\04	.84937\05	.56614\05
15	.12928\04	.76073\05	.46290\05	.29012\05	.18668\05

DIRICHLET D INTEGRAL
1/A = 3 R = 2

M	B= 1	B= 2	B= 3	B= 4	B= 5
1	.93750000	.88800000	.84722222	.81269709	.78286743
2	.84375000	.73440000	.65277778	.58911678	.53784943
3	.73828125	.57888000	.47222222	.39620882	.33951616
4	.63281250	.44064000	.32638889	.25247862	.20169353
5	.53393555	.32659200	.21788194	.15440412	.11439769
6	.44494629	.23701248	.14149306	.09139235	.06254070
7	.36708069	.16908134	.08984375	.05267067	.03317595
8	.30033875	.11891681	.05598958	.02968529	.01716032
9	.24402523	.08263711	.03434245	.01641635	.00868732
10	.19709730	.05683821	.02077908	.00893114	.00431687
11	.15838176	.03874673	.01242405	.00479002	.00211050
12	.12670541	.02620846	.00735135	.00253689	.00101709
13	.10096837	.01760582	.00430976	.00132865	.00048392
14	.08018077	.01175462	.00250583	.00068892	.00022761
15	.06347644	.00780507	.00144619	.00035400	.00010595

M	B= 6	B= 7	B= 8	B= 9	B=10
1	.75668795	.73342320	.71253642	.69362411	.67637552
2	.49554137	.45994608	.42952559	.40318794	.38013438
3	.29575375	.26104853	.23292050	.20971055	.19026833
4	.16518046	.13798382	.11714334	.10079699	.08772332
5	.08770347	.06908471	.05562988	.04562054	.03799202
6	.04472989	.03312725	.02523762	.01968160	.01565326
7	.02207223	.01533179	.01102739	.00816277	.00618984
8	.01059422	.00688753	.00466841	.00327485	.00236437
9	.00496616	.00301624	.00192355	.00127696	.00087667
10	.00228073	.00129200	.00077414	.00048573	.00031675
11	.00102879	.00054278	.00030517	.00018077	.00011186
12	.00045674	.00022413	.00011811	.65986\04	.38709\04
13	.00019992	.91139\04	.44966\04	.23671\04	.13153\04
14	.86395\04	.36549\04	.16867\04	.83590\05	.43963\05
15	.36907\04	.14474\04	.62422\05	.29100\05	.14476\05

M	B=11	B=12	B=13	B=14	B=15
1	.66054638	.64594136	.63240189	.61979755	.60801988
2	.35976584	.34162346	.32534929	.31065969	.29732671
3	.17377214	.15962006	.14736153	.13665291	.12722774
4	.07709268	.06832469	.06100299	.05482230	.04955445
5	.03205763	.02735924	.02358223	.02050505	.01796824
6	.01265971	.01038770	.00863150	.00725213	.00615333
7	.00479114	.00377469	.00302004	.00244916	.00201013
8	.00174930	.00132177	.00101718	.00079545	.00063096
9	.00061928	.00044832	.00033155	.00024981	.00019136
10	.00021342	.00014789	.00010502	.76181\04	.56316\04
11	.71824\04	.47605\04	.32433\04	.22636\04	.16138\04
12	.23666\04	.14991\04	.97929\05	.65716\05	.45159\05
13	.76512\05	.46289\05	.28974\05	.18684\05	.12369\05
14	.24315\05	.14040\05	.84163\06	.52125\06	.33227\06
15	.76071\06	.41902\06	.24040\06	.14293\06	.87687\07

DIRICHLET D INTEGRAL
1/A = 2 R = 2

M	B= 1	B= 2	B= 3	B= 4	B= 5
1	.88888889	.81250000	.75520000	.70987654	.67271290
2	.74074074	.59375000	.49792000	.43004115	.37922489
3	.59259259	.40625000	.30208000	.23662551	.19220369
4	.46090535	.26562500	.17305600	.12185642	.09053842
5	.35116598	.16796875	.09502720	.05974699	.04039173
6	.26337449	.10351563	.05049549	.02819692	.01727155
7	.19509221	.06250000	.02613576	.01290454	.00713665
8	.14306762	.03710938	.01323827	.00575793	.00286638
9	.10404918	.02172852	.00658506	.00251486	.00112399
10	.07514663	.01257324	.00322542	.00107851	.00043177
11	.05395143	.00720215	.00155894	.00045526	.00016292
12	.03853673	.00408936	.00074477	.00018953	.60514\04
13	.02740390	.00230408	.00035219	.77939\04	.22166\04
14	.01941110	.00128937	.00016503	.31702\04	.80188\05
15	.01370195	.00071716	.76708\04	.12769\04	.28685\05

M	B= 6	B= 7	B= 8	B= 9	B=10
1	.64143372	.61457894	.59115904	.57047441	.55201340
2	.33964157	.30787135	.28176819	.25991374	.24133110
3	.16035843	.13657310	.11823181	.10372262	.09200223
4	.06997013	.05572710	.04545121	.03779092	.03192580
5	.02879107	.02136669	.01636830	.01286487	.01032734
6	.01131567	.00780241	.00560029	.00415187	.00316112
7	.00428520	.00273863	.00183793	.00128295	.00092498
8	.00157348	.00093009	.00058258	.00038230	.00026065
9	.00056285	.00030716	.00017928	.00011045	.71121\04
10	.00019686	.99020\04	.53781\04	.31065\04	.18873\04
11	.67514\04	.31255\04	.15776\04	.85348\05	.48871\05
12	.22758\04	.96836\05	.45372\05	.22966\05	.12383\05
13	.75543\05	.29509\05	.12821\05	.60660\06	.30774\06
14	.24733\05	.88593\06	.35657\06	.15756\06	.75152\07
15	.79980\06	.26243\06	.97758\07	.40311\07	.18063\07

M	B=11	B=12	B=13	B=14	B=15
1	.53539132	.52031222	.50654390	.49390110	.48223385
2	.22532441	.21138397	.19912748	.18826236	.17856072
3	.08236790	.07433032	.06753919	.06173764	.05673339
4	.02733419	.02367135	.02070196	.01826093	.01622960
5	.00843862	.00700012	.00588274	.00499991	.00429194
6	.00246105	.00195266	.00157472	.00128804	.00106671
7	.00068517	.00051931	.00040143	.00031567	.00025197
8	.00018347	.00013269	.98221\04	.74186\04	.57030\04
9	.47522\04	.32764\04	.23203\04	.16820\04	.12443\04
10	.11959\04	.78528\05	.53165\05	.36959\05	.26296\05
11	.29341\05	.18336\05	.11858\05	.79008\06	.54027\06
12	.70384\06	.41828\06	.25825\06	.16480\06	.10825\06
13	.16547\06	.93452\07	.55046\07	.33627\07	.21206\07
14	.38199\07	.20489\07	.11508\07	.67260\08	.40703\08
15	.86740\08	.44161\08	.23637\08	.13211\08	.76686\09

DIRICHLET D INTEGRAL
A = 1 R = 2

M	B= 1	B= 2	B= 3	B= 4	B= 5
1	.75000000	.62962963	.55468750	.50208000	.46244856
2	.50000000	.33333333	.25000000	.20000000	.16666667
3	.31250000	.16049383	.10058594	.07020800	.05242627
4	.18750000	.07270233	.03759766	.02269440	.01508440
5	.10937500	.03155007	.01333618	.00691456	.00406998
6	.06250000	.01326017	.00454712	.00201411	.00104548
7	.03515625	.00543616	.00150299	.00056617	.00025826
8	.01953125	.00218463	.00048447	.00015460	.61787\04
9	.01074219	.00086369	.00015295	.41210\04	.14391\04
10	.00585938	.00033682	.47445\04	.10762\04	.32762\05
11	.00317383	.00012984	.14499\04	.27613\05	.73128\06
12	.00170898	.49551\04	.43735\05	.69773\06	.16044\06
13	.00091553	.18747\04	.13043\05	.17394\06	.34668\07
14	.00048828	.70389\05	.38510\06	.42846\07	.73905\08
15	.00025940	.26251\05	.11269\06	.10442\07	.15565\08

M	B= 6	B= 7	B= 8	B= 9	B=10
1	.43116267	.40562725	.38425730	.36602157	.35021564
2	.14285714	.12500000	.11111111	.10000000	.09090909
3	.04100142	.03316420	.02752047	.02330108	.02005112
4	.01070762	.00797202	.00615378	.00488674	.00397004
5	.00261158	.00178376	.00127727	.00094903	.00072639
6	.00060440	.00037768	.00025029	.00017366	.00012501
7	.00013413	.76503\04	.46828\04	.30289\04	.20476\04
8	.28759\04	.14941\04	.84326\05	.50770\05	.32191\05
9	.59901\05	.28295\05	.14702\05	.82284\06	.48875\06
10	.12171\05	.52190\06	.24931\06	.12955\06	.72011\07
11	.24206\06	.94083\07	.41267\07	.19888\07	.10335\07
12	.47244\07	.16622\07	.66867\08	.29858\08	.14493\08
13	.90684\08	.28846\08	.10632\08	.43943\09	.19907\09
14	.17150\08	.49267\09	.16619\09	.63530\10	.26840\10
15	.32003\09	.82941\10	.25585\10	.90380\11	.35583\11

M	B=11	B=12	B=13	B=14	B=15
1	.33633947	.32402676	.31300210	.30305382	.29401614
2	.08333333	.07692308	.07142857	.06666667	.06250000
3	.01748637	.01542115	.01372967	.01232402	.01114113
4	.00328628	.00276321	.00235447	.00202922	.00176630
5	.00056969	.00045598	.00037131	.00030686	.00025688
6	.92751\04	.70571\04	.54853\04	.43422\04	.34922\04
7	.14353\04	.10370\04	.76853\05	.58219\05	.44947\05
8	.21294\05	.14592\05	.10303\05	.74622\06	.55259\06
9	.30478\06	.19792\06	.13301\06	.92046\07	.65335\07
10	.42294\07	.26005\07	.16622\07	.10982\07	.74671\08
11	.57119\08	.33229\08	.20186\08	.12726\08	.82837\09
12	.75312\09	.41422\09	.23901\09	.14368\09	.89493\10
13	.97196\10	.50508\10	.27665\10	.15850\10	.94416\11
14	.12304\10	.60374\11	.31373\11	.17123\11	.97497\12
15	.15307\11	.70876\12	.34924\12	.18148\12	.98733\13

DIRICHLET D INTEGRAL
A = 2 R = 2

M	B= 1	B= 2	B= 3	B= 4	B= 5
1	.55555556	.42400000	.35526864	.31148707	.28053282
2	.25925926	.14240000	.09680490	.07286039	.05820604
3	.11111111	.04256000	.02306012	.01472412	.01034434
4	.04526749	.01184000	.00505013	.00271261	.00166557
5	.01783265	.00313600	.00104305	.00046823	.00025004
6	.00685871	.00080128	.00020625	.76963\04	.35599\04
7	.00259107	.00019917	.39426\04	.12174\04	.48608\05
8	.00096530	.48435\04	.73347\05	.18669\05	.64149\06
9	.00035564	.11571\04	.13343\05	.27901\06	.82291\07
10	.00012984	.27238\05	.23824\06	.40803\07	.10305\07
11	.47042\04	.63324\06	.41864\07	.58570\08	.12641\08
12	.16935\04	.14565\06	.72557\08	.82728\09	.15227\09
13	.60632\05	.33194\07	.12425\08	.11521\09	.18054\10
14	.21604\05	.75039\08	.21054\09	.15844\10	.21106\11
15	.76661\06	.16843\08	.35343\10	.21547\11	.24364\12

M	B= 6	B= 7	B= 8	B= 9	B=10
1	.25718225	.23876971	.22377533	.21126147	.20061444
2	.04835316	.04129232	.03599339	.03187530	.02858606
3	.00773542	.00604411	.00487876	.00403799	.00340920
4	.00111577	.00079459	.00059203	.00045671	.00036216
5	.00014948	.96719\04	.66345\04	.47599\04	.35383\04
6	.18931\04	.11100\04	.69949\05	.46584\05	.32410\05
7	.22931\05	.12157\05	.70240\06	.43351\06	.28188\06
8	.26781\06	.12812\06	.67757\07	.38700\07	.23488\07
9	.30338\07	.13073\07	.63190\08	.33356\08	.18875\08
10	.33485\08	.12976\08	.57248\09	.27896\09	.14702\09
11	.36138\09	.12576\09	.50577\10	.22726\10	.11145\10
12	.38243\10	.11935\10	.43706\11	.18091\11	.82483\12
13	.39775\11	.11118\11	.37034\12	.14109\12	.59754\13
14	.40734\12	.10187\12	.30835\13	.10802\13	.42467\14
15	.41142\13	.91961\14	.25271\14	*	*

M	B=11	B=12	B=13	B=14	B=15
1	.19141363	.18335993	.17623408	.16987122	.16414472
2	.02590030	.02366720	.02178211	.02017019	.01877653
3	.00292516	.00254362	.00223686	.00198605	.00177802
4	.00029367	.00024257	.00020350	.00017299	.00014875
5	.27070\04	.21208\04	.16951\04	.13780\04	.11368\04
6	.23361\05	.17338\05	.13187\05	.10242\05	.80986\06
7	.19118\06	.13426\06	.97088\07	.71972\07	.54506\07
8	.14973\07	.99406\08	.68276\08	.48271\08	.34987\08
9	.11298\08	.70843\09	.46181\09	.31116\09	.21570\09
10	.82562\10	.48853\10	.30203\10	.19381\10	.12842\10
11	.58664\11	.32733\11	.19179\11	.11714\11	.74143\12
12	.40663\12	.21380\12	.11865\12	.68934\13	.41659\13
13	.27571\13	.13651\13	.71710\14	.39610\14	.22844\14
14	.18326\14	*	*	*	*
15	*	*	*	*	*

DIRICHLET D INTEGRAL
A = 3 R = 2

M	B= 1	B= 2	B= 3	B= 4	B= 5
1	.43750000	.31778426	.26020000	.22501098	.20076609
2	.15625000	.07788421	.05068000	.03718005	.02919859
3	.05078125	.01683822	.00856000	.00526809	.00361207
4	.01562500	.00337444	.00132400	.00067799	.00040344
5	.00463867	.00064235	.00019270	.81577\04	.41927\04
6	.00134277	.00011778	.26812\04	.93337\05	.41267\05
7	.00038147	.20989\04	.36028\05	.10267\05	.38917\06
8	.00010681	.36570\05	.47080\06	.10941\06	.35447\07
9	.29564\04	.62559\06	.60130\07	.11356\07	.31366\08
10	.81062\05	.10541\06	.75340\08	.11529\08	.27083\09
11	.22054\05	.17536\07	.92872\09	.11484\09	.22897\10
12	.59605\06	.28855\08	.11289\09	.11254\10	.19006\11
13	.16019\06	.47035\09	.13555\10	.10870\11	.15524\12
14	.42841\07	.76039\10	.16102\11	.10367\12	.12500\13
15	.11409\07	.12204\10	.18946\12	.97756\14	*

M	B= 6	B= 7	B= 8	B= 9	B=10
1	.18280181	.16882404	.15755866	.14823517	.14035719
2	.02395796	.02026740	.01753490	.01543417	.01377112
3	.00265431	.00204667	.00163489	.00134170	.00112480
4	.00026428	.00018505	.00013606	.00010384	.81611\04
5	.24392\04	.15463\04	.10439\04	.73927\05	.54369\05
6	.21255\05	.12168\05	.75254\06	.49366\06	.33920\06
7	.17698\06	.91284\07	.51626\07	.31319\07	.20078\07
8	.14198\07	.65854\08	.34001\08	.19048\08	.11379\08
9	.11042\08	.45975\09	.21638\09	.11180\09	.62162\10
10	.83642\10	.31210\10	.13372\10	.63647\11	.32904\11
11	.61928\11	.20680\11	.80556\12	.35283\12	.16945\12
12	.44946\12	.13414\12	.47454\13	.19107\13	.85168\14
13	.32052\13	.85389\14	.27405\14	.10135\14	*
14	.22503\14	*	*	*	*
15	*	*	*	*	*

M	B=11	B=12	B=13	B=14	B=15
1	.13358887	.12769392	.12250072	.11788125	.11373785
2	.01242331	.01130980	.01037501	.00957955	.00889475
3	.00095934	.00082992	.00072657	.00064255	.00057323
4	.65680\04	.53901\04	.44965\04	.38036\04	.32562\04
5	.41221\05	.32046\05	.25441\05	.20561\05	.16873\05
6	.24194\06	.17796\06	.13431\06	.10360\06	.81427\07
7	.13456\07	.93541\08	.67046\08	.49319\08	.37097\08
8	.71577\09	.46981\09	.31952\09	.22396\09	.16110\09
9	.36665\10	.22702\10	.14639\10	.97704\11	.67164\11
10	.18182\11	.10611\11	.64828\12	.41171\12	.27031\12
11	.87639\13	.48176\13	.27866\13	.16829\13	.10547\13
12	.41200\14	.21317\14	.11667\14	*	*
13	*	*	*	*	*
14	*	*	*	*	*
15	*	*	*	*	*

DIRICHLET D INTEGRAL
A = 4 R = 2

M	B= 1	B= 2	B= 3	B= 4	B= 5
1	.36000000	.25377229	.20506985	.17598462	.15622347
2	.10400000	.04892547	.03106172	.02246991	.01748417
3	.02720000	.00828126	.00406252	.00245054	.00165841
4	.00672000	.00129648	.00048557	.00024228	.00014177
5	.00160000	.00019256	.54549\04	.22371\04	.11265\04
6	.00037120	.27528\04	.58538\05	.19627\05	.84715\06
7	.84480\04	.38227\05	.60637\06	.16547\06	.61011\07
8	.18944\04	.51882\06	.61060\07	.13509\07	.42423\08
9	.41984\05	.69118\07	.60077\08	.10740\08	.28649\09
10	.92160\06	.90677\08	.57977\09	.83497\10	.18875\10
11	.20070\06	.11743\08	.55036\10	.63682\11	.12174\11
12	.43418\07	.15041\09	.51509\11	.47772\12	.77081\13
13	.93389\08	.19083\10	.47617\12	.35320\13	.48018\14
14	.19988\08	.24009\11	.43543\13	.25781\14	*
15	.42598\09	.29987\12	.39438\14	*	*

M	B= 6	B= 7	B= 8	B= 9	B=10
1	.14172107	.13051632	.12153505	.11413442	.10790368
2	.01425163	.01199611	.01033792	.00907029	.00807141
3	.00120741	.00092452	.00073447	.00060009	.00050123
4	.91777\04	.63695\04	.46508\04	.35296\04	.27612\04
5	.64602\05	.40516\05	.27123\05	.19080\05	.13954\05
6	.42904\06	.24254\06	.14855\06	.96686\07	.66006\07
7	.27214\07	.13836\07	.77385\08	.46529\08	.29609\08
8	.16626\08	.75878\09	.38689\09	.21459\09	.12714\09
9	.98442\10	.40257\10	.18686\10	.95482\11	.52608\11
10	.56756\11	.20764\11	.87617\12	.41199\12	.21088\12
11	.31979\12	.10451\12	.40043\13	.17308\13	.82227\14
12	.17661\13	.51493\14	.17893\14	*	*
13	*	*	*	*	*
14	*	*	*	*	*
15	*	*	*	*	*

M	B=11	B=12	B=13	B=14	B=15
1	.10256677	.09793057	.09385545	.09023767	.08699841
2	.00726500	.00660098	.00604512	.00557330	.00516801
3	.00042617	.00036770	.00032116	.00028344	.00025241
4	.22135\04	.18105\04	.15060\04	.12706\04	.10853\04
5	.10531\05	.81540\06	.64511\06	.51978\06	.42541\06
6	.46825\07	.34283\07	.25770\07	.19809\07	.15521\07
7	.19723\08	.13639\08	.97313\09	.71300\09	.53443\09
8	.79426\10	.51828\10	.35071\10	.24474\10	.17536\10
9	.30795\11	.18945\11	.12148\11	.80685\12	.55227\12
10	.11556\12	.66969\13	.40665\13	.25689\13	.16787\13
11	.42146\14	.22991\14	.13211\14	*	*
12	*	*	*	*	*
13	*	*	*	*	*
14	*	*	*	*	*
15	*	*	*	*	*

DIRICHLET D INTEGRAL
A = 5 R = 2

M	B= 1	B= 2	B= 3	B= 4	B= 5
1	.30555556	.21111946	.16915894	.14446288	.12783023
2	.07407407	.03353596	.02095795	.01502716	.01162555
3	.01620370	.00466312	.00223613	.00133194	.00089408
4	.00334362	.00059891	.00021777	.00010690	.61904\04
5	.00066444	.72920\04	.19918\04	.80074\05	.39813\05
6	.00012860	.85418\05	.17395\05	.56967\06	.24224\06
7	.24410\04	.97161\06	.14659\06	.38933\07	.14110\07
8	.45645\05	.10799\06	.12006\07	.25761\08	.79340\09
9	.84345\06	.11780\07	.96068\09	.16595\09	.43320\10
10	.15436\06	.12653\08	.75383\10	.10453\10	.23073\11
11	.28023\07	.13414\09	.58181\11	.64585\12	.12029\12
12	.50533\08	.14065\10	.44268\12	.39246\13	.61558\14
13	.90603\09	.14605\11	.33267\13	.23503\14	*
14	.16164\09	.15040\12	.24728\14	*	*
15	.28712\10	.15375\13	*	*	*

M	B= 6	B= 7	B= 8	B= 9	B=10
1	.11569635	.10636243	.09890591	.09277821	.08763058
2	.00943733	.00791917	.00680792	.00596136	.00529617
3	.00064720	.00049343	.00039068	.00031833	.00026529
4	.39781\04	.27458\04	.19963\04	.15098\04	.11778\04
5	.22630\05	.14099\05	.93900\06	.65782\06	.47948\06
6	.12141\06	.68106\07	.41463\07	.26858\07	.18263\07
7	.62192\08	.31342\08	.17410\08	.10411\08	.65955\09
8	.30678\09	.13863\09	.70144\10	.38667\10	.22794\10
9	.14664\10	.59310\11	.27296\11	.13854\11	.75908\12
10	.68245\12	.24666\12	.10311\12	.48128\13	.24485\13
11	.31035\13	.10010\13	.37962\14	.16277\14	*
12	.13832\14	*	*	*	*
13	*	*	*	*	*
14	*	*	*	*	*
15	*	*	*	*	*

M	B=11	B=12	B=13	B=14	B=15
1	.08322960	.07941250	.07606195	.07309102	.07043375
2	.00476042	.00432016	.00395227	.00364046	.00337299
3	.00022513	.00019392	.00016914	.00014909	.00013262
4	.94194\05	.76886\05	.63841\05	.53782\05	.45874\05
5	.36080\06	.27870\06	.22003\06	.17696\06	.14459\06
6	.12913\07	.94278\08	.70695\08	.54226\08	.42409\08
7	.43767\09	.30169\09	.21467\09	.15691\09	.11736\09
8	.14181\10	.92203\11	.62200\11	.43289\11	.30944\11
9	.44228\12	.27102\12	.17320\12	.11470\12	.78302\13
10	.13350\13	.77032\14	.46602\14	.29344\14	.19122\14
11	*	*	*	*	*
12	*	*	*	*	*
13	*	*	*	*	*
14	*	*	*	*	*
15	*	*	*	*	*

DIRICHLET D INTEGRAL
1/A = 5 R = 3

M	B= 1	B= 2	B= 3	B= 4	B= 5
1	.99537037	.99095615	.98673344	.98268263	.97878737
2	.98379630	.96877152	.95475614	.94161702	.92924719
3	.96450617	.93270782	.90394169	.87771074	.85363193
4	.93771433	.88413720	.83737541	.79608061	.75925883
5	.90422454	.82559225	.75969874	.70357338	.65512757
6	.86515311	.76007766	.67591736	.60703510	.54966142
7	.82174041	.69060254	.59063382	.51228386	.44949522
8	.77522680	.61989019	.50762949	.42364961	.35907237
9	.72677512	.55022284	.42970683	.34391208	.28076752
10	.67742619	.48338343	.35870591	.27447828	.21528328
11	.62807727	.42066209	.29562325	.21567261	.16213511
12	.57947605	.36290217	.24077767	.16705174	.12010859
13	.53222487	.31056680	.19398579	.12769112	.08763129
14	.48679104	.26381255	.15472391	.09641785	.06304193
15	.44352072	.22256164	.12226439	.07198292	.04476405

M	B= 6	B= 7	B= 8	B= 9	B=10
1	.97503381	.97141008	.96790592	.76451235	.96122149
2	.91755915	.90648026	.89594937	.88591448	.87633094
3	.83140328	.81078223	.79157084	.77360537	.75674879
4	.72615681	.69619063	.66889892	.64391119	.62092574
5	.61284354	.57558809	.54249586	.51289331	.48624745
6	.50117557	.45969674	.42383914	.39255876	.36505389
7	.39824741	.35577218	.32010466	.28981410	.26383566
8	.30828441	.26758405	.23444415	.20708843	.18423706
9	.23301741	.19608708	.16697687	.14365422	.12470291
10	.17232805	.14032080	.11593345	.09699466	.08204292
11	.12492382	.09825318	.07863796	.06389123	.05259248
12	.08891120	.06743532	.05220943	.04114077	.03291884
13	.06221721	.04543816	.03398472	.02594238	.02015634
14	.04286071	.03009857	.02172093	.01604474	.01209310
15	.02910013	.01962441	.01364906	.00974638	.00711958

M	B=11	B=12	B=13	B=14	B=15
1	.95802635	.95492068	.95189892	.94895601	.94608742
2	.86716007	.85836810	.84992538	.84180569	.83398575
3	.74088524	.72591586	.71175560	.69833079	.68557716
4	.59969385	.58000830	.56169472	.54460521	.52861326
5	.46213045	.44019457	.42015398	.40177139	.38484803
6	.34069867	.31899733	.29955213	.28204026	.26619706
7	.24136166	.22176930	.20457124	.18938119	.17588935
8	.16494791	.14851439	.13439774	.12218114	.11153788
9	.10911214	.09614523	.08525527	.07602961	.06815228
10	.07006781	.06035431	.05238600	.04578310	.04026176
11	.04379125	.03683506	.03126573	.02675509	.02306385
12	.02669577	.02190729	.01816849	.01521097	.01284395
13	.01590476	.01272232	.01030082	.00843122	.00696872
14	.00927652	.00722713	.00570853	.00456486	.00369093
15	.00530486	.00402224	.00309727	.00241819	.00191157

DIRICHLET D INTEGRAL
1/A = 4 R = 3

M	B= 1	B= 2	B= 3	B= 4	B= 5
1	.99200000	.98456790	.97761501	.97107387	.96489157
2	.97280000	.94855967	.92668871	.90676343	.88846783
3	.94208000	.89300412	.85060213	.81341165	.78039896
4	.90112000	.82213077	.75720379	.70266935	.65607817
5	.85196800	.74135040	.65537030	.58654602	.53017929
6	.79691776	.65599756	.55299499	.47476552	.41356923
7	.73819750	.57064472	.45606731	.37383554	.31259620
8	.67779953	.48881437	.36848116	.28717509	.22970654
9	.61740155	.41294518	.29225974	.21575491	.16457559
10	.55834575	.34450234	.22796531	.15887763	.11524872
11	.50165218	.28415488	.17514389	.11488665	.07905239
12	.44805099	.23196972	.13272189	.08171310	.05321183
13	.39802321	.18759301	.09931909	.05724664	.03520611
14	.35184372	.15040427	.07347235	.03955388	.02292764
15	.30962247	.11963832	.05377996	.02698291	.01471544

M	B= 6	B= 7	B= 8	B= 9	B=10
1	.95902567	.95344142	.94810994	.94300690	.93811157
2	.87155905	.85584594	.84117506	.82742117	.81448062
3	.75080594	.72406007	.69971853	.67743191	.65691980
4	.61572115	.58036301	.54908510	.52118785	.49612773
5	.48315961	.44334041	.40918979	.37958379	.35367830
6	.36455749	.32454249	.29134526	.26342793	.23967563
7	.26563705	.22876260	.19923099	.17518361	.15532150
8	.18762100	.15590474	.13143025	.11216430	.09673869
9	.12886521	.10308872	.08395590	.06941813	.05814967
10	.08630655	.06633366	.05209633	.04166871	.03385366
11	.05649777	.04164245	.03148792	.02432827	.01914600
12	.03622303	.02556075	.01858136	.01384986	.01054583
13	.02278636	.01537005	.01072724	.00770437	.00566994
14	.01408560	.00906911	.00606935	.00419558	.00298132
15	.00856798	.00525870	.00337065	.00224035	.00153571

M	B=11	B=12	B=13	B=14	B=15
1	.93340611	.92887502	.92450475	.92028337	.91620031
2	.80226659	.79070555	.77973468	.76929983	.75935401
3	.63795385	.62034578	.60393858	.58860001	.57421775
4	.47347517	.45288558	.43407888	.41682484	.40093222
5	.33082662	.31052468	.29237362	.27605364	.26130526
6	.21926138	.20155961	.18608884	.17247285	.16041374
7	.13871293	.12467450	.11269510	.10238589	.09344657
8	.08420544	.07389105	.06530648	.05808977	.05196846
9	.04926416	.04215221	.03638456	.03165235	.02772916
10	.02787963	.02323373	.01956577	.01663099	.01425472
11	.01530955	.01241336	.01018921	.00845512	.00708478
12	.00817930	.00644657	.00515324	.00417135	.00341455
13	.00426136	.00326189	.00253733	.00200205	.00159990
14	.00216934	.00161141	.00121886	.00093683	.00073042
15	.00108097	.00077862	.00057228	.00042821	.00032554

DIRICHLET D INTEGRAL
1/A = 3 R = 3

M	B= 1	B= 2	B= 3	B= 4	B= 5
1	.98437500	.97056000	.95814043	.94683737	.93645221
2	.94921875	.90720000	.87143133	.84035558	.81293612
3	.89648438	.81734400	.75414738	.70211341	.65828043
4	.83056641	.71228160	.62442130	.55630419	.50180254
5	.75640869	.60279552	.49746817	.42102055	.36318859
6	.67854309	.49724099	.38328270	.30625252	.25138329
7	.60067749	.40109230	.28682002	.21522460	.16738386
8	.52559280	.31724587	.20921947	.14675778	.10773747
9	.45520091	.24662138	.14921287	.09744294	.06730203
10	.39067501	.18879959	.10430871	.06318695	.04093897
11	.33260170	.14257054	.07162626	.04011579	.02431646
12	.28112762	.10634888	.04840050	.02498811	.01413646
13	.23608781	.07845820	.03223504	.01529915	.00805994
14	.19711105	.05730615	.02118796	.00922135	.00451466
15	.16370240	.04147784	.01376060	.00547902	.00248814

M	B= 6	B= 7	B= 8	B= 9	B=10
1	.92683760	.91788060	.90949249	.90160208	.89415126
2	.78844611	.76635548	.74626540	.72786812	.71092116
3	.62069407	.58800163	.55923116	.53366281	.51075002
4	.45712421	.41978541	.38808570	.36081883	.33710368
5	.31803569	.28189215	.25236961	.22784819	.20719152
6	.21071826	.17962760	.15525268	.13574314	.11985411
7	.13382960	.10938843	.09103616	.07690764	.06580126
8	.08190912	.06402633	.05119103	.04170292	.03451444
9	.04852123	.03618743	.02774070	.02175447	.01738931
10	.02791998	.01982617	.01454606	.01096365	.00845256
11	.01565277	.01056321	.00740534	.00535696	.00397838
12	.00857170	.00548791	.00367075	.00254523	.00181874
13	.00459501	.00278661	.00177594	.00117891	.00080969
14	.00241582	.00138569	.00084038	.00053349	.00035183
15	.00124769	.00067597	.00038965	.00023631	.00014950

M	B=11	B=12	B=13	B=14	B=15
1	.88709191	.88038365	.87399226	.86788844	.86204687
2	.69523001	.68063631	.66700940	.65424024	.64223690
3	.49006897	.47128509	.45413018	.43838628	.42387416
4	.31628073	.29784591	.28140698	.26665390	.25333826
5	.18957957	.17440630	.16121449	.14965295	.13944770
6	.10672079	.09572582	.08641807	.07846116	.07159978
7	.05691434	.04969401	.04374937	.03879768	.03463034
8	.02895353	.02457401	.02107090	.01823042	.01589935
9	.01412858	.01164199	.00971150	.00818909	.00697177
10	.00664330	.00530867	.00430389	.00353385	.00293437
11	.00302131	.00233901	.00184131	.00147093	.00119039
12	.00133324	.00099904	.00076302	.00059257	.00046705
13	.00057240	.00041480	.00030712	.00023172	.00017776
14	.00023964	.00016781	.00012037	.88167\04	.65790\04
15	.98036\04	.66289\04	.46030\04	.32714\04	.23731\04

DIRICHLET D INTEGRAL
1/A = 2 R = 3

M	B= 1	B= 2	B= 3	B= 4	B= 5
1	.96296296	.93359375	.90918400	.88827446	.86997900
2	.88888889	.81054688	.75089920	.70326075	.66394041
3	.79012346	.66113281	.57229312	.50673011	.45602903
4	.68038409	.51220703	.40859238	.33848530	.28798923
5	.57064472	.38037109	.27654554	.21258351	.16989618
6	.46822131	.27270508	.17906532	.12686762	.09473901
7	.37717828	.18981934	.11170585	.07252845	.05038151
8	.29914139	.12884521	.06750585	.03996875	.02572694
9	.23411065	.08558655	.03969145	.02133711	.01268315
10	.18112265	.05579376	.02278581	.01107853	.00606280
11	.13873224	.03577805	.01280818	.00561270	.00282011
12	.10533374	.02261162	.00706636	.00278212	.00128021
13	.07935712	.01410675	.00383400	.00135230	.00056858
14	.05937511	.00869942	.00204921	.00064580	.00024758
15	.04415073	.00530910	.00108049	.00030350	.00010588

M	B= 6	B= 7	B= 8	B= 9	B=10
1	.85371546	.83907932	.82577729	.81358961	.80234711
2	.63069211	.60205195	.57701557	.55486581	.53507390
3	.41546924	.38217543	.35428597	.33053727	.31003865
4	.24994938	.22030680	.19658821	.17720128	.16107499
5	.13973779	.11750217	.10055447	.08728902	.07667702
6	.07352875	.05877629	.04809211	.04010059	.03396348
7	.03676447	.02785366	.02173520	.01737021	.01415755
8	.01759597	.01260214	.00935783	.00715405	.00560186
9	.00810806	.00547676	.00386236	.00281988	.00211819
10	.00361369	.00229738	.00153601	.00106933	.00076952
11	.00156370	.00093389	.00059101	.00039178	.00026977
12	.00065900	.00036910	.00022077	.00013918	.91597\04
13	.00027120	.00014222	.80291\04	.48080\04	.30213\04
14	.00010923	.53554\04	.28500\04	.16194\04	.97063\05
15	.43139\04	.19747\04	.98944\05	.53288\05	.30439\05

M	B=11	B=12	B=13	B=14	B=15
1	.79191666	.78219157	.77308496	.76452509	.75645207
2	.51723959	.50105314	.48627038	.47269564	.46016995
3	.29214263	.27636625	.26234130	.24978177	.23846187
4	.14746277	.13582857	.12577771	.11701327	.10930770
5	.06803189	.06087964	.05488365	.04979888	.04544321
6	.02914593	.02529341	.02216324	.01958472	.01743488
7	.01173100	.00985771	.00838426	.00720642	.00625149
8	.00447527	.00363663	.00299872	.00250435	.00211493
9	.00162938	.00127890	.00102132	.00082796	.00068010
10	.00056929	.00043114	.00033314	.00026193	.00020911
11	.00019174	.00013997	.00010456	.79670\04	.61770\04
12	.62481\04	.43930\04	.31697\04	.23389\04	.17600\04
13	.19761\04	.13370\04	.93117\05	.66495\05	.48533\05
14	.60822\05	.39569\05	.26581\05	.18358\05	.12989\05
15	.18259\05	.11414\05	.73908\06	.49337\06	.33820\06

DIRICHLET D INTEGRAL
A = 1 R = 3

M	B= 1	B= 2	B= 3	B= 4	B= 5
1	.87500000	.80246914	.75207520	.71385344	.68329368
2	.68750000	.54732510	.46395874	.40740915	.36596371
3	.50000000	.33333333	.25000000	.20000000	.16666667
4	.34375000	.18701417	.12216949	.08808417	.06755134
5	.22656250	.09861302	.05543613	.03572206	.02505420
6	.14453125	.04953513	.02372813	.01357170	.00866076
7	.08984375	.02392928	.00968659	.00488919	.00282640
8	.05468750	.01119409	.00380209	.00168494	.00087900
9	.03271484	.00509746	.00144367	.00055922	.00026237
10	.01928711	.00226867	.00053281	.00017968	.75582\04
11	.01123047	.00098997	.00019185	.56120\04	.21107\04
12	.00646973	.00042463	.67600\04	.17097\04	.57346\05
13	.00369263	.00017941	.23369\04	.50946\05	.15203\05
14	.00209045	.74795\04	.79416\05	.14883\05	.39430\06
15	.00117493	.30811\04	.26579\05	.42708\06	.10025\06

M	B= 6	B= 7	B= 8	B= 9	B=10
1	.65797823	.63646441	.61782457	.60142829	.58682842
2	.33399034	.30840585	.28736437	.26968567	.25457601
3	.14285714	.12500000	.11111111	.10000000	.09090909
4	.05403932	.04457793	.03764074	.03237047	.02825192
5	.01861482	.01441958	.01152794	.00944662	.00789621
6	.00595234	.00431564	.00325779	.00253791	.00202766
7	.00179069	.00121177	.00086171	.00063691	.00048554
8	.00051182	.00032247	.00021557	.00015090	.00010960
9	.00014002	.81956\04	.51409\04	.34030\04	.23514\04
10	.36881\04	.20013\04	.11760\04	.73502\05	.48261\05
11	.93959\05	.47183\05	.25931\05	.15283\05	.95244\06
12	.23241\05	.10782\05	.55340\06	.30719\06	.18150\06
13	.55988\06	.23958\06	.11469\06	.59893\07	.33517\07
14	.13171\06	.51907\07	.23148\07	.11359\07	.60152\08
15	.30323\07	.10992\07	.45608\08	.21010\08	.10518\08

M	B=11	B=12	B=13	B=14	B=15
1	.57369686	.56178605	.55090485	.54090273	.53165918
2	.24147970	.22999482	.21982270	.21073624	.20255934
3	.08333333	.07692308	.07142857	.06666667	.06250000
4	.02495861	.02227446	.02005134	.01818457	.01659830
5	.00670864	.00577775	.00503374	.00442915	.00393077
6	.00165384	.00137241	.00115561	.00098530	.00084925
7	.00037962	.00030313	.00024639	.00020335	.00017006
8	.82032\04	.62955\04	.49346\04	.39384\04	.31928\04
9	.16828\04	.12399\04	.93628\05	.72202\05	.56698\05
10	.32987\05	.23312\05	.16944\05	.12615\05	.95887\06
11	.62113\06	.42063\06	.29404\06	.21120\06	.15529\06
12	.11283\06	.73158\07	.49148\07	.34033\07	.24190\07
13	.19843\07	.12309\07	.79415\08	.52983\08	.36385\08
14	.33888\08	.20096\08	.12443\08	.79936\09	.53006\09
15	.56343\09	.31920\09	.18956\09	.11719\09	.74998\10

DIRICHLET D INTEGRAL
A = 2 R = 3

M	B= 1	B= 2	B= 3	B= 4	B= 5
1	.70370370	.59552000	.53310999	.49065872	.45913744
2	.40740741	.27296000	.21076617	.17417952	.14979327
3	.20987654	.10707200	.06980865	.05108383	.03997100
4	.10013717	.03784960	.02052235	.01316544	.00930519
5	.04526749	.01241344	.00553102	.00308600	.00195887
6	.01966164	.00384625	.00139449	.00067221	.00038134
7	.00828126	.00113961	.00033340	.00013806	.69700\04
8	.00340395	.00032565	.76318\04	.27011\04	.12091\04
9	.00137174	.90313\04	.16847\04	.50734\05	.20068\05
10	.00054380	.24423\04	.36057\05	.92023\06	.32067\06
11	.00021263	.64638\05	.75146\06	.16194\06	.49573\07
12	.82167\04	.16792\05	.15302\06	.27750\07	.74438\08
13	.31431\04	.42919\06	.30528\07	.46449\08	.10892\08
14	.11917\04	.10814\06	.59812\08	.76137\09	.15573\09
15	.44835\05	.26900\07	.11530\08	.12247\09	.21806\10

M	B= 6	B= 7	B= 8	B= 9	B=10
1	.43441232	.41427374	.39741365	.38299922	.37047030
2	.13223662	.11891653	.10841922	.09990457	.09284012
3	.03267088	.02753511	.02373900	.02082643	.01852560
4	.00700314	.00550725	.00447345	.00372511	.00316347
5	.00135143	.00098812	.00075401	.00059449	.00048097
6	.00024027	.00016284	.00011644	.86745\04	.66739\04
7	.39974\04	.25047\04	.16747\04	.11766\04	.85946\05
8	.62935\05	.36371\05	.22694\05	.15011\05	.10396\05
9	.94552\06	.50291\06	.29231\06	.18177\06	.11920\06
10	.13643\06	.66656\07	.36030\07	.21034\07	.13044\07
11	.19004\07	.85129\08	.42730\08	.23388\08	.13702\08
12	.25659\08	.10521\08	.48970\09	.25100\09	.13876\09
13	.33697\09	.12627\09	.54428\10	.26095\10	.13600\10
14	.43163\10	.14760\10	.58845\11	.26362\11	.12940\11
15	.54056\11	.16846\11	.62045\12	.25945\12	.11985\12

M	B=11	B=12	B=13	B=14	B=15
1	.35943374	.34960391	.34076738	.33276086	.32545689
2	.08687112	.08175152	.07730501	.07340169	.06994365
3	.01666496	.01513107	.01384609	.01275486	.01181729
4	.00272959	.00238636	.00210942	.00188220	.00169308
5	.00039733	.00033394	.00028473	.00024577	.00021439
6	.52703\04	.42522\04	.34929\04	.29134\04	.24620\04
7	.64788\05	.50119\05	.39620\05	.31901\05	.26095\05
8	.74717\06	.55359\06	.42073\06	.32673\06	.25849\06
9	.81580\07	.57831\07	.42218\07	.31596\07	.24158\07
10	.84930\08	.57550\08	.40321\08	.29060\08	.21458\08
11	.84780\09	.54867\09	.36865\09	.25568\09	.18222\09
12	.81523\10	.50348\10	.32417\10	.21622\10	.14864\10
13	.75799\11	.44638\11	.27523\11	.17643\11	.11693\11
14	.68365\12	.38362\12	.22637\12	.13938\12	.89004\13
15	.59973\13	.32045\13	.18085\13	.10690\13	.65739\14

DIRICHLET D INTEGRAL
A = 3 R = 3

M	B= 1	B= 2	B= 3	B= 4	B= 5
1	.57812500	.46665080	.40797100	.36994881	.34258683
2	.26171875	.15958487	.11780470	.09472278	.07993686
3	.10351563	.04588589	.02803078	.01972850	.01503330
4	.03759766	.01178931	.00587274	.00358365	.00244901
5	.01287842	.00279680	.00112282	.00058948	.00035927
6	.00422668	.00062495	.00020023	.89853\04	.48608\04
7	.00134277	.00013327	.33793\04	.12888\04	.61628\05
8	.00041580	.27369\04	.54526\05	.17586\05	.74056\06
9	.00012612	.54494\05	.84748\06	.23011\06	.85053\07
10	.37611\04	.10571\05	.12761\06	.29053\07	.93965\08
11	.11057\04	.20058\06	.18697\07	.35562\08	.10037\08
12	.32112\05	.37338\07	.26752\08	.42366\09	.10407\09
13	.92294\06	.68356\08	.37486\09	.49278\10	.10511\10
14	.26287\06	.12332\08	.51565\10	.56108\11	.10369\11
15	.74273\07	.21960\09	.69769\11	.62671\12	.10015\12

M	B= 6	B= 7	B= 8	B= 9	B=10
1	.32159893	.30479261	.29091104	.27917364	.26906587
2	.06958564	.06189605	.05593587	.05116635	.04725360
3	.01205095	.01000486	.00852194	.00740216	.00652922
4	.00179843	.00138784	.00111045	.00091329	.00076756
5	.00024067	.00017204	.00012895	.00010019	.80064\04
6	.29594\04	.19541\04	.13687\04	.10025\04	.76048\05
7	.33992\05	.20678\05	.13506\05	.93093\06	.66930\06
8	.36899\06	.20631\06	.12541\06	.81217\07	.55265\07
9	.38182\07	.19580\07	.11058\07	.67183\08	.43212\08
10	.37915\08	.17799\08	.93241\09	.53069\09	.32226\09
11	.36321\09	.15580\09	.75593\10	.40255\10	.23053\10
12	.33708\10	.13190\10	.59191\11	.29457\11	.15892\11
13	.30413\11	.10840\11	.44929\12	.20872\12	.10598\12
14	.26755\12	.86726\13	.33162\13	.14365\13	.68586\14
15	.23004\13	.67725\14	.23863\14	*	*

M	B=11	B=12	B=13	B=14	B=15
1	.26023245	.25241889	.24543731	.23914544	.23343329
2	.04397922	.04119404	.03879258	.03669805	.03485314
3	.00583118	.00526127	.00478786	.00438884	.00404827
4	.00065641	.00056945	.00049995	.00044341	.00039671
5	.65450\04	.54509\04	.46108\04	.39519\04	.34256\04
6	.59336\05	.47378\05	.38566\05	.31909\05	.26773\05
7	.49777\06	.38060\06	.29782\06	.23765\06	.19284\06
8	.39128\07	.28619\07	.21508\07	.16538\07	.12968\07
9	.29094\08	.20336\08	.14664\08	.10857\08	.82211\09
10	.20611\09	.13755\09	.95094\10	.67738\10	.49500\10
11	.13993\10	.89079\11	.59000\11	.40408\11	.28478\11
12	.91464\12	.55499\12	.35190\12	.23157\12	.15730\12
13	.57784\13	.33395\13	.20257\13	.12801\13	.83764\14
14	.35399\14	.19471\14	.11292\14	*	*
15	*	*	*	*	*

DIRICHLET D INTEGRAL
A = 4 R = 3

M	B= 1	B= 2	B= 3	B= 4	B= 5
1	.48800000	.38220800	.32945284	.29619069	.27266276
2	.18080000	.10402095	.07480998	.05922992	.04946408
3	.05792000	.02357552	.01387538	.00955636	.00717725
4	.01696000	.00475318	.00225663	.00133953	.00089881
5	.00467200	.00088266	.00033413	.00016965	.00010115
6	.00123136	.00015415	.46077\04	.19882\04	.10484\04
7	.00031386	.25665\04	.60071\05	.21904\05	.10173\05
8	.77926\04	.41123\05	.74819\06	.22940\06	.93489\07
9	.18944\04	.63846\06	.89716\07	.23027\07	.82074\08
10	.45261\05	.96542\07	.10418\07	.22292\08	.69281\09
11	.10658\05	.14274\07	.11767\08	.20916\09	.56523\10
12	.24789\06	.20699\08	.12976\09	.19095\10	.44754\11
13	.57049\07	.29514\09	.14010\10	.17016\11	.34508\12
14	.13009\07	.41464\10	.14847\11	.14841\12	.25983\13
15	.29426\08	.57492\11	.15473\12	.12695\13	.19150\14

M	B= 6	B= 7	B= 8	B= 9	B=10
1	.25483320	.24068573	.22908412	.21933190	.21097474
2	.04272972	.03778304	.03398253	.03096289	.02850035
3	.00569314	.00468848	.00396788	.00342829	.00301056
4	.00065139	.00049763	.00039500	.00032275	.00026977
5	.66696\04	.47111\04	.34978\04	.26967\04	.21412\04
6	.62671\05	.40812\05	.28276\05	.20528\05	.15458\05
7	.54956\06	.32911\06	.21233\06	.14490\06	.10332\06
8	.45512\07	.25006\07	.14994\07	.96035\08	.64751\08
9	.35911\08	.18065\08	.10050\08	.60320\09	.38409\09
10	.27181\09	.12494\09	.64388\10	.36165\10	.21722\10
11	.19840\10	.83187\11	.39651\11	.20816\11	.11781\11
12	.14026\11	.53554\12	.23577\12	.11555\12	.61553\13
13	.96376\13	.33458\13	.13587\13	.62092\14	.31104\14
14	.64555\14	.20347\14	*	*	*
15	*	*	*	*	*

M	B=11	B=12	B=13	B=14	B=15
1	.20370166	.19729151	.19158202	.18645101	.18180444
2	.02644993	.02471340	.02322181	.02192521	.02078654
3	.00267848	.00240872	.00218562	.00199831	.00183899
4	.00022963	.00019840	.00017357	.00015346	.00013691
5	.17408\04	.14429\04	.12155\04	.10380\04	.89679\05
6	.11986\05	.95185\06	.77116\06	.63540\06	.53115\06
7	.76301\07	.57990\07	.45139\07	.35852\07	.28972\07
8	.45490\08	.33051\08	.24694\08	.18891\08	.14746\08
9	.25641\09	.17792\09	.12749\09	.93860\10	.70723\10
10	.13766\10	.91141\11	.62579\11	.44308\11	.32206\11
11	.70800\12	.44689\12	.29381\12	.19992\12	.14009\12
12	.35050\13	.21075\13	.13258\13	.86643\14	.58495\14
13	.16768\14	*	*	*	*
14	*	*	*	*	*
15	*	*	*	*	*

DIRICHLET D INTEGRAL
A = 5 R = 3

M	B= 1	B= 2	B= 3	B= 4	B= 5
1	.42129630	.32319576	.27599076	.24674394	.22628001
2	.13194444	.07299269	.05161088	.04046003	.03356554
3	.03549383	.01364595	.00784045	.00532605	.00396374
4	.00870199	.00226317	.00104177	.00060767	.00040308
5	.00200403	.00034519	.00012584	.62559\04	.36788\04
6	.00044117	.49469\04	.14144\04	.59543\05	.30898\05
7	.93871\04	.67544\05	.15021\05	.53246\06	.24281\06
8	.19449\04	.88714\06	.15233\06	.45243\07	.18065\07
9	.39444\05	.11287\06	.14867\07	.36833\08	.12834\08
10	.78602\06	.13982\07	.14048\08	.28913\09	.87654\10
11	.15436\06	.16932\08	.12909\09	.21992\10	.57847\11
12	.29937\07	.20109\09	.11579\10	.16274\11	.37044\12
13	.57445\08	.23479\10	.10168\11	.11753\12	.23097\13
14	.10921\08	.27008\11	.87632\13	.83061\14	.14062\14
15	.20595\09	.30660\12	.74265\14	*	*

M	B= 6	B= 7	B= 8	B= 9	B=10
1	.21088975	.19874715	.18883405	.18053139	.17343794
2	.02885589	.02542060	.02279566	.02071924	.01903209
3	.00312345	.00255933	.00215732	.00185784	.00162698
4	.00028973	.00021996	.00017373	.00014138	.00011777
5	.24021\04	.16842\04	.12431\04	.95385\05	.75437\05
6	.18262\05	.11791\05	.81145\06	.58591\06	.43919\06
7	.12950\06	.76804\07	.49176\07	.33355\07	.23662\07
8	.86686\08	.47119\08	.28016\08	.17822\08	.11949\08
9	.55271\09	.27476\09	.15145\09	.90222\10	.57095\10
10	.33796\10	.15335\10	.78235\11	.43587\11	.26005\11
11	.19925\11	.82378\12	.38839\12	.20211\12	.11356\12
12	.11375\12	.42782\13	.18615\13	.90372\14	.47771\14
13	.63113\14	.21558\14	*	*	*
14	*	*	*	*	*
15	*	*	*	*	*

M	B=11	B=12	B=13	B=14	B=15
1	.16728052	.16186572	.15705217	.15273378	.14882914
2	.01763164	.01644875	.01543506	.01455570	.01378487
3	.00144411	.00129602	.00117388	.00107158	.00098474
4	.99960\04	.86152\04	.75206\04	.66364\04	.59105\04
5	.61124\05	.50518\05	.42447\05	.36166\05	.31185\05
6	.33923\06	.26851\06	.21691\06	.17827\06	.14868\06
7	.17399\07	.13175\07	.10222\07	.80961\08	.65259\08
8	.83546\09	.60453\09	.45008\09	.34323\09	.26718\09
9	.37918\10	.26194\10	.18696\10	.13718\10	.10305\10
10	.16387\11	.10797\11	.73828\12	.52081\12	.37732\12
11	.67836\13	.42596\13	.27880\13	.18896\13	.13194\13
12	.27026\14	.16160\14	.10117\14	*	*
13	*	*	*	*	*
14	*	*	*	*	*
15	*	*	*	*	*

DIRICHLET D INTEGRAL
1/A = 5 R = 4

M	B= 1	B= 2	B= 3	B= 4	B= 5
1	.99922840	.99847245	.99773116	.99700364	.99628908
2	.99665638	.99342111	.99028507	.98724047	.98428056
3	.99129801	.98301312	.97510099	.96752485	.96025367
4	.98236740	.96591427	.95048636	.93595972	.92223292
5	.96934359	.94142098	.91580163	.89215175	.87020658
6	.95197851	.90945863	.87141662	.83708858	.80588866
7	.93027216	.87049911	.81852767	.77281578	.73222244
8	.90443126	.82543650	.75890816	.70203023	.65279692
9	.87482191	.77544949	.69464703	.62767849	.57130340
10	.84192262	.72186935	.62791251	.55261778	.49110906
11	.80628173	.66606533	.56076239	.47936853	.41497901
12	.76848078	.60935321	.49500939	.40996470	.34494307
13	.72910480	.55292877	.43214058	.34589189	.28228219
14	.68871917	.49782475	.37328399	.28809502	.22759954
15	.64785276	.44488839	.31921224	.23703402	.18094124

M	B= 6	B= 7	B= 8	B= 9	B=10
1	.99558680	.99489615	.99421655	.99354750	.99288850
2	.98139943	.97859189	.97585330	.97317953	.97056686
3	.95326102	.94652405	.94002290	.93374014	.92766035
4	.90922163	.89685484	.88507204	.87382121	.86305725
5	.84975213	.83061270	.81264211	.79571734	.77973378
6	.77735737	.75112757	.72690127	.70443332	.68351962
7	.69588088	.66311819	.63340226	.60630572	.58148058
8	.60973507	.57173510	.53794341	.50769140	.48044716
9	.52322477	.48176665	.44567628	.41399786	.38598914
10	.44005628	.39710910	.36056264	.32915124	.30191643
11	.36305821	.32051398	.28517229	.25546440	.23023289
12	.29409330	.25356746	.22074643	.19379512	.17139563
13	.23411458	.19682975	.16742678	.14386610	.12472312
14	.18331192	.15006037	.12455682	.10463560	.08882684
15	.14129971	.11246756	.09098383	.07464150	.06198620

M	B=11	B=12	B=13	B=14	B=15
1	.99223912	.99159895	.99096763	.99034479	.98973013
2	.96801194	.96551170	.96306336	.96066437	.95831239
3	.92176983	.91605633	.91050884	.90511740	.89987301
4	.85274072	.84283695	.83331524	.82414829	.81531169
5	.76460172	.75024362	.73659195	.72358756	.71117834
6	.66398856	.64569443	.62851260	.61233558	.59707015
7	.55864013	.53754565	.51799651	.49982267	.48287890
8	.45578176	.43334513	.41284853	.39405151	.37675213
9	.36106483	.33875708	.31868744	.30054648	.28407875
10	.27811973	.25718351	.23864982	.22215124	.20738982
11	.20860721	.18992142	.17365849	.15941161	.14685689
12	.15258100	.13662846	.12298872	.11123826	.10104618
13	.10897911	.09589043	.08490434	.07560335	.06766756
14	.07610678	.06574613	.05721502	.05012164	.04417170
15	.05203204	.04409413	.03768631	.03245683	.02814685

DIRICHLET D INTEGRAL
1/A = 4 R = 4

M	B= 1	B= 2	B= 3	B= 4	B= 5
1	.99840000	.99685642	.99536361	.99391689	.99251232
2	.99328000	.98693035	.98090339	.97516127	.96967324
3	.98304000	.96743383	.95296482	.93946805	.92681430
4	.96665600	.93695067	.91017804	.88582502	.86350523
5	.94371840	.89545970	.85337745	.81619489	.78299079
6	.91435827	.84408994	.78507556	.73461519	.69084719
7	.87912612	.78476444	.70877652	.64612435	.59349761
8	.83886080	.71983406	.62831853	.55580724	.49698798
9	.79456895	.65175919	.54735038	.46810771	.40619751
10	.74732431	.58286660	.46898162	.38644241	.32448982
11	.69818988	.51518696	.39560062	.31307803	.25371290
12	.64816210	.45036569	.32883029	.24920011	.19442147
13	.59813433	.38963333	.26958103	.19509556	.14620290
14	.54887620	.33381938	.21816172	.15038246	.10801670
15	.50102546	.28339447	.17441578	.11423896	.07849181

M	B= 6	B= 7	B= 8	B= 9	B=10
1	.99114656	.98981672	.98852026	.98725496	.98601886
2	.96441382	.95936160	.95449836	.94980841	.94527813
3	.91490002	.90364063	.89296599	.88281719	.87314416
4	.84291962	.82383144	.80604976	.78941823	.77380703
5	.75307806	.72593042	.70113592	.67836605	.65735481
6	.65243681	.61839680	.58797807	.56060004	.53580451
7	.54861983	.50986789	.47605000	.44626945	.41983744
8	.44836426	.40753731	.37280522	.34292647	.31697388
9	.35669526	.31635575	.28295744	.25493106	.23113844
10	.27667734	.23893114	.20856424	.18374090	.16316942
11	.20957950	.17588266	.14957920	.12866065	.11175732
12	.15526697	.12640028	.10456718	.08769518	.07441583
13	.11266235	.08882168	.07137302	.05828581	.04826307
14	.08017107	.06111602	.04763765	.03783634	.03053950
15	.05601718	.04123113	.03113515	.02402456	.01888329

M	B=11	B=12	B=13	B=14	B=15
1	.98481019	.98362738	.98246899	.98133374	.98022044
2	.94089559	.93665025	.93253278	.92853483	.92464892
3	.86390394	.85505933	.84657793	.83843126	.83059417
4	.75910718	.74522622	.73208501	.71961529	.70775775
5	.63788383	.61977174	.60286635	.58703873	.57217881
6	.51322423	.49256082	.47356898	.45604488	.43981757
7	.39621541	.37497574	.35577430	.33833082	.32241457
8	.29424100	.27418011	.25636010	.24043702	.22613318
9	.21073527	.19308297	.17769025	.16417345	.15222896
10	.14591779	.13129882	.11879623	.10801560	.09865116
11	.09790877	.08642527	.07680105	.06865841	.06171069
12	.06379742	.05518870	.04812394	.04226334	.03735454
13	.04044811	.03425854	.02928842	.02524859	.02192900
14	.02499632	.02071071	.01734589	.01466774	.01251005
15	.01508118	.01221384	.01001405	.00830061	.00694787

DIRICHLET D INTEGRAL
1/A = 3 R = 4

M	B= 1	B= 2	B= 3	B= 4	B= 5
1	.99609375	.99244800	.98902011	.98577847	.98269873
2	.98437500	.97032960	.95753910	.94577640	.93487481
3	.96240234	.93014784	.90189841	.87677562	.85416758
4	.92944336	.87219763	.82430329	.78333340	.74768956
5	.88618469	.79962163	.73088611	.67461251	.62747216
6	.83427429	.71712885	.62938687	.56095980	.50597387
7	.77587509	.62984256	.52727004	.45140720	.39313423
8	.71330452	.54247028	.43056341	.35239537	.29518687
9	.64877862	.45882379	.34338265	.26752308	.21478756
10	.58425272	.38163348	.26795832	.19794325	.15184398
11	.52133996	.31257085	.20495383	.14304263	.10453910
12	.46128688	.25239460	.15389770	.10114581	.07023710
13	.40498711	.20115182	.11361049	.07010019	.04614018
14	.35301810	.15838664	.08256144	.04769099	.02968585
15	.30568917	.12332764	.05913101	.03189256	.01873412

M	B= 6	B= 7	B= 8	B= 9	B=10
1	.97976160	.97695142	.97425527	.97166231	.96916335
2	.92470756	.91517548	.90619929	.89771438	.88966733
3	.83362907	.81482509	.79749673	.78143942	.76648861
4	.71626260	.68825143	.66305891	.64022830	.61940273
5	.58726837	.55248194	.52202274	.49508479	.47105731
6	.46075312	.42286859	.39064551	.36288888	.33872157
7	.34705345	.30976444	.27901841	.25326902	.23141836
8	.25180368	.21796988	.19097778	.16903572	.15091465
9	.17653235	.14783869	.12572905	.10831009	.09432854
10	.11992911	.09695121	.07987865	.06686141	.05671910
11	.07915534	.06164473	.04911883	.03988827	.03291521
12	.05087329	.03809670	.02931038	.02306041	.01848773
13	.03190392	.02293397	.01701201	.01295066	.01007585
14	.01955859	.01347468	.00962371	.00708029	.00534019
15	.01174033	.00774035	.00531589	.00377548	.00275782

M	B=11	B=12	B=13	B=14	B=15
1	.96675048	.96441685	.96215648	.75996410	.95783504
2	.88201334	.87471444	.86773813	.86105634	.85464464
3	.75250980	.73939164	.72704082	.71537838	.70433688
4	.60029823	.58268535	.56637617	.55121495	.53707128
5	.44946769	.42994367	.41218751	.39595780	.38105650
6	.31748431	.29867144	.28188832	.26682219	.25322186
7	.21266571	.19641366	.18220762	.16969572	.15860150
8	.13574580	.12289917	.11190780	.10241892	.09416156
9	.08292656	.07350013	.06561349	.05894555	.05325535
10	.04867016	.04218072	.03687617	.03248748	.02881753
11	.02753560	.02330967	.01993759	.01720960	.01497571
12	.01506159	.01244162	.01040236	.00879036	.00749855
13	.00798584	.00643128	.00525188	.00434152	.00362803
14	.00411377	.00322717	.00257200	.00207850	.00170044
15	.00206306	.00157528	.00122442	.00096667	.00077377

DIRICHLET D INTEGRAL
1/A = 2 R = 4

M	B= 1	B= 2	B= 3	B= 4	B= 5
1	.98765432	.97705078	.96769638	.95929366	.95164609
2	.95473251	.91870117	.88872141	.86304529	.84059876
3	.89986283	.82714844	.77067551	.72490825	.68670032
4	.82670325	.71362305	.63243846	.57060884	.52157652
5	.74135040	.59185791	.49353431	.42360959	.37118679
6	.65030377	.47372437	.36821619	.29855920	.24940622
7	.55926434	.36734009	.26398098	.20097625	.15929946
8	.47255669	.27693176	.18267413	.12989199	.09728500
9	.39307468	.20360374	.12249164	.08096360	.05708864
10	.32242400	.14637852	.07985710	.04885696	.03232551
11	.26119341	.10314751	.05076278	.02863582	.01772478
12	.20924019	.07138443	.03154097	.01634764	.00944017
13	.16594583	.04860353	.01919684	.00911194	.00489647
14	.13042226	.03260684	.01146606	.00496920	.00247901
15	.10166508	.02158237	.00673180	.00265630	.00122752

M	B= 6	B= 7	B= 8	B= 9	B=10
1	.94461559	.93810076	.93202455	.92632679	.92095947
2	.82067047	.80276320	.78651490	.77165338	.75796859
3	.65409471	.62579355	.60089357	.57874202	.55885261
4	.48152550	.44806156	.41959568	.39502619	.37356206
5	.33035527	.29761644	.27076089	.24832172	.22928527
6	.21302580	.18511215	.16308343	.14530094	.13067648
7	.13007978	.10868118	.09246958	.07984939	.06980382
8	.07569374	.06063757	.04970902	.04151858	.03521764
9	.04219823	.03233299	.02548500	.02055175	.01688875
10	.02263922	.01655475	.01252260	.00973455	.00773902
11	.01173312	.00817166	.00592210	.00443117	.00340372
12	.00589334	.00390203	.00270501	.00194557	.00144223
13	.00287685	.00180773	.00119700	.00082654	.00059066
14	.00136814	.00081460	.00051451	.00034068	.00023445
15	.00063521	.00035783	.00021531	.00013656	.90420\04

M	B=11	B=12	B=13	B=14	B=15
1	.91588358	.91106693	.90648260	.90210782	.89792316
2	.74529492	.73349945	.72247378	.71212839	.70238844
3	.54085392	.52445628	.50942968	.49558867	.48278170
4	.35461876	.33775387	.32262565	.30896541	.29655868
5	.21292782	.19871840	.18625834	.17524233	.16543249
6	.11846030	.10811977	.09926656	.09161121	.08493376
7	.06165764	.05494682	.04934317	.04460876	.04056737
8	.03026363	.02629626	.02306852	.02040638	.01818431
9	.01409977	.01193075	.01021302	.00883117	.00770419
10	.00626925	.00516029	.00430616	.00363648	.00310322
11	.00267244	.00213763	.00173734	.00143169	.00119420
12	.00109639	.00085144	.00067341	.00054111	.00044088
13	.00043433	.00032719	.00025163	.00019702	.00015670
14	.00016661	.00012165	.90911\04	.69315\04	.53783\04
15	.62045\04	.43877\04	.31839\04	.23624\04	.17873\04

DIRICHLET D INTEGRAL
A = 1 R = 4

M	B= 1	B= 2	B= 3	B= 4	B= 5
1	.93750000	.89666209	.86619568	.84189039	.82168474
2	.81250000	.71132449	.64424896	.59512323	.55692874
3	.65625000	.50810344	.42219067	.36498811	.32367936
4	.50000000	.33333333	.25000000	.20000000	.16666667
5	.36328125	.20411861	.13641471	.10010009	.07787331
6	.25390625	.11813353	.06960723	.04651000	.03359635
7	.17187500	.06523584	.03358439	.02030689	.01355736
8	.11328125	.03462724	.01545299	.00840936	.00516795
9	.07299805	.01776951	.00682621	.00332707	.00187528
10	.04614258	.00885640	.00291040	.00126491	.00065175
11	.02868652	.00430308	.00120284	.00046431	.00021804
12	.01757813	.00204441	.00048360	.00016519	.70503\04
13	.01063538	.00095219	.00018971	.57146\04	.22110\04
14	.00636292	.00043568	.72791\04	.19277\04	.67448\05
15	.00376892	.00019619	.27377\04	.63550\05	.20064\05

M	B= 6	B= 7	B= 8	B= 9	B=10
1	.80440881	.78933259	.77596922	.76397743	.75310863
2	.52601872	.50027173	.47835278	.45937142	.44270648
3	.29220038	.26727476	.24696259	.23003534	.21567376
4	.14285714	.12500000	.11111111	.10000000	.09090909
5	.06305304	.05255872	.04478877	.03883460	.03414541
6	.02558862	.02025070	.01649740	.01374781	.01166694
7	.00967840	.00725084	.00563335	.00450261	.00368156
8	.00344703	.00243842	.00180288	.00137962	.00108517
9	.00116535	.00077661	.00054539	.00039893	.00030144
10	.00037637	.00023579	.00015701	.00010961	.79465\04
11	.00011673	.68616\04	.43253\04	.28781\04	.19995\04
12	.34915\04	.19223\04	.11454\04	.72551\05	.48247\05
13	.10108\04	.52035\05	.29266\05	.17626\05	.11208\05
14	.28410\05	.13654\05	.72391\06	.41406\06	.25154\06
15	.77726\06	.34821\06	.17382\06	.94324\07	.54691\07

M	B=11	B=12	B=13	B=14	B=15
1	.74317609	.73403593	.72557498	.71770256	.71034493
2	.42790884	.41464421	.40265758	.39175047	.38176557
3	.20330865	.19253114	.18303930	.17460486	.16705179
4	.08333333	.07692308	.07142857	.06666667	.06250000
5	.03036931	.02727181	.02469108	.02251208	.02065097
6	.01005003	.00876587	.00772704	.00687339	.00616233
7	.00306677	.00259458	.00222408	.00192804	.00168775
8	.00087297	.00071552	.00059581	.00050288	.00042943
9	.00023389	.00018553	.00014992	.00012310	.00010247
10	.59406\04	.45557\04	.35697\04	.28488\04	.23099\04
11	.14387\04	.10657\04	.80899\05	.62710\05	.49495\05
12	.33379\05	.23863\05	.17537\05	.13195\05	.10131\05
13	.74496\06	.51358\06	.36513\06	.26649\06	.19894\06
14	.16047\06	.10661\06	.73271\07	.51841\07	.37605\07
15	.33463\07	.21406\07	.14214\07	.97436\08	.68645\08

DIRICHLET D INTEGRAL

A = 2 R = 4

M	B= 1	B= 2	B= 3	B= 4	B= 5
1	.80246914	.71942400	.66824914	.63190387	.60404016
2	.53909465	.40648960	.33811138	.29492168	.26457404
3	.31961591	.19464448	.14181654	.11246842	.09370470
4	.17329675	.08277453	.05199532	.03710585	.02850008
5	.08794391	.03219712	.01721885	.01096713	.00771989
6	.04242239	.01168173	.00526315	.00297129	.00190729
7	.01966164	.00400773	.00150747	.00074980	.00043700
8	.00882318	.00131310	.00040906	.00017831	.93988\04
9	.00385555	.00041394	.00010604	.40313\04	.19151\04
10	.00164772	.00012627	.26431\04	.87248\05	.37234\05
11	.00069099	.37443\04	.63668\05	.18176\05	.69474\06
12	.00028511	.10832\04	.14883\05	.36608\06	.12498\06
13	.00011599	.30662\05	.33876\06	.71551\07	.21762\07
14	.46608\04	.85139\06	.75297\07	.13612\07	.36795\08
15	.18525\04	.23237\06	.16382\07	.25273\08	.60576\09

M	B= 6	B= 7	B= 8	B= 9	B=10
1	.58162472	.56298291	.54709729	.53330585	.52115497
2	.24179136	.22389721	.20937314	.19728621	.18702794
3	.08062847	.07096809	.06352380	.05760124	.05277014
4	.02295886	.01912208	.01632308	.01419923	.01253745
5	.00579616	.00455179	.00369461	.00307559	.00261183
6	.00132959	.00098153	.00075567	.00060073	.00048977
7	.00028191	.00019511	.00014217	.00010773	.84191\04
8	.55941\04	.36212\04	.24923\04	.17973\04	.13445\04
9	.10488\04	.63359\05	.41117\05	.28175\05	.20148\05
10	.18715\05	.10531\05	.64328\06	.41826\06	.28558\06
11	.31976\06	.16726\06	.96032\07	.59171\07	.38531\07
12	.52561\07	.25515\07	.13748\07	.80177\08	.49742\08
13	.83455\08	.37534\08	.18955\08	.10450\08	.61708\09
14	.12842\08	.53430\09	.25256\09	.13149\09	.73832\10
15	.19207\09	.73815\10	.32619\10	.16021\10	.85462\11

M	B=11	B=12	B=13	B=14	B=15
1	.51032113	.50056579	.49170845	.48360937	.47615830
2	.17818258	.17045527	.16363038	.15754616	.15207858
3	.04874939	.04534737	.04242887	.03989567	.03767470
4	.01120480	.01011433	.00920687	.00844088	.00778635
5	.00225402	.00197125	.00174325	.00155628	.00140072
6	.00040752	.00034482	.00029589	.00025694	.00022542
7	.67450\04	.55151\04	.45872\04	.38710\04	.33075\04
8	.10358\04	.81757\05	.65850\05	.53954\05	.44862\05
9	.14910\05	.11349\05	.88440\06	.70303\06	.56846\06
10	.20280\06	.14872\06	.11204\06	.86342\07	.67847\07
11	.26230\07	.18517\07	.13475\07	.10061\07	.76779\08
12	.32431\08	.22022\08	.15469\08	.11182\08	.82830\09
13	.38499\09	.25126\09	.17025\09	.11908\09	.85570\10
14	.44042\10	.27606\10	.18032\10	.12196\10	.84977\11
15	.48705\11	.29300\11	.18438\11	.12053\11	.81384\12

DIRICHLET D INTEGRAL
A = 3 R = 4

M	B= 1	B= 2	B= 3	B= 4	B= 5
1	.68359375	.58671010	.53258172	.49615374	.46920166
2	.36718750	.25308818	.20129108	.17071014	.15014091
3	.16943359	.09009903	.06149911	.04685693	.03796967
4	.07055664	.02811498	.01622183	.01099713	.00815440
5	.02729797	.00796501	.00383749	.00229663	.00154964
6	.00999451	.00209497	.00083413	.00043775	.00026749
7	.00350571	.00051941	.00016937	.77486\04	.42698\04
8	.00118828	.00012271	.32512\04	.12897\04	.63845\05
9	.00039166	.27848\04	.59520\05	.20376\05	.90299\06
10	.00012612	.61081\05	.10463\05	.30778\06	.12171\06
11	.39823\04	.13010\05	.17759\06	.44703\07	.15728\07
12	.12364\04	.27016\06	.29226\07	.62724\08	.19580\08
13	.37833\05	.54857\07	.46803\08	.85345\09	.23577\09
14	.11430\05	.10921\07	.73150\09	.11297\09	.27552\10
15	.34145\06	.21361\08	.11186\09	.14586\10	.31334\11

M	B= 6	B= 7	B= 8	B= 9	B=10
1	.44807222	.43084627	.41640002	.40402303	.39323981
2	.13517552	.12369930	.11456050	.10707363	.10080292
3	.03200040	.02771193	.02447955	.02195407	.01992505
4	.00640001	.00522341	.00438636	.00376410	.00328548
5	.00112786	.00086464	.00068836	.00056393	.00047246
6	.00017988	.00012916	.97255\04	.75913\04	.60944\04
7	.26444\04	.17739\04	.12606\04	.93589\05	.71889\05
8	.36311\05	.22701\05	.15198\05	.10714\05	.78635\06
9	.47033\06	.27348\06	.17217\06	.11509\06	.80606\07
10	.57917\07	.31258\07	.18475\07	.11694\07	.78067\08
11	.68217\08	.34111\08	.18900\08	.11312\08	.71905\09
12	.77240\09	.35722\09	.18527\09	.10474\09	.63323\10
13	.84421\10	.36051\10	.17479\10	.93226\11	.53554\11
14	.89374\11	.35188\11	.15927\11	.80061\12	.43660\12
15	.91920\12	.33317\12	.14062\12	.66546\13	.34419\13

M	B=11	B=12	B=13	B=14	B=15
1	.38371768	.37521550	.36755335	.36059371	.35422926
2	.09545687	.09083247	.08678354	.08320193	.08000568
3	.01825815	.01686357	.01567899	.01465982	.01377330
4	.00290721	.00260157	.00235003	.00213977	.00196168
5	.00040302	.00034891	.00030581	.00027084	.00024203
6	.50044\04	.41859\04	.35557\04	.30600\04	.26630\04
7	.56750\05	.45812\05	.37677\05	.31478\05	.26656\05
8	.59606\06	.46390\06	.36905\06	.29909\06	.24626\06
9	.58605\07	.43930\07	.33777\07	.26532\07	.21227\07
10	.54386\08	.39231\08	.29129\08	.22162\08	.17217\08
11	.47953\09	.33258\09	.23829\09	.17549\09	.13230\09
12	.40389\10	.26911\10	.18594\10	.13246\10	.96851\11
13	.32641\11	.20879\11	.13902\11	.95735\12	.67856\12
14	.25408\12	.15591\12	.99969\13	.66517\13	.45678\13
15	.19110\13	.11242\13	.69374\14	.44574\14	.29642\14

DIRICHLET D INTEGRAL

A = 4 R = 4

M	B= 1	B= 2	B= 3	B= 4	B= 5
1	.59040000	.49206445	.44026313	.40645581	.38193280
2	.26272000	.17097258	.13237842	.11044864	.09605083
3	.09888000	.04831846	.03174420	.02364155	.01886246
4	.03334400	.01188741	.00653011	.00430105	.00312783
5	.01040640	.00264505	.00120040	.00069391	.00045747
6	.00306637	.00054513	.00020230	.00010196	.60650\04
7	.00086436	.00010573	.31798\04	.13891\04	.74250\05
8	.00023521	.19521\04	.47198\05	.17778\05	.85062\06
9	.62198\04	.34591\05	.66759\06	.21578\06	.92100\07
10	.16060\04	.59203\06	.90617\07	.25025\07	.94977\08
11	.40647\05	.98357\07	.11869\07	.27892\08	.93854\09
12	.10113\05	.15923\07	.15069\08	.30021\09	.89312\10
13	.24789\06	.25202\08	.18609\09	.31323\10	.82177\11
14	.59985\07	.39095\09	.22423\10	.31784\11	.73358\12
15	.14351\07	.59575\10	.26429\11	.31453\12	.63717\13

M	B= 6	B= 7	B= 8	B= 9	B=10
1	.36297803	.34769116	.33498108	.32416850	.31480432
2	.08575306	.07795762	.07181309	.06682119	.06266938
3	.01571615	.01348945	.01183082	.01054732	.00952437
4	.00242022	.00195370	.00162622	.00138540	.00120184
5	.00032739	.00024776	.00019521	.00015857	.00013190
6	.40002\04	.28299\04	.21058\04	.16279\04	.12963\04
7	.44989\05	.29679\05	.20814\05	.15286\05	.11636\05
8	.47212\06	.28976\06	.19117\06	.13317\06	.96766\07
9	.46703\07	.26610\07	.16486\07	.10877\07	.75361\08
10	.43892\08	.23172\08	.13460\08	.83999\09	.55422\09
11	.39439\09	.19256\09	.10471\09	.61730\10	.38745\10
12	.34052\10	.15350\10	.78031\11	.43403\11	.25888\11
13	.28371\11	.11789\11	.55944\12	.29327\12	.16607\12
14	.22890\12	.87536\13	.38731\13	.19114\13	.10266\13
15	.17936\13	.63037\14	.25973\14	.12055\14	*

M	B=11	B=12	B=13	B=14	B=15
1	.30657756	.29926473	.29270034	.28675871	.28134235
2	.05915095	.05612327	.05348447	.05115971	.04909264
3	.00868972	.00799555	.00740896	.00690660	.00647141
4	.00105786	.00094230	.00084774	.00076911	.00070281
5	.00011183	.96297\04	.84005\04	.74090\04	.65960\04
6	.10571\04	.87887\05	.74261\05	.63608\05	.55122\05
7	.91146\06	.73087\06	.59757\06	.49666\06	.41861\06
8	.72729\07	.56188\07	.44413\07	.35789\07	.29318\07
9	.54288\08	.40370\08	.30824\08	.24063\08	.19146\08
10	.38227\09	.27338\09	.20147\09	.15227\09	.11759\09
11	.25564\10	.17567\10	.12486\10	.91303\11	.68399\11
12	.16325\11	.10771\11	.73786\12	.52167\12	.37889\12
13	% .999999\13	.63302\13	.41766\13	.28534\13	.20082\13
14	.58984\14	.35797\14	.22734\14	.15000\14	.10224\14
15	*	*	*	*	*

DIRICHLET D INTEGRAL
A = 5 R = 4

M	B= 1	B= 2	B= 3	B= 4	B= 5
1	.51774691	.42263297	.37445024	.34362140	.32153859
2	.19624486	.12276203	.09337587	.07708444	.06655123
3	.06228567	.02873959	.01841528	.01351538	.01067724
4	.01763260	.00583253	.00310342	.00200675	.00144151
5	.00460879	.00106807	.00046635	.00026369	.00017132
6	.00113578	.00018090	.64157\04	.31515\04	.18433\04
7	.00026752	.28809\04	.82248\05	.34895\05	.18299\05
8	.60794\04	.43640\05	.99500\06	.36271\06	.16989\06
9	.13419\04	.63417\06	.11465\06	.35738\07	.14900\07
10	.28916\05	.88981\07	.12673\07	.33632\08	.12441\08
11	.61056\06	.12115\07	.13513\08	.30410\09	.99515\10
12	.12671\06	.16071\08	.13963\09	.26545\10	.76637\11
13	.25905\07	.20837\09	.14032\10	.22458\11	.57053\12
14	.52276\08	.26477\10	.13756\11	.18476\12	.41201\13
15	.10429\08	.33044\11	.13189\12	.14821\13	.28946\14

M	B= 6	B= 7	B= 8	B= 9	B=10
1	.30462175	.29107064	.27986421	.27037278	.26218320
2	.05909808	.05350144	.04911807	.04557540	.04264171
3	.00883208	.00753841	.00658183	.00584599	.00526244
4	.00110546	.00088626	.00073366	.00062219	.00053770
5	.00012132	.91069\04	.71295\04	.57607\04	.47707\04
6	.12011\04	.84191\05	.62192\05	.47788\05	.37864\05
7	.10937\05	.71407\06	.49671\06	.36235\06	.27428\06
8	.92873\07	.56348\07	.36841\07	.25475\07	.18398\07
9	.74304\08	.41807\08	.25647\08	.16786\08	.11552\08
10	.56460\09	.29402\09	.16897\09	.10453\09	.68473\10
11	.41004\10	.19727\10	.10604\10	.61932\11	.38572\11
12	.28608\11	.12694\11	.63734\12	.35098\12	.20762\12
13	.19257\12	.78676\13	.36847\13	.19112\13	.10728\13
14	.12550\13	.47139\14	.20567\14	.10036\14	*
15	*	*	*	*	*

M	B=11	B=12	B=13	B=14	B=15
1	.25501119	.24865352	.24296041	.23781855	.23314037
2	.04016477	.03804014	.03619362	.03457092	.03313133
3	.00478830	.00439540	.00406445	.00378182	.00353760
4	.00047174	.00041901	.00037602	.00034038	.00031042
5	.40294\04	.34583\04	.30082\04	.26463\04	.23505\04
6	.30744\05	.25467\05	.21448\05	.18319\05	.15834\05
7	.21383\06	.17076\06	.13911\06	.11525\06	.96864\07
8	.13756\07	.10579\07	.83294\08	.66886\08	.54622\08
9	.82746\09	.61233\09	.46553\09	.36206\09	.28710\09
10	.46941\10	.33394\10	.24496\10	.18439\10	.14189\10
11	.25283\11	.17277\11	.12219\11	.88965\12	.66391\12
12	.13001\12	.85267\13	.58106\13	.40893\13	.29579\13
13	.64118\14	.40330\14	.26462\14	.17991\14	.12607\14
14	*	*	*	*	*
15	*	*	*	*	*

DIRICHLET D INTEGRAL
1/A = 5 R = 5

M	B= 1	B= 2	B= 3	B= 4	B= 5
1	.99987140	.99974396	.99961765	.99949241	.99936822
2	.99933556	.99868086	.99803541	.99739877	.99677053
3	.99799597	.99603638	.99411840	.99223953	.99039753
4	.99539121	.99092911	.98660206	.98240002	.97831420
5	.99104994	.98249149	.97428676	.96640378	.95881523
6	.98453803	.96997343	.95620207	.94313837	.93071087
7	.97549372	.95281785	.93172159	.91200453	.89350345
8	.96364998	.93070518	.90062214	.87297948	.84744312
9	.94884530	.90356860	.86309045	.82660056	.79347458
10	.93102485	.87158453	.81968637	.77388528	.73310099
11	.91023433	.83514452	.77127486	.71620936	.66819886
12	.88660874	.79481462	.71893963	.65516492	.60081724
13	.86035808	.75128839	.66389081	.59241888	.53298979
14	.83175160	.70533838	.60737632	.52958679	.46657957
15	.80110179	.65777018	.55060446	.46813231	.40316764

M	B= 6	B= 7	B= 8	B= 9	B=10
1	.99924503	.99912282	.99900155	.99888119	.99876173
2	.99615035	.99553788	.99493282	.99433488	.99374381
3	.98859041	.98681636	.98507376	.98336109	.98167700
4	.97433687	.97046117	.96668095	.96299071	.95938545
5	.95149750	.94443001	.93759467	.93097545	.92455809
6	.91885912	.90753141	.89668307	.88627521	.87627369
7	.87608336	.85963112	.84405086	.82926050	.81518919
8	.82374331	.80165885	.78100598	.76163027	.74340068
9	.76322108	.73544627	.70982983	.68610772	.66405987
10	.69650583	.66345250	.63342598	.60601032	.58086525
11	.62594066	.58844173	.55493000	.52479507	.49754726
12	.55396419	.51317156	.47735099	.44566085	.41743957
13	.48288313	.44013534	.40329357	.37125987	.34318931
14	.41469914	.37138267	.33478299	.30353715	.27661904
15	.35099972	.30842422	.27319364	.24369020	.21872324

M	B=11	B=12	B=13	B=14	B=15
1	.99864313	.99852536	.99840842	.99829227	.99817691
2	.99315935	.99258128	.99200939	.99144347	.99088335
3	.98002023	.97838962	.97678409	.97520265	.97364437
4	.95586065	.95241218	.94903626	.94572943	.94248848
5	.91832980	.91227907	.90639548	.90066960	.89509279
6	.86664839	.85737253	.84842220	.83977595	.83141446
7	.80177532	.78896490	.77671039	.76496963	.75370512
8	.72620500	.70994643	.69454086	.67991472	.66600335
9	.64350097	.62427362	.60624304	.58929297	.57332254
10	.55770920	.53630688	.51645983	.49799937	.48078107
11	.47278868	.45019240	.42948705	.41044537	.39287547
12	.39215970	.36939537	.34879887	.33008348	.31301061
13	.31842128	.29643205	.27680124	.25918768	.24331173
14	.25324250	.23279588	.21479682	.19886020	.18467526
15	.19739947	.17903724	.16310863	.14919929	.13698012

DIRICHLET D INTEGRAL
1/A = 4 R = 5

M	B= 1	B= 2	B= 3	B= 4	B= 5
1	.99968000	.99936573	.99905681	.99875291	.99845375
2	.99840000	.99684531	.99533211	.99385716	.99241767
3	.99532800	.99085188	.98655094	.98240816	.97840928
4	.98959360	.97979997	.97053983	.96175005	.95337921
5	.98041856	.96238858	.94566900	.93007450	.91545875
6	.96720650	.93778707	.91112272	.88675552	.86433566
7	.94959043	.90571756	.86698681	.83240817	.80125167
8	.92744450	.86644802	.81418110	.76872184	.72870550
9	.90086939	.82071979	.75428438	.69814995	.64998751
10	.87016037	.76963525	.68930364	.62354224	.56867207
11	.83576628	.71452890	.62143340	.54779206	.48816713
12	.79824544	.65684068	.55283987	.47354851	.41138134
13	.75822322	.59800412	.48549092	.40301801	.34053014
14	.71635382	.53935707	.42104089	.33785925	.27707196
15	.67328814	.48207778	.36076945	.27916006	.22174737

M	B= 6	B= 7	B= 8	B= 9	B=10
1	.99815907	.99786865	.99758226	.99729973	.99702088
2	.99101122	.98963567	.98828914	.98696994	.98567656
3	.97454216	.97079637	.96716284	.96363362	.96020167
4	.94538469	.93773066	.93038658	.92332621	.91652675
5	.90170373	.88871257	.87640459	.86471176	.85357609
6	.84358909	.82429664	.80628001	.78939202	.77350966
7	.77296284	.74711031	.72335173	.70141081	.68106135
8	.69312892	.66123396	.63243446	.60626840	.58236564
9	.60814334	.57140526	.53886130	.50981026	.48370292
10	.52217414	.48226024	.44762293	.41728255	.39048964
11	.43897462	.39775401	.36275914	.33271631	.30667460
12	.36153424	.32082202	.28705325	.25867334	.23455074
13	.29191784	.25327469	.22199812	.19629439	.17489160
14	.23126547	.19587454	.16795720	.14554546	.12728102
15	.17990612	.14852779	.12443166	.10555450	.09051140

M	B=11	B=12	B=13	B=14	B=15
1	.99674556	.99647361	.99620492	.99593935	.99567679
2	.98440762	.98316189	.98193824	.98073562	.97955309
3	.95686075	.95360525	.95043014	.94733086	.94430329
4	.90996822	.90363305	.89750562	.89157201	.88581974
5	.84294769	.83278328	.82304506	.81369975	.80471794
6	.75852901	.74436145	.73093073	.71817080	.70602400
7	.66211587	.64441727	.62783262	.61224852	.59756744
8	.56042543	.54020040	.52148479	.50410589	.48791738
9	.46010220	.43865543	.41907451	.40112156	.38459826
10	.36666052	.34533349	.32613838	.30877477	.29299624
11	.28390984	.26386097	.24608667	.23023514	.21602262
12	.21384342	.19591281	.18026648	.16651910	.15436515
13	.15686571	.14153121	.12837015	.11698488	.10706567
14	.11220115	.09960764	.08898394	.07994112	.07218176
15	.07834539	.06837844	.06011963	.05320668	.04736768

DIRICHLET D INTEGRAL
1/A = 3 R = 5

M	B= 1	B= 2	B= 3	B= 4	B= 5
1	.99902344	.99808512	.99718050	.99630598	.99545862
2	.99536133	.99100109	.98687767	.98295946	.97922164
3	.98712158	.97535232	.96448902	.95438445	.94492748
4	.97270203	.94860564	.92700763	.90742393	.88950435
5	.95107269	.90963524	.87375276	.84216526	.81400175
6	.92187309	.85881857	.80633364	.76165570	.72296583
7	.88537359	.79780846	.72809380	.67092766	.62299515
8	.84235632	.72913179	.64332732	.57580102	.52114134
9	.79396190	.65574423	.55650712	.48182080	.42362221
10	.74153460	.58063124	.47167603	.39352466	.33505697
11	.68648594	.50650402	.39206282	.31408516	.25822004
12	.63018618	.43560592	.31991819	.24527176	.19418561
13	.57388641	.36962165	.25652610	.18763074	.14269567
14	.51866933	.30966895	.20233106	.14077663	.10260187
15	.46542429	.25634789	.15712451	.10370880	.07227722

M	B= 6	B= 7	B= 8	B= 9	B=10
1	.99463598	.99383601	.99305693	.99229723	.99155555
2	.97564423	.97221080	.96890762	.96572306	.96264715
3	.93603157	.92762768	.91965962	.91208094	.90485268
4	.87298591	.85766540	.84338202	.83000613	.81743147
5	.78863313	.76558926	.74450910	.72510923	.70716291
6	.68900315	.65885832	.63185465	.60747554	.58531807
7	.58209602	.54670184	.51571207	.48831079	.46387828
8	.47591701	.43783633	.40530498	.37717657	.35260408
9	.37703844	.33894596	.30724831	.28048533	.25760853
10	.28988164	.25407360	.22509493	.20123545	.18130298
11	.21664027	.18474846	.15968488	.13958728	.12319665
12	.15763157	.13054647	.10990458	.09380400	.08099890
13	.11184374	.08979471	.07352118	.06118835	.05163277
14	.07749763	.06021894	.04788445	.03881223	.03197087
15	.05251422	.03943308	.03041232	.02397967	.01926279

M	B=11	B=12	B=13	B=14	B=15
1	.99083073	.99012171	.98942756	.98874743	.98808057
2	.95967128	.95678791	.95399043	.95127295	.94863027
3	.89794178	.89131991	.88496258	.87884842	.87295868
4	.80556979	.79434688	.78369973	.77357431	.76392394
5	.69048590	.67492644	.66035801	.64667408	.63378406
6	.56506234	.54645035	.52927121	.51335053	.49854256
7	.44193418	.42209954	.40407089	.38760197	.37249055
8	.33094713	.31171202	.29451165	.27903796	.26504260
9	.23784556	.22061491	.20547047	.19206437	.18012136
10	.16444256	.15002608	.13758241	.12675174	.11725491
11	.10963472	.09827237	.08864857	.08041868	.07332056
12	.07064464	.06215174	.05509837	.04917612	.04415513
13	.04408854	.03803516	.03310924	.02905097	.02567079
14	.02670155	.02256867	.01927562	.01661523	.01443754
15	.01572144	.01300831	.01089293	.00921809	.00787394

DIRICHLET D INTEGRAL
1/A = 2 R = 5

M	B= 1	B= 2	B= 3	B= 4	B= 5
1	.99588477	.99217224	.98876994	.98561724	.98267150
2	.98216735	.96688080	.95342761	.94137397	.93043174
3	.95473251	.91830063	.88776436	.86147341	.83839776
4	.91205609	.84640312	.79441415	.75165378	.71553213
5	.85515419	.75604725	.68247489	.62498854	.57844955
6	.78687192	.65477991	.56353442	.49616009	.44409576
7	.71100273	.55061364	.44825027	.37717814	.32494882
8	.63152071	.45043671	.34438176	.27546717	.22747535
9	.55203870	.35920942	.25624694	.19391033	.15290901
10	.47550047	.27981335	.18514240	.13196608	.09903503
11	.40406478	.21331754	.13020524	.08706889	.06199249
12	.33912325	.15943666	.08932633	.05583338	.03760814
13	.28139745	.11701741	.05989895	.03487652	.02216595
14	.23107238	.08445854	.03932968	.02126453	.01272054
15	.18793662	.06002538	.02532659	.01267784	.00712182

M	B= 6	B= 7	B= 8	B= 9	B=10
1	.97990119	.97728217	.97479542	.97242559	.97016011
2	.92039783	.91112275	.90249272	.89441884	.88683014
3	.81784719	.79933476	.78250329	.76708253	.75286284
4	.68440451	.65716172	.63302099	.61140983	.59189738
5	.53977507	.50698231	.47872765	.45406273	.43229591
6	.40250435	.36842586	.33993786	.31573279	.29488777
7	.28495868	.25337161	.22780413	.20669632	.18898403
8	.19235310	.16566949	.14479537	.12807739	.11442672
9	.12430267	.10344253	.08769885	.07548489	.06579299
10	.07718488	.06192180	.05082644	.04249997	.03608680
11	.04620584	.03566208	.02829101	.02294724	.01895676
12	.02674604	.01982193	.01517362	.01192226	.00957076
13	.01500936	.01066262	.00786441	.00597820	.00465835
14	.00818503	.00556451	.00394901	.00290077	.00219180
15	.00434649	.00282347	.00192549	.00136522	.00099930

M	B=11	B=12	B=13	B=14	B=15
1	.96798851	.96590193	.96389282	.96195468	.96008185
2	.87966891	.87288755	.86644626	.86031138	.85445417
3	.73967821	.72739482	.71590322	.70511268	.69494718
4	.57415153	.55791115	.54296742	.52915083	.51632206
5	.41290897	.39550479	.37977314	.36546760	.35238966
6	.27673139	.26076265	.24659955	.23394545	.22256620
7	.17391626	.16094788	.14967356	.13978572	.13104670
8	.10309875	.09356814	.08545428	.07847510	.07241745
9	.05795611	.05151719	.04615383	.04163286	.03778201
10	.03103917	.02699292	.02369807	.02097836	.01870651
11	.01590262	.01351605	.01161772	.01008434	.00882900
12	.00782232	.00649159	.00545838	.00464224	.00398780
13	.00370581	.00300023	.00246588	.00205337	.00172956
14	.00169557	.00133812	.00107423	.00087527	.00072246
15	.00075112	.00057738	.00045244	.00036048	.00029141

DIRICHLET D INTEGRAL
A = 1 R = 5

M	B= 1	B= 2	B= 3	B= 4	B= 5
1	.96875000	.94655286	.92912057	.91469064	.90234554
2	.89062500	.82360413	.77565653	.73858613	.70853300
3	.77343750	.65696286	.58208142	.52846882	.48753019
4	.63671875	.48428142	.39733199	.34013505	.29922255
5	.50000000	.33333333	.25000000	.20000000	.16666667
6	.37695313	.21631564	.14670953	.10887523	.08547660
7	.27441406	.13345079	.08108938	.05547073	.04083679
8	.19384766	.07880691	.04255367	.02668393	.01834531
9	.13342285	.04480014	.02134140	.01220640	.00780828
10	.08978271	.02463208	.01028406	.00534099	.00316834
11	.05923462	.01314986	.00478307	.00224627	.00123192
12	.03840637	.00683845	.00215517	.00091176	.00046097
13	.02452087	.00347383	.00094378	.00035840	.00016661
14	.01544189	.00172780	.00040277	.00013684	.58343\04
15	.00960541	.00084311	.00016790	.50877\04	.19849\04

M	B= 6	B= 7	B= 8	B= 9	B=10
1	.89154071	.88192420	.87325442	.86535785	.85810540
2	.68337082	.66180464	.64298826	.62633904	.61143915
3	.45489119	.42804947	.40545238	.38607687	.36921690
4	.26829031	.24396154	.22425138	.20791070	.19411109
5	.14285714	.12500000	.11111111	.10000000	.09090909
6	.06974143	.05851850	.05015683	.04371369	.03861421
7	.03160152	.02535317	.02090278	.01760529	.01508433
8	.01342180	.01026935	.00812693	.00660293	.00547899
9	.00538556	.00392078	.00297264	.00232608	.00186668
10	.00205481	.00142041	.00102995	.00077506	.00060078
11	.00074946	.00049097	.00033994	.00024568	.00018373
12	.00026248	.00016267	.00010738	.74440\04	.53651\04
13	.88610\04	.51860\04	.32596\04	.21649\04	.15022\04
14	.28926\04	.15963\04	.95404\05	.60638\05	.40470\05
15	.91569\05	.47577\05	.27004\05	.16408\05	.10523\05

M	B=11	B=12	B=13	B=14	B=15
1	.85139828	.84515913	.83932618	.83384925	.82868706
2	.59797910	.58572345	.57448911	.56413099	.55453223
3	.35436688	.34115371	.32929508	.31857281	.30881516
4	.18227982	.17200732	.16299233	.15500798	.14787989
5	.08333333	.07692308	.07142857	.06666667	.06250000
6	.03448929	.03109180	.02825045	.02584304	.02378018
7	.01310740	.01152412	.01023344	.00916526	.00826960
8	.00462550	.00396153	.00343439	.00300858	.00265947
9	.00152922	.00127445	.00107764	.00092258	.00079835
10	.00047707	.00038651	.00031849	.00026627	.00022540
11	.00014128	.00011117	.89201\04	.72771\04	.60225\04
12	.39911\04	.30478\04	.23795\04	.18930\04	.15307\04
13	.10801\04	.79985\05	.60718\05	.47077\05	.37170\05
14	.28103\05	.20166\05	.14875\05	.11233\05	.86562\06
15	.70517\06	.48998\06	.35098\06	.25801\06	.19394\06

DIRICHLET D INTEGRAL
A = 2 R = 5

M	B= 1	B= 2	B= 3	B= 4	B= 5
1	.86831276	.80683264	.76702715	.73782810	.71490239
2	.64883402	.52892723	.46206534	.41762651	.38520448
3	.42935528	.29506202	.23172450	.19395569	.16851502
4	.25864960	.14546299	.10107449	.07753998	.06298981
5	.14484581	.06510799	.03952368	.02756405	.02081415
6	.07656353	.02697138	.01415294	.00891143	.00622444
7	.03862894	.01048670	.00471299	.00266321	.00171330
8	.01875843	.00386714	.00147642	.00074473	.00043757
9	.00882318	.00136348	.00043896	.00019669	.00010614
10	.00403954	.00046254	.00012473	.49429\04	.24309\04
11	.00180718	.00015173	.34065\04	.11889\04	.53127\05
12	.00079246	.48331\04	.89825\05	.27506\05	.11136\05
13	.00034148	.14999\04	.22955\05	.61453\06	.22485\06
14	.00014490	.45475\05	.57038\06	.13304\06	.43882\07
15	.60650\04	.13503\05	.13817\06	.27993\07	.83035\08

M	B= 6	B= 7	B= 8	B= 9	B=10
1	.69610888	.68023516	.66652951	.65449451	.64378438
2	.36013068	.33994898	.32322423	.30905275	.29683262
3	.15004059	.13592131	.12472323	.11558896	.10797211
4	.05311024	.04596397	.04055371	.03631435	.03290192
5	.01654526	.01363138	.01153052	.00995221	.00872784
6	.00464731	.00363420	.00294009	.00244102	.00206847
7	.00119757	.00088659	.00068454	.00054573	.00044616
8	.00028681	.00020055	.00014750	.00011273	.88797\04
9	.64476\04	.42492\04	.29718\04	.21743\04	.16480\04
10	.13713\04	.85010\05	.56445\05	.39480\05	.28762\05
11	.27768\05	.16163\05	.10173\05	.67938\06	.47521\06
12	.53814\06	.29360\06	.17493\06	.11141\06	.74743\07
13	.10025\06	.51182\07	.28828\07	.17489\07	.11243\07
14	.18017\07	.85947\08	.45705\08	.26384\08	.16238\08
15	.31339\08	.13947\08	.69942\09	.38380\09	.22594\09

M	B=11	B=12	B=13	B=14	B=15
1	.63414933	.62540319	.61740366	.61003956	.60322247
2	.28614489	.27668756	.26823650	.26062132	.25370981
3	.10150688	.09593842	.09108336	.08680613	.08300418
4	.03009516	.02774529	.02574862	.02403063	.02253644
5	.00775334	.00696124	.00630602	.00575593	.00528821
6	.00178190	.00155601	.00137428	.00122554	.00110200
7	.00037223	.00031576	.00027163	.00023644	.00020791
8	.71661\04	.58996\04	.49384\04	.41924\04	.36024\04
9	.12852\04	.10258\04	.83484\05	.69071\05	.57958\05
10	.21651\05	.16740\05	.13236\05	.10665\05	.87334\06
11	.34498\06	.25818\06	.19817\06	.15540\06	.12412\06
12	52282\07	.37841\07	.28179\07	.21493\07	.16734\07
13	.75715\08	.52962\08	.38235\08	.28350\08	.21504\08
14	.10520\08	.71064\09	.49708\09	.35807\09	.26449\09
15	.14070\09	.91729\10	.62130\10	.43458\10	.31245\10

DIRICHLET D INTEGRAL
A = 3 R = 5

M	B= 1	B= 2	B= 3	B= 4	B= 5
1	.76269531	.68149672	.63406200	.60122694	.57643059
2	.46606445	.34921565	.29204497	.25665970	.23202630
3	.24359131	.14703914	.10826948	.08703355	.07349513
4	.11381531	.05373985	.03431999	.02499554	.01958414
5	.04892731	.01764893	.00965744	.00632362	.00457213
6	.01972771	.00532983	.00247312	.00144634	.00096052
7	.00756121	.00150414	.00058648	.00030455	.00018501
8	.00278151	.00040137	.00013044	.59830\04	.33121\04
9	.00098912	.00010217	.27472\04	.11077\04	.55689\05
10	.00034187	.24982\04	.55203\05	.19479\05	.88657\06
11	.00011534	.58990\05	.10647\05	.32743\06	.13451\06
12	.38107\04	.13511\05	.19803\06	.52880\07	.19553\07
13	.12364\04	.30123\06	.35668\07	.82398\08	.27352\08
14	.39483\05	.65571\07	.62415\08	.12432\08	.36956\09
15	.12432\05	.13970\07	.10641\08	.18218\09	.48383\10

M	B= 6	B= 7	B= 8	B= 9	B=10
1	.55667723	.54036052	.52652434	.51455600	.50404067
2	.21361412	.19917879	.18746627	.17771403	.16942843
3	.06404489	.05703768	.05161293	.04727523	.04371842
4	.01607148	.01361608	.01180699	.01042068	.00932550
5	.00351898	.00282701	.00234294	.00198812	.00171850
6	.00069092	.00052487	.00041481	.00033778	.00028156
7	.00012399	.88845\04	.66823\04	.52131\04	.41845\04
8	.20621\04	.13908\04	.99367\05	.74152\05	.57243\05
9	.32128\05	.20351\05	.13789\05	.98294\06	.72887\06
10	.47279\06	.28074\06	.18011\06	.12248\06	.87143\07
11	.66156\07	.36759\07	.22297\07	.14446\07	.98516\08
12	.88502\08	.45936\08	.26308\08	.16221\08	.10592\08
13	.11371\08	.55043\09	.29723\09	.17422\09	.10883\09
14	.14084\09	.63487\10	.32284\10	.17970\10	.10728\10
15	.16872\10	.70720\11	.33825\11	.17861\11	.10182\11

M	B=11	B=12	B=13	B=14	B=15
1	.49468516	.48627508	.47864917	.47168329	.46527994
2	.16227397	.15601364	.15047461	.14552753	.14107335
3	.04074268	.03821180	.03602960	.03412612	.03244921
4	.00843906	.00770722	.00709299	.00657025	.00612007
5	.00150766	.00133888	.00120114	.00108688	.00099076
6	.00023914	.00020625	.00018018	.00015911	.00014182
7	.34365\04	.28754\04	.24436\04	.21042\04	.18323\04
8	.45404\05	.36821\05	.30415\05	.25517\05	.21695\05
9	.55780\06	.43801\06	.35139\06	.28703\06	.23810\06
10	.64282\07	.48836\07	.38021\07	.30218\07	.24443\07
11	.69984\08	.51399\08	.38807\08	.29989\08	.23640\08
12	.72398\09	.51363\09	.37583\09	.28224\09	.21671\09
13	.71512\10	.48973\10	.34705\10	.25313\10	.18921\10
14	.67722\11	.44737\11	.30686\11	.21725\11	.15801\11
15	.61700\12	.39291\12	.26071\12	.17907\12	.12667\12

DIRICHLET D INTEGRAL
A = 4 R = 5

M	B= 1	B= 2	B= 3	B= 4	B= 5
1	.67232000	.58417861	.53581247	.50343616	.47951659
2	.34464000	.24414083	.19871453	.17172479	.15342275
3	.14803200	.08222018	.05824181	.04572986	.03799382
4	.05628160	.02380988	.01446795	.01022851	.00785297
5	.01958144	.00616332	.00317472	.00200595	.00141575
6	.00636938	.00146230	.00063202	.00035461	.00022903
7	.00196536	.00032352	.00011627	.57597\04	.33905\04
8	.00058124	.67575\04	.20031\04	.87154\05	.46586\05
9	.00016601	.13450\04	.32643\05	.12415\05	.60054\06
10	.46050\04	.25692\05	.50710\06	.16784\06	.73241\07
11	.12462\04	.47365\06	.75559\07	.21675\07	.85071\08
12	.33013\05	.84653\07	.10852\07	.26878\08	.94622\09
13	.85858\06	.14722\07	.15086\08	.32144\09	.10123\09
14	.21970\06	.24987\08	.20368\09	.37210\10	.10458\10
15	.55422\07	.41499\09	.26784\10	.41820\11	.10464\11

M	B= 6	B= 7	B= 8	B= 9	B=10
1	.46076185	.44545830	.43260808	.42158261	.41196216
2	.13999734	.12962145	.12129833	.11443314	.10864650
3	.03270957	.02885514	.02590975	.02357955	.02168588
4	.00634839	.00531592	.00456622	.00399851	.00355447
5	.00107050	.00084827	.00069529	.00058462	.00050143
6	.00016143	.00012073	.94220\04	.75932\04	.62742\04
7	.22208\04	.15637\04	.11597\04	.89435\05	.71095\05
8	.28279\05	.18706\05	.13160\05	.96969\06	.74064\06
9	.33697\06	.20897\06	.13923\06	.97887\07	.71751\07
10	.37895\07	.21991\07	.13854\07	.92818\08	.65222\08
11	.40498\08	.21952\08	.13058\08	.83261\09	.56027\09
12	.41355\09	.20904\09	.11724\09	.71067\10	.45750\10
13	.40540\10	.19078\10	.10075\10	.57996\11	.35686\11
14	.38298\11	.16754\11	.83211\12	.45438\12	.26698\12
15	.34982\12	.14205\12	.66272\13	.34294\13	.19225\13

M	B=11	B=12	B=13	B=14	B=15
1	.40345354	.39584444	.38897653	.38272884	.37700699
2	.10368379	.09936705	.09556767	.09219019	.08916203
3	.02011366	.01878537	.01764679	.01665880	.01579250
4	.00319810	.00290604	.00266250	.00245642	.00227986
5	.00043698	.00038581	.00034434	.00031015	.00028155
6	.52890\04	.45318\04	.39361\04	.34580\04	.30678\04
7	.57902\05	.48098\05	.40615\05	.34775\05	.30128\05
8	.58213\06	.46836\06	.38419\06	.32034\06	.27084\06
9	.54371\07	.42331\07	.33705\07	.27348\07	.22551\07
10	.47603\08	.35834\08	.27674\08	.21838\08	.17551\08
11	.39351\09	.28618\09	.21422\09	.16429\09	.12863\09
12	.30896\10	.21691\10	.15728\10	.11716\10	.89311\11
13	.23152\11	.15681\11	.11006\11	.79590\12	.59041\12
14	.16627\12	.10857\12	.73717\13	.51724\13	.37320\13
15	.11486\13	.72251\14	.47431\14	.32273\14	.22639\14

DIRICHLET D INTEGRAL

A = 5 R = 5

M	B= 1	B= 2	B= 3	B= 4	B= 5
1	.59812243	.50919713	.46240486	.43175637	.40942885
2	.26322445	.17926123	.14326674	.12243923	.10855030
3	.09577546	.05023260	.03468186	.02681764	.02204905
4	.03065641	.01203457	.00707802	.00490899	.00372035
5	.00895006	.00256910	.00127221	.00078568	.00054608
6	.00243816	.00050171	.00020708	.00011315	.71805\04
7	.00062929	.91244\04	.31109\04	.14955\04	.86301\05
8	.00015554	.15653\04	.43726\05	.18397\05	.96194\06
9	.37107\04	.25571\05	.58099\06	.21293\06	.10053\06
10	.85941\05	.40070\06	.73552\07	.23377\07	.99351\08
11	.19412\05	.60577\07	.89275\08	.24506\08	.93474\09
12	.42912\06	.88754\08	.10442\08	.24661\09	.84188\10
13	.93107\07	.12650\08	.11817\09	.23927\10	.72916\11
14	.19874\07	.17594\09	.12986\10	.22465\11	.60964\12
15	.41814\08	.23939\10	.13897\11	.20476\12	.49362\13

M	B= 6	B= 7	B= 8	B= 9	B=10
1	.39209780	.37806458	.36635345	.35635629	.34767053
2	.09848228	.09077092	.08462925	.07959303	.07536893
3	.01883596	.01651626	.01475798	.01337614	.01225938
4	.00297905	.00247621	.00211438	.00184241	.00163098
5	.00040832	.00032077	.00026111	.00021830	.00018634
6	.49969\04	.37004\04	.28654\04	.22944\04	.18856\04
7	.55725\05	.38809\05	.28533\05	.21847\05	.17262\05
8	.57474\06	.37561\06	.26173\06	.19135\06	.14519\06
9	.55438\07	.33930\07	.22371\07	.15595\07	.11350\07
10	.50445\08	.28859\08	.17977\08	.11934\08	.83214\09
11	.43602\09	.23274\09	.13678\09	.86360\10	.57637\10
12	.36001\10	.17901\10	.99106\11	.59448\11	.37938\11
13	.28528\11	.13192\11	.68718\12	.39118\12	.23847\12
14	.21780\12	.93527\13	.45781\13	.24706\13	.14375\13
15	.16075\13	.64007\14	.29406\14	.15029\14	*

M	B=11	B=12	B=13	B=14	B=15
1	.34001692	.33319452	.32705424	.32148271	.31639184
2	.07176157	.06863529	.06589263	.06346156	.06128760
3	.01133659	.01056017	.00989705	.00932349	.00882203
4	.00146219	.00132449	.00121012	.00111368	.00103132
5	.00016172	.00014226	.00012657	.00011368	.00010293
6	.15821\04	.13501\04	.11685\04	.10233\04	.90523\05
7	.13987\05	.11567\05	.97289\06	.83008\06	.71691\06
8	.11348\06	.90854\07	.74209\07	.61640\07	.51939\07
9	.85484\08	.66204\08	.52470\08	.42401\08	.34835\08
10	.60340\09	.45166\09	.34709\09	.27269\09	.21831\09
11	.40200\10	.29060\10	.21638\10	.16518\10	.12879\10
12	.25430\11	.17740\11	.12791\11	.94815\12	.71959\12
13	.15351\12	.10327\12	.72052\13	.51835\13	.38273\13
14	.88789\14	.57562\14	.38842\14	.27104\14	.19461\14
15	*	*	*	*	*

DIRICHLET D INTEGRAL
1/A = 5 R = 6

M	B= 1	B= 2	B= 3	B= 4	B= 5
1	.99997857	.99995722	.99993596	.99991479	.99989369
2	.99987140	.99974366	.99961676	.99949066	.99936536
3	.99955883	.99912220	.99868995	.99826191	.99783794
4	.99886422	.99774562	.99664335	.99555667	.99448489
5	.99756184	.99517559	.99283814	.99054675	.98829893
6	.99539121	.99091593	.98656451	.98232850	.97820045
7	.99207496	.98445330	.97710878	.97001892	.96316422
8	.98733746	.97529934	.96382193	.95285175	.94234331
9	.98092210	.96302931	.94618063	.93026035	.91517206
10	.97260589	.94731367	.92383474	.90193808	.88143535
11	.96221063	.92794045	.89663732	.86787073	.84129780
12	.94961032	.90482778	.86465609	.82833992	.79529240
13	.93473495	.87802653	.82816708	.78389993	.74427395
14	.91757105	.84771434	.78763309	.73533303	.68934592
15	.89815951	.81418249	.74367062	.68359021	.63177166

M	B= 6	B= 7	B= 8	B= 9	B=10
1	.99987268	.99985175	.99983089	.99981011	.99978940
2	.99924082	.99911704	.99899398	.99887163	.99874997
3	.99741789	.99700163	.99658905	.99618003	.99577447
4	.99342738	.99238357	.99135292	.99033494	.98932917
5	.98609246	.98392530	.98179562	.97970170	.97764200
6	.97417373	.97024240	.96640112	.96264505	.95896978
7	.95652761	.95009402	.94385009	.93778386	.93188458
8	.93225750	.92256032	.91322206	.90421648	.89552032
9	.90083436	.88717777	.87414240	.86167621	.84973366
10	.86217066	.84401332	.82685253	.81059348	.79515434
11	.81664062	.79367062	.77219738	.75206055	.73312376
12	.76504833	.73723266	.71153867	.68771238	.66554118
13	.70855263	.67615457	.64661338	.61954960	.59465075
14	.64856233	.61212574	.57936267	.54973523	.52280794
15	.58661731	.54692184	.51175721	.48039632	.45226084

M	B=11	B=12	B=13	B=14	B=15
1	.99976877	.99974821	.99972772	.99970730	.99968694
2	.99862900	.99850868	.99838902	.99826999	.99815158
3	.99537227	.99497334	.99457758	.99418492	.99379527
4	.98833518	.98735258	.98638099	.98542005	.98446943
5	.97561507	.97361958	.97165430	.96971806	.96780980
6	.95537129	.95184589	.94839019	.94500105	.94167558
7	.92614255	.92054895	.91509578	.90977569	.90458197
8	.88711281	.87897533	.87109109	.86344489	.85602294
9	.83827457	.82726337	.81666835	.80646110	.79661609
10	.78046393	.76645998	.75308760	.74029822	.72804858
11	.71527010	.69839853	.68242116	.66726105	.65285049
12	.64484538	.62547183	.60728899	.59018310	.57405518
13	.57165673	.55034898	.53054223	.51207829	.49482110
14	.49822401	.47568798	.45495290	.43581060	.41808424
15	.42688463	.40388756	.38295642	.36383071	.34629191

DIRICHLET D INTEGRAL
1/A = 4 R = 6

M	B= 1	B= 2	B= 3	B= 4	B= 5
1	.99993600	.99987259	.99980974	.99974744	.99968565
2	.99962880	.99926306	.99890247	.99854677	.99819571
3	.99876864	.99756452	.99638573	.99523060	.99409768
4	.99693363	.99396452	.99108423	.98828561	.98556251
5	.99363062	.98753946	.98169726	.97607985	.97066693
6	.98834579	.97736923	.96698526	.95712557	.94773422
7	.98059472	.96264824	.94594510	.93031752	.91563160
8	.96996468	.94277569	.91794471	.89510371	.87396714
9	.95614562	.91741575	.88280088	.85156632	.82315910
10	.93894857	.88652437	.84079627	.80041056	.76438043
11	.91831211	.85034489	.79264207	.74287535	.69940402
12	.89429878	.80937811	.73939879	.68058974	.63038689
13	.86708367	.76433383	.68237130	.61540282	.55963472
14	.83693770	.71607191	.62299414	.54921294	.48938270
15	.80420779	.66553987	.56272114	.48381726	.42161879

M	B= 6	B= 7	B= 8	B= 9	B=10
1	.99962437	.99956357	.99950324	.99944335	.99938391
2	.99784907	.99750665	.99716828	.99683380	.99650304
3	.99298567	.99189342	.99081989	.98976414	.98872533
4	.98290958	.98032215	.97779606	.97532762	.97291350
5	.96544122	.96038782	.95549376	.95074767	.94613948
6	.93876456	.93017721	.92193849	.91401933	.90639441
7	.90177834	.88866759	.87622373	.86438263	.85308932
8	.85430835	.83594397	.81872331	.80252087	.78723092
9	.79715169	.77320633	.75105130	.73046466	.71126270
10	.73196404	.70258972	.67580777	.65125823	.62864870
11	.66103061	.62685612	.59618975	.56849020	.54332621
12	.58697600	.54903180	.51556062	.48580089	.45915803
13	.51246855	.47205975	.43706013	.40645998	.37948734
14	.43996425	.39851905	.36331142	.33307223	.30685205
15	.37151866	.33043941	.29625017	.26743095	.24286997

M	B=11	B=12	B=13	B=14	B=15
1	.99932489	.99926627	.99920806	.99915024	.99909280
2	.99617586	.99585214	.99553176	.99521460	.99490055
3	.98770266	.98669544	.98570299	.98472471	.98376004
4	.97055071	.96823655	.96596853	.96374441	.96156210
5	.94166025	.93730195	.93305737	.92891998	.92488387
6	.89904152	.89194103	.88507550	.87842933	.87198852
7	.84229628	.83196211	.82205048	.81252934	.80337023
8	.77276348	.75904127	.74599744	.73357372	.72171901
9	.69329163	.67642144	.66054117	.64555544	.63138163
10	.60773856	.58832757	.57024741	.55335536	.53752741
11	.52034920	.49927385	.47986402	.46192234	.44528237
12	.43516027	.41342809	.39365233	.37557845	.35899478
13	.35554145	.33414748	.31492507	.29756578	.28181688
14	.28392659	.26373386	.24583127	.22986565	.21555184
15	.22173639	.20339768	.18736438	.17325229	.16075592

DIRICHLET D INTEGRAL
1/A = 3 R = 6

M	B= 1	B= 2	B= 3	B= 4	B= 5
1	.99975586	.99951741	.99928415	.99905568	.99883163
2	.99865723	.99736278	.99611136	.99489869	.99372125
3	.99577332	.99176820	.98795472	.98430961	.98081425
4	.99000549	.98073844	.97208025	.96394226	.95625583
5	.98027229	.96246049	.94619409	.93120770	.91730305
6	.96567249	.93564615	.90894949	.88491804	.86307389
7	.94559777	.89974054	.86020536	.82554513	.79476063
8	.91978741	.85498106	.80102217	.75509613	.71534179
9	.88833103	.80232700	.73344708	.67670758	.62896054
10	.85163192	.74329959	.66017292	.59410212	.54018224
11	.81034543	.67977653	.58415874	.51104529	.45331650
12	.76530561	.61378073	.50828355	.43090204	.37192278
13	.71745081	.54729251	.43507890	.35634470	.29854354
14	.66775544	.48210287	.36655981	.28922006	.23464897
15	.61717265	.41971513	.30415164	.23055223	.18074102

M	B= 6	B= 7	B= 8	B= 9	B=10
1	.99861172	.99839567	.99818325	.99797426	.99776851
2	.99257606	.99146060	.99037269	.98931042	.98827212
3	.97745340	.97421436	.97108639	.96806030	.96512813
4	.94896639	.94202955	.93540865	.92907292	.92299622
5	.90432718	.89215909	.88070108	.86987297	.85960796
6	.84306023	.82460268	.80748529	.79153471	.77660953
7	.76713117	.74211901	.71931199	.69838707	.67908633
8	.68046492	.64953002	.62183973	.59686096	.57417754
9	.58809968	.55265225	.52155201	.49400481	.46940509
10	.49526148	.45721143	.42453736	.39615576	.37126032
11	.40658726	.36800183	.33561731	.30806443	.28434965
12	.32562222	.28841119	.25792698	.23255228	.21114391
13	.25461601	.22030574	.19290485	.17061489	.15219851
14	.19456905	.16418869	.14056537	.12180476	.10663935
15	.14544441	.11951593	.09990834	.08472199	.07272177

M	B=11	B=12	B=13	B=14	B=15
1	.99756584	.99736609	.99716913	.99697484	.99678310
2	.98725628	.98626160	.98528686	.98433100	.98339302
3	.96228294	.95951862	.95682976	.95421153	.95165960
4	.91715614	.91153323	.90611054	.90087316	.89580787
5	.84984974	.84055034	.83166851	.82316851	.81501909
6	.76259274	.74938640	.73690763	.72508564	.71385948
7	.66120056	.64455773	.62901472	.61445117	.60076497
8	.55345880	.53443789	.51689664	.50065456	.48556077
9	.44728181	.42726227	.40904719	.39239310	.37709966
10	.34923759	.32961194	.31200852	.29612762	.28172659
11	.26373420	.24565720	.22968504	.21547721	.20276258
12	.19287173	.17711898	.16341827	.15140926	.14080989
13	.13677891	.12371900	.11254580	.10290149	.09451089
14	.09419329	.08384466	.07514130	.06774777	.06141072
15	.06307601	.05520790	.04870707	.04327508	.03869068

DIRICHLET D INTEGRAL
1/A = 2 R = 6

M	B= 1	B= 2	B= 3	B= 4	B= 5
1	.99862826	.99735069	.99614960	.99501262	.99393060
2	.99314129	.98697376	.98133906	.97613294	.97128190
3	.98033836	.96342182	.94850028	.93511123	.92294470
4	.95757761	.92298365	.89371557	.86833599	.84593282
5	.92343647	.86483228	.81758429	.77817977	.74451621
6	.87791495	.79105598	.72475092	.67181260	.62819842
7	.82227754	.70589674	.62230341	.55869429	.50832706
8	.75869193	.61465764	.51790362	.44803526	.39503508
9	.68980752	.52266176	.41827551	.34703423	.29542132
10	.61837184	.43448546	.32832759	.26012511	.21307303
11	.54693615	.35353599	.25089987	.18906175	.14855162
12	.47766519	.28193787	.18696595	.13350103	.10033206
13	.41224261	.22064091	.13607925	.09175604	.06578297
14	.35185253	.16965294	.09688354	.06149117	.04195020
15	.29721389	.12831569	.06757017	.04024602	.02606587

M	B= 6	B= 7	B= 8	B= 9	B=10
1	.99289651	.99190477	.99095086	.99003105	.98914219
2	.96673155	.96244020	.95837499	.95450947	.95082191
3	.91178074	.90145660	.89184791	.88285715	.87440624
4	.82588703	.80775744	.79121764	.77601888	.76196700
5	.71523242	.68939591	.66633989	.64557119	.62671492
6	.59141849	.55983663	.53232389	.50807091	.48647932
7	.46726157	.43301920	.40395228	.37891741	.35709326
8	.35336313	.31968980	.29188608	.26852365	.24860674
9	.25638120	.22587038	.20140574	.18137879	.16470225
10	.17891305	.15314093	.13310413	.11714560	.10418051
11	.12038498	.09990705	.08449298	.07256335	.06311766
12	.07829142	.06287589	.05165744	.04323047	.03673431
13	.04932175	.03826376	.03049393	.02483563	.02059309
14	.03016079	.02256609	.01742033	.01379132	.01114722
15	.01793707	.01292287	.00965092	.00741859	.00583950

M	B=11	B=12	B=13	B=14	B=15
1	.98828161	.98744703	.98663646	.98584814	.98508054
2	.94729423	.94391119	.94065977	.93752876	.93450843
3	.86643155	.85888049	.85170898	.84487971	.83836078
4	.74890738	.73671481	.72528641	.71453657	.70439331
5	.60947964	.59363439	.57899316	.56540401	.55274131
6	.46709534	.44956742	.43361822	.41902546	.40560857
7	.33787284	.32079682	.30551048	.29173509	.27924828
8	.23141886	.21643078	.20324276	.19154698	.18110247
9	.15061519	.13856939	.12816009	.11908209	.11110100
10	.09347096	.08449885	.07689051	.07037017	.06473022
11	.05549510	.04924379	.04404560	.03967084	.03595010
12	.03161740	.02751270	.02416810	.02140567	.01909686
13	.01733403	.01477873	.01273984	.01108815	.00973232
14	.00916795	.00765223	.00646868	.00552884	.00477150
15	.00468855	.00382826	.00317128	.00266017	.00225606

DIRICHLET D INTEGRAL
A = 1 R = 6

M	B= 1	B= 2	B= 3	B= 4	B= 5
1	.98437500	.97256516	.96291925	.95470307	.94751464
2	.93750000	.89547664	.86365252	.83802118	.81657067
3	.85546875	.77100374	.71238912	.66808405	.63280621
4	.74609375	.62010835	.54089073	.48508852	.44302186
5	.62304688	.46793090	.38050450	.32348978	.28298395
6	.50000000	.33333333	.25000000	.20000000	.16666667
7	.38720703	.22557878	.15459817	.11564622	.09137678
8	.29052734	.14586781	.09060369	.06302182	.04702205
9	.21197510	.09059020	.05062383	.03258014	.02287071
10	.15087891	.05427292	.02710503	.01606702	.01057631
11	.10505676	.03148592	.01396777	.00759445	.00467348
12	.07173157	.01774598	.00695363	.00345458	.00198178
13	.04812622	.00974444	.00335509	.00151756	.00080941
14	.03178406	.00522572	.00157332	.00064574	.00031942
15	.02069473	.00274276	.00071879	.00026685	.00012214

M	B= 6	B= 7	B= 8	B= 9	B=10
1	.94110640	.93531360	.93002042	.92514195	.92061392
2	.79813941	.78199340	.76763822	.75472459	.74299639
3	.60370808	.57908566	.55784179	.53923057	.52272308
4	.40983784	.38279381	.36020521	.34097179	.32434003
5	.25253002	.22869097	.20945711	.19356937	.18019602
6	.14285714	.12500000	.11111111	.10000000	.09090909
7	.07495664	.06318495	.05437544	.04756064	.04214816
8	.03679738	.02980607	.02478082	.02102734	.01813715
9	.01702568	.01322168	.01059988	.00871169	.00730375
10	.00747053	.00555048	.00428358	.00340498	.00277128
11	.00312481	.00221704	.00164449	.00126261	.00099647
12	.00125155	.00084646	.00060257	.00044632	.00034120
13	.00048182	.00031012	.00021159	.00015102	.00011172
14	.00017888	.00010940	.71445\04	.49083\04	.35105\04
15	.64229\04	.37270\04	.23269\04	.15372\04	.10620\04

M	B=11	B=12	B=13	B=14	B=15
1	.91638637	.91241962	.90868162	.90514601	.90179091
2	.73226016	.72236615	.71319605	.70465476	.69666467
3	.50793142	.49456321	.48239291	.47124315	.46097210
4	.30977403	.29688078	.28536456	.27499801	.26560304
5	.16876408	.15886506	.15019941	.14254209	.13572059
6	.08333333	.07692308	.07142857	.06666667	.06250000
7	.03775624	.03412843	.03108647	.02850279	.02628387
8	.01585597	.01401825	.01251202	.01125922	.01020388
9	.00622394	.00537629	.00469773	.00414541	.00368934
10	.00229946	.00193886	.00165713	.00143286	.00125144
11	.00080428	.00066138	.00055249	.00046778	.00040068
12	.00026764	.00021447	.00017498	.00014498	.00012172
13	.85093\04	.66398\04	.52874\04	.42842\04	.35236\04
14	.25943\04	.19697\04	.15300\04	.12117\04	.97575\05
15	.76083\05	.56172\05	.42533\05	.32905\05	.25932\05

DIRICHLET D INTEGRAL

A = 2 R = 6

M	B= 1	B= 2	B= 3	B= 4	B= 5
1	.91220850	.86769854	.83772020	.81515320	.79709249
2	.73662551	.63409993	.57336863	.53138532	.49985365
3	.53177869	.39877337	.33058093	.28766709	.25759436
4	.34969263	.22183866	.16596198	.13421181	.11356176
5	.21312808	.11170877	.07447781	.05552202	.04413603
6	.12208505	.05183151	.03047376	.02080083	.01545979
7	.06644764	.02246261	.01153996	.00716993	.00496149
8	.03465483	.00918858	.00409129	.00230164	.00147734
9	.01743373	.00357717	.00137029	.00069466	.00041217
10	.00850427	.00133411	.00043671	.00019862	.00010860
11	.00403954	.00047920	.00013321	.54134\04	.27194\04
12	.00187482	.00016651	.39082\04	.14136\04	.65066\05
13	.00085260	.56169\04	.11072\04	.35519\05	.14941\05
14	.00038080	.18453\04	.30395\05	.86185\06	.33049\06
15	.00016737	.59189\05	.81085\06	.20258\06	.70653\07

M	B= 6	B= 7	B= 8	B= 9	B=10
1	.78206192	.76920783	.75799194	.74805302	.73913696
2	.47490439	.45444144	.43721092	.42240772	.40948649
3	.23506558	.21740129	.20308457	.19118507	.18109692
4	.09897090	.08806613	.07957887	.07276733	.06716771
5	.03657484	.03120288	.02719647	.02409719	.02163023
6	.01213725	.00989890	.00830244	.00711398	.00619929
7	.00367856	.00286093	.00230428	.00190612	.00161019
8	.00103148	.00076327	.00058928	.00046987	.00038427
9	.00027030	.00018992	.00014031	.00010769	.85163\04
10	.66732\04	.44435\04	.31367\04	.23143\04	.17678\04
11	.15623\04	.98416\05	.66282\05	.46957\05	.34609\05
12	.34875\05	.20748\05	.13314\05	.90461\06	.64270\06
13	.74565\06	.41831\06	.25541\06	.16626\06	.11376\06
14	.15329\06	.80967\07	.46985\07	.29270\07	.19272\07
15	.30400\07	.15097\07	.83163\08	.49534\08	.31355\08

M	B=11	B=12	B=13	B=14	B=15
1	.73105827	.72367742	.71688688	.71060200	.70475502
2	.39806209	.38785328	.37864891	.37028675	.36263971
3	.17240671	.16482158	.15812756	.15216432	.14680903
4	.06247456	.05847821	.05502973	.05202031	.04936851
5	.01962113	.01795392	.01654860	.01534824	.01431121
6	.00547628	.00489220	.00441170	.00401031	.00367056
7	.00138341	.00120523	.00106229	.00094559	.00084888
8	.00032073	.00027223	.00023431	.00020408	.00017957
9	.68985\04	.56992\04	.47865\04	.40763\04	.35131\04
10	.13884\04	.11155\04	.91349\05	.76015\05	.64131\05
11	.26330\05	.20559\05	.16404\05	.13330\05	.11002\05
12	.47326\06	.35884\06	.27878\06	.22109\06	.17845\06
13	.81013\07	.59608\07	.45065\07	.34860\07	.27500\07
14	.13262\07	.94629\08	.69578\08	.52472\08	.40436\08
15	.20837\08	.14409\08	.10298\08	.75670\09	.56941\09

DIRICHLET D INTEGRAL
A = 3 R = 6

M	B= 1	B= 2	B= 3	B= 4	B= 5
1	.82202148	.75547843	.71518923	.68666501	.66476935
2	.55505371	.44191202	.38325393	.34560538	.31869710
3	.32145691	.21315508	.16593565	.13871248	.12070231
4	.16572571	.08888749	.06119821	.04698176	.03832297
5	.07812691	.03310548	.01991902	.01393842	.01060188
6	.03432751	.01126528	.00586351	.00371599	.00262379
7	.01425278	.00356024	.00158871	.00090671	.00059194
8	.00564933	.00105780	.00040143	.00020527	.00012345
9	.00215418	.00029824	.95547\04	.43567\04	.24059\04
10	.00079495	.80376\04	.21592\04	.87411\05	.44186\05
11	.00028524	.20827\04	.46621\05	.16689\05	.76998\06
12	.99890\04	.52132\05	.96681\06	.30486\06	.12803\06
13	.34246\04	.12655\05	.19339\06	.53524\07	.20407\07
14	.11523\04	.29891\06	.37447\07	.90661\08	.31305\08
15	.38130\05	.68885\07	.70411\08	.14864\08	.46373\09

M	B= 6	B= 7	B= 8	B= 9	B=10
1	.64710295	.63235732	.61974307	.60874872	.59902465
2	.29816649	.28179708	.26832518	.25696947	.24721686
3	.10776205	.09793762	.09017867	.08386671	.07861209
4	.03248685	.02827995	.02509899	.02260624	.02059788
5	.00849974	.00706489	.00602836	.00524726	.00463908
6	.00198271	.00156940	.00128471	.00107873	.00092396
7	.00042035	.00031611	.00024777	.00020038	.00016605
8	.82161\04	.58575\04	.43886\04	.34133\04	.27332\04
9	.14969\04	.10097\04	.72193\05	.53929\05	.41682\05
10	.25639\05	.16334\05	.11128\05	.79743\06	.59427\06
11	.41578\06	.24973\06	.16190\06	.11116\06	.79790\07
12	.64203\07	.36296\07	.22361\07	.14693\07	.10150\07
13	.94851\08	.50393\08	.29465\08	.18510\08	.12293\08
14	.13461\08	.67111\09	.37197\09	.22318\09	.14238\09
15	.18415\09	.86030\10	.45147\10	.25846\10	.15827\10

M	B=11	B=12	B=13	B=14	B=15
1	.59032176	.58245653	.57529005	.56871475	.56264569
2	.23871428	.23120955	.22451696	.21849625	.21303918
3	.07415617	.07032003	.06697559	.06402860	.06140800
4	.01894357	.01755604	.01637466	.01535595	.01446792
5	.00415307	.00375636	.00342680	.00314893	.00291166
6	.00080410	.00070899	.00063196	.00056850	.00051546
7	.00014032	.00012048	.00010483	.92250\04	.81959\04
8	.22403\04	.18715\04	.15884\04	.13662\04	.11886\04
9	.33105\05	.26884\05	.22238\05	.18684\05	.15908\05
10	.45694\06	.36042\06	.29038\06	.23815\06	.19832\06
11	.59344\07	.45432\07	.35626\07	.28506\07	.23205\07
12	.72958\08	.54173\08	.41320\08	.32237\08	.25640\08
13	.85341\09	.61417\09	.45539\09	.34623\09	.26893\09
14	.95385\10	.66489\10	.47897\10	.35471\10	.26894\10
15	.10224\10	.68989\11	.48257\11	.34792\11	.25738\11

DIRICHLET D INTEGRAL

A = 4 R = 6

M	B= 1	B= 2	B= 3	B= 4	B= 5
1	.73785600	.66060656	.61681040	.58689622	.56447283
2	.42328320	.31920070	.26947458	.23890507	.21766015
3	.20308224	.12412743	.09294840	.07585903	.06491289
4	.08564173	.04123061	.02700179	.02010972	.01606331
5	.03279350	.01214903	.00687861	.00464139	.00344359
6	.01165421	.00325706	.00157848	.00095901	.00065802
7	.00390313	.00080872	.00033252	.00018089	.00011434
8	.00124562	.00018841	.65200\04	.31599\04	.18335\04
9	.00038193	.41595\04	.12026\04	.51681\05	.27436\05
10	.00011323	.87682\05	.21036\05	.79818\06	.38651\06
11	.32614\04	.17756\05	.35130\06	.11721\06	.51622\07
12	.91640\05	.34713\06	.56307\07	.16456\07	.65742\08
13	.25197\05	.65780\07	.87003\08	.22194\08	.80217\09
14	.67970\06	.12123\07	.13008\08	.28866\09	.94158\10
15	.18028\06	.21791\08	.18878\09	.36323\10	.10668\10

M	B= 6	B= 7	B= 8	B= 9	B=10
1	.54669192	.53204908	.51965818	.50895569	.49956227
2	.20177515	.18930518	.17916973	.17071411	.16351538
3	.05722769	.05149488	.04703052	.04344053	.04048095
4	.01340590	.01152766	.01012947	.00904773	.00818553
5	.00271039	.00222070	.00187296	.00161457	.00141573
6	.00048687	.00037912	.00030628	.00025437	.00021586
7	.79298\04	.58553\04	.45226\04	.36133\04	.29637\04
8	.11887\04	.83058\05	.61232\05	.46995\05	.37212\05
9	.16588\05	.10946\05	.76905\06	.56625\06	.43237\06
10	.21742\06	.13526\06	.90424\07	.63796\07	.46925\07
11	.26958\07	.15783\07	.10027\07	.67705\08	.47926\08
12	.31807\08	.17497\08	.10549\08	.68098\09	.46347\09
13	.35886\09	.18520\09	.10583\09	.65244\10	.42656\10
14	.38876\10	.18794\10	.10166\10	.59800\11	.37525\11
15	.40581\11	.18352\11	.93868\12	.52627\12	.31670\12

M	B=11	B=12	B=13	B=14	B=15
1	.49121109	.48370772	.47690636	.47069506	.46498608
2	.15728649	.15182476	.14698247	.14264903	.13873976
3	.03799220	.03586526	.03402290	.03240884	.03098099
4	.00748184	.00689635	.00640137	.00597723	.00560959
5	.00125842	.00113113	.00102619	.00093832	.00086374
6	.00018636	.00016317	.00014455	.00012931	.00011666
7	.24824\04	.21150\04	.18279\04	.15987\04	.14126\04
8	.30207\05	.25024\05	.21081\05	.18013\05	.15578\05
9	.33984\06	.27349\06	.22443\06	.18722\06	.15838\06
10	.35680\07	.27872\07	.22264\07	.18122\07	.14988\07
11	.35223\08	.26688\08	.20738\08	.16460\08	.13303\08
12	.32897\09	.24160\09	.18251\09	.14118\09	.11144\09
13	.29218\10	.20785\10	.15256\10	.11495\10	.88572\11
14	.24787\11	.17069\11	.12165\11	.89242\12	.67094\12
15	.20159\12	.13429\12	.92890\13	.66311\13	.48623\13

DIRICHLET D INTEGRAL
A = 5 R = 6

M	B= 1	B= 2	B= 3	B= 4	B= 5
1	.66510202	.58380974	.53972064	.51030417	.48858849
2	.33020405	.23949812	.19848165	.17396378	.15722790
3	.13484689	.07785620	.05679875	.04562948	.03862162
4	.04802149	.02146034	.01359591	.00992758	.00782367
5	.01546197	.00522621	.00284305	.00187380	.00136828
6	.00460879	.00115510	.00053428	.00031591	.00021284
7	.00129254	.00023606	.92025\04	.48549\04	.30065\04
8	.00034504	.45214\04	.14737\04	.69020\05	.39149\05
9	.88430\04	.81994\05	.22181\05	.91797\06	.47535\06
10	.21900\04	.14189\05	.31643\06	.11522\06	.54304\07
11	.52676\05	.23576\06	.43074\07	.13743\07	.58787\08
12	.12355\05	.37803\07	.56253\08	.15668\08	.60659\09
13	.28351\06	.58737\08	.70799\09	.17152\09	.59949\10
14	.63814\07	.88737\09	.86197\10	.18102\10	.56979\11
15	.14120\07	.13072\09	.10184\10	.18481\11	.52265\12

M	B= 6	B= 7	B= 8	B= 9	B=10
1	.47155824	.45765265	.44596598	.43592894	.42716193
2	.14487406	.13527082	.12752637	.12110709	.11567175
3	.03377288	.03019583	.02743467	.02523029	.02342402
4	.00646445	.00551556	.00481606	.00427917	.00385410
5	.00106442	.00086422	.00072358	.00061997	.00054083
6	.00015540	.00011977	.95960\04	.79154\04	.66783\04
7	.20542\04	.14995\04	.11476\04	.91000\05	.74164\05
8	.24968\05	.17227\05	.12573\05	.95701\06	.75252\06
9	.28228\06	.18374\06	.12769\06	.93178\07	.70614\07
10	.29958\07	.18363\07	.12133\07	.84780\08	.61858\08
11	.30064\08	.17324\08	.10868\08	.72632\09	.50973\09
12	.28698\09	.15521\09	.92324\10	.58951\10	.39757\10
13	.26186\10	.13272\10	.74767\11	.45564\11	.29503\11
14	.22937\11	.10879\11	.57968\12	.33682\12	.20921\12
15	.19355\12	.85780\13	.43185\13	.23902\13	.14230\13

M	B=11	B=12	B=13	B=14	B=15
1	.41940001	.41245145	.40617334	.40045647	.39521566
2	.11099072	.10690302	.10329205	.10007099	.09717370
3	.02191301	.02062752	.01951848	.01855033	.01769662
4	.00350915	.00322357	.00298319	.00277802	.00260079
5	.00047860	.00042851	.00038742	.00035316	.00032418
6	.57370\04	.50013\04	.44134\04	.39347\04	.35389\04
7	.61779\05	.52386\05	.45082\05	.39282\05	.34592\05
8	.60724\06	.50043\06	.41965\06	.35712\06	.30772\06
9	.55146\07	.44130\07	.36036\07	.29930\07	.25221\07
10	.46711\08	.36270\08	.28821\08	.23349\08	.19231\08
11	.37187\09	.27997\09	.21634\09	.17086\09	.13747\09
12	.28000\10	.20425\10	.15339\10	.11803\10	.92727\11
13	.20043\11	.14157\11	.10326\11	.77378\12	.59327\12
14	.13700\12	.93637\13	.66301\13	.48357\13	.36168\13
15	.89758\14	.59327\14	.40756\14	.28918\14	.21090\14

DIRICHLET D INTEGRAL
1/A = 5 R = 7

M	B= 1	B= 2	B= 3	B= 4	B= 5
1	.99999643	.99999286	.99998930	.99998575	.99998220
2	.99997559	.99995125	.99992699	.99990281	.99987869
3	.99990613	.99981271	.99971973	.99962719	.99953506
4	.99973248	.99946688	.99920314	.99894121	.99868104
5	.99937071	.99874790	.99813134	.99752080	.99691609
6	.99870746	.99743343	.99617707	.99493764	.99371442
7	.99760204	.99525048	.99294278	.99067669	.98845014
8	.99589128	.99188730	.98798126	.98416715	.98043958
9	.99339642	.98700980	.98082354	.97482316	.96899595
10	.98993133	.98028072	.97101080	.76208960	.95348947
11	.98531121	.97137967	.95812712	.94548786	.93340604
12	.97936106	.96002235	.94182983	.92465683	.90839762
13	.97192338	.94597731	.92187487	.89938516	.87831917
14	.96286466	.92907946	.89813530	.86963196	.84324851
15	.95208047	.90923925	.87061167	.83553459	.80348776

M	B= 6	B= 7	B= 8	B= 9	B=10
1	.99997867	.99997513	.99997161	.99996808	.99996457
2	.99985464	.99983066	.99980676	.99978291	.99975914
3	.99944335	.99935204	.99926113	.99917061	.99908047
4	.99842259	.99816581	.99791065	.99765708	.99740506
5	.99631700	.99572337	.99513503	.79455181	.99397358
6	.99250678	.99131413	.99013591	.98897161	.98782076
7	.98626127	.98410836	.98198986	.97990431	.97785038
8	.97679374	.97322526	.96973019	.96630490	.96294611
9	.96333063	.95781716	.95244651	.94721053	.94210185
10	.94518626	.93715871	.92938798	.92185726	.91455150
11	.92183362	.91072896	.90005561	.88978147	.87987806
12	.89296277	.87827591	.86427112	.85089109	.83808564
13	.85851990	.83985531	.82221318	.80549725	.78962427
14	.81872330	.79583997	.77441749	.75430277	.73536515
15	.77405526	.74689928	.72174167	.69835075	.67653154

M	B=11	B=12	B=13	B=14	B=15
1	.99996106	.99995756	.99995406	.99995057	.99994708
2	.99973542	.99971178	.99968819	.99966467	.99964121
3	.99899070	.99890130	.99881226	.99872357	.99863523
4	.99715455	.99690553	.99665796	.99641180	.99616703
5	.99340020	.99283153	.99226744	.99170783	.99115258
6	.98668291	.98555763	.98444454	.98334326	.98225344
7	.97582682	.97383249	.97186630	.96992726	.96801441
8	.95965075	.95641603	.95323935	.95011831	.94705065
9	.93711374	.93224006	.92747517	.92281388	.91825141
10	.90745710	.90056178	.89385434	.88732459	.88096319
11	.87032001	.86108458	.85215127	.84350159	.83511872
12	.82581051	.81402646	.80269851	.79179536	.78128882
13	.77452172	.76012599	.74638101	.73323704	.72064976
14	.71749223	.70058658	.68456323	.66934758	.65487385
15	.65611842	.63696961	.61896280	.60199183	.58596397

DIRICHLET D INTEGRAL
1/A = 4 R = 7

M	B= 1	B= 2	B= 3	B= 4	B= 5
1	.99998720	.99997446	.99996178	.99994916	.99993660
2	.99991552	.99983169	.99974849	.99966589	.99958387
3	.99968614	.99937597	.99906930	.99876598	.99846587
4	.99913564	.99828605	.99745036	.99662780	.99581770
5	.99803464	.99611638	.99424188	.99240819	.99061277
6	.99609687	.99232083	.98866097	.98510800	.98165392
7	.99299644	.98629407	.97986217	.97367517	.96771151
8	.98839009	.97742349	.96702316	.95712666	.94768236
9	.98194119	.96514376	.94943240	.93466873	.92074099
10	.97334267	.94898715	.92656598	.90579715	.88645852
11	.96233656	.92862402	.89814458	.87036382	.84487165
12	.94872900	.90389015	.86416486	.82860270	.79649438
13	.93239993	.87479929	.82490358	.78111475	.74227476
14	.91330749	.84154152	.78089662	.72881700	.68350705
15	.89148755	.80446882	.73289734	.67286520	.62171474

M	B= 6	B= 7	B= 8	B= 9	B=10
1	.99992409	.99991163	.99989923	.99988687	.99987457
2	.99950242	.99942151	.99934114	.99926128	.99918192
3	.99816883	.99787476	.99758354	.99729507	.99700925
4	.99501944	.99423247	.99345627	.99269039	.99193440
5	.98885334	.98712787	.98543454	.98377170	.98213787
6	.97829175	.97501538	.97181938	.96869889	.96564957
7	.96195275	.95638292	.95098808	.94575596	.94067569
8	.93864695	.92998362	.92166072	.91365080	.90592985
9	.90755731	.89504102	.88312733	.87176089	.86089396
10	.86837140	.85138955	.83539158	.82027543	.80595441
11	.82134578	.79952782	.77920725	.76021008	.74239080
12	.76729584	.74058034	.71600690	.69329870	.67222799
13	.70751786	.67618031	.64774249	.62179039	.59798911
14	.64366014	.60829810	.57667139	.54819428	.52240129
15	.57756585	.53904573	.50512565	.47501771	.44810723

M	B=11	B=12	B=13	B=14	B=15
1	.99986232	.99985011	.99983795	.99982584	.99981377
2	.99910305	.99902466	.99894673	.99886925	.99879221
3	.99672600	.99644524	.99616689	.99589088	.99561714
4	.99118791	.99045056	.98972199	.98900191	.98829002
5	.98053166	.97895184	.97739724	.97586680	.97435953
6	.96266749	.95974908	.95689110	.95409058	.95134479
7	.93573757	.93093288	.92625378	.92169316	.91724455
8	.89847671	.89127262	.88430085	.87754641	.87099578
9	.85048503	.84049776	.83090011	.82166365	.81276307
10	.79235417	.77941043	.76706722	.75527544	.74399181
11	.72562641	.70981202	.69485738	.68068431	.66722463
12	.65260516	.63427072	.61708927	.60094493	.58573782
13	.57606417	.55578807	.53697026	.51944975	.50308936
14	.49891712	.47743533	.45770289	.43950876	.42267540
15	.42390693	.40202508	.38214293	.36399823	.34737310

DIRICHLET D INTEGRAL
1/A = 3 R = 7

M	B= 1	B= 2	B= 3	B= 4	B= 5
1	.99993896	.99987878	.99981940	.99976078	.99970286
2	.99961853	.99924536	.99887980	.99852131	.99816938
3	.99865723	.99735747	.99609641	.99487053	.99367687
4	.99649429	.99314692	.98993863	.98685413	.98388098
5	.99243879	.98534141	.97864052	.97228443	.96623218
6	.98574722	.97264442	.96049518	.94915393	.93850889
7	.97570986	.95392675	.93415314	.91603451	.89930723
8	.96172924	.92838956	.89885735	.87236486	.84836069
9	.94337969	.89567276	.85457242	.81857150	.78662681
10	.92044275	.85589784	.80199607	.75601596	.71614600
11	.89291842	.80964777	.74246560	.68679989	.63973039
12	.86101522	.75789625	.67778931	.61348308	.56056726
13	.82512412	.70190359	.61004011	.53877517	.48182254
14	.78578195	.64309758	.54134656	.46525389	.40631091
15	.74362962	.58295548	.47370993	.39514630	.33627147

M	B= 6	B= 7	B= 8	B= 9	B=10
1	.99964561	.99958901	.99953302	.99947761	.99942276
2	.99782359	.99748358	.99714902	.99681962	.99649510
3	.99251292	.99137652	.99026577	.98917904	.98811483
4	.98100882	.97822891	.97553378	.97291697	.97037282
5	.96045056	.95491214	.94959390	.94447629	.93954250
6	.92847174	.91897121	.90994871	.90135537	.89314990
7	.88376886	.86925992	.85565218	.84284073	.83073854
8	.82643378	.80626858	.78761714	.77028076	.75409763
9	.75798465	.73208205	.70848715	.68686129	.66693400
10	.68111997	.65001878	.62215492	.59700133	.57414553
11	.59928552	.56407652	.53309155	.50557291	.48094008
12	.51617079	.47833360	.44566635	.41715369	.39203463
13	.43524279	.39643284	.36359944	.33546488	.31109308
14	.35939310	.32122730	.28962441	.26306510	.24046210
15	.29073288	.25461345	.22537257	.20129408	.18117934

M	B=11	B=12	B=13	B=14	B=15
1	.99936844	.99931465	.99926134	.99920852	.99915616
2	.99617524	.99585981	.99554862	.99524148	.99493822
3	.98707185	.98604892	.98504496	.98405901	.98309020
4	.96789637	.96548321	.96312939	.96083140	.95858604
5	.93477795	.93016987	.92570704	.92137948	.91717831
6	.88529704	.87776639	.87053153	.86356929	.85685929
7	.81927247	.80838041	.79800912	.78811258	.77865071
8	.73893409	.72467841	.71123623	.69852705	.68648168
9	.64848584	.63133638	.61533548	.60035692	.58629361
10	.55325904	.53407633	.51637993	.49998966	.48475474
11	.45873957	.43861125	.42026491	.40346378	.38801244
12	.36972650	.34977491	.33181988	.31557221	.30079674
13	.28978194	.27099420	.25431156	.23940338	.22600476
14	.22101716	.20413121	.18934612	.17630570	.16472900
15	.16416746	.14962519	.13707710	.12615980	.11659115

DIRICHLET D INTEGRAL
1/A = 2 R = 7

M	B= 1	B= 2	B= 3	B= 4	B= 5
1	.99954275	.99910772	.99869146	.99829144	.99790573
2	.99740893	.99500340	.99274841	.99061944	.98859838
3	.99171874	.98426738	.97745876	.97116872	.96530924
4	.98033836	.96331741	.94823744	.93466065	.92229050
5	.96137106	.92944491	.90216695	.87833038	.85715681
6	.93355236	.88154025	.83891587	.80290950	.77182529
7	.89646075	.82032990	.76081168	.71240003	.67190510
8	.85053781	.74815906	.67221288	.61295913	.56507597
9	.79696105	.66847652	.57851428	.51143975	.45922267
10	.73743131	.58520994	.48513002	.41408128	.36094196
11	.67393293	.50218877	.39669924	.32564291	.27471847
12	.60851035	.42271180	.31662344	.24905210	.20276612
13	.54308777	.34929484	.24692985	.18548168	.14534914
14	.47934269	.28358728	.18838377	.13469590	.10134274
15	.41863309	.22641937	.14074845	.09550403	.06882906

M	B= 6	B= 7	B= 8	B= 9	B=10
1	.99753276	.99717129	.99682027	.99647880	.99614613
2	.98667127	.98482706	.98305680	.98135309	.97970972
3	.95981493	.95463553	.94973132	.94507026	.94062601
4	.91091485	.90037570	.89055164	.88134703	.87268500
5	.83811111	.82080791	.80495958	.79034527	.77679147
6	.74454558	.72029349	.69850661	.67876453	.66074472
7	.63731615	.60728651	.58087196	.55738614	.53631541
8	.52536220	.49175580	.46285835	.43768250	.41550819
9	.41726418	.38272011	.35372743	.32900991	.30766133
10	.31966191	.28665480	.25965470	.23715724	.21812337
11	.23655275	.20696389	.18340663	.16424509	.14838120
12	.16936505	.14429775	.12489940	.10951368	.09706109
13	.11751716	.09732809	.08216175	.07044487	.06118269
14	.07915445	.06362060	.05230652	.04380143	.03724070
15	.05183644	.04037130	.03228437	.02637502	.02193061

M	B=11	B=12	B=13	B=14	B=15
1	.99582162	.99550468	.99519481	.99489157	.99459456
2	.97812141	.97658364	.97509247	.97364446	.97223659
3	.93637662	.93230355	.92839096	.92462521	.92099442
4	.86450269	.85674798	.84937708	.84235282	.83564334
5	.76415922	.75233538	.74122656	.73075476	.72085407
6	.64419437	.62891163	.61473286	.60152357	.58917192
7	.51726601	.49992990	.48406193	.46946403	.45597409
8	.39579602	.37813281	.36219596	.34772940	.33452694
9	.28901898	.27258607	.25798233	.24491134	.23313844
10	.20181178	.18767909	.17531774	.16441596	.15473114
11	.13505130	.12370834	.11395056	.10547649	.09805563
12	.08681034	.07824985	.07101191	.06482599	.05948902
13	.05371932	.04760695	.04253073	.03826369	.03463858
14	.03206984	.02791961	.02453624	.02174049	.01940285
15	.01850697	.01581576	.01366337	.01191595	.01047861

DIRICHLET D INTEGRAL
A = 1 R = 7

M	B= 1	B= 2	B= 3	B= 4	B= 5
1	.99218750	.98599280	.98077352	.97622464	.97217187
2	.96484375	.93947901	.91940109	.90270135	.88836887
3	.91015625	.85231140	.80972917	.77615135	.74852206
4	.82812500	.73132657	.66571266	.61696806	.57868404
5	.72558594	.59311767	.51123076	.45425939	.41172844
6	.61279297	.45584630	.36819400	.31140780	.27127181
7	.50000000	.33333333	.25000000	.20000000	.16666667
8	.39526367	.23292415	.16089433	.12107750	.09612890
9	.30361938	.15619348	.09864420	.06948234	.05236691
10	.22724915	.10090589	.05787801	.03798767	.02708858
11	.16615295	.06302254	.03263060	.01987407	.01336837
12	.11894226	.03817252	.01774007	.00998821	.00631991
13	.08353424	.02248399	.00932968	.00483860	.00287235
14	.05765915	.01290967	.00475947	.00226608	.00125898
15	.03917694	.00724110	.00236097	.00102872	.00053365

M	B= 6	B= 7	B= 8	B= 9	B=10
1	.96850445	.96514677	.96204459	.95915744	.95645421
2	.87579667	.86458916	.85447301	.84525089	.83677540
3	.72511423	.70485522	.68703296	.67115068	.65684811
4	.54746311	.52129936	.49891575	.47945248	.46230506
5	.37844686	.35150808	.32914068	.31019538	.29388968
6	.24122331	.21778577	.19893451	.18340541	.17036595
7	.14285714	.12500000	.11111111	.10000000	.09090909
8	.07917155	.06696758	.05780401	.05069446	.04503308
9	.04132481	.03371398	.02820562	.02406636	.02086189
10	.02043022	.01604311	.01298714	.01076569	.00909548
11	.00961358	.00725277	.00567229	.00456220	.00375245
12	.00432413	.00312873	.00236065	.00183998	.00147184
13	.00186607	.00129284	.00093980	.00070908	.00055110
14	.00077515	.00051344	.00035914	.00026203	.00019769
15	.00031083	.00019656	.00013214	.93129\04	.68146\04

M	B=11	B=12	B=13	B=14	B=15
1	.95391039	.95150631	.94922587	.94705573	.94498468
2	.82893332	.82163574	.81481147	.80840258	.80236130
3	.64385580	.63196704	.62101981	.61088487	.60145746
4	.44703354	.43330857	.42087766	.40954340	.39914876
5	.27966998	.26713216	.25597347	.24596203	.23691686
6	.15924409	.14963302	.14123505	.13382701	.12723813
7	.08333333	.07692308	.07142857	.06666667	.06250000
8	.04042834	.03661655	.03341400	.03068897	.02834471
9	.01832037	.01626386	.01457149	.01315859	.01196427
10	.00780494	.00678498	.00596341	.00529085	.00473253
11	.00314340	.00267360	.00230344	.00200648	.00176452
12	.00120253	.00099993	.00084387	.00072126	.00062324
13	.00043879	.00035644	.00029447	.00024680	.00020944
14	.00015327	.00012154	.98232\04	.80691\04	.67215\04
15	.51408\04	.39772\04	.31429\04	.25289\04	.20668\04

DIRICHLET D INTEGRAL
A = 2 R = 7

M	B= 1	B= 2	B= 3	B= 4	B= 5
1	.94147234	.90973208	.88764180	.87064961	.85683114
2	.80490779	.72060323	.66816667	.63074442	.60197024
3	.62282172	.49852628	.43041003	.38566089	.35329646
4	.44073566	.30654416	.24263553	.20427181	.17829900
5	.28899727	.17063493	.12229664	.09596764	.07935715
6	.17772246	.08733975	.05609594	.04075743	.03177852
7	.10353925	.04163259	.02374669	.01588312	.01162923
8	.05761630	.01867014	.00938146	.00574677	.00393679
9	.03082792	.00794160	.00348991	.00194874	.00124490
10	.01594549	.00322555	.00123135	.00062407	.00037063
11	.00800819	.00125779	.00041454	.00018993	.00010456
12	.00391928	.00047303	.00013382	.55217\04	.28106\04
13	.00187482	.00017223	.41597\04	.15403\04	.72303\05
14	.00087881	.60906\04	.12495\04	.41383\05	.17871\05
15	.00040451	.20977\04	.36381\05	.10743\05	.42581\06

M	B= 6	B= 7	B= 8	B= 9	B=10
1	.84518473	.83512049	.82626101	.81834995	.81120509
2	.57877710	.55946228	.54298708	.52867355	.51605604
3	.32844149	.30855187	.29215093	.27831403	.26642821
4	.15936060	.14483674	.13328352	.12383493	.11593761
5	.06789466	.05949097	.05305497	.04796081	.04382351
6	.02593236	.02184437	.01883565	.01653433	.01472045
7	.00902349	.00728940	.00606501	.00516127	.00447080
8	.00289657	.00223912	.00179451	.00147825	.00124429
9	.00086637	.00063961	.00049296	.00039256	.00032072
10	.00024341	.00017131	.00012678	.97477\04	.77218\04
11	.64665\04	.43310\04	.30736\04	.22790\04	.17487\04
12	.16334\04	.10394\04	.70641\05	.50458\05	.37468\05
13	.39410\05	.23792\05	.15466\05	.10631\05	.76329\06
14	.91195\06	.52154\06	.32389\06	.21404\06	.14846\06
15	.20307\06	.10987\06	.65110\07	.41327\07	.27671\07

M	B=11	B=12	B=13	B=14	B=15
1	.80469224	.79870964	.79317832	.78803571	.78323143
2	.50480170	.49466479	.48545906	.47704017	.46929410
3	.25606873	.24693052	.23878799	.23147015	.22484471
4	.10922022	.10342348	.09836051	.09389285	.08991562
5	.04039309	.03750009	.03502546	.03288311	.03100919
6	.01325600	.01205015	.01104082	.01018417	.00944841
7	.00392854	.00349297	.00313649	.00284007	.00259023
8	.00106573	.00092593	.00081414	.00072313	.00064789
9	.00026749	.00022690	.00019520	.00016995	.00014950
10	.62654\04	.51846\04	.43611\04	.37197\04	.32105\04
11	.13792\04	.11125\04	.91442\05	.76358\05	.64633\05
12	.28702\05	.22554\05	.18102\05	.14791\05	.12272\05
13	.56747\06	.43408\06	.34001\06	.27171\06	.22087\06
14	.10704\06	.79660\07	.60861\07	.47540\07	.37845\07
15	.19335\07	.13990\07	.10420\07	.79522\08	.61967\08

DIRICHLET D INTEGRAL
A = 3 R = 7

M	B= 1	B= 2	B= 3	B= 4	B= 5
1	.86651611	.81280768	.77930134	.75512789	.73631678
2	.63291931	.52757668	.47033647	.43250269	.40488113
3	.39932251	.28470122	.23132409	.19926845	.17742772
4	.22412491	.13279859	.09691717	.07749060	.06519890
5	.11462641	.05514743	.03573025	.02632298	.02082102
6	.05440223	.02083631	.01186641	.00800517	.00592686
7	.02429014	.00727947	.00361226	.00221921	.00153205
8	.01030953	.00238065	.00102122	.00056853	.00036470
9	.00419301	.00073578	.00027088	.00013603	.80823\04
10	.00164447	.00021654	.67962\04	.30655\04	.16820\04
11	.00062505	.61053\04	.16235\04	.65515\05	.33103\05
12	.00023118	.16574\04	.37123\05	.13353\05	.61967\06
13	.83479\04	.43500\05	.81626\06	.26077\06	.11087\06
14	.29512\04	.11077\05	.17324\06	.48990\07	.19037\07
15	.10238\04	.27445\06	.35604\07	.88843\08	.31480\08

M	B= 6	B= 7	B= 8	B= 9	B=10
1	.72097663	.70806084	.69693072	.68716846	.67848617
2	.38345408	.36613707	.35172172	.33945053	.32882017
3	.16137152	.14895137	.13898669	.13076950	.12384673
4	.05666527	.05036374	.04550125	.04162354	.03845101
5	.01722543	.01469704	.01282424	.01138220	.01023800
6	.00465230	.00380079	.00319644	.00274780	.00240299
7	.00113774	.00088800	.00071846	.00059729	.00050719
8	.00025557	.00019018	.00014778	.00011865	.97714\04
9	.53316\04	.37755\04	.28136\04	.21788\04	.17384\04
10	.10422\04	.70106\05	.50031\05	.37324\05	.28822\05
11	.19224\05	.12264\05	.83703\06	.60091\06	.44868\06
12	.33662\06	.20335\06	.13256\06	.91480\07	.65986\07
13	.56229\07	.32116\07	.19972\07	.13236\07	.92155\08
14	.89978\08	.48521\08	.28751\08	.18281\08	.12275\08
15	.13843\08	.70381\09	.39695\09	.24192\09	.15655\09

M	B=11	B=12	B=13	B=14	B=15
1	.67067726	.66358862	.65710369	.65113177	.64560089
2	.31948074	.31118004	.30373085	.29699075	.29084907
3	.11791358	.11275662	.10822147	.10419337	.10058504
4	.03580183	.03355241	.03161568	.02992847	.02844378
5	.00930813	.00853756	.00788858	.00733448	.00685582
6	.00213056	.00191043	.00172922	.00157769	.00144928
7	.00043804	.00038362	.00033986	.00030404	.00027428
8	.82125\04	.70180\04	.60807\04	.53303\04	.47193\04
9	.14205\04	.11835\04	.10022\04	.86026\05	.74709\05
10	.22877\05	.18571\05	.15358\05	.12902\05	.10984\05
11	.34566\06	.27318\06	.22050\06	.18118\06	.15115\06
12	.49303\07	.37909\07	.29848\07	.23975\07	.19589\07
13	.66730\08	.49886\08	.38293\08	.30054\08	.24039\08
14	.86082\09	.62532\09	.46769\09	.35849\09	.28057\09
15	.10624\09	.74948\10	.54590\10	.40846\10	.31269\10

DIRICHLET D INTEGRAL
A = 4 R = 7

M	B= 1	B= 2	B= 3	B= 4	B= 5
1	.79028480	.72360627	.68475724	.65777239	.63729941
2	.49668352	.39303439	.34124537	.30852112	.28532615
3	.26180250	.17242750	.13478887	.11333715	.09920757
4	.12087388	.06442481	.04472362	.03467691	.02855808
5	.05040957	.02125077	.01295484	.00919845	.00709296
6	.01940528	.00634451	.00336381	.00217420	.00156327
7	.00700356	.00174545	.00079794	.00046699	.00031193
8	.00239721	.00044842	.00017538	.92483\04	.57194\04
9	.00078499	.00010868	.36105\04	.17078\04	.97473\05
10	.00024758	.25050\04	.70211\05	.29664\05	.15580\05
11	.75607\04	.55256\05	.12986\05	.48815\06	.23528\06
12	.22453\04	.11726\05	.22974\06	.76547\07	.33768\07
13	.65062\05	.24043\06	.39056\07	.11494\07	.46291\08
14	.18450\05	.47803\07	.64049\08	.16593\08	.60867\09
15	.51322\06	.92442\08	.10167\08	.23112\09	.77042\10

M	B= 6	B= 7	B= 8	B= 9	B=10
1	.62091254	.60731464	.59573413	.58567639	.57680618
2	.26771578	.25371805	.24222092	.23254226	.22423672
3	.08907014	.08137363	.07529055	.07033602	.06620550
4	.02442403	.02143399	.01916461	.01737933	.01593540
5	.00575776	.00484001	.00417242	.00366594	.00326906
6	.00120076	.00096450	.00080009	.00067997	.00058889
7	.00022608	.00017315	.00013797	.00011326	.95146\04
8	.39017\04	.28437\04	.21731\04	.17207\04	.14006\04
9	.62441\05	.43227\05	.31634\05	.24132\05	.19012\05
10	.93515\06	.61388\06	.42960\06	.31537\06	.24024\06
11	.13204\06	.82057\07	.54842\07	.38700\07	.28480\07
12	.17684\07	.10388\07	.66221\08	.44872\08	.31873\08
13	.22579\08	.12518\08	.76027\09	.49419\09	.33853\09
14	.27603\09	.14423\09	.83358\10	.51929\10	.34277\10
15	.32428\10	.15948\10	.87612\11	.52260\11	.33214\11

M	B=11	B=12	B=13	B=14	B=15
1	.56888625	.56174287	.55524513	.54929208	.54380432
2	.21699926	.21061277	.20491787	.19979444	.19514997
3	.06269731	.05967212	.05703027	.05469847	.05262146
4	.01474153	.01373651	.01287777	.01213475	.01148491
5	.00294996	.00268797	.00246914	.00228365	.00212448
6	.00051775	.00046084	.00041440	.00037588	.00034346
7	.81416\04	.70723\04	.62206\04	.55293\04	.49592\04
8	.11653\04	.98705\05	.84856\05	.73867\05	.64991\05
9	.15366\05	.12682\05	.10649\05	.90726\06	.78261\06
10	.18848\06	.15145\06	.12413\06	.10345\06	.87441\07
11	.21670\07	.16941\07	.13545\07	.11035\07	.91363\08
12	.23503\08	.17865\08	.13926\08	.11086\08	.89852\09
13	.24175\09	.17856\09	.13561\09	.10544\09	.83625\10
14	.23688\10	.16991\10	.12567\10	.95379\11	.73992\11
15	.22198\11	.15453\11	.11125\11	.82383\12	.62486\12

DIRICHLET D INTEGRAL
A = 5 R = 7

M	B= 1	B= 2	B= 3	B= 4	B= 5
1	.72091835	.64772943	.60702201	.57944237	.55885862
2	.39532310	.30110126	.25662133	.22933113	.21035281
3	.17825959	.11096844	.08449111	.06990680	.06050697
4	.06972784	.03451687	.02317156	.01760689	.01429957
5	.02450628	.00943108	.00552169	.00382756	.00290343
6	.00792504	.00232520	.00117607	.00073939	.00052175
7	.00239796	.00052719	.00022840	.00012955	.84735\04
8	.00068720	.00011146	.41043\04	.20903\04	.12630\04
9	.00018822	.22210\04	.69011\05	.31416\05	.17480\05
10	.49621\04	.42053\05	.10953\05	.44382\06	.22675\06
11	.12660\04	.76159\06	.16524\06	.59365\07	.27773\07
12	.31396\05	.13263\06	.23832\07	.75630\08	.32314\08
13	.75952\06	.22308\07	.33016\08	.92224\09	.35898\09
14	.17977\06	.36372\08	.44110\09	.10809\09	.38239\10
15	.41728\07	.57665\09	.57025\10	.12220\10	.39200\11

M	B= 6	B= 7	B= 8	B= 9	B=10
1	.54257880	.52919419	.51788016	.50811473	.49954772
2	.19614018	.18496161	.17585734	.16824650	.16175403
3	.05386659	.04888430	.04498338	.04183075	.03921961
4	.01210328	.01053540	.00935769	.00843907	.00770140
5	.00232844	.00193884	.00165863	.00144801	.00128424
6	.00039527	.00031411	.00025832	.00021797	.00018762
7	.60477\04	.45770\04	.36123\04	.29419\04	.24548\04
8	.84710\05	.60938\05	.46083\05	.36175\05	.29231\05
9	.10993\05	.75030\06	.54288\06	.41029\06	.32069\06
10	.13341\06	.86246\07	.59625\07	.43334\07	.32733\07
11	.15255\07	.93263\08	.61524\08	.42953\08	.31327\08
12	.16539\08	.95470\09	.60023\09	.40212\09	.28292\09
13	.17088\09	.92998\10	.55657\10	.35746\10	.24242\10
14	.16899\10	.86586\11	.49271\11	.30308\11	.19795\11
15	.16055\11	.77344\12	.41801\12	.24605\12	.15465\12

M	B=11	B=12	B=13	B=14	B=15
1	.49193346	.48509332	.47889362	.47323181	.46802765
2	.15612535	.15118070	.14678897	.14285196	.13929445
3	.03701436	.03512206	.03347673	.03203013	.03074611
4	.00709521	.00658764	.00615600	.00578409	.00546005
5	.00115343	.00104666	.00095793	.00088306	.00081907
6	.00016409	.00014538	.00013020	.00011767	.00010716
7	.20885\04	.18050\04	.15805\04	.13993\04	.12505\04
8	.24168\05	.20359\05	.17418\05	.15097\05	.13231\05
9	.25746\06	.21123\06	.17645\06	.14964\06	.12854\06
10	.25495\07	.20358\07	.16593\07	.13760\07	.11580\07
11	.23654\08	.18369\08	.14600\08	.11833\08	.97512\09
12	.20693\09	.15619\09	.12099\09	.95792\10	.77257\10
13	.17163\10	.12583\10	.94946\11	.73393\11	.57907\11
14	.13557\11	.96484\12	.70877\12	.53467\12	.41253\12
15	.10239\12	.70694\13	.50532\13	.37183\13	.28043\13

DIRICHLET D INTEGRAL
1/A = 5 R = 8

M	B= 1	B= 2	B= 3	B= 4	B= 5
1	.99999940	.99999881	.99999822	.99999762	.99999703
2	.99999544	.99999088	.99998633	.99998178	.99997724
3	.99998055	.99996115	.99994178	.99992247	.99990319
4	.99993921	.99987862	.99981824	.99975805	.99969806
5	.99984446	.99968969	.99953567	.99938240	.99922985
6	.99965496	.99931234	.99897208	.99863412	.99829840
7	.99931280	.99863226	.99795815	.99729025	.99662837
8	.99874255	.99750146	.99627605	.99506570	.99386985
9	.99785153	.99573979	.99366300	.99161950	.98960782
10	.99653149	.99313945	.98981946	.98656755	.98338015
11	.99466145	.98947198	.98442152	.97950119	.97470306
12	.99211138	.98449740	.97713656	.97001019	.96310186
13	.98874672	.97797476	.96764089	.95770807	.94814424
14	.98443304	.96967331	.95563808	.94225776	.92947302
15	.97904094	.95938348	.94087656	.92339540	.90683542

M	B= 6	B= 7	B= 8	B= 9	B=10
1	.99999644	.99999584	.99999525	.99999466	.99999407
2	.99997271	.99996818	.99996366	.99995915	.99995464
3	.99988395	.99986476	.99984560	.99982649	.99980742
4	.99963827	.99957867	.99951925	.99946002	.99940097
5	.99907800	.99892685	.99877639	.99862659	.99847744
6	.99796485	.99763343	.99730408	.99697675	.99665141
7	.99597232	.99532193	.99467704	.99403750	.99340316
8	.99268796	.99151955	.99036415	.98922134	.98809072
9	.98762658	.98567452	.98375047	.98185338	.97998221
10	.98025402	.97718622	.97417404	.97121501	.96830683
11	.97002003	.96544572	.96097433	.95660058	.95231964
12	.95639706	.94988284	.94354760	.93738087	.93137319
13	.93892144	.93001501	.92140314	.91306633	.90498715
14	.91723282	.90549277	.89421403	.88336232	.87290724
15	.89110793	.87613692	.86185665	.84820984	.83514620

M	B=11	B=12	B=13	B=14	B=15
1	.99999348	.99999289	.99999230	.99999171	.99999112
2	.99995014	.99994564	.99994115	.99993667	.99993219
3	.99978838	.99976938	.99975042	.99973150	.99971262
4	.99934209	.99928340	.99922487	.99916652	.99910834
5	.99832894	.99818107	.99803382	.99788718	.99774114
6	.99632800	.99600649	.99568683	.99536899	.99505293
7	.99277388	.99214954	.99153001	.99091517	.99030492
8	.98697190	.98586454	.98476829	.98368285	.98260791
9	.97813605	.97631402	.97451530	.97273911	.97098473
10	.96544738	.96263470	.95986696	.95714245	.95445957
11	.94812709	.94401886	.93999117	.93604055	.93216376
12	.92551595	.91980127	.91422195	.90877135	.90344338
13	.89714984	.88954017	.88214519	.87495312	.86795313
14	.86282166	.85308124	.84366406	.83455031	.82572197
15	.82262132	.81059576	.79903433	.78790545	.77718068

DIRICHLET D INTEGRAL
1/A = 4 R = 8

M	B= 1	B= 2	B= 3	B= 4	B= 5
1	.99999744	.99999489	.99999234	.99998980	.99998726
2	.99998106	.99996219	.99994340	.99992468	.99990603
3	.99992207	.99984463	.99976766	.99969115	.99961508
4	.99976479	.99953174	.99930077	.99907181	.99884477
5	.99941876	.99884514	.99827877	.99771933	.99716650
6	.99875438	.99753134	.99632954	.99514781	.99398510
7	.99760279	.99526401	.99297955	.99074581	.98855964
8	.99576025	.99165617	.98767644	.98381139	.98005267
9	.99299644	.98628083	.97982505	.97360558	.96760239
10	.98906568	.97869796	.96883270	.95941735	.95040810
11	.98371986	.96848457	.95415824	.94063211	.92781730
12	.97672169	.95526477	.93535934	.91679467	.89940234
13	.96785734	.93873702	.91213269	.88765945	.86501667
14	.95694737	.91869608	.88434335	.85322208	.82482384
15	.94385540	.89504818	.85204048	.81372954	.77929640

M	B= 6	B= 7	B= 8	B= 9	B=10
1	.99998473	.99998221	.99997969	.99997718	.99997468
2	.99988745	.99986894	.99985050	.99983212	.99981381
3	.99953944	.99946422	.99938941	.99931499	.99924096
4	.99861959	.99839622	.99817459	.99795466	.99773636
5	.99662002	.99607963	.99554509	.99501619	.99449274
6	.99284046	.99171304	.99060204	.98950676	.98842653
7	.98641824	.98431912	.98226000	.98023886	.97825384
8	.97639296	.97282580	.96934546	.96594680	.96262522
9	.96179827	.95617826	.95072923	.94543957	.94029897
10	.94176789	.93346499	.92547197	.91776491	.91032275
11	.91564003	.90403815	.89295872	.88235618	.87219100
12	.88304514	.86760954	.85300033	.83913674	.82594961
13	.84396442	.82430774	.80588585	.78856441	.77222996
14	.79875151	.77468855	.75237841	.73161012	.71220804
15	.74811565	.71969881	.69365703	.66967584	.64749738

M	B=11	B=12	B=13	B=14	B=15
1	.99997218	.99996968	.99996719	.99996471	.99996223
2	.99979556	.99977737	.99975924	.99974118	.99972317
3	.99916732	.99909404	.99902112	.99894856	.99887635
4	.99751966	.99730451	.99709087	.99687869	.99666795
5	.99397454	.99346143	.99295324	.99244982	.99195102
6	.98736076	.98630888	.98527036	.98424473	.98323153
7	.97630323	.97438549	.97249918	.97064298	.96881565
8	.95937653	.95619693	.95308296	.95003143	.94703943
9	.93529817	.93042886	.92568353	.92105533	.91653806
10	.90312688	.89616071	.88940939	.88285957	.87649917
11	.86242859	.85303848	.84399368	.83527013	.82684631
12	.81337918	.80137341	.78988671	.77887884	.76831413
13	.75678574	.74214849	.72824606	.71501549	.70240148
14	.69402441	.67693366	.66082826	.64561535	.63121427
15	.62690768	.60772734	.58980449	.57300955	.55723109

DIRICHLET D INTEGRAL
1/A = 3 R = 8

M	B= 1	B= 2	B= 3	B= 4	B= 5
1	.99998474	.99996961	.99995461	.99993972	.99992494
2	.99989319	.99978779	.99968372	.99958089	.99947926
3	.99958420	.99917656	.99877649	.99838345	.99799701
4	.99881172	.99765676	.99653210	.99543520	.99436393
5	.99721849	.99454448	.99196622	.98947422	.98706059
6	.99435067	.98899320	.98388984	.97901090	.97433236
7	.98969047	.98007406	.97104591	.96252601	.95445142
8	.98270016	.96687838	.95227592	.93870192	.92601033
9	.97287004	.94862397	.92668365	.90663770	.88818066
10	.95976322	.92474782	.89376649	.86600622	.84088235
11	.94305202	.89497257	.85349417	.81712661	.78483386
12	.92254282	.85933952	.80632592	.76094491	.72148114
13	.89818814	.81820738	.75316875	.69893359	.65282809
14	.87008659	.77222031	.69528845	.63293323	.58120631
15	.83847235	.72225255	.63419020	.56496605	.50902763

M	B= 6	B= 7	B= 8	B= 9	B=10
1	.99991027	.99989571	.99988125	.99986688	.99985261
2	.99937875	.99927932	.99918092	.99908350	.99898703
3	.99761675	.99724234	.99687347	.99650985	.99615123
4	.99331645	.99229117	.99128668	.99030174	.98933523
5	.98471871	.98244290	.98022826	.97807049	.97596584
6	.96983445	.96550055	.96131654	.95727029	.95335125
7	.94677136	.93944413	.93243489	.92571416	.91925672
8	.91408624	.90283723	.89218757	.88207428	.87244433
9	.87107833	.85514651	.84023734	.82623004	.81302454
10	.81795867	.79690018	.77744357	.75937783	.74253109
11	.75586485	.72965706	.70577797	.68388781	.66371477
12	.68672810	.65580549	.62805197	.60295859	.58012577
13	.61303028	.57824676	.54752891	.52016225	.49559658
14	.53750506	.50003385	.46750836	.43898254	.41374237
15	.46283944	.42403374	.39096040	.36243175	.33756977

M	B=11	B=12	B=13	B=14	B=15
1	.99983843	.99982433	.99981032	.99979640	.99978256
2	.99889147	.99879678	.99870294	.99860992	.99851768
3	.99579738	.99544809	.99510318	.99476245	.99442575
4	.98838616	.98745363	.98653680	.98563494	.98474736
5	.97391094	.97190280	.96993874	.96801632	.96613333
6	.94955018	.94585892	.94227023	.93877764	.93537532
7	.91304079	.90704738	.90125980	.89566332	.89024483
8	.86325254	.85446006	.84603319	.83794252	.83016215
9	.80053690	.78869592	.77744066	.76671853	.75648378
10	.72676145	.71195028	.69799743	.68481748	.67233706
11	.64503805	.62767586	.61147679	.59631347	.58207771
12	.55923436	.54002571	.52228749	.50584341	.49054565
13	.47340014	.45322864	.43480362	.41789711	.40232039
14	.39123768	.37103701	.35279676	.33623960	.32113901
15	.31571059	.29634248	.27906434	.26355716	.24956394

DIRICHLET D INTEGRAL

1/A = 2 R = 8

M	B= 1	B= 2	B= 3	B= 4	B= 5
1	.99984758	.99970044	.99955790	.99941943	.99928464
2	.99903470	.99811878	.99724453	.99640629	.99559968
3	.99659605	.99343772	.99047912	.98768794	.98504024
4	.99117682	.98321649	.97592894	.96918682	.96289926
5	.98124157	.96488510	.95031567	.93714262	.92509836
6	.96534517	.93632918	.91129601	.88925193	.86954572
7	.94238370	.89637225	.85810207	.82539255	.79688033
8	.91176840	.84503209	.79176615	.74772887	.71039732
9	.87349928	.78351857	.71492116	.66024760	.61528893
10	.82814329	.71401467	.63129413	.56794352	.51753780
11	.77673984	.63932481	.54506581	.47592618	.42282641
12	.72066334	.56248272	.46026039	.38867124	.33572641
13	.66147148	.48639586	.38027902	.30954225	.25928062
14	.60076188	.41357552	.30762786	.24060769	.19495729
15	.54005229	.34597267	.24383651	.18270412	.14287345

M	B= 6	B= 7	B= 8	B= 9	B=10
1	.99915317	.99902475	.99889915	.99877615	.99865558
2	.99482124	.99406815	.99333807	.99262901	.99193929
3	.98251752	.98010513	.97779116	.97556577	.97342070
4	.95699836	.95143170	.94615774	.94114283	.93635926
5	.91398978	.90367199	.89403300	.88498414	.87645381
6	.85172407	.83545716	.82049665	.80665048	.79376672
7	.77165196	.74906381	.72864456	.71003837	.69296987
8	.67814778	.64987405	.62478922	.60231406	.58201033
9	.57744848	.54501847	.51682016	.49200856	.46995908
10	.47628490	.44178100	.41241822	.38707507	.36494248
11	.38065049	.34627722	.31768598	.29350679	.27277577
12	.29498958	.26268750	.23645864	.21474820	.19649024
13	.22188570	.19307779	.17027058	.15181210	.13659965
14	.16217199	.13766867	.11877609	.10384056	.09178814
15	.11530702	.09534593	.08037431	.06882406	.05970520

M	B=11	B=12	B=13	B=14	B=15
1	.99853728	.99842112	.99830697	.99819472	.99808426
2	.99126744	.99061219	.98997243	.98934715	.98873546
3	.97134891	.96934437	.96740184	.96551672	.96368498
4	.93178387	.92739707	.92318209	.91912447	.91521160
5	.86838322	.86072346	.85343331	.84647766	.83982638
6	.78172299	.77041915	.75977214	.74971228	.74018047
7	.67722153	.66261844	.64901793	.63630203	.62437214
8	.56353875	.54663157	.53107397	.51669111	.50333894
9	.45019827	.43235953	.41615382	.40134966	.38775911
10	.34542032	.32805296	.31248757	.29844622	.28570669
11	.25479386	.23904098	.22512178	.21273006	.20162486
12	.18092935	.16751515	.15583715	.14558295	.13651062
13	.12387014	.11307942	.10382952	.09582294	.08883307
14	.08189361	.07365116	.06669817	.06076851	.05566287
15	.05236583	.04636143	.04137981	.03719614	.03364493

DIRICHLET D INTEGRAL
A = 1 R = 8

M	B= 1	B= 2	B= 3	B= 4	B= 5
1	.99609375	.99287681	.99009728	.98762865	.98539592
2	.98046875	.96557996	.95336985	.94294608	.93381408
3	.94531250	.90735113	.87806361	.85417034	.83398025
4	.88671875	.81632693	.76568230	.72642054	.69454876
5	.80615234	.70020442	.62973066	.57809444	.53797953
6	.70947266	.57230271	.48865687	.43103341	.38834729
7	.60473633	.44645952	.35870772	.30215416	.26234546
8	.50000000	.33333333	.25000000	.20000000	.16666667
9	.40180969	.23893478	.16607213	.12556119	.10006412
10	.31452942	.16495521	.10555260	.07508615	.05703855
11	.24034119	.11001678	.06441790	.04293910	.03099909
12	.17964172	.07108548	.03787268	.02356580	.01612369
13	.13158798	.04461145	.02151327	.01245193	.00805349
14	.09462357	.02725597	.01183852	.00635271	.00387453
15	.06690025	.01624548	.00632599	.00313733	.00180027

M	B= 6	B= 7	B= 8	B= 9	B=10
1	.98335006	.98145690	.97969144	.97803475	.97647211
2	.92566673	.91829847	.91156413	.90535692	.89959577
3	.81649860	.80108773	.78731322	.77486522	.76351469
4	.66784869	.64496155	.62499577	.60733550	.59153805
5	.50555907	.47859780	.45568547	.43587989	.41852315
6	.35515605	.32843492	.30635187	.28772454	.27175168
7	.23264138	.20953783	.19100057	.17576299	.16299289
8	.14285714	.12500000	.11111111	.10000000	.09090909
9	.08267103	.07011524	.06066269	.05331200	.04744656
10	.04530731	.03717039	.03124926	.02677864	.02330299
11	.02361956	.01870973	.01526093	.01273551	.01082438
12	.01175940	.00897856	.00709546	.00575955	.00477637
13	.00561100	.00412284	.00315256	.00248636	.00200996
14	.00257394	.00181737	.00134297	.00102805	.00080940
15	.00113835	.00077125	.00055014	.00040836	.00031286

M	B=11	B=12	B=13	B=14	B=15
1	.97499174	.97358412	.97224137	.97095689	.96972513
2	.89421753	.88917194	.88441833	.87992324	.87565880
3	.75308743	.74344775	.73448788	.72612077	.71827505
4	.57727492	.56429581	.55240588	.54145056	.53130527
5	.40313927	.38937388	.37695667	.36567721	.35536884
6	.25786875	.24566517	.23483439	.22514233	.21640686
7	.15211967	.14273833	.13455301	.12734224	.12093681
8	.08333333	.07692308	.07142857	.06666667	.06250000
9	.04266701	.03870385	.03536898	.03252733	.03007951
10	.02053589	.01828910	.01643429	.01488125	.01356492
11	.00933903	.00815885	.00720360	.00641810	.00576333
12	.00403097	.00345181	.00299247	.00262171	.00231791
13	.00165789	.00139056	.00118292	.00101850	.00088612
14	.00065201	.00053530	.00044656	.00037766	.00032318
15	.00024595	.00019752	.00016150	.00013409	.00011281

DIRICHLET D INTEGRAL
A = 2 R = 8

M	B= 1	B= 2	B= 3	B= 4	B= 5
1	.96098156	.93859302	.92256926	.91001292	.89966021
2	.85693238	.78958142	.74595028	.71397133	.68889365
3	.70085861	.58964853	.52526591	.48144061	.44890654
4	.52744331	.39423000	.32616239	.28342003	.25351482
5	.36847929	.23915257	.18153316	.14837013	.12659259
6	.24130807	.13337334	.09193945	.07021608	.05687992
7	.14946219	.06916572	.04291590	.03045354	.02332809
8	.08823160	.03367372	.01865853	.01224040	.00883511
9	.04996248	.01551343	.00762143	.00460117	.00311931
10	.02728449	.00680756	.00294576	.00162973	.00103464
11	.01443362	.00286102	.00108379	.00054733	.00032449
12	.00742406	.00115689	.00038147	.00017520	.96748\04
13	.00372457	.00045186	.00012900	.53695\04	.27550\04
14	.00182739	.00017104	.42061\04	.15816\04	.75219\05
15	.00087881	.62924\04	.13266\04	.44919\05	.19759\05

M	B= 6	B= 7	B= 8	B= 9	B=10
1	.89083916	.88314762	.87632495	.87019223	.86462111
2	.66836347	.65104621	.63611387	.62301836	.61137868
3	.42340326	.40264795	.38528729	.37045883	.35758223
4	.23113922	.21361276	.19941981	.18763183	.17764464
5	.11108620	.09942272	.09029505	.08293439	.07685759
6	.04786871	.04137388	.03647029	.03263625	.02955546
7	.01877138	.01563070	.01334674	.01161754	.01026665
8	.00677935	.00542699	.00448117	.00378863	.00326318
9	.00227687	.00174891	.00139431	.00114350	.00095887
10	.00071681	.00052738	.00040537	.00032211	.00026269
11	.00021293	.00014981	.00011087	.85262\04	.67564\04
12	.60014\04	.40315\04	.28692\04	.21332\04	.16410\04
13	.16125\04	.10327\04	.70594\05	.50691\05	.37825\05
14	.41467\05	.25284\05	.16582\05	.11489\05	.83092\06
15	.10242\05	.59379\06	.37321\06	.24929\06	.17460\06

M	B=11	B=12	B=13	B=14	B=15
1	.85951631	.85480507	.85043057	.84634754	.84251933
2	.60091961	.59143611	.58277144	.57480324	.56743419
3	.34625001	.33616612	.32710933	.31891041	.31143727
4	.16904631	.16154521	.15492844	.14903652	.14374733
5	.07174504	.06737646	.06359482	.06028501	.05736062
6	.02702511	.02490919	.02311310	.02156904	.02022709
7	.00918450	.00829966	.00756367	.00694258	.00641193
8	.00285301	.00252531	.00225841	.00203744	.00185194
9	.00081854	.00070908	.00062184	.00055102	.00049263
10	.00021876	.00018532	.00015926	.00013853	.00012175
11	.54843\04	.45403\04	.38211\04	.32607\04	.28159\04
12	.12974\04	.10489\04	.86399\05	.72295\05	.61313\05
13	.29105\05	.22966\05	.18504\05	.15175\05	.12633\05
14	.62185\06	.47859\06	.37700\06	.30286\06	.24741\06
15	.12701\06	.95285\07	.73345\07	.57692\07	.46225\07

DIRICHLET D INTEGRAL
A = 3 R = 8

M	B= 1	B= 2	B= 3	B= 4	B= 5
1	.89988708	.85701256	.82956704	.80944177	.79359526
2	.69966125	.60437880	.55052835	.51404845	.48693735
3	.47440720	.35826993	.30115905	.26568451	.24092029
4	.28669548	.18402539	.14071703	.11622622	.10023925
5	.15764368	.08406197	.05780730	.04438258	.03622297
6	.08021259	.03484861	.02134385	.01514088	.01164466
7	.03827076	.01331742	.00720283	.00469594	.00339073
8	.01729984	.00474870	.00225064	.00134209	.00090676
9	.00746972	.00159517	.00065784	.00035721	.00022514
10	.00310078	.00050868	.00018134	.89301\04	.52352\04
11	.00124398	.00015496	.47461\04	.21114\04	.11483\04
12	.00048438	.45324\04	.11859\04	.47488\05	.23896\05
13	.00018370	.12784\04	.28421\05	.10209\05	.47416\06
14	.68060\04	.34899\05	.65588\06	.21063\06	.90093\07
15	.24693\04	.92493\06	.14624\06	.41856\07	.16451\07

M	B= 6	B= 7	B= 8	B= 9	B=10
1	.78055344	.76949013	.75989597	.75143499	.74387401
2	.46561280	.44818263	.43353419	.42096174	.40999163
3	.22236801	.20779459	.19595001	.18607302	.17767011
4	.08886842	.08030495	.07358686	.06815251	.06365060
5	.03072950	.02677206	.02378066	.02143675	.01954831
6	.00942292	.00789520	.00678439	.00594245	.00528349
7	.00260998	.00209856	.00174145	.00147995	.00128135
8	.00066227	.00051001	.00040806	.00033602	.00028299
9	.00015566	.00011461	.88282\04	.70357\04	.57575\04
10	.34194\04	.24028\04	.17795\04	.13710\04	.10892\04
11	.70704\05	.47345\05	.33670\05	.25050\05	.19302\05
12	.13845\05	.88214\06	.60167\06	.43184\06	.32245\06
13	.25803\06	.15621\06	.10207\06	.70605\07	.51048\07
14	.45968\07	.26406\07	.16509\07	.10997\07	.76927\08
15	.78566\08	.42767\08	.25559\08	.16379\08	.11078\08

M	B=11	B=12	B=13	B=14	B=15
1	.73704469	.73082157	.72510869	.71983103	.71492886
2	.40029170	.39162071	.38379835	.37668656	.37017749
3	.17040548	.16404161	.15840515	.15336619	.14882522
4	.05984940	.05658943	.05375710	.05126915	.04906307
5	.01799263	.01668765	.01557633	.01461783	.01378208
6	.00475439	.00432062	.00395882	.00365262	.00339024
7	.00112606	.00100174	.00090025	.00081601	.00074512
8	.00024264	.00021112	.00018595	.00016548	.00014856
9	.48123\04	.40922\04	.35302\04	.30824\04	.27193\04
10	.88667\05	.73638\05	.62177\05	.53238\05	.46130\05
11	.15293\05	.12396\05	.10238\05	.85919\06	.73085\06
12	.24847\06	.19643\06	.15862\06	.13040\06	.10884\06
13	.38229\07	.29459\07	.23245\07	.18710\07	.15318\07
14	.55951\08	.42002\08	.32368\08	.25498\08	.20466\08
15	.78199\09	.57157\09	.42996\09	.33134\09	.26064\09

DIRICHLET D INTEGRAL

A = 4 R = 8

M	B= 1	B= 2	B= 3	B= 4	B= 5
1	.83222784	.77530537	.74135162	.71742344	.69908033
2	.56379238	.46352252	.41150731	.37787780	.35364671
3	.32220047	.22534333	.18232706	.15700869	.13994511
4	.16113920	.09326948	.06785740	.05434099	.04585903
5	.07255550	.03398500	.02199232	.01626840	.01293836
6	.03003532	.01116770	.00636809	.00432640	.00322973
7	.01160991	.00336882	.00167865	.00104206	.00072760
8	.00423975	.00094548	.00040857	.00023066	.00015015
9	.00147594	.00024945	.92831\04	.47457\04	.28714\04
10	.00049325	.62379\04	.19861\04	.91566\05	.51352\05
11	.00015914	.14883\04	.40296\05	.16690\05	.86524\06
12	.49789\04	.34062\05	.77974\06	.28909\06	.13819\06
13	.15163\04	.75118\06	.14461\06	.47826\07	.21029\07
14	.45086\05	.16023\06	.25806\07	.75887\08	.30623\08
15	.13123\05	.33164\07	.44470\08	.11591\08	.42831\09

M	B= 6	B= 7	B= 8	B= 9	B=10
1	.68427985	.67191791	.66133212	.65209484	.64391452
2	.33501419	.32005001	.30765177	.29713612	.28805296
3	.12748340	.11788483	.11020599	.10388622	.09856921
4	.03999460	.03567285	.03234082	.02968393	.02750950
5	.01076517	.00923635	.00810239	.00722761	.00653202
6	.00255614	.00210474	.00178317	.00154350	.00135854
7	.00054631	.00043082	.00035192	.00029518	.00025273
8	.00010670	.80434\04	.63255\04	.51348\04	.42720\04
9	.19268\04	.13861\04	.10479\04	.82229\05	.66413\05
10	.32471\05	.22253\05	.16153\05	.12239\05	.95872\06
11	.51453\06	.33542\06	.23346\06	.17064\06	.12952\06
12	.77139\07	.47763\07	.31840\07	.22426\07	.16481\07
13	.10999\07	.64595\08	.41194\08	.27934\08	.19861\08
14	.14981\08	.83337\09	.50788\09	.33127\09	.22768\09
15	.19567\09	.10296\09	.59902\10	.37549\10	.24930\10

M	B=11	B=12	B=13	B=14	B=15
1	.63658374	.62994992	.62389762	.61833751	.61319907
2	.28009161	.27302943	.26670197	.26098462	.25578093
3	.09401651	.09006186	.08658541	.08349830	.08073313
4	.02569264	.02414869	.02281811	.02165779	.02063563
5	.00596544	.00549483	.00509754	.00475754	.00446316
6	.00121182	.00109280	.00099445	.00091190	.00084169
7	.00021997	.00019403	.00017306	.00015582	.00014141
8	.36246\04	.31249\04	.27300\04	.24117\04	.21510\04
9	.54884\05	.46212\05	.39517\05	.34235\05	.29990\05
10	.77109\06	.63365\06	.53005\06	.45007\06	.38705\06
11	.10131\06	.81197\07	.66404\07	.55234\07	.46611\07
12	.12528\07	.97871\08	.78209\08	.63696\08	.52721\08
13	.14661\08	.11157\08	.87073\09	.69402\09	.56320\09
14	.16312\09	.12085\09	.92063\10	.71783\10	.57088\10
15	.17322\10	.12488\10	.92813\11	.70762\11	.55131\11

DIRICHLET D INTEGRAL
A = 5 R = 8

M	B= 1	B= 2	B= 3	B= 4	B= 5
1	.76743196	.70226236	.66522109	.63979325	.62063691
2	.45734124	.36226302	.31573705	.28656608	.26596427
3	.22477320	.14870657	.11718263	.09927191	.08747143
4	.09556874	.05141354	.03616430	.02837169	.02360401
5	.03635002	.01555477	.00966435	.00697642	.00545592
6	.01266254	.00422842	.00229942	.00151883	.00111227
7	.00410872	.00105263	.00049691	.00029882	.00020421
8	.00125745	.00024339	.98989\04	.53947\04	.34293\04
9	.00036643	.52840\04	.18386\04	.90416\05	.53305\05
10	.00010242	.10863\04	.32129\05	.14199\05	.77422\06
11	.27620\04	.21291\05	.53203\06	.21051\06	.10587\06
12	.72197\05	.40009\06	.83980\07	.29641\07	.13717\07
13	.18362\05	.72413\07	.12698\07	.39844\08	.16924\08
14	.45584\06	.12672\07	.18470\08	.51351\09	.19974\09
15	.11075\06	.21511\08	.25932\09	.63687\10	.22636\10

M	B= 6	B= 7	B= 8	B= 9	B=10
1	.60537605	.59275533	.58203440	.57274165	.56455907
2	.25035068	.23795069	.22776916	.21919802	.21184140
3	.07899247	.07254249	.06743378	.06326381	.05977991
4	.02036652	.01801320	.01621854	.01480035	.01364848
5	.00448308	.00380887	.00331470	.00293718	.00263945
6	.00086787	.00070673	.00059340	.00050982	.00044588
7	.00015092	.00011756	.95082\04	.79097\04	.67250\04
8	.23948\04	.17811\04	.13856\04	.11147\04	.92041\05
9	.35098\05	.24882\05	.18592\05	.14449\05	.11574\05
10	.47965\06	.32359\06	.23194\06	.17393\06	.13505\06
11	.61593\07	.39482\07	.27114\07	.19600\07	.14739\07
12	.74793\08	.45489\08	.29894\08	.20810\08	.15144\08
13	.86340\09	.49753\09	.31253\09	.20932\09	.14729\09
14	.95177\10	.51893\10	.31126\10	.20039\10	.13623\10
15	.10057\10	.51817\11	.29646\11	.18331\11	.12032\11

M	B=11	B=12	B=13	B=14	B=15
1	.55726257	.55068850	.54471397	.53924444	.53420567
2	.20542871	.19976767	.19471726	.19017134	.18604819
3	.05681473	.05425261	.05201080	.05002836	.04825935
4	.01269229	.01188436	.01119159	.01059018	.01006253
5	.00239863	.00219983	.00203291	.00189074	.00176818
6	.00039554	.00035497	.00032164	.00029380	.00027024
7	.58178\04	.51047\04	.45317\04	.40629\04	.36734\04
8	.77582\05	.66503\05	.57804\05	.50834\05	.45153\05
9	.94979\06	.79481\06	.67597\06	.58277\06	.50828\06
10	.10781\06	.88013\07	.73198\07	.61832\07	.52928\07
11	.11437\07	.91029\08	.73990\08	.61208\08	.51399\08
12	.11414\08	.88519\09	.70279\09	.56909\09	.46861\09
13	.10776\09	.81379\10	.63078\10	.49974\10	.40335\10
14	.96693\11	.71063\11	.53750\11	.41645\11	.32933\11
15	.82788\12	.59182\12	.43658\12	.33067\12	.25610\12

DIRICHLET D INTEGRAL
1/A = 5 R = 9

M	B= 1	B= 2	B= 3	B= 4	B= 5
1	.99999990	.99999980	.99999970	.99999960	.99999950
2	.99999916	.99999831	.99999747	.99999663	.99999579
3	.99999606	.99999212	.99998818	.99998425	.99998032
4	.99998658	.99997318	.99995981	.99994645	.99993312
5	.99996289	.99992588	.99988895	.99985210	.99981534
6	.99991157	.99982344	.99973562	.99964809	.99956084
7	.99981178	.99962446	.99943803	.99925246	.99906775
8	.99963357	.99926955	.99890788	.99854849	.99819134
9	.99933656	.99867896	.99802701	.99738054	.99673938
10	.99886905	.99775115	.99664581	.99555258	.99447105
11	.99816778	.99636282	.99458392	.99282998	.99109999
12	.99715838	.99437041	.99163332	.98894463	.98630208
13	.99575644	.99161306	.98756391	.98360366	.97972751
14	.99386921	.98791702	.98213121	.97650100	.97101677
15	.99139783	.98310109	.97508590	.96733143	.95981941

M	B= 6	B= 7	B= 8	B= 9	B=10
1	.99999941	.99999931	.99999921	.99999911	.99999901
2	.99999495	.99999411	.99999327	.99999243	.99999159
3	.99997640	.99997248	.99996856	.99996465	.99996074
4	.99991981	.99990651	.99989324	.99987999	.99986676
5	.99977867	.99974208	.99970557	.99966914	.99963279
6	.99947388	.99938720	.99930080	.99921467	.99912880
7	.99888387	.99870080	.99851854	.99833707	.99815637
8	.99783636	.99748351	.99713274	.99678400	.99643726
9	.99610340	.99547245	.99484639	.99422509	.99360845
10	.99340082	.99234153	.99129284	.99025444	.98922601
11	.98939303	.98770823	.98604480	.98440198	.98277908
12	.98370358	.98114724	.97863130	.97615412	.97371421
13	.97593112	.97221053	.96856214	.96498261	.96146891
14	.96566992	.96045270	.95535809	.95037970	.94551169
15	.95253364	.94545967	.93858457	.93189665	.92538534

M	B=11	B=12	B=13	B=14	B=15
1	.99999891	.99999881	.99999871	.99999861	.99999852
2	.99999075	.99998992	.99998908	.99998824	.99998741
3	.99995684	.99995294	.99994904	.99994515	.99994126
4	.99985355	.99984035	.99982718	.99981403	.99980089
5	.99959652	.99956033	.99952422	.99948818	.99945222
6	.99904320	.99895786	.99887277	.99878794	.99870335
7	.99797643	.99779724	.99761878	.99744104	.99726402
8	.99609246	.99574958	.99540857	.99506939	.99473202
9	.99299635	.99238867	.99178531	.99118619	.99059119
10	.98820728	.98719798	.98619786	.98520666	.98422417
11	.98117544	.97959044	.97802351	.97647411	.97494170
12	.97131016	.96894065	.96660446	.96430044	.96202750
13	.95801819	.95462785	.95129545	.94801872	.94479554
14	.94074871	.93608582	.93151849	.92704251	.92265395
15	.91904101	.91285490	.90681894	.90092573	.89516845

DIRICHLET D INTEGRAL
1/A = 4 R = 9

M	B= 1	B= 2	B= 3	B= 4	B= 5
1	.99999949	.99999898	.99999847	.99999796	.99999745
2	.99999580	.99999161	.99998743	.99998326	.99997910
3	.99998106	.99996218	.99994336	.99992460	.99990590
4	.99993780	.99987591	.99981433	.99975303	.99969202
5	.99983399	.99966918	.99950551	.99934296	.99918148
6	.99961807	.99923998	.99886557	.99849470	.99812724
7	.99921501	.99844080	.99767683	.99692264	.99617780
8	.99852406	.99707515	.99565169	.99425229	.99287571
9	.99741854	.99489883	.99243665	.99002830	.98767053
10	.99574797	.99162633	.98762466	.98373401	.97994657
11	.99334235	.98694195	.98077498	.97482134	.96906380
12	.99001821	.98051483	.97143893	.96274837	.95440767
13	.98558604	.97201664	.95918932	.94702096	.93544286
14	.97985830	.96114043	.94365073	.92723402	.91176454
15	.97265772	.94761897	.92452776	.90310829	.88314164

M	B= 6	B= 7	B= 8	B= 9	B=10
1	.99999694	.99999643	.99999592	.99999542	.99999491
2	.99997494	.99997080	.99996666	.99996253	.99995840
3	.99988726	.99986867	.99985014	.99983167	.99981325
4	.99963129	.99957083	.99951064	.99945070	.99939103
5	.99902106	.99886166	.99870324	.99854579	.99838929
6	.99776307	.99740208	.99704417	.99668924	.99633720
7	.99544191	.99471462	.99399559	.99328452	.99258114
8	.99152083	.99018662	.98887215	.98757658	.98629912
9	.98536042	.98309538	.98087306	.97869132	.97654822
10	.97625550	.97265473	.96913884	.96570298	.96234276
11	.96348744	.95807924	.95282770	.94772263	.94275490
12	.94638659	.93865904	.93120233	.92399655	.91702409
13	.92439742	.91383570	.90371566	.89400089	.88465957
14	.89713861	.88326951	.87008384	.85751878	.84552011
15	.86445058	.84688929	.83033615	.81468851	.79985885

M	B=11	B=12	B=13	B=14	B=15
1	.99999440	.99999390	.99999339	.99999289	.99999239
2	.99995429	.99995018	.99994608	.99994199	.99993791
3	.99979488	.99977657	.99975830	.99974009	.99972193
4	.99933160	.99927241	.99921347	.99915476	.99909628
5	.99823370	.99807900	.99792518	.99777222	.99762009
6	.99598796	.99564145	.99529759	.99495631	.99461755
7	.99188516	.99119636	.99051448	.98983933	.98917070
8	.98503904	.98379567	.98256839	.98135662	.98015981
9	.97444197	.97237095	.97033364	.96832864	.96635466
10	.95905421	.95583368	.95267786	.94958371	.94654841
11	.93791632	.93319950	.92859772	.92410488	.91971537
12	.91026929	.90371815	.89735808	.89117770	.88516670
13	.87566367	.86698840	.85861166	.85051367	.84267663
14	.83404065	.82303903	.81247878	.80232752	.79255640
15	.78577191	.77236241	.75957337	.74735471	.73566219

DIRICHLET D INTEGRAL
1/A = 3 R = 9

M	B= 1	B= 2	B= 3	B= 4	B= 5
1	.99999619	.99999239	.99998861	.99998486	.99998112
2	.99997044	.99994111	.99991201	.99988313	.99985446
3	.99987388	.99974928	.99962612	.99950432	.99938382
4	.99960834	.99922350	.99884502	.99847250	.99810559
5	.99901088	.99804580	.99710279	.99618019	.99527654
6	.99784582	.99576263	.99374346	.99178258	.98987517
7	.99580699	.99179645	.98794733	.98424273	.98066871
8	.99253028	.98548010	.97879312	.97242524	.96634100
9	.98761522	.97610978	.96534749	.95522482	.94566060
10	.98065222	.96300881	.94676812	.93170794	.91765851
11	.97125217	.94559340	.92240423	.90124334	.88178286
12	.95907483	.92343149	.89187936	.86359991	.83800277
13	.94385316	.89628765	.85514517	.81899706	.78684457
14	.92541152	.86414943	.81250000	.76810166	.72935696
15	.90367673	.82723343	.76457174	.71197482	.66701397

M	B= 6	B= 7	B= 8	B= 9	B=10
1	.99997739	.99997369	.99997000	.99996632	.99996267
2	.99982600	.99979772	.99976964	.99974173	.99971401
3	.99926456	.99914649	.99902954	.99891369	.99879889
4	.99774397	.99738737	.99703554	.99668826	.99634532
5	.99439062	.99352132	.99266768	.99182884	.99100400
6	.98801711	.98620484	.98443525	.98270561	.98101347
7	.97721360	.97386751	.97062188	.96746931	.96440328
8	.96051131	.95491185	.94952204	.94432425	.93930319
9	.93658956	.92795826	.91972216	.91184370	.90429081
10	.90448646	.89208468	.88036554	.86925629	.85869570
11	.86377134	.84701101	.83134307	.81663790	.80278809
12	.81464646	.79319132	.77336998	.75496792	.73781026
13	.75796046	.73179799	.70793554	.68604095	.66584792
14	.69513467	.66460475	.63714053	.61225774	.58957472
15	.62801884	.59379434	.56345805	.53634141	.51192691

M	B=11	B=12	B=13	B=14	B=15
1	.99995902	.99995539	.99995178	.99994818	.99994459
2	.99968645	.99965905	.99963182	.99960475	.99957783
3	.99868510	.99857228	.99846040	.99834944	.99823936
4	.99600654	.99567174	.99534076	.99501347	.99468973
5	.99019248	.98939363	.98860686	.98783164	.98706748
6	.97935667	.97773323	.97614139	.97457953	.97304620
7	.96141805	.95850848	.95566999	.95289846	.95019015
8	.93444555	.92973958	.92517489	.92074221	.91643325
9	.89703584	.89005476	.88332649	.87683247	.87055620
10	.84863174	.83901969	.82982083	.82100139	.81253167
11	.78970357	.77730800	.76553603	.75433130	.74364479
12	.72175247	.70667366	.69247167	.67905937	.66636184
13	.64713967	.62973740	.61349197	.59827768	.58398767
14	.56878564	.54964189	.53193885	.51550618	.50020077
15	.48980653	.46965353	.45120265	.43423593	.41857236

DIRICHLET D INTEGRAL
1/A = 2 R = 9

M	B= 1	B= 2	B= 3	B= 4	B= 5
1	.99994919	.99989965	.99985123	.99980382	.99975735
2	.99964436	.99930178	.99897056	.99864940	.99833728
3	.99862826	.99732799	.99608809	.99490036	.99375849
4	.99614445	.99256367	.98920720	.98603932	.98303338
5	.99117682	.98318625	.97584993	.96904715	.96269103
6	.98256627	.96724964	.95352752	.94106286	.92962214
7	.96917208	.94304160	.92028716	.90009958	.88194050
8	.95003752	.90941523	.87516948	.84558374	.81956387
9	.92452477	.86601487	.81845596	.77856565	.74434466
10	.89239761	.81335535	.75166084	.70157066	.65975030
11	.85384502	.75275455	.67729683	.61816706	.57022918
12	.80945113	.68615121	.59849372	.53238905	.48046227
13	.76012458	.61585679	.51856742	.44816076	.39468819
14	.70700368	.54429241	.44062670	.36883484	.31622778
15	.65135322	.47375301	.36727704	.29690628	.24725532

M	B= 6	B= 7	B= 8	B= 9	B=10
1	.99971173	.99966691	.99962284	.99957946	.99953674
2	.99803335	.99773690	.99744735	.99716419	.99688697
3	.99265744	.99159312	.99056214	.98956160	.98858907
4	.98016879	.97742912	.97480102	.97227345	.96983708
5	.95671629	.95107227	.94571864	.94062262	.93575713
6	.91903530	.90917382	.89993773	.89124743	.88303836
7	.86543062	.85029119	.83631057	.82332385	.81119977
8	.79636547	.77545736	.75644633	.73903277	.72298289
9	.71448180	.68806982	.66445476	.64314972	.62378261
10	.62409563	.59319690	.56606560	.54198337	.52041283
11	.53037054	.49657456	.46746732	.44207508	.41968542
12	.43842861	.40360756	.37422647	.34906125	.32723662
13	.35261869	.31861560	.29053953	.26695224	.24684944
14	.27606721	.24443857	.21891100	.19789493	.18030721
15	.21053450	.18239197	.16021312	.14233644	.12765786

M	B=11	B=12	B=13	B=14	B=15
1	.99949463	.99945311	.99941215	.99937171	.99933178
2	.99661530	.99634883	.99608727	.99583033	.99557778
3	.98764241	.98671977	.98581952	.98494022	.98408058
4	.96748400	.96520739	.96300134	.96086064	.95878073
5	.93109948	.92663042	.92233344	.91819426	.91420044
6	.87525729	.86785973	.86080808	.85407025	.84761855
7	.79983197	.78913298	.77902990	.76946133	.76037497
8	.70811066	.69426551	.68132389	.66918307	.65775677
9	.60606273	.58975884	.57468403	.56068514	.54763521
10	.50094223	.48324966	.46707908	.45222377	.43851466
11	.39976323	.38189768	.36576731	.35111652	.33773917
12	.30810855	.29119168	.27611265	.26257918	.25035894
13	.22950767	.21439204	.20109822	.18931473	.17879760
14	.16538442	.15257312	.14146240	.13174087	.12316829
15	.11541645	.10507164	.09622940	.08859610	.08194872

DIRICHLET D INTEGRAL
A = 1 R = 9

M	B= 1	B= 2	B= 3	B= 4	B= 5
1	.99804688	.99638858	.99492569	.99360581	.99239682
2	.98925781	.98070190	.97347980	.96718024	.96156608
3	.96728516	.94317941	.92385975	.90765219	.89365222
4	.92700195	.87796484	.84093864	.81121522	.78642053
5	.86657715	.78610408	.72929956	.68587788	.65102040
6	.78802490	.67503063	.60103339	.54742901	.50615638
7	.69638062	.55564503	.47078312	.41279364	.37010755
8	.59819031	.43890341	.35112056	.29478932	.25526915
9	.50000000	.33333333	.25000000	.20000000	.16666667
10	.40726471	.24397227	.17042890	.12934541	.10339363
11	.32380295	.17251212	.11157102	.08000526	.06116445
12	.25172234	.11813138	.07033973	.04747880	.03462031
13	.19165516	.07851897	.04282018	.02711198	.01880778
14	.14313936	.05076643	.02523261	.01493700	.00983439
15	.10501981	.03199092	.01442508	.00795907	.00496221

M	B= 6	B= 7	B= 8	B= 9	B=10
1	.99127721	.99023169	.98924891	.98832013	.98743848
2	.95648569	.95183519	.94753975	.94354346	.93980322
3	.88130899	.87025964	.86025125	.85109997	.84266759
4	.76518164	.74663048	.73018241	.71542460	.70205453
5	.62208790	.59748207	.57616413	.55742245	.54074916
6	.47304548	.44568217	.42255454	.40265889	.38529825
7	.33708826	.31062218	.28883308	.27051534	.25485503
8	.22586057	.20303937	.18476501	.16976976	.15722233
9	.14285714	.12500000	.11111111	.10000000	.09090909
10	.08563798	.07278865	.06309445	.05554172	.04950503
11	.04884248	.04025192	.03397311	.02921413	.02550160
12	.02659750	.02121724	.01741225	.01460941	.01247714
13	.01387421	.01069634	.00852430	.00697085	.00581931
14	.00695316	.00517310	.00399860	.00318378	.00259571
15	.00335676	.00240675	.00180230	.00139591	.00111056

M	B=11	B=12	B=13	B=14	B=15
1	.98659842	.98579540	.98502565	.98428598	.98357368
2	.93628507	.93296165	.92981060	.92681336	.92395433
3	.83484743	.82755533	.82072362	.81429705	.80822989
4	.68984390	.67861614	.66823189	.65857923	.64956694
5	.52576978	.51220066	.49982212	.48846073	.47797735
6	.36997088	.35630485	.34401763	.33289008	.32274924
7	.24128064	.22937775	.21883782	.20942583	.20095934
8	.14655373	.13736073	.12934915	.12229911	.11604277
9	.08333333	.07692308	.07142857	.06666667	.06250000
10	.04457855	.04048800	.03704164	.03410162	.03156634
11	.02253681	.02012276	.01812477	.01644790	.01502350
12	.01081205	.00948337	.00840370	.00751267	.00676743
13	.00494062	.00425390	.00370631	.00326216	.00289655
14	.00215752	.00182231	.00156016	.00135127	.00118211
15	.00090307	.00074779	.00062875	.00053560	.00046143

DIRICHLET D INTEGRAL
A = 2 R = 9

M	B= 1	B= 2	B= 3	B= 4	B= 5
1	.97398771	.95832500	.94683899	.93769132	.93005734
2	.89595082	.84332342	.80802729	.78156016	.76045099
3	.76588935	.66969628	.61136954	.57045454	.53940141
4	.60692532	.48033874	.41172574	.36697772	.33478975
5	.44796130	.31383461	.24956618	.21082776	.18451984
6	.31019248	.18872788	.13783718	.10964839	.09164099
7	.20303895	.10550035	.07015118	.05225712	.04154229
8	.12650072	.05530317	.03322066	.02305961	.01737573
9	.07547523	.02738902	.01475890	.00950370	.00676652
10	.04334807	.01289712	.00619410	.00368493	.00247189
11	.02407177	.00580551	.00247015	.00135247	.00085248
12	.01297330	.00250964	.00094069	.00047233	.00027904
13	.00680748	.00104595	.00034356	.00015766	.87090\04
14	.00348742	.00042170	.00012078	.50493\04	.26020\04
15	.00174835	.00016496	.40998\04	.15567\04	.74677\05

M	B= 6	B= 7	B= 8	B= 9	B=10
1	.92349010	.91771823	.91256381	.90790351	.90364812
2	.74293726	.72800088	.71500072	.70350710	.69321815
3	.51463129	.49418204	.47686886	.46192510	.44882811
4	.31017711	.29055200	.27441863	.26084368	.24921035
5	.16528925	.15051061	.13873373	.12908715	.12101316
6	.07909049	.06981312	.06265801	.05695997	.05230714
7	.03443975	.02939920	.02564243	.02273735	.02042540
8	.01380092	.01136964	.00962102	.00830953	.00729329
9	.00513687	.00407654	.00334163	.00280763	.00240511
10	.00178973	.00136582	.00108306	.00088421	.00073854
11	.00058748	.00043045	.00032977	.00026132	.00021263
12	.00018268	.00012832	.94864\04	.72893\04	.57728\04
13	.54069\04	.36360\04	.25907\04	.19285\04	.14853\04
14	.15293\04	.98321\05	.67450\05	.48594\05	.36372\05
15	.41482\05	.25465\05	.16802\05	.11706\05	.85080\06

M	B=11	B=12	B=13	B=14	B=15
1	.89973094	.89610073	.89271724	.88954822	.88656746
2	.68391382	.67542879	.66763580	.66043480	.65374578
3	.43720672	.42678853	.41736818	.40878725	.40092121
4	.23909231	.23018413	.22226041	.21515064	.20872313
5	.11413689	.10819632	.10300231	.09841464	.09432692
6	.04843055	.04514691	.04232687	.03987647	.03772579
7	.01854262	.01698017	.01566300	.01453774	.01356540
8	.00648505	.00582839	.00528535	.00482949	.00444187
9	.00209271	.00184441	.00164308	.00147711	.00133830
10	.00062830	.00054262	.00047456	.00041949	.00037420
11	.00017672	.00014945	.00012824	.00011140	.97794\04
12	.46839\04	.38765\04	.32617\04	.27830\04	.24031\04
13	.11757\04	.95164\05	.78474\05	.65736\05	.55809\05
14	.28067\05	.22206\05	.17937\05	.14745\05	.12303\05
15	.63967\06	.49441\06	.39100\06	.31527\06	.25844\06

DIRICHLET D INTEGRAL
A = 3 R = 9

M	B= 1	B= 2	B= 3	B= 4	B= 5
1	.92491531	.89097329	.86874853	.85221660	.83906357
2	.75597477	.67168968	.62242009	.58833048	.56260571
3	.54479909	.43104776	.37246645	.33502256	.30833998
4	.35122138	.24073504	.19125534	.16222989	.14277901
5	.20603810	.11962599	.08640372	.06863026	.05747368
6	.11166897	.05388041	.03504738	.02591453	.02056488
7	.05662031	.02232540	.01296966	.00888137	.00665476
8	.02712996	.00861117	.00443443	.00279920	.00197403
9	.01238478	.00312129	.00141504	.00081989	.00054256
10	.00542178	.00107141	.00042489	.00022507	.00013938
11	.00228843	.00035048	.00012086	.58312\04	.33704\04
12	.00093539	.00010984	.32749\04	.14341\04	.77179\05
13	.00037163	.33121\04	.84936\05	.33647\05	.16819\05
14	.00014395	.96467\05	.21170\05	.75618\06	.35033\06
15	.54508\04	.27224\05	.50881\06	.16338\06	.70003\07

M	B= 6	B= 7	B= 8	B= 9	B=10
1	.82815045	.81883145	.81070482	.80350353	.79704110
2	.54212936	.52522919	.51090956	.49853243	.48766578
3	.28802838	.27186383	.25858136	.24740039	.23780914
4	.12866033	.11785029	.10925121	.10221177	.09631887
5	.04977319	.04411267	.03976083	.03630095	.03347766
6	.01706302	.01459652	.01276668	.01135561	.01023439
7	.00527595	.00434716	.00368324	.00318724	.00280385
8	.00149181	.00118171	.00096843	.00081424	.00069842
9	.00038998	.00029647	.00023468	.00019149	.00016000
10	.95088\04	.69266\04	.52889\04	.41839\04	.34020\04
11	.21783\04	.15181\04	.11168\04	.85562\05	.67650\05
12	.47166\05	.31403\05	.22233\05	.16481\05	.12660\05
13	.97025\06	.61632\06	.41944\06	.30056\06	.22415\06
14	.19045\06	.11527\06	.75329\07	.52134\07	.37718\07
15	.35803\07	.20621\07	.12928\07	.86341\08	.60555\08

M	B=11	B=12	B=13	B=14	B=15
1	.79118211	.78582518	.78089241	.77632266	.77206710
2	.47800437	.46932500	.46145984	.45427971	.44768317
3	.22945596	.22208999	.21552671	.20962689	.20428319
4	.09129673	.08695355	.08315141	.07978835	.07678721
5	.03112542	.02913209	.02741887	.02592873	.02461933
6	.00932198	.00856492	.00792655	.00738088	.00690899
7	.00249939	.00225222	.00204787	.00187631	.00173037
8	.00060873	.00053754	.00047987	.00043234	.00039259
9	.00013623	.00011780	.00010318	.91359\04	.81642\04
10	.28277\04	.23927\04	.20550\04	.17872\04	.15710\04
11	.54848\05	.45388\05	.38202\05	.32617\05	.28189\05
12	.10005\05	.80920\06	.66711\06	.55890\06	.47472\06
13	.17256\06	.13632\06	.11002\06	.90408\07	.75438\07
14	.28267\07	.21800\07	.17216\07	.13870\07	.11365\07
15	.44152\08	.33225\08	.25662\08	.20260\08	.16296\08

DIRICHLET D INTEGRAL
A = 4 R = 9

M	B= 1	B= 2	B= 3	B= 4	B= 5
1	.86578227	.81759271	.78824720	.76730154	.75109764
2	.62419036	.52932801	.47850906	.44500087	.42051554
3	.38259845	.28113171	.23399174	.20548030	.18588849
4	.20543105	.12731739	.09628560	.07918534	.06818176
5	.09913061	.05069340	.03446690	.02637302	.02151536
6	.04385438	.01815907	.01099965	.00778743	.00599619
7	.01805881	.00595460	.00318759	.00207772	.00150480
8	.00700356	.00181139	.00085065	.00050819	.00034513
9	.00258146	.00051653	.00021135	.00011524	.73178\04
10	.00091089	.00013923	.49323\04	.24451\04	.14477\04
11	.00030949	.35712\04	.10889\04	.48893\05	.26925\05
12	.00010173	.87658\05	.22873\05	.92705\06	.47368\06
13	.32476\04	.20685\05	.45947\06	.16753\06	.79236\07
14	.10102\04	.47109\06	.88629\07	.28979\07	.12659\07
15	.30702\05	.10389\06	.16476\07	.48160\08	.19390\08

M	B= 6	B= 7	B= 8	B= 9	B=10
1	.73793052	.72686952	.71735201	.70901246	.70160032
2	.40148163	.38605939	.37318647	.36219830	.35265380
3	.17136449	.16004066	.15088877	.14329037	.13684837
4	.06042539	.05461918	.05008351	.04642588	.04340278
5	.01826856	.01593986	.01418444	.01281131	.01170613
6	.00486396	.00408743	.00352344	.00309605	.00276141
7	.00116319	.00093958	.00078333	.00066876	.00058158
8	.00025363	.00019658	.00015827	.00013113	.00011107
9	.51020\04	.37878\04	.29413\04	.23621\04	.19471\04
10	.95567\05	.67857\05	.50756\05	.39471\05	.31633\05
11	.16796\05	.11389\05	.81963\06	.61661\06	.48006\06
12	.27873\06	.18025\06	.12466\06	.90635\07	.68496\07
13	.43906\07	.27040\07	.17952\07	.12604\07	.92392\08
14	.65946\08	.38629\08	.24594\08	.16659\08	.11836\08
15	.94808\09	.52757\09	.32178\09	.21011\09	.14459\09

M	B=11	B=12	B=13	B=14	B=15
1	.69493654	.68888884	.68335680	.67826246	.67354407
2	.34424643	.33675525	.33001619	.32390435	.31832261
3	.13129468	.12644092	.12215030	.11832082	.11487463
4	.04085461	.03867212	.03677781	.03511507	.03364154
5	.01079618	.01003302	.00938310	.00882243	.00833340
6	.00249252	.00227187	.00208762	.00193149	.00179754
7	.00051327	.00045846	.00041362	.00037632	.00034485
8	.95762\04	.83760\04	.74142\04	.66293\04	.59785\04
9	.16386\04	.14025\04	.12173\04	.10690\04	.94834\05
10	.25966\05	.21733\05	.18485\05	.15937\05	.13900\05
11	.38408\06	.31416\06	.26172\06	.22141\06	.18978\06
12	.53377\07	.42644\07	.34777\07	.28855\07	.24297\07
13	.70082\08	.54654\08	.43611\08	.35474\08	.29331\08
14	.87337\09	.66451\09	.51856\09	.41334\09	.33547\09
15	.10373\09	.76956\10	.58704\10	.45835\10	.36501\10

DIRICHLET D INTEGRAL
A = 5 R = 9

M	B= 1	B= 2	B= 3	B= 4	B= 5
1	.80619330	.74864756	.71531028	.69215939	.67457455
2	.51548325	.42165667	.37430879	.34406765	.32242705
3	.27322488	.19010425	.15411085	.13311873	.11902951
4	.12517809	.07217151	.05276496	.04250085	.03606692
5	.05115470	.02391946	.01561169	.01165339	.00934540
6	.01907790	.00710117	.00410121	.00282186	.00213065
7	.00660358	.00192436	.00097568	.00061582	.00043634
8	.00214847	.00048282	.00021334	.00012298	.81523\04
9	.00066344	.00011339	.43374\04	.22740\04	.14063\04
10	.00019592	.25146\04	.82736\05	.39299\05	.22613\05
11	.55671\04	.53022\05	.14917\05	.63960\06	.34155\06
12	.15295\04	.10692\05	.25576\06	.98641\07	.48768\07
13	.40793\05	.20715\06	.41910\07	.14491\07	.66179\08
14	.10598\05	.38721\07	.65919\08	.20370\08	.85734\09
15	.26891\06	.70061\08	.99883\09	.27499\09	.10644\09

M	B= 6	B= 7	B= 8	B= 9	B=10
1	.66047655	.64875748	.63875951	.63006134	.62237751
2	.30585877	.29259176	.28162283	.27233384	.26431958
3	.10876058	.10085820	.09453821	.08933646	.08495880
4	.03161584	.02833125	.02579464	.02376833	.02210688
5	.00783403	.00676680	.00597220	.00535696	.00486601
6	.00170335	.00141518	.00120866	.00105385	.00093376
7	.00033186	.00026470	.00021843	.00018490	.00015963
8	.58854\04	.44988\04	.35820\04	.29402\04	.24710\04
9	.96168\05	.70334\05	.53962\05	.42905\05	.35067\05
10	.14619\05	.10214\05	.75417\06	.58025\06	.46084\06
11	.20835\06	.13887\06	.98565\07	.73311\07	.56531\07
12	.28021\07	.17791\07	.12125\07	.87105\08	.65163\08
13	.35753\08	.21596\08	.14118\08	.97867\09	.70975\09
14	.43480\09	.24954\09	.15631\09	.10447\09	.73395\10
15	.50596\10	.27555\10	.16522\10	.10638\10	.72349\11

M	B=11	B=12	B=13	B=14	B=15
1	.61550604	.60929897	.60364479	.59845747	.59366931
2	.25730132	.25107995	.24550876	.24047680	.23589836
3	.08120880	.07794967	.07508295	.07253568	.07025258
4	.02071608	.01953202	.01850978	.01761679	.01682882
5	.00446478	.00413043	.00384732	.00360433	.00339336
6	.00083802	.00075999	.00069523	.00064065	.00059405
7	.00014000	.00012436	.00011165	.00010113	.92310\04
8	.21159\04	.18397\04	.16199\04	.14415\04	.12945\04
9	.29297\05	.24915\05	.21503\05	.18789\05	.16591\05
10	.37534\06	.31201\06	.26380\06	.22622\06	.19635\06
11	.44855\07	.36426\07	.30153\07	.25365\07	.21630\07
12	.50338\08	.39911\08	.32331\08	.26666\08	.22332\08
13	.53345\09	.41271\09	.32700\09	.26433\09	.21733\09
14	.53638\10	.40471\10	.31350\10	.24826\10	.20030\10
15	.51381\11	.37789\11	.28605\11	.22182\11	.17557\11

DIRICHLET D INTEGRAL
1/A = 5 R = 10

M	B= 1	B= 2	B= 3	B= 4	B= 5
1	.99999998	.99999997	.99999995	.99999993	.99999992
2	.99999985	.99999969	.99999954	.99999938	.99999923
3	.99999921	.99999843	.99999764	.99999686	.99999607
4	.99999711	.99999422	.99999133	.99998845	.99998556
5	.99999141	.99998282	.99997425	.99996568	.99995713
6	.99997810	.99995624	.99993441	.99991262	.99989086
7	.99995038	.99990088	.99985149	.99980222	.99975306
8	.99989758	.99979549	.99969374	.99959232	.99949121
9	.99980408	.99960903	.99941484	.99922150	.99902898
10	.99964824	.99929858	.99895096	.99860535	.99826170
11	.99940150	.99880767	.99821839	.99763354	.99705298
12	.99902764	.99806509	.99711201	.99616810	.99523308
13	.99848244	.99698433	.99550490	.99404345	.99259934
14	.99771357	.99546387	.99324925	.99106816	.98891923
15	.99666095	.99338833	.99017868	.98702885	.98393598

M	B= 6	B= 7	B= 8	B= 9	B=10
1	.99999990	.99999988	.99999987	.99999985	.99999983
2	.99999907	.99999892	.99999877	.99999861	.99999846
3	.99999529	.99999451	.99999372	.99999294	.99999216
4	.99998268	.99997981	.99997693	.99997405	.99997118
5	.99994858	.99994005	.99993152	.99992300	.99991450
6	.99986914	.99984746	.99982581	.99980419	.99978261
7	.99970402	.99965508	.99960625	.99955754	.99950893
8	.99939043	.99928995	.99918979	.99908992	.99899036
9	.99883728	.99864637	.99845625	.99826689	.99807830
10	.99791996	.99758009	.99724205	.99690582	.99657134
11	.99647662	.99590435	.99533607	.99477168	.99421110
12	.99430667	.99338863	.99247872	.99157671	.99068240
13	.99117195	.98976072	.98836514	.98698470	.98561893
14	.98680115	.98471273	.98265286	.98062052	.97861474
15	.98089746	.97791090	.97497407	.97208494	.96924161

M	B=11	B=12	B=13	B=14	B=15
1	.99999982	.99999980	.99999979	.99999977	.99999975
2	.99999830	.99999815	.99999800	.99999784	.99999769
3	.99999138	.99999059	.99998981	.99998903	.99998825
4	.99996831	.99996545	.99996258	.99995972	.99995685
5	.99990600	.99989751	.99988903	.99988056	.99987209
6	.99976106	.99973954	.99971806	.99969661	.99967519
7	.99946042	.99941202	.99936372	.99931552	.99926743
8	.99889108	.99879211	.99869341	.99859500	.99849687
9	.99789045	.99770333	.99751693	.99733124	.99714625
10	.99623860	.99590754	.99557816	.99525040	.99492425
11	.99365424	.99310101	.99255135	.99200517	.99146240
12	.98979559	.98891608	.98804370	.98717827	.98631963
13	.98426741	.98292971	.98160545	.98029425	.97899577
14	.97663460	.97467928	.97274796	.97083990	.96895438
15	.96644232	.96368541	.96096935	.95829270	.95565409

DIRICHLET D INTEGRAL

1/A = 4 R = 10

M	B= 1	B= 2	B= 3	B= 4	B= 5
1	.99999990	.99999980	.99999969	.99999959	.99999949
2	.99999908	.99999816	.99999724	.99999632	.99999540
3	.99999547	.99999096	.99998645	.99998194	.99997745
4	.99998394	.99996792	.99995195	.99993601	.99992012
5	.99995395	.99990808	.99986239	.99981687	.99977151
6	.99988677	.99977418	.99966219	.99955080	.99943999
7	.99975242	.99950675	.99926291	.99902086	.99878053
8	.99950675	.99901863	.99853544	.99805698	.99758308
9	.99908911	.99819078	.99730442	.99642947	.99556546
10	.99842088	.99687011	.99534609	.99384740	.99237277
11	.99740517	.99486995	.99239042	.98996313	.98758505
12	.99592778	.99197339	.98812788	.98438348	.98073342
13	.99385943	.98793949	.98222090	.97668725	.97132438
14	.99105921	.98251203	.97431923	.96644797	.95887036
15	.98737891	.97543034	.96407827	.95326029	.94292423

M	B= 6	B= 7	B= 8	B= 9	B=10
1	.99999939	.99999928	.99999918	.99999908	.99999898
2	.99999449	.99999357	.99999266	.99999174	.99999083
3	.99997296	.99996848	.99996401	.99995955	.99995509
4	.99990427	.99988846	.99987268	.99985694	.99984125
5	.99972633	.99968130	.99963643	.99959172	.99954717
6	.99932975	.99922005	.99911090	.99900227	.99889416
7	.99854187	.99830484	.99806939	.99783548	.99760307
8	.99711359	.99664834	.99618722	.99573007	.99527680
9	.99471192	.99386844	.99303464	.99221017	.99139470
10	.99092103	.98949114	.98808214	.98669315	.98532334
11	.98525343	.98296583	.98072003	.97851400	.97634589
12	.97717171	.97369300	.97029253	.96696600	.96370949
13	.96611996	.96106313	.95614430	.95135487	.94668716
14	.95156243	.94450339	.93767500	.93106117	.92464761
15	.93302594	.92352760	.91439645	.90560392	.89712489

M	B=11	B=12	B=13	B=14	B=15
1	.99999888	.99999878	.99999867	.99999857	.99999847
2	.99998992	.99998901	.99998810	.99998719	.99998628
3	.99995064	.99994620	.99994176	.99993733	.99993291
4	.99982559	.99980996	.99979437	.99977882	.99976330
5	.99950276	.99945850	.99941438	.99937041	.99932658
6	.99878655	.99867944	.99857281	.99846665	.99836097
7	.99737212	.99714260	.99691447	.99668770	.99646226
8	.99482727	.99438139	.99393905	.99350016	.99306463
9	.99058791	.98978954	.98899929	.98821694	.98744223
10	.98397198	.98263836	.98132183	.98002179	.97873767
11	.97421401	.97211680	.97005281	.96802069	.96601922
12	.96051947	.95739268	.95432616	.95131715	.94836313
13	.94213422	.93768975	.93334803	.92910382	.92495231
14	.91842154	.91237147	.90648703	.90075882	.89517828
15	.88893710	.88102076	.87335811	.86593322	.85873166

DIRICHLET D INTEGRAL
1/A = 3 R = 10

M	B= 1	B= 2	B= 3	B= 4	B= 5
1	.99999705	.99999810	.99999715	.99999620	.99999526
2	.99999189	.99998383	.99997580	.99996781	.99995986
3	.99996239	.99992506	.99988800	.99985119	.99981464
4	.99987388	.99974912	.99962566	.99950344	.99938239
5	.99965813	.99932148	.99898973	.99866260	.99833984
6	.99920505	.99842676	.99766390	.99691540	.99618034
7	.99835553	.99675718	.99520088	.99368323	.99220134
8	.99689922	.99391218	.99102717	.98823459	.98552641
9	.99457822	.98941142	.98446912	.97972697	.97516498
10	.99109672	.98272047	.97479883	.96727533	.96010469
11	.98613558	.97328700	.96129505	.95003893	.93942402
12	.97937040	.96058315	.94331148	.92731452	.91240824
13	.97049109	.94414933	.92034341	.89862236	.87865047
14	.95922119	.92363434	.89208436	.86377300	.83812222
15	.94533508	.89882782	.85846552	.82290535	.79120271

M	B= 6	B= 7	B= 8	B= 9	B=10
1	.99999432	.99999338	.99999245	.99999152	.99999059
2	.99995194	.99994406	.99993621	.99992839	.99992060
3	.99977832	.99974224	.99970638	.99967074	.99963530
4	.99926248	.99914364	.99902585	.99890906	.99879323
5	.99802122	.99770654	.99739563	.99708833	.99678447
6	.99545792	.99474743	.99404821	.99335971	.99268140
7	.99075272	.98933516	.98794675	.98658578	.98525070
8	.98289581	.98033692	.97784464	.97541448	.97304248
9	.97076649	.96651738	.96240554	.95842049	.95455303
10	.95324985	.94667994	.94036885	.93429430	.92843702
11	.92937440	.91982794	.91073297	.90204596	.89372978
12	.89844807	.88531791	.87292282	.86118398	.85003512
13	.86016986	.84297746	.82691012	.81183449	.79764005
14	.81469835	.79316688	.77326386	.75477715	.73753349
15	.76266631	.73677423	.71312250	.69139190	.67132575

M	B=11	B=12	B=13	B=14	B=15
1	.99998966	.99998874	.99998782	.99998690	.99998598
2	.99991285	.99990512	.99989743	.99988976	.99988212
3	.99960008	.99956505	.99953021	.99949557	.99946110
4	.99867834	.99856436	.99845124	.99833898	.99822754
5	.99648392	.99618656	.99589227	.99560093	.99531246
6	.99201280	.99135349	.99070307	.99006117	.98942746
7	.98394013	.98265283	.98138765	.98014355	.97891958
8	.97072508	.96845912	.96624174	.96407033	.96194255
9	.95079507	.94713943	.94357968	.94011007	.93672538
10	.92278021	.91730913	.91201074	.90687344	.90188684
11	.88575249	.87808633	.87070701	.86359313	.85672572
12	.83941989	.82928992	.81960335	.81032366	.80141878
13	.78423407	.77153791	.75948424	.74801496	.73707956
14	.72138949	.70622507	.69193861	.67844339	.66566473
15	.65271461	.63538534	.61919322	.60401606	.58974981

DIRICHLET D INTEGRAL
1/A = 2 R = 10

M	B= 1	B= 2	B= 3	B= 4	B= 5
1	.99998306	.99996643	.99995007	.99993397	.99991810
2	.99987016	.99974376	.99962042	.99949984	.99938180
3	.99945620	.99893290	.99842746	.99793784	.99746240
4	.99835228	.99679022	.99530068	.99387391	.99250238
5	.99596046	.99220165	.98867318	.98533909	.98217243
6	.99149573	.98376032	.97663410	.97000782	.96380218
7	.98405451	.96993942	.95721964	.94561033	.93491196
8	.97271551	.94932185	.92877061	.91040740	.89379051
9	.95665193	.92083324	.89026489	.86359500	.83994632
10	.93523383	.88392785	.84154405	.80553293	.77430759
11	.90810423	.83869222	.78338746	.73774835	.69912917
12	.87521986	.78585319	.71742833	.66272888	.61765059
13	.83685477	.72669769	.64594246	.58356890	.53360669
14	.79357107	.66292704	.57156247	.50355638	.45071189
15	.74616512	.59647601	.49697587	.42578452	.37222207

M	B= 6	B= 7	B= 8	B= 9	B=10
1	.99990244	.99988700	.99987174	.99985667	.99984178
2	.99926607	.99915251	.99904095	.99893127	.99882336
3	.99699982	.99654900	.99610900	.99567901	.99525833
4	.99118008	.98990211	.98866438	.98746345	.98629635
5	.97915226	.97626183	.97348748	.97081788	.96824344
6	.95795722	.95242613	.94717150	.94216278	.93737466
7	.92497788	.91569644	.90698024	.89875932	.89097663
8	.87860509	.86461756	.85164912	.83955943	.82823612
9	.81871340	.79945902	.78185631	.76565397	.75065441
10	.74680922	.72229468	.70022175	.68018250	.66186221
11	.66582615	.63667842	.61086029	.58776425	.56693085
12	.57964669	.54703474	.51864941	.49365284	.47142429
13	.49249188	.45794469	.42842878	.40286573	.38047328
14	.40832968	.37350248	.34432608	.31949653	.29808845
15	.33041791	.29686390	.26932892	.24632374	.22681498

M	B=11	B=12	B=13	B=14	B=15
1	.99982705	.99981247	.99979805	.99978377	.99976963
2	.99871712	.99861245	.99850927	.99840752	.99830713
3	.99484635	.99444252	.99404637	.99365746	.99327540
4	.98516054	.98405377	.98297409	.98191976	.98088920
5	.96575602	.96334857	.96101497	.95874986	.95654849
6	.93278584	.92837823	.92413626	.92004643	.91609696
7	.88358495	.87654467	.86982219	.86338872	.85721937
8	.81758762	.80753822	.79802458	.78899306	.78039786
9	.73669929	.72365969	.71142919	.69991886	.68905365
10	.64501303	.62943626	.61497023	.60148163	.58885931
11	.54800460	.53070514	.51480767	.50012936	.48651968
12	.45149222	.43349090	.41713154	.40218252	.38845561
13	.36066844	.34300670	.32714256	.31280289	.29976866
14	.27942550	.26300111	.24842779	.23540379	.22369037
15	.21006252	.19552193	.18278365	.17153353	.16152651

DIRICHLET D INTEGRAL
A = 1 R = 10

M	B= 1	B= 2	B= 3	B= 4	B= 5
1	.99902344	.99817337	.99741037	.99671272	.99606673
2	.99414063	.98930517	.98512484	.98141218	.97805530
3	.98071289	.96580073	.95346815	.94287843	.93356056
4	.95385742	.92085247	.89492073	.87349008	.85520048
5	.91021729	.85158937	.80813005	.77371783	.74532413
6	.84912109	.76041368	.69880282	.65226449	.61525507
7	.77275085	.65416759	.57752827	.52253964	.48051857
8	.68547058	.54193816	.45620533	.39801812	.35541321
9	.59273529	.43265530	.34488024	.28875643	.24949153
10	.50000000	.33333333	.25000000	.20000000	.16666667
11	.41190147	.24827431	.17416177	.13259580	.10625920
12	.33181190	.17911787	.11687556	.08436803	.06484205
13	.26173353	.12541427	.07572768	.05165183	.03797640
14	.20243645	.08538449	.04747919	.03050026	.02140216
15	.15372813	.05662610	.02886484	.01741045	.01163390

M	B= 6	B= 7	B= 8	B= 9	B=10
1	.99546304	.99489486	.99435707	.99384571	.99335759
2	.97498074	.97213713	.96948682	.96700126	.96465817
3	.92521885	.91765374	.91072332	.90432268	.89837177
4	.83923731	.82507103	.81233677	.80077178	.79018029
5	.72122254	.70033409	.68193917	.66553385	.65075157
6	.58477443	.55902316	.53684115	.51743910	.50025715
7	.44701516	.41947240	.39629888	.37644375	.35918067
8	.32259956	.29639473	.27488916	.25686032	.24148548
9	.22033931	.19776038	.17970992	.16491993	.15256029
10	.14285714	.12500000	.11111111	.10000000	.09090909
11	.08819583	.07509677	.06519659	.05747138	.05128829
12	.05200667	.04301990	.03642741	.03141464	.02749300
13	.02937621	.02357058	.01944150	.01638488	.01404934
14	.01593797	.01238635	.00994001	.00817853	.00686496
15	.00832633	.00625906	.00488148	.00391748	.00321640

M	B=11	B=12	B=13	B=14	B=15
1	.99289015	.99244125	.99200911	.99159220	.99118923
2	.96243976	.96033161	.95832177	.95640027	.95455867
3	.89280800	.88758140	.88265136	.87798443	.87355266
4	.78041233	.77135036	.76290049	.75498650	.74754564
5	.63731751	.62502035	.61369423	.60320658	.59344984
6	.48488466	.47101211	.45840084	.44686319	.43624916
7	.34398894	.33048398	.31837467	.30743589	.29749046
8	.22818856	.21655295	.20626910	.19710181	.18886878
9	.14206365	.13302844	.12516205	.11824601	.11211371
10	.08333333	.07692308	.07142857	.06666667	.06250000
11	.04623606	.04203635	.03849437	.03546989	.03285949
12	.02435320	.02179071	.01966537	.01787814	.01635724
13	.01221834	.01075206	.00955669	.00856722	.00773736
14	.00585724	.00506584	.00443199	.00391576	.00348922
15	.00269042	.00228556	.00196714	.00171208	.00150455

DIRICHLET D INTEGRAL
A = 2 R = 10

M	B= 1	B= 2	B= 3	B= 4	B= 5
1	.98265847	.97177118	.96361400	.95702328	.95146360
2	.92485337	.88444650	.85651894	.83515631	.81786464
3	.81887735	.73789400	.68679823	.65001342	.62155561
4	.67757600	.56145053	.49525379	.45065630	.41780402
5	.52449953	.39115929	.32318383	.28056203	.25077306
6	.38162816	.25140817	.19263744	.15852300	.13597833
7	.26256869	.15029881	.10588754	.08214496	.06735778
8	.17185671	.08422733	.05414448	.03940675	.03078370
9	.10760239	.04455344	.02595335	.01764383	.01309035
10	.06476617	.02238022	.01173900	.00742459	.00521695
11	.03763657	.01073176	.00503886	.00295392	.00196062
12	.02119439	.00493469	.00206261	.00111684	.00069850
13	.01160311	.00218437	.00080857	.00040306	.00023699
14	.00619265	.00093399	.00030467	.00013938	.76871\04
15	.00322978	.00038689	.00011070	.46335\04	.23922\04

M	B= 6	B= 7	B= 8	B= 9	B=10
1	.94663968	.94236984	.93853372	.93504714	.93184875
2	.80334935	.79085011	.77988165	.77011515	.76131740
3	.59851072	.57924828	.56276767	.54841257	.53573037
4	.39220972	.37148681	.35422895	.33954442	.32683582
5	.22850303	.21107112	.19696284	.18525096	.17533254
6	.11984462	.10765849	.09808776	.09034582	.08393667
7	.05725164	.04989881	.04430288	.03989702	.03633498
8	.02516812	.02123928	.01834567	.01613066	.01438338
9	.01027085	.00837705	.00702899	.00602675	.00525604
10	.00391994	.00308488	.00251108	.00209718	.00178715
11	.00140803	.00106752	.00084193	.00068420	.00056922
12	.00047857	.00034905	.00026642	.00021047	.00017080
13	.00015463	.00010835	.79947\04	.61338\04	.48518\04
14	.47690\04	.32062\04	.22845\04	.17008\04	.13103\04
15	.14088\04	.90760\05	.62388\05	.45033\05	.33767\05

M	B=11	B=12	B=13	B=14	B=15
1	.92889238	.92614236	.92357052	.92115418	.91887484
2	.75331694	.74598408	.73921827	.73294005	.72708548
3	.52439640	.51417026	.50486925	.49635145	.48850452
4	.31568529	.30579027	.29692538	.28891868	.28163632
5	.16679669	.15935267	.15278831	.14694472	.14170031
6	.07853122	.07390204	.06988664	.06636565	.06324928
7	.03339328	.03092114	.02881318	.02699345	.02540584
8	.01297151	.01180799	.01083329	.01000535	.00929368
9	.00464716	.00415541	.00375093	.00341303	.00312700
10	.00154785	.00135858	.00120579	.00108032	.00097577
11	.00048256	.00041548	.00036236	.00031950	.00028436
12	.00014163	.00011954	.00010240	.88809\04	.77850\04
13	.39327\04	.32521\04	.27344\04	.23317\04	.20124\04
14	.10375\04	.84012\05	.69306\05	.58081\05	.49332\05
15	.26103\05	.20686\05	.16735\05	.13777\05	.11512\05

DIRICHLET D INTEGRAL
A = 3 R = 10

M	B= 1	B= 2	B= 3	B= 4	B= 5
1	.94368649	.91699061	.89915420	.88571555	.87492361
2	.80290270	.72965094	.68555012	.65446693	.63069479
3	.60932499	.50089289	.44277764	.40470063	.37707922
4	.41574728	.30096884	.24685648	.21409747	.19164394
5	.25846540	.16118582	.12131892	.09914884	.08484351
6	.14836808	.07822547	.05352984	.04098599	.03339306
7	.07955725	.03488254	.02152657	.01536423	.01187686
8	.04023678	.01445560	.00798635	.00528958	.00386743
9	.01934778	.00561900	.00276062	.00168974	.00116516
10	.00890328	.00206436	.00089633	.00050507	.00032759
11	.00394214	.00072137	.00027520	.00014224	.86560\04
12	.00168708	.00024104	.80348\04	.37964\04	.21625\04
13	.00070049	.77355\04	.22415\04	.96500\05	.51336\05
14	.00028309	.23936\04	.59992\05	.23460\05	.11631\05
15	.00011165	.71642\05	.15459\05	.54748\06	.25241\06

M	B= 6	B= 7	B= 8	B= 9	B=10
1	.86590421	.85815647	.85136618	.84532308	.83987945
2	.61157433	.59565842	.58207582	.57026303	.55983555
3	.35576002	.33860138	.32436787	.31228822	.30185175
4	.17505871	.16217872	.15181030	.14323506	.13599184
5	.07475738	.06721418	.06133046	.05659422	.05268718
6	.02829556	.02463101	.02186543	.01970122	.01795936
7	.00965283	.00811867	.00700000	.00614992	.00548301
8	.00300783	.00243982	.00204017	.00174558	.00152051
9	.00086531	.00067581	.00054733	.00045555	.00038732
10	.00023185	.00017407	.00013637	.00011030	.91460\04
11	.58275\04	.42002\04	.31792\04	.24965\04	.20171\04
12	.13824\04	.95516\05	.69777\05	.53146\05	.41810\05
13	.31109\05	.20579\05	.14494\05	.10698\05	.81887\06
14	.66704\06	.42191\06	.28620\06	.20455\06	.15223\06
15	.13680\06	.82635\07	.53934\07	.37296\07	.26968\07

M	B=11	B=12	B=13	B=14	B=15
1	.83492751	.83038619	.82619295	.82229854	.81866348
2	.55051986	.54211474	.53446811	.52746231	.52100459
3	.29270453	.28459216	.27732627	.27076383	.26479398
4	.12976955	.12434983	.11957437	.11532526	.11151266
5	.04940048	.04659104	.04415738	.04202538	.04013957
6	.01652567	.01532388	.01430107	.01341938	.01265096
7	.00494636	.00450552	.00413714	.00382482	.00355675
8	.00134358	.00120124	.00108452	.00098725	.00090506
9	.00033497	.00029374	.00026059	.00023344	.00021087
10	.77355\04	.66492\04	.57927\04	.51040\04	.45409\04
11	.16673\04	.14039\04	.12004\04	.10398\04	.91073\05
12	.33751\05	.27824\05	.23339\05	.19866\05	.17123\05
13	.64518\06	.52042\06	.42805\06	.35788\06	.30342\06
14	.11699\06	.92291\07	.74396\07	.61070\07	.50913\07
15	.20205\07	.15580\07	.12303\07	.99117\08	.81223\08

DIRICHLET D INTEGRAL
A = 4 R = 10

M	B= 1	B= 2	B= 3	B= 4	B= 5
1	.89262582	.85209548	.82695332	.80880208	.79464479
2	.67787745	.58970427	.54112147	.50853251	.48442465
3	.44165425	.33820486	.28823545	.25729317	.23567209
4	.25267569	.16588493	.12960698	.10899910	.09545012
5	.12983963	.07152562	.05070275	.03991711	.03326969
6	.06105143	.02770483	.01766196	.01294648	.01023183
7	.02665733	.00980351	.00557668	.00378790	.00282931
8	.01093432	.00321095	.00161824	.00101411	.00071376
9	.00425203	.00098361	.00043629	.00025124	.00016616
10	.00157912	.00028417	.00011026	.58123\04	.36028\04
11	.00056341	.77956\04	.26304\04	.12649\04	.73306\05
12	.00019407	.20422\04	.59600\05	.26055\05	.14085\05
13	.64794\04	.51333\05	.12890\05	.51061\06	.25689\06
14	.21040\04	.12430\05	.26721\06	.95623\07	.44677\07
15	.66643\05	.29093\06	.53293\07	.17177\07	.74378\08

M	B= 6	B= 7	B= 8	B= 9	B=10
1	.78306772	.77329234	.76484470	.75741512	.75079030
2	.46550531	.45005719	.43707892	.42593900	.41621543
3	.21943533	.20664286	.19621282	.18748766	.18004139
4	.08574011	.07837345	.07255387	.06781549	.06386602
5	.02873515	.02542887	.02290197	.02090190	.01927543
6	.00847354	.00724369	.00633566	.00563780	.00508466
7	.00224119	.00184725	.00156666	.00135751	.00119607
8	.00053960	.00042808	.00035155	.00029628	.00025477
9	.00011964	.91214\04	.72440\04	.59317\04	.49737\04
10	.24660\04	.18040\04	.13838\04	.10999\04	.89853\05
11	.47611\05	.33373\05	.24700\05	.19037\05	.15140\05
12	.86651\06	.58121\06	.41457\06	.30958\06	.23951\06
13	.14946\06	.95804\07	.65793\07	.47560\07	.35768\07
14	.24542\07	.15016\07	.99183\08	.69349\08	.50663\08
15	.38517\08	.22468\08	.14260\08	.96365\09	.68341\09

M	B=11	B=12	B=13	B=14	B=15
1	.74481717	.73938224	.73439906	.72980031	.72553254
2	.40761318	.39991852	.39297196	.38665160	.38086229
3	.17358421	.16791103	.16287209	.15835506	.15427380
4	.06051207	.05762012	.05509474	.05286575	.05088029
5	.01792400	.01678125	.01580080	.01494924	.01420184
6	.00463533	.00426300	.00394934	.00368141	.00344981
7	.00106797	.00096401	.00087807	.00080591	.00074451
8	.00022262	.00019709	.00017639	.00015931	.00014501
9	.42499\04	.36877\04	.32409\04	.28789\04	.25809\04
10	.75025\05	.63766\05	.54999\05	.48028\05	.42385\05
11	.12345\05	.10271\05	.86899\06	.74564\06	.64750\06
12	.19057\06	.15513\06	.12868\06	.10845\06	.92632\07
13	.27756\07	.22094\07	.17960\07	.14860\07	.12480\07
14	.38320\08	.29812\08	.23738\08	.19274\08	.15911\08
15	.50354\09	.38268\09	.29834\09	.23763\09	.19274\09

DIRICHLET D INTEGRAL
A = 5 R = 10

M	B= 1	B= 2	B= 3	B= 4	B= 5
1	.83849442	.78801247	.75826483	.73739245	.72142152
2	.56931845	.47835087	.43119600	.40058774	.37843184
3	.32257381	.23417709	.19442013	.17070421	.15453104
4	.15807738	.09664718	.07300109	.06012070	.05187766
5	.06897515	.03477812	.02366424	.01817016	.01488294
6	.02739411	.01118598	.00679826	.00484159	.00375099
7	.01006867	.00327596	.00176470	.00116031	.00084758
8	.00346851	.00088600	.00042007	.00025392	.00017437
9	.00113095	.00022374	.92759\04	.51341\04	.33057\04
10	.00035176	.53223\04	.19176\04	.96823\05	.58301\05
11	.00010502	.12010\04	.37389\05	.17161\05	.96403\06
12	.30249\04	.25862\05	.69185\06	.28770\06	.15042\06
13	.84409\05	.53402\06	.12212\06	.45860\07	.22267\07
14	.22900\05	.10617\06	.20652\07	.69821\08	.31417\08
15	.60575\06	.20397\07	.33587\08	.10192\08	.42414\09

M	B= 6	B= 7	B= 8	B= 9	B=10
1	.70854476	.69779169	.68858256	.68054423	.67342281
2	.36131848	.34751653	.33603663	.32626473	.31779572
3	.14259782	.13332313	.12584375	.11964371	.11439329
4	.04608358	.04175269	.03837192	.03564628	.03339338
5	.01268743	.01111221	.00992371	.00899294	.00824280
6	.00306059	.00258599	.00224039	.00197778	.00177161
7	.00066038	.00053730	.00045092	.00038730	.00033870
8	.00012945	.00010126	.82206\04	.68624\04	.58533\04
9	.23337\04	.17521\04	.13744\04	.11139\04	.92587\05
10	.39066\05	.28111\05	.21280\05	.16729\05	.13539\05
11	.61203\06	.42152\06	.30760\06	.23433\06	.18452\06
12	.90324\07	.59463\07	.41784\07	.30819\07	.23595\07
13	.12626\07	.79351\08	.53638\08	.38272\08	.28468\08
14	.16795\08	.10064\08	.65378\09	.45091\09	.32563\09
15	.21344\09	.12180\09	.75969\10	.50606\10	.35459\10

M	B=11	B=12	B=13	B=14	B=15
1	.66703801	.66125730	.65598056	.65113033	.64664549
2	.31034952	.30372505	.29777354	.29238205	.28746304
3	.10987072	.10592060	.10243041	.09931638	.09651471
4	.03149396	.02986652	.02845331	.02721223	.02611176
5	.00762430	.00710484	.00666181	.00627908	.00594478
6	.00160552	.00146886	.00135447	.00125731	.00117375
7	.00030046	.00026967	.00024439	.00022328	.00020542
8	.50788\04	.44686\04	.39774\04	.35747\04	.32394\04
9	.78519\05	.67683\05	.59134\05	.52254\05	.46622\05
10	.11214\05	.94646\06	.81127\06	.70450\06	.61860\06
11	.14918\06	.12320\06	.10355\06	.88328\07	.76297\07
12	.18606\07	.15028\07	.12380\07	.10369\07	.88072\08
13	.21884\08	.17277\08	.13943\08	.11461\08	.95697\09
14	.24387\09	.18810\09	.14864\09	.11987\09	.98355\10
15	.25858\10	.19475\10	.15064\10	.11914\10	.96020\11

TABLE E (A > 1)

This table gives the value of the incomplete Dirichlet integral of type 2

$$D_A^{(B)}(R, M) = \frac{\Gamma(M+BR)}{\Gamma^B(R)\Gamma(M)} \int_A^\infty \cdots \int_A^\infty \frac{\prod_{i=1}^{B} x_i^{R-1}\, dx_i}{\left(1 + \sum_{i=1}^{B} x_i\right)^{M+BR}}$$

for

$M = R$
$A = 3(1)10$ and $A^{-1} = .40(.10).60(.05).80$,
$B = 1(1)10$

The common value of M and R increases from 1 by ones until we reach the line that contains .01; this line is completed.

Note: The symbols M, B, R and A which are all capital here in the table correspond to the same lower case symbols in the text.

D-INTEGRAL PROBABILITY M=R, B= 1

M=R	1	2	3	4	5
			A=10		
0+	.0909090909	.0232907588	.0065258831	.0019074600	.0005715525
			A= 9		
0+	.1000000000	.0280000000	.0085600000	.0027280000	.0008909200
			A= 8		
0+	.1111111111	.0342935528	.0115327948	.0040395411	.0014492806
			A= 7		
0+	.1250000000	.0429687500	.0160522461	.0062389374	.0024822801
			A= 6		
0+	.1428571429	.0553935860	.0232641161	.0101500468	.0045297314
			A= 5		
0+	.1666666667	.0740740741	.0354938272	.0176326017	.0089500616
			A= 4		
0+	.2000000000	.1040000000	.0579200000	.0333440000	.0195814400
5+	.0116542054	.0070035612	.0042397497	.0025814628	.0015791205
			A= 3		
0+	.2500000000	.1562500000	.1035156250	.0705566406	.0489273071
5+	.0343275070	.0242901444	.0172998384	.0123847794	.0089032793
			A=1/.40		
0+	.2857142857	.1982507289	.1447016124	.1082736421	.0822536632
5+	.0631369441	.0488319161	.0379884547	.0296898873	.0232918989
10+	.0183301935	.0144639296	.0114393014	.0090651819	.0071963415
			A= 2		
0+	.3333333333	.2592592593	.2098765432	.1732967535	.1448458060
5+	.1220850480	.1035392452	.0882315984	.0754752261	.0647661728
10+	.0557229722	.0480499536	.0415136784	.0359271184	.0311386385
15+	.0270240927	.0234810117	.0204242358	.0177825777	.0154962304
20+	.0135147293	.0117953316	.0103017133	.0090029148	.0078724791
			A=1/.60		
0+	.3750000000	.3164062500	.2752075195	.2430210114	.2166180164
5+	.1943404893	.1751957396	.1585295511	.1438815340	.1309119354
10+	.1193608867	.1090240107	.0997369737	.0913652457	.0837970540
15+	.0769383803	.0707093114	.0650413168	.0598751758	.0551593728
20+	.0508488341	.0469039214	.0432896193	.0399748722	.0369320379
25+	.0341364339	.0315659566	.0292007606	.0270229852	.0250165218
30+	.0231668134	.0214606812	.0198861744	.0184324395	.0170896052
35+	.0158486825	.0147014753	.0136405025	.0126589281	.0117504999
40+	.0109094940	.0101306662	.0094092074	.0087407046	.0081211050
			A=1/.65		
0+	.3939393939	.3432952111	.3070211407	.2781528876	.2540297376
5+	.2332957904	.2151448603	.1990487974	.1846377218	.1716396697
10+	.1598471302	.1490971068	.1392585485	.1302240596	.1219042234
15+	.1142235910	.1071177718	.1005312726	.0944158618	.0887293070
20+	.0834343882	.0784981134	.0738910884	.0695870047	.0655622187
25+	.0617954034	.0582672569	.0549602570	.0518584533	.0489472896
30+	.0462134511	.0436447329	.0412299256	.0389587152	.0368215961
35+	.0348097941	.0329151984	.0311303015	.0294481455	.0278622745
40+	.0263666918	.0249558217	.0236244745	.0223678161	.0211813392
45+	.0200608382	.0190023860	.0180023127	.0170571868	.0161637975
50+	.0153191385	.0145203936	.0137649232	.0130502520	.0123740572
55+	.0117341587	.0111285090	.0105551845	.0100123769	.0094983864
			A=1/.70		
0+	.4117647059	.3690209648	.3379614989	.3128846522	.2916257337
5+	.2730885453	.2566252984	.2418140727	.2283609699	.2160509411
10+	.2047206032	.1942420901	.1845129054	.1754492519	.1669814787
15+	.1590508768	.1516073620	.1446077613	.1380145196	.1317947068
20+	.1259192436	.1203622887	.1151007491	.1101138830	.1053829748
25+	.1008910675	.0966227393	.0925639164	.0887017146	.0850243051

D-INTEGRAL PROBABILITY M=R,B= 1

M=R	1	2	3	4	5
			A=1/.70		
30+	.0815207983	.0781811458	.0749960532	.0719569065	.0690557061
35+	.0662850094	.0636378809	.0611078470	.0586888562	.0563752442
40+	.0541617020	.0520432473	.0500151997	.0480731567	.0462129740
45+	.0444307460	.0427227886	.0410856241	.0395159666	.0380107094
50+	.0365669125	.0351817921	.0338527107	.0325771675	.0313527898
55+	.0301773256	.0290486359	.0279646882	.0269235500	.0259233832
60+	.0249624387	.0240390511	.0231516342	.0222986769	.0214787384
65+	.0206904452	.0199324868	.0192036129	.0185026300	.0178283984
70+	.0171798295	.0165558835	.0159555664	.0153779279	.0148220596
75+	.0142870925	.0137721952	.0132765721	.0127994617	.0123401351
80+	.0118978942	.0114720704	.0110620231	.0106671388	.0102868291
85+	.0099205305	.0095677024	.0092278266	.0089004062	.0085849644
			A=1/.75		
0+	.4285714286	.3935860058	.3678824299	.3468999190	.3289149096
5+	.3130587381	.2988205433	.2858691824	.2739750754	.2629710037
10+	.2527304798	.2431549250	.2341656286	.2256984735	.2177003444
15+	.2101266059	.2029392826	.1961057171	.1895975595	.1833899935
20+	.1774611346	.1717915552	.1663639058	.1611626093	.1561736105
25+	.1513841717	.1467827014	.1423586121	.1381021996	.1340045415
30+	.1300574095	.1262531955	.1225848462	.1190458079	.1156299775
35+	.1123316595	.1091455293	.1060665993	.1030901900	.1002119041
40+	.0974276030	.0947333864	.0921255732	.0896006852	.0871554320
45+	.0847866968	.0824915249	.0802671116	.0781107926	.0760200344
50+	.0739924255	.0720256693	.0701175761	.0682660571	.0664691181
55+	.0647248537	.0630314425	.0613871420	.0597902844	.0582392722
60+	.0567325747	.0552687241	.0538463122	.0524639878	.0511204531
65+	.0498144615	.0485448147	.0473103606	.0461099911	.0449426398
70+	.0438072800	.0427029232	.0416286169	.0405834433	.0395665176
75+	.0385769866	.0376140274	.0366768458	.0357646753	.0348767760
80+	.0340124331	.0331709565	.0323516791	.0315539563	.0307771650
85+	.0300207028	.0292839868	.0285664534	.0278675573	.0271867705
90+	.0265235824	.0258774982	.0252480391	.0246347413	.0240371553
95+	.0234548460	.0228873915	.0223343831	.0217954244	.0212701313
100+	.0207581314	.0202590634	.0197725769	.0192983321	.0188359993
105+	.0183852585	.0179457992	.0175173201	.0170995287	.0166921411
110+	.0162948814	.0159074820	.0155296829	.0151612315	.0148018826
115+	.0144513978	.0141095457	.0137761012	.0134508458	.0131335670
120+	.0128240583	.0125221189	.0122275538	.0119401732	.0116597926
125+	.0113862327	.0111193191	.0108588821	.0106047567	.0103567824
130+	.0101148031	.0098786669	.0096482260	.0094233366	.0092038588
			A=1/.80		
0+	.4444444444	.4170096022	.3966874968	.3799614842	.3655069053
5+	.3526583908	.3410259908	.3303578286	.3204799006	.3112659211
10+	.3026207058	.2944703345	.2867559913	.2794299293	.2724527274
15+	.2657913660	.2594178411	.2533081441	.2474414940	.2417997513
20+	.2363669619	.2311289993	.2260732777	.2211885226	.2164645824
25+	.2118922750	.2074632592	.2031699288	.1990053226	.1949630484
30+	.1910372184	.1872223934	.1835135359	.1799059677	.1763953349
35+	.1729775760	.1696488946	.1664057347	.1632447603	.1601628352
40+	.1571570070	.1542244917	.1513626602	.1485690262	.1458412354
45+	.1431770557	.1405743681	.1380311587	.1355455117	.1331156023
50+	.1307396909	.1284161174	.1261432962	.1239197115	.1217439131
55+	.1196145122	.1175301779	.1154896341	.1134916559	.1115350671
60+	.1096187373	.1077415796	.1059025482	.1041006361	.1023348735
65+	.1006043256	.0989080908	.0972452994	.0956151117	.0940167169
70+	.0924493315	.0909121983	.0894045847	.0879257823	.0864751052
75+	.0850518896	.0836554921	.0822852897	.0809406783	.0796210722

D-INTEGRAL PROBABILITY M=R,B= 1

M=R	1	2	3	4	5
			A=1/.80		
80+	.0783259032	.0770546202	.0758066880	.0745815872	.0733788134
85+	.0721978764	.0710383000	.0698996214	.0687813904	.0676831694
90+	.0666045325	.0655450656	.0645043654	.0634820394	.0624777055
95+	.0614909915	.0605215349	.0595689823	.0586329897	.0577132213
100+	.0568093502	.0559210574	.0550480317	.0541899699	.0533465757
105+	.0525175604	.0517026420	.0509015454	.0501140019	.0493397490
110+	.0485785308	.0478300969	.0470942029	.0463706099	.0456590847
115+	.0449593991	.0442713303	.0435946603	.0429291761	.0422746695
120+	.0416309365	.0409977781	.0403749994	.0397624097	.0391598224
125+	.0385670550	.0379839290	.0374102695	.0368459054	.0362906692
130+	.0357443969	.0352069281	.0346781053	.0341577748	.0336457856
135+	.0331419902	.0326462437	.0321584046	.0316783338	.0312058954
140+	.0307409561	.0302833851	.0298330544	.0293898385	.0289536143
145+	.0285242612	.0281016611	.0276856979	.0272762580	.0268732299
150+	.0264765043	.0260859739	.0257015338	.0253230806	.0249505133
155+	.0245837325	.0242226410	.0238671432	.0235171453	.0231725554
160+	.0228332833	.0224992404	.0221703397	.0218464959	.0215276254
165+	.0212136459	.0209044767	.0206000387	.0203002540	.0200050463
170+	.0197143408	.0194280637	.0191461428	.0188685072	.0185950871
175+	.0183258142	.0180606211	.0177994419	.0175422117	.0172888669
180+	.0170393448	.0167935841	.0165515242	.0163131059	.0160782710
185+	.0158469621	.0156191229	.0153946983	.0151736339	.0149558763
190+	.0147413730	.0145300724	.0143219240	.0141168780	.0139148853
195+	.0137158978	.0135198684	.0133267505	.0131364983	.0129490672
200+	.0127644127	.0125824916	.0124032612	.0122266795	.0120527053
205+	.0118812981	.0117124178	.0115460254	.0113820822	.0112205505
210+	.0110613927	.0109045724	.0107500535	.0105978004	.0104477783
215+	.0102999529	.0101542905	.0100107579	.0098693224	.0097299520

D-INTEGRAL PROBABILITY M=R,B= 2

M=R	1	2	3	4	5
			A=10		
0+	.0476190476	.0096718960	.0022034125	.0005314833	.0001327190
			A= 9		
0+	.0526315789	.0117479148	.0029355901	.0007759756	.0002122404
			A= 8		
0+	.0588235294	.0145711857	.0040311311	.0011783936	.0003562083
			A= 7		
0+	.0666666667	.0185481481	.0057460412	.0018782243	.0006343391
			A= 6		
0+	.0769230769	.0244039074	.0085864818	.0031817223	.0012168408
			A= 5		
0+	.0909090909	.0335359607	.0136459520	.0058325317	.0025690989
			A= 4		
0+	.1111111111	.0489254687	.0235755239	.0118874138	.0061633213
			A= 3		
0+	.1428571429	.0778842149	.0458858857	.0281149836	.0176489310
5+	.0112652752	.0072794733	.0047486960	.0031212925	.0020643631
			A=1/.40		
0+	.1666666667	.1030092593	.0681463799	.0466648453	.0326460430
5+	.0231789795	.0166382164	.0120446997	.0087786843	.0064340112
			A= 2		
0+	.2000000000	.1424000000	.1070720000	.0827745280	.0651079885
5+	.0518315089	.0416325871	.0336737249	.0273890162	.0223802173
10+	.0183585136	.0151095715	.0124714246	.0103198858	.0085585953
			A=1/.60		
0+	.2307692308	.1824515948	.1504380892	.1267281244	.1081964742
5+	.0932431231	.0809193259	.0706082512	.0618828404	.0544336805
10+	.0480287965	.0424895913	.0376756276	.0334745647	.0297952565
15+	.0265628725	.0237153573	.0212008063	.0189754818	.0170022890
20+	.0152495885	.0136902578	.0123009434	.0110614563	.0099542812
			A=1/.65		
0+	.2452830189	.2023806914	.1731140303	.1508310926	.1329420090
5+	.1181235778	.1055914958	.0948351573	.0854999773	.0773278675
10+	.0701241566	.0637378555	.0580492095	.0529614873	.0483953620
15+	.0442849442	.0405749092	.0372183666	.0341752522	.0314110901
20+	.0288960286	.0266040759	.0245124902	.0226012846	.0208528243
25+	.0192514933	.0177834185	.0164362385	.0151989092	.0140615386
30+	.0130152465	.0120520442	.0111647310	.0103468044	.0095923833
			A=1/.70		
0+	.2592592593	.2221149667	.1961533385	.1759309934	.1593356357
5+	.1452912622	.1331614670	.1225324325	.1131168525	.1047054491
10+	.0971400964	.0902978258	.0840807627	.0784095049	.0732186013
15+	.0684533688	.0640675899	.0600218109	.0562820577	.0528188505
20+	.0496064360	.0466221806	.0438460850	.0412603928	.0388492693
25+	.0365985377	.0344954600	.0325285533	.0306874345	.0289626894
30+	.0273457603	.0258288495	.0244048369	.0230672071	.0218099877
35+	.0206276935	.0195152793	.0184680968	.0174818576	.0165526000
40+	.0156766592	.0148506411	.0140713988	.0133360114	.0126417651
45+	.0119861360	.0113667749	.0107814931	.0102282496	.0097051403
			A=1/.75		
0+	.2727272727	.2415818592	.2193690916	.2017396096	.1870104038
5+	.1743266738	.1631841953	.1532558424	.1443149041	.1361964530
10+	.1287759690	.1219566408	.1156613912	.1098276403	.1044037367
15+	.0993464490	.0946191561	.0901905111	.0860334352	.0821243479
20+	.0784425666	.0749698351	.0716899453	.0685884329	.0656523293
25+	.0628699565	.0602307579	.0577251563	.0553444345	.0530806336
30+	.0509264671	.0488752464	.0469208171	.0450575032	.0432800595
35+	.0415836291	.0399637064	.0384161042	.0369369257	.0355225379

D-INTEGRAL PROBABILITY M=R,B= 2

M=R	1	2	3	4	5
			A=1/.75		
40+	.0341695496	.0328747907	.0316352939	.0304482787	.0293111362
45+	.0282214164	.0271768158	.0261751663	.0252144259	.0242926693
50+	.0234080797	.0225589414	.0217436328	.0209606203	.0202084524
55+	.0194857542	.0187912229	.0181236229	.0174817818	.0168645865
60+	.0162709795	.0156999560	.0151505603	.0146218831	.0141130591
65+	.0136232642	.0131517132	.0126976579	.0122603847	.0118392132
70+	.0114334940	.0110426072	.0106659608	.0103029895	.0099531529
			A=1/.80		
0+	.2857142857	.2607246980	.2426017828	.2279965209	.2156157066
5+	.2048041227	.1951762158	.1864824319	.1785501948	.1712542061
10+	.1645000318	.1582143635	.1523389110	.1468263974	.1416378321
15+	.1367405986	.1321070751	.1277136189	.1235398019	.1195678267
20+	.1157820724	.1121687370	.1087155529	.1054115567	.1022469023
25+	.0992127072	.0963009251	.0935042395	.0908159741	.0882300173
30+	.0857407572	.0833430268	.0810320563	.0788034315	.0766530584
35+	.0745771317	.0725721074	.0706346786	.0687617544	.0669504406
40+	.0651980232	.0635019531	.0618598330	.0602694052	.0587285406
45+	.0572352292	.0557875711	.0543837684	.0530221184	.0517010062
50+	.0504188995	.0491743424	.0479659507	.0467924074	.0456524580
55+	.0445449072	.0434686145	.0424224918	.0414054996	.0404166443
60+	.0394549760	.0385195856	.0376096025	.0367241928	.0358625573
65+	.0350239293	.0342075733	.0334127832	.0326388807	.0318852142
70+	.0311511572	.0304361074	.0297394850	.0290607322	.0283993118
75+	.0277547062	.0271264171	.0265139637	.0259168826	.0253347268
80+	.0247670649	.0242134808	.0236735724	.0231469515	.0226332432
85+	.0221320850	.0216431267	.0211660297	.0207004663	.0202461199
90+	.0198026838	.0193698613	.0189473653	.0185349176	.0181322490
95+	.0177390986	.0173552136	.0169803490	.0166142674	.0162567387
100+	.0159075396	.0155664537	.0152332711	.0149077879	.0145898065
105+	.0142791351	.0139755874	.0136789827	.0133891454	.0131059049
110+	.0128290958	.0125585571	.0122941325	.0120356701	.0117830224
115+	.0115360459	.0112946010	.0110585522	.0108277676	.0106021190
120+	.0103814816	.0101657341	.0099547584	.0097484398	.0095466664

D-INTEGRAL PROBABILITY M=R, B= 3

M=R	1	2	3	4	5
A=10					
0+	.0322580645	.0058419825	.0012013818	.0002637167	.0000602674
A= 9					
0+	.0357142857	.0071240778	.0016104202	.0003881968	.0000973640
A= 8					
0+	.0400000000	.0088794112	.0022281602	.0005954863	.0001654675
A= 7					
0+	.0454545455	.0113729502	.0032064447	.0009612782	.0002993526
A= 6					
0+	.0526315789	.0150846150	.0048512616	.0016555421	.0005861312
A= 5					
0+	.0625000000	.0209579468	.0078404517	.0031034231	.0012722148
A= 4					
0+	.0769230769	.0310617221	.0138753765	.0065301103	.0031747188
A= 3					
0+	.1000000000	.0506800000	.0280307800	.0162218289	.0096574366
A=1/.40					
0+	.1176470588	.0682688468	.0427876443	.0279108886	.0186687482
5+	.0127067344	.0087616368	.0061026350	.0042852037	.0030292097
A= 2					
0+	.1428571429	.0968049027	.0698086543	.0519953226	.0395236764
5+	.0304737588	.0237466873	.0186585304	.0147588983	.0117389979
10+	.0093806598	.0075261323	.0060592410	.0048931538	.0039621733
A=1/.60					
0+	.1666666667	.1269433013	.1014658191	.0831626611	.0692491826
5+	.0583071498	.0495037205	.0423038453	.0363419718	.0313569668
10+	.0271559104	.0235924987	.0205534389	.0179494982	.0157093986
15+	.0137755282	.0121008514	.0106466352	.0093807443	.0082763425
A=1/.65					
0+	.1780821918	.1423445259	.1186065956	.1009874518	.0871681935
5+	.0759647359	.0666786522	.0588584227	.0521930895	.0464583892
10+	.0414868173	.0371497865	.0333463896	.0299960047	.0270332468
15+	.0244044161	.0220649334	.0199774468	.0181104074	.0164369791
20+	.0149341923	.0135822781	.0123641377	.0112649166	.0102716583
25+	.0093730205	.0085590416	.0078209459	.0071509816	.0065422834
A=1/.70					
0+	.1891891892	.1578545193	.1364110684	.1200513318	.1068795467
5+	.0959273357	.0866227267	.C785950293	.0715880280	.0654160709
10+	.0599397002	.0550511442	.0506651923	.0467131941	.0431389661
15+	.0398959129	.0369449490	.0342529642	.0317916686	.0295367078
20+	.0274669748	.0255640683	.0238118600	.0221961459	.0207043627
25+	.0193253542	.0180491786	.0168669467	.0157706868	.0147532304
30+	.0138081144	.0129294981	.0121120917	.0113510946	.0106421416
35+	.0099812568	.0093648124	.0087894934	.0082522662	.0077503505
A=1/.75					
0+	.2000000000	.1734044444	.1547352810	.1401585599	.1281629814
5+	.1179770765	.1091453066	.1013723584	.0944539317	.0882417646
10+	.0826242452	.0775148814	.0728450577	.0685592768	.0646119166
15+	.0609649530	.0575863159	.0544486776	.0515285403	.0488055383
20+	.0462618954	.0438819975	.0416520514	.0395598106	.0375943505
25+	.0357458851	.0340056143	.0323655963	.0308186410	.0293582188
30+	.0279783838	.0266737075	.0254392221	.0242703708	.0231629658
35+	.0221131504	.0211173664	.0201723256	.0192749842	.0184225206
40+	.0176123151	.0168419326	.0161091062	.0154117234	.0147478133
45+	.0141155349	.0135131670	.0129390985	.0123918205	.0118699178
50+	.0113720626	.0108970077	.0104435810	.0100106798	.0095972662
A=1/.80					
0+	.2105263158	.1889359028	.1734479383	.1611176240	.1507845154

D-INTEGRAL PROBABILITY M=R,B= 3

M=R	1	2	3	4	5
			A=1/.80		
5+	.1418568055	.1339853663	.1269440330	.1205764161	.1147690420
10+	.1094364754	.1045124796	.0999444670	.0956898575	.0917135989
15+	.0879864266	.0844836091	.0811840234	.0780694603	.0751240927
20+	.0723340626	.0696871565	.0671725465	.0647805811	.0625026157
25+	.0603308731	.0582583278	.0562786096	.0543859224	.0525749757
30+	.0508409261	.0491793275	.0475860876	.0460574312	.0445898674
35+	.0431801617	.0418253112	.0405225226	.0392691934	.0380628945
40+	.0369013555	.0357824507	.0347041876	.0336646957	.0326622171
45+	.0316950974	.0307617783	.0298607899	.0289907447	.0281503314
50+	.0273383097	.0265535052	.0257948052	.0250611544	.0243515511
55+	.0236650441	.0230007289	.0223577454	.0217352747	.0211325372
60+	.0205487893	.0199833225	.0194354603	.0189045569	.0183899956
65+	.0178911868	.0174075668	.0169385962	.0164837588	.0160425604
70+	.0156145273	.0151992057	.0147961603	.0144049739	.0140252457
75+	.0136565914	.0132986416	.0129510417	.0126134509	.0122855414
80+	.0119669983	.0116575184	.0113568101	.0110645928	.0107805963
85+	.0105045603	.0102362341	.0099753762	.0097217538	.0094751425

D-INTEGRAL PROBABILITY M=R,B= 4

M=R	1	2	3	4	5
			A=10		
0+	.0243902439	.0041108658	.0007924205	.0001639063	.0000354288
			A= 9		
0+	.0270270270	.0050238829	.0010658393	.0002423832	.0000575650
			A= 8		
0+	.0303030303	.0062784995	.0014809000	.0003739240	.0000985241
			A= 7		
0+	.0344827586	.0080689859	.0021424904	.0006079630	.0001798443
			A= 6		
0+	.0400000000	.0107501363	.0032642320	.0010569195	.0003562708
			A= 5		
0+	.0476190476	.0150271574	.0053260503	.0020067475	.0007856815
			A= 4		
0+	.0588235294	.0224699099	.0095563644	.0043010528	.0020059542
			A= 3		
0+	.0769230769	.0371800500	.0197284981	.0109971311	.0063236234
			A=1/.40		
0+	.0909090909	.0506183530	.0306078791	.0193326247	.0125521520
5+	.0083084420	.0055791665	.0037887301	.0025962534	.0017924515
			A= 2		
0+	.1111111111	.0728603928	.0510838264	.0371058462	.0275640538
5+	.0208008298	.0158831188	.0122404004	.0095037023	.0074245944
			A=1/.60		
0+	.1304347826	.0968858910	.0758152486	.0609845708	.0499212125
5+	.0413712457	.0346039739	.0291542994	.0247076541	.0210416883
10+	.0179939062	.0154424234	.0132938951	.0114756099	.0099301252
			A=1/.65		
0+	.1397849462	.1093610946	.0894955981	.0750015103	.0638123136
5+	.0548735863	.0475659942	.0414914668	.0363776869	.0320296783
10+	.0283029175	.0250873281	.0222972208	.0198646875	.0177351051
15+	.0158639815	.0142146855	.0127567776	.0114647591	.0103171183
20+	.0092955932	.0083845946	.0075707480	.0068425279	.0061899618
			A=1/.70		
0+	.1489361702	.1220539797	.1039026091	.0902471882	.0793953561

D-INTEGRAL PROBABILITY M=R, B= 4

M=R	1	2	3	4	5
			A=1/.70		
5+	.0704809082	.0629929632	.0566014204	.0510790071	.0462617753
10+	.0420271654	.0382809487	.0349490182	.0319719961	.0293015595
15+	.0268978601	.0247276646	.0227629840	.0209800438	.0193584976
20+	.0178808157	.0165318061	.0152982318	.0141685047	.0131324352
25+	.0121810287	.0113063167	.0105012163	.0097594130	.0090752613
			A=1/.75		
0+	.1578947368	.1349067945	.1189285743	.1065897394	.0965408059
5+	.0880901309	.0808290753	.0744929775	.0688990659	.0639150045
10+	.0594414331	.0554015776	.0517347205	.0483919123	.0453330505
15+	.0425248276	.0399392527	.0375525609	.0353443944	.0332971751
20+	.0313956189	.0296263523	.0279776075	.0264389771	.0250012138
25+	.0236560667	.0223961459	.0212148087	.0201060649	.0190644969
30+	.0180851908	.0171636782	.0162958866	.0154780954	.0147068988
35+	.0139791728	.0132920470	.0126428792	.0120292334	.0114488605
40+	.0108996805	.0103797677	.0098873365	.0094207297	.0089784068
			A=1/.80		
0+	.1666666667	.1478688432	.1344713155	.1238922464	.1150963489
5+	.1075527531	.1009476589	.0950776859	.0898022399	.0850193860
10+	.0806524587	.0766420975	.0729412435	.0695118541	.0663226662
15+	.0633476254	.0605647548	.0579553205	.0555032052	.0531944288
20+	.0510167767	.0489595063	.0470131147	.0451691503	.0434200603
25+	.0417590649	.0401800540	.0386774999	.0372463853	.0358821410
30+	.0345805942	.0333379230	.0321506189	.0310154524	.0299294449
35+	.0288898430	.0278940970	.0269398408	.0260248754	.0251471535
40+	.0243047661	.0234959306	.0227189798	.0219723530	.0212545868
45+	.0205643075	.0199002242	.0192611228	.0186458597	.0180533568
50+	.0174825971	.0169326201	.0164025178	.0158914314	.0153985478
55+	.0149230965	.0144643471	.0140216064	.0135942164	.0131815519
60+	.0127830184	.0123980505	.0120261099	.0116666842	.0113192848
65+	.0109834462	.0106587242	.0103446950	.0100409537	.0097471138

D-INTEGRAL PROBABILITY M=R, B= 5

M=R	1	2	3	4	5
			A=10		
0+	.0196078431	.0031420301	.0005784562	.0001147004	.0000238327
			A= 9		
0+	.0217391304	.0038451889	.0007797749	.0001701294	.0000388698
			A= 8		
0+	.0243902439	.0048137047	.0010864144	.0002634377	.0000668376
			A= 7		
0+	.0277777778	.0061999933	.0015772611	.0004303534	.0001227281
			A= 6		
0+	.0322580645	.0082838923	.0024140966	.0007528051	.0002450147
			A= 5		
0+	.0384615385	.0116255457	.0039637398	.0014415087	.0005460800
			A= 4		
0+	.0476190476	.0174841715	.0071772535	.0031278323	.0014157544
			A= 3		
0+	.0625000000	.0291985869	.0150332972	.0081544047	.0045721255
			A=1/.40		
0+	.0740740741	.0400332398	.0235809324	.0145467207	.0092414258
			A= 2		
0+	.0909090909	.0582060372	.0399709997	.0285000801	.0208141474
5+	.0154597948	.0116292278	.0088351058	.0067665172	.0052169503
			A=1/.60		
0+	.1071428571	.0781350402	.0601846870	.0477386308	.0385828335
5+	.0315982648	.0261368267	.0217890116	.0182800373	.0154172702
10+	.0130611084	.0111077219	.0094782635	.0081118310	.0069607133
			A=1/.65		
0+	.1150442478	.0885957284	.0715322213	.0592384892	.0498591422
5+	.0424479999	.0364512191	.0315145044	.0273967939	.0239264911
10+	.0209771710	.0184531381	.0162803662	.0144005618	.0127671337
15+	.0113423730	.0100954306	.0090008339	.0080373790	.0071872896
			A=1/.70		
0+	.1228070175	.0993137995	.0835976146	.0718953358	.0626855980
5+	.0551883790	.0489442013	.0436569372	.0391233691	.0351974433
10+	.0317704111	.0287590108	.0260980429	.0237354966	.0216292340
15+	.0197446643	.0180530706	.0165303781	.0151562316	.0139132903
20+	.0127866833	.0117635816	.0108328586	.0099848178	.0092109719
			A=1/.75		
0+	.1304347826	.1102413069	.0962990218	.0856191012	.0769883584
5+	.0697828483	.0636337634	.0583024446	.0536243336	.0494805183
10+	.0457819195	.0424598771	.0394602326	.0367394436	.0342619379
15+	.0319982561	.0299237140	.0280174169	.0262615198	.0246406616
20+	.0231415266	.0217525001	.0204633937	.0192652246	.0181500366
25+	.0171107530	.0161410552	.0152352816	.0143883426	.0135956494
30+	.0128530532	.0121567923	.0115034511	.0108899159	.0103133473
35+	.0097711482	.0092609400	.0087805397	.0083279410	.0079012970
			A=1/.80		
0+	.1379310345	.1213353155	.1095565234	.1003110047	.0926688994
5+	.0861511218	.0804740037	.0754536245	.0709628813	.0669096866
10+	.0632248492	.0598548583	.0567573479	.0538981175	.0512491036
15+	.0487869548	.0464920062	.0443475228	.0423391315	.0404543872
20+	.0386824353	.0370137470	.0354399082	.0339534497	.0325477088
25+	.0312167163	.0299551024	.0287580189	.0276210732	.0265402731
30+	.0255119795	.0245328661	.0235998851	.0227102365	.0218613430
35+	.0210508270	.0202764906	.0195362988	.0188283636	.0181509309
40+	.0175023683	.0168811543	.0162858690	.0157151856	.0151678626
45+	.0146427368	.0141387174	.0136547802	.0131899626	.0127433591
50+	.0123141167	.0119014319	.0115045461	.0111227436	.0107553479
55+	.0104017192	.0100612522	.0097333736	.0094175402	.0091132369

D-INTEGRAL PROBABILITY M=R, B= 6

M=R	1	2	3	4	5
			A=10		
0+	.0163934426	.0025288831	.0004495609	.0000863170	.0000174036
			A= 9		
0+	.0181818182	.0030978419	.0006069804	.0001283069	.0000284615
			A= 8		
0+	.0204081633	.0038828117	.0008473312	.0001992106	.0000491053
			A= 7		
0+	.0232558140	.0050087304	.0012332350	.0003265424	.0000905540
			A= 6		
0+	.0270270270	.0067058446	.0018937543	.0005737741	.0001818000
			A= 5		
0+	.0322580645	.0094373251	.0031234540	.0011054599	.0004083179
			A= 4		
0+	.0400000000	.0142516320	.0056931407	.0024202162	.0010705005
			A= 3		
0+	.0526315789	.0239579583	.0120509455	.0064000094	.0035189782
			A=1/.40		
0+	.0625000000	.0330151442	.0190552713	.0115409156	.0072087096
			A= 2		
0+	.0769230769	.0483531636	.0326708763	.0229588589	.0165452622
5+	.0121372537	.0090234852	.0067793502	.0051368666	.0039199364
			A=1/.60		
0+	.0909090909	.0653590841	.0497232627	.0390077657	.0312107277
5+	.0253229958	.0207630184	.0171655240	.0142869861	.0119578166
10+	.0100559434	.0084911891	.0071955368	.0061168027	.0052143779
			A=1/.65		
0+	.0977443609	.0743562180	.0594014101	.0487319401	.0406667265
5+	.0343487558	.0292778002	.0251351147	.0217047940	.0188338828
10+	.0164102481	.0143494418	.0125864722	.0110704229	.0097608079
			A=1/.70		
0+	.1044776119	.0836222688	.0697656969	.0595302095	.0515361000
5+	.0450748592	.0397295639	.0352320922	.0313989520	.0280986757
10+	.0252337010	.0227295903	.0205282622	.0185835532	.0168582000
15+	.0153217275	.0139489294	.0127187536	.0116134665	.0106180173
20+	.0097195471	.0089070034	.0081708361	.0075027528	.0068955223
			A=1/.75		
0+	.1111111111	.0931170350	.0807526587	.0713405096	.0637804601
5+	.0575049593	.0521783958	.0475837527	.0435715961	.0400341274
10+	.0368907501	.0340794731	.0315515074	.0292677183	.0271962102
15+	.0253106332	.0235889626	.0220126016	.0205657077	.0192346781
20+	.0180077516	.0168746951	.0158265556	.0148554595	.0139544512
25+	.0131173595	.0123386886	.0116135267	.0109374697	.0103065566
30+	.0097172155	.0091662165	.0086506323	.0081678038	.0077153103
			A=1/.80		
0+	.1176470588	.1028040320	.0922988363	.0840907566	.0773373735
5+	.0716028018	.0666285663	.0622470256	.0583423599	.0548307096
10+	.0516491206	.0487489562	.0460917532	.0436465001	.0413877833
15+	.0392944852	.0373483467	.0355357766	.0338423328	.0322573254
20+	.0307710099	.0293748450	.0280613003	.0268237012	.0256561035
25+	.0245531901	.0235101854	.0225227844	.0215870932	.0206995782
30+	.0198570239	.0190564958	.0182953091	.0175710018	.0168813111
35+	.0162241529	.0155976039	.0149998861	.0144293526	.0138844759
40+	.0133638365	.0128661139	.0123900775	.0119345791	.0114985463
45+	.0110809759	.0106809285	.0102975237	.0099299352	.0095773869

D-INTEGRAL PROBABILITY M=R,B= 7

M=R	1	2	3	4	5
			A=10		
0+	.0140845070	.0021084789	.0003645105	.0000682099	.0000134263
			A= 9		
0+	.0156250000	.0025847376	.0004927392	.0001015586	.0000220028
			A= 8		
0+	.0175438596	.0032426273	.0006888778	.0001580037	.0000380598
			A= 7		
0+	.0200000000	.0041877467	.0010045197	.0002596716	.0000704156
			A= 6		
0+	.0232558140	.0056152475	.0015463981	.0004578390	.0001419794
			A= 5		
0+	.0277777778	.0079191651	.0025593340	.0008862562	.0003207736
			A= 4		
0+	.0344827586	.0119961116	.0046884777	.0019536953	.0008482671
			A= 3		
0+	.0454545455	.0202673950	.0100048563	.0052234057	.0028270147
			A=1/.40		
0+	.0540540541	.0280374431	.0159182147	.0094984160	.0058519742
			A= 2		
0+	.0666666667	.0412923236	.0275351134	.0191220807	.0136313753
5+	.0098988978	.0072893978	.0054269927	.0040765408	.0030848753
			A=1/.60		
0+	.0789473684	.0561125104	.0422590425	.0328540354	.0260711640
5+	.0209916649	.0170882665	.0140314378	.0116026566	.0096505895
			A=1/.65		
0+	.0849673203	.0640009461	.0506871536	.0412632479	.0341930322
5+	.0286934653	.0243085773	.0207487833	.0178186691	.0153803527
10+	.0133331724	.0116016405	.0101279093	.0088668536	.0077827459
			A=1/.70		
0+	.0909090909	.0721574515	.0597634488	.0506671018	.0436067505
5+	.0379335401	.0332659729	.0293591602	.0260459043	.0232067297
10+	.0207532250	.0186181337	.0167491367	.0151047779	.0136516985
15+	.0123627016	.0112153628	.0101910120	.0092739715	.0084509767
			A=1/.75		
0+	.0967741935	.0805478576	.0694378104	.0610229122	.0542973346
5+	.0487407114	.0440452104	.0400119031	.0365039712	.0334228631
10+	.0306950201	.0282639677	.0260853437	.0241236351	.0223499574
15+	.0207404972	.0192753925	.0179379081	.0167138176	.0155909326
20+	.0145587370	.0136081014	.0127310553	.0119206043	.0111705829
25+	.0104755329	.0098306035	.0092314689	.0086742587	.0081554993
			A=1/.80		
0+	.1025641026	.0891408505	.0796590654	.0722777057	.0662272646
5+	.0611080426	.0566826998	.0527972567	.0493453581	.0462500408
10+	.0434535725	.0409113911	.0385882913	.0364559193	.0344910657
15+	.0326744677	.0309399456	.0294237677	.0279641726	.0266010047
20+	.0253254314	.0241297213	.0230070668	.0219514426	.0209574895
25+	.0200204202	.0191359411	.0183001869	.0175096660	.0167612144
30+	.0160519568	.0153792726	.0147407680	.0141342504	.0135577074
35+	.0130092883	.0124872873	.0119901297	.0115163590	.0110646260
40+	.0106336787	.0102223538	.0098295690	.0094543156	.0090956528

D-INTEGRAL PROBABILITY M=R,B= 8

M=R	1	2	3	4	5
			A=10		
0+	.0123456790	.0018035129	.0003047026	.0000558203	.0000107718
			A= 9		
0+	.0136986301	.0022121469	.0004122826	.0000832205	.0000176819
			A= 8		
0+	.0153846154	.0027771685	.0005770739	.0001296841	.0000306485
			A= 7		
0+	.0175438596	.0035898682	.0008427519	.0002135715	.0000568522
			A= 6		
0+	.0204081633	.0048193241	.0012999373	.0003775865	.0001150265
			A= 5		
0+	.0243902439	.0068079202	.0021573165	.0007336588	.0002611117
			A= 4		
0+	.0303030303	.0103379177	.0039678839	.0016262214	.0006952872
			A= 3		
0+	.0400000000	.0175348975	.0085219435	.0043863629	.0023429413
			A=1/.40		
0+	.0476190476	.0243315962	.0136262481	.0080303122	.0048910192
			A= 2		
0+	.0588235294	.0359933920	.0237390015	.0163230792	.0115305168
5+	.0083024386	.0060650070	.0044811681	.0033416341	.0025110826
			A=1/.60		
0+	.0697674419	.0491196447	.0366801992	.0283009802	.0223027182
5+	.0178420224	.0144365729	.0117862094	.0096927952	.0080197608
			A=1/.65		
0+	.0751445087	.0561402689	.0441389365	.0356995880	.0294078401
5+	.0245427445	.0206853285	.0175702573	.0150190606	.0129062467
10+	.0111405264	.0096536890	.0083936543	.0073199447	.0064006289
			A=1/.70		
0+	.0804597701	.0634225241	.0522077964	.0440207809	.0376992621
5+	.0326446795	.0285053498	.0250558801	.0221426978	.0196563158
10+	.0175159129	.0156601640	.0140414884	.0126222850	.0113723778
15+	.0102672307	.0092866676	.0084139353	.0076350035	.0069380329
			A=1/.75		
0+	.0857142857	.0709369796	.0608465951	.0532356874	.0471779180
5+	.0421927988	.0379959751	.0344037896	.0312900558	.0285639987
10+	.0261579642	.0240200941	.0221097205	.0203943446	.0188475802
15+	.0174477133	.0161766657	.0150192333	.0139625155	.0129954800
20+	.0121086260	.0112937210	.0105435906	.0098519512	.0092132738
			A=1/.80		
0+	.0909090909	.0786564611	.0700134979	.0633053489	.0578239486
5+	.0532002052	.0492146794	.0457249643	.0426327186	.0398668141
10+	.0373739319	.0351129496	.0330514087	.0311631925	.0294269448
15+	.0278249591	.0263423791	.0249666100	.0236868764	.0224933847
20+	.0213795614	.0203368472	.0193595343	.0184421340	.0175797706
25+	.0167680938	.0160032063	.0152816039	.0146001250	.0139559081
30+	.0133463555	.0127691029	.0122219928	.0117030517	.0112104702
35+	.0107425863	.0102978701	.0098749107	.0094724048	.0090891464

D-INTEGRAL PROBABILITY M=R,B= 9

M=R	1	2	3	4	5
			A=10		
0+	.0109890110	.0015728372	.0002606224	.0000468949	.0000088988
			A= 9		
0+	.0121951220	.0019300930	.0003529119	.0000699890	.0000146273
			A= 8		
0+	.0136986301	.0024244596	.0004944472	.0001092107	.0000253967
			A= 7		
0+	.0156250000	.0031362353	.0007229704	.0001801605	.0000472116
			A= 6		
0+	.0181818182	.0042144026	.0011169808	.0003192312	.0000957921
			A= 5		
0+	.0217391304	.0059613602	.0018578352	.0006221910	.0002182992
			A= 4		
0+	.0270270270	.0090702922	.0034282946	.0013854016	.0005846156
			A= 3		
0+	.0357142857	.0154341708	.0074021642	.0037641020	.0019881247
			A=1/.40		
0+	.0425531915	.0214699938	.0118842370	.0069297326	.0041795059
			A= 2		
0+	.0526315789	.0318753018	.0208264257	.0141992317	.0099522061
			A=1/.60		
0+	.0625000000	.0436513305	.0323609004	.0248060061	.0194321394
5+	.0154596117	.0124438773	.0101092988	.0082746476	.0068155183
			A=1/.65		
0+	.0673575130	.0499747553	.0390468220	.0314049405	.0257383747
5+	.0213789385	.0179389878	.0151736326	.0129186197	.0110588016
10+	.0095106750	.0082120433	.0071155696	.0061845791	.0053902324
			A=1/.70		
0+	.0721649485	.0565509314	.0463069450	.0388623270	.0331396088
5+	.0285831211	.0248665673	.0217811140	.0191847259	.0169763461
10+	.0150815374	.0134439450	.0120199450	.0107751432	.0096820014
			A=1/.75		
0+	.0769230769	.0633542037	.0541086508	.0471593905	.0416478577
5+	.0371276393	.0333344430	.0300976476	.0273001288	.0248577282
10+	.0227078049	.0208024111	.0191040019	.0175826197	.0162139800
15+	.0149781300	.0138584846	.0128411182	.0119142339	.0110677598
20+	.0102930368	.0095825733	.0089298516	.0083291724	.0077755288
			A=1/.80		
0+	.0816326531	.0703607416	.0624173388	.0562676807	.0512560660
5+	.0470395754	.0434140920	.0402471166	.0374471659	.0349481016
10+	.0327003754	.0306658008	.0288142608	.0271215435	.0255678681
15+	.0241368496	.0228147539	.0215899489	.0204524927	.0193938191
20+	.0184064943	.0174840251	.0166207069	.0158115018	.0150519388
25+	.0143380330	.0136662181	.0130332909	.0124363643	.0118728281
30+	.0113403153	.0108366736	.0103599411	.0099083250	.0094801835

D-INTEGRAL PROBABILITY M=R, B=10

M=R	1	2	3	4	5
			A=10		
0+	.0099009901	.0013926329	.0002269402	.0000402065	.0000075199
			A= 9		
0+	.0109890110	.0017096073	.0003075010	.0000600610	.0000123749
			A= 8		
0+	.0123456790	.0021485167	.0004311703	.0000938239	.0000215166
			A= 7		
0+	.0140845070	.0027809633	.0006310944	.0001549984	.0000400710
			A= 6		
0+	.0163934426	.0037399898	.0009763539	.0002751632	.0000814982
			A= 5		
0+	.0196078431	.0052961660	.0016269754	.0005376958	.0001863369
			A= 4		
0+	.0243902439	.0080714087	.0030105616	.0012018352	.0005014315
			A= 3		
0+	.0322580645	.0137711224	.0065292235	.0032854815	.0017185028
			A=1/.40		
0+	.0384615385	.0191963872	.0105188723	.0060772686	.0036342456
			A= 2		
0+	.0476190476	.0285860580	.0185256017	.0125374490	.0087278377
			A=1/.60		
0+	.0566037736	.0392612129	.0289229378	.0220447360	.0171791954
5+	.0136011350	.0108981726	.0088154379	.0071859391	.0058954556
			A=1/.65		
0+	.0610328638	.0450125218	.0349788206	.0279958887	.0228421748
5+	.0188948435	.0157930977	.0133095019	.0112918908	.0096338746
			A=1/.70		
0+	.0654205607	.0510067977	.0415758364	.0347487073	.0295209569
5+	.0253738935	.0220030785	.0192139044	.0168742222	.0148901776
10+	.0131927712	.0117298691	.0104611968	.0093550661	.0083861581
			A=1/.75		
0+	.0697674419	.0572214389	.0486872455	.0422919535	.0372354078
5+	.0331006222	.0296406291	.0266960555	.0241575859	.0219467534
10+	.0200052222	.0182883963	.0167614013	.0153964492	.0141710453
15+	.0130667337	.0120681958	.0111625876	.0103390445	.0095883039
			A=1/.80		
0+	.0740740741	.0636353478	.0562841512	.0506052085	.0459879293
5+	.0421120079	.0387865800	.0358877154	.0333298609	.0310511898
10+	.0290054093	.0271568652	.0254774580	.0239446152	.0225399105
15+	.0212480936	.0200563926	.0189539999	.0179316864	.0169815077
20+	.0160965752	.0152708782	.0144991404	.0137767066	.0130994492
25+	.0124636929	.0118661515	.0113038762	.0107742116	.0102747592
30+	.0098033459	.0093579977	.0089369170	.0085384623	.0081611320

TABLE F (A > 1)

This table gives the value of the incomplete Dirichlet integral of type 2

$$D_A^{(B)}(R, M) = \frac{\Gamma(M+BR)}{\Gamma^B(R)\Gamma(M)} \int_A^\infty \cdots \int_A^\infty \frac{\prod_{i=1}^{B} x_i^{R-1}\, dx_i}{\left(1 + \sum_{i=1}^{B} x_i\right)^{M+BR}}$$

for

$$M = R+1$$
$$A = 3(1)10 \quad \text{and} \quad A^{-1} = .40(.10).60(.05).80,$$
$$B = 1(1)10$$

The value of R increases from 1 by ones at least until we reach the the line that contains .01; the last line is completed.

Note: The symbols M, B, R and A which are all capital here in the table correspond to the same lower case symbols in the text.

D-INTEGRAL PROBABILITY M=R+1,B= 1

R	1	2	3	4	5
			A=10		
0+	.0082644628	.0028003552	.0008811438	.0002746842	.0000857680
			A= 9		
0+	.0100000000	.0037000000	.0012700000	.0004316500	.0001469026
			A= 8		
0+	.0123456790	.0050297211	.0018986115	.0007092062	.0002651615
			A= 7		
0+	.0156250000	.0070800781	.0029678345	.0012300611	.0005100351
			A= 6		
0+	.0204081633	.0104123282	.0049044191	.0022816052	.0010611939
			A= 5		
0+	.0277777778	.0162037037	.0087019890	.0046087915	.0024381565
			A= 4		
0+	.0400000000	.0272000000	.0169600000	.0104064000	.0063693824
5+	.0039031316	.0023972087	.0014759383	.0009108924	.0005634137
			A= 3		
0+	.0625000000	.0507812500	.0375976562	.0272979736	.0197277069
5+	.0142527819	.0103095323	.0074697204	.0054217792	.0039421417
			A=1/.40		
0+	.0816326531	.0733027905	.0597030149	.0475603581	.0376479852
5+	.0297585456	.0235305062	.0186251308	.0147612476	.0117145864
10+	.0093089111	.0074064638	.0058996893	.0047045543	.0037553020
			A= 2		
0+	.1111111111	.1111111111	.1001371742	.0879439110	.0765635320
5+	.0664476395	.0576163049	.0499624815	.0433480662	.0376365711
10+	.0327039162	.0284411281	.0247539985	.0215616785	.0187950012
15+	.0163948495	.0143106842	.0124992615	.0109235356	.0095517272
20+	.0083565361	.0073144768	.0064053179	.0056116077	.0049182738
			A=1/.60		
0+	.1406250000	.1516113281	.1464614868	.1374090314	.1275079083
5+	.1177614902	.1085309858	.0999374824	.0920031398	.0847077406
10+	.0780133828	.0718758627	.0662500615	.0610924790	.0563623592
15+	.0520221048	.0480373328	.0443767529	.0410119637	.0379172179
20+	.0350691834	.0324467131	.0300306307	.0278035350	.0257496219
25+	.0238545249	.0221051723	.0204896589	.0189971318	.0176176883
30+	.0163422845	.0151626540	.0140712346	.0130611025	.0121259144
35+	.0112598538	.0104575839	.0097142049	.0090252152	.0083864766
40+	.0077941828	.0072448311	.0067351960	.0062623061	.0058234231
			A=1/.65		
0+	.1551882461	.1722888793	.1709279473	.1644294661	.1562839869
5+	.1477271198	.1392634208	.1311108695	.1233611905	.1160462694
10+	.1091684483	.1027153318	.0966673868	.0910019745	.0856955278
15+	.0807247289	.0760671331	.0717014785	.0676078178	.0637675470
20+	.0601633784	.0567792813	.0536004081	.0506130137	.0478043752
25+	.0451627125	.0426771144	.0403374681	.0381343957	.0360591939
30+	.0341037798	.0322606411	.0305227907	.0288837252	.0273373867
35+	.0258781287	.0245006843	.0232001374	.0219718965	.0208116705
40+	.0197154468	.0186794708	.0177002277	.0167744250	.0158989774
45+	.0150709920	.0142877548	.0135467193	.0128454942	.0121818335
50+	.0115536269	.0109588905	.0103957587	.0098624766	.0093573929
55+	.0088789531	.0084256932	.0079962347	.0075892784	.0072036001
			A=1/.70		
0+	.1695501730	.1930173250	.1958593676	.1924174475	.1865816659
5+	.1797968126	.1726950198	.1655798232	.1586041403	.1518456931
10+	.1453423619	.1391100436	.1331522020	.1274652037	.1220414013
15+	.1168709599	.1119429582	.1072460584	.1027689138	.0985004149
20+	.0944298325	.0905468978	.0868418410	.0833054032	.0799283335
25+	.0767038744	.0736227428	.0706781066	.0678630603	.0651711003

D-INTEGRAL PROBABILITY M=R+1,B= 1

R	1	2	3	4	5
			A=1/.70		
30+	.0625961004	.0601322881	.0577742220	.0555167705	.0533550917
35+	.0512846146	.0493010218	.0474002326	.0455783884	.0438318382
40+	.0421571257	.0405509771	.0390102897	.0375321215	.0361136816
45+	.0347523207	.0334455230	.0321908986	.0309861755	.0298291935
50+	.0287178973	.0276503310	.0266246322	.0256390272	.0246918259
55+	.0237814174	.0229062655	.0220649051	.0212559383	.0204780309
60+	.0197299088	.0190103557	.0183182092	.0176523589	.0170117433
65+	.0163953477	.0158022014	.0152313763	.0146819842	.0141531751
70+	.0136441354	.0131540863	.0126822818	.0122280076	.0117905793
75+	.0113693409	.0109636643	.0105729468	.0101966111	.0098341033
80+	.0094848925	.0091484693	.0088243450	.0085120508	.0082111368
85+	.0079211713	.0076417397	.0073724442	.0071129028	.0068627485
			A=1/.75		
0+	.1836734694	.2136609746	.2210048534	.2210048534	.2179217090
5+	.2133913742	.2081610169	.2026104337	.1969465732	.1912873364
10+	.1857015960	.1802298504	.1748955427	.1697115702	.1646841748
15+	.1598153429	.1551043242	.1505486139	.1461445975	.1418879808
20+	.1377740787	.1337980100	.1299548297	.1262396179	.1226475388
25+	.1191738799	.1158140762	.1125637248	.1094185925	.1063746178
30+	.1034279110	.1005747505	.0978115783	.0951349947	.0925417520
35+	.0900287479	.0875930190	.0852317343	.0829421887	.0807217967
40+	.0785680865	.0764786940	.0744513575	.0724839123	.0705742860
45+	.0687204933	.0669206320	.0651728787	.0634754847	.0618267725
50+	.0602251319	.0586690171	.0571569433	.0556874838	.0542592673
55+	.0528709752	.0515213392	.0502091388	.0489331993	.0476923896
60+	.0464856202	.0453118413	.0441700413	.0430592449	.0419785115
65+	.0409269338	.0399036364	.0389077743	.0379385319	.0369951214
70+	.0360767821	.0351827789	.0343124016	.0334649635	.0326398010
75+	.0318362721	.0310537561	.0302916524	.0295493798	.0288263760
80+	.0281220967	.0274360147	.0267676197	.0261164174	.0254819291
85+	.0248636908	.0242612531	.0236741804	.0231020503	.0225444536
90+	.0220009933	.0214712845	.0209549540	.0204516396	.0199609901
95+	.0194826646	.0190163324	.0185616724	.0181183730	.0176861319
100+	.0172646553	.0168536581	.0164528634	.0160620023	.0156808136
105+	.0153090436	.0149464456	.0145927804	.0142478150	.0139113234
110+	.0135830856	.0132628882	.0129505233	.0126457891	.0123484892
115+	.0120584329	.0117754344	.0114993133	.0112298941	.0109670059
120+	.0107104827	.0104601630	.0102158895	.0099775092	.0097448735
125+	.0095178374	.0092962600	.0090800043	.0088689367	.0086629273
130+	.0084618496	.0082655806	.0080740003	.0078869920	.0077044423
			A=1/.80		
0+	.1975308642	.2341106539	.2461533830	.2498702747	.2498702747
5+	.2479667910	.2450125307	.2414564766	.2375540853	.2334589834
10+	.2292673638	.2250412453	.2208214337	.2166351125	.2125004744
15+	.2084296424	.2044305680	.2005082934	.1966658091	.1929046472
20+	.1892252979	.1856275053	.1821104816	.1786730613	.1753138150
25+	.1720311327	.1688232859	.1656884732	.1626248549	.1596305777
30+	.1567037940	.1538426753	.1510454225	.1483102728	.1456355049
35+	.1430194426	.1404604562	.1379569644	.1355074343	.1331103815
40+	.1307643692	.1284680081	.1262199544	.1240189094	.1218636179
45+	.1197528669	.1176854838	.1156603356	.1136763271	.1117323996
50+	.1098275294	.1079607268	.1061310342	.1043375256	.1025793047
55+	.1008555039	.0991652835	.0975078303	.0958823565	.0942880990
60+	.0927243180	.0911902967	.0896853397	.0882087729	.0867599421
65+	.0853382125	.0839429679	.0825736103	.0812295586	.0799102486
70+	.0786151321	.0773436762	.0760953630	.0748696889	.0736661642
75+	.0724843126	.0713236706	.0701837871	.0690642232	.0679645514

D-INTEGRAL PROBABILITY M=R+1,B= 1

A=1/.80

R	1	2	3	4	5
80+	.0668843556	.0658232305	.0647807811	.0637566228	.0627503806
85+	.0617616892	.0607901923	.0598355425	.0588974011	.0579754378
90+	.0570693302	.0561787638	.0553034317	.0544430344	.0535972796
95+	.0527658817	.0519485620	.0511450482	.0503550746	.0495783814
100+	.0488147147	.0480638266	.0473254748	.0465994225	.0458854380
105+	.0451832951	.0444927723	.0438136534	.0431457265	.0424887848
110+	.0418426255	.0412070507	.0405818664	.0399668829	.0393619146
115+	.0387667798	.0381813005	.0376053027	.0370386159	.0364810732
120+	.0359325110	.0353927695	.0348616917	.0343391241	.0338249162
125+	.0333189208	.0328209934	.0323309926	.0318487797	.0313742189
130+	.0309071769	.0304475234	.0299951303	.0295498724	.0291116265
135+	.0286802722	.0282556913	.0278377679	.0274263884	.0270214412
140+	.0266228171	.0262304088	.0258441111	.0254638209	.0250894369
145+	.0247208599	.0243579923	.0240007386	.0236490050	.0233026994
150+	.0229617314	.0226260124	.0222954552	.0219699746	.0216494865
155+	.0213339088	.0210231605	.0207171624	.0204158365	.0201191063
160+	.0198268968	.0195391342	.0192557460	.0189766611	.0187018098
165+	.0184311233	.0181645343	.0179019766	.0176433852	.0173886963
170+	.0171378470	.0168907759	.0166474222	.0164077266	.0161716305
175+	.0159390766	.0157100084	.0154843704	.0152621082	.0150431682
180+	.0148274978	.0146150453	.0144057598	.0141995913	.0139964908
185+	.0137964100	.0135993014	.0134051184	.0132138151	.0130253465
190+	.0128396683	.0126567368	.0124765093	.0122989436	.0121239984
195+	.0119516329	.0117818070	.0116144815	.0114496176	.0112871772
200+	.0111271229	.0109694180	.0108140261	.0106609117	.0105100398
205+	.0103613758	.0102148861	.0100705371	.0099282961	.0097881310
210+	.0096500099	.0095139017	.0093797757	.0092476017	.0091173500
215+	.0089889913	.0088624970	.0087378387	.0086149886	.0084939193

D-INTEGRAL PROBABILITY M=R+1,B= 2

R	1	2	3	4	5
			A=10		
0+	.0022675737	.0007103154	.0001935169	.0000515818	.0000137379
			A= 9		
0+	.0027700831	.0009527083	.0002846128	.0000831296	.0000242501
			A= 8		
0+	.0034602076	.0013191469	.0004361437	.0001408626	.0000454133
			A= 7		
0+	.0044444444	.0019002469	.0007032040	.0002539228	.0000914624
			A= 6		
0+	.0059171598	.0028791278	.0012093415	.0004949396	.0002018725
			A= 5		
0+	.0082644628	.0046631191	.0022631719	.0010680710	.0005017088
			A= 4		
0+	.0123456790	.0082812579	.0047531843	.0026450490	.0014622976
			A= 3		
0+	.0204081633	.0168382222	.0117893055	.0079650058	.0053298263
5+	.0035602435	.0023806502	.0015951663	.0010714087	.0007213749
			A=1/.40		
0+	.0277777778	.0258487654	.0202838774	.0153036912	.0114089645
5+	.0084764781	.0062965097	.0046823906	.0034877743	.0026026839
			A= 2		
0+	.0400000000	.0425600000	.0378496000	.0321971200	.0269713842
5+	.0224626128	.0186701367	.0155134282	.0128971250	.0107317641
10+	.0089396142	.0074552860	.0062245876	.0052029163	.0043536674
			A=1/.60		
0+	.0532544379	.0624654922	.0607393711	.0562213248	.0510818565
5+	.0460362080	.0413335726	.0370495288	.0331898502	.0297319268
10+	.0266425570	.0238858984	.0214270324	.0192334853	.0172757713
15+	.0155274666	.0139650653	.0125677419	.0113170847	.0101968307
20+	.0091926162	.0082917515	.0074830177	.0067564884	.0061033724
			A=1/.65		
0+	.0601637593	.0733864748	.0739189517	.0707133443	.0663001699
5+	.0615902657	.0569523674	.0525407890	.0484158396	.0445940357
10+	.0410709621	.0378323403	.0348596141	.0321328320	.0296321407
15+	.0273385399	.0252342377	.0233027866	.0215290999	.0198994053
20+	.0184011661	.0170229890	.0157545278	.0145863898	.0135100470
25+	.0125177532	.0116024690	.0107577926	.0099778984	.0092574806
30+	.0085917029	.0079761540	.0074068065	.0068799809	.0063923126
			A=1/.70		
0+	.0672153635	.0848362178	.0880904908	.0866807188	.0834713635
5+	.0795548983	.0754117619	.0712706810	.0672443546	.0633876880
10+	.0597252483	.0562651437	.0530064187	.0499431542	.0470668085
15+	.0443675786	.0418352017	.0394594250	.0372302783	.0351382253
20+	.0331742409	.0313298427	.0295970955	.0279685985	.0264374626
25+	.0249972833	.0236421106	.0223664185	.0211650751	.0200333136
30+	.0189667055	.0179611345	.0170127726	.0161180586	.0152736772
35+	.0144765405	.0137237706	.0130126837	.0123407756	.0117057082
40+	.0111052974	.0105375018	.0100004119	.0094922412	.0090113170
45+	.0085560723	.0081250388	.0077168395	.0073301827	.0069638562
			A=1/.75		
0+	.0743801653	.0967333329	.1031259022	.1039631576	.1024207150
5+	.0997596074	.0965637988	.0931310971	.0896230669	.0861306957
10+	.0827060801	.0793788765	.0761653448	.0730735502	.0701064662
15+	.0672638811	.0645435965	.0619421958	.0594555429	.0570791095
20+	.0548081904	.0526380456	.0505639931	.0485814685	.0466860618
25+	.0448735396	.0431398562	.0414811576	.0398937816	.0383742531
30+	.0369192792	.0355257414	.0341906881	.0329113262	.0316850131
35+	.0305092484	.0293816657	.0283000253	.0272622069	.0262662020

D-INTEGRAL PROBABILITY M=R+1,B= 2

R	1	2	3	4	5
			A=1/.75		
40+	.0253101080	.0243921216	.0235105332	.0226637212	.0218501466
45+	.0210683489	.0203169407	.0195946041	.0189000863	.0182321964
50+	.0175898011	.0169718225	.0163772339	.0158050579	.0152543633
55+	.0147242624	.0142139090	.0137224962	.0132492539	.0127934474
60+	.0123543750	.0119313669	.0115237829	.0111310116	.0107524683
65+	.0103875941	.0100358544	.0096967378	.0093697548	.0090544370
70+	.0087503360	.0084570222	.0081740841	.0079011277	.0076377751
			A=1/.80		
0+	.0816326531	.1090022015	.1188984667	.1223899231	.1229442784
5+	.1219798823	.1201768588	.1178969686	.1153469334	.1126510799
10+	.1098869776	.1071042829	.1043353079	.1016012373	.0989159307
15+	.0962883262	.0937240005	.0912262067	.0887965772	.0864356085
20+	.0841430010	.0819178980	.0797590580	.0776649779	.0756339831
25+	.0736642922	.0717540647	.0699014361	.0681045427	.0663615401
30+	.0646706160	.0630299995	.0614379674	.0598928479	.0583930237
35+	.0569369323	.0555230673	.0541499774	.0528162655	.0515205882
40+	.0502616537	.0490382203	.0478490951	.0466931317	.0455692289
45+	.0444763285	.0434134138	.0423795077	.0413736712	.0403950017
50+	.0394426311	.0385157248	.0376134800	.0367351242	.0358799140
55+	.0350471338	.0342360946	.0334461325	.0326766085	.0319269062
60+	.0311964319	.0304846129	.0297908971	.0291147518	.0284556628
65+	.0278131342	.0271866868	.0265758584	.0259802020	.0253992860
70+	.0248326935	.0242800211	.0237408792	.0232148906	.0227016910
75+	.0222009274	.0217122585	.0212353540	.0207698939	.0203155686
80+	.0198720780	.0194391317	.0190164481	.0186037544	.0182007862
85+	.0178072871	.0174230086	.0170477096	.0166811565	.0163231223
90+	.0159733872	.0156317375	.0152979660	.0149718715	.0146532587
95+	.0143419379	.0140377250	.0137404410	.0134499122	.0131659696
100+	.0128884493	.0126171917	.0123520419	.0120928492	.0118394671
105+	.0115917531	.0113495688	.0111127795	.0108812540	.0106548650
110+	.0104334884	.0102170036	.0100052930	.0097982424	.0095957405
115+	.0093976791	.0092039527	.0090144587	.0088290970	.0086477704
120+	.0084703840	.0082968455	.0081270649	.0079609545	.0077984290

D-INTEGRAL PROBABILITY M=R+1,B= 3

R	1	2	3	4	5
			A=10		
0+	.0010405827	.0003243716	.0000833076	.0000207029	.0000051263
			A= 9		
0+	.0012755102	.0004375353	.0001234835	.0000336949	.0000091565
			A= 8		
0+	.0016000000	.0006100787	.0001910570	.0000577925	.0000173987
			A= 7		
0+	.0020661157	.0008866474	.0003118144	.0001057856	.0000356897
			A= 6		
0+	.0027700831	.0013590350	.0005448160	.0002103452	.0000806750
			A= 5		
0+	.0039062500	.0022361279	.0010417650	.0004663476	.0002070769
			A= 4		
0+	.0059171598	.0040625183	.0022566310	.0012004035	.0006320151
			A= 3		
0+	.0100000000	.0085600000	.0058727440	.0038374875	.0024731169
			A=1/.40		
0+	.0138408304	.0134953110	.0104705673	.0077050856	.0055767355
5+	.0040154634	.0028888989	.0020804529	.0015008827	.0010849991
			A= 2		
0+	.0204081633	.0230601195	.0205223538	.0172188485	.0141529383
5+	.0115399620	.0093814633	.0076214271	.0061941048	.0050388557
10+	.0041040129	.0033470078	.0027333091	.0022351078	.0018300766
			A=1/.60		
0+	.0277777778	.0350175278	.0344341591	.0317423292	.0285565370
5+	.0254154699	.0225058527	.0198829025	.0175494264	.0154873979
10+	.0136714214	.0120746523	.0106713765	.0094380479	.0083536043
15+	.0073994453	.0065592636	.0058188185	.0051656996	.0045891007
			A=1/.65		
0+	.0317132670	.0418023288	.0427834707	.0409370885	.0381577694
5+	.0351398571	.0321649269	.0293497235	.0267387180	.0243429527
10+	.0221577357	.0201711722	.0183684292	.0167339038	.0152523193
15+	.0139092564	.0126913833	.0115865235	.0105836382	.0096727652
20+	.0088449382	.0080920996	.0074070133	.0067831823	.0062147720
25+	.0056965409	.0052237767	.0047922409	.0043981172	.0040379675
			A=1/.70		
0+	.0357925493	.0490709409	.0520036891	.0513883339	.0493797968
5+	.0468192251	.0440793271	.0413373707	.0386800056	.0361487247
10+	.0337614102	.0315232422	.0294325035	.0274837849	.0256698021
15+	.0239824418	.0224133682	.0209543712	.0195975610	.0183354706
20+	.0171611022	.0160679399	.0150499423	.0141015238	.0132175295
25+	.0123932072	.0116241785	.0109064106	.0102361887	.0096100907
30+	.0090249623	.0084778957	.0079662081	.0074874236	.0070392555
35+	.0066195909	.0062264758	.0058581027	.0055127979	.0051890110
			A=1/.75		
0+	.0400000000	.0567822222	.0620390590	.0630418153	.0621867465
5+	.0604520338	.0582961855	.0559541016	.0535531773	.0511648416
10+	.0488296319	.0465702779	.0443989133	.0423212270	.0403389383
15+	.0384513137	.0366561174	.0349502171	.0333299747	.0317914999
20+	.0303308153	.0289439634	.0276270750	.0263764109	.0251883877
25+	.0240595904	.0229867782	.0219668843	.0209970131	.0200744342
30+	.0191965758	.0183610170	.0175654800	.0168078219	.0160860269
35+	.0153981983	.0147425519	.0141174079	.0135211852	.0129523946
40+	.0124096331	.0118915783	.0113969836	.0109246727	.0104735359
45+	.0100425254	.0096306515	.0092369791	.0088606243	.0085007511
50+	.0081565688	.0078273286	.0075123219	.0072108772	.0069223584
			A=1/.80		
0+	.0443213296	.0648952881	.0728274893	.0758250762	.0765084571

D-INTEGRAL PROBABILITY M=R+1,B= 3

R	1	2	3	4	5
			A=1/.80		
5+	.0759834342	.0747886330	.0732129748	.0714222557	.0695163200
10+	.0675574020	.0655852180	.0636254754	.0616948940	.0598042850
15+	.0579604993	.0561676927	.0544281671	.0527429401	.0511121369
20+	.0495352640	.0480114014	.0465393409	.0451176836	.0437449115
25+	.0424194384	.0411396471	.0399039159	.0387106376	.0375582336
30+	.0364451627	.0353699276	.0343310793	.0333272193	.0323570006
35+	.0314191283	.0305123591	.0296355004	.0287874089	.0279669894
40+	.0271731931	.0264050159	.0256614966	.0249417151	.0242447908
45+	.0235698807	.0229161776	.0222829089	.0216693346	.0210747459
50+	.0204984638	.0199398375	.0193982434	.0188730836	.0183637846
55+	.0178697963	.0173905911	.0169256624	.0164745239	.0160367087
60+	.0156117631	.0151992713	.0147988040	.0144099681	.0140323806
65+	.0136656734	.0133094920	.0129634956	.0126273559	.0123007570
70+	.0119833944	.0116749750	.0113752164	.0110838463	.0108006023
75+	.0105252314	.0102574895	.0099971414	.0097439599	.0094977258
80+	.0092582277	.0090252613	.0087986295	.0085781417	.0083636141
85+	.0081548689	.0079517344	.0077540445	.0075616389	.0073743623

D-INTEGRAL PROBABILITY M=R+1,B= 4

R	1	2	3	4	5
			A=10		
0+	.0005948840	.0001881318	.0000468193	.0000111663	.0000026464
			A= 9		
0+	.0007304602	.0002545422	.0000696965	.0000182731	.0000047582
			A= 8		
0+	.0009182736	.0003562679	.0001084097	.0000315537	.0000091150
			A= 7		
0+	.0011890606	.0005202730	.0001781247	.0000582530	.0000188911
			A= 6		
0+	.0016000000	.0008025186	.0003139847	.0001171343	.0000432813
			A= 5		
0+	.0022675737	.0013319395	.0006076683	.0002636903	.0001131519
			A= 4		
0+	.0034602076	.0024505441	.0013395278	.0006939056	.0003546108
			A= 3		
0+	.0059171598	.0052680926	.0035836500	.0022966293	.0014463410
			A=1/.40		
0+	.0082644628	.0084329070	.0065236277	.0047327206	.0033638282
5+	.0023748624	.0016742480	.0011812715	.0008349345	.0005914302
			A= 2		
0+	.0123456790	.0147241180	.0131654433	.0109671293	.0089114269
5+	.0071699259	.0057467667	.0046011714	.0036849350	.0029539158
			A=1/.60		
0+	.0170132325	.0228148691	.0226952634	.0209018399	.0186985544
5+	.0165129976	.0144937502	.0126846246	.0110878101	.0096891315
10+	.0084687816	.0074059983	.0064810659	.0056760924	.0049752188
			A=1/.65		
0+	.0195398312	.0274981366	.0285601280	.0273805064	.0254458792
5+	.0233103426	.0211996052	.0192070479	.0173679213	.0156906742
10+	.0141712447	.0127999362	.0115648357	.0104535356	.0094539907
15+	.0085549219	.0077459811	.0070177885	.0063619044	.0057707691
20+	.0052376294	.0047564613	.0043218952	.0039291451	.0035739451
			A=1/.70		
0+	.0221819828	.0325809680	.0351438677	.0348888340	.0335088923

D-INTEGRAL PROBABILITY M=R+1,B= 4

R	1	2	3	4	5
			A=1/.70		
5+	.0316784546	.0296986334	.0277128450	.0257909819	.0239662481
10+	.0222526984	.0206541341	.0191688353	.0177921675	.0165180524
15+	.0153398100	.0142506406	.0132438988	.0123132420	.0114527060
20+	.0106567357	.0099201891	.0092383267	.0086067918	.0080215861
25+	.0074790450	.0069758106	.0065088076	.0060752188	.0056724631
			A=1/.75		
0+	.0249307479	.0380417833	.0424242246	.0434189669	.0429087409
5+	.0416804337	.0401068186	.0383797533	.0366033925	.0348360311
10+	.0331106039	.0314454276	.0298501363	.0283291015	.0268834706
15+	.0255124144	.0242139074	.0229852233	.0218232523	.0207247071
20+	.0196862544	.0187046004	.0177765430	.0168990046	.0160690503
25+	.0152838959	.0145409114	.0138376187	.0131716880	.0125409315
30+	.0119432971	.0113768609	.0108398202	.0103304856	.0098472742
35+	.0093887029	.0089533814	.0085400065	.0081473558	.0077742829
40+	.0074197115	.0070826314	.0067620936	.0064572063	.0061671312
			A=1/.80		
0+	.0277777778	.0438579041	.0503718726	.0529463662	.0536378226
5+	.0533373529	.0524847580	.0513176502	.0499729055	.0485330863
10+	.0470496524	.0455553957	.0440714785	.0426116012	.0411845606
15+	.0397958723	.0384488237	.0371451746	.0358856312	.0346701715
20+	.0334982729	.0323690709	.0312814735	.0302342412	.0292260461
25+	.0282555126	.0273212481	.0264218642	.0255559915	.0247222907
30+	.0239194595	.0231462373	.0224014083	.0216838026	.0209922975
35+	.0203258164	.0196833291	.0190638498	.0184664364	.0178901889
40+	.0173342477	.0167977920	.0162800383	.0157802385	.0152976785
45+	.0148316765	.0143815813	.0139467712	.0135266524	.0131206573
50+	.0127282437	.0123488935	.0119821110	.0116274224	.0112843744
55+	.0109525332	.0106314837	.0103208285	.0100201869	.0097291943
60+	.0094475015	.0091747735	.0089106893	.0086549411	.0084072336
65+	.0081672835	.0079348187	.0077095783	.0074913115	.0072797776

D-INTEGRAL PROBABILITY M=R+1,B= 5

R	1	2	3	4	5
			A=10		
0+	.0003844675	.0001241528	.0000303146	.0000070337	.0000016177
			A= 9		
0+	.0004725898	.0001683052	.0000452523	.0000115514	.0000029212
			A= 8		
0+	.0005948840	.0002361355	.0000706299	.0000200346	.0000056259
			A= 7		
0+	.0007716049	.0003458987	.0001165575	.0000371944	.0000117390
			A= 6		
0+	.0010405827	.0005357086	.0002066408	.0000753409	.0000271346
			A= 5		
0+	.0014792899	.0008940775	.0004030836	.0001713201	.0000718053
			A= 4		
0+	.0022675737	.0016584088	.0008988109	.0004574733	.0002290319
			A= 3		
0+	.0039062500	.0036120731	.0024490054	.0015496389	.0009605241
			A=1/.40		
0+	.0054869684	.0058402102	.0045206290	.0032499297	.0022813225
			A= 2		
0+	.0082644628	.0103443393	.0093051914	.0077198906	.0062244447
5+	.0049614914	.0039367924	.0031193091	.0024718888	.0019606152
			A=1/.60		
0+	.0114795918	.0162471225	.0163402239	.0150555527	.0134210394
5+	.0117891292	.0102828914	.0089387896	.0077589006	.0067318584
10+	.0058417381	.0050719058	.0044066475	.0038317869	.0033348378
			A=1/.65		
0+	.0132351789	.0197099306	.0207446865	.0199380798	.0184997336
5+	.0168875260	.0152889499	.0137816246	.0123947826	.0111353762
10+	.0100000078	.0089806639	.0080675565	.0072505452	.0065198366
15+	.0058663089	.0052816372	.0047583158	.0042896280	.0038695895
			A=1/.70		
0+	.0150815636	.0235013822	.0257453144	.0256737193	.0246638797
5+	.0232744239	.0217565726	.0202302481	.0187538427	.0173549918
10+	.0160452890	.0148277631	.0137008556	.0126606086	.0117018955
15+	.0108191227	.0100066279	.0092589044	.0085707206	.0079371789
20+	.0073537370	.0068162089	.0063207519	.0058638488	.0054422854
			A=1/.75		
0+	.0170132325	.0276098930	.0313362434	.0322759873	.0319589253
5+	.0310378071	.0298247772	.0284811898	.0270947688	.0257145838
10+	.0243682891	.0230711916	.0218312709	.0206520760	.0195344510
15+	.0184775912	.0174797018	.0165384156	.0156510595	.0148148267
20+	.0140268856	.0132844506	.0125848241	.0119254230	.0113037923
25+	.0107176109	.0101646933	.0096429868	.0091505675	.0086856344
30+	.0082465033	.0078316000	.0074394538	.0070686907	.0067180271
35+	.0063862637	.0060722800	.0057750281	.0054935287	.0052268652
			A=1/.80		
0+	.0190249703	.0320225191	.0375052654	.0397445261	.0404075384
5+	.0402337244	.0395922053	.0386836606	.0376240915	.0364835852
10+	.0353058417	.0341186874	.0329400486	.0317814959	.0306504255
15+	.0295514418	.0284872574	.0274592903	.0264680684	.0255135072
20+	.0245951032	.0237120691	.0228634306	.0220480948	.0212648986
25+	.0205126444	.0197901251	.0190961412	.0184295140	.0177890938
30+	.0171737657	.0165824535	.0160141214	.0154677756	.0149424639
35+	.0144372758	.0139513414	.0134838304	.0130339510	.0126009482
40+	.0121841022	.0117827276	.0113961709	.0110238097	.0106650510
45+	.0103193295	.0099861068	.0096648695	.0093551283	.0090564164
50+	.0087682889	.0084903211	.0082221078	.0079632624	.0077134155
55+	.0074722145	.0072393225	.0070144176	.0067971922	.0065873519

D-INTEGRAL PROBABILITY M=R+1,B= 6

R	1	2	3	4	5
			A=10		
0+	.0002687450	.0000888031	.0000214174	.0000048715	.0000010959
			A= 9		
0+	.0003305785	.0001205478	.0000320339	.0000080207	.0000019849
			A= 8		
0+	.0004164931	.0001694164	.0000501206	.0000139548	.0000038374
			A= 7		
0+	.0005408329	.0002487012	.0000829690	.0000260112	.0000080461
			A= 6		
0+	.0007304602	.0003862698	.0001476955	.0000529664	.0000187179
			A= 5		
0+	.0010405827	.0006472009	.0002897279	.0001213168	.0000499686
			A= 4		
0+	.0016000000	.0012074125	.0006513891	.0003273907	.0001614307
			A= 3		
0+	.0027700831	.0026543123	.0017984346	.0011278553	.0006909184
			A=1/.40		
0+	.0039062500	.0043224720	.0033534590	.0023959373	.0016665034
			A= 2		
0+	.0059171598	.0077354182	.0070031359	.0057961629	.0046473106
5+	.0036785586	.0028965728	.0022768743	.0017897305	.0014080257
			A=1/.60		
0+	.0082644628	.0122694366	.0124665572	.0114989078	.0102264419
5+	.0089477622	.0077677273	.0067176363	.0057995642	.0050041938
10+	.0043183754	.0037283901	.0032213212	.0027855649	.0024109486
			A=1/.65		
0+	.0095539601	.0149551892	.0159305624	.0153526027	.0142329245
5+	.0129599097	.0116934446	.0104998416	.0094040917	.0084121660
10+	.0075212257	.0067245280	.0060138514	.0053807036	.0048169119
			A=1/.70		
0+	.0109155714	.0179147562	.0198965766	.0199257078	.0191529099
5+	.0180525853	.0168395139	.0156164838	.0144335521	.0133143694
10+	.0122688150	.0112994291	.0104048359	.0095816237	.0088254019
15+	.0081313995	.0074948052	.0069109554	.0063754347	.0058841229
20+	.0054332114	.0050192014	.0046388921	.0042893628	.0039679533
			A=1/.75		
0+	.0123456790	.0211420132	.0243672082	.0252418955	.0250420275
5+	.0243222214	.0233499380	.0222639954	.0211399886	.0200202032
10+	.0189284373	.0178778245	.0168751735	.0159234741	.0150233912
15+	.0141741777	.0133742440	.0126215180	.0119136744	.0112482821
20+	.0106228975	.0100351238	.0094826468	.0089632566	.0084748579
25+	.0080154752	.0075832528	.0071764520	.0067934472	.0064327203
30+	.0060928551	.0057725311	.0054705172	.0051856662	.0049169090
			A=1/.80		
0+	.0138408304	.0246295250	.0293396522	.0313135229	.0319379893
5+	.0318410331	.0313393270	.0306059965	.0297414292	.0288063769
10+	.0278387392	.0268626412	.0258936071	.0249416373	.0240131040
15+	.0231119551	.0222404969	.0213999147	.0205906248	.0198125158
20+	.0190651170	.0183477157	.0176594412	.0169993235	.0163663357
25+	.0157594243	.0151775305	.0146196054	.0140846203	.0135715742
30+	.0130794983	.0126074592	.0121545605	.0117199432	.0113027861
35+	.0109023052	.0105177528	.0101484163	.0097936172	.0094527098
40+	.0091250795	.0088101416	.0085073401	.0082161458	.0079360556
45+	.0076665905	.0074072950	.0071577356	.0069174995	.0066861938

D-INTEGRAL PROBABILITY M=R+1,B= 7

R	1	2	3	4	5
			A=10		
0+	.0001983733	.0000671025	.0000160496	.0000035956	.0000007950
			A= 9		
0+	.0002441406	.0000911820	.0000240408	.0000059315	.0000014434
			A= 8		
0+	.0003077870	.0001283073	.0000376837	.0000103446	.0000027986
			A= 7		
0+	.0004000000	.0001886561	.0000625277	.0000193406	.0000058899
			A= 6		
0+	.0005408329	.0002936327	.0001116520	.0000395414	.0000137688
			A= 5		
0+	.0007716049	.0004934303	.0002199590	.0000910689	.0000370043
			A= 4		
0+	.0011890606	.0009245220	.0004976337	.0002477561	.0001207260
			A= 3		
0+	.0020661157	.0020466743	.0013878365	.0008646419	.0005248677
			A=1/.40		
0+	.0029218408	.0033510371	.0026079633	.0018550891	.0012812187
			A= 2		
0+	.0044444444	.0060441113	.0055072458	.0045517935	.0036341982
5+	.0028609305	.0022391189	.0017489125	.0013658240	.0010675176
			A=1/.60		
0+	.0062326870	.0096590215	.0099078181	.0091516921	.0081257670
5+	.0070883327	.0061306166	.0052800729	.0045387884	.0038989736
			A=1/.65		
0+	.0072194455	.0118162456	.0127253637	.0122971175	.0113956014
5+	.0103569081	.0093201206	.0083430594	.0074475615	.0066388973
10+	.0059146777	.0052691335	.0046952276	.0041857038	.0037335963
			A=1/.70		
0+	.0082644628	.0142051186	.0159723609	.0160594800	.0154481536
5+	.0145491241	.0135495131	.0125391069	.0115616612	.0106378376
10+	.0097762519	.0089791081	.0082451975	.0075715461	.0069543375
15+	.0063894362	.0058726821	.0054000532	.0049677510	.0045722404
			A=1/.75		
0+	.0093652445	.0168226650	.0196559391	.0204671942	.0203426940
5+	.0197626203	.0189603152	.0180573458	.0171200239	.0161854544
10+	.0152745042	.0143986679	.0135638830	.0127727354	.0120257748
15+	.0113223177	.0106609482	.0100398342	.0094569282	.0089100961
20+	.0083971983	.0079161411	.0074649075	.0070415752	.0066443256
25+	.0062714476	.0059213366	.0055924928	.0052835162	.0049931023
			A=1/.80		
0+	.0105193951	.0196647276	.0237786603	.0255394391	.0261243278
5+	.0260765717	.0256726154	.0250643512	.0243401067	.0235534178
10+	.0227377030	.0219142334	.0210966864	.0202938604	.0195113491
15+	.0187526058	.0180196356	.0173134562	.0166344087	.0159823716
20+	.0153569091	.0147573748	.0141829856	.0136328738	.0131061237
25+	.0126017987	.0121189594	.0116566771	.0112140427	.0107901732
30+	.0103842153	.0099953480	.0096227840	.0092657703	.0089235878
35+	.0085955508	.0082810065	.0079793334	.0076899408	.0074122670
40+	.0071457781	.0068899672	.0066443524	.0064084760	.0061819032

D-INTEGRAL PROBABILITY M=R+1, B= 8

R	1	2	3	4	5
			A=10		
0+	.0001524158	.0000527604	.0000125469	.0000027775	.0000006056
			A= 9		
0+	.0001876525	.0000717497	.0000188160	.0000045888	.0000011016
			A= 8		
0+	.0002366864	.0001010619	.0000295367	.0000080181	.0000021409
			A= 7		
0+	.0003077870	.0001487818	.0000491002	.0000150271	.0000045189
			A= 6		
0+	.0004164931	.0002319538	.0000878895	.0000308203	.0000106049
			A= 5		
0+	.0005948840	.0003906761	.0001737303	.0000712952	.0000286542
			A= 4		
0+	.0009182736	.0007344715	.0003949982	.0001952130	.0000942196
			A= 3		
0+	.0016000000	.0016348876	.0011104462	.0006883587	.0004148095
			A=1/.40		
0+	.0022675737	.0026882437	.0020996282	.0014886646	.0010223572
			A= 2		
0+	.0034602076	.0048787574	.0044734477	.0036946116	.0029400934
5+	.0023042839	.0017945081	.0013943081	.0010830593	.0008419336
			A=1/.60		
0+	.0048674959	.0078432510	.0081166079	.0075090101	.0066596118
5+	.0057954736	.0049972472	.0042894045	.0036740345	.0031445154
			A=1/.65		
0+	.0056466972	.0096227485	.0104675967	.0101423621	.0093975654
5+	.0085286635	.0076585361	.0068383994	.0060876506	.0054110296
10+	.0048065070	.0042690746	.0037926142	.0033708232	.0029976620
			A=1/.70		
0+	.0064737746	.0116010576	.0131912159	.0133125695	.0128164991
5+	.0120641247	.0112209195	.0103664882	.0095396628	.0087587815
10+	.0080314895	.0073597443	.0067424821	.0061770814	.0056601826
15+	.0051881512	.0047573374	.0043642199	.0040054799	.0036780347
			A=1/.75		
0+	.0073469388	.0137770934	.0162970198	.0170500240	.0169761512
5+	.0164974271	.0158203754	.0150529694	.0142542039	.0134571061
10+	.0126802434	.0119338311	.0112231339	.0105504326	.0099161977
15+	.0093198063	.0087599882	.0082351077	.0077433420	.0072827954
20+	.0068515711	.0064478164	.0060697496	.0057156756	.0053839933
			A=1/.80		
0+	.0082644628	.0161487292	.0197907025	.0213775276	.0219249716
5+	.0219099801	.0215772798	.0210619504	.0204427500	.0197674836
10+	.0190660124	.0183573438	.0176536943	.0169629141	.0162899860
15+	.0156379780	.0150086632	.0144029331	.0138210757	.0132629673
20+	.0127282052	.0122162007	.0117262446	.0112575540	.0108093050
25+	.0103806566	.0099707668	.0095788046	.0092039575	.0088454375
30+	.0085024842	.0081743669	.0078603857	.0075598719	.0072721878
35+	.0069967260	.0067329088	.0064801872	.0062380394	.0060059702

D-INTEGRAL PROBABILITY M=R+1,B= 9

R	1	2	3	4	5
			A=10		
0+	.0001207584	.0000427495	.0000101256	.0000022198	.0000004785
			A= 9		
0+	.0001487210	.0000581725	.0000151993	.0000036719	.0000008716
			A= 8		
0+	.0001876525	.0000820025	.0000238875	.0000064260	.0000016972
			A= 7		
0+	.0002441406	.0001208445	.0000397689	.0000120671	.0000035910
			A= 6		
0+	.0003305785	.0001886500	.0000713280	.0000248140	.0000084543
			A= 5		
0+	.0004725898	.0003183271	.0001413808	.0000576075	.0000229440
			A= 4		
0+	.0007304602	.0006000894	.0003227500	.0001585698	.0000759319
			A= 3		
0+	.0012755102	.0013417025	.0009132948	.0005639282	.0003377825
			A=1/.40		
0+	.0018107741	.0022137991	.0017356262	.0012275962	.0008391740
			A= 2		
0+	.0027700831	.0040379918	.0037251067	.0030755905	.0024410246
			A=1/.60		
0+	.0039062500	.0065232436	.0068064031	.0063073059	.0055892950
5+	.0048545749	.0041753600	.0035737361	.0030517821	.0026037894
			A=1/.65		
0+	.0045370346	.0080222278	.0088077395	.0085561643	.0079282374
5+	.0071870057	.0064423917	.0057403325	.0050982800	.0045205554
10+	.0040054190	.0035484693	.0031443171	.0027874135	.0024724485
			A=1/.70		
0+	.0052077798	.0096940255	.0111364052	.0112780553	.0108673151
5+	.0102256439	.0095011429	.0087652397	.0080528167	.0073803551
10+	.0067547303	.0061777107	.0056483556	.0051643315	.0047226480
			A=1/.75		
0+	.0059171598	.0115387192	.0138032119	.0145038825	.0144651677
5+	.0140625406	.0134809378	.0128173641	.0121249070	.0114333049
10+	.0107592742	.0101120055	.0094962278	.0089139833	.0083656854
15+	.0078507660	.0073680774	.0069161464	.0064933345	.0060979407
20+	.0057282655	.0053826515	.0050595070	.0047573199	.0044746641
			A=1/.80		
0+	.0066638900	.0135554788	.0168156904	.0182582599	.0187713679
5+	.0187788469	.0184998361	.0180558008	.0175177335	.0169287854
10+	.0163159265	.0156963339	.0150810340	.0144771132	.0138890626
15+	.0133196415	.0127704392	.0122422489	.0117353194	.0112495290
20+	.0107845050	.0103397084	.0099144926	.0095081459	.0091199210
25+	.0087490559	.0083947893	.0080563704	.0077330664	.0074241670
30+	.0071289879	.0068468719	.0065771904	.0063193434	.0060727591

D-INTEGRAL PROBABILITY M=R+1,B=10

R	1	2	3	4	5
			A=10		
0+	.0000980296	.0000354620	.0000083762	.0000018215	.0000003888
			A= 9		
0+	.0001207584	.0000482811	.0000125833	.0000030163	.0000007092
			A= 8		
0+	.0001524158	.0000681033	.0000197955	.0000052854	.0000013831
			A= 7		
0+	.0001983733	.0001004450	.0000329977	.0000099415	.0000029324
			A= 6		
0+	.0002687450	.0001569770	.0000592814	.0000204877	.0000069223
			A= 5		
0+	.0003844675	.0002652864	.0001177726	.0000477074	.0000188559
			A= 4		
0+	.0005948840	.0005012303	.0002697661	.0001319023	.0000627423
			A= 3		
0+	.0010405827	.0011248000	.0007675598	.0004724625	.0002815636
			A=1/.40		
0+	.0014792899	.0018612487	.0014648870	.0010341963	.0007042386
			A= 2		
0+	.0022675737	.0034091978	.0031634741	.0026118287	.0020684725
			A=1/.60		
0+	.0032039872	.0055298445	.0058145238	.0053972229	.0047800576
5+	.0041450104	.0035574117	.0030373925	.0025870111	.0022012869
			A=1/.65		
0+	.0037250105	.0068140379	.0075458433	.0073485795	.0068104455
5+	.0061680839	.0055208318	.0049103124	.0043523895	.0038510431
			A=1/.70		
0+	.0042798498	.0082501200	.0095677574	.0097211851	.0093754988
5+	.0088197997	.0081879602	.0075447037	.0069216763	.0063338413
10+	.0057874479	.0052841142	.0048230041	.0044020220	.0040184812
			A=1/.75		
0+	.0048674959	.0098389071	.0118916631	.0125456641	.0125319998
5+	.0121881233	.0116812498	.0110993583	.0104906725	.0098822241
10+	.0092892088	.0087199778	.0081788241	.0076675975	.0071866689
15+	.0067355186	.0063131029	.0059180843	.0055489778	.0052042437
			A=1/.80		
0+	.0054869684	.0115804147	.0145261480	.0158473854	.0163293982
5+	.0163526251	.0161151658	.0157272517	.0152534557	.0147330780
10+	.0141907001	.0136419741	.0130969549	.0125620956	.0120414826
15+	.0115376231	.0110519572	.0105851992	.0101375683	.0097089463
20+	.0092989879	.0089071967	.0085329798	.0081756857	.0078346313
25+	.0075091215	.0071984622	.0069019698	.0066189778	.0063488407
30+	.0060909372	.0058446713	.0056094732	.0053847998	.0051701338

TABLE G (A < 1)

This table gives the value of the incomplete Dirichlet integral of type 2

$$D_A^{(B)}(R, M) = \frac{\Gamma(M+BR)}{\Gamma^B(R)\Gamma(M)} \int_A^{\infty} \cdots \int_A^{\infty} \frac{\prod_{i=1}^{B} x_i^{R-1}\, dx_i}{\left(1 + \sum_{i=1}^{B} x_i\right)^{M+BR}}$$

for

$$M = R$$
$$A^{-1} = 3(1)10 \quad \text{and} \quad A = .40(.10).60(.05).80,$$
$$B = 1(1)10$$

The common value of M and R increases from 1 by ones until after we reach the line containing .99; the last line is then completed.

<u>Note</u>: The symbols M, B, R and A which are all capital here in the table correspond to the same lower case symbols in the text.

D-INTEGRAL PROBABILITY M=R, B= 1

R	1	2	3	4	5
			A=1/10		
0+	.9090909091	.9767092412	.9934741169	.9980925400	.9994284475
			A=1/ 9		
0+	.9000000000	.9720000000	.9914400000	.9972720000	.9991090800
			A=1/ 8		
0+	.8888888889	.9657064472	.9884672052	.9959604589	.9985507194
			A=1/ 7		
0+	.8750000000	.9570312500	.9839477539	.9937610626	.9975177199
			A=1/ 6		
0+	.8571428571	.9446064140	.9767358839	.9898499532	.9954702686
			A=1/ 5		
0+	.8333333333	.9259259259	.9645061728	.9823673983	.9910499384
5+	.9953912085	.9976020404	.9987425490	.9993365639	.9996482384
			A=1/ 4		
0+	.8000000000	.8960000000	.9420800000	.9666560000	.9804185600
5+	.9883457946	.9929964388	.9957602503	.9974185372	.9984208795
			A=1/ 3		
0+	.7500000000	.8437500000	.8964843750	.9294433594	.9510726929
5+	.9656724930	.9757098556	.9827001616	.9876152206	.9910967207
10+	.9935772895	.9953531513	.9966295519	.9975500332	.9982157383
			A= .40		
0+	.7142857143	.8017492711	.8552983876	.8917263579	.9177463368
5+	.9368630559	.9511680839	.9620115453	.9703101127	.9767081011
10+	.9816698065	.9855360704	.9885606986	.9909348181	.9928036585
15+	.9942783897	.9954446312	.9963686641	.9971020236	.9976849301
			A=1/ 2		
0+	.6666666667	.7407407407	.7901234568	.8267032465	.8551541940
5+	.8779149520	.8964607548	.9117684016	.9245247739	.9352338272
10+	.9442770278	.9519500464	.9584863216	.9640728816	.9688613615
15+	.9729759073	.9765189883	.9795757642	.9822174223	.9845037696
20+	.9864852707	.9882046684	.9896982867	.9909970852	.9921275209
25+	.9931122560	.9939707430	.9947197111	.9953735721	.9959447611
			A= .60		
0+	.6250000000	.6835937500	.7247924805	.7569789886	.7833819836
5+	.8056595107	.8248042604	.8414704489	.8561184660	.8690880646
10+	.8806391133	.8909759893	.9002630263	.9086347543	.9162029460
15+	.9230616197	.9292906886	.9349586832	.9401248242	.9448406272
20+	.9491511659	.9530960786	.9567103807	.9600251278	.9630679621
25+	.9658635661	.9684340434	.9707992394	.9729770148	.9749834782
30+	.9768331866	.9785393188	.9801138256	.9815675605	.9829103948
35+	.9841513175	.9852985247	.9863594975	.9873410719	.9882495001
40+	.9890905060	.9898693338	.9905907926	.9912592954	.9918788950
45+	.9924533155	.9929859812	.9934800429	.9939384008	.9943637266
			A= .65		
0+	.6060606061	.6567047889	.6929788593	.7218471124	.7459702624
5+	.7667042096	.7848551397	.8009512026	.8153622782	.8283603303
10+	.8401528698	.8509028932	.8607414515	.8697759404	.8780957766
15+	.8857764090	.8928822282	.8994687274	.9055841382	.9112706930
20+	.9165656118	.9215018866	.9261089116	.9304129953	.9344377813
25+	.9382045966	.9417327431	.9450397430	.9481415467	.9510527104
30+	.9537865489	.9563552671	.9587700744	.9610412848	.9631784039
35+	.9651902059	.9670848016	.9688696985	.9705518545	.9721377255
40+	.9736333082	.9750441783	.9763755255	.9776321839	.9788186608
45+	.9799391618	.9809976140	.9819976873	.9829428132	.9838362025
50+	.9846808615	.9854796064	.9862350768	.9869497480	.9876259428
55+	.9882658413	.9888714910	.9894448155	.9899876231	.9905016136
60+	.9909883865	.9914494464	.9918862098	.9923000104	.9926921046

D-INTEGRAL PROBABILITY M=R, B=1

M=R	1	2	3	4	5
			A= .70		
0+	.5882352941	.6309790352	.6620385011	.6871153478	.7083742663
5+	.7269114547	.7433747016	.7581859273	.7716390301	.7839490589
10+	.7952793968	.8057579099	.8154870946	.8245507481	.8330185213
15+	.8409491232	.8483926380	.8553922387	.8619854804	.8682052932
20+	.8740807564	.8796377113	.8848992509	.8898861170	.8946170252
25+	.8991089325	.9033772607	.9074360836	.9112982854	.9149756949
30+	.9184792017	.9218188542	.9250039468	.9280430935	.9309442939
35+	.9337149906	.9363621191	.9388921530	.9413111438	.9436247558
40+	.9458382980	.9479567527	.9499848003	.9519268433	.9537870260
45+	.9555692540	.9572772114	.9589143759	.9604840334	.9619892906
50+	.9634330875	.9648182079	.9661472893	.9674228325	.9686472102
55+	.9698226744	.9709513641	.9720353118	.9730764500	.9740766168
60+	.9750375613	.9759609489	.9768483658	.9777013231	.9785212616
65+	.9793095548	.9800675132	.9807963871	.9814973700	.9821716016
70+	.9828201705	.9834441165	.9840444336	.9846220721	.9851779404
75+	.9857129075	.9862278048	.9867234279	.9872005383	.9876598649
80+	.9881021058	.9885279296	.9889379769	.9893328612	.9897131709
85+	.9900794695	.9904322976	.9907721734	.9910995938	.9914150356
90+	.9917189560	.9920117940	.9922939704	.9925658895	.9928279392
			A= .75		
0+	.5714285714	.6064139942	.6321175701	.6531000810	.6710850904
5+	.6869412619	.7011794567	.7141308176	.7260249246	.7370289963
10+	.7472695202	.7568450750	.7658343714	.7743015265	.7822996556
15+	.7898733941	.7970607174	.8038942829	.8104024405	.8166100065
20+	.8225388654	.8282084448	.8336360942	.8388373907	.8438263895
25+	.8486158283	.8532172986	.8576413879	.8618978004	.8659954585
30+	.8699425905	.8737468045	.8774151538	.8809541921	.8843700225
35+	.8876683405	.8908544707	.8939334007	.8969098100	.8997880959
40+	.9025723970	.9052666136	.9078744268	.9103993148	.9128445680
45+	.9152133032	.9175084751	.9197328884	.9218892074	.9239799656
50+	.9260075745	.9279743307	.9298824239	.9317339429	.9335308819
55+	.9352751463	.9369685575	.9386128580	.9402097156	.9417607278
60+	.9432674253	.9447312759	.9461536878	.9475360122	.9488795469
65+	.9501855385	.9514551853	.9526896394	.9538900089	.9550573602
70+	.9561927200	.9572970768	.9583713831	.9594165567	.9604334824
75+	.9614230134	.9623859726	.9633231542	.9642353247	.9651232240
80+	.9659875669	.9668290435	.9676483209	.9684460437	.9692228350
85+	.9699792972	.9707160132	.9714335466	.9721324427	.9728132295
90+	.9734764176	.9741225018	.9747519609	.9753652587	.9759628447
95+	.9765451540	.9771126085	.9776656169	.9782045756	.9787298687
100+	.9792418686	.9797409366	.9802274231	.9807016679	.9811640007
105+	.9816147415	.9820542008	.9824826799	.9829004713	.9833078589
110+	.9837051186	.9840925180	.9844703171	.9848387685	.9851981174
115+	.9855486022	.9858904543	.9862238988	.9865491542	.9868664330
120+	.9871759417	.9874778811	.9877724462	.9880598268	.9883402074
125+	.9886137673	.9888806809	.9891411179	.9893952433	.9896432176
130+	.9898851969	.9901213331	.9903517740	.9905766634	.9907961412
135+	.9910103436	.9912194030	.9914234485	.9916226056	.9918169965
			A= .80		
0+	.5555555556	.5829903978	.6033125032	.6200385158	.6344930947
5+	.6473416092	.6589740092	.6696421714	.6795200994	.6887340789
10+	.6973792942	.7055296655	.7132440087	.7205700707	.7275472726
15+	.7342086340	.7405821589	.7466918559	.7525585060	.7582002487
20+	.7636330381	.7688710007	.7739267223	.7788114774	.7835354176
25+	.7881077250	.7925367408	.7968300712	.8009946774	.8050369516
30+	.8089627816	.8127776066	.8164864641	.8200940323	.8236046651
35+	.8270224240	.8303511054	.8335942653	.8367552397	.8398371648

D-INTEGRAL PROBABILITY M=R, B=1

A= .80

M=R	1	2	3	4	5
40+	.8428429930	.8457755083	.8486373398	.8514309738	.8541587646
45+	.8568229443	.8594256319	.8619688413	.8644544883	.8668843977
50+	.8692603091	.8715838826	.8738567038	.8760802885	.8782560869
55+	.8803854878	.8824698221	.8845103659	.8865083441	.8884649329
60+	.8903812627	.8922584204	.8940974518	.8958993639	.8976651265
65+	.8993956744	.9010919092	.9027547006	.9043848883	.9059832831
70+	.9075506685	.9090878017	.9105954153	.9120742177	.9135248948
75+	.9149481104	.9163445079	.9177147103	.9190593217	.9203789278
80+	.9216740968	.9229453798	.9241933120	.9254184128	.9266211866
85+	.9278021236	.9289617000	.9301003786	.9312186096	.9323168306
90+	.9333954675	.9344549344	.9354956346	.9365179606	.9375222945
95+	.9385090085	.9394784651	.9404310177	.9413670103	.9422867787
100+	.9431906498	.9440789426	.9449519683	.9458100301	.9466534243
105+	.9474824396	.9482973580	.9490984546	.9498859981	.9506602510
110+	.9514214692	.9521699031	.9529057971	.9536293901	.9543409153
115+	.9550406009	.9557286697	.9564053397	.9570708239	.9577253305
120+	.9583690635	.9590022219	.9596250006	.9602375903	.9608401776
125+	.9614329450	.9620160710	.9625897305	.9631540946	.9637093308
130+	.9642556031	.9647930719	.9653218947	.9658422252	.9663542144
135+	.9668580098	.9673537563	.9678415954	.9683216662	.9687941046
140+	.9692590439	.9697166149	.9701669456	.9706101615	.9710463857
145+	.9714757388	.9718983389	.9723143021	.9727237420	.9731267701
150+	.9735234957	.9739140261	.9742984662	.9746769194	.9750494867
155+	.9754162675	.9757773590	.9761328568	.9764828547	.9768274446
160+	.9771667167	.9775007596	.9778296603	.9781535041	.9784723746
165+	.9787863541	.9790955233	.9793999613	.9796997460	.9799949537
170+	.9802856592	.9805719363	.9808538572	.9811314928	.9814049129
175+	.9816741858	.9819393789	.9822005581	.9824577883	.9827111331
180+	.9829606552	.9832064159	.9834484758	.9836868941	.9839217290
185+	.9841530379	.9843808771	.9846053017	.9848263661	.9850441237
190+	.9852586270	.9854699276	.9856780760	.9858831220	.9860851147
195+	.9862841022	.9864801316	.9866732495	.9868635017	.9870509328
200+	.9872355873	.9874175084	.9875967388	.9877733205	.9879472947
205+	.9881187019	.9882875822	.9884539746	.9886179178	.9887794495
210+	.9889386073	.9890954276	.9892499465	.9894021996	.9895522217
215+	.9897000471	.9898457095	.9899892421	.9901306776	.9902700480
220+	.9904073850	.9905427195	.9906760822	.9908075032	.9909370119
225+	.9910646375	.9911904087	.9913143534	.9914364996	.9915568743

D-INTEGRAL PROBABILITY M=R,B= 2

R	1	2	3	4	5
			A=1/10		
0+	.8333333333	.9548611111	.9871117863	.9962051966	.9988594904
			A=1/ 9		
0+	.8181818182	.9460419370	.9831518900	.9945832054	.9982240843
			A=1/ 8		
0+	.8000000000	.9344000000	.9774080000	.9920020480	.9971159859
			A=1/ 7		
0+	.7777777778	.9186099680	.9687679347	.9877024694	.9950744007
			A=1/ 6		
0+	.7500000000	.8964843750	.9551925659	.9801371098	.9910563984
5+	.9959271708	.9981290922	.9991346144	.9995974564	.9998118753
			A=1/ 5		
0+	.7142857143	.8642232403	.9327078246	.9659142738	.9824914892
5+	.9909159298	.9952504773	.9975014561	.9986789612	.9992985761
			A=1/ 4		
0+	.6666666667	.8148148148	.8930041152	.9369506681	.9623885812
5+	.9773692345	.9862940741	.9916561675	.9948988272	.9968701078
			A=1/ 3		
0+	.6000000000	.7344000000	.8173440000	.8721976320	.9096352358
5+	.9356461472	.9539267509	.9668783824	.9761097814	.9827204657
10+	.9874723955	.9908989597	.9933764145	.9951717762	.9964754616
			A= .40		
0+	.5555555556	.6744398720	.7538466170	.8109594113	.8534432823
5+	.8856487354	.9103581685	.9294746557	.9443545399	.9559908995
10+	.9651244687	.9723151867	.9779905591	.9824794841	.9860365264
15+	.9888596764	.9911035387	.9928892434	.9943119634	.9954466589
			A=1/ 2		
0+	.5000000000	.5937500000	.6611328125	.7136230469	.7560472488
5+	.7910559773	.8203298971	.8450320950	.8660148718	.8839278531
10+	.8992803004	.9124797901	.9238576601	.9336865443	.9421929179
15+	.9495663510	.9559665081	.9615285542	.9663674023	.9705810991
20+	.9742535572	.9774567821	.9802527026	.9826946877	.9848288108
25+	.9866949088	.9883274743	.9897564099	.9910076681	.9921037962
30+	.9930644012	.9939065480	.9946451000	.9952930132	.9958615879
			A= .60		
0+	.4545454545	.5242128270	.5763086392	.6187592686	.6547519890
5+	.6859661823	.7134303145	.7378374606	.7596864693	.7793539334
10+	.7971345163	.8132652298	.8279409024	.8413244982	.8535542681
15+	.8647488681	.8750111242	.8844308691	.8930871231	.9010498037
20+	.9083810845	.9151364941	.9213658137	.9271138193	.9324209026
25+	.9373235937	.9418550059	.9460452161	.9499215935	.9535090840
30+	.9568304584	.9599065298	.9627563454	.9653973556	.9678455637
35+	.9701156603	.9722211419	.9741744179	.9759869067	.9776691218
40+	.9792307495	.9806807193	.9820272674	.9832779948	.9844399189
45+	.9855195219	.9865227938	.9874552718	.9883220770	.9891279467
50+	.9898772650	.9905740902	.9912221805	.9918250166	.9923858236
			A= .65		
0+	.4347826087	.4933158472	.5375894164	.5741675022	.6056523817
5+	.6333949814	.6582103224	.6806397628	.7010677649	.7197811127
10+	.7370018951	.7529072303	.7676417393	.7813257990	.7940612106
15+	.8059352175	.8170234313	.8273920119	.8370993263	.8461972334
20+	.8547320948	.8627455814	.8702753265	.8773554591	.8840170459
25+	.8902884591	.8961956872	.9017625986	.9070111664	.9119616631
30+	.9166328282	.9210420156	.9252053214	.9291376974	.9328530511
35+	.9363643348	.9396836252	.9428221938	.9457905714	.9485986055
40+	.9512555121	.9537699231	.9561499293	.9584031190	.9605366139
45+	.9625571020	.9644708669	.9662838160	.9680015050	.9696291618
50+	.9711717078	.9726337775	.9740197373	.9753337023	.9765795520

D-INTEGRAL PROBABILITY M=R,B= 2

R	1	2	3	4	5
			A= .65		
55+	.9777609452	.9788813335	.9799439741	.9809519415	.9819081384
60+	.9828153064	.9836760353	.9844927718	.9852678287	.9860033918
65+	.9867015278	.9873641914	.9879932308	.9885903950	.9891573384
70+	.9896956268	.9902067420	.9906920870	.9911529896	.9915907076
75+	.9920064318	.9924012902	.9927763513	.9931326272	.9934710772
			A= .70		
0+	.4166666667	.4647714120	.5014246874	.5319957183	.5585922321
5+	.5822970381	.6037562176	.6233940547	.6415084436	.6583191169
10+	.6739944153	.6886672365	.7024450797	.7154166621	.7276564409
15+	.7392278025	.7501853706	.7605767172	.7704436546	.7798232294
20+	.7887484998	.7972491503	.8053519858	.8130813317	.8204593629
25+	.8275063755	.8342410134	.8406804602	.8468406004	.8527361583
30+	.8583808167	.8637873195	.8689675611	.8739326648	.8786930505
35+	.8832584959	.8876381895	.8918407788	.8958744120	.8997467766
40+	.9034651337	.9070363486	.9104669187	.9137629992	.9169304257
45+	.9199747356	.9229011870	.9257147767	.9284202561	.9310221459
50+	.9335247503	.9359321695	.9382483113	.9404769019	.9426214966
55+	.9446854883	.9466721171	.9485844779	.9504255284	.9521980959
60+	.9539048842	.9555484797	.9571313575	.9586558867	.9601243357
65+	.9615388771	.9629015920	.9642144749	.9654794372	.9666983115
70+	.9678728551	.9690047534	.9700956231	.9711470155	.9721604194
75+	.9731372638	.9740789204	.9749867064	.9758618869	.9767056768
80+	.9775192431	.9783037072	.9790601466	.9797895968	.9804930530
85+	.9811714721	.9818257737	.9824568424	.9830655286	.9836526502
90+	.9842189939	.9847653164	.9852923458	.9858007826	.9862913007
95+	.9867645487	.9872211508	.9876617077	.9880867979	.9884969778
100+	.9888927835	.9892747310	.9896433170	.9899990200	.9903423007
105+	.9906736028	.9909933536	.9913019648	.9915998329	.9918873399
			A= .75		
0+	.4000000000	.4384000000	.4677760000	.4924241920	.5140199066
5+	.5334178786	.5511243374	.5674691230	.5826819257	.5969307441
10+	.6103431845	.6230191231	.6350386610	.6464673516	.6573597637
15+	.6677619875	.6777134431	.6872482164	.6963960651	.7051831896
20+	.7136328334	.7217657568	.7296006150	.7371542640	.7444420103
25+	.7514778161	.7582744705	.7648437322	.7711964502	.7773426664
30+	.7832917030	.7890522375	.7946323680	.8000396693	.8052812425
35+	.8103637578	.8152934927	.8200763655	.8247179650	.8292235770
40+	.8335982080	.8378466066	.8419732822	.8459825225	.8498784087
45+	.8536648300	.8573454960	.8609239485	.8644035722	.8677876043
50+	.8710791436	.8742811586	.8773964948	.8804278822	.8833779410
55+	.8862491885	.8890440436	.8917648327	.8944137941	.8969930824
60+	.8995047727	.9019508649	.9043332864	.9066538964	.9089144885
65+	.9111167938	.9132624839	.9153531734	.9173904223	.9193757384
70+	.9213105795	.9231963557	.9250344309	.9268261252	.9285727164
75+	.9302754413	.9319354983	.9335540479	.9351322145	.9366710880
80+	.9381717248	.9396351491	.9410623543	.9424543036	.9438119316
85+	.9451361453	.9464278247	.9476878239	.9489169723	.9501160749
90+	.9512859137	.9524272480	.9535408155	.9546273328	.9556874963
95+	.9567219826	.9577314494	.9587165358	.9596778634	.9606160363
100+	.9615316420	.9624252518	.9632974214	.9641486911	.9649795868
105+	.9657906198	.9665822879	.9673550752	.9681094529	.9688458797
110+	.9695648020	.9702666541	.9709518591	.9716208286	.9722739637
115+	.9729116547	.9735342816	.9741422147	.9747358144	.9753154319
120+	.9758814093	.9764340795	.9769737673	.9775007888	.9780154521
125+	.9785180572	.9790088967	.9794882555	.9799564114	.9804136351
130+	.9808601904	.9812963343	.9817223174	.9821383842	.9825447726
135+	.9829417149	.9833294374	.9837081607	.9840781000	.9844394651

D-INTEGRAL PROBABILITY M=R,B= 2

R	1	2	3	4	5
			A= .75		
140+	.9847924604	.9851372854	.9854741347	.9858031979	.9861246600
145+	.9864387014	.9867454980	.9870452216	.9873380394	.9876241147
150+	.9879036069	.9881766713	.9884434594	.9887041192	.9889587949
155+	.9892076273	.9894507536	.9896883079	.9899204208	.9901472200
160+	.9903688299	.9905853719	.9907969646	.9910037237	.9912057620
165+	.9914031897	.9915961142	.9917846406	.9919688712	.9921489059
			A= .80		
0+	.3846153846	.4140261195	.4365541420	.4555150342	.4721959746
5+	.4872507753	.5010648702	.5138881583	.5258937578	.5372076001
10+	.5479247970	.5581193554	.5678502728	.5771655358	.5861048422
15+	.5947015112	.6029838614	.6109762253	.6186997127	.6261727952
20+	.6334117594	.6404310643	.6472436266	.6538610498	.6602938117
25+	.6665514182	.6726425305	.6785750717	.6843563162	.6899929661
30+	.6954912150	.7008568041	.7060950696	.7112109838	.7162091915
35+	.7210940412	.7258696124	.7305397402	.7351080365	.7395779091
40+	.7439525787	.7482350939	.7524283447	.7565350747	.7605578920
45+	.7644992793	.7683616024	.7721471185	.7758579837	.7794962596
50+	.7830639193	.7865628533	.7899948741	.7933617215	.7966650667
55+	.7999065162	.8030876155	.8062098526	.8092746612	.8122834234
60+	.8152374725	.8181380958	.8209865364	.8237839959	.8265316362
65+	.8292305815	.8318819197	.8344867048	.8370459578	.8395606685
70+	.8420317969	.8444602743	.8468470046	.8491928656	.8514987098
75+	.8537653657	.8559936387	.8581843118	.8603381469	.8624558852
80+	.8645382480	.8665859380	.8685996391	.8705800179	.8725277237
85+	.8744433897	.8763276331	.8781810556	.8800042445	.8817977725
90+	.8835621988	.8852980689	.8870059157	.8886862594	.8903396082
95+	.8919664584	.8935672952	.8951425926	.8966928139	.8982184121
100+	.8997198303	.9011975015	.9026518494	.9040832887	.9054922247
105+	.9068790545	.9082441663	.9095879405	.9109107492	.9122129569
110+	.9134949204	.9147569893	.9159995059	.9172228056	.9184272170
115+	.9196130621	.9207806563	.9219303088	.9230623227	.9241769952
120+	.9252746174	.9263554750	.9274198480	.9284680110	.9295002333
125+	.9305167791	.9315179075	.9325038727	.9334749241	.9344313065
130+	.9353732599	.9363010200	.9372148181	.9381148810	.9390014317
135+	.9398746888	.9407348671	.9415821773	.9424168263	.9432390174
140+	.9440489502	.9448468207	.9456328212	.9464071408	.9471699652
145+	.9479214768	.9486618547	.9493912749	.9501099103	.9508179308
150+	.9515155034	.9522027921	.9528799580	.9535471595	.9542045524
155+	.9548522895	.9554905212	.9561193954	.9567390572	.9573496496
160+	.9579513128	.9585441849	.9591284015	.9597040961	.9602713998
165+	.9608304415	.9613813480	.9619242440	.9624592522	.9629864931
170+	.9635060853	.9640181453	.9645227880	.9650201260	.9655102705
175+	.9659933304	.9664694131	.9669386243	.9674010679	.9678568460
180+	.9683060591	.9687488063	.9691851847	.9696152902	.9700392169
185+	.9704570576	.9708689033	.9712748440	.9716749678	.9720693617
190+	.9724581112	.9728413004	.9732190121	.9735913280	.9739583281
195+	.9743200914	.9746766957	.9750282174	.9753747317	.9757163129
200+	.9760530337	.9763849660	.9767121805	.9770347468	.9773527332
205+	.9776662073	.9779752354	.9782798828	.9785802139	.9788762919
210+	.9791681793	.9794559374	.9797396266	.9800193065	.9802950357
215+	.9805668718	.9808348716	.9810990912	.9813595855	.9816164088
220+	.9818696146	.9821192554	.9823653830	.9826080483	.9828473018
225+	.9830831927	.9833157699	.9835450813	.9837711742	.9839940950
230+	.9842138897	.9844306033	.9846442804	.9848549648	.9850626995
235+	.9852675270	.9854694893	.9856686275	.9858649822	.9860585934
240+	.9862495006	.9864377424	.9866233573	.9868063827	.9869868559
245+	.9871648133	.9873402910	.9875133244	.9876839485	.9878521977

D-INTEGRAL PROBABILITY M=R, B= 2

R	1	2	3	4	5
			A= .80		
250+	.9880181059	.9881817065	.9883430324	.9885021161	.9886589894
255+	.9888136840	.9889662308	.9891166603	.9892650026	.9894112875
260+	.9895555441	.9896978012	.9898380873	.9899764303	.9901128576
265+	.9902473965	.9903800738	.9905109157	.9906399482	.9907671970
270+	.9908926873	.9910164438	.9911384911	.9912588533	.9913775542

D-INTEGRAL PROBABILITY M=R, B= 3

M=R	1	2	3	4	5
			A=1/10		
0+	.7692307692	.9343025589	.9809043911	.9943374246	.9982930918
			A=1/ 9		
0+	.7500000000	.9218750000	.9751180013	.9919322214	.9973448951
			A=1/ 8		
0+	.7272727273	.9056510050	.9667838420	.9881209081	.9956953872
			A=1/ 7		
0+	.7000000000	.8839600000	.9543700600	.9818125022	.9926684276
5+	.9970164822	.9987762275	.9994946924	.9997901866	.9999124703
			A=1/ 6		
0+	.6666666667	.8541380887	.9351379865	.9708217146	.9867511654
5+	.9939367607	.9972065480	.9987055897	.9993972374	.9997181176
			A=1/ 5		
0+	.6250000000	.8117675781	.9039416909	.9504863556	.9742867588
5+	.9865645085	.9929427841	.9962760467	.9980270088	.9989509631
			A=1/ 4		
0+	.5714285714	.7492116380	.8506021283	.9101780441	.9456689996
5+	.9669852568	.9798621651	.9876764379	.9924366456	.9953460892
			A=1/ 3		
0+	.5000000000	.6527777778	.7541473765	.8243032931	.8737527650
5+	.9089494921	.9341531432	.9522759246	.9653474861	.9747988272
10+	.9816465021	.9866164702	.9902292063	.9928590147	.9947757606
			A= .40		
0+	.4545454545	.5847526560	.6776642591	.7474444560	.8010419907
5+	.8427057839	.8753285194	.9009948327	.9212566815	.9372925955
10+	.9500090407	.9601092783	.9681422563	.9745383904	.9796363051
15+	.9837031181	.9869499913	.9895441481	.9916182140	.9932775106
			A=1/ 2		
0+	.4000000000	.4979200000	.5722931200	.6324384563	.6824748856
5+	.7247509163	.7608116769	.7917661491	.8184559573	.8415440518
10+	.8615662111	.8789633805	.8941032790	.9072955798	.9188030156
15+	.9288497734	.9376280091	.9453030138	.9520173797	.9578944082
20+	.9630409275	.9675496453	.9715011263	.9749654660	.9780037132
25+	.9806690853	.9830080090	.9850610152	.9868635093	.9884464371
30+	.9898368598	.9910584527	.9921319379	.9930754591	.9939049088
			A= .60		
0+	.3571428571	.4270913905	.4817549955	.5276672820	.5675368236
5+	.6028113697	.6343885584	.6628813655	.6887374989	.7123003602
10+	.7338431928	.7535896172	.7717266988	.7884136587	.8037879087
15+	.8179693735	.8310636769	.8431645487	.8543556838	.8647122077
20+	.8743018493	.8831858969	.8914199866	.8990547604	.9061364229
25+	.9127072156	.9188058250	.9244677370	.9297255458	.9346092253
30+	.9391463696	.9433624056	.9472807836	.9509231466	.9543094843
35+	.9574582699	.9603865853	.9631102335	.9656438416	.9680009536
40+	.9701941157	.9722349542	.9741342461	.9759019842	.9775474372
45+	.9790792042	.9805052648	.9818330257	.9830693634	.9842206630

D-INTEGRAL PROBABILITY M=R,B= 3

R	1	2	3	4	5
			A= .60		
50+	.9852928550	.9862914482	.9872215606	.9880879482	.9888950309
55+	.9896469170	.9903474257	.9910001078	.9916082652	.9921749683
			A= .65		
0+	.3389830508	.3967429844	.4421558404	.4806843239	.5145547402
5+	.5449345591	.5725327077	.5978229683	.6211441065	.6427507918
10+	.6628419826	.6815778945	.6990907202	.7154917123	.7308760400
15+	.7453262268	.7589146498	.7717054004	.7837556984	.7951169868
20+	.8058357938	.8159544205	.8255114990	.8345424485	.8430798544
25+	.8511537851	.8587920605	.8660204804	.8728630213	.8793420062
30+	.8854782536	.8912912075	.8967990523	.9020188151	.9069664564
35+	.9116569518	.9161043652	.9203219153	.9243220352	.9281164272
40+	.9317161127	.9351314775	.9383723138	.9414478584	.9443668282
45+	.9471374529	.9497675049	.9522643277	.9546348617	.9568856681
50+	.9590229519	.9610525823	.9629801126	.9648107983	.9665496142
55+	.9682012702	.9697702268	.9712607083	.9726767166	.9740220434
60+	.9753002816	.9765148364	.9776689355	.9787656389	.9798078477
65+	.9807983129	.9817396433	.9826343133	.9834846701	.9842929400
70+	.9850612355	.9857915607	.9864858176	.9871458108	.9877732533
75+	.9883697706	.9889369060	.9894761243	.9899888160	.9904763014
80+	.9909398340	.9913806041	.9917997417	.9921983203	.9925773590
			A= .70		
0+	.3225806452	.3692984120	.4061076151	.4375167425	.4653434267
5+	.4905295952	.5136395892	.5350450342	.5550075098	.5737204984
10+	.5913327129	.6079620051	.6237041255	.6386384900	.6528321193
15+	.6663424120	.6792191515	.6915059898	.7032415676	.7144603742
20+	.7251934190	.7354687625	.7453119417	.7547463150	.7637933450
25+	.7724728319	.7808031087	.7888012054	.7964829880	.8038632775
30+	.8109559520	.8177740347	.8243297714	.8306346975	.8366996969
35+	.8425350549	.8481505037	.8535552648	.8587580852	.8637672712
40+	.8685907182	.8732359381	.8777100837	.8820199713	.8861721014
45+	.8901726774	.8940276229	.8977425976	.9013230120	.9047740411
50+	.9081006370	.9113075405	.9143992918	.9173802415	.9202545589
55+	.9230262419	.9256991249	.9282768863	.9307630566	.9331610245
60+	.9354740441	.9377052404	.9398576154	.9419340539	.9439373278
65+	.9458701020	.9477349382	.9495342999	.9512705562	.9529459859
70+	.9545627815	.9561230523	.9576288281	.9590820628	.9604846366
75+	.9618383600	.9631449758	.9644061622	.9656235354	.9667986516
80+	.9679330102	.9690280550	.9700851772	.9711057171	.9720909661
85+	.9730421683	.9739605230	.9748471857	.9757032703	.9765298502
90+	.9773279603	.9780985981	.9788427253	.9795612690	.9802551232
95+	.9809251499	.9815721801	.9821970153	.9828004284	.9833831649
100+	.9839459436	.9844894579	.9850143767	.9855213450	.9860109851
105+	.9864838974	.9869406608	.9873818342	.9878079565	.9882195477
110+	.9886171097	.9890011264	.9893720652	.9897303766	.9900764958
115+	.9904108424	.9907338215	.9910458241	.9913472273	.9916383954
			A= .75		
0+	.3076923077	.3444395666	.3733636681	.3981007449	.4201065934
5+	.4401303551	.4586173525	.4758582639	.4920557167	.5073580255
10+	.5218779264	.5357037299	.5489063341	.5615438344	.5736646661
15+	.5853098133	.5965144033	.6073088826	.6177199025	.6277709977
20+	.6374831129	.6468750192	.6559636456	.6647643471	.6732911229
25+	.6815567962	.6895731635	.6973511192	.7049007610	.7122314785
30+	.7193520302	.7262706081	.7329948945	.7395321108	.7458890603
35+	.7520721654	.7580875009	.7639408223	.7696375923	.7751830034
40+	.7805819982	.7858392880	.7909593691	.7959465378	.8008049036
45+	.8055384015	.8101508033	.8146457271	.8190266471	.8232969018
50+	.8274597015	.8315181358	.8354751800	.8393337011	.8430964634

D-INTEGRAL PROBABILITY M=R,B= 3

R	1	2	3	4	5
			A= .75		
55+	.8467661338	.8503452865	.8538364078	.8572418996	.8605640841
60+	.8638052069	.8669674406	.8700528880	.8730635850	.8760015037
65+	.8788685547	.8816665898	.8843974042	.8870627391	.8896642834
70+	.8922036759	.8946825072	.8971023212	.8994646174	.9017708518
75+	.9040224393	.9062207541	.9083671321	.9104628716	.9125092348
80+	.9145074490	.9164587079	.9183641722	.9202249715	.9220422044
85+	.9238169403	.9255502196	.9272430552	.9288964330	.9305113130
90+	.9320886298	.9336292935	.9351341905	.9366041841	.9380401153
95+	.9394428035	.9408130467	.9421516227	.9434592893	.9447367853
100+	.9459848303	.9472041260	.9483953562	.9495591876	.9506962703
105+	.9518072378	.9528927080	.9539532834	.9549895515	.9560020852
110+	.9569914434	.9579581709	.9589027995	.9598258475	.9607278209
115+	.9616092130	.9624705052	.9633121673	.9641346573	.9649384225
120+	.9657238990	.9664915127	.9672416788	.9679748029	.9686912806
125+	.9693914980	.9700758320	.9707446503	.9713983122	.9720371679
130+	.9726615596	.9732718212	.9738682786	.9744512502	.9750210466
135+	.9755779711	.9761223199	.9766543819	.9771744396	.9776827685
140+	.9781796378	.9786653103	.9791400425	.9796040850	.9800576826
145+	.9805010741	.9809344929	.9813581669	.9817723186	.9821771654
150+	.9825729195	.9829597884	.9833379744	.9837076753	.9840690844
155+	.9844223903	.9847677773	.9851054253	.9854355104	.9857582042
160+	.9860736744	.9863820851	.9866835964	.9869783645	.9872665425
165+	.9875482794	.9878237212	.9880930102	.9883562858	.9886136839
170+	.9888653373	.9891113758	.9893519263	.9895871127	.9898170559
175+	.9900418744	.9902616836	.9904765965	.9906867234	.9908921722
180+	.9910930481	.9912894542	.9914814909	.9916692567	.9918528474
			A= .80		
0+	.2941176471	.3218810063	.3436568883	.3622706213	.3788481442
5+	.3939664694	.4079671013	.4210718855	.4334346533	.4451673557
10+	.4563545596	.4670620684	.4773423377	.4872380315	.4967844435
15+	.5060111982	.5149434766	.5236029195	.5320083076	.5401760806
20+	.5481207413	.5558551729	.5633908929	.5707382566	.5779066236
25+	.5849044945	.5917396236	.5984191136	.6049494949	.6113367926
30+	.6175865843	.6237040483	.6296940065	.6355609604	.6413091231
35+	.6469424470	.6524646480	.6578792266	.6631894876	.6683985558
40+	.6735093917	.6785248044	.6834474634	.6882799095	.6930245643
45+	.6976837389	.7022596416	.7067543850	.7111699925	.7155084044
50+	.7197714826	.7239610163	.7280787257	.7321262669	.7361052350
55+	.7400171679	.7438635494	.7476458124	.7513653413	.7550234748
60+	.7586215082	.7621606954	.7656422512	.7690673532	.7724371432
65+	.7757527295	.7790151877	.7822255629	.7853848703	.7884940973
70+	.7915542038	.7945661242	.7975307679	.8004490203	.8033217442
75+	.8061497802	.8089339478	.8116750460	.8143738544	.8170311336
80+	.8196476260	.8222240563	.8247611324	.8272595458	.8297199720
85+	.8321430714	.8345294894	.8368798573	.8391947923	.8414748984
90+	.8437207663	.8459329743	.8481120883	.8502586625	.8523732395
95+	.8544563506	.8565085164	.8585302467	.8605220413	.8624843897
100+	.8644177721	.8663226588	.8681995112	.8700487816	.8718709137
105+	.8736663425	.8754354949	.8771787896	.8788966376	.8805894419
110+	.8822575984	.8839014954	.8855215142	.8871180290	.8886914074
115+	.8902420103	.8917701920	.8932763005	.8947606777	.8962236593
120+	.8976655752	.8990867494	.9004875005	.9018681411	.9032289789
125+	.9045703159	.9058924490	.9071956702	.9084802664	.9097465196
130+	.9109947070	.9122251013	.9134379704	.9146335779	.9158121830
135+	.9169740405	.9181194010	.9192485110	.9203616130	.9214589453
140+	.9225407425	.9236072355	.9246586512	.9256952128	.9267171402
145+	.9277246494	.9287179531	.9296972608	.9306627782	.9316147081

D-INTEGRAL PROBABILITY M=R, B= 3

R	1	2	3	4	5
		A= .80			
150+	.9325532498	.9334785998	.9343909510	.9352904938	.9361774151
155+	.9370518991	.9379141273	.9387642780	.9396025269	.9404290469
160+	.9412440084	.9420475789	.9428399235	.9436212046	.9443915822
165+	.9451512138	.9459002544	.9466388569	.9473671715	.9480853464
170+	.9487935273	.9494918578	.9501804794	.9508595314	.9515291508
175+	.9521894728	.9528406304	.9534827546	.9541159745	.9547404173
180+	.9553562081	.9559634703	.9565623254	.9571528930	.9577352911
185+	.9583096357	.9588760413	.9594346206	.9599854845	.9605287426
190+	.9610645024	.9615928703	.9621139508	.9626278469	.9631346602
195+	.9636344907	.9641274370	.9646135961	.9650930639	.9655659346
200+	.9660323010	.9664922548	.9669458862	.9673932839	.9678345358
205+	.9682697279	.9686989455	.9691222723	.9695397910	.9699515829
210+	.9703577283	.9707583063	.9711533947	.9715430704	.9719274090
215+	.9723064851	.9726803723	.9730491430	.9734128685	.9737716194
220+	.9741254648	.9744744733	.9748187123	.9751582480	.9754931460
225+	.9758234709	.9761492861	.9764706545	.9767876379	.9771002970
230+	.9774086921	.9777128822	.9780129258	.9783088803	.9786008024
235+	.9788887480	.9791727723	.9794529295	.9797292731	.9800018560
240+	.9802707301	.9805359468	.9807975566	.9810556094	.9813101543
245+	.9815612397	.9818089135	.9820532226	.9822942136	.9825319322
250+	.9827664235	.9829977319	.9832259015	.9834509753	.9836729961
255+	.9838920058	.9841080459	.9843211572	.9845313801	.9847387542
260+	.9849433187	.9851451123	.9853441729	.9855405382	.9857342450
265+	.9859253300	.9861138291	.9862997777	.9864832109	.9866641632
270+	.9868426685	.9870187603	.9871924719	.9873638356	.9875328838
275+	.9876996480	.9878641596	.9880264494	.9881865478	.9883444846
280+	.9885002896	.9886539918	.9888056200	.9889552025	.9891027672
285+	.9892483417	.9893919532	.9895336284	.9896733939	.9898112755
290+	.9899472992	.9900814900	.9902138732	.9903444732	.9904733145
295+	.9906004209	.9907258161	.9908495234	.9909715658	.9910919659

D-INTEGRAL PROBABILITY M=R,B= 4

R	1	2	3	4	5
			A=1/10		
0+	.7142857143	.9149036541	.9748440523	.9924887058	.9977292159
			A=1/ 9		
0+	.6923076923	.8992916438	.9673224911	.9893177424	.9964713994
			A=1/ 8		
0+	.6666666667	.8791152263	.9565610532	.9843134983	.9942885337
5+	.9979027528	.9992239259	.9997108868	.9998916786	.9999592184
			A=1/ 7		
0+	.6363636364	.8524848989	.9406777793	.9760807111	.9902983085
5+	.9960387055	.9983721518	.9993271520	.9997204591	.9998833435
			A=1/ 6		
0+	.6000000000	.8164800000	.9163856160	.9618697579	.9825481260
5+	.9919753109	.9962922726	.9982789426	.9991977064	.9996245603
			A=1/ 5		
0+	.5555555556	.7664538491	.8777107423	.9359597168	.9664037783
5+	.9823284996	.9906766899	.9950657016	.9973805362	.9986053516
			A=1/ 4		
0+	.5000000000	.6948242187	.8134116530	.8858250161	.9300744956
5+	.9571255718	.9736751678	.9838113941	.9900282990	.9938474013
10+	.9961974741	.9976460802	.9985405472	.9990937769	.9994365026
			A=1/ 3		
0+	.4285714286	.5891167796	.7021134124	.7833334012	.8421652586
5+	.8849180388	.9160345087	.9387019191	.9552248206	.9672753259
10+	.9760686353	.9824886215	.9871784932	.9906065145	.9931137230
15+	.9949486046	.9962922978	.9972769163	.9979988728	.9985285687
			A= .40		
0+	.3846153846	.5176970955	.6178062225	.6956824319	.7570986163
5+	.8058494967	.8446798603	.8756691375	.9004303598	.9202307472
10+	.9360729519	.9487535801	.9589071063	.9670396470	.9735553250
15+	.9787770359	.9829628948	.9863193064	.9890113728	.9911711932
20+	.9929044870	.9942958821	.9954131402	.9963105312	.9970315312
			A=1/ 2		
0+	.3333333333	.4300411523	.5067301097	.5706088374	.6249885412
5+	.6718126029	.7124000319	.7477288697	.7785656544	.8055329298
10+	.8291482712	.8498488142	.8680077713	.8839462019	.8979417942
15+	.9102356648	.9210377756	.9305313504	.9388765366	.9462134826
20+	.9526649509	.9583385529	.9633286737	.9677181347	.9715796373
25+	.9749770183	.9779663448	.9805968696	.9829118671	.9849493647
30+	.9867427837	.9883215002	.9897113379	.9909350001	.9920124483
35+	.9929612351	.9937967963	.9945327073	.9951809084	.9957519025
			A= .60		
0+	.2941176471	.3613174550	.4156473584	.4623602837	.5036853641
5+	.5408226394	.5745203360	.6052931048	.6335204311	.6594967896
10+	.6834596186	.7056060491	.7261034827	.7450965980	.7627121749
15+	.7790625323	.7942480507	.8083590732	.8214773731	.8336773095
20+	.8450267572	.8555878659	.8654176909	.8745687244	.8830893473
25+	.8910242200	.8984146212	.9052987461	.9117119693	.9176870800
30+	.9232544911	.9284424286	.9332771022	.9377828600	.9419823297
35+	.9458965472	.9495450741	.9529461064	.9561165731	.9590722278
40+	.9618277332	.9643967384	.9667919511	.9690252041	.9711075162
45+	.9730491502	.9748596647	.9765479639	.9781223426	.9795905286
50+	.9809597220	.9822366315	.9834275081	.9845381773	.9855740674
55+	.9865402376	.9874414026	.9882819570	.9890659965	.9897973386
60+	.9904795417	.9911159229	.9917095740	.9922633774	.9927800200
			A= .65		
0+	.2777777778	.3326127459	.3770253874	.4154803101	.4498387175
5+	.4810817664	.5098061957	.5364113239	.5611832894	.5843379445
10+	.6060447419	.6264409814	.6456407798	.6637409751	.6808251568

D-INTEGRAL PROBABILITY M=R,B= 4

R	1	2	3	4	5
			A= .65		
15+	.6969665051	.7122298457	.7266731721	.7403487971	.7533042409
20+	.7655829265	.7772247348	.7882664512	.7987421321	.8086834078
25+	.8181197365	.8270786187	.8355857806	.8436653321	.8513399041
30+	.8586307685	.8655579444	.8721402918	.8783955958	.8843406418
35+	.8899912838	.8953625056	.9004684773	.9053226063	.9099375846
40+	.9143254320	.9184975361	.9224646890	.9262371221	.9298245374
45+	.9332361377	.9364806541	.9395663718	.9425011546	.9452924673
50+	.9479473977	.9504726756	.9528746929	.9551595203	.9573329246
55+	.9594003845	.9613671052	.9632380327	.9650178669	.9667110744
60+	.9683219003	.9698543792	.9713123463	.9726994470	.9740191468
65+	.9752747403	.9764693594	.9776059820	.9786874393	.9797164234
70+	.9806954939	.9816270846	.9825135102	.9833569714	.9841595611
75+	.9849232698	.9856499904	.9863415229	.9869995795	.9876257883
80+	.9882216982	.9887887819	.9893284405	.9898420064	.9903307472
85+	.9907958683	.9912385165	.9916597825	.9920607040	.9924422678
			A= .70		
0+	.2631578947	.3070329898	.3425005335	.3732954998	.4009587334
5+	.4262933008	.4497808955	.4717386130	.4923895078	.5118984995
10+	.5303923598	.5479716280	.5647181083	.5806997977	.5959742455
15+	.6105909141	.6245928811	.6380180956	.6509003233	.6632698696
20+	.6751541431	.6865780985	.6975645915	.7081346651	.7183077831
25+	.7281020234	.7375342379	.7466201875	.7553746556	.7638115451
30+	.7719439617	.7797842858	.7873442347	.7946349170	.8016668810
35+	.8084501565	.8149942927	.8213083916	.8274011379	.8332808263
40+	.8389553856	.8444324011	.8497191345	.8548225425	.8597492935
45+	.8645057833	.8690981493	.8735322835	.8778138451	.8819482717
50+	.8859407898	.8897964250	.8935200108	.8971161979	.9005894618
55+	.9039441106	.9071842924	.9103140019	.9133370871	.9162572550
60+	.9190780778	.9218029980	.9244353338	.9269782842	.9294349333
65+	.9318082550	.9341011175	.9363162869	.9384564314	.9405241250
70+	.9425218509	.9444520050	.9463168993	.9481187646	.9498597540
75+	.9515419453	.9531673442	.9547378863	.9562554404	.9577218103
80+	.9591387375	.9605079031	.9618309304	.9631093867	.9643447850
85+	.9655385867	.9666922026	.9678069952	.9688842803	.9699253285
90+	.9709313670	.9719035807	.9728431144	.9737510734	.9746285256
95+	.9754765020	.9762959989	.9770879783	.9778533695	.9785930704
100+	.9793079481	.9799988401	.9806665556	.9813118764	.9819355575
105+	.9825383284	.9831208939	.9836839346	.9842281084	.9847540506
110+	.9852623751	.9857536749	.9862285232	.9866874734	.9871310604
115+	.9875598010	.9879741944	.9883747230	.9887618529	.9891360345
120+	.9894977027	.9898472781	.9901851669	.9905117615	.9908274412
125+	.9911325724	.9914275092	.9917125938	.9919881567	.9922545174
			A= .75		
0+	.2500000000	.2841720581	.3116625398	.3355169728	.3569836042
5+	.3767091230	.3950787260	.4123438815	.4286797625	.4442144541
10+	.4590452046	.4732481098	.4868842053	.5000034700	.5126475546
15+	.5248516976	.5366461069	.5480569786	.5591072623	.5698172465
20+	.5802050134	.5902867968	.6000772671	.6095897610	.6188364688
25+	.6278285872	.6365764475	.6450896210	.6533770086	.6614469164
30+	.6693071199	.6769649195	.6844271874	.6917004094	.6987907203
35+	.7057039353	.7124455772	.7190209012	.7254349157	.7316924015
40+	.7377979291	.7437558733	.7495704272	.7552456144	.7607852998
45+	.7661931999	.7714728918	.7766278215	.7816613111	.7865765664
50+	.7913766826	.7960646503	.8006433611	.8051156123	.8094841115
55+	.8137514808	.8179202611	.8219929154	.8259718320	.8298593286
60+	.8336576543	.8373689928	.8409954653	.8445391325	.8480019972
65+	.8513860065	.8546930535	.8579249799	.8610835773	.8641705892

D-INTEGRAL PROBABILITY M=R, B= 4

R	1	2	3	4	5
			A= .75		
70+	.8671877127	.8701365999	.8730188596	.8758360584	.8785897226
75+	.8812813390	.8839123565	.8864841870	.8889982067	.8914557573
80+	.8938581466	.8962066501	.8985025116	.9007469439	.9029411304
85+	.9050862251	.9071833543	.9092336164	.9112380836	.9131978020
90+	.9151137926	.9169870520	.9188185528	.9206092445	.9223600541
95+	.9240718863	.9257456249	.9273821324	.9289822513	.9305468040
100+	.9320765939	.9335724056	.9350350052	.9364651410	.9378635441
105+	.9392309285	.9405679914	.9418754144	.9431538629	.9444039871
110+	.9456264224	.9468217893	.9479906943	.9491337298	.9502514747
115+	.9513444948	.9524133427	.9534585586	.9544806703	.9554801936
120+	.9564576325	.9574134797	.9583482165	.9592623135	.9601562305
125+	.9610304169	.9618853119	.9627213449	.9635389355	.9643384938
130+	.9651204207	.9658851081	.9666329389	.9673642877	.9680795203
135+	.9687789944	.9694630599	.9701320586	.9707863246	.9714261847
140+	.9720519581	.9726639571	.9732624869	.9738478458	.9744203256
145+	.9749802115	.9755277821	.9760633102	.9765870621	.9770992983
150+	.9776002737	.9780902371	.9785694322	.9790380968	.9794964638
155+	.9799447607	.9803832100	.9808120291	.9812314309	.9816416231
160+	.9820428092	.9824351880	.9828189538	.9831942967	.9835614025
165+	.9839204530	.9842716257	.9846150945	.9849510291	.9852795958
170+	.9856009569	.9859152712	.9862226940	.9865233771	.9868174691
175+	.9871051151	.9873864570	.9876616338	.9879307811	.9881940318
180+	.9884515155	.9887033593	.9889496872	.9891906206	.9894262783
185+	.9896567762	.9898822279	.9901027443	.9903184340	.9905294032
190+	.9907357555	.9909375926	.9911350136	.9913281155	.9915169934
			A= .80		
0+	.2380952381	.2636801657	.2841161104	.3017906244	.3176778616
5+	.3322808595	.3458982386	.3587241997	.3708934245	.3825038896
10+	.3936295549	.4043279201	.4146447774	.4246173332	.4342763359
15+	.4436475689	.4527529279	.4616112138	.4702387297	.4786497364
20+	.4868568067	.4948711040	.5027026039	.5103602730	.5178522145
25+	.5251857871	.5323677042	.5394041162	.5463006794	.5530626149
30+	.5596947582	.5662016020	.5725873325	.5788558615	.5850108535
35+	.5910557499	.5969937901	.6028280298	.6085613571	.6141965074
40+	.6197360759	.6251825292	.6305382151	.6358053722	.6409861381
45+	.6460825563	.6510965836	.6560300956	.6608848924	.6656627035
50+	.6703651926	.6749939615	.6795505539	.6840364592	.6884531151
55+	.6928019113	.6970841915	.7013012565	.7054543658	.7095447406
60+	.7135735651	.7175419887	.7214511275	.7253020660	.7290958586
65+	.7328335310	.7365160814	.7401444816	.7437196786	.7472425951
70+	.7507141309	.7541351636	.7575065495	.7608291248	.7641037058
75+	.7673310899	.7705120564	.7736473673	.7767377673	.7797839851
80+	.7827867334	.7857467099	.7886645976	.7915410650	.7943767670
85+	.7971723453	.7999284284	.8026456324	.8053245612	.8079658069
90+	.8105699503	.8131375608	.8156691972	.8181654076	.8206267301
95+	.8230536925	.8254468134	.8278066015	.8301335566	.8324281694
100+	.8346909220	.8369222878	.8391227320	.8412927118	.8434326763
105+	.8455430670	.8476243179	.8496768554	.8517010989	.8536974608
110+	.8556663465	.8576081546	.8595232774	.8614121004	.8632750031
115+	.8651123586	.8669245342	.8687118911	.8704747847	.8722135648
120+	.8739285757	.8756201561	.8772886394	.8789343539	.8805576226
125+	.8821587635	.8837380897	.8852959095	.8868325263	.8883482389
130+	.8898433416	.8913181241	.8927728718	.8942078656	.8956233824
135+	.8970196947	.8983970709	.8997557756	.9010960691	.9024182082
140+	.9037224455	.9050090301	.9062782073	.9075302189	.9087653029
145+	.9099836941	.9111856235	.9123713191	.9135410052	.9146949031
150+	.9158332307	.9169562028	.9180640312	.9191569244	.9202350880

D-INTEGRAL PROBABILITY M=R,B= 4

A= .80

R	1	2	3	4	5
155+	.9212987247	.9223480342	.9233832134	.9244044561	.9254119538
160+	.9264058948	.9273864650	.9283538475	.9293082229	.9302497691
165+	.9311786615	.9320950731	.9329991744	.9338911335	.9347711160
170+	.9356392853	.9364958026	.9373408265	.9381745136	.9389970182
175+	.9398084926	.9406090868	.9413989487	.9421782242	.9429470571
180+	.9437055892	.9444539604	.9451923086	.9459207697	.9466394777
185+	.9473485650	.9480481619	.9487383968	.9494193967	.9500912865
190+	.9507541895	.9514082272	.9520535196	.9526901849	.9533183397
195+	.9539380990	.9545495761	.9551528830	.9557481299	.9563354256
200+	.9569148775	.9574865912	.9580506711	.9586072202	.9591563400
205+	.9596981306	.9602326907	.9607601177	.9612805077	.9617939554
210+	.9623005542	.9628003963	.9632935726	.9637801728	.9642602853
215+	.9647339972	.9652013948	.9656625627	.9661175848	.9665665436
220+	.9670095205	.9674465958	.9678778489	.9683033577	.9687231996
225+	.9691374504	.9695461852	.9699494780	.9703474019	.9707400286
230+	.9711274294	.9715096743	.9718868323	.9722589717	.9726261596
235+	.9729884625	.9733459457	.9736986738	.9740467104	.9743901185
240+	.9747289599	.9750632958	.9753931864	.9757186913	.9760398691
245+	.9763567778	.9766694744	.9769780152	.9772824559	.9775828512
250+	.9778792552	.9781717213	.9784603022	.9787450496	.9790260150
255+	.9793032487	.9795768007	.9798467201	.9801130554	.9803758545
260+	.9806351647	.9808910325	.9811435038	.9813926240	.9816384379
265+	.9818809895	.9821203224	.9823564795	.9825895033	.9828194354
270+	.9830463172	.9832701892	.9834910918	.9837090644	.9839241461
275+	.9841363755	.9843457906	.9845524289	.9847563274	.9849575227
280+	.9851560507	.9853519469	.9855452466	.9857359841	.9859241938
285+	.9861099092	.9862931636	.9864739897	.9866524199	.9868284860
290+	.9870022197	.9871736518	.9873428131	.9875097337	.9876744436
295+	.9878369720	.9879973481	.9881556005	.9883117574	.9884658468
300+	.9886178961	.9887679324	.9889159826	.9890620730	.9892062297
305+	.9893484785	.9894888447	.9896273533	.9897640290	.9898988962
310+	.9900319790	.9901633010	.9902928856	.9904207560	.9905469349
315+	.9906714449	.9907943080	.9909155461	.9910351809	.9911532337

D-INTEGRAL PROBABILITY M=R,B= 5

R	1	2	3	4	5
			A=1/10		
0+	.6666666667	.8965530864	.9689235396	.9906585475	.9971678278
			A=1/ 9		
0+	.6428571429	.8781180825	.9597510809	.9867385431	.9956034887
			A=1/ 8		
0+	.6153846154	.8545124979	.9467099846	.9805765572	.9928950561
5+	.9973849388	.9990311509	.9996388491	.9998646453	.9999490323
			A=1/ 7		
0+	.5833333333	.8237164030	.9276259405	.9704977191	.9879626602
5+	.9950689372	.9979699462	.9991600510	.9996508356	.9998542415
			A=1/ 6		
0+	.5454545455	.7826952409	.8987828962	.9532518360	.9784414944
5+	.9900416648	.9953860317	.9978546250	.9989988536	.9995312012
			A=1/ 5		
0+	.5000000000	.7268000000	.8536319300	.9222329190	.9588152266
5+	.9782004498	.9884501408	.9938698499	.9967393833	.9982616967
			A=1/ 4		
0+	.4444444444	.6488321375	.7803989607	.8635052336	.9154587455
5+	.9477342167	.9677115138	.9800526714	.9876705261	.9923727668
10+	.9952767712	.9970717374	.9981823772	.9988703968	.9992971380
			A=1/ 3		
0+	.3750000000	.5378494263	.6582804290	.7476895649	.8140017544
5+	.8630738949	.8993072303	.9260103265	.9456605959	.9601046902
10+	.9707134861	.9785012321	.9842163132	.9884097971	.9914868252
15+	.9937449148	.9954023749	.9966193104	.9975131053	.9981698124
			A= .40		
0+	.3333333333	.4654288066	.5692279564	.6523914497	.7194616077
5+	.7736615779	.8174733345	.8528740142	.8814603001	.9045281563
10+	.9231308067	.9381238838	.9502017681	.9599272585	.9677559179
15+	.9740560624	.9791251484	.9832031796	.9864836474	.9891224398
20+	.9912450790	.9929525901	.9943262534	.9954314447	.9963207367
			A=1/ 2		
0+	.2857142857	.3792248856	.4560290271	.5215765185	.5784495481
5+	.6281984229	.6719051040	.7103973202	.7443446615	.7743075856
10+	.8007651815	.8241323948	.8447715988	.8630009144	.8791005381
15+	.8933177720	.9058711595	.9169539682	.9267371757	.9353720606
20+	.9429924696	.9497168116	.9556498196	.9608841097	.9655015635
25+	.9695745545	.9731670365	.9763355096	.9791298774	.9815942078
30+	.9837674074	.9856838197	.9873737546	.9888639586	.9901780296
35+	.9913367854	.9923585888	.9932596355	.9940542087	.9947549036
			A= .60		
0+	.2500000000	.3136515625	.3665432937	.4128901931	.4545149978
5+	.4924007630	.5271605896	.5592172105	.5888840604	.6164064340
10+	.6419842824	.6657857220	.6879554871	.7086204636	.7278934550
15+	.7458758348	.7626594723	.7783281703	.7929587672	.8066219999
20+	.8193831973	.8313028460	.8424370624	.8528379915	.8625541506
25+	.8716307276	.8801098439	.8880307882	.8954302258	.9023423882
30+	.9087992446	.9148306594	.9204645356	.9257269481	.9306422657
35+	.9352332649	.9395212344	.9435260734	.9472663826	.9507595485
40+	.9540218226	.9570683956	.9599134656	.9625703029	.9650513101
45+	.9673680784	.9695314399	.9715515173	.9734377697	.9751990355
50+	.9768435731	.9783790983	.9798128199	.9811514724	.9824013471
55+	.9835683210	.9846578838	.9856751633	.9866249489	.9875117139
60+	.9883396363	.9891126179	.9898343025	.9905080931	.9911371674
65+	.9917244927	.9922728401	.9927847968	.9932627788	.9937090417
			A= .65		
0+	.2352941176	.2867815908	.3294923689	.3670853915	.4011163233
5+	.4324069459	.4614561195	.4885968831	.5140675420	.5380478802

D-INTEGRAL PROBABILITY M=R,B= 5

R	1	2	3	4	5
			A= .65		
10+	.5606792899	.5820767350	.6023362503	.6215398487	.6397588502
15+	.6570562111	.6734881966	.6891056101	.7039547165	.7180779479
20+	.7315144523	.7443005278	.7564699703	.7680543567	.7790832776
25+	.7895845316	.7995842893	.8091072328	.8181766758	.8268146685
30+	.8350420887	.8428787238	.8503433420	.8574537580	.8642268914
35+	.8706788194	.8768248263	.8826794474	.8882565106	.8935691745
40+	.8986299634	.9034508008	.9080430399	.9124174925	.9165844561
45+	.9205537397	.9243346874	.9279362011	.9313667625	.9346344526
50+	.9377469714	.9407116559	.9435354976	.9462251586	.9487869873
55+	.9512270334	.9535510616	.9557645654	.9578727795	.9598806923
60+	.9617930571	.9636144034	.9653490471	.9670011008	.9685744832
65+	.9700729282	.9714999934	.9728590685	.9741533832	.9753860146
70+	.9765598945	.9776778160	.9787424400	.9797563017	.9807218163
75+	.9816412844	.9825168979	.9833507447	.9841448137	.9849009996
80+	.9856211072	.9863068554	.9869598819	.9875817463	.9881739343
85+	.9887378609	.9892748739	.9897862569	.9902732326	.9907369654
90+	.9911785644	.9915990858	.9919995355	.9923808717	.9927440065
			A= .70		
0+	.2222222222	.2630949377	.2968245109	.3265219347	.3534976680
5+	.3784373962	.4017521113	.4237117317	.4445055938	.4642733121
10+	.4831219691	.5011363569	.5183854129	.5349264413	.5508079842
15+	.5660718328	.5807544726	.5948881466	.6085016513	.6216209441
20+	.6342696128	.6464692453	.6582397219	.6695994515	.6805655633
25+	.6911540630	.7013799638	.7112573940	.7207996891	.7300194686
30+	.7389287026	.7475387680	.7558604984	.7639042262	.7716798206
35+	.7791967204	.7864639631	.7934902113	.8002837754	.8068526353
40+	.8132044587	.8193466182	.8252862076	.8310300555	.8365847385
45+	.8419565938	.8471517296	.8521760362	.8570351951	.8617346883
50+	.8662798065	.8706756572	.8749271720	.8790391135	.8830160822
55+	.8868625223	.8905827281	.8941808492	.8976608961	.9010267452
60+	.9042821436	.9074307139	.9104759583	.9134212632	.9162699030
65+	.9190250440	.9216897483	.9242669770	.9267595943	.9291703700
70+	.9315019833	.9337570254	.9359380028	.9380473400	.9400873821
75+	.9420603977	.9439685810	.9458140549	.9475988727	.9493250208
80+	.9509944207	.9526089313	.9541703507	.9556804186	.9571408176
85+	.9585531757	.9599190676	.9612400168	.9625174968	.9637529333
90+	.9649477054	.9661031472	.9672205492	.9683011597	.9693461865
95+	.9703567976	.9713341233	.9722792564	.9731932546	.9740771406
100+	.9749319037	.9757585012	.9765578587	.9773308717	.9780784064
105+	.9788013006	.9795003647	.9801763826	.9808301125	.9814622878
110+	.9820736178	.9826647885	.9832364637	.9837892852	.9843238736
115+	.9848408296	.9853407338	.9858241479	.9862916152	.9867436611
120+	.9871807937	.9876035044	.9880122688	.9884075463	.9887897817
125+	.9891594049	.9895168318	.9898624646	.9901966921	.9905198907
130+	.9908324239	.9911346438	.9914268904	.9917094928	.9919827693
			A= .75		
0+	.2105263158	.2421327966	.2680102979	.2907273910	.3113597534
5+	.3304678361	.3483858340	.3653317457	.3814571717	.3968727247
10+	.4116621925	.4258909861	.4396114534	.4528663717	.4656913250
15+	.4781163731	.4901672526	.5018662623	.5132329267	.5242845029
20+	.5350363741	.5455023573	.5556949499	.5656255259	.5753044970
25+	.5847414433	.5939452220	.6029240577	.6116856180	.6202370777
30+	.6285851725	.6367362461	.6446962895	.6524709757	.6600656891
35+	.6674855521	.6747354472	.6818200374	.6887437835	.6955109600
40+	.7021256688	.7085918515	.7149133007	.7210936695	.7271364810
45+	.7330451356	.7388229191	.7444730089	.7499984800	.7554023105
50+	.7606873869	.7658565085	.7709123918	.7758576742	.7806949176

D-INTEGRAL PROBABILITY M=R,B= 5

R	1	2	3	4	5
			A= .75		
55+	.7854266123	.7900551792	.7945829735	.7990122869	.8033453506
60+	.8075843369	.8117313620	.8157884880	.8197577247	.8236410314
65+	.8274403186	.8311574501	.8347942437	.8383524735	.8418338708
70+	.8452401254	.8485728871	.8518337670	.8550243380	.8581461368
75+	.8612006641	.8641893862	.8671137356	.8699751121	.8727748837
80+	.8755143871	.8781949292	.8808177871	.8833842094	.8858954169
85+	.8883526030	.8907569345	.8931095524	.8954115725	.8976640858
90+	.8998681595	.9020248370	.9041351392	.9062000642	.9082205887
95+	.9101976677	.9121322356	.9140252064	.9158774741	.9176899136
100+	.9194633804	.9211987119	.9228967270	.9245582270	.9261839961
105+	.9277748013	.9293313932	.9308545061	.9323448586	.9338031538
110+	.9352300796	.9366263092	.9379925013	.9393293005	.9406373374
115+	.9419172291	.9431695798	.9443949802	.9455940086	.9467672311
120+	.9479152014	.9490384613	.9501375411	.9512129599	.9522652254
125+	.9532948346	.9543022737	.9552880188	.9562525355	.9571962795
130+	.9581196969	.9590232242	.9599072883	.9607723074	.9616186904
135+	.9624468377	.9632571409	.9640499833	.9648257402	.9655847785
140+	.9663274577	.9670541291	.9677651369	.9684608178	.9691415011
145+	.9698075094	.9704591581	.9710967560	.9717206052	.9723310015
150+	.9729282341	.9735125863	.9740843350	.9746437516	.9751911013
155+	.9757266437	.9762506330	.9767633178	.9772649415	.9777557421
160+	.9782359527	.9787058013	.9791655109	.9796152999	.9800553820
165+	.9804859663	.9809072572	.9813194551	.9817227557	.9821173509
170+	.9825034282	.9828811711	.9832507593	.9836123684	.9839661706
175+	.9843123340	.9846510234	.9849823999	.9853066212	.9856238415
180+	.9859342118	.9862378799	.9865349903	.9868256845	.9871101007
185+	.9873883746	.9876606386	.9879270223	.9881876526	.9884426537
190+	.9886921471	.9889362516	.9891750834	.9894087565	.9896373822
195+	.9898610693	.9900799246	.9902940523	.9905035544	.9907085310
200+	.9909090796	.9911052959	.9912972734	.9914851037	.9916688763
			A= .80		
0+	.2000000000	.2235101389	.2425673185	.2592049013	.2742707074
5+	.2882055180	.3012716450	.3136397894	.3254283030	.3367232135
10+	.3475893869	.3580771850	.3682266563	.3780702901	.3876348935
15+	.3969429092	.4060133655	.4148625761	.4235046660	.4319519730
20+	.4402153601	.4483044597	.4562278697	.4639933099	.4716077501
25+	.4790775142	.4864083676	.4936055885	.5006740293	.5076181670
30+	.5144421471	.5211498208	.5277447766	.5342303681	.5406097379
35+	.5468858383	.5530614495	.5591391957	.5651215592	.5710108926
40+	.5768094302	.5825192975	.5881425203	.5936810321	.5991366817
45+	.6045112391	.6098064012	.6150237975	.6201649939	.6252314978
50+	.6302247616	.6351461858	.6399971231	.6447788807	.6494927230
55+	.6541398743	.6587215213	.6632388144	.6676928706	.6720847745
60+	.6764155803	.6806863136	.6848979721	.6890515276	.6931479270
65+	.6971880932	.7011729266	.7051033057	.7089800884	.7128041124
70+	.7165761966	.7202971414	.7239677295	.7275887269	.7311608831
75+	.7346849319	.7381615921	.7415915678	.7449755490	.7483142120
80+	.7516082200	.7548582237	.7580648609	.7612287581	.7643505297
85+	.7674307791	.7704700987	.7734690705	.7764282659	.7793482465
90+	.7822295643	.7850727615	.7878783713	.7906469179	.7933789166
95+	.7960748745	.7987352900	.8013606537	.8039514481	.8065081479
100+	.8090312205	.8115211257	.8139783161	.8164032374	.8187963282
105+	.8211580205	.8234887396	.8257889044	.8280589273	.8302992147
110+	.8325101669	.8346921780	.8368456366	.8389709253	.8410684212
115+	.8431384958	.8451815153	.8471978405	.8491878272	.8511518257
120+	.8530901817	.8550032356	.8568913232	.8587547754	.8605939186
125+	.8624090744	.8642005599	.8659686879	.8677137668	.8694361006

D-INTEGRAL PROBABILITY M=R, B= 5

R	1	2	3	4	5
			A= .80		
130+	.8711359891	.8728137280	.8744696090	.8761039195	.8777169432
135+	.8793089598	.8808802453	.8824310717	.8839617075	.8854724173
140+	.8869634625	.8884351005	.8898875855	.8913211682	.8927360960
145+	.8941326127	.8955109592	.8968713728	.8982140878	.8995393354
150+	.9008473437	.9021383375	.9034125390	.9046701672	.9059114381
155+	.9071365651	.9083457585	.9095392260	.9107171725	.9118798001
160+	.9130273084	.9141598940	.9152777514	.9163810722	.9174700455
165+	.9185448580	.9196056940	.9206527352	.9216861609	.9227061484
170+	.9237128721	.9247065047	.9256872162	.9266551746	.9276105456
175+	.9285534928	.9294841775	.9304027592	.9313093949	.9322042400
180+	.9330874476	.9339591687	.9348195527	.9356687466	.9365068958
185+	.9373341438	.9381506321	.9389565004	.9397518865	.9405369266
190+	.9413117550	.9420765043	.9428313052	.9435762870	.9443115772
195+	.9450373015	.9457535840	.9464605475	.9471583128	.9478469994
200+	.9485267251	.9491976062	.9498597575	.9505132924	.9511583227
205+	.9517949588	.9524233097	.9530434829	.9536555846	.9542597197
210+	.9548559914	.9554445019	.9560253520	.9565986410	.9571644673
215+	.9577229277	.9582741177	.9588181319	.9593550634	.9598850041
220+	.9604080449	.9609242754	.9614337839	.9619366579	.9624329834
225+	.9629228455	.9634063281	.9638835142	.9643544855	.9648193228
230+	.9652781056	.9657309128	.9661778218	.9666189094	.9670542512
235+	.9674839218	.9679079949	.9683265433	.9687396388	.9691473521
240+	.9695497533	.9699469115	.9703388946	.9707257701	.9711076043
245+	.9714844627	.9718564101	.9722235102	.9725858261	.9729434200
250+	.9732963532	.9736446865	.9739884795	.9743277914	.9746626804
255+	.9749932041	.9753194192	.9756413817	.9759591470	.9762727697
260+	.9765823036	.9768878020	.9771893173	.9774869014	.9777806054
265+	.9780704798	.9783565745	.9786389385	.9789176206	.9791926685
270+	.9794641296	.9797320506	.9799964776	.9802574560	.9805150307
275+	.9807692460	.9810201457	.9812677729	.9815121703	.9817533798
280+	.9819914431	.9822264010	.9824582940	.9826871621	.9829130445
285+	.9831359803	.9833560077	.9835731647	.9837874887	.9839990165
290+	.9842077845	.9844138288	.9846171849	.9848178876	.9850159717
295+	.9852114712	.9854044197	.9855948507	.9857827968	.9859682904
300+	.9861513635	.9863320476	.9865103740	.9866863732	.9868600757
305+	.9870315113	.9872007097	.9873677000	.9875325109	.9876951710
310+	.9878557081	.9880141501	.9881705242	.9883248574	.9884771763
315+	.9886275071	.9887758759	.9889223081	.9890668290	.9892094636
320+	.9893502365	.9894891719	.9896262938	.9897616258	.9898951914
325+	.9900270135	.9901571149	.9902855180	.9904122451	.9905373179
330+	.9906607580	.9907825867	.9909028252	.9910214940	.9911386137

D-INTEGRAL PROBABILITY M=R,B= 6

R	1	2	3	4	5
			A=1/10		
0+	.6250000000	.8791547269	.9631361960	.9888464801	.9966088937
			A=1/ 9		
0+	.6000000000	.8582072889	.9523908419	.9841934713	.9947410586
			A=1/ 8		
0+	.5714285714	.8316117070	.9372043611	.9769070711	.9915146044
5+	.9968696391	.9988388632	.9995669064	.9998376306	.9999388498
			A=1/ 7		
0+	.5384615385	.7972802277	.9151584068	.9650550659	.9856601964
5+	.9941069797	.9975695802	.9989933846	.9995813151	.9998251641
			A=1/ 6		
0+	.5000000000	.7521580826	.8822032237	.9449422959	.9744260455
5+	.9881347513	.9944876045	.9974325911	.9988006693	.9994380383
			A=1/ 5		
0+	.4545454545	.6917258306	.8314032809	.9092216340	.9514974991
5+	.9741737298	.9862612679	.9926879629	.9961034000	.9979199556
			A=1/ 4		
0+	.4000000000	.6093184000	.7508059362	.8429196241	.9017037267
5+	.9387645605	.9619527509	.9763929559	.9853604228	.9909210306
10+	.9943670519	.9965026590	.9978267296	.9986482253	.9991583528
			A=1/ 3		
0+	.3333333333	.4955413677	.6206940666	.7162625981	.7886331346
5+	.8430602275	.8837688589	.9140862420	.9365895647	.9532498522
10+	.9655599845	.9746423195	.9813358477	.9862649805	.9898928561
15+	.9925620988	.9945256752	.9959700768	.9970326369	.9978144091
			A= .40		
0+	.2941176471	.4234072715	.5288315584	.6154627863	.6866966114
5+	.7451665994	.7930456173	.8321583430	.8640396600	.8899760803
10+	.9110405735	.9281235299	.9419604468	.9531564297	.9622075796
15+	.9695194205	.9754225974	.9801861224	.9840284638	.9871267690
20+	.9896244922	.9916376756	.9932601013	.9945675029	.9956210007
			A=1/ 2		
0+	.2500000000	.3396415710	.4154692422	.4815255002	.5397750667
5+	.5914184902	.6373161520	.6781477805	.7144818029	.7468092168
10+	.7755619165	.8011236323	.8238371231	.8440093503	.8619154988
15+	.8778022921	.8918908415	.9043791623	.9154444312	.9252450296
20+	.9339224026	.9416027513	.9483985739	.9544100674	.9597264011
25+	.9644268703	.9685819401	.9722541882	.9754991545	.9783661047
30+	.9808987174	.9831356993	.9851113363	.9868559863	.9883965199
35+	.9897567125	.9909575952	.9920177657	.9929536662	.9937798298
			A= .60		
0+	.2173913043	.2774344506	.3284815308	.3739270530	.4152623537
5+	.4532896090	.4885068265	.5212566858	.5517937514	.5803184207
10+	.6069955576	.6319654029	.6553503015	.6772590330	.6977897060
15+	.7170317571	.7350673746	.7519725422	.7678178249	.7826689780
20+	.7965874293	.8096306719	.8218525908	.8333037403	.8440315840
25+	.8540807061	.8634929999	.8723078381	.8805622279	.8882909542
30+	.8955267118	.9023002282	.9086403787	.9145742943	.9201274630
35+	.9253238251	.9301858629	.9347346859	.9389901103	.9429707351
40+	.9466940136	.9501763202	.9534330151	.9564785038	.9593262943
45+	.9619890507	.9644786441	.9668062002	.9689821445	.9710162450
50+	.9729176522	.9746949367	.9763561252	.9779087338	.9793597996
55+	.9807159106	.9819832332	.9831675391	.9842742293	.9853083581
60+	.9862746543	.9871775421	.9880211601	.9888093797	.9895458215
65+	.9902338720	.9908766978	.9914772602	.9920383280	.9925624899
			A= .65		
0+	.2040816327	.2523232218	.2931464056	.3295725991	.3629095744
5+	.3938479711	.4228050790	.4500579710	.4758039962	.5001916006

D-INTEGRAL PROBABILITY M=R, B= 6

R	1	2	3	4	5
			A= .65		
10+	.5233374187	.5453363928	.5662680793	.5862007486	.6051941454
15+	.6233014011	.6405703951	.6570447437	.6727645351	.6877668835
20+	.7020863559	.7157553046	.7288041310	.7412614977	.7531545007
25+	.7645088123	.7753488001	.7856976281	.7955773432	.8050089502
30+	.8140124773	.8226070345	.8308108657	.8386413952	.8461152705
35+	.8532484017	.8600559965	.8665525947	.8727520981	.8786678005
40+	.8843124145	.8896980973	.8948364750	.8997386661	.9044153028
45+	.9088765523	.9131321364	.9171913508	.9210630829	.9247558295
50+	.9282777131	.9316364977	.9348396047	.9378941264	.9408068412
55+	.9435842259	.9462324694	.9487574846	.9511649204	.9534601728
60+	.9556483962	.9577345135	.9597232262	.9616190241	.9634261945
65+	.9651488308	.9667908412	.9683559568	.9698477392	.9712695882
70+	.9726247487	.9739163177	.9751472509	.9763203686	.9774383622
75+	.9785037998	.9795191315	.9804866948	.9814087198	.9822873339
80+	.9831245664	.9839223527	.9846825392	.9854068867	.9860970744
85+	.9867547038	.9873813022	.9879783258	.9885471633	.9890891388
90+	.9896055148	.9900974949	.9905662267	.9910128042	.9914382704
95+	.9918436196	.9922297994	.9925977133	.9929482226	.9932821480
			A= .70		
0+	.1923076923	.2303720989	.2623287073	.2907933287	.3168895047
5+	.3412067542	.3640977263	.3857931396	.4064541105	.4261989092
10+	.4451178714	.4632822762	.4807499158	.4975687422	.5137793391
15+	.5294166485	.5445112071	.5590900508	.5731773903	.5867951230
20+	.5999632266	.6127000676	.6250226434	.6369467771	.6484872732
25+	.6596580458	.6704722231	.6809422340	.6910798808	.7008964000
30+	.7104025143	.7196084761	.7285241061	.7371588253	.7455216841
35+	.7536213869	.7614663139	.7690645412	.7764238572	.7835517791
40+	.7904555660	.7971422322	.8036185583	.8098911019	.8159662072
45+	.8218500141	.8275484662	.8330673183	.8384121444	.8435883434
50+	.8486011459	.8534556204	.8581566783	.8627090797	.8671174383
55+	.8713862261	.8755197784	.8795222976	.8833978580	.8871504094
60+	.8907837812	.8943016860	.8977077233	.9010053830	.9041980485
65+	.9072890002	.9102814184	.9131783863	.9159828933	.9186978373
70+	.9213260275	.9238701875	.9263329573	.9287168962	.9310244850
75+	.9332581283	.9354201568	.9375128297	.9395383366	.9414987996
80+	.9433962754	.9452327572	.9470101765	.9487304053	.9503952573
85+	.9520064902	.9535658071	.9550748582	.9565352423	.9579485085
90+	.9593161576	.9606396436	.9619203751	.9631597166	.9643589899
95+	.9655194756	.9666424139	.9677290062	.9687804162	.9697977709
100+	.9707821619	.9717346464	.9726562482	.9735479590	.9744107390
105+	.9752455181	.9760531967	.9768346469	.9775907129	.9783222123
110+	.9790299367	.9797146525	.9803771019	.9810180034	.9816380525
115+	.9822379227	.9828182660	.9833797135	.9839228763	.9844483458
120+	.9849566948	.9854484774	.9859242301	.9863844724	.9868297070
125+	.9872604203	.9876770835	.9880801522	.9884700678	.9888472573
130+	.9892121339	.9895650977	.9899065357	.9902368226	.9905563209
135+	.9908653816	.9911643442	.9914535373	.9917332790	.9920038768
			A= .75		
0+	.1818181818	.2110923083	.2354141416	.2569724077	.2767024261
5+	.2950939483	.3124391644	.3289284111	.3446937425	.3598312334
10+	.3744134317	.3884967901	.4021263422	.4153387734	.4281645111
15+	.4406291900	.4527547056	.4645599893	.4760615890	.4872741113
20+	.4982105641	.5088826250	.5193008539	.5294748638	.5394134593
25+	.5491247495	.5586162409	.5678949149	.5769672923	.5858394877
30+	.5945172553	.6030060284	.6113109524	.6194369142	.6273885666
35+	.6351703500	.6427865115	.6502411209	.6575380856	.6646811630
40+	.6716739720	.6785200027	.6852226260	.6917851005	.6982105807

D-INTEGRAL PROBABILITY M=R, B= 6

R	1	2	3	4	5
			A= .75		
45+	.7045021227	.7106626902	.7166951598	.7226023257	.7283869039
50+	.7340515361	.7395987936	.7450311801	.7503511353	.7555610371
55+	.7606632044	.7656598996	.7705533306	.7753456531	.7800389720
60+	.7846353439	.7891367780	.7935452383	.7978626446	.8020908742
65+	.8062317630	.8102871067	.8142586623	.8181481487	.8219572482
70+	.8256876073	.8293408374	.8329185163	.8364221886	.8398533666
75+	.8432135311	.8465041324	.8497265908	.8528822974	.8559726145
80+	.8589988767	.8619623915	.8648644394	.8677062750	.8704891275
85+	.8732142010	.8758826754	.8784957065	.8810544270	.8835599465
90+	.8860133524	.8884157100	.8907680635	.8930714356	.8953268287
95+	.8975352251	.8996975872	.9018148581	.9038879618	.9059178041
100+	.9079052722	.9098512358	.9117565467	.9136220400	.9154485339
105+	.9172368299	.9189877137	.9207019550	.9223803081	.9240235121
110+	.9256322911	.9272073547	.9287493984	.9302591032	.9317371368
115+	.9331841532	.9346007933	.9359876850	.9373454435	.9386746716
120+	.9399759598	.9412498867	.9424970193	.9437179130	.9449131118
125+	.9460831490	.9472285467	.9483498167	.9494474602	.9505219682
130+	.9515738219	.9526034925	.9536114417	.9545981217	.9555639757
135+	.9565094375	.9574349323	.9583408766	.9592276783	.9600957369
140+	.9609454438	.9617771825	.9625913284	.9633882494	.9641683057
145+	.9649318502	.9656792285	.9664107791	.9671268337	.9678277168
150+	.9685137467	.9691852347	.9698424859	.9704857990	.9711154668
155+	.9717317758	.9723350066	.9729254340	.9735033273	.9740689500
160+	.9746225602	.9751644108	.9756947492	.9762138180	.9767218544
165+	.9772190909	.9777057552	.9781820702	.9786482541	.9791045207
170+	.9795510792	.9799881346	.9804158874	.9808345343	.9812442676
175+	.9816452756	.9820377427	.9824218497	.9827977732	.9831656864
180+	.9835257588	.9838781564	.9842230417	.9845605737	.9848909082
185+	.9852141977	.9855305915	.9858402358	.9861432736	.9864398452
190+	.9867300875	.9870141351	.9872921192	.9875641686	.9878304093
195+	.9880909647	.9883459555	.9885954999	.9888397136	.9890787100
200+	.9893126000	.9895414920	.9897654925	.9899847054	.9901992327
205+	.9904091739	.9906146267	.9908156866	.9910124472	.9912049999
210+	.9913934345	.9915778386	.9917582981	.9919348970	.9921077176
			A= .80		
0+	.1724137931	.1940811794	.2118622406	.2275072688	.2417609941
5+	.2550130126	.2674956721	.2793602359	.2907115263	.3016256576
10+	.3121599410	.3223588013	.3322575025	.3418845985	.3512636028
15+	.3604141615	.3693528979	.3780940352	.3866498633	.3950310964
20+	.4032471498	.4113063567	.4192161423	.4269831622	.4346134156
25+	.4421123381	.4494848780	.4567355602	.4638685394	.4708876453
30+	.4777964209	.4845981550	.4912959105	.4978925482	.5043907482
35+	.5107930274	.5171017563	.5233191722	.5294473918	.5354884219
40+	.5414441692	.5473164487	.5531069909	.5588174494	.5644494062
45+	.5700043775	.5754838189	.5808891289	.5862216541	.5914826918
50+	.5966734938	.6017952695	.6068491881	.6118363819	.6167579475
55+	.6216149490	.6264084192	.6311393615	.6358087516	.6404175391
60+	.6449666485	.6494569811	.6538894154	.6582648089	.6625839986
65+	.6668478022	.6710570189	.6752124304	.6793148014	.6833648803
70+	.6873634000	.6913110788	.6952086203	.6990567146	.7028560383
75+	.7066072554	.7103110174	.7139679643	.7175787241	.7211439140
80+	.7246641404	.7281399991	.7315720761	.7349609475	.7383071797
85+	.7416113301	.7448739470	.7480955702	.7512767307	.7544179514
90+	.7575197473	.7605826251	.7636070844	.7665936170	.7695427074
95+	.7724548331	.7753304647	.7781700658	.7809740935	.7837429983
100+	.7864772244	.7891772098	.7918433862	.7944761796	.7970760099
105+	.7996432913	.8021784325	.8046818364	.8071539007	.8095950177

D-INTEGRAL PROBABILITY M=R, B= 6

R	1	2	3	4	5
			A= .80		
110+	.8120055744	.8143859526	.8167365293	.8190576763	.8213497605
115+	.8236131443	.8258481850	.8280552355	.8302346440	.8323867542
120+	.8345119056	.8366104332	.8386826675	.8407289352	.8427495584
125+	.8447448555	.8467151407	.8486607241	.8505819122	.8524790072
130+	.8543523080	.8562021094	.8580287027	.8598323753	.8616134113
135+	.8633720912	.8651086919	.8668234868	.8685167462	.8701887368
140+	.8718397220	.8734699621	.8750797140	.8766692315	.8782387653
145+	.8797885631	.8813188693	.8828299254	.8843219701	.8857952389
150+	.8872499645	.8886863769	.8901047031	.8915051673	.8928879911
155+	.8942533932	.8956015898	.8969327943	.8982472175	.8995450679
160+	.9008265510	.9020918701	.9033412260	.9045748168	.9057928384
165+	.9069954843	.9081829455	.9093554108	.9105130667	.9116560973
170+	.9127846845	.9138990079	.9149992452	.9160855716	.9171581602
175+	.9182171822	.9192628066	.9202952001	.9213145278	.9223209524
180+	.9233146347	.9242957338	.9252644065	.9262208079	.9271650911
185+	.9280974073	.9290179060	.9299267348	.9308240393	.9317099636
190+	.9325846499	.9334482388	.9343008689	.9351426772	.9359737993
195+	.9367943688	.9376045177	.9384043766	.9391940743	.9399737380
200+	.9407434935	.9415034649	.9422537750	.9429945447	.9437258938
205+	.9444479405	.9451608014	.9458645918	.9465594257	.9472454155
210+	.9479226722	.9485913056	.9492514240	.9499031345	.9505465428
215+	.9511817532	.9518088690	.9524279918	.9530392224	.9536426601
220+	.9542384029	.9548265478	.9554071905	.9559804255	.9565463462
225+	.9571050448	.9576566123	.9582011387	.9587387128	.9592694222
230+	.9597933537	.9603105929	.9608212241	.9613253308	.9618229955
235+	.9623142996	.9627993234	.9632781463	.9637508468	.9642175022
240+	.9646781890	.9651329827	.9655819579	.9660251882	.9664627464
245+	.9668947043	.9673211328	.9677421020	.9681576810	.9685679381
250+	.9689729408	.9693727558	.9697674487	.9701570847	.9705417277
255+	.9709214412	.9712962878	.9716663291	.9720316263	.9723922396
260+	.9727482284	.9730996515	.9734465670	.9737890321	.9741271033
265+	.9744608366	.9747902871	.9751155093	.9754365569	.9757534831
270+	.9760663404	.9763751804	.9766800544	.9769810128	.9772781056
275+	.9775713819	.9778608903	.9781466790	.9784287952	.9787072858
280+	.9789821970	.9792535745	.9795214633	.9797859078	.9800469522
285+	.9803046396	.9805590130	.9808101147	.9810579864	.9813026694
290+	.9815442043	.9817826314	.9820179905	.9822503206	.9824796604
295+	.9827060483	.9829295219	.9831501185	.9833678749	.9835828273
300+	.9837950117	.9840044634	.9842112174	.9844153082	.9846167700
305+	.9848156363	.9850119404	.9852057150	.9853969926	.9855858051
310+	.9857721842	.9859561609	.9861377661	.9863170301	.9864939830
315+	.9866686544	.9868410734	.9870112690	.9871792697	.9873451037
320+	.9875087986	.9876703820	.9878298810	.9879873222	.9881427322
325+	.9882961369	.9884475620	.9885970331	.9887445752	.9888902130
330+	.9890339710	.9891758733	.9893159438	.9894542060	.9895906832
335+	.9897253982	.9898583737	.9899896321	.9901191954	.9902470855
340+	.9903733237	.9904979315	.9906209296	.9907423389	.9908621797
345+	.9909804723	.9910972366	.9912124921	.9913262583	.9914385545

D-INTEGRAL PROBABILITY M=R, B= 7

R	1	2	3	4	5
			A=1/10		
0+	.5882352941	.8626249590	.9574758738	.9870520560	.9960523807
			A=1/ 9		
0+	.5625000000	.8394337390	.9452300188	.9816814413	.9938840084
5+	.9979490944	.9993080592	.9997651234	.9999198308	.9999725044
			A=1/ 8		
0+	.5333333333	.8102199648	.9280207607	.9733022459	.9901468458
5+	.9963568139	.9986470581	.9994950582	.9998106345	.9999286710
			A=1/ 7		
0+	.5000000000	.7728725914	.9032263997	.9597450806	.9833897174
5+	.9931526453	.9971710243	.9988271482	.9995118971	.9997961111
			A=1/ 6		
0+	.4615384615	.7243780971	.8665402227	.9369185811	.9704970362
5+	.9862535757	.9935967821	.9970127971	.9986031444	.9993450696
			A=1/ 5		
0+	.4166666667	.6604216008	.8107822322	.8968548372	.9444300118
5+	.9702424036	.9841083633	.9915195494	.9954724451	.9975800876
			A=1/ 4		
0+	.3636363636	.5749271628	.7240600661	.8238314390	.8887125684
5+	.9301772081	.9563829180	.9728257957	.9830953848	.9894911421
10+	.9934678925	.9959386741	.9974735355	.9984272347	.9990201357
			A=1/ 3		
0+	.3000000000	.4599460800	.5880016337	.6882514265	.7655892555
5+	.8246026826	.8692599199	.9028372289	.9279582607	.9466799356
10+	.9605903219	.9709016216	.9785311880	.9841686645	.9883298597
15+	.9913990317	.9936615520	.9953288439	.9965572539	.9974622360
			A= .40		
0+	.2631578947	.3888062620	.4945926039	.5834657695	.6578000993
5+	.7196640565	.7709094029	.8131839976	.8479342895	.8764126285
10+	.8996904647	.9186754090	.9341300002	.9466905632	.9568851222
15+	.9651497841	.9718433289	.9772599470	.9816401965	.9851803166
20+	.9880400730	.9903493167	.9922134331	.9937178475	.9949317341
			A=1/ 2		
0+	.2222222222	.3078713475	.3821754262	.4480615576	.5069823065
5+	.5598366323	.6072865148	.6498740540	.6880698239	.7222946765
10+	.7529303951	.7803254080	.8047982512	.8266399956	.8461162031
15+	.8634686663	.8789170481	.8926604627	.9048790106	.9157352641
20+	.9253756961	.9339320448	.9415226055	.9482534455	.9542195403
25+	.9595058277	.9641881833	.9683343166	.9720045927	.9752527820
30+	.9781267428	.9806690398	.9829175045	.9849057408	.9866635807
35+	.9882174938	.9895909554	.9908047758	.9918773959	.9928251507
			A= .60		
0+	.1923076923	.2489358897	.2980287448	.3423320826	.3830688835
5+	.4208924427	.4562044847	.4892795687	.5203212710	.5494903893
10+	.5769202612	.6027256008	.6270078471	.6498585283	.6713614523
15+	.6915941742	.7106290085	.7285337462	.7453721772	.7612044824
20+	.7760875374	.7900751540	.8032182801	.8155651685	.8271615245
25+	.8380506377	.8482735015	.8578689248	.8668736353	.8753223785
30+	.8832480111	.8906815905	.8976524609	.9041883358	.9103153776
35+	.9160582740	.9214403112	.9264834453	.9312083699	.9356345811
40+	.9397804406	.9436632353	.9472992349	.9507037466	.9538911679
45+	.9568750361	.9596680764	.9622822472	.9647287833	.9670182371
50+	.9691605175	.9711649271	.9730401974	.9747945217	.9764355874
55+	.9779706050	.9794063371	.9807491248	.9820049128	.9831792735
60+	.9842774295	.9853042744	.9862643936	.9871620824	.9880013641
65+	.9887860068	.9895195392	.9902052652	.9908462781	.9914454734
70+	.9920055617	.9925290796	.9930184014	.9934757484	.9939031994

D-INTEGRAL PROBABILITY M=R, B= 7

M=R	1	2	3	4	5
			A= .65		
0+	.1801801802	.2254325697	.2643825362	.2995454584	.3320284851
5+	.3624150561	.3910551808	.4181793273	.4439503854	.4684901404
10+	.4918939194	.5142392238	.5355910772	.5560054764	.5755316929
15+	.5942138516	.6120920396	.6292031023	.6455812243	.6612583619
20+	.6762645686	.6906282454	.7043763344	.7175344719	.7301271101
25+	.7421776168	.7537083559	.7647407556	.7752953658	.7853919074
30+	.7950493150	.8042857748	.8131187587	.8215650548	.8296407956
35+	.8373614838	.8447420167	.8517967087	.8585393132	.8649830427
40+	.8711405884	.8770241394	.8826453998	.8880156069	.8931455476
45+	.8980455746	.9027256219	.9071952204	.9114635122	.9155392651
50+	.9194308864	.9231464360	.9266936396	.9300799015	.9333123163
55+	.9363976809	.9393425063	.9421530278	.9448352167	.9473947898
60+	.9498372199	.9521677453	.9543913792	.9565129187	.9585369532
65+	.9604678734	.9623098789	.9640669862	.9657430362	.9673417014
70+	.9688664930	.9703207675	.9717077331	.9730304563	.9742918673
75+	.9754947664	.9766418288	.9777356108	.9787785540	.9797729909
80+	.9807211493	.9816251566	.9824870448	.9833087539	.9840921366
85+	.9848389615	.9855509171	.9862296154	.9868765951	.9874933249
90+	.9880812064	.9886415774	.9891757146	.9896848364	.9901701051
95+	.9906326300	.9910734694	.9914936330	.9918940840	.9922757412
			A= .70		
0+	.1694915254	.2050250960	.2352978093	.2625304998	.2876949707
5+	.3113021357	.3336567295	.3549571320	.3753410633	.3949090016
10+	.4137372383	.4318856459	.4494025439	.4663278764	.4826953613
15+	.4985339850	.5138690694	.5287230490	.5431160474	.5570663125
20+	.5705905490	.5837041769	.5964215330	.6087560316	.6207202922
25+	.6323262435	.6435852069	.6545079660	.6651048231	.6753856472
30+	.6853599133	.6950367361	.7044248985	.7135328759	.7223688568
35+	.7309407615	.7392562575	.7473227732	.7551475106	.7627374558
40+	.7700993886	.7772398917	.7841653581	.7908819986	.7973958485
45+	.8037127733	.8098384747	.8157784957	.8215382255	.8271229042
50+	.8325376274	.8377873498	.8428768895	.8478109319	.8525940332
55+	.8572306236	.8617250110	.8660813841	.8703038155	.8743962649
60+	.8783625815	.8822065076	.8859316808	.8895416370	.8930398126
65+	.8964295476	.8997140879	.9028965874	.9059801108	.9089676358
70+	.9118620551	.9146661791	.9173827376	.9200143820	.9225636876
75+	.9250331554	.9274252140	.9297422218	.9319864686	.9341601775
80+	.9362655069	.9383045520	.9402793464	.9421918641	.9440440211
85+	.9458376766	.9475746350	.9492566472	.9508854121	.9524625780
90+	.9539897441	.9554684618	.9569002362	.9582865271	.9596287505
95+	.9609282799	.9621864471	.9634045440	.9645838233	.9657254996
100+	.9668307509	.9679007192	.9689365118	.9699392021	.9709098309
105+	.9718494071	.9727589087	.9736392836	.9744914507	.9753163006
110+	.9761146964	.9768874748	.9776354464	.9783593968	.9790600874
115+	.9797382559	.9803946171	.9810298636	.9816446665	.9822396760
120+	.9828155219	.9833728145	.9839121451	.9844340862	.9849391927
125+	.9854280020	.9859010345	.9863587944	.9868017702	.9872304348
130+	.9876452462	.9880466482	.9884350704	.9888109289	.9891746267
135+	.9895265540	.9898670886	.9901965966	.9905154322	.9908239385
140+	.9911224477	.9914112815	.9916907512	.9919611585	.9922227952
			A= .75		
0+	.1600000000	.1872091874	.2101004332	.2305580177	.2494028161
5+	.2670665939	.2838069874	.2997915302	.3151361326	.3299248381
10+	.3442208712	.3580732353	.3715208633	.3845953445	.3973227768
15+	.4097250662	.4218208585	.4336262245	.4451551723	.4564200368
20+	.4674317809	.4782002304	.4887342612	.4990419472	.5091306818
25+	.5190072752	.5286780352	.5381488342	.5474251641	.5565121833

D-INTEGRAL PROBABILITY M=R, B= 7

M=R	1	2	3	4	5
			A= .75		
30+	.5654147558	.5741374846	.5826847397	.5910606827	.5992692874
35+	.6073143580	.6151995448	.6229283574	.6305041769	.6379302663
40+	.6452097794	.6523457690	.6593411943	.6661989269	.6729217563
45+	.6795123955	.6859734852	.6923075977	.6985172408	.7046048609
50+	.7105728463	.7164235295	.7221591901	.7277820565	.7332943086
55+	.7386980791	.7439954558	.7491884826	.7542791616	.7592694540
60+	.7641612817	.7689565282	.7736570399	.7782646273	.7827810656
65+	.7872080958	.7915474258	.7958007308	.7999696545	.8040558095
70+	.8080607782	.8119861133	.8158333389	.8196039504	.8232994157
75+	.8269211756	.8304706441	.8339492092	.8373582335	.8406990544
80+	.8439729846	.8471813130	.8503253044	.8534062008	.8564252210
85+	.8593835618	.8622823977	.8651228817	.8679061456	.8706333005
90+	.8733054368	.8759236250	.8784889158	.8810023405	.8834649111
95+	.8858776214	.8882414463	.8905573428	.8928262502	.8950490902
100+	.8972267676	.8993601699	.9014501685	.9034976183	.9055033580
105+	.9074682111	.9093929851	.9112784727	.9131254516	.9149346847
110+	.9167069206	.9184428939	.9201433251	.9218089212	.9234403755
115+	.9250383685	.9266035675	.9281366272	.9296381898	.9311088852
120+	.9325493311	.9339601337	.9353418873	.9366951749	.9380205682
125+	.9393186280	.9405899041	.9418349359	.9430542522	.9442483717
130+	.9454178030	.9465630446	.9476845857	.9487829057	.9498584747
135+	.9509117538	.9519431948	.9529532410	.9539423266	.9549108778
140+	.9558593120	.9567880386	.9576974589	.9585879664	.9594599466
145+	.9603137777	.9611498301	.9619684672	.9627700450	.9635549124
150+	.9643234116	.9650758778	.9658126396	.9665340191	.9672403319
155+	.9679318875	.9686089889	.9692719333	.9699210118	.9705565099
160+	.9711787072	.9717878777	.9723842899	.9729682070	.9735398868
165+	.9740995819	.9746475400	.9751840036	.9757092103	.9762233931
170+	.9767267802	.9772195951	.9777020568	.9781743800	.9786367748
175+	.9790894474	.9795325994	.9799664287	.9803911288	.9808068895
180+	.9812138967	.9816123325	.9820023753	.9823841998	.9827579771
185+	.9831238751	.9834820578	.9838326862	.9841759179	.9845119073
190+	.9848408055	.9851627607	.9854779179	.9857864192	.9860884037
195+	.9863840078	.9866733650	.9869566060	.9872338588	.9875052490
200+	.9877708992	.9880309299	.9882854588	.9885346014	.9887784705
205+	.9890171769	.9892508290	.9894795329	.9897033925	.9899225096
210+	.9901369839	.9903469130	.9905523924	.9907535158	.9909503749
215+	.9911430592	.9913316569	.9915162538	.9916969343	.9918737810
			A= .80		
0+	.1515151515	.1715751862	.1882118845	.2029476687	.2164428404
5+	.2290448294	.2409612331	.2523272358	.2632364580	.2737568006
10+	.2839393124	.2938234927	.3034406342	.3128160215	.3219704297
15+	.3309211769	.3396828823	.3482680245	.3566873604	.3649502450
20+	.3730648799	.3810385074	.3888775657	.3965878130	.4041744286
25+	.4116420956	.4189950690	.4262372324	.4333721458	.4404030853
30+	.4473330773	.4541649274	.4609012453	.4675444661	.4740968688
35+	.4805605923	.4869376497	.4932299400	.4994392597	.5055673114
40+	.5116157128	.5175860039	.5234796537	.5292980662	.5350425855
45+	.5407145004	.5463150491	.5518454226	.5573067685	.5627001937
50+	.5680267676	.5732875247	.5784834663	.5836155634	.5886847583
55+	.5936919663	.5986380775	.6035239580	.6083504516	.6131183808
60+	.6178285481	.6224817370	.6270787131	.6316202246	.6361070038
65+	.6405397671	.6449192164	.6492460395	.6535209107	.6577444911
70+	.6619174298	.6660403639	.6701139191	.6741387100	.6781153409
75+	.6820444057	.6859264887	.6897621645	.6935519988	.6972965481
80+	.7009963607	.7046519764	.7082639269	.7118327361	.7153589204
85+	.7188429886	.7222854426	.7256867770	.7290474797	.7323680320

D-INTEGRAL PROBABILITY M=R, B= 7

M=R	1	2	3	4	5
			A= .80		
90+	.7356489085	.7388905778	.7420935020	.7452581372	.7483849336
95+	.7514743357	.7545267823	.7575427064	.7605225359	.7634666931
100+	.7663755952	.7692496542	.7720892771	.7748948658	.7776668174
105+	.7804055244	.7831113743	.7857847502	.7884260305	.7910355891
110+	.7936137957	.7961610153	.7986776091	.8011639336	.8036203415
115+	.8060471813	.8084447972	.8108135300	.8131537160	.8154656880
120+	.8177497748	.8200063015	.8222355895	.8244379565	.8266137167
125+	.8287631805	.8308866549	.8329844434	.8350568461	.8371041598
130+	.8391266776	.8411246896	.8430984826	.8450483399	.8469745420
135+	.8488773659	.8507570857	.8526139722	.8544482933	.8562603139
140+	.8580502958	.8598184980	.8615651764	.8632905842	.8649949717
145+	.8666785862	.8683416725	.8699844725	.8716072254	.8732101676
150+	.8747935330	.8763575528	.8779024555	.8794284672	.8809358113
155+	.8824247088	.8838953780	.8853480350	.8867828933	.8882001639
160+	.8896000557	.8909827750	.8923485257	.8936975097	.8950299263
165+	.8963459726	.8976458437	.8989297323	.9001978287	.9014503214
170+	.9026873967	.9039092385	.9051160288	.9063079477	.9074851728
175+	.9086478801	.9097962434	.9109304345	.9120506232	.9131569775
180+	.9142496634	.9153288449	.9163946843	.9174473419	.9184869761
185+	.9195137438	.9205277997	.9215292969	.9225183867	.9234952188
190+	.9244599410	.9254126994	.9263536384	.9272829009	.9282006280
195+	.9291069591	.9300020321	.9308859834	.9317589476	.9326210578
200+	.9334724456	.9343132410	.9351435726	.9359635675	.9367733510
205+	.9375730475	.9383627793	.9391426679	.9399128330	.9406733928
210+	.9414244645	.9421661637	.9428986046	.9436219001	.9443361618
215+	.9450415001	.9457380239	.9464258410	.9471050576	.9477757791
220+	.9484381094	.9490921512	.9497380059	.9503757740	.9510055544
225+	.9516274452	.9522415430	.9528479436	.9534467415	.9540380300
230+	.9546219013	.9551984467	.9557677563	.9563299190	.9568850228
235+	.9574331546	.9579744003	.9585088447	.9590365717	.9595576641
240+	.9600722037	.9605802714	.9610819471	.9615773097	.9620664373
245+	.9625494068	.9630262944	.9634971753	.9639621239	.9644212136
250+	.9648745169	.9653221054	.9657640500	.9662004207	.9666312865
255+	.9670567157	.9674767757	.9678915333	.9683010542	.9687054035
260+	.9691046454	.9694988435	.9698880603	.9702723579	.9706517974
265+	.9710264393	.9713963434	.9717615686	.9721221732	.9724782147
270+	.9728297501	.9731768356	.9735195265	.9738578778	.9741919436
275+	.9745217774	.9748474321	.9751689598	.9754864120	.9757998399
280+	.9761092935	.9764148227	.9767164765	.9770143034	.9773083514
285+	.9775986676	.9778852989	.9781682914	.9784476907	.9787235418
290+	.9789958892	.9792647769	.9795302482	.9797923459	.9800511125
295+	.9803065898	.9805588190	.9808078409	.9810536958	.9812964236
300+	.9815360635	.9817726543	.9820062344	.9822368416	.9824645134
305+	.9826892867	.9829111979	.9831302832	.9833465780	.9835601176
310+	.9837709366	.9839790693	.9841845497	.9843874110	.9845876864
315+	.9847854085	.9849806095	.9851733211	.9853635749	.9855514018
320+	.9857368325	.9859198972	.9861006259	.9862790480	.9864551927
325+	.9866290887	.9868007646	.9869702484	.9871375678	.9873027502
330+	.9874658226	.9876268117	.9877857439	.9879426453	.9880975416
335+	.9882504580	.9884014199	.9885504518	.9886975782	.9888428234
340+	.9889862111	.9891277649	.9892675081	.9894054636	.9895416541
345+	.9896761021	.9898088295	.9899398583	.9900692100	.9901969059
350+	.9903229670	.9904474142	.9905702679	.9906915483	.9908112755
355+	.9909294691	.9910461488	.9911613336	.9912750427	.9913872949

D-INTEGRAL PROBABILITY M=R,B= 8

R	1	2	3	4	5
			A=1/10		
0+	.5555555556	.8468905797	.9519368800	.9852748473	.9954982569
			A=1/ 9		
0+	.5294117647	.8216893979	.9382578829	.9792014282	.9930322412
5+	.9976596429	.9992097844	.9997316636	.9999083936	.9999685791
			A=1/ 8		
0+	.5000000000	.7901746258	.9191381941	.9697594847	.9887914637
5+	.9958464248	.9984557309	.9994233039	.9997836569	.9999184959
			A=1/ 7		
0+	.4666666667	.7502436544	.8917872181	.9545607782	.9811501011
5+	.9922057550	.9967742500	.9986613372	.9994425807	.9997670824
			A=1/ 6		
0+	.4285714286	.6989641719	.8517035502	.9291607195	.9666501417
5+	.9843972111	.9927133671	.9965952009	.9984062698	.9992522933
			A=1/ 5		
0+	.3846153846	.6322660161	.7915708392	.8850720372	.9375946697
5+	.9664011244	.9819898586	.9903641513	.9948463850	.9972420538
			A=1/ 4		
0+	.3333333333	.5446682456	.6997185284	.8060497646	.8764045909
5+	.9219384871	.9509880780	.9693454556	.9808730616	.9880821410
10+	.9925789014	.9953796232	.9971227300	.9982073987	.9988824760
			A=1/ 3		
0+	.2727272727	.4295255899	.5592311610	.6630589057	.7445091044
5+	.8074852915	.8556521762	.8921875663	.9197221641	.9403688542
10+	.9557892548	.9672702512	.9757971619	.9821178434	.9867960917
15+	.9902547023	.9928094195	.9946952726	.9960867602	.9971131797
			A= .40		
0+	.2380952381	.3597670282	.4651237764	.5553881917	.6320431568
5+	.6966335945	.7506955510	.7956903587	.8329608791	.8637092605
10+	.8889905719	.9097167671	.9266667061	.9404992437	.9517674154
15+	.9609324907	.9683771786	.9744176140	.9793139739	.9832797136
20+	.9864894931	.9890859050	.9911851385	.9928817122	.9942524082
			A=1/ 2		
0+	.2000000000	.2817681920	.3542859684	.4195956838	.4787276473
5+	.5323238864	.5808719628	.6247892167	.6644547570	.7002217533
10+	.7324219407	.7613670616	.7873492130	.8106409269	.8314953210
15+	.8501464327	.8668097528	.8816829354	.8949466438	.9067654928
20+	.9172890503	.9266528661	.9349795037	.9423795543	.9489526186
25+	.9547882470	.9599668290	.9645604288	.9686335652	.9722439340
30+	.9754430744	.9782769813	.9807866648	.9830086605	.9849754936
35+	.9867160986	.9882561998	.9896186539	.9908237588	.9918895315
40+	.9928319567	.9936652104	.9944018597	.9950530415	.9956286220
			A= .60		
0+	.1724137931	.2258976594	.2730584551	.3161252228	.3561029361
5+	.3935223967	.4287050790	.4618679730	.4931709333	.5227402915
10+	.5506815268	.5770864505	.6020374529	.6256100966	.6478747411
15+	.6688975831	.6887413335	.7074656673	.7251275290	.7417813438
20+	.7574791702	.7722708137	.7862039167	.7993240337	.8116746972
25+	.8232974784	.8342320461	.8445162233	.8541860443	.8632758119
30+	.8718181547	.8798440849	.8873830564	.8944630230	.9011104961
35+	.9073506030	.9132071433	.9187026456	.9238584231	.9286946279
40+	.9332303038	.9374834386	.9414710143	.9452090559	.9487126787
45+	.9519961340	.9550728532	.9579554898	.9606559606	.9631854843
50+	.9655546192	.9677732985	.9698508650	.9717961032	.9736172707
55+	.9753221275	.9769179642	.9784116289	.9798095524	.9811177723
60+	.9823419560	.9834874219	.9845591606	.9855618535	.9864998917
65+	.9873773932	.9881982188	.9889659883	.9896840946	.9903557176
70+	.9909838369	.9915712446	.9921205561	.9926342215	.9931145356

D-INTEGRAL PROBABILITY M=R,B= 8

R	1	2	3	4	5
			A= .65		
0+	.1612903226	.2038402885	.2410092607	.2749070601	.3064774497
5+	.3362159195	.3644169150	.3912721361	.4169155730	.4414464536
10+	.4649419061	.4874643887	.5090662633	.5297927231	.5496837244
15+	.5687752966	.5871004487	.6046898086	.6215720808	.6377743797
20+	.6533224738	.6682409678	.6825534398	.6962825450	.7094500968
25+	.7220771284	.7341839433	.7457901543	.7569147166	.7675759549
30+	.7777915867	.7875787422	.7969539829	.8059333174	.8145322172
35+	.8227656302	.8306479950	.8381932539	.8454148653	.8523258167
40+	.8589386369	.8652654079	.8713177773	.8771069700	.8826437999
45+	.8879386817	.8930016427	.8978423338	.9024700409	.9068936963
50+	.9111218896	.9151628782	.9190245987	.9227146765	.9262404368
55+	.9296089142	.9328268629	.9359007662	.9388368461	.9416410723
60+	.9443191713	.9468766353	.9493187305	.9516505057	.9538768000
65+	.9560022509	.9580313021	.9599682104	.9618170532	.9635817354
70+	.9652659960	.9668734148	.9684074184	.9698712867	.9712681582
75+	.9726010361	.9738727938	.9750861799	.9762438232	.9773482383
80+	.9784018295	.9794068957	.9803656350	.9812801485	.9821524446
85+	.9829844429	.9837779780	.9845348028	.9852565925	.9859449473
90+	.9866013961	.9872273995	.9878243524	.9883935871	.9889363763
95+	.9894539350	.9899474236	.9904179501	.9908665722	.9912943000
100+	.9917020974	.9920908849	.9924615409	.9928149039	.9931517741
			A= .70		
0+	.1515151515	.1847945236	.2135108222	.2395663817	.2638088066
5+	.2866843435	.3084581841	.3293019388	.3493338949	.3686396932
10+	.3872838603	.4053166649	.4227784070	.4397022120	.4561159152
15+	.4720433686	.4875053686	.5025203297	.5171047810	.5312737385
20+	.5450409886	.5584193052	.5714206193	.5840561512	.5963365163
25+	.6082718083	.6198716670	.6311453324	.6421016896	.6527493046
30+	.6630964546	.6731511529	.6829211691	.6924140470	.7016371189
35+	.7105975182	.7193021897	.7277578990	.7359712406	.7439486440
40+	.7516963807	.7592205688	.7665271784	.7736220357	.7805108266
45+	.7871991010	.7936922758	.7999956376	.8061143466	.8120534384
50+	.8178178275	.8234123090	.8288415617	.8341101501	.8392225271
55+	.8441830355	.8489959110	.8536652838	.8581951811	.8625895286
60+	.8668521532	.8709867846	.8749970573	.8788865127	.8826586011
65+	.8863166833	.8898640327	.8933038371	.8966392007	.8998731455
70+	.9030086138	.9060484691	.9089954986	.9118524144	.9146218556
75+	.9173063897	.9199085143	.9224306589	.9248751863	.9272443943
80+	.9295405171	.9317657271	.9339221361	.9360117969	.9380367051
85+	.9399987997	.9418999653	.9437420331	.9455267824	.9472559418
90+	.9489311903	.9505541592	.9521264326	.9536495492	.9551250032
95+	.9565542455	.9579386849	.9592796892	.9605785862	.9618366651
100+	.9630551771	.9642353367	.9653783226	.9664852787	.9675573152
105+	.9685955092	.9696009060	.9705745196	.9715173339	.9724303034
110+	.9733143540	.9741703837	.9749992638	.9758018391	.9765789291
115+	.9773313285	.9780598079	.9787651148	.9794479737	.9801090873
120+	.9807491367	.9813687825	.9819686650	.9825494048	.9831116038
125+	.9836558451	.9841826942	.9846926991	.9851863909	.9856642844
130+	.9861268784	.9865746565	.9870080872	.9874276245	.9878337083
135+	.9882267649	.9886072075	.9889754362	.9893318387	.9896767908
140+	.9900106564	.9903337881	.9906465274	.9909492054	.9912421424
145+	.9915256489	.9918000258	.9920655644	.9923225467	.9925712460
			A= .75		
0+	.1428571429	.1682498426	.1898472705	.2092868363	.2272952224
5+	.2442562470	.2603989165	.2758718479	.2907775448	.3051900482
10+	.3191648208	.3327446535	.3459633823	.3588483282	.3714219548
15+	.3837030284	.3957074508	.4074488693	.4189391329	.4301886384

D-INTEGRAL PROBABILITY M=R,B= 8

R	1	2	3	4	5
			A= .75		
20+	.4412065973	.4520012443	.4625800014	.4729496094	.4831162336
25+	.4930855497	.5028628139	.5124529213	.5218604536	.5310897194
30+	.5401447881	.5490295182	.5577475813	.5663024831	.5746975804
35+	.5829360967	.5910211347	.5989556881	.6067426510	.6143848262
40+	.6218849332	.6292456140	.6364694395	.6435589139	.6505164793
45+	.6573445199	.6640453650	.6706212925	.6770745315	.6834072645
50+	.6896216302	.6957197246	.7017036039	.7075752852	.7133367482
55+	.7189899367	.7245367598	.7299790928	.7353187784	.7405576273
60+	.7456974197	.7507399055	.7556868050	.7605398103	.7653005851
65+	.7699707657	.7745519616	.7790457559	.7834537057	.7877773431
70+	.7920181749	.7961776834	.8002573271	.8042585406	.8081827352
75+	.8120312993	.8158055987	.8195069770	.8231367557	.8266962349
80+	.8301866935	.8336093893	.8369655593	.8402564205	.8434831695
85+	.8466469833	.8497490194	.8527904160	.8557722923	.8586957490
90+	.8615618680	.8643717134	.8671263314	.8698267501	.8724739807
95+	.8750690170	.8776128358	.8801063974	.8825506457	.8849465082
100+	.8872948964	.8895967065	.8918528186	.8940640980	.8962313947
105+	.8983555439	.9004373663	.9024776681	.9044772413	.9064368640
110+	.9083573005	.9102393017	.9120836050	.9138909346	.9156620021
115+	.9173975059	.9190981323	.9207645549	.9223974355	.9239974238
120+	.9255651575	.9271012631	.9286063556	.9300810385	.9315259047
125+	.9329415360	.9343285036	.9356873682	.9370186800	.9383229793
130+	.9396007963	.9408526513	.9420790551	.9432805088	.9444575044
135+	.9456105245	.9467400427	.9478465238	.9489304239	.9499921904
140+	.9510322623	.9520510705	.9530490374	.9540265777	.9549840982
145+	.9559219978	.9568406679	.9577404925	.9586218483	.9594851046
150+	.9603306238	.9611587612	.9619698656	.9627642786	.9635423357
155+	.9643043656	.9650506909	.9657816276	.9664974861	.9671985702
160+	.9678851783	.9685576028	.9692161302	.9698610418	.9704926132
165+	.9711111146	.9717168109	.9723099620	.9728908225	.9734596420
170+	.9740166654	.9745621326	.9750962787	.9756193345	.9761315259
175+	.9766330745	.9771241975	.9776051077	.9780760138	.9785371204
180+	.9789886278	.9794307327	.9798636274	.9802875008	.9807025379
185+	.9811089198	.9815068243	.9818964254	.9822778938	.9826513966
190+	.9830170978	.9833751578	.9837257339	.9840689805	.9844050485
195+	.9847340860	.9850562379	.9853716466	.9856804511	.9859827879
200+	.9862787908	.9865685907	.9868523159	.9871300921	.9874020427
205+	.9876682881	.9879289467	.9881841343	.9884339644	.9886785482
210+	.9889179944	.9891524100	.9893818993	.9896065647	.9898265066
215+	.9900418232	.9902526106	.9904589633	.9906609736	.9908587318
220+	.9910523266	.9912418448	.9914273713	.9916089895	.9917867810
			A= .80		
0+	.1351351351	.1537961081	.1694158194	.1833310281	.1961321989
5+	.2081317111	.2195165457	.2304084805	.2408917922	.2510275285
10+	.2608615120	.2704291313	.2797583635	.2888717628	.2977878165
15+	.3065218966	.3150869458	.3234939822	.3317524778	.3398706479
20+	.3478556746	.3557138828	.3634508797	.3710716666	.3785807305
25+	.3859821176	.3932794954	.4004762031	.4075752945	.4145795741
30+	.4214916269	.4283138450	.4350484494	.4416975092	.4482629581
35+	.4547466083	.4611501637	.4674752302	.4737233252	.4798958867
40+	.4859942796	.4920198035	.4979736974	.5038571457	.5096712825
45+	.5154171956	.5210959305	.5267084936	.5322558551	.5377389519
50+	.5431586898	.5485159461	.5538115710	.5590463902	.5642212057
55+	.5693367980	.5743939272	.5793933343	.5843357421	.5892218567
60+	.5940523682	.5988279516	.6035492677	.6082169636	.6128316738
65+	.6173940206	.6219046146	.6263640556	.6307729325	.6351318245
70+	.6394413010	.6437019221	.6479142391	.6520787950	.6561961243

D-INTEGRAL PROBABILITY M=R,B= 8

R	1	2	3	4	5
			A= .80		
75+	.6602667538	.6642912027	.6682699827	.6722035987	.6760925485
80+	.6799373233	.6837384080	.6874962810	.6912114149	.6948842762
85+	.6985153257	.7021050186	.7056538047	.7091621283	.7126304288
90+	.7160591401	.7194486915	.7227995071	.7261120066	.7293866047
95+	.7326237117	.7358237332	.7389870706	.7421141209	.7452052768
100+	.7482609268	.7512814552	.7542672423	.7572186646	.7601360943
105+	.7630198999	.7658704461	.7686880938	.7714732001	.7742261186
110+	.7769471990	.7796367876	.7822952272	.7849228570	.7875200127
115+	.7900870268	.7926242283	.7951319428	.7976104926	.8000601969
120+	.8024813715	.8048743292	.8072393795	.8095768287	.8118869801
125+	.8141701341	.8164265877	.8186566354	.8208605681	.8230386744
130+	.8251912395	.8273185459	.8294208734	.8314984986	.8335516957
135+	.8355807358	.8375858874	.8395674162	.8415255854	.8434606553
140+	.8453728835	.8472625253	.8491298330	.8509750567	.8527984436
145+	.8546002386	.8563806839	.8581400195	.8598784826	.8615963082
150+	.8632937288	.8649709745	.8666282730	.8682658496	.8698839274
155+	.8714827271	.8730624671	.8746233636	.8761656304	.8776894793
160+	.8791951197	.8806827588	.8821526018	.8836048516	.8850397090
165+	.8864573727	.8878580392	.8892419032	.8906091572	.8919599914
170+	.8932945944	.8946131526	.8959158504	.8972028703	.8984743929
175+	.8997305967	.9009716584	.9021977528	.9034090529	.9046057298
180+	.9057879527	.9069558890	.9081097042	.9092495624	.9103756254
185+	.9114880537	.9125870058	.9136726384	.9147451069	.9158045645
190+	.9168511632	.9178850530	.9189063824	.9199152983	.9209119460
195+	.9218964691	.9228690097	.9238297083	.9247787039	.9257161340
200+	.9266421345	.9275568398	.9284603829	.9293528952	.9302345068
205+	.9311053462	.9319655406	.9328152157	.9336544959	.9344835040
210+	.9353023617	.9361111891	.9369101052	.9376992274	.9384786720
215+	.9392485539	.9400089868	.9407600829	.9415019533	.9422347079
220+	.9429584552	.9436733027	.9443793564	.9450767213	.9457655011
225+	.9464457985	.9471177147	.9477813502	.9484368040	.9490841740
230+	.9497235573	.9503550494	.9509787451	.9515947381	.9522031207
235+	.9528039844	.9533974197	.9539835160	.9545623615	.9551340435
240+	.9556986485	.9562562617	.9568069675	.9573508493	.9578879894
245+	.9584184692	.9589423694	.9594597695	.9599707481	.9604753829
250+	.9609737509	.9614659279	.9619519891	.9624320085	.9629060596
255+	.9633742148	.9638365457	.9642931230	.9647440169	.9651892963
260+	.9656290297	.9660632845	.9664921275	.9669156247	.9673338412
265+	.9677468414	.9681546890	.9685574469	.9689551773	.9693479416
270+	.9697358006	.9701188141	.9704970415	.9708705413	.9712393716
275+	.9716035894	.9719632513	.9723184131	.9726691301	.9730154567
280+	.9733574470	.9736951540	.9740286305	.9743579284	.9746830992
285+	.9750041935	.9753212615	.9756343528	.9759435163	.9762488005
290+	.9765502530	.9768479213	.9771418518	.9774320908	.9777186838
295+	.9780016757	.9782811112	.9785570340	.9788294876	.9790985148
300+	.9793641582	.9796264594	.9798854600	.9801412006	.9803937217
305+	.9806430632	.9808892645	.9811323645	.9813724016	.9816094138
310+	.9818434387	.9820745134	.9823026744	.9825279579	.9827503998
315+	.9829700352	.9831868991	.9834010260	.9836124498	.9838212043
320+	.9840273226	.9842308376	.9844317816	.9846301867	.9848260845
325+	.9850195063	.9852104830	.9853990449	.9855852224	.9857690450
330+	.9859505422	.9861297430	.9863066761	.9864813698	.9866538522
335+	.9868241507	.9869922928	.9871583054	.9873222151	.9874840482
340+	.9876438307	.9878015884	.9879573464	.9881111300	.9882629637
345+	.9884128721	.9885608792	.9887070090	.9888512848	.9889937301
350+	.9891343677	.9892732203	.9894103104	.9895456600	.9896792910
355+	.9898112250	.9899414833	.9900700869	.9901970567	.9903224132

D-INTEGRAL PROBABILITY M=R,B= 8

R	1	2	3	4	5
			A= .80		
360+	.9904461767	.9905683672	.9906890045	.9908081081	.9909256973
365+	.9910417912	.9911564086	.9912695682	.9913812882	.9914915868

D-INTEGRAL PROBABILITY M=R,B= 9

R	1	2	3	4	5
			A=1/10		
0+	.5263157895	.8318871274	.9465139290	.9835144450	.9949464912
			A=1/ 9		
0+	.5000000000	.8048806087	.9314646077	.9767524627	.9921856637
5+	.9973710473	.9991116504	.9996982270	.9998969603	.9999646544
			A=1/ 8		
0+	.4705882353	.7713372960	.9105377628	.9662763666	.9874481561
5+	.9953384345	.9982648770	.9993516429	.9997566977	.9999083245
			A=1/ 7		
0+	.4375000000	.7291855552	.8808032329	.9494957727	.9789402960
5+	.9912661381	.9963792298	.9984959472	.9993733652	.9997380780
			A=1/ 6		
0+	.4000000000	.6755997098	.8376157574	.9216509192	.9628814020
5+	.9825647915	.9918371718	.9961797620	.9982100368	.9991597075
			A=1/ 5		
0+	.3571428571	.6067729185	.7736053691	.8738212133	.9309754532
5+	.9626450513	.9799043085	.9892213405	.9942250937	.9969058172
			A=1/ 4		
0+	.3076923077	.5177991684	.6774319097	.7894182290	.8647117606
5+	.9140193269	.9457559625	.9659468040	.9786913209	.9866931455
10+	.9916997154	.9948253572	.9967742520	.9979886920	.9987453634
			A=1/ 3		
0+	.2500000000	.4031879441	.5336628148	.6402283031	.7251092317
5+	.7915347118	.8428407332	.8820742822	.9118437048	.9342943043
10+	.9511435962	.9637404372	.9731292030	.9801098404	.9852899861
15+	.9891281960	.9919687445	.9940690523	.9956209751	.9967671351
			A= .40		
0+	.2173913043	.3350130522	.4394379134	.5304895905	.6088802048
5+	.6756780237	.7321174153	.7794719344	.8189728939	.8517619334
10+	.8788672870	.9011956854	.9195340387	.9345568322	.9468364975
15+	.9568549781	.9650153745	.9716530059	.9770455349	.9814219954
20+	.9849706916	.9878460095	.9901742244	.9920584084	.9935825459
			A=1/ 2		
0+	.1818181818	.2599137436	.3305372662	.3950261928	.4540627343
5+	.5080709070	.5573861446	.6023140610	.6431497249	.6801825001
10+	.7136956657	.7439644010	.7712535575	.7958157595	.8178899985
15+	.8377007296	.8554574136	.8713544290	.8855712765	.8982730054
20+	.9096108013	.9197226852	.9287342831	.9367596333	.9439020080
25+	.9502547276	.9559019561	.9609194659	.9653753658	.9693307870
30+	.9728405255	.9759536391	.9787140004	.9811608042	.9833290347
35+	.9852498907	.9869511733	.9884576386	.9897913160	.9909717968
40+	.9920164944	.9929408780	.9937586833	.9944821013	.9951219478
			A= .60		
0+	.1562500000	.2068697918	.2521786996	.2939900378	.3331304857
5+	.3700296435	.4049419759	.4380358366	.4694336795	.4992319716
10+	.5275117380	.5543444224	.5797952541	.6039252240	.6267922572
15+	.6484519076	.6689577657	.6883616905	.7067139362	.7240632144
20+	.7404567201	.7559401373	.7705576361	.7843518662	.7973639524
25+	.8096334931	.8211985642	.8320957279	.8423600473	.8520251052

D-INTEGRAL PROBABILITY M=R, B= 9

R	1	2	3	4	5
			A= .60		
30+	.8611230286	.8696845166	.8777388726	.8853140390	.8924366346
35+	.8991319944	.9054242102	.9113361735	.9168896177	.9221051624
40+	.9270023561	.9315997202	.9359147917	.9399641659	.9437635380
45+	.9473277442	.9506708012	.9538059456	.9567456710	.9595017651
50+	.9620853446	.9645068894	.9667762753	.9689028054	.9708952404
55+	.9727618274	.9745103277	.9761480430	.9776818410	.9791181790
60+	.9804631272	.9817223903	.9829013281	.9840049755	.9850380608
65+	.9860050238	.9869100323	.9877569979	.9885495914	.9892912568
70+	.9899852248	.9906345256	.9912420006	.9918103141	.9923419640
			A= .65		
0+	.1459854015	.1861065585	.2216136038	.2542876692	.2849382646
5+	.3139882095	.3416858141	.3681904060	.3936116858	.4180297820
10+	.4415062908	.4640907240	.4858244531	.5067432103	.5268787231
15+	.5462598052	.5649131001	.5828635943	.6001349752	.6167498840
20+	.6327300939	.6480966373	.6628698963	.6770696666	.6907152038
25+	.7038252557	.7164180850	.7285114860	.7401227965	.7512689057
30+	.7619662611	.7722308733	.7820783204	.7915237517	.8005818920
35+	.8092670449	.8175930975	.8255735247	.8332213944	.8405493721
40+	.8475697276	.8542943403	.8607347061	.8669019441	.8728068038
45+	.8784596725	.8838705827	.8890492204	.8940049328	.8987467365
50+	.9032833258	.9076230811	.9117740772	.9157440915	.9195406128
55+	.9231708492	.9266417369	.9299599477	.9331318981	.9361637565
60+	.9390614514	.9418306797	.9444769134	.9470054081	.9494212097
65+	.9517291620	.9539339137	.9560399251	.9580514750	.9599726673
70+	.9618074371	.9635595575	.9652326448	.9668301652	.9683554399
75+	.9698116508	.9712018459	.9725289447	.9737957425	.9750049161
80+	.9761590280	.9772605310	.9783117727	.9793149994	.9802723608
85+	.9811859133	.9820576241	.9828893750	.9836829653	.9844401162
90+	.9851624729	.9858516089	.9865090280	.9871361679	.9877344028
95+	.9883050459	.9888493522	.9893685210	.9898636981	.9903359782
100+	.9907864072	.9912159840	.9916256630	.9920163554	.9923889318
			A= .70		
0+	.1369863014	.1682620163	.1955550157	.2205076890	.2438649330
5+	.2660193076	.2872031077	.3075656302	.3272089266	.3462062049
10+	.3646120988	.3824687694	.3998097154	.4166622521	.4330491763
15+	.4489899184	.4645013568	.4795984055	.4942944455	.5086016464
20+	.5225312088	.5360935485	.5492984395	.5621551241	.5746723989
25+	.5868586834	.5987220731	.6102703824	.6215111789	.6324518103
30+	.6430994265	.6534609974	.6635433270	.6733530650	.6828967163
35+	.6921806487	.7012110988	.7099941779	.7185358755	.7268420633
40+	.7349184979	.7427708236	.7504045743	.7578251754	.7650379458
45+	.7720480986	.7788607435	.7854808872	.7919134348	.7981631913
50+	.8042348623	.8101330552	.8158622805	.8214269524	.8268313907
55+	.8320798211	.8371763768	.8421250995	.8469299405	.8515947623
60+	.8561233393	.8605193591	.8647864241	.8689280525	.8729476796
65+	.8768486592	.8806342647	.8843076906	.8878720539	.8913303952
70+	.8946856804	.8979408013	.9010985781	.9041617597	.9071330256
75+	.9100149870	.9128101884	.9155211087	.9181501627	.9206997022
80+	.9231720176	.9255693388	.9278938372	.9301476261	.9323327626
85+	.9344512486	.9365050321	.9384960085	.9404260214	.9422968645
90+	.9441102823	.9458679712	.9475715808	.9492227152	.9508229337
95+	.9523737523	.9538766444	.9553330422	.9567443374	.9581118825
100+	.9594369915	.9607209411	.9619649718	.9631702882	.9643380608
105+	.9654694262	.9665654883	.9676273189	.9686559590	.9696524193
110+	.9706176809	.9715526965	.9724583908	.9733356613	.9741853792
115+	.9750083901	.9758055145	.9765775489	.9773252659	.9780494153
120+	.9787507245	.9794298993	.9800876246	.9807245643	.9813413630

D-INTEGRAL PROBABILITY M=R, B= 9

R	1	2	3	4	5
			A= .70		
125+	.9819386455	.9825170181	.9830770687	.9836193674	.9841444672
130+	.9846529044	.9851451989	.9856218548	.9860833612	.9865301919
135+	.9869628064	.9873816503	.9877871555	.9881797406	.9885598115
140+	.9889277616	.9892839720	.9896288125	.9899626413	.9902858054
145+	.9905986412	.9909014750	.9911946225	.9914783899	.9917530740
			A= .75		
0+	.1290322581	.1528253881	.1732584610	.1917661831	.2089968550
5+	.2252942418	.2408634001	.2558371126	.2703066562	.2843376909
10+	.2979791710	.3112686730	.3242357492	.3369041277	.3492932067
15+	.3614190990	.3732953815	.3849336430	.3963438933	.4075348737
20+	.4185142951	.4292890240	.4398652288	.4502484965	.4604439261
25+	.4704562045	.4802896685	.4899483553	.4994360448	.5087562945
30+	.5179124686	.5269077625	.5357452238	.5444277694	.5529582004
35+	.5613392152	.5695734200	.5776633382	.5856114186	.5934200418
40+	.6010915269	.6086281360	.6160320793	.6233055186	.6304505707
45+	.6374693106	.6443637742	.6511359602	.6577878324	.6643213213
50+	.6707383258	.6770407145	.6832303269	.6893089745	.6952784418
55+	.7011404869	.7068968428	.7125492174	.7180992945	.7235487345
60+	.7288991746	.7341522291	.7393094903	.7443725287	.7493428929
65+	.7542221105	.7590116882	.7637131120	.7683278475	.7728573401
70+	.7773030152	.7816662787	.7859485168	.7901510965	.7942753656
75+	.7983226531	.8022942692	.8061915054	.8100156350	.8137679131
80+	.8174495765	.8210618445	.8246059185	.8280829822	.8314942022
85+	.8348407278	.8381236913	.8413442080	.8445033767	.8476022794
90+	.8506419819	.8536235338	.8565479686	.8594163039	.8622295417
95+	.8649886684	.8676946550	.8703484573	.8729510161	.8755032573
100+	.8780060922	.8804604174	.8828671153	.8852270539	.8875410875
105+	.8898100563	.8920347868	.8942160921	.8963547719	.8984516127
110+	.9005073880	.9025228586	.9044987723	.9064358646	.9083348586
115+	.9101964653	.9120213834	.9138103002	.9155638908	.9172828190
120+	.9189677374	.9206192870	.9222380979	.9238247895	.9253799702
125+	.9269042379	.9283981799	.9298623735	.9312973857	.9327037735
130+	.9340820841	.9354328550	.9367566141	.9380538801	.9393251622
135+	.9405709606	.9417917665	.9429880624	.9441603219	.9453090102
140+	.9464345839	.9475374915	.9486181733	.9496770614	.9507145803
145+	.9517311464	.9527271688	.9537030488	.9546591804	.9555959503
150+	.9565137382	.9574129166	.9582938510	.9591569004	.9600024167
155+	.9608307457	.9616422264	.9624371914	.9632159674	.9639788745
160+	.9647262273	.9654583339	.9661754970	.9668780133	.9675661741
165+	.9682402650	.9689005663	.9695473528	.9701808941	.9708014548
170+	.9714092942	.9720046669	.9725878223	.9731590053	.9737184558
175+	.9742664094	.9748030970	.9753287449	.9758435751	.9763478056
180+	.9768416496	.9773253167	.9777990121	.9782629372	.9787172892
185+	.9791622618	.9795980447	.9800248238	.9804427816	.9808520970
190+	.9812529451	.9816454980	.9820299239	.9824063882	.9827750526
195+	.9831360760	.9834896139	.9838358187	.9841748400	.9845068243
200+	.9848319152	.9851502534	.9854619770	.9857672211	.9860661183
205+	.9863587984	.9866453889	.9869260143	.9872007970	.9874698569
210+	.9877333113	.9879912753	.9882438617	.9884911810	.9887333415
215+	.9889704493	.9892026083	.9894299205	.9896524857	.9898704018
220+	.9900837645	.9902926678	.9904972036	.9906974623	.9908935321
225+	.9910854995	.9912734494	.9914574648	.9916376271	.9918140161
			A= .80		
0+	.1219512195	.1393900505	.1541064953	.1672841006	.1794549061
5+	.1909018427	.2017945403	.2122434867	.2223250880	.2320946200
10+	.2415935038	.2508536664	.2599002931	.2687536392	.2774302650
15+	.2859439036	.2943060853	.3025265968	.3106138267	.3185750281

D-INTEGRAL PROBABILITY M=R,B= 9

A= .80

R	1	2	3	4	5
20+	.3264165222	.3341438581	.3417619398	.3492751276	.3566873209
25+	.3640020253	.3712224084	.3783513458	.3853914596	.3923451511
30+	.3992146279	.4060019274	.4127089370	.4193374107	.4258889846
35+	.4323651890	.4387674602	.4450971497	.4513555332	.4575438177
40+	.4636631483	.4697146139	.4756992525	.4816180558	.4874719731
45+	.4932619152	.4989887573	.5046533425	.5102564838	.5157989668
50+	.5212815519	.5267049758	.5320699537	.5373771808	.5426273334
55+	.5478210704	.5529590348	.5580418541	.5630701418	.5680444980
60+	.5729655105	.5778337553	.5826497974	.5874141912	.5921274816
65+	.5967902038	.6014028844	.6059660416	.6104801855	.6149458188
70+	.6193634367	.6237335277	.6280565734	.6323330492	.6365634244
75+	.6407481624	.6448877208	.6489825520	.6530331029	.6570398155
80+	.6610031265	.6649234682	.6688012681	.6726369492	.6764309301
85+	.6801836249	.6838954439	.6875667931	.6911980744	.6947896861
90+	.6983420224	.7018554739	.7053304274	.7087672662	.7121663700
95+	.7155281150	.7188528741	.7221410166	.7253929085	.7286089127
100+	.7317893888	.7349346929	.7380451785	.7411211953	.7441630906
105+	.7471712082	.7501458889	.7530874708	.7559962887	.7588726748
110+	.7617169581	.7645294650	.7673105188	.7700604402	.7727795469
115+	.7754681539	.7781265737	.7807551156	.7833540865	.7859237907
120+	.7884645295	.7909766017	.7934603037	.7959159290	.7983437687
125+	.8007441110	.8031172420	.8054634450	.8077830008	.8100761877
130+	.8123432815	.8145845557	.8168002812	.8189907264	.8211561575
135+	.8232968380	.8254130292	.8275049902	.8295729773	.8316172448
140+	.8336380447	.8356356264	.8376102373	.8395621224	.8414915245
145+	.8433986839	.8452838390	.8471472258	.8489890783	.8508096279
150+	.8526091043	.8543877347	.8561457443	.8578833562	.8596007913
155+	.8612982685	.8629760045	.8646342140	.8662731097	.8678929021
160+	.8694937998	.8710760094	.8726397354	.8741851805	.8757125451
165+	.8772220281	.8787138261	.8801881339	.8816451444	.8830850486
170+	.8845080356	.8859142927	.8873040052	.8886773567	.8900345290
175+	.8913757018	.8927010535	.8940107603	.8953049967	.8965839357
180+	.8978477482	.8990966037	.9003306698	.9015501124	.9027550957
185+	.9039457824	.9051223334	.9062849079	.9074336637	.9085687567
190+	.9096903415	.9107985707	.9118935957	.9129755663	.9140446305
195+	.9151009349	.9161446247	.9171758435	.9181947332	.9192014346
200+	.9201960867	.9211788273	.9221497925	.9231091170	.9240569344
205+	.9249933765	.9259185740	.9268326559	.9277357501	.9286279830
210+	.9295094798	.9303803642	.9312407587	.9320907844	.9329305611
215+	.9337602075	.9345798407	.9353895768	.9361895305	.9369798155
220+	.9377605440	.9385318270	.9392937745	.9400464951	.9407900964
225+	.9415246846	.9422503651	.9429672417	.9436754174	.9443749939
230+	.9450660719	.9457487510	.9464231295	.9470893049	.9477473734
235+	.9483974303	.9490395698	.9496738849	.9503004678	.9509194095
240+	.9515308002	.9521347289	.9527312837	.9533205517	.9539026190
245+	.9544775707	.9550454912	.9556064636	.9561605702	.9567078926
250+	.9572485112	.9577825055	.9583099543	.9588309354	.9593455258
255+	.9598538014	.9603558374	.9608517084	.9613414876	.9618252479
260+	.9623030610	.9627749981	.9632411293	.9637015240	.9641562509
265+	.9646053778	.9650489718	.9654870993	.9659198257	.9663472159
270+	.9667693340	.9671862432	.9675980061	.9680046846	.9684063400
275+	.9688030325	.9691948221	.9695817678	.9699639279	.9703413603
280+	.9707141218	.9710822690	.9714458576	.9718049426	.9721595784
285+	.9725098191	.9728557176	.9731973266	.9735346980	.9738678832
290+	.9741969330	.9745218976	.9748428264	.9751597685	.9754727723
295+	.9757818857	.9760871560	.9763886299	.9766863536	.9769803729
300+	.9772707327	.9775574776	.9778406519	.9781202989	.9783964617

D-INTEGRAL PROBABILITY M=R, B= 9

R	1	2	3	4	5
			A= .80		
305+	.9786691828	.9789385044	.9792044678	.9794671141	.9797264839
310+	.9799826172	.9802355537	.9804853324	.9807319920	.9809755708
315+	.9812161065	.9814536363	.9816881972	.9819198256	.9821485574
320+	.9823744282	.9825974731	.9828177270	.9830352240	.9832499980
325+	.9834620826	.9836715109	.9838783154	.9840825286	.9842841824
330+	.9844833082	.9846799373	.9848741004	.9850658279	.9852551499
335+	.9854420961	.9856266958	.9858089780	.9859889712	.9861667039
340+	.9863422039	.9865154988	.9866866159	.9868555822	.9870224243
345+	.9871871685	.9873498407	.9875104667	.9876690717	.9878256809
350+	.9879803190	.9881330104	.9882837793	.9884326496	.9885796447
355+	.9887247881	.9888681027	.9890096112	.9891493360	.9892872995
360+	.9894235233	.9895580293	.9896908387	.9898219726	.9899514521
365+	.9900792975	.9902055293	.9903301676	.9904532323	.9905747429
370+	.9906947189	.9908131793	.9909301432	.9910456290	.9911596554

D-INTEGRAL PROBABILITY M=R, B=10

R	1	2	3	4	5
			A=1/10		
0+	.5000000000	.8175575370	.9412021017	.9817704577	.9943970533
5+	.9982743232	.9994655955	.9998335087	.9999478397	.9999835784
			A=1/ 9		
0+	.4736842105	.7889256483	.9248411654	.9743336266	.9913441854
5+	.9970832988	.9990136565	.9996648136	.9998855307	.9999607303
			A=1/ 8		
0+	.4444444444	.7535892598	.9022023774	.9628506302	.9861166351
5+	.9948328070	.9980744919	.9992800747	.9997297570	.9998981567
			A=1/ 7		
0+	.4117647059	.7095235885	.8702410879	.9445442052	.9767593139
5+	.9903336317	.9959859372	.9983309738	.9993042500	.9997090978
			A=1/ 6		
0+	.3750000000	.6540250849	.8242098985	.9143732438	.9591871783
5+	.9807555058	.9909680181	.9957664417	.9980144371	.9990673104
			A=1/ 5		
0+	.3333333333	.5835554481	.7567487916	.8630572486	.9245580911
5+	.9589697806	.9778503761	.9880907158	.9936084516	.9965713424
			A=1/ 4		
0+	.2857142857	.4937507149	.6569197967	.7738070326	.8535760896
5+	.9063944109	.9406756945	.9626252220	.9765482184	.9853233427
10+	.9908299963	.9942757359	.9964280435	.9977710910	.9986087883
			A=1/ 3		
0+	.2307692308	.3801343755	.5107500214	.6194027308	.7071629109
5+	.7766095255	.8307385438	.8724443323	.9042908131	.9284370198
10+	.9466418321	.9603053287	.9705232496	.9781422550	.9838101279
15+	.9880186807	.9911390385	.9934498969	.9951597311	.9964240045
			A= .40		
0+	.2000000000	.3136364809	.4168107762	.5082133141	.5878927072
5+	.6564872980	.7149477966	.7643636178	.8058511430	.8404850684
10+	.8692594417	.8930686105	.9127010888	.9288414553	.9420769277
15+	.9529063710	.9617502713	.9689607577	.9748311206	.9796045319
20+	.9834818309	.9866283474	.9891797964	.9912473129	.9929217136
25+	.9942770781	.9953737399	.9962607705	.9969780361	.9975578917
			A=1/ 2		
0+	.1666666667	.2413311008	.3100386480	.3735620562	.4322959135
5+	.4864793194	.5363154140	.5820103312	.6237826078	.6618622150

D-INTEGRAL PROBABILITY M=R, B=10

R	1	2	3	4	5
			A=1/ 2		
10+	.6964863221	.7278945039	.7563244022	.7820081674	.8051697235
15+	.8260227850	.8447695182	.8615997319	.8766904914	.8902060627
20+	.9022981074	.9131060627	.9227576540	.9313694965	.9390477521
25+	.9458888151	.9519800071	.9574002645	.9622208078	.9665057853
30+	.9703128848	.9736939101	.9766953203	.9793587317	.9817213813
35+	.9838165548	.9856739778	.9873201743	.9887787927	.9900709026
40+	.9912152633	.9922285669	.9931256575	.9939197291	.9946225031
			A= .60		
0+	.1428571429	.1908772172	.2344377292	.2750142303	.3132864595
5+	.3495999243	.3841529112	.4170725837	.4484493713	.4783538758
10+	.5068456978	.5339782301	.5598013193	.5843627465	.6077090370
15+	.6298858788	.6509383114	.6709107815	.6898471235	.7077904988
20+	.7247833164	.7408671481	.7560826464	.7704694683	.7840662102
25+	.7969103518	.8090382121	.8204849153	.8312843669	.8414692388
30+	.8510709632	.8601197334	.8686445113	.8766730408	.8842318657
35+	.8913463522	.8980407142	.9043380422	.9102603341	.9158285276
40+	.9210625346	.9259812756	.9306027161	.9349439018	.9390209951
45+	.9428493107	.9464433516	.9498168437	.9529827706	.9559534070
50+	.9587403522	.9613545614	.9638063774	.9661055602	.9682613164
55+	.9702823268	.9721767740	.9739523677	.9756163696	.9771756177
60+	.9786365485	.9800052189	.9812873273	.9824882327	.9836129745
65+	.9846662894	.9856526293	.9865761771	.9874408620	.9882503746
70+	.9890081800	.9897175314	.9903814826	.9910028992	.9915844701
75+	.9921287181	.9926380094	.9931145632	.9935604606	.9939776528
			A= .65		
0+	.1333333333	.1712725837	.2052415478	.2367520076	.2665021141
5+	.2948541555	.3220178748	.3481254194	.3732659854	.3975034771
10+	.4208862061	.4434525299	.4652342783	.4862589070	.5065508876
15+	.5261326222	.5450250540	.5632480778	.5808208182	.5977618164
20+	.6140891553	.6298205418	.6449733595	.6595646998	.6736113797
25+	.6871299494	.7001366930	.7126476253	.7246784864	.7362447337
30+	.7473615342	.7580437567	.7683059640	.7781624062	.7876270143
35+	.7967133952	.8054348280	.8138042599	.8218343050	.8295372426
40+	.8369250170	.8440092383	.8508011836	.8573117992	.8635517033
45+	.8695311897	.8752602315	.8807484859	.8860052986	.8910397095
50+	.8958604583	.9004759902	.9048944622	.9091237494	.9131714515
55+	.9170448993	.9207511618	.9242970525	.9276891368	.9309337381
60+	.9340369454	.9370046196	.9398424007	.9425557139	.9451497769
65+	.9476296059	.9500000227	.9522656603	.9544309698	.9565002260
70+	.9584775340	.9603668343	.9621719093	.9638963884	.9655437537
75+	.9671173452	.9686203661	.9700558877	.9714268544	.9727360884
80+	.9739862944	.9751800639	.9763198798	.9774081202	.9784470631
85+	.9794388894	.9803856878	.9812894575	.9821521124	.9829754841
90+	.9837613256	.9845113142	.9852270545	.9859100817	.9865618643
95+	.9871838067	.9877772521	.9883434846	.9888837325	.9893991696
100+	.9898909183	.9903600515	.9908075946	.9912345274	.9916417865
105+	.9920302667	.9924008230	.9927542724	.9930913953	.9934129370
			A= .70		
0+	.1250000000	.1544912422	.1804871918	.2044155942	.2269355216
5+	.2483945343	.2689973190	.2888744253	.3081142424	.3267794946
10+	.3449164507	.3625603938	.3797390324	.3964747138	.4127859086
15+	.4286882328	.4441951680	.4593185789	.4740690907	.4884563678
20+	.5024893238	.5161762783	.5295250777	.5425431873	.5552377622
25+	.5676157021	.5796836940	.5914482452	.6029157089	.6140923042
30+	.6249841312	.6355971828	.6459373542	.6560104491	.6658221850
35+	.6753781971	.6846840409	.6937451939	.7025670568	.7111549545
40+	.7195141362	.7276497756	.7355669707	.7432707436	.7507660403

D-INTEGRAL PROBABILITY M=R,B=10

R	1	2	3	4	5
			A= .70		
45+	.7580577300	.7651506052	.7720493808	.7787586942	.7852831048
50+	.7916270938	.7977950640	.8037913397	.8096201667	.8152857124
55+	.8207920656	.8261432371	.8313431596	.8363956884	.8413046014
60+	.8460735996	.8507063080	.8552062758	.8595769770	.8638218114
65+	.8679441050	.8719471110	.8758340105	.8796079133	.8832718588
70+	.8868288172	.8902816900	.8936333111	.8968864481	.9000438027
75+	.9031080125	.9060816514	.9089672307	.9117672008	.9144839514
80+	.9171198131	.9196770583	.9221579026	.9245645054	.9268989711
85+	.9291633505	.9313596415	.9334897906	.9355556933	.9375591959
90+	.9395020959	.9413861435	.9432130426	.9449844514	.9467019840
95+	.9483672109	.9499816602	.9515468186	.9530641325	.9545350086
100+	.9559608148	.9573428818	.9586825031	.9599809365	.9612394049
105+	.9624590968	.9636411676	.9647867402	.9658969057	.9669727245
110+	.9680152269	.9690254139	.9700042578	.9709527033	.9718716678
115+	.9727620426	.9736246930	.9744604595	.9752701582	.9760545813
120+	.9768144984	.9775506562	.9782637797	.9789545729	.9796237187
125+	.9802718803	.9808997010	.9815078055	.9820967996	.9826672713
130+	.9832197913	.9837549131	.9842731737	.9847750941	.9852611798
135+	.9857319210	.9861877932	.9866292578	.9870567621	.9874707400
140+	.9878716122	.9882597867	.9886356593	.9889996136	.9893520216
145+	.9896932440	.9900236305	.9903435201	.9906532416	.9909531136
150+	.9912434449	.9915245349	.9917966739	.9920601433	.9923152155
			A= .75		
0+	.1176470588	.1400260416	.1594111899	.1770687701	.1935811444
5+	.2092582727	.2242849777	.2387806622	.2528271160	.2664829168
10+	.2797915173	.2927860839	.3054925428	.3179315776	.3301199850
15+	.3420716227	.3537980875	.3653092114	.3766134314	.3877180690
20+	.3986295448	.4093535462	.4198951577	.4302589659	.4404491425
25+	.4504695121	.4603236067	.4700147108	.4795458982	.4889200629
30+	.4981399443	.5072081484	.5161271659	.5248993871	.5335271144
35+	.5420125732	.5503579211	.5585652553	.5666366193	.5745740089
40+	.5823793764	.5900546353	.5976016632	.6050223053	.6123183768
45+	.6194916652	.6265439321	.6334769146	.6402923272	.6469918628
50+	.6535771932	.6600499708	.6664118288	.6726643819	.6788092271
55+	.6848479436	.6907820938	.6966132233	.7023428611	.7079725201
60+	.7135036970	.7189378727	.7242765121	.7295210646	.7346729641
65+	.7397336288	.7447044614	.7495868495	.7543821651	.7590917649
70+	.7637169904	.7682591678	.7727196082	.7770996072	.7814004456
75+	.7856233889	.7897696875	.7938405767	.7978372768	.8017609932
80+	.8056129161	.8093942211	.8131060687	.8167496048	.8203259602
85+	.8238362514	.8272815801	.8306630333	.8339816838	.8372385896
90+	.8404347947	.8435713287	.8466492069	.8496694307	.8526329874
95+	.8555408503	.8583939791	.8611933197	.8639398042	.8666343515
100+	.8692778668	.8718712423	.8744153568	.8769110762	.8793592532
105+	.8817607279	.8841163276	.8864268671	.8886931486	.8909159620
110+	.8930960849	.8952342831	.8973313100	.8993879076	.9014048057
115+	.9033827230	.9053223665	.9072244319	.9090896037	.9109185553
120+	.9127119494	.9144704376	.9161946610	.9178852501	.9195428252
125+	.9211679960	.9227613624	.9243235139	.9258550305	.9273564823
130+	.9288284297	.9302714238	.9316860061	.9330727090	.9344320559
135+	.9357645611	.9370707300	.9383510594	.9396060374	.9408361437
140+	.9420418496	.9432236181	.9443819042	.9455171549	.9466298093
145+	.9477202986	.9487890466	.9498364695	.9508629759	.9518689675
150+	.9528548383	.9538209758	.9547677600	.9556955645	.9566047557
155+	.9574956938	.9583687323	.9592242181	.9600624921	.9608838886
160+	.9616887361	.9624773569	.9632500676	.9640071786	.9647489949
165+	.9654758156	.9661879346	.9668856400	.9675692148	.9682389365

D-INTEGRAL PROBABILITY M=R,B=10

R	1	2	3	4	5
170+	.9688950776	.9695379055	.9701676825	.9707846661	.9713891087
175+	.9719812584	.9725613581	.9731296466	.9736863578	.9742317214
180+	.9747659628	.9752893028	.9758019583	.9763041420	.9767960624
185+	.9772779242	.9777499280	.9782122709	.9786651457	.9791087420
190+	.9795432455	.9799688382	.9803856989	.9807940027	.9811939215
195+	.9815856237	.9819692745	.9823450360	.9827130670	.9830735233
200+	.9834265575	.9837723196	.9841109562	.9844426115	.9847674264
205+	.9850855395	.9853970863	.9857021999	.9860010107	.9862936465
210+	.9865802327	.9868608921	.9871357450	.9874049096	.9876685016
215+	.9879266344	.9881794191	.9884269648	.9886693783	.9889067642
220+	.9891392251	.9893668618	.9895897726	.9898080542	.9900218014
225+	.9902311068	.9904360615	.9906367545	.9908332733	.9910257032
230+	.9912141284	.9913986309	.9915792913	.9917561884	.9919293998

A= .80

R	1	2	3	4	5
0+	.1111111111	.1274763107	.1413881990	.1539023524	.1655013915
5+	.1764432409	.1868827766	.1969208815	.2066272664	.2160522977
10+	.2252336544	.2342003212	.2429751109	.2515763260	.2600188901
15+	.2683151443	.2764754187	.2845084544	.2924217190	.3002216476
20+	.3079138291	.3155031525	.3229939227	.3303899536	.3376946432
25+	.3449110354	.3520418703	.3590896268	.3660565569	.3729447159
30+	.3797559869	.3864921023	.3931546617	.3997451473	.4062649379
35+	.4127153196	.4190974966	.4254125996	.4316616935	.4378457844
40+	.4439658249	.4500227200	.4560173314	.4619504814	.4678229572

A= .80

R	1	2	3	4	5
45+	.4736355131	.4793888745	.4850837397	.4907207823	.4963006540
50+	.5018239853	.5072913881	.5127034569	.5180607702	.5233638916
55+	.5286133710	.5338097456	.5389535411	.5440452720	.5490854427
60+	.5540745481	.5590130744	.5639014993	.5687402929	.5735299180
65+	.5782708302	.5829634792	.5876083081	.5922057546	.5967562505
70+	.6012602228	.6057180934	.6101302795	.6144971936	.6188192442
75+	.6230968355	.6273303677	.6315202374	.6356668371	.6397705562
80+	.6438317804	.6478508922	.6518282706	.6557642917	.6596593286
85+	.6635137512	.6673279265	.6711022187	.6748369893	.6785325968
90+	.6821893974	.6858077442	.6893879879	.6929304768	.6964355563
95+	.6999035697	.7033348574	.7067297577	.7100886063	.7134117366
100+	.7166994796	.7199521638	.7231701157	.7263536591	.7295031157
105+	.7326188050	.7357010441	.7387501477	.7417664285	.7447501969
110+	.7477017610	.7506214267	.7535094978	.7563662758	.7591920599
115+	.7619871475	.7647518334	.7674864104	.7701911693	.7728663985
120+	.7755123845	.7781294114	.7807177613	.7832777143	.7858095482
125+	.7883135388	.7907899596	.7932390824	.7956611765	.7980565094
130+	.8004253464	.8027679506	.8050845834	.8073755038	.8096409689
135+	.8118812338	.8140965514	.8162871728	.8184533468	.8205953205
140+	.8227133387	.8248076445	.8268784787	.8289260803	.8309506863
145+	.8329525317	.8349318496	.8368888711	.8388238253	.8407369395
150+	.8426284390	.8444985471	.8463474854	.8481754734	.8499827287
155+	.8517694673	.8535359029	.8552822477	.8570087118	.8587155037
160+	.8604028298	.8620708948	.8637199015	.8653500511	.8669615428
165+	.8685545741	.8701293406	.8716860363	.8732248534	.8747459822
170+	.8762496114	.8777359280	.8792051172	.8806573625	.8820928457
175+	.8835117470	.8849142448	.8863005159	.8876707355	.8890250769
180+	.8903637121	.8916868113	.8929945431	.8942870745	.8955645708
185+	.8968271961	.8980751124	.8993084805	.9005274596	.9017322073
190+	.9029228796	.9040996311	.9052626149	.9064119826	.9075478842
195+	.9086704684	.9097798822	.9108762714	.9119597803	.9130305517
200+	.9140887269	.9151344460	.9161678476	.9171890688	.9181982456
205+	.9191955124	.9201810022	.9211548470	.9221171771	.9230681216
210+	.9240078084	.9249363640	.9258539135	.9267605809	.9276564889

D-INTEGRAL PROBABILITY M=R,B=10

R	1	2	3	4	5
215+	.9285417589	.9294165110	.9302808642	.9311349361	.9319788433
220+	.9328127009	.9336366231	.9344507228	.9352551116	.9360499002
225+	.9368351978	.9376111127	.9383777521	.9391352219	.9398836270
230+	.9406230711	.9413536569	.9420754859	.9427886588	.9434932748
235+	.9441894324	.9448772288	.9455567605	.9462281225	.9468914093
240+	.9475467139	.9481941287	.9488337448	.9494656525	.9500899411
245+	.9507066989	.9513160133	.9519179706	.9525126564	.9531001551
250+	.9536805505	.9542539253	.9548203612	.9553799391	.9559327392
255+	.9564788406	.9570183215	.9575512594	.9580777310	.9585978118
260+	.9591115768	.9596191002	.9601204551	.9606157141	.9611049486
265+	.9615882298	.9620656274	.9625372110	.9630030489	.9634632090
270+	.9639177582	.9643667627	.9648102882	.9652483993	.9656811601
275+	.9661086340	.9665308836	.9669479707	.9673599567	.9677669020
280+	.9681688664	.9685659092	.9689580889	.9693454633	.9697280896
285+	.9701060243	.9704793234	.9708480420	.9712122349	.9715719560
290+	.9719272588	.9722781960	.9726248198	.9729671818	.9733053331
295+	.9736393240	.9739692044	.9742950235	.9746168301	.9749346723
300+	.9752485977	.9755586533	.9758648857	.9761673407	.9764660640
305+	.9767611003	.9770524940	.9773402891	.9776245289	.9779052563
310+	.9781825137	.9784563429	.9787267854	.9789938820	.9792576732
315+	.9795181990	.9797754989	.9800296119	.9802805766	.9805284312
320+	.9807732133	.9810149602	.9812537088	.9814894953	.9817223558
325+	.9819523259	.9821794405	.9824037345	.9826252421	.9828439973
330+	.9830600336	.9832733839	.9834840812	.9836921577	.9838976453
335+	.9841005756	.9843009799	.9844988889	.9846943331	.9848873427
340+	.9850779473	.9852661765	.9854520592	.9856356241	.9858168996
345+	.9859959137	.9861726942	.9863472683	.9865196632	.9866899056
350+	.9868580218	.9870240379	.9871879797	.9873498727	.9875097421
355+	.9876676126	.9878235090	.9879774553	.9881294757	.9882795938
360+	.9884278331	.9885742166	.9887187672	.9888615075	.9890024598
365+	.9891416461	.9892790883	.9894148077	.9895488258	.9896811635
370+	.9898118414	.9899408802	.9900683001	.9901941211	.9903183629
375+	.9904410450	.9905621869	.9906818074	.9907999255	.9909165598
380+	.9910317285	.9911454500	.9912577420	.9913686224	.9914781085

TABLE H (A > 1)

This table gives the value of the incomplete Dirichlet integral of type 2

$$D_A^{(B)}(R, M) = \frac{\Gamma(M+BR)}{\Gamma^R(R)\Gamma(M)} \int_A^{\infty} \cdots \int_A^{\infty} \frac{\prod_{i=1}^{B} x_i^{R-1} dx_i}{\left(1 + \sum_{i=1}^{B} x_i\right)^{M+BR}}$$

for

$$M = R+1$$
$$A^{-1} = 3(1)10 \quad \text{and} \quad A = .40(.10).60(.05).80,$$
$$B = 1(1)10$$

The value of R increases from 1 by ones at least until we reach the line containing .99; the last line is completed.

<u>Note</u>: The symbols M, B, R and A which are all capital here in the table correspond to the same lower case symbols in the text.

D-INTEGRAL PROBABILITY M=R+1, B= 1

R	1	2	3	4	5
			A=1/10		
0+	.8264462810	.9562188375	.9878293776	.9964597641	.9989426629
			A=1/ 9		
0+	.8100000000	.9477000000	.9841500000	.9949756500	.9983650626
			A=1/ 8		
0+	.7901234568	.9364426155	.9788330219	.9926301239	.9973666003
			A=1/ 7		
0+	.7656250000	.9211425781	.9708633423	.9887521863	.9955454748
			A=1/ 6		
0+	.7346938776	.8996251562	.9583761868	.9819815116	.9920017311
			A=1/ 5		
0+	.6944444444	.8680555556	.9377143347	.9693435881	.9845380333
5+	.9920749605	.9958912776	.9978515267	.9988690522	.9994014961
			A=1/ 4		
0+	.6400000000	.8192000000	.9011200000	.9437184000	.9672065024
5+	.9805947208	.9883900864	.9929964388	.9957479667	.9974051726
			A=1/ 3		
0+	.5625000000	.7382812500	.8305664062	.8861846924	.9218730927
5+	.9455977678	.9617292434	.9728700437	.9806522204	.9861355831
10+	.9900255660	.9928003500	.9947885895	.9962186228	.9972504658
			A= .40		
0+	.5102040816	.6768013328	.7702997901	.8310130740	.8731406588
5+	.9034846574	.9258666739	.9426482213	.9553814729	.9651307887
10+	.9726485240	.9784786045	.9830210865	.9865741905	.9893626190
15+	.9915571595	.9932885545	.9946574921	.9957419083	.9966023894
			A=1/ 2		
0+	.4444444444	.5925925926	.6803840878	.7413504039	.7868719199
5+	.8222775435	.8505378145	.8734992846	.8923976140	.9081042255
10+	.9212579718	.9323412210	.9417266417	.9497074417	.9565177243
15+	.9623466641	.9673486609	.9716507898	.9753583801	.9785592665
20+	.9813270775	.9837238137	.9858018913	.9876057780	.9891733156
25+	.9905367949	.9917238387	.9927581280	.9936600053	.9944469767
			A= .60		
0+	.3906250000	.5187988281	.5960464478	.6513670087	.6942718755
5+	.7290805115	.7581395067	.7828783802	.8042400719	.8228838698
10+	.8392916094	.8538278413	.8667761140	.8783619877	.8887682512
15+	.8981453441	.9066187099	.9142941194	.9212616121	.9275984724
20+	.9333715153	.9386388703	.9434513921	.9478537906	.9518855461
25+	.9555816571	.9589732591	.9620881378	.9649511615	.9675846446
30+	.9700086576	.9722412916	.9742988857	.9761962236	.9779467040
35+	.9795624887	.9810546332	.9824331999	.9837073590	.9848854769
40+	.9859751948	.9869834986	.9879167811	.9887808970	.9895812131
45+	.9903226526	.9910097347	.9916466110	.9922370977	.9927847046
			A= .65		
0+	.3673094582	.4856984571	.5568856660	.6081236910	.6482245118
5+	.6811355390	.7089737002	.7330132747	.7540857470	.7727669301
10+	.7894741878	.8045211182	.8181502899	.8305538553	.8418870810
15+	.8522775469	.8618315896	.8706389333	.8787760942	.8863089330
20+	.8932946020	.8997830545	.9058182313	.9114390044	.9166799378
25+	.9215719057	.9261426006	.9304169542	.9344174892	.9381646148
30+	.9416768776	.9449711752	.9480629396	.9509662949	.9536941946
35+	.9562585406	.9586702875	.9609395344	.9630756056	.9650871216
40+	.9669820632	.9687678275	.9704512786	.9720387928	.9735362990
45+	.9749493155	.9762829829	.9775420939	.9787311205	.9798542385
50+	.9809153500	.9819181033	.9828659122	.9837619727	.9846092785
55+	.9854106356	.9861686752	.9868858658	.9875645245	.9882068274
60+	.9888148184	.9893904190	.9899354353	.9904515665	.9909404112

D-INTEGRAL PROBABILITY M=R+1, B= 1

R	1	2	3	4	5
			A= .70		
0+	.3460207612	.4549753954	.5199363697	.5666481430	.6033301984
5+	.6336197220	.6594444231	.6819516777	.7018822005	.7197438108
10+	.7359011554	.7506258635	.7641263912	.7765667000	.7880784440
15+	.7987692063	.8087282341	.8180305357	.8267398746	.8349110012
20+	.8425913453	.8498223205	.8566403428	.8630776373	.8691628839
25+	.8749217394	.8803772642	.8855502738	.8904596310	.8951224902
30+	.8995545038	.9037699966	.9077821155	.9116029574	.9152436796
35+	.9187145958	.9220252600	.9251845387	.9282006760	.9310813497
40+	.9338337217	.9364644824	.9389798904	.9413858081	.9436877335
45+	.9458908287	.9479999459	.9500196505	.9519542423	.9538077748
50+	.9555840724	.9572867467	.9589192107	.9604846923	.9619862464
55+	.9634267662	.9648089937	.9661355288	.9674088384	.9686312644
60+	.9698050314	.9709322535	.9720149407	.9730550052	.9740542665
65+	.9750144573	.9759372279	.9768241505	.9776767242	.9784963784
70+	.9792844763	.9800423193	.9807711491	.9814721518	.9821464600
75+	.9827951559	.9834192739	.9840198027	.9845976876	.9851538330
80+	.9856891041	.9862043285	.9867002987	.9871777733	.9876374786
85+	.9880801103	.9885063349	.9889167910	.9893120903	.9896928196
90+	.9900595412	.9904127941	.9907530954	.9910809412	.9913968071
			A= .75		
0+	.3265306122	.4264889629	.4852399935	.5272050154	.5600918897
5+	.5872738980	.6105199304	.6308720690	.6489964224	.6653453290
10+	.6802406365	.6939200005	.7065642855	.7183146232	.7292834859
15+	.7395621311	.7492257590	.7583371797	.7669494784	.7751079938
20+	.7828518095	.7902148997	.7972270180	.8039143994	.8103003178
25+	.8164055365	.8222486733	.8278465006	.8332141932	.8383655349
30+	.8433130920	.8480683596	.8526418860	.8570433789	.8612817971
35+	.8653654288	.8693019604	.8730985358	.8767618087	.8802979886
40+	.8837128804	.8870119212	.8902002111	.8932825419	.8962634220
45+	.8991470996	.9019375822	.9046386555	.9072538995	.9097867038
50+	.9122402809	.9146176785	.9169217910	.9191553695	.9213210311
55+	.9234212678	.9254584542	.9274348548	.9293526305	.9312138451
60+	.9330204707	.9347743931	.9364774168	.9381312693	.9397376053
65+	.9412980109	.9428140071	.9442870531	.9457185496	.9471098418
70+	.9484622220	.9497769326	.9510551679	.9522980770	.9535067658
75+	.9546822988	.9558257012	.9569379607	.9580200292	.9590728241
80+	.9600972305	.9610941017	.9620642615	.9630085048	.9639275990
85+	.9648222853	.9656932796	.9665412736	.9673669358	.9681709125
90+	.9689538285	.9697162880	.9704588757	.9711821571	.9718866795
95+	.9725729726	.9732415493	.9738929062	.9745275242	.9751458692
100+	.9757483924	.9763355313	.9769077096	.9774653380	.9780088150
105+	.9785385266	.9790548472	.9795581401	.9800487575	.9805270412
110+	.9809933229	.9814479242	.9818911575	.9823233261	.9827447241
115+	.9831556372	.9835563430	.9839471109	.9843282025	.9846998719
120+	.9850623662	.9854159252	.9857607819	.9860971629	.9864252883
125+	.9867453719	.9870576218	.9873622401	.9876594233	.9879493625
130+	.9882322434	.9885082468	.9887775483	.9890403188	.9892967246
135+	.9895469274	.9897910844	.9900293487	.9902618693	.9904887909
			A= .80		
0+	.3086419753	.4000914495	.4527783893	.4899473063	.5188564640
5+	.5426500094	.5629605490	.5807408194	.5965942841	.6109271411
10+	.6240259521	.6361005763	.6473094511	.6577752539	.6675950195
15+	.6768469104	.6855948857	.6938920051	.7017828210	.7093051447
20+	.7164913740	.7233695068	.7299639262	.7362960162	.7423846501
25+	.7482465828	.7538967676	.7593486157	.7646142097	.7697044809
30+	.7746293573	.7793978885	.7840183508	.7884983374	.7928448351
35+	.7970642905	.8011626671	.8051454949	.8090179138	.8127847112

D-INTEGRAL PROBABILITY M=R+1, B= 1

R	1	2	3	4	5
			A= .80		
40+	.8164503553	.8200190247	.8234946340	.8268808571	.8301811471
45+	.8333987555	.8365367477	.8395980182	.8425853037	.8455011950
50+	.8483481477	.8511284920	.8538444419	.8564981026	.8590914785
55+	.8616264796	.8641049277	.8665285621	.8688990448	.8712179648
60+	.8734868435	.8757071375	.8778802434	.8800075007	.8820901950
65+	.8841295613	.8861267863	.8880830115	.8899993352	.8918768148
70+	.8937164690	.8955192796	.8972861935	.8990181243	.9007159537
75+	.9023805335	.9040126863	.9056132076	.9071828665	.9087224070
80+	.9102325491	.9117139901	.9131674051	.9145934483	.9159927538
85+	.9173659364	.9187135922	.9200362997	.9213346203	.9226090990
90+	.9238602651	.9250886325	.9262947009	.9274789556	.9286418685
95+	.9297838986	.9309054922	.9320070836	.9330890953	.9341519387
100+	.9351960142	.9362217119	.9372294114	.9382194828	.9391922866
105+	.9401481743	.9410874883	.9420105626	.9429177228	.9438092867
110+	.9446855639	.9455468569	.9463934607	.9472256631	.9480437452
115+	.9488479815	.9496386399	.9504159821	.9511802636	.9519317343
120+	.9526706380	.9533972132	.9541116928	.9548143047	.9555052715
125+	.9561848108	.9568531355	.9575104536	.9581569689	.9587928805
130+	.9594183830	.9600336673	.9606389197	.9612343228	.9618200553
135+	.9623962919	.9629632039	.9635209588	.9640697207	.9646096503
140+	.9651409049	.9656636386	.9661780023	.9666841440	.9671822084
145+	.9676723374	.9681546701	.9686293429	.9690964891	.9695562397
150+	.9700087229	.9704540645	.9708923877	.9713238134	.9717484600
155+	.9721664438	.9725778786	.9729828761	.9733815459	.9737739955
160+	.9741603302	.9745406534	.9749150666	.9752836693	.9756465590
165+	.9760038315	.9763555809	.9767018993	.9770428772	.9773786036
170+	.9777091655	.9780346484	.9783551365	.9786707121	.9789814562
175+	.9792874482	.9795887662	.9798854866	.9801776847	.9804654344
180+	.9807488082	.9810278772	.9813027114	.9815733795	.9818399489
185+	.9821024859	.9823610555	.9826157217	.9828665474	.9831135940
190+	.9833569224	.9835965920	.9838326612	.9840651877	.9842942279
195+	.9845198372	.9847420703	.9849609806	.9851766209	.9853890429
200+	.9855982975	.9858044347	.9860075037	.9862075526	.9864046291
205+	.9865987797	.9867900504	.9869784863	.9871641316	.9873470301
210+	.9875272244	.9877047569	.9878796688	.9880520009	.9882217934
215+	.9883890855	.9885539159	.9887163229	.9888763437	.9890340153
220+	.9891893739	.9893424550	.9894932938	.9896419246	.9897883814
225+	.9899326975	.9900749056	.9902150381	.9903531267	.9904892025

D-INTEGRAL PROBABILITY M=R+1,B= 2

R	1	2	3	4	5
			A=1/10		
0+	.6944444444	.9162808642	.9760880612	.9929718548	.9978920355
			A=1/ 9		
0+	.6694214876	.9007767726	.9690044006	.9900518624	.9967452180
			A=1/ 8		
0+	.6400000000	.8806400000	.9588736000	.9854648320	.9947696071
			A=1/ 7		
0+	.6049382716	.8538840624	.9439075046	.9779495970	.9911863172
			A=1/ 6		
0+	.5625000000	.8173828125	.9209461212	.9650134742	.9842791291
5+	.9928555871	.9967243892	.9984874773	.9992975074	.9996721390
			A=1/ 5		
0+	.5102040816	.7660498602	.8841371975	.9414209774	.9699734287
5+	.9844532964	.9918872994	.9957397937	.9977511513	.9988076713
			A=1/ 4		
0+	.4444444444	.6913580247	.8221307727	.8954597030	.9377870733
5+	.9626482352	.9774234862	.9862808273	.9916263293	.9948699494
			A=1/ 3		
0+	.3600000000	.5788800000	.7122816000	.7996216320	.8588185657
5+	.8997405429	.9283895554	.9486239681	.9630088123	.9732869966
10+	.9806608171	.9859687724	.9898005000	.9925733271	.9945841662
			A= .40		
0+	.3086419753	.5017019763	.6270076878	.7154655039	.7804753081
5+	.8293478464	.8666175533	.8953174876	.9175745882	.9349275525
10+	.9485136794	.9591866132	.9675944344	.9742334581	.9794863692
15+	.9836498498	.9869549322	.9895821919	.9916732083	.9933392828
			A=1/ 2		
0+	.2500000000	.4062500000	.5122070312	.5918579102	.6547799110
5+	.7058967352	.7481590621	.7835185701	.8133553532	.8386922238
10+	.8603133953	.8788354780	.8947526381	.9084667893	.9203086188
15+	.9305527323	.9394288739	.9471304353	.9538210357	.9596396960
20+	.9647049634	.9691182386	.9729664867	.9763244645	.9792565648
25+	.9818183522	.9840578491	.9860166186	.9877306785	.9892312772
30+	.9905455535	.9916970991	.9927064402	.9935914499	.9943677025
			A= .60		
0+	.2066115702	.3314167562	.4163432585	.4816479609	.5349602713
5+	.5799912140	.6188548902	.6528897889	.6830093953	.7098728401
10+	.7339764139	.7557064545	.7753717878	.7932246218	.8094745605
15+	.8242983342	.8378467585	.8502498407	.8616206127	.8720580655
20+	.8816494372	.8904720226	.8985946284	.9060787549	.9129795689
25+	.9193467118	.9252249779	.9306548877	.9356731767	.9403132150
30+	.9446053694	.9485773182	.9522543260	.9556594844	.9588139245
35+	.9617370044	.9644464758	.9669586327	.9692884432	.9714496690
40+	.9734549709	.9753160056	.9770435116	.9786473871	.9801367614
45+	.9815200583	.9828050546	.9839989329	.9851083299	.9861393802
50+	.9870977566	.9879887065	.9888170855	.9895873883	.9903037762
			A= .65		
0+	.1890359168	.3003258217	.3753449295	.4330405349	.4804034559
5+	.5207670405	.5559928533	.5872366091	.6152722594	.6406483688
10+	.6637712401	.6849525183	.7044381154	.7224266536	.7390817066
15+	.7545402050	.7689183786	.7823160638	.7948198976	.8065057304
20+	.8174404835	.8276835993	.8372881908	.8463019650	.8547679714
25+	.8627252177	.8702091792	.8772522250	.8838839776	.8901316183
30+	.8960201485	.9015726146	.9068103031	.9117529107	.9164186927
35+	.9208245944	.9249863667	.9289186689	.9326351604	.9361485827
40+	.9394708330	.9426130301	.9455855749	.9483982041	.9510600394
45+	.9535796324	.9559650057	.9582236895	.9603627567	.9623888536
50+	.9643082290	.9661267608	.9678499804	.9694830953	.9710310101

D-INTEGRAL PROBABILITY M=R+1,B= 2

R	1	2	3	4	5
			A= .65		
55+	.9724983459	.9738894579	.9752084527	.9764592031	.9776453628
60+	.9787703799	.9798375093	.9808498242	.9818102270	.9827214596
65+	.9835861128	.9844066351	.9851853411	.9859244190	.9866259384
70+	.9872918565	.9879240252	.9885241963	.9890940277	.9896350888
75+	.9901488648	.9906367621	.9911001123	.9915401765	.9919581490
			A= .70		
0+	.1736111111	.2727744020	.3385800011	.3888787557	.4301457812
5+	.4654198402	.4963714755	.5240215984	.5490432229	.5719060079
10+	.5929525555	.6124418047	.6305752163	.6475133530	.6633868072
15+	.6783036488	.6923546520	.7056170548	.7181573240	.7300332277
20+	.7412954159	.7519886448	.7621527377	.7718233501	.7810325840
25+	.7898094876	.7981804658	.8061696193	.8137990282	.8210889904
30+	.8280582233	.8347240359	.8411024767	.8472084611	.8530558817
35+	.8586577046	.8640260543	.8691722874	.8741070589	.8788403798
40+	.8833816693	.8877398014	.8919231460	.8959396069	.8997966549
45+	.9035013591	.9070604141	.9104801656	.9137666333	.9169255320
50+	.9199622908	.9228820709	.9256897822	.9283900979	.9309874687
55+	.9334861353	.9358901411	.9382033424	.9404294190	.9425718843
60+	.9446340932	.9466192516	.9485304231	.9503705374	.9521423962
65+	.9538486800	.9554919544	.9570746752	.9585991944	.9600677644
70+	.9614825436	.9628456002	.9641589167	.9654243940	.9666438549
75+	.9678190477	.9689516499	.9700432708	.9710954552	.9721096858
80+	.9730873861	.9740299230	.9749386092	.9758147056	.9766594236
85+	.9774739269	.9782593338	.9790167193	.9797471163	.9804515179
90+	.9811308791	.9817861178	.9824181170	.9830277257	.9836157606
95+	.9841830074	.9847302218	.9852581311	.9857674350	.9862588070
100+	.9867328950	.9871903227	.9876316907	.9880575770	.9884685378
105+	.9888651090	.9892478065	.9896171270	.9899735488	.9903175327
			A= .75		
0+	.1600000000	.2483200000	.3056896000	.3490140160	.3843351876
5+	.4144511528	.4408805743	.4645399371	.4860259064	.5057497043
10+	.5240076029	.5410207775	.5569592165	.5719567755	.5861210432
15+	.5995400294	.6122868325	.6244229786	.6360008628	.6470655701
20+	.6576562557	.6678072080	.6775486786	.6869075387	.6959078031
25+	.7045710542	.7129167877	.7209626964	.7287249064	.7362181739
30+	.7434560505	.7504510240	.7572146377	.7637575927	.7700898362
35+	.7762206376	.7821586551	.7879119936	.7934882553	.7988945848
40+	.8041377084	.8092239694	.8141593591	.8189495449	.8235998952
45+	.8281155019	.8325012007	.8367615888	.8409010424	.8449237308
50+	.8488336308	.8526345384	.8563300809	.8599237270	.8634187964
55+	.8668184684	.8701257906	.8733436857	.8764749589	.8795223039
60+	.8824883094	.8853754640	.8881861618	.8909227070	.8935873182
65+	.8961821329	.8987092113	.9011705397	.9035680343	.9059035442
70+	.9081788542	.9103956883	.9125557118	.9146605338	.9167117100
75+	.9187107447	.9206590929	.9225581625	.9244093159	.9262138719
80+	.9279731079	.9296882607	.9313605286	.9329910729	.9345810189
85+	.9361314577	.9376434470	.9391180127	.9405561499	.9419588240
90+	.9433269717	.9446615021	.9459632977	.9472332152	.9484720863
95+	.9496807190	.9508598978	.9520103848	.9531329205	.9542282244
100+	.9552969956	.9563399137	.9573576393	.9583508145	.9593200635
105+	.9602659935	.9611891949	.9620902418	.9629696928	.9638280910
110+	.9646659652	.9654838294	.9662821842	.9670615165	.9678223003
115+	.9685649967	.9692900549	.9699979118	.9706889931	.9713637131
120+	.9720224751	.9726656720	.9732936864	.9739068907	.9745056480
125+	.9750903116	.9756612258	.9762187260	.9767631390	.9772947832
130+	.9778139687	.9783209978	.9788161649	.9792997572	.9797720543
135+	.9802333288	.9806838465	.9811238663	.9815536407	.9819734157

D-INTEGRAL PROBABILITY M=R+1,B= 2

R	1	2	3	4	5
			A= .75		
140+	.9823834314	.9827839214	.9831751140	.9835572313	.9839304902
145+	.9842951021	.9846512730	.9849992039	.9853390909	.9856711251
150+	.9859954929	.9863123761	.9866219521	.9869243938	.9872198700
155+	.9875085453	.9877905802	.9880661313	.9883353515	.9885983897
160+	.9888553915	.9891064987	.9893518497	.9895915797	.9898258203
165+	.9900547003	.9902783452	.9904968773	.9907104163	.9909190787
			A= .80		
0+	.1479289941	.2265730838	.2763051816	.3131965225	.3429137591
5+	.3680502743	.3899981512	.4095886062	.4273572274	.4436694205
10+	.4587858314	.4728991259	.4861559372	.4986706126	.5105341776
15+	.5218203831	.5325899050	.5428933346	.5527733557	.5622663601
20+	.5714036673	.5802124599	.5887165102	.5969367526	.6048917389
25+	.6125980046	.6200703667	.6273221667	.6343654727	.6412112460
30+	.6478694815	.6543493260	.6606591785	.6668067755	.6727992643
35+	.6786432662	.6843449314	.6899099861	.6953437742	.7006512939
40+	.7058372293	.7109059788	.7158616803	.7207082329	.7254493176
45+	.7300884139	.7346288167	.7390736499	.7434258794	.7476883250
50+	.7518636705	.7559544733	.7599631734	.7638921008	.7677434832
55+	.7715194521	.7752220493	.7788532321	.7824148785	.7859087919
60+	.7893367054	.7927002860	.7960011378	.7992408060	.8024207798
65+	.8055424951	.8086073379	.8116166461	.8145717125	.8174737867
70+	.8203240773	.8231237536	.8258739479	.8285757567	.8312302428
75+	.8338384362	.8364013363	.8389199124	.8413951055	.8438278295
80+	.8462189719	.8485693952	.8508799381	.8531514158	.8553846214
85+	.8575803267	.8597392827	.8618622210	.8639498537	.8660028746
90+	.8680219600	.8700077689	.8719609437	.8738821110	.8757718821
95+	.8776308531	.8794596059	.8812587086	.8830287155	.8847701683
100+	.8864835956	.8881695142	.8898284286	.8914608323	.8930672073
105+	.8946480249	.8962037459	.8977348211	.8992416911	.9007247871
110+	.9021845310	.9036213355	.9050356047	.9064277338	.9077981101
115+	.9091471124	.9104751119	.9117824719	.9130695485	.9143366902
120+	.9155842387	.9168125286	.9180218879	.9192126380	.9203850939
125+	.9215395644	.9226763521	.9237957538	.9248980606	.9259835577
130+	.9270525250	.9281052370	.9291419628	.9301629666	.9311685074
135+	.9321588395	.9331342122	.9340948703	.9350410541	.9359729991
140+	.9368909369	.9377950944	.9386856948	.9395629568	.9404270954
145+	.9412783215	.9421168423	.9429428614	.9437565784	.9445581897
150+	.9453478879	.9461258624	.9468922991	.9476473808	.9483912868
155+	.9491241935	.9498462741	.9505576989	.9512586351	.9519492470
160+	.9526296962	.9533001415	.9539607387	.9546116413	.9552530000
165+	.9558849629	.9565076757	.9571212815	.9577259210	.9583217327
170+	.9589088525	.9594874143	.9600575494	.9606193873	.9611730551
175+	.9617186779	.9622563786	.9627862782	.9633084956	.9638231478
180+	.9643303498	.9648302148	.9653228542	.9658083774	.9662868921
185+	.9667585043	.9672233182	.9676814364	.9681329597	.9685779874
190+	.9690166172	.9694489451	.9698750657	.9702950720	.9707090555
195+	.9711171063	.9715193131	.9719157630	.9723065419	.9726917343
200+	.9730714232	.9734456906	.9738146169	.9741782814	.9745367622
205+	.9748901359	.9752384781	.9755818633	.9759203647	.9762540544
210+	.9765830032	.9769072811	.9772269569	.9775420981	.9778527715
215+	.9781590427	.9784609762	.9787586357	.9790520838	.9793413821
220+	.9796265913	.9799077712	.9801849806	.9804582774	.9807277188
225+	.9809933609	.9812552591	.9815134677	.9817680405	.9820190302
230+	.9822664890	.9825104680	.9827510177	.9829881878	.9832220272
235+	.9834525841	.9836799060	.9839040397	.9841250311	.9843429257
240+	.9845577681	.9847696022	.9849784715	.9851844187	.9853874857
245+	.9855877139	.9857851443	.9859798169	.9861717713	.9863610465

D-INTEGRAL PROBABILITY M=R+1,B= 2

R	1	2	3	4	5
			A= .80		
250+	.9865476811	.9867317126	.9869131786	.9870921156	.9872685599
255+	.9874425470	.9876141122	.9877832899	.9879501143	.9881146189
260+	.9882768369	.9884368007	.9885945425	.9887500940	.9889034862
265+	.9890547500	.9892039155	.9893510125	.9894960705	.9896391184
270+	.9897801847	.9899192974	.9900564844	.9901917727	.9903251894

D-INTEGRAL PROBABILITY M=R+1,B= 3

R	1	2	3	4	5
			A=1/10		
0+	.5917159763	.8796701920	.9647476062	.9895344948	.9968479983
			A=1/ 9		
0+	.5625000000	.8583984375	.9545059204	.9852241785	.9951400935
			A=1/ 8		
0+	.5289256198	.8311946357	.9399994773	.9784920713	.9922077462
			A=1/ 7		
0+	.4900000000	.7957600000	.9188530960	.9675566703	.9869176840
5+	.9946804182	.9978205823	.9991012281	.9996272640	.9998446752
			A=1/ 6		
0+	.4444444444	.7486663618	.8870172481	.9489795433	.9768113366
5+	.9893913793	.9951167838	.9977397709	.9989487100	.9995089157
			A=1/ 5		
0+	.3906250000	.6847381592	.8373754099	.9158016269	.9562020682
5+	.9771087815	.9879812638	.9936629978	.9966458110	.9982183928
			A=1/ 4		
0+	.3265306122	.5962141625	.7572037860	.8533774474	.9111227244
5+	.9459450962	.9670231629	.9798250535	.9876246644	.9923904156
			A=1/ 3		
0+	.2500000000	.4722222222	.6244212963	.7308861073	.8063336404
5+	.8602053627	.8988573465	.9266836516	.9467681211	.9612950501
10+	.9718209542	.9794597378	.9850111212	.9890507322	.9919937887
			A= .40		
0+	.2066115702	.3932266515	.5285688859	.6304942600	.7088556709
5+	.7698027496	.8175428035	.8551132629	.8847779913	.9082578176
10+	.9268774947	.9416656059	.9534255648	.9627876390	.9702478830
15+	.9761977106	.9809465971	.9847396417	.9877712332	.9901957298
			A=1/ 2		
0+	.1600000000	.3020800000	.4085923840	.4935343145	.5635344205
5+	.6223034080	.6722128809	.7149211639	.7516608402	.7833874603
10+	.8108640433	.8347128963	.8554494489	.8735055190	.8892460118
15+	.9029813364	.9149769096	.9254606035	.9346286959	.9426506990
20+	.9496733281	.9558237964	.9612125755	.9659357213	.9700768479
25+	.9737088083	.9768951324	.9796912607	.9821456057	.9843004676
30+	.9861928255	.9878550220	.9893153568	.9905986025	.9917264515
			A= .60		
0+	.1275510204	.2356054875	.3159701959	.3810424273	.4361771901
5+	.4841323196	.5265363338	.5644496484	.5986142207	.6295781868
10+	.6577638534	.6835074719	.7070838617	.7287223849	.7486177310
15+	.7669374495	.7838273701	.7994156057	.8138155793	.8271283589
20+	.8394444941	.8508454843	.8614049701	.8711897132	.8802604131
25+	.8886723940	.8964761892	.9037180426	.9104403417	.9166819955
30+	.9224787651	.9278635558	.9328666749	.9375160621	.9418374952
35+	.9458547741	.9495898873	.9530631613	.9562933966	.9592979902
40+	.9620930482	.9646934869	.9671131265	.9693647756	.9714603092
45+	.9734107397	.9752262829	.9769164175	.9784899405	.9799550181

D-INTEGRAL PROBABILITY M=R+1,B= 3

R	1	2	3	4	5
			A= .60		
50+	.9813192324	.9825896248	.9837727353	.9848746397	.9859009830
55+	.9868570112	.9877475999	.9885772814	.9893502690	.9900704802
			A= .65		
0+	.1149095088	.2092779592	.2784184270	.3341878480	.3815614649
5+	.4230379739	.4600555545	.4935262678	.5240714536	.5521382406
10+	.5780625707	.6021057487	.6244768588	.6453471607	.6648596966
15+	.6831359186	.7002803893	.7163842011	.7315275176	.7457815000
20+	.7592097923	.7718696857	.7838130434	.7950870464	.8057348003
25+	.8157958362	.8253065277	.8343004410	.8428086323	.8508599019
30+	.8584810125	.8656968797	.8725307371	.8790042823	.8851378051
35+	.8909503019	.8964595771	.9016823347	.9066342604	.9113300952
40+	.9157837033	.9200081325	.9240156704	.9278178951	.9314257221
45+	.9348494470	.9380987854	.9411829088	.9441104793	.9468896799
50+	.9495282442	.9520334831	.9544123100	.9566712640	.9588165319
55+	.9608539685	.9627891157	.9646272202	.9663732504	.9680319118
60+	.9696076618	.9711047236	.9725270988	.9738785801	.9751627621
65+	.9763830525	.9775426823	.9786447149	.9796920555	.9806874596
70+	.9816335407	.9825327779	.9833875234	.9842000088	.9849723516
75+	.9857065613	.9864045450	.9870681126	.9876989821	.9882987843
80+	.9888690674	.9894113012	.9899268810	.9904171320	.9908833121
			A= .70		
0+	.1040582726	.1865885219	.2457959900	.2930683826	.3330746829
5+	.3681168091	.3994959157	.4280219621	.4542362000	.4785204727
10+	.5011559366	.5223568989	.5422914398	.5610945736	.5788769754
15+	.5957309599	.6117346934	.6269552332	.6414507692	.6552723076
20+	.6684649575	.6810689289	.6931203168	.7046517266	.7156927764
25+	.7262705065	.7364097154	.7461332384	.7554621802	.7644161110
30+	.7730132331	.7812705219	.7892038488	.7968280861	.8041571989
35+	.8112043248	.8179818446	.8245014444	.8307741701	.8368104768
40+	.8426202727	.8482129578	.8535974598	.8587822655	.8637754505
45+	.8685847047	.8732173570	.8776803967	.8819804943	.8861240193
50+	.8901170577	.8939654276	.8976746941	.9012501822	.9046969900
55+	.9080200001	.9112238906	.9143131452	.9172920630	.9201647669
60+	.9229352129	.9256071970	.9281843633	.9306702107	.9330680995
65+	.9353812579	.9376127873	.9397656686	.9418427670	.9438468368
70+	.9457805268	.9476463842	.9494468593	.9511843091	.9528610016
75+	.9544791192	.9560407623	.9575479527	.9590026365	.9604066877
80+	.9617619102	.9630700417	.9643327553	.9655516627	.9667283162
85+	.9678642114	.9689607891	.9700194374	.9710414940	.9720282480
90+	.9729809416	.9739007720	.9747888934	.9756464180	.9764744181
95+	.9772739276	.9780459431	.9787914254	.9795113010	.9802064634
100+	.9808777740	.9815260636	.9821521335	.9827567563	.9833406775
105+	.9839046157	.9844492645	.9849752927	.9854833454	.9859740452
110+	.9864479925	.9869057667	.9873479266	.9877750115	.9881875419
115+	.9885860196	.9889709293	.9893427384	.9897018980	.9900488435
			A= .75		
0+	.0946745562	.1669608224	.2174625446	.2571406286	.2904007080
5+	.3193827698	.3452799872	.3688251837	.3905014657	.4106449633
10+	.4294996802	.4472489151	.4640343071	.4799679264	.4951402516
15+	.5096256073	.5234859771	.5367737411	.5495336856	.5618045045
20+	.5736199417	.5850096716	.5959999882	.6066143510	.6168738210
25+	.6267974146	.6364023909	.6457044896	.6547181276	.6634565639
30+	.6719320374	.6801558851	.6881386412	.6958901236	.7034195070
35+	.7107353871	.7178458362	.7247584514	.7314803978	.7380184452
40+	.7443790020	.7505681442	.7565916416	.7624549814	.7681633892
45+	.7737218474	.7791351128	.7844077317	.7895440534	.7945482437
50+	.7994242958	.8041760408	.8088071579	.8133211824	.8177215145

D-INTEGRAL PROBABILITY M=R+1, B= 3

R	1	2	3	4	5
			A= .75		
55+	.8220114265	.8261940697	.8302724807	.8342495875	.8381282147
60+	.8419110889	.8456008432	.8492000217	.8527110836	.8561364072
65+	.8594782932	.8627389683	.8659205886	.8690252420	.8720549516
70+	.8750116780	.8778973219	.8807137264	.8834626797	.8861459163
75+	.8887651200	.8913219250	.8938179183	.8962546411	.8986335904
80+	.9009562207	.9032239457	.9054381392	.9076001367	.9097112369
85+	.9117727026	.9137857620	.9157516099	.9176714088	.9195462897
90+	.9213773533	.9231656712	.9249122860	.9266182131	.9282844410
95+	.9299119323	.9315016245	.9330544306	.9345712399	.9360529189
100+	.9375003115	.9389142403	.9402955065	.9416448911	.9429631552
105+	.9442510407	.9455092705	.9467385496	.9479395651	.9491129869
110+	.9502594683	.9513796460	.9524741413	.9535435598	.9545884922
115+	.9556095145	.9566071887	.9575820628	.9585346715	.9594655364
120+	.9603751663	.9612640575	.9621326944	.9629815495	.9638110840
125+	.9646217477	.9654139797	.9661882085	.9669448521	.9676843187
130+	.9684070065	.9691133043	.9698035914	.9704782383	.9711376063
135+	.9717820483	.9724119089	.9730275243	.9736292228	.9742173249
140+	.9747921434	.9753539840	.9759031447	.9764399170	.9769645851
145+	.9774774266	.9779787128	.9784687083	.9789476717	.9794158555
150+	.9798735061	.9803208643	.9807581654	.9811856388	.9816035088
155+	.9820119946	.9824113100	.9828016640	.9831832606	.9835562992
160+	.9839209745	.9842774766	.9846259914	.9849667002	.9852997804
165+	.9856254050	.9859437432	.9862549603	.9865592177	.9868566729
170+	.9871474802	.9874317900	.9877097493	.9879815017	.9882471876
175+	.9885069441	.9887609052	.9890092017	.9892519615	.9894893095
180+	.9897213679	.9899482559	.9901700900	.9903869841	.9905990495
			A= .80		
0+	.0865051903	.1499167128	.1928391701	.2258355552	.2530756951
5+	.2765585172	.2973856880	.3162265215	.3335176563	.3495597785
10+	.3645689370	.3787057715	.3920931328	.4048272129	.4169848520
15+	.4286284977	.4398096653	.4505714129	.4609501501	.4709769842
20+	.4806787398	.4900787424	.4991974289	.5080528292	.5166609497
25+	.5250360834	.5331910605	.5411374556	.5488857574	.5564455114
30+	.5638254384	.5710335353	.5780771601	.5849631047	.5916976577
35+	.5982866574	.6047355392	.6110493754	.6172329113	.6232905956
40+	.6292266081	.6350448833	.6407491325	.6463428621	.6518293907
45+	.6572118646	.6624932705	.6676764485	.6727641027	.6777588111
50+	.6826630347	.6874791252	.6922093332	.6968558141	.7014206347
55+	.7059057788	.7103131522	.7146445874	.7189018481	.7230866332
60+	.7272005802	.7312452692	.7352222253	.7391329219	.7429787838
65+	.7467611888	.7504814709	.7541409219	.7577407939	.7612823008
70+	.7647666202	.7681948953	.7715682360	.7748877207	.7781543977
75+	.7813692862	.7845333777	.7876476373	.7907130043	.7937303936
80+	.7967006968	.7996247825	.8025034977	.8053376686	.8081281009
85+	.8108755809	.8135808763	.8162447366	.8188678937	.8214510626
90+	.8239949422	.8265002154	.8289675498	.8313975983	.8337909995
95+	.8361483779	.8384703449	.8407574986	.8430104245	.8452296959
100+	.8474158741	.8495695089	.8516911387	.8537812912	.8558404833
105+	.8578692217	.8598680030	.8618373138	.8637776316	.8656894244
110+	.8675731511	.8694292619	.8712581985	.8730603940	.8748362737
115+	.8765862546	.8783107463	.8800101504	.8816848616	.8833352671
120+	.8849617470	.8865646747	.8881444169	.8897013337	.8912357785
125+	.8927480990	.8942386362	.8957077255	.8971556963	.8985828723
130+	.8999895714	.9013761064	.9027427843	.9040899071	.9054177717
135+	.9067266697	.9080168881	.9092887087	.9105424089	.9117782611
140+	.9129965336	.9141974900	.9153813893	.9165484867	.9176990329
145+	.9188332745	.9199514542	.9210538106	.9221405785	.9232119889

D-INTEGRAL PROBABILITY M=R+1,B= 3

A= .80

R	1	2	3	4	5
150+	.9242682691	.9253096425	.9263363293	.9273485458	.9283465051
155+	.9293304168	.9303004871	.9312569190	.9321999124	.9331296638
160+	.9340463668	.9349502119	.9358413865	.9367200753	.9375864598
165+	.9384407191	.9392830291	.9401135631	.9409324920	.9417399835
170+	.9425362033	.9433213141	.9440954763	.9448588480	.9456115844
175+	.9463538388	.9470857620	.9478075024	.9485192063	.9492210177
180+	.9499130784	.9505955281	.9512685043	.9519321426	.9525865764
185+	.9532319372	.9538683543	.9544959555	.9551148662	.9557252102
190+	.9563271095	.9569206840	.9575060521	.9580833303	.9586526334
195+	.9592140745	.9597677650	.9603138147	.9608523316	.9613834225
200+	.9619071922	.9624237441	.9629331803	.9634356010	.9639311053
205+	.9644197905	.9649017528	.9653770868	.9658458858	.9663082417
210+	.9667642450	.9672139850	.9676575496	.9680950255	.9685264981
215+	.9689520515	.9693717687	.9697857314	.9701940202	.9705967145
220+	.9709938925	.9713856313	.9717720069	.9721530942	.9725289671
225+	.9728996982	.9732653594	.9736260212	.9739817533	.9743326245
230+	.9746787024	.9750200536	.9753567440	.9756888384	.9760164006
235+	.9763394937	.9766581796	.9769725196	.9772825740	.9775884022
240+	.9778900629	.9781876138	.9784811119	.9787706133	.9790561733
245+	.9793378465	.9796156868	.9798897470	.9801600796	.9804267359
250+	.9806897668	.9809492225	.9812051521	.9814576045	.9817066276
255+	.9819522688	.9821945745	.9824335909	.9826693633	.9829019364
260+	.9831313541	.9833576601	.9835808971	.9838011072	.9840183323
265+	.9842326133	.9844439907	.9846525044	.9848581938	.9850610976
270+	.9852612542	.9854587011	.9856534757	.9858456146	.9860351539
275+	.9862221292	.9864065758	.9865885284	.9867680209	.9869450873
280+	.9871197606	.9872920737	.9874620589	.9876297480	.9877951724
285+	.9879583632	.9881193508	.9882781654	.9884348367	.9885893939
290+	.9887418660	.9888922814	.9890406682	.9891870541	.9893314665
295+	.9894739321	.9896144777	.9897531294	.9898899129	.9900248538

D-INTEGRAL PROBABILITY M=R+1, B= 4

R	1	2	3	4	5
			A=1/10		
0+	.5102040816	.8459637202	.9537824527	.9861460120	.9958104362
			A=1/ 9		
0+	.4792899408	.8199033134	.9406042588	.9804884745	.9935493339
			A=1/ 8		
0+	.4444444444	.7870370370	.9221065013	.9717009505	.9896798250
5+	.9962111278	.9985991533	.9994786920	.9998048904	.9999266163
			A=1/ 7		
0+	.4049586777	.7449760152	.8954716274	.9575423150	.9827351593
5+	.9929475976	.9971033792	.9988037900	.9995035234	.9997930195
			A=1/ 6		
0+	.3600000000	.6903360000	.8560543968	.9337827603	.9695800852
5+	.9859943308	.9935283606	.9969975774	.9986015066	.9993461558
			A=1/ 5		
0+	.3086419753	.6183659146	.7960806116	.8921517505	.9431383690
5+	.9700189238	.9841671493	.9916195034	.9955525818	.9976335358
			A=1/ 4		
0+	.2500000000	.5227050781	.7026693523	.8161948890	.8867555213
5+	.9303175191	.9571266570	.9736055766	.9837340104	.9899631312
10+	.9937981435	.9961622707	.9976217478	.9985240893	.9990828094
			A=1/ 3		
0+	.1836734694	.3962088197	.5563041874	.6746125137	.7616556957
5+	.8255451286	.8723648577	.9066376965	.9317079386	.9500389311
10+	.9634399287	.9732369660	.9804004423	.9856397611	.9894731820
15+	.9922791850	.9943341350	.9958398447	.9969437159	.9977534436
			A= .40		
0+	.1479289941	.3200892381	.4565884529	.5649237840	.6513478907
5+	.7204836192	.7758635120	.8202516862	.8558387880	.8843723138
10+	.9072507318	.9255948536	.9403034889	.9520975145	.9615550513
15+	.9691396551	.9752229489	.9801028045	.9840179461	.9871596781
20+	.9896812975	.9917056451	.9933311602	.9946367348	.9956856019
			A=1/ 2		
0+	.1111111111	.2366255144	.3384853046	.4236095946	.4961600946
5+	.5586942859	.6129591285	.6602476017	.7015706615	.7377483469
10+	.7694624142	.7972891631	.8217213229	.8431835060	.8620436558
15+	.8786218665	.8931973938	.9060143659	.9172865220	.9272012034
20+	.9359227481	.9435954043	.9503458445	.9562853462	.9615116887
25+	.9661108080	.9701582436	.9737204043	.9768556778	.9796154040
30+	.9820447289	.9841833541	.9860661955	.9877239617	.9891836627
35+	.9904690561	.9916010400	.9925979982	.9934761036	.9942495844
			A= .60		
0+	.0865051903	.1788099193	.2524208972	.3144741421	.3685964030
5+	.4167597562	.4601641769	.4996084788	.5356626358	.5687558620
10+	.5992253081	.6273447604	.6533424722	.6774127492	.6997237698
15+	.7204230451	.7396413449	.7574955984	.7740910892	.7895231529
20+	.8038785182	.8172363842	.8296693017	.8412439043	.8520215244
25+	.8620587170	.8714077120	.8801168065	.8882307101	.8957908494
30+	.9028356397	.9094007277	.9155192104	.9212218328	.9265371674
35+	.9314917776	.9361103668	.9404159154	.9444298063	.9481719399
40+	.9516608406	.9549137548	.9579467415	.9607747558	.9634117266
45+	.9658706281	.9681635467	.9703017423	.9722957059	.9741552124
50+	.9758893703	.9775066678	.9790150147	.9804217830	.9817338432
55+	.9829575989	.9840990186	.9851636656	.9861567255	.9870830324
60+	.9879470921	.9887531054	.9895049880	.9902063906	.9908607167
			A= .65		
0+	.0771604938	.1566608748	.2188380462	.2708796905	.3162708912
5+	.3568492228	.3936989602	.4275186155	.4587896766	.4878621656
10+	.5150016306	.5404166709	.5642759266	.5867190329	.6078639469

D-INTEGRAL PROBABILITY M=R+1, B= 4

R	1	2	3	4	5
			A= .65		
15+	.6278120048	.6466515094	.6644603367	.6813078707	.6972564682
20+	.7123625852	.7266776598	.7402488114	.7531194048	.7653295087
25+	.7769162742	.7879142504	.7983556494	.8082705725	.8176872037
30+	.8266319770	.8351297224	.8432037942	.8508761841	.8581676223
35+	.8650976669	.8716847860	.8779464300	.8838990989	.8895584023
40+	.8949391155	.9000552302	.9049200011	.9095459899	.9139451052
45+	.9181286397	.9221073050	.9258912641	.9294901612	.9329131505
50+	.9361689224	.9392657283	.9422114040	.9450133918	.9476787608
55+	.9502142269	.9526261704	.9549206540	.9571034388	.9591799996
60+	.9611555401	.9630350058	.9648230978	.9665242850	.9681428155
65+	.9696827281	.9711478623	.9725418690	.9738682191	.9751302132
70+	.9763309895	.9774735325	.9785606798	.9795951303	.9805794503
75+	.9815160805	.9824073422	.9832554430	.9840624824	.9848304575
80+	.9855612675	.9862567189	.9869185301	.9875483354	.9881476893
85+	.9887180707	.9892608860	.9897774733	.9902691054	.9907369928
			A= .70		
0+	.0692520776	.1379299536	.1903032234	.2335533001	.2710393326
5+	.3044952669	.3349244869	.3629604738	.3890305552	.4134382802
10+	.4364083825	.4581129806	.4786876740	.4982418768	.5168656988
15+	.5346346748	.5516131020	.5678564528	.5834131545	.5983259262
20+	.6126328005	.6263679146	.6395621324	.6522435386	.6644378364
25+	.6761686701	.6874578898	.6983257708	.7087911959	.7188718093
30+	.7285841465	.7379437455	.7469652418	.7556624505	.7640484383
35+	.7721355850	.7799356394	.7874597665	.7947185915	.8017222370
40+	.8084803581	.8150021726	.8212964888	.8273717309	.8332359614
45+	.8388969025	.8443619546	.8496382141	.8547324896	.8596513166
50+	.8644009712	.8689874832	.8734166478	.8776940368	.8818250086
55+	.8858147185	.8896681276	.8933900108	.8969849660	.9004574204
60+	.9038116388	.9070517297	.9101816520	.9132052208	.9161261138
65+	.9189478760	.9216739256	.9243075586	.9268519536	.9293101764
70+	.9316851844	.9339798302	.9361968664	.9383389485	.9404086393
75+	.9424084116	.9443406519	.9462076638	.9480116705	.9497548180
80+	.9514391780	.9530667504	.9546394658	.9561591885	.9576277182
85+	.9590467925	.9604180895	.9617432295	.9630237772	.9642612433
90+	.9654570872	.9666127180	.9677294967	.9688087377	.9698517105
95+	.9708596412	.9718337142	.9727750735	.9736848239	.9745640328
100+	.9754137312	.9762349149	.9770285460	.9777955539	.9785368362
105+	.9792532602	.9799456639	.9806148565	.9812616200	.9818867100
110+	.9824908561	.9830747635	.9836391134	.9841845641	.9847117513
115+	.9852212894	.9857137721	.9861897728	.9866498457	.9870945261
120+	.9875243314	.9879397613	.9883412989	.9887294107	.9891045477
125+	.9894671455	.9898176251	.9901563933	.9904838430	.9908003542
			A= .75		
0+	.0625000000	.1220035553	.1660207247	.2016779058	.2322106910
5+	.2592614371	.2837672883	.3063120408	.3272842073	.3469558987
10+	.3655256034	.3831429745	.3999239850	.4159606135	.4313272694
15+	.4460851905	.4602855371	.4739716203	.4871805400	.4999444123
20+	.5122913030	.5242459487	.5358303216	.5470640765	.5579649081
25+	.5685488410	.5788304653	.5888231313	.5985391101	.6079897289
30+	.6171854834	.6261361339	.6348507871	.6433379653	.6516056669
35+	.6596614179	.6675123175	.6751650769	.6826260540	.6899012838
40+	.6969965051	.7039171843	.7106685360	.7172555426	.7236829706
45+	.7299553859	.7360771674	.7420525193	.7478854823	.7535799436
50+	.7591396461	.7645681970	.7698690750	.7750456381	.7801011294
55+	.7850386834	.7898613312	.7945720060	.7991735476	.8036687066
60+	.8080601487	.8123504585	.8165421429	.8206376344	.8246392943
65+	.8285494156	.8323702255	.8361038884	.8397525078	.8433181291

D-INTEGRAL PROBABILITY M=R+1,B= 4

R	1	2	3	4	5
			A= .75		
70+	.8468027411	.8502082788	.8535366248	.8567896112	.8599690215
75+	.8630765922	.8661140141	.8690829342	.8719849570	.8748216456
80+	.8775945233	.8803050747	.8829547471	.8855449511	.8880770624
85+	.8905524223	.8929723390	.8953380883	.8976509148	.8999120327
90+	.9021226266	.9042838524	.9063968379	.9084626840	.9104824649
95+	.9124572293	.9143880006	.9162757781	.9181215372	.9199262305
100+	.9216907876	.9234161167	.9251031044	.9267526166	.9283654988
105+	.9299425770	.9314846576	.9329925287	.9344669597	.9359087024
110+	.9373184910	.9386970430	.9400450592	.9413632242	.9426522070
115+	.9439126612	.9451452253	.9463505233	.9475291650	.9486817459
120+	.9498088484	.9509110413	.9519888806	.9530429095	.9540736590
125+	.9550816480	.9560673835	.9570313613	.9579740658	.9588959703
130+	.9597975377	.9606792203	.9615414601	.9623846892	.9632093301
135+	.9640157956	.9648044893	.9655758057	.9663301304	.9670678404
140+	.9677893042	.9684948820	.9691849260	.9698597805	.9705197820
145+	.9711652596	.9717965348	.9724139222	.9730177291	.9736082562
150+	.9741857972	.9747506395	.9753030639	.9758433450	.9763717512
155+	.9768885452	.9773939834	.9778883168	.9783717908	.9788446452
160+	.9793071146	.9797594281	.9802018102	.9806344799	.9810576516
165+	.9814715349	.9818763347	.9822722514	.9826594809	.9830382149
170+	.9834086405	.9837709410	.9841252954	.9844718790	.9848108628
175+	.9851424145	.9854666976	.9857838724	.9860940954	.9863975196
180+	.9866942948	.9869845673	.9872684803	.9875461737	.9878177845
185+	.9880834465	.9883432906	.9885974447	.9888460340	.9890891809
190+	.9893270051	.9895596237	.9897871510	.9900096991	.9902273773
			A= .80		
0+	.0566893424	.1083900766	.1453116421	.1744885054	.1990318860
5+	.2204975676	.2397627529	.2573681058	.2736698397	.2889150154
10+	.3032821012	.3169043370	.3298839430	.3423011569	.3542202014
15+	.3656933520	.3767637881	.3874676380	.3978354775	.4078934491
20+	.4176641102	.4271670868	.4364195835	.4454367862	.4542321835
25+	.4628178260	.4712045374	.4794020873	.4874193350	.4952643480
30+	.5029445027	.5104665687	.5178367802	.5250608975	.5321442590
35+	.5390918270	.5459082265	.5525977792	.5591645335	.5656122902
40+	.5719446256	.5781649119	.5842763346	.5902819089	.5961844935
45+	.6019868035	.6076914216	.6133008083	.6188173111	.6242431727
50+	.6295805384	.6348314632	.6399979176	.6450817933	.6500849085
55+	.6550090121	.6598557887	.6646268617	.6693237977	.6739481090
60+	.6785012573	.6829846563	.6873996741	.6917476359	.6960298260
65+	.7002474902	.7044018372	.7084940410	.7125252420	.7164965489
70+	.7204090403	.7242637656	.7280617465	.7318039783	.7354914309
75+	.7391250498	.7427057573	.7462344532	.7497120157	.7531393026
80+	.7565171513	.7598463804	.7631277899	.7663621619	.7695502613
85+	.7726928366	.7757906199	.7788443280	.7818546627	.7848223112
90+	.7877479467	.7906322286	.7934758034	.7962793045	.7990433530
95+	.8017685578	.8044555163	.8071048143	.8097170267	.8122927174
100+	.8148324401	.8173367381	.8198061448	.8222411841	.8246423702
105+	.8270102084	.8293451948	.8316478168	.8339185536	.8361578757
110+	.8383662456	.8405441179	.8426919395	.8448101496	.8468991802
115+	.8489594559	.8509913943	.8529954061	.8549718952	.8569212589
120+	.8588438881	.8607401671	.8626104743	.8644551819	.8662746562
125+	.8680692574	.8698393403	.8715852540	.8733073421	.8750059427
130+	.8766813889	.8783340082	.8799641234	.8815720522	.8831581073
135+	.8847225967	.8862658237	.8877880870	.8892896806	.8907708942
140+	.8922320131	.8936733183	.8950950865	.8964975903	.8978810983
145+	.8992458749	.9005921809	.9019202728	.9032304038	.9045228229
150+	.9057977756	.9070555040	.9082962464	.9095202376	.9107277092

D-INTEGRAL PROBABILITY M=R+1, B= 4

R	1	2	3	4	5
		A= .80			
155+	.9119188891	.9130940022	.9142532700	.9153969107	.9165251394
160+	.9176381683	.9187362062	.9198194591	.9208881300	.9219424190
165+	.9229825233	.9240086373	.9250209527	.9260196583	.9270049403
170+	.9279769822	.9289359651	.9298820672	.9308154645	.9317363303
175+	.9326448355	.9335411486	.9344254358	.9352978607	.9361585850
180+	.9370077678	.9378455662	.9386721348	.9394876262	.9402921911
185+	.9410859777	.9418691323	.9426417993	.9434041208	.9441562371
190+	.9448982865	.9456304055	.9463527284	.9470653881	.9477685152
195+	.9484622387	.9491466858	.9498219820	.9504882509	.9511456146
200+	.9517941933	.9524341057	.9530654688	.9536883980	.9543030072
205+	.9549094084	.9555077125	.9560980287	.9566804645	.9572551263
210+	.9578221187	.9583815450	.9589335072	.9594781057	.9600154396
215+	.9605456068	.9610687036	.9615848251	.9620940652	.9625965163
220+	.9630922698	.9635814156	.9640640426	.9645402382	.9650100890
225+	.9654736801	.9659310955	.9663824181	.9668277298	.9672671112
230+	.9677006419	.9681284003	.9685504638	.9689669089	.9693778109
235+	.9697832440	.9701832816	.9705779959	.9709674584	.9713517393
240+	.9717309082	.9721050333	.9724741823	.9728384218	.9731978175
245+	.9735524343	.9739023361	.9742475860	.9745882463	.9749243782
250+	.9752560425	.9755832989	.9759062063	.9762248228	.9765392058
255+	.9768494120	.9771554970	.9774575161	.9777555235	.9780495729
260+	.9783397170	.9786260082	.9789084977	.9791872365	.9794622746
265+	.9797336613	.9800014455	.9802656752	.9805263979	.9807836603
270+	.9810375087	.9812879886	.9815351450	.9817790221	.9820196637
275+	.9822571130	.9824914125	.9827226043	.9829507298	.9831758298
280+	.9833979447	.9836171142	.9838333778	.9840467739	.9842573410
285+	.9844651166	.9846701380	.9848724418	.9850720643	.9852690412
290+	.9854634076	.9856551985	.9858444480	.9860311900	.9862154579
295+	.9863972846	.9865767027	.9867537441	.9869284405	.9871008232
300+	.9872709229	.9874387700	.9876043944	.9877678258	.9879290933
305+	.9880882257	.9882452514	.9884001984	.9885530943	.9887039665
310+	.9888528418	.9889997469	.9891447078	.9892877504	.9894289003
315+	.9895681826	.9897056221	.9898412433	.9899750704	.9901071271

D-INTEGRAL PROBABILITY M=R+1, B= 5

R	1	2	3	4	5
			A=1/10		
0+	.4444444444	.8148104252	.9431695274	.9828048302	.9947792378
			A=1/ 9		
0+	.4132653061	.7847571787	.9272549568	.9758409242	.9919726003
			A=1/ 8		
0+	.3786982249	.7473324227	.9051051042	.9650815816	.9871847243
5+	.9952799558	.9982520090	.9993489577	.9997562286	.9999082931
			A=1/ 7		
0+	.3402777778	.7001901216	.8735736070	.9478791376	.9786346994
5+	.9912339179	.9963906528	.9985074039	.9993800313	.9997414230
			A=1/ 6		
0+	.2975206612	.6401698654	.8276350341	.9193412648	.9625692442
5+	.9826612171	.9919584673	.9962607634	.9982558696	.9991838538
			A=1/ 5		
0+	.2500000000	.5631500000	.7592588275	.8702065818	.9307108715
5+	.9631642213	.9804395834	.9896078185	.9944710464	.9970529827
			A=1/ 4		
0+	.1975308642	.4642605879	.6560781731	.7829907907	.8643356609
5+	.9156315974	.9476823574	.9676023882	.9799465714	.9875850451
10+	.9923094496	.9952320028	.9970409846	.9981616549	.9988566204
			A=1/ 3		
0+	.1406250000	.3395161629	.5018025365	.6274721631	.7229658342
5+	.7947606345	.8483606916	.8881806553	.9176585060	.9394240239
10+	.9554648742	.9672708522	.9759518519	.9823310760	.9870170884
15+	.9904586885	.9929862671	.9948427364	.9962065339	.9972086594
			A= .40		
0+	.1111111111	.2678013717	.4015892367	.5125487081	.6038725721
5+	.6786981362	.7397967323	.7895485856	.8299685274	.8627449368
10+	.8892816776	.9107388356	.9280702580	.9420570169	.9533365599
15+	.9624276821	.9697516456	.9756498684	.9803986241	.9842211817
20+	.9872977828	.9897738059	.9917664263	.9933700334	.9946606270
			A=1/ 2		
0+	.0816326531	.1922036858	.2879892346	.3711867864	.4440957556
5+	.5083270593	.5650759698	.6152889325	.6597502976	.6991291700
10+	.7340065612	.7648924059	.7922370964	.8164398855	.8378553962
15+	.8567989129	.8735508361	.8883605263	.9014496732	.9130152803
20+	.9232323249	.9322561359	.9402245246	.9472596945	.9534699522
25+	.9589512408	.9637885114	.9680569501	.9718230735	.9751457065
30+	.9780768541	.9806624775	.9829431850	.9849548460	.9867291365
35+	.9882940233	.9896741940	.9908914370	.9919649797	.9929117867
			A= .60		
0+	.0625000000	.1418376953	.2088315431	.2671938041	.3193136764
5+	.3665690021	.4098214355	.4496551706	.4864938382	.5206617698
10+	.5524182646	.5819778627	.6095229271	.6352117784	.6591841493
15+	.6815649609	.7024670150	.7219929674	.7402368108	.7572850165
20+	.7732174340	.7881080145	.8020254040	.8150334406	.8271915749
25+	.8385552345	.8491761415	.8591025930	.8683797127	.8770496761
30+	.8851519156	.8927233077	.8997983441	.9064092902	.9125863305
35+	.9183577045	.9237498322	.9287874313	.9334936261	.9378900499
40+	.9419969398	.9458332259	.9494166147	.9527636669	.9558898708
45+	.9588097105	.9615367298	.9640835929	.9664621402	.9686834412
50+	.9707578439	.9726950214	.9745040150	.9761932750	.9777706989
55+	.9792436671	.9806190760	.9819033698	.9831025696	.9842223011
60+	.9852678196	.9862440352	.9871555344	.9880066014	.9888012382
65+	.9895431828	.9902359262	.9908827293	.9914866371	.9920504937
			A= .65		
0+	.0553633218	.1230137932	.1788318561	.2269778393	.2698817746
5+	.3088909369	.3448181054	.3781932135	.4093844228	.4386610772

D-INTEGRAL PROBABILITY M=R+1,B= 5

R	1	2	3	4	5
			A= .65		
10+	.4662287945	.4922501990	.5168577802	.5401621979	.5622578361
15+	.5832266275	.6031407591	.6220646278	.6400562854	.6571685252
20+	.6734497134	.6889444361	.7036940088	.7177368843	.7311089839
25+	.7438439681	.7559734623	.7675272459	.7785334131	.7890185104
30+	.7990076565	.8085246461	.8175920421	.8262312574	.8344626279
35+	.8423054785	.8497781827	.8568982175	.8636822135	.8701460004
40+	.8763046501	.8821725163	.8877632713	.8930899402	.8981649338
45+	.9030000784	.9076066447	.9119953744	.9161765060	.9201597989
50+	.9239545554	.9275696436	.9310135165	.9342942325	.9374194735
55+	.9403965623	.9432324798	.9459338804	.9485071076	.9509582083
60+	.9532929464	.9555168161	.9576350545	.9596526533	.9615743703
65+	.9634047401	.9651480847	.9668085234	.9683899817	.9698962009
70+	.9713307463	.9726970155	.9739982463	.9752375238	.9764177881
75+	.9775418402	.9786123495	.9796318591	.9806027921	.9815274571
80+	.9824080538	.9832466777	.9840453251	.9848058979	.9855302082
85+	.9862199818	.9868768630	.9875024182	.9880981393	.9886654475
90+	.9892056966	.9897201761	.9902101142	.9906766808	.9911209901
			A= .70		
0+	.0493827160	.1073134067	.1537613253	.1931836587	.2280177649
5+	.2595809684	.2886519740	.3157281015	.3411470757	.3651497783
10+	.3879150281	.4095800555	.4302531725	.4500219453	.4689586542
15+	.4871240532	.5045700259	.5213415051	.5374778878	.5530140968
20+	.5679813906	.5824079885	.5963195613	.6097396193	.6226898240
25+	.6351902394	.6472595380	.6589151703	.6701735054	.6810499487
30+	.6915590419	.7017145468	.7115295174	.7210163625	.7301868990
35+	.7390523986	.7476236292	.7559108906	.7639240464	.7716725528
40+	.7791654835	.7864115527	.7934191358	.8001962882	.8067507616
45+	.8130900202	.8192212548	.8251513957	.8308871248	.8364348870
50+	.8418009008	.8469911674	.8520114804	.8568674342	.8615644318
55+	.8661076926	.8705022593	.8747530047	.8788646385	.8828417127
60+	.8866886277	.8904096380	.8940088570	.8974902623	.9008577004
65+	.9041148911	.9072654323	.9103128038	.9132603714	.9161113910
70+	.9188690120	.9215362810	.9241161455	.9266114567	.9290249731
75+	.9313593635	.9336172096	.9358010095	.9379131799	.9399560590
80+	.9419319091	.9438429189	.9456912061	.9474788197	.9492077418
85+	.9508798906	.9524971216	.9540612302	.9555739536	.9570369722
90+	.9584519121	.9598203465	.9611437973	.9624237371	.9636615905
95+	.9648587358	.9660165065	.9671361925	.9682190420	.9692662624
100+	.9702790217	.9712584500	.9722056405	.9731216508	.9740075039
105+	.9748641895	.9756926651	.9764938569	.9772686607	.9780179433
110+	.9787425431	.9794432710	.9801209115	.9807762234	.9814099409
115+	.9820227737	.9826154088	.9831885105	.9837427213	.9842786626
120+	.9847969357	.9852981220	.9857827840	.9862514655	.9867046928
125+	.9871429746	.9875668032	.9879766547	.9883729893	.9887562525
130+	.9891268749	.9894852730	.9898318497	.9901669945	.9904910844
			A= .75		
0+	.0443213296	.0941296907	.1327477456	.1648180121	.1927553501
5+	.2178393136	.2408156203	.2621545225	.2821713515	.3010880367
10+	.3190669672	.3362308198	.3526747760	.3684743770	.3836907570
15+	.3983742388	.4125668693	.4263042500	.4396168842	.4525311880
20+	.4650702590	.4772544710	.4891019394	.5006288892	.5118499497
25+	.5227783929	.5334263271	.5438048569	.5539242148	.5637938717
30+	.5734226295	.5828186987	.5919897651	.6009430466	.6096853416
35+	.6182230711	.6265623149	.6347088434	.6426681454	.6504454519
40+	.6580457581	.6654738417	.6727342800	.6798314646	.6867696150
45+	.6935527900	.7001848990	.7066697111	.7130108643	.7192118730
50+	.7252761352	.7312069392	.7370074695	.7426808121	.7482299596

D-INTEGRAL PROBABILITY M=R+1,B= 5

R	1	2	3	4	5
			A= .75		
55+	.7536578158	.7589671999	.7641608505	.7692414290	.7742115231
60+	.7790736498	.7838302587	.7884837340	.7930363977	.7974905115
65+	.8018482793	.8061118493	.8102833156	.8143647206	.8183580563
70+	.8222652660	.8260882462	.8298288479	.8334888777	.8370700998
75+	.8405742365	.8440029702	.8473579439	.8506407628	.8538529951
80+	.8569961731	.8600717942	.8630813217	.8660261861	.8689077853
85+	.8717274862	.8744866247	.8771865075	.8798284116	.8824135863
90+	.8849432527	.8874186055	.8898408128	.8922110170	.8945303357
95+	.8967998619	.8990206648	.9011937903	.9033202615	.9054010793
100+	.9074372228	.9094296501	.9113792983	.9132870845	.9151539057
105+	.9169806399	.9187681458	.9205172639	.9222288165	.9239036083
110+	.9255424266	.9271460418	.9287152079	.9302506625	.9317531277
115+	.9332233099	.9346619004	.9360695758	.9374469982	.9387948153
120+	.9401136613	.9414041567	.9426669086	.9439025114	.9451115466
125+	.9462945833	.9474521785	.9485848772	.9496932131	.9507777080
130+	.9518388729	.9528772078	.9538932021	.9548873346	.9558600740
135+	.9568118790	.9577431984	.9586544716	.9595461284	.9604185897
140+	.9612722671	.9621075637	.9629248738	.9637245834	.9645070701
145+	.9652727037	.9660218459	.9667548507	.9674720645	.9681738265
150+	.9688604685	.9695323152	.9701896843	.9708328870	.9714622275
155+	.9720780037	.9726805071	.9732700229	.9738468303	.9744112025
160+	.9749634068	.9755037049	.9760323527	.9765496008	.9770556943
165+	.9775508733	.9780353725	.9785094217	.9789732457	.9794270645
170+	.9798710936	.9803055435	.9807306206	.9811465265	.9815534587
175+	.9819516105	.9823411709	.9827223249	.9830952536	.9834601341
180+	.9838171399	.9841664406	.9845082021	.9848425870	.9851697542
185+	.9854898592	.9858030543	.9861094884	.9864093072	.9867026533
190+	.9869896664	.9872704828	.9875452364	.9878140577	.9880770748
195+	.9883344128	.9885861942	.9888325389	.9890735641	.9893093847
200+	.9895401128	.9897658584	.9899867290	.9902028298	.9904142637
			A= .80		
0+	.0400000000	.0829879132	.1150734989	.1410039114	.1631492655
5+	.1827434802	.2004968343	.2168526637	.2321055144	.2464608724
10+	.2600677943	.2730378986	.2854570040	.2973925781	.3088986798
15+	.3200193418	.3307909493	.3412439508	.3514041154	.3612934723
20+	.3709310251	.3803333029	.3895147901	.3984882668	.4072650802
25+	.4158553644	.4242682183	.4325118524	.4405937100	.4485205685
30+	.4562986235	.4639335608	.4714306168	.4787946303	.4860300869
35+	.4931411574	.5001317305	.5070054420	.5137656994	.5204157044
40+	.5269584718	.5333968468	.5397335199	.5459710402	.5521118277
45+	.5581581832	.5641122986	.5699762648	.5757520798	.5814416552
50+	.5870468229	.5925693401	.5980108953	.6033731120	.6086575538
55+	.6138657276	.6189990876	.6240590382	.6290469372	.6339640985
60+	.6388117943	.6435912580	.6483036859	.6529502390	.6575320455
65+	.6620502018	.6665057744	.6708998015	.6752332940	.6795072368
70+	.6837225904	.6878802916	.6919812544	.6960263716	.7000165150
75+	.7039525367	.7078352696	.7116655284	.7154441101	.7191717948
80+	.7228493462	.7264775124	.7300570261	.7335886054	.7370729543
85+	.7405107631	.7439027088	.7472494554	.7505516548	.7538099466
90+	.7570249587	.7601973078	.7633275995	.7664164287	.7694643799
95+	.7724720275	.7754399360	.7783686602	.7812587459	.7841107295
100+	.7869251386	.7897024923	.7924433010	.7951480670	.7978172845
105+	.8004514401	.8030510123	.8056164724	.8081482843	.8106469046
110+	.8131127830	.8155463623	.8179480785	.8203183613	.8226576334
115+	.8249663118	.8272448068	.8294935228	.8317128585	.8339032063
120+	.8360649532	.8381984804	.8403041638	.8423823735	.8444334747
125+	.8464578272	.8484557856	.8504276995	.8523739137	.8542947679

D-INTEGRAL PROBABILITY M=R+1,B= 5

A= .80

R	1	2	3	4	5
130+	.8561905971	.8580617318	.8599084977	.8617312158	.8635302031
135+	.8653057716	.8670582296	.8687878807	.8704950245	.8721799564
140+	.8738429680	.8754843466	.8771043757	.8787033349	.8802815003
145+	.8818391438	.8833765338	.8848939353	.8863916094	.8878698138
150+	.8893288027	.8907688271	.8921901344	.8935929687	.8949775709
155+	.8963441788	.8976930268	.8990243463	.9003383656	.9016353101
160+	.9029154019	.9041788604	.9054259022	.9066567406	.9078715866
165+	.9090706480	.9102541302	.9114222355	.9125751639	.9137131125
170+	.9148362761	.9159448467	.9170390138	.9181189644	.9191848834
175+	.9202369527	.9212753522	.9223002593	.9233118493	.9243102949
180+	.9252957667	.9262684331	.9272284602	.9281760121	.9291112505
185+	.9300343353	.9309454240	.9318446723	.9327322338	.9336082600
190+	.9344729006	.9353263032	.9361686137	.9369999757	.9378205313
195+	.9386304207	.9394297822	.9402187522	.9409974656	.9417660553
200+	.9425246526	.9432733871	.9440123867	.9447417777	.9454616845
205+	.9461722302	.9468735362	.9475657223	.9482489068	.9489232064
210+	.9495887363	.9502456103	.9508939405	.9515338379	.9521654118
215+	.9527887701	.9534040194	.9540112648	.9546106103	.9552021582
220+	.9557860097	.9563622646	.9569310215	.9574923777	.9580464292
225+	.9585932707	.9591329958	.9596656968	.9601914649	.9607103900
230+	.9612225610	.9617280654	.9622269899	.9627194197	.9632054392
235+	.9636851316	.9641585790	.9646258626	.9650870622	.9655422569
240+	.9659915246	.9664349424	.9668725861	.9673045307	.9677308503
245+	.9681516179	.9685669055	.9689767844	.9693813248	.9697805961
250+	.9701746666	.9705636040	.9709474750	.9713263452	.9717002799
255+	.9720693429	.9724335977	.9727931068	.9731479318	.9734981335
260+	.9738437721	.9741849069	.9745215964	.9748538983	.9751818699
265+	.9755055672	.9758250459	.9761403609	.9764515663	.9767587155
270+	.9770618612	.9773610556	.9776563499	.9779477950	.9782354410
275+	.9785193371	.9787995322	.9790760744	.9793490113	.9796183897
280+	.9798842560	.9801466559	.9804056344	.9806612361	.9809135049
285+	.9811624842	.9814082168	.9816507449	.9818901104	.9821263542
290+	.9823595171	.9825896392	.9828167601	.9830409188	.9832621538
295+	.9834805034	.9836960050	.9839086957	.9841186121	.9843257904
300+	.9845302662	.9847320747	.9849312506	.9851278283	.9853218416
305+	.9855133238	.9857023080	.9858888267	.9860729121	.9862545957
310+	.9864339090	.9866108827	.9867855475	.9869579333	.9871280699
315+	.9872959865	.9874617122	.9876252754	.9877867044	.9879460270
320+	.9881032705	.9882584622	.9884116288	.9885627966	.9887119918
325+	.9888592400	.9890045666	.9891479967	.9892895550	.9894292659
330+	.9895671535	.9897032416	.9898375536	.9899701128	.9901009418

D-INTEGRAL PROBABILITY M=R+1, B= 6

R	1	2	3	4	5
			A=1/10		
0+	.3906250000	.7859161124	.9328879140	.9795094609	.9937542956
			A=1/ 9		
0+	.3600000000	.7525230933	.9144184899	.9712779681	.9904095689
			A=1/ 8		
0+	.3265306122	.7114181392	.8889176230	.9586249486	.9847213914
5+	.9943549459	.9979060612	.9992194567	.9997076126	.9998899790
			A=1/ 7		
0+	.2899408284	.6603759709	.8529999911	.9385428392	.9746125864
5+	.9895388069	.9956823144	.9982120560	.9992567857	.9996898851
			A=1/ 6		
0+	.2500000000	.5965470679	.8014195373	.9055851743	.9557644692
5+	.9793890867	.9904064939	.9955292023	.9979117726	.9990220039
			A=1/ 5		
0+	.2066115702	.5164946532	.7261568098	.8497521275	.9188591232
5+	.9565276065	.9767937406	.9876265755	.9934008160	.9964766228
			A=1/ 4		
0+	.1600000000	.4167270400	.6157211458	.7530780564	.8435890914
5+	.9017783400	.9386469519	.9617982725	.9762554988	.9852534301
10+	.9908431095	.9943124709	.9964653730	.9978016915	.9986316166
			A=1/ 3		
0+	.1111111111	.2957537463	.4571242060	.5872683726	.6890031523
5+	.7671311657	.8264337768	.8710783252	.9044864606	.9293744024
10+	.9478525636	.9615367822	.9716513306	.9791166896	.9846209556
15+	.9886762713	.9916625584	.9938609439	.9954790556	.9966700205
			A= .40		
0+	.0865051903	.2287728175	.3581654580	.4696107900	.5638412734
5+	.6426715966	.7081286788	.7621750073	.8066015743	.8429899362
10+	.8727073459	.8969182943	.9166036974	.9325828784	.9455356565
15+	.9560231088	.9645063181	.9713628565	.9769010097	.9813718818
20+	.9849795888	.9878897716	.9902366586	.9921288952	.9936543330
			A=1/ 2		
0+	.0625000000	.1603584290	.2499493832	.3303552676	.4025043543
5+	.4672615718	.5253622009	.5774484786	.6240956330	.6658261540
10+	.7031175665	.7364070250	.7660944410	.7925449697	.8160912389
15+	.8370354817	.8556516291	.8721873721	.8868661795	.8998892570
20+	.9114374270	.9216729175	.9307410498	.9387718174	.9458813540
25+	.9521732913	.9577400074	.9626637694	.9670177765	.9708671079
30+	.9742695813	.9772765292	.9799334994	.9822808842	.9843544864
35+	.9861860261	.9878035947	.9892320605	.9904934310	.9916071759
			A= .60		
0+	.0472589792	.1161574604	.1772171675	.2319029098	.2817200956
5+	.3276032479	.3701531501	.4097836451	.4467990194	.4814358121
10+	.5138864724	.5443133647	.5728574004	.5996435501	.6247844769
15+	.6483830024	.6705338262	.6913247583	.7108376249	.7291489536
20+	.7463305041	.7624496922	.7775699360	.7917509469	.8050489791
25+	.8175170482	.8292051265	.8401603192	.8504270268	.8600470947
30+	.8690599528	.8775027472	.8854104632	.8928160425	.8997504940
35+	.9062429988	.9123210100	.9180103478	.9233352897	.9283186561
40+	.9329818925	.9373451462	.9414273407	.9452462453	.9488185413
45+	.9521598853	.9552849686	.9582075737	.9609406283	.9634962550
50+	.9658858201	.9681199783	.9702087159	.9721613905	.9739867698
55+	.9756930672	.9772879756	.9787786997	.9801719857	.9814741498
60+	.9826911048	.9838283852	.9848911712	.9858843100	.9868123377
65+	.9876794979	.9884897611	.9892468413	.9899542123	.9906151234
			A= .65		
0+	.0416493128	.0999526696	.1502849722	.1948111493	.2352097541
5+	.2724656807	.3071858388	.3397696066	.3704956900	.3995685866

D-INTEGRAL PROBABILITY M=R+1, B= 6

R	1	2	3	4	5
			A= .65		
10+	.4271448884	.4533489529	.4782826571	.5020316993	.5246698029
15+	.5462615980	.5668646461	.5865308918	.6053077251	.6232387713
20+	.6403644882	.6567226252	.6723485797	.6872756795	.7015354082
25+	.7151575879	.7281705296	.7406011582	.7524751178	.7638168613
30+	.7746497284	.7849960132	.7948770238	.8043131362	.8133238420
35+	.8219277923	.8301428379	.8379860662	.8454738357	.8526218077
40+	.8594449772	.8659577005	.8721737226	.8781062026	.8837677381
45+	.8891703880	.8943256948	.8992447059	.9039379938	.9084156752
50+	.9126874303	.9167625200	.9206498034	.9243577546	.9278944780
55+	.9312677239	.9344849034	.9375531019	.9404790932	.9432693523
60+	.9459300680	.9484671550	.9508862652	.9531927994	.9553919171
65+	.9574885479	.9594874003	.9613929719	.9632095581	.9649412609
70+	.9665919975	.9681655081	.9696653635	.9710949730	.9724575907
75+	.9737563232	.9749941353	.9761738568	.9772981883	.9783697068
80+	.9793908713	.9803640283	.9812914162	.9821751708	.9830173296
85+	.9838198362	.9845845444	.9853132228	.9860075578	.9866691582
90+	.9872995580	.9879002202	.9884725400	.9890178476	.9895374117
95+	.9900324416	.9905040907	.9909534584	.9913815928	.9917894933
			A= .70		
0+	.0369822485	.0865797674	.1280693765	.1641032315	.1964615304
5+	.2261545345	.2537917751	.2797663700	.3043465291	.3277239135
10+	.3500408576	.3714065569	.3919071519	.4116122605	.4305793543
15+	.4488567745	.4664858638	.4835025049	.4999382532	.5158211848
20+	.5311765405	.5460272215	.5603941769	.5742967082	.5877527130
25+	.6007788795	.6133908462	.6256033308	.6374302381	.6488847483
30+	.6599793916	.6707261113	.6811363166	.6912209282	.7009904173
35+	.7104548395	.7196238640	.7285067991	.7371126154	.7454499650
40+	.7535272001	.7613523882	.7689333271	.7762775575	.7833923747
45+	.7902848398	.7969617892	.8034298439	.8096954180	.8157647264
50+	.8216437923	.8273384540	.8328543715	.8381970324	.8433717578
55+	.8483837078	.8532378866	.8579391477	.8624921982	.8669016040
60+	.8711717936	.8753070625	.8793115772	.8831893791	.8869443882
65+	.8905804069	.8941011231	.8975101138	.9008108485	.9040066922
70+	.9071009082	.9100966617	.9129970220	.9158049657	.9185233794
75+	.9211550620	.9237027278	.9261690082	.9285564551	.9308675423
80+	.9331046682	.9352701582	.9373662664	.9393951780	.9413590112
85+	.9432598193	.9450995924	.9468802594	.9486036899	.9502716956
90+	.9518860326	.9534484024	.9549604539	.9564237851	.9578399443
95+	.9592104315	.9605367003	.9618201591	.9630621722	.9642640613
100+	.9654271072	.9665525501	.9676415918	.9686953962	.9697150909
105+	.9707017678	.9716564848	.9725802664	.9734741047	.9743389606
110+	.9751757649	.9759854186	.9767687946	.9775267380	.9782600672
115+	.9789695745	.9796560274	.9803201687	.9809627178	.9815843712
120+	.9821858032	.9827676666	.9833305936	.9838751960	.9844020662
125+	.9849117776	.9854048854	.9858819269	.9863434224	.9867898754
130+	.9872217733	.9876395878	.9880437757	.9884347788	.9888130251
135+	.9891789286	.9895328900	.9898752971	.9902065254	.9905269381
			A= .75		
0+	.0330578512	.0754573414	.1096675830	.1386765875	.1643127995
5+	.1875883133	.2091049717	.2292464973	.2482718415	.2663639192
10+	.2836568277	.3002519457	.3162279264	.3316471562	.3465600760
15+	.3610081593	.3750260167	.3886429183	.4018839148	.4147706796
20+	.4273221492	.4395550192	.4514841311	.4631227789	.4744829543
25+	.4855755443	.4964104917	.5069969279	.5173432812	.5274573682
30+	.5373464696	.5470173944	.5564765338	.5657299079	.5747832050
35+	.5836418158	.5923108623	.6007952240	.6090995594	.6172283257
40+	.6251857956	.6329760727	.6406031040	.6480706923	.6553825062

D-INTEGRAL PROBABILITY M=R+1,B= 6

R	1	2	3	4	5
			A= .75		
45+	.6625420893	.6695528688	.6764181629	.6831411872	.6897250609
50+	.6961728123	.7024873836	.7086716357	.7147283519	.7206602419
55+	.7264699450	.7321600335	.7377330155	.7431913376	.7485373870
60+	.7537734945	.7589019360	.7639249349	.7688446635	.7736632451
65+	.7783827556	.7830052249	.7875326381	.7919669375	.7963100231
70+	.8005637544	.8047299513	.8088103952	.8128068298	.8167209628
75+	.8205544658	.8243089762	.8279860974	.8315873999	.8351144219
80+	.8385686704	.8419516216	.8452647217	.8485093877	.8516870078
85+	.8547989423	.8578465241	.8608310593	.8637538277	.8666160836
90+	.8694190561	.8721639495	.8748519444	.8774841975	.8800618425
95+	.8825859905	.8850577305	.8874781297	.8898482339	.8921690683
100+	.8944416374	.8966669260	.8988458990	.9009795021	.9030686623
105+	.9051142879	.9071172693	.9090784790	.9109987720	.9128789864
110+	.9147199436	.9165224483	.9182872895	.9200152400	.9217070575
115+	.9233634843	.9249852479	.9265730613	.9281276230	.9296496178
120+	.9311397164	.9325985762	.9340268414	.9354251431	.9367941000
125+	.9381343180	.9394463910	.9407309008	.9419884175	.9432194997
130+	.9444246948	.9456045389	.9467595574	.9478902651	.9489971663
135+	.9500807550	.9511415155	.9521799217	.9531964385	.9541915209
140+	.9551656148	.9561191570	.9570525755	.9579662894	.9588607094
145+	.9597362378	.9605932685	.9614321875	.9622533730	.9630571953
150+	.9638440172	.9646141940	.9653680737	.9661059973	.9668282987
155+	.9675353050	.9682273363	.9689047066	.9695677231	.9702166867
160+	.9708518922	.9714736283	.9720821776	.9726778172	.9732608181
165+	.9738314460	.9743899608	.9749366174	.9754716651	.9759953482
170+	.9765079059	.9770095725	.9775005773	.9779811449	.9784514954
175+	.9789118440	.9793624018	.9798033753	.9802349667	.9806573740
180+	.9810707911	.9814754079	.9818714103	.9822589804	.9826382963
185+	.9830095326	.9833728601	.9837284462	.9840764547	.9844170459
190+	.9847503768	.9850766013	.9853958698	.9857083297	.9860141253
195+	.9863133978	.9866062855	.9868929239	.9871734455	.9874479801
200+	.9877166547	.9879795937	.9882369188	.9884887493	.9887352019
205+	.9889763908	.9892124279	.9894434226	.9896694822	.9898907115
210+	.9901072132	.9903190880	.9905264342	.9907293482	.9909279244
			A= .80		
0+	.0297265161	.0661385070	.0943557901	.1175944923	.1376939411
5+	.1556505024	.1720489747	.1872585819	.2015264516	.2150258496
10+	.2278828907	.2401922327	.2520267650	.2634438433	.2744894449
15+	.2852010231	.2956095201	.3057408179	.3156168074	.3252561877
20+	.3346750759	.3438874767	.3529056498	.3617404011	.3704013148
25+	.3788969412	.3872349490	.3954222517	.4034651103	.4113692207
30+	.4191397856	.4267815757	.4342989821	.4416960600	.4489765668
35+	.4561439949	.4632015998	.4701524245	.4769993213	.4837449701
40+	.4903918951	.4969424789	.5033989758	.5097635228	.5160381496
45+	.5222247880	.5283252795	.5343413826	.5402747792	.5461270809
50+	.5518998334	.5575945219	.5632125751	.5687553692	.5742242312
55+	.5796204427	.5849452420	.5901998275	.5953853598	.6005029641
60+	.6055537321	.6105387242	.6154589708	.6203154744	.6251092108
65+	.6298411304	.6345121599	.6391232030	.6436751418	.6481688377
70+	.6526051322	.6569848483	.6613087908	.6655777471	.6697924883
75+	.6739537697	.6780623310	.6821188977	.6861241809	.6900788783
80+	.6939836746	.6978392416	.7016462393	.7054053156	.7091171072
85+	.7127822398	.7164013283	.7199749774	.7235037817	.7269883261
90+	.7304291861	.7338269277	.7371821084	.7404952766	.7437669726
95+	.7469977281	.7501880670	.7533385052	.7564495511	.7595217054
100+	.7625554616	.7655513063	.7685097186	.7714311713	.7743161302
105+	.7771650546	.7799783976	.7827566057	.7855001195	.7882093735

D-INTEGRAL PROBABILITY M=R+1,B= 6

R	1	2	3	4	5
			A= .80		
110+	.7908847963	.7935268107	.7961358338	.7987122770	.8012565465
115+	.8037690429	.8062501615	.8087002926	.8111198210	.8135091270
120+	.8158685856	.8181985670	.8204994368	.8227715555	.8250152795
125+	.8272309603	.8294189449	.8315795761	.8337131923	.8358201275
130+	.8379007115	.8399552701	.8419841248	.8439875933	.8459659891
135+	.8479196220	.8498487978	.8517538184	.8536349822	.8554925838
140+	.8573269139	.8591382600	.8609269059	.8626931317	.8644372143
145+	.8661594271	.8678600401	.8695393200	.8711975303	.8728349312
150+	.8744517797	.8760483296	.8776248317	.8791815337	.8807186802
155+	.8822365130	.8837352707	.8852151892	.8866765013	.8881194371
160+	.8895442239	.8909510863	.8923402460	.8937119220	.8950663307
165+	.8964036860	.8977241989	.8990280781	.9003155295	.9015867569
170+	.9028419611	.9040813408	.9053050923	.9065134093	.9077064833
175+	.9088845034	.9100476565	.9111961270	.9123300973	.9134497475
180+	.9145552554	.9156467968	.9167245453	.9177886724	.9188393474
185+	.9198767376	.9209010084	.9219123231	.9229108428	.9238967269
190+	.9248701328	.9258312160	.9267801299	.9277170264	.9286420551
195+	.9295553642	.9304570998	.9313474064	.9322264266	.9330943012
200+	.9339511696	.9347971691	.9356324356	.9364571032	.9372713044
205+	.9380751701	.9388688297	.9396524106	.9404260393	.9411898401
210+	.9419439361	.9426884490	.9434234987	.9441492039	.9448656816
215+	.9455730477	.9462714163	.9469609003	.9476416112	.9483136591
220+	.9489771528	.9496321997	.9502789059	.9509173762	.9515477141
225+	.9521700218	.9527844003	.9533909493	.9539897674	.9545809517
230+	.9551645984	.9557408023	.9563096572	.9568712557	.9574256892
235+	.9579730479	.9585134211	.9590468968	.9595735620	.9600935026
240+	.9606068034	.9611135483	.9616138200	.9621077002	.9625952696
245+	.9630766080	.9635517941	.9640209056	.9644840193	.9649412111
250+	.9653925559	.9658381275	.9662779990	.9667122427	.9671409296
255+	.9675641302	.9679819139	.9683943493	.9688015042	.9692034455
260+	.9696002393	.9699919508	.9703786445	.9707603839	.9711372320
265+	.9715092508	.9718765016	.9722390448	.9725969403	.9729502470
270+	.9732990233	.9736433267	.9739832140	.9743187413	.9746499640
275+	.9749769370	.9752997141	.9756183487	.9759328936	.9762434008
280+	.9765499217	.9768525069	.9771512066	.9774460702	.9777371466
285+	.9780244841	.9783081302	.9785881320	.9788645359	.9791373878
290+	.9794067329	.9796726160	.9799350812	.9801941721	.9804499318
295+	.9807024027	.9809516268	.9811976455	.9814404998	.9816802300
300+	.9819168761	.9821504774	.9823810728	.9826087008	.9828333992
305+	.9830552054	.9832741565	.9834902889	.9837036387	.9839142414
310+	.9841221322	.9843273457	.9845299162	.9847298776	.9849272632
315+	.9851221059	.9853144384	.9855042927	.9856917006	.9858766935
320+	.9860593021	.9862395572	.9864174887	.9865931266	.9867665001
325+	.9869376384	.9871065700	.9872733233	.9874379261	.9876004061
330+	.9877607904	.9879191060	.9880753794	.9882296367	.9883819039
335+	.9885322065	.9886805697	.9888270184	.9889715771	.9891142702
340+	.9892551216	.9893941549	.9895313935	.9896668604	.9898005784
345+	.9899325700	.9900628573	.9901914623	.9903184066	.9904437114

D-INTEGRAL PROBABILITY M=R+1,B= 7

R	1	2	3	4	5
			A=1/10		
0+	.3460207612	.7590319820	.9229185797	.9762584962	.9927355057
			A=1/ 9		
0+	.3164062500	.7228395983	.9020595368	.9667962866	.9888599297
5+	.9962594258	.9987381739	.9995719731	.9998540271	.9999499778
			A=1/ 8		
0+	.2844444444	.6787604574	.8734762246	.9523228000	.9822888343
5+	.9934359788	.9975612958	.9990901875	.9996590422	.9998716738
			A=1/ 7		
0+	.2500000000	.6247335825	.8336158971	.9295117364	.9706653905
5+	.9878617244	.9949782789	.9979177327	.9991337843	.9996384057
			A=1/ 6		
0+	.2130177515	.5582548983	.7771307569	.8924542585	.9491529258
5+	.9761752267	.9888718686	.9948027735	.9975691901	.9988606008
			A=1/ 5		
0+	.1736111111	.4765563323	.6961906300	.8306126975	.9075314109
5+	.9500940225	.9732252610	.9856745153	.9923415275	.9959043513
			A=1/ 4		
0+	.1322314050	.3773477717	.5803630063	.7259304245	.8242966427
5+	.8886675858	.9299836216	.9561782622	.9726547262	.9829658317
10+	.9893981313	.9934032741	.9958947511	.9974441336	.9984077714
			A=1/ 3		
0+	.0900000000	.2610485280	.4197854081	.5524819447	.6588583217
5+	.7421190101	.8062685806	.8551465135	.8920846796	.9198279376
10+	.9405670383	.9560139773	.9674868865	.9759896875	.9822807998
15+	.9869296395	.9903616885	.9928937074	.9947608440	.9961372754
			A= .40		
0+	.0692520776	.1986610363	.3229982491	.4336847614	.5295174369
5+	.6111747257	.6799954845	.7375269098	.7853171042	.8248149410
10+	.8573250473	.8839929549	.9058068628	.9236081779	.9381061890
15+	.9498941323	.9594650663	.9672266964	.9735147348	.9786046491
20+	.9827218152	.9860501748	.9887395440	.9909117316	.9926656293
			A=1/ 2		
0+	.0493827160	.1365730969	.2203068028	.2976164387	.3684258634
5+	.4330191971	.4917557981	.5450184675	.5931968985	.6366784211
10+	.6758413703	.7110501702	.7426518280	.7709736135	.7963217076
15+	.8189806174	.8392131814	.8572610140	.8733452653	.8876675980
20+	.9004113035	.9117424954	.9218113368	.9307532639	.9386901849
25+	.9457316323	.9519758598	.9575108751	.9624154042	.9667597860
30+	.9706067982	.9740124146	.9770264980	.9796934311	.9820526892
35+	.9841393593	.9859846083	.9876161052	.9890584013	.9903332705
			A= .60		
0+	.0369822485	.0974498780	.1533235573	.2045727767	.2520655439
5+	.2964042781	.3379888557	.3770987470	.4139419828	.4486827733
10+	.4814573475	.5123833007	.5415652684	.5690984588	.5950709007
15+	.6195649026	.6426580178	.6644236953	.6849317286	.7042485730
20+	.7224375773	.7395591610	.7556709537	.7708279127	.7850824247
25+	.7984843990	.8110813543	.8229185023	.8340388295	.8444831773
30+	.8542903213	.8634970501	.8721382433	.8802469487	.8878544581
35+	.8949903830	.9016827278	.9079579623	.9138410923	.9193557284
40+	.9245241526	.9293673838	.9339052398	.9381563985	.9421384561
45+	.9458679832	.9493605791	.9526309233	.9556928252	.9585592714
50+	.9612424713	.9637538998	.9661043387	.9683039160	.9703621430
55+	.9722879497	.9740897183	.9757753151	.9773521209	.9788270593
60+	.9802066239	.9814969037	.9827036076	.9838320870	.9848873576
65+	.9858741193	.9867967759	.9876594530	.9884660149	.9892200810
70+	.9899250406	.9905840674	.9912001325	.9917760175	.9923143262

D-INTEGRAL PROBABILITY M=R+1, B= 7

R	1	2	3	4	5
			A= .65		
0+	.0324648973	.0833273634	.1289880289	.1702704787	.2083095034
5+	.2438192661	.2772496028	.3088985784	.3389744173	.3676297832
10+	.3949815265	.4211225526	.4461292307	.4700661755	.4929894226
15+	.5149485887	.5359883735	.5561496233	.5754700967	.5939850235
20+	.6117275193	.6287288960	.6450188977	.6606258822	.6755769622
25+	.6898981166	.7036142793	.7167494115	.7293265609	.7413679121
30+	.7528948279	.7639278870	.7744869157	.7845910169	.7942585963
35+	.8035073868	.8123544710	.8208163023	.8289087251	.8366469946
40+	.8440457945	.8511192562	.8578809751	.8643440286	.8705209920
45+	.8764239551	.8820645379	.8874539059	.8926027855	.8975214787
50+	.9022198772	.9067074769	.9109933917	.9150863665	.9189947905
55+	.9227267101	.9262898413	.9296915816	.9329390216	.9360389571
60+	.9389978993	.9418220864	.9445174933	.9470898423	.9495446126
65+	.9518870497	.9541221750	.9562547941	.9582895060	.9602307107
70+	.9620826179	.9638492543	.9655344711	.9671419512	.9686752163
75+	.9701376333	.9715324210	.9728626561	.9741312791	.9753411002
80+	.9764948049	.9775949589	.9786440135	.9796443105	.9805980869
85+	.9815074790	.9823745274	.9832011807	.9839892997	.9847406609
90+	.9854569608	.9861398186	.9867907804	.9874113219	.9880028515
95+	.9885667136	.9891041912	.9896165088	.9901048347	.9905702837
			A= .70		
0+	.0287273772	.0717713338	.1091257258	.1422170998	.1723475273
5+	.2002973223	.2265470837	.2514094870	.2750985067	.2977670377
10+	.3195284237	.3404693904	.3606581557	.3801497052	.3989893334
15+	.4172150871	.4348594908	.4519507903	.4685138666	.4845709163
20+	.5001419673	.5152452737	.5298976215	.5441145688	.5579106348
25+	.5712994510	.5842938832	.5969061289	.6091477986	.6210299813
30+	.6325632987	.6437579514	.6546237558	.6651701770	.6754063554
35+	.6853411300	.6949830579	.7043404319	.7134212949	.7222334534
40+	.7307844883	.7390817656	.7471324452	.7549434891	.7625216687
45+	.7698735714	.7770056072	.7839240137	.7906348617	.7971440602
50+	.8034573609	.8095803624	.8155185143	.8212771216	.8268613479
55+	.8322762189	.8375266266	.8426173318	.8475529678	.8523380433
60+	.8569769456	.8614739434	.8658331898	.8700587249	.8741544787
65+	.8781242739	.8819718281	.8857007567	.8893145753	.8928167020
70+	.8962104602	.8994990805	.9026857030	.9057733801	.9087650781
75+	.9116636795	.9144719855	.9171927175	.9198285195	.9223819601
80+	.9248555341	.9272516650	.9295727063	.9318209435	.9339985959
85+	.9361078187	.9381507040	.9401292831	.9420455278	.9439013522
90+	.9456986139	.9474391160	.9491246084	.9507567892	.9523373059
95+	.9538677575	.9553496950	.9567846233	.9581740022	.9595192480
100+	.9608217342	.9620827932	.9633037172	.9644857594	.9656301352
105+	.9667380230	.9678105656	.9688488710	.9698540137	.9708270351
110+	.9717689450	.9726807224	.9735633161	.9744176459	.9752446036
115+	.9760450532	.9768198322	.9775697524	.9782956005	.9789981387
120+	.9796781059	.9803362179	.9809731682	.9815896291	.9821862517
125+	.9827636668	.9833224858	.9838633007	.9843866854	.9848931954
130+	.9853833691	.9858577281	.9863167773	.9867610061	.9871908884
135+	.9876068830	.9880094347	.9883989737	.9887759172	.9891406689
140+	.9894936198	.9898351486	.9901656220	.9904853950	.9907948115
			A= .75		
0+	.0256000000	.0622327818	.0928296965	.1192452408	.1428787940
5+	.1645413473	.1847248101	.2037460506	.2218200904	.2390992353
10+	.2556952528	.2716926214	.2871568119	.3021396678	.3166830189
15+	.3308211782	.3445827108	.3579917139	.3710687621	.3838316158
20+	.3962957616	.4084748295	.4203809184	.4320248547	.4434163972
25+	.4545644032	.4654769633	.4761615114	.4866249159	.4968735548

D-INTEGRAL PROBABILITY M=R+1, B= 7

R	1	2	3	4	5
			A= .75		
30+	.5069133791	.5167499653	.5263885606	.5358341204	.5450913406
35+	.5541646856	.5630584118	.5717765882	.5803231141	.5887017347
40+	.5969160547	.6049695496	.6128655765	.6206073831	.6281981158
45+	.6356408266	.6429384799	.6500939576	.6571100644	.6639895324
50+	.6707350251	.6773491409	.6838344165	.6901933299	.6964283030
55+	.7025417042	.7085358505	.7144130096	.7201754020	.7258252021
60+	.7313645407	.7367955055	.7421201433	.7473404607	.7524584255
65+	.7574759679	.7623949812	.7672173232	.7719448167	.7765792506
70+	.7811223807	.7855759304	.7899415914	.7942210244	.7984158599
75+	.8025276989	.8065581129	.8105086453	.8143808116	.8181760996
80+	.8218959707	.8255418597	.8291151754	.8326173017	.8360495970
85+	.8394133957	.8427100080	.8459407203	.8491067960	.8522094758
90+	.8552499778	.8582294980	.8611492110	.8640102701	.8668138074
95+	.8695609348	.8722527438	.8748903060	.8774746736	.8800068795
100+	.8824879377	.8849188437	.8873005748	.8896340901	.8919203313
105+	.8941602226	.8963546714	.8985045680	.9006107865	.9026741845
110+	.9046956039	.9066758709	.9086157963	.9105161757	.9123777899
115+	.9142014050	.9159877728	.9177376310	.9194517034	.9211307002
120+	.9227753179	.9243862404	.9259641380	.9275096689	.9290234784
125+	.9305061995	.9319584536	.9333808496	.9347739853	.9361384468
130+	.9374748090	.9387836358	.9400654802	.9413208848	.9425503813
135+	.9437544916	.9449337272	.9460885898	.9472195714	.9483271544
140+	.9494118120	.9504740080	.9515141971	.9525328254	.9535303302
145+	.9545071400	.9554636752	.9564003479	.9573175621	.9582157137
150+	.9590951911	.9599563749	.9607996381	.9616253465	.9624338587
155+	.9632255260	.9640006929	.9647596971	.9655028694	.9662305343
160+	.9669430096	.9676406070	.9683236318	.9689923832	.9696471547
165+	.9702882337	.9709159018	.9715304351	.9721321042	.9727211742
170+	.9732979047	.9738625504	.9744153608	.9749565802	.9754864482
175+	.9760051993	.9765130636	.9770102663	.9774970282	.9779735654
180+	.9784400900	.9788968095	.9793439272	.9797816425	.9802101506
185+	.9806296427	.9810403061	.9814423244	.9818358774	.9822211412
190+	.9825982884	.9829674880	.9833289056	.9836827034	.9840290402
195+	.9843680717	.9846999502	.9850248252	.9853428428	.9856541463
200+	.9859588760	.9862571693	.9865491609	.9868349825	.9871147633
205+	.9873886298	.9876567057	.9879191125	.9881759689	.9884273912
210+	.9886734934	.9889143871	.9891501816	.9893809839	.9896068989
215+	.9898280291	.9900444750	.9902563352	.9904637061	.9906666819
			A= .80		
0+	.0229568411	.0542954731	.0793870413	.1003906203	.1187552435
5+	.1352979535	.1505073520	.1646952467	.1780717420	.1907847945
10+	.2029424078	.2146257948	.2258975588	.2368069846	.2473935807
15+	.2576895193	.2677213623	.2775113081	.2870781109	.2964377709
20+	.3056040595	.3145889252	.3234028117	.3320549090	.3405533549
25+	.3489053969	.3571175250	.3651955798	.3731448428	.3809701110
30+	.3886757597	.3962657953	.4037439004	.4111134717	.4183776532
35+	.4255393639	.4326013227	.4395660690	.4464359814	.4532132939
40+	.4599001096	.4664984135	.4730100835	.4794368996	.4857805531
45+	.4920426540	.4982247377	.5043282714	.5103546594	.5163052482
50+	.5221813306	.5279841500	.5337149040	.5393747475	.5449647960
55+	.5504861276	.5559397864	.5613267841	.5666481020	.5719046935
60+	.5770974850	.5822273780	.5872952505	.5923019582	.5972483356
65+	.6021351974	.6069633393	.6117335394	.6164465583	.6211031407
70+	.6257040157	.6302498978	.6347414873	.6391794710	.6435645229
75+	.6478973047	.6521784658	.6564086448	.6605884689	.6647185548
80+	.6687995092	.6728319286	.6768164004	.6807535026	.6846438043
85+	.6884878661	.6922862401	.6960394704	.6997480933	.7034126373

D-INTEGRAL PROBABILITY M=R+1, B= 7

A= .80

R	1	2	3	4	5
90+	.7070336235	.7106115659	.7141469713	.7176403396	.7210921642
95+	.7245029316	.7278731224	.7312032104	.7344936638	.7377449445
100+	.7409575086	.7441318065	.7472682831	.7503673776	.7534295240
105+	.7564551508	.7594446813	.7623985340	.7653171221	.7682008538
110+	.7710501328	.7738653577	.7766469225	.7793952166	.7821106250
115+	.7847935280	.7874443017	.7900633176	.7926509431	.7952075414
120+	.7977334714	.8002290882	.8026947425	.8051307812	.8075375472
125+	.8099153796	.8122646136	.8145855807	.8168786085	.8191440211
130+	.8213821389	.8235932785	.8257777534	.8279358731	.8300679440
135+	.8321742689	.8342551472	.8363108750	.8383417452	.8403480472
140+	.8423300674	.8442880888	.8462223915	.8481332522	.8500209448
145+	.8518857398	.8537279050	.8555477051	.8573454018	.8591212540
150+	.8608755175	.8626084455	.8643202883	.8660112932	.8676817051
155+	.8693317658	.8709617147	.8725717883	.8741622207	.8757332431
160+	.8772850842	.8788179703	.8803321251	.8818277695	.8833051224
165+	.8847643999	.8862058158	.8876295815	.8890359059	.8904249957
170+	.8917970552	.8931522865	.8944908893	.8958130610	.8971189970
175+	.8984088901	.8996829315	.9009413096	.9021842111	.9034118205
180+	.9046243200	.9058218900	.9070047088	.9081729524	.9093267952
185+	.9104664094	.9115919651	.9127036308	.9138015728	.9148859556
190+	.9159569419	.9170146923	.9180593658	.9190911195	.9201101087
195+	.9211164868	.9221104057	.9230920152	.9240614638	.9250188978
200+	.9259644623	.9268983004	.9278205536	.9287313619	.9296308636
205+	.9305191953	.9313964922	.9322628878	.9331185141	.9339635016
210+	.9347979792	.9356220743	.9364359130	.9372396197	.9380333174
215+	.9388171279	.9395911712	.9403555662	.9411104302	.9418558794
220+	.9425920283	.9433189902	.9440368773	.9447458001	.9454458680
225+	.9461371892	.9468198704	.9474940173	.9481597341	.9488171240
230+	.9494662888	.9501073292	.9507403448	.9513654337	.9519826932
235+	.9525922193	.9531941068	.9537884494	.9543753399	.9549548697
240+	.9555271292	.9560922079	.9566501941	.9572011749	.9577452366
245+	.9582824645	.9588129426	.9593367542	.9598539815	.9603647055
250+	.9608690066	.9613669640	.9618586560	.9623441600	.9628235525
255+	.9632969090	.9637643042	.9642258117	.9646815044	.9651314543
260+	.9655757325	.9660144093	.9664475540	.9668752352	.9672975207
265+	.9677144775	.9681261715	.9685326682	.9689340321	.9693303270
270+	.9697216157	.9701079607	.9704894232	.9708660640	.9712379432
275+	.9716051200	.9719676529	.9723255998	.9726790178	.9730279633
280+	.9733724920	.9737126592	.9740485191	.9743801254	.9747075314
285+	.9750307894	.9753499512	.9756650680	.9759761903	.9762833681
290+	.9765866506	.9768860867	.9771817243	.9774736111	.9777617940
295+	.9780463193	.9783272329	.9786045800	.9788784054	.9791487530
300+	.9794156666	.9796791892	.9799393634	.9801962311	.9804498339
305+	.9807002127	.9809474081	.9811914599	.9814324078	.9816702907
310+	.9819051471	.9821370151	.9823659323	.9825919358	.9828150623
315+	.9830353479	.9832528284	.9834675392	.9836795151	.9838887905
320+	.9840953996	.9842993759	.9845007525	.9846995624	.9848958377
325+	.9850896106	.9852809126	.9854697748	.9856562281	.9858403029
330+	.9860220292	.9862014367	.9863785546	.9865534120	.9867260373
335+	.9868964588	.9870647043	.9872308014	.9873947773	.9875566587
340+	.9877164723	.9878742441	.9880300000	.9881837655	.9883355658
345+	.9884854259	.9886333703	.9887794233	.9889236088	.9890659505
350+	.9892064719	.9893451959	.9894821454	.9896173429	.9897508106
355+	.9898825704	.9900126442	.9901410531	.9902678184	.9903929610

D-INTEGRAL PROBABILITY M=R+1,B= 8

R	1	2	3	4	5
			A=1/10		
0+	.3086419753	.7339459207	.9132441454	.9730506021	.9917227675
			A=1/ 9		
0+	.2802768166	.6954046529	.8901463826	.9623927765	.9873233860
5+	.9957339268	.9985593539	.9995110622	.9998332128	.9999428382
			A=1/ 8		
0+	.2500000000	.6489239894	.8587212752	.9461675568	.9798861163
5+	.9925229400	.9972176986	.9989611482	.9996105171	.9998533777
			A=1/ 7		
0+	.2177777778	.5926291604	.8153058860	.9207663762	.9667899370
5+	.9862021597	.9942784645	.9976244210	.9990110250	.9995869843
			A=1/ 6		
0+	.1836734694	.5243664924	.7545395599	.8798961713	.9427230668
5+	.9730171348	.9873540540	.9940813621	.9972280977	.9986996394
			A=1/ 5		
0+	.1479289941	.4419879614	.6688989199	.8126421123	.8966830717
5+	.9438500931	.9697301859	.9837504748	.9912928405	.9953360689
			A=1/ 4		
0+	.1111111111	.3442206433	.5490851001	.7011359157	.8062800098
5+	.8762237290	.9216607229	.9507292298	.9691388407	.9807200279
10+	.9879736057	.9925040413	.9953289675	.9970889196	.9981850600
			A=1/ 3		
0+	.0743801653	.2329204999	.3880857049	.5220227409	.6318546545
5+	.7193119918	.7876171378	.8402373426	.8803655424	.9107329685
10+	.9335779574	.9506846733	.9634481449	.9729440235	.9799931034
15+	.9852167493	.9890824715	.9919403396	.9940515008	.9956101936
			A= .40		
0+	.0566893424	.1748111851	.2939328470	.4031276115	.4996848422
5+	.5833226502	.6547606085	.7151500531	.7657929042	.8079924494
10+	.8429747460	.8718505797	.8956012359	.9150781019	.9310099611
15+	.9440141768	.9546094242	.9632285723	.9702309141	.9759133290
20+	.9805202020	.9842520693	.9872730459	.9897171356	.9916935442
			A=1/ 2		
0+	.0400000000	.1182318080	.1965882138	.2707608944	.3399378642
5+	.4039522752	.4628583462	.5168201632	.5660655979	.6108602909
10+	.6514904731	.6882508817	.7214361098	.7513344468	.7782235880
15+	.8023677604	.8240159236	.8434007832	.8607384123	.8762283195
20+	.8900538421	.9023827650	.9133680924	.9231489131	.9318513186
25+	.9395893379	.9464658671	.9525735759	.9579957775	.9628072557
30+	.9670750422	.9708591428	.9742132105	.9771851673	.9798177744
35+	.9821491539	.9842132649	.9860403352	.9876572531	.9890879214
40+	.9903535766	.9914730763	.9924631579	.9933386699	.9941127801
			A= .60		
0+	.0297265161	.0833146401	.1346834229	.1827968685	.2280578057
5+	.2708176924	.3113208351	.3497404590	.3862070021	.4208252856
10+	.4536845507	.4848643281	.5144378927	.5424743012	.5690395881
15+	.5941974551	.6180096539	.6405361853	.6618353877	.6819639629
20+	.7009769676	.7189277886	.7358681133	.7518479023	.7669153666
25+	.7811169534	.7944973399	.8070994357	.8189643933	.8301316259
30+	.8406388327	.8505220290	.8598155826	.8685522534	.8767632369
35+	.8844782110	.8917253839	.8985315447	.9049221141	.9109211965
40+	.9165516316	.9218350464	.9267919068	.9314415681	.9358023248
45+	.9398914597	.9437252916	.9473192213	.9506877768	.9538446567
50+	.9568027722	.9595742878	.9621706600	.9646026745	.9668804824
55+	.9690136341	.9710111124	.9728813635	.9746323269	.9762714639
60+	.9778057843	.9792418725	.9805859117	.9818437071	.9830207075
65+	.9841220268	.9851524630	.9861165173	.9870184115	.9878621047
70+	.9886513091	.9893895048	.9900799537	.9907257131	.9913296476

D-INTEGRAL PROBABILITY M=R+1,B= 8

R	1	2	3	4	5
			A= .65		
0+	.0260145682	.0708709208	.1125499515	.1509584753	.1868312503
5+	.2206774035	.2528256316	.2834947415	.3128371137	.3409638434
10+	.3679595310	.3938912586	.4188142083	.4427752808	.4658154803
15+	.4879715208	.5092769248	.5297627868	.5494583091	.5683911816
20+	.5865878531	.6040737260	.6208732968	.6370102582	.6525075721
25+	.6673875247	.6816717649	.6953813339	.7085366871	.7211577106
30+	.7332637341	.7448735416	.7560053804	.7666769689	.7769055044
35+	.7867076699	.7960996419	.8050970973	.8137152213	.8219687151
40+	.8298718047	.8374382487	.8446813477	.8516139533	.8582484778
45+	.8645969035	.8706707928	.8764812984	.8820391730	.8873547802
50+	.8924381044	.8972987613	.9019460081	.9063887541	.9106355710
55+	.9146947028	.9185740761	.9222813099	.9258237259	.9292083577
60+	.9324419606	.9355310209	.9384817650	.9413001687	.9439919656
65+	.9465626558	.9490175146	.9513616001	.9535997616	.9557366472
70+	.9577767112	.9597242217	.9615832672	.9633577640	.9650514626
75+	.9666679540	.9682106764	.9696829208	.9710878368	.9724284389
80+	.9737076110	.9749281124	.9760925826	.9772035462	.9782634175
85+	.9792745053	.9802390170	.9811590631	.9820366611	.9828737391
90+	.9836721404	.9844336261	.9851598793	.9858525080	.9865130486
95+	.9871429688	.9877436706	.9883164932	.9888627155	.9893835590
100+	.9898801900	.9903537225	.9908052198	.9912356973	.9916461243
			A= .70		
0+	.0229568411	.0607592511	.0946435950	.1251865799	.1533353740
5+	.1796950740	.2046465337	.2284395291	.2512452614	.2731858208
10+	.2943513492	.3148104645	.3346168413	.3538135085	.3724357404
15+	.3905130523	.4080706079	.4251302328	.4417111553	.4578305569
20+	.4735039868	.4887456769	.5035687843	.5179855796	.5320075944
25+	.5456457370	.5589103847	.5718114573	.5843584763	.5965606123
30+	.6084267238	.6199653885	.6311849291	.6420934344	.6526987767
35+	.6630086265	.6730304640	.6827715900	.6922391334	.7014400595
40+	.7103811752	.7190691351	.7275104454	.7357114685	.7436784265
45+	.7514174041	.7589343518	.7662350885	.7733253043	.7802105621
50+	.7868963005	.7933878356	.7996903631	.8058089602	.8117485877
55+	.8175140916	.8231102053	.8285415512	.8338126426	.8389278856
60+	.8438915808	.8487079251	.8533810136	.8579148413	.8623133049
65+	.8665802047	.8707192462	.8747340419	.8786281133	.8824048923
70+	.8860677234	.8896198648	.8930644910	.8964046936	.8996434837
75+	.9027837933	.9058284770	.9087803137	.9116420083	.9144161931
80+	.9171054299	.9197122111	.9222389614	.9246880395	.9270617396
85+	.9293622927	.9315918686	.9337525765	.9358464674	.9378755350
90+	.9398417171	.9417468970	.9435929051	.9453815199	.9471144695
95+	.9487934328	.9504200409	.9519958779	.9535224826	.9550013497
100+	.9564339303	.9578216337	.9591658285	.9604678431	.9617289674
105+	.9629504534	.9641335165	.9652793364	.9663890580	.9674637923
110+	.9685046177	.9695125804	.9704886956	.9714339483	.9723492941
115+	.9732356599	.9740939452	.9749250221	.9757297368	.9765089100
120+	.9772633374	.9779937909	.9787010189	.9793857474	.9800486799
125+	.9806904990	.9813118662	.9819134228	.9824957908	.9830595729
130+	.9836053536	.9841336991	.9846451586	.9851402643	.9856195318
135+	.9860834612	.9865325368	.9869672281	.9873879901	.9877952637
140+	.9881894760	.9885710410	.9889403599	.9892978211	.9896438012
145+	.9899786650	.9903027659	.9906164462	.9909200375	.9912138613
			A= .75		
0+	.0204081633	.0524645931	.0800684053	.1042785265	.1261732720
5+	.1464093709	.1653932135	.1833891726	.2005775372	.2170862788
10+	.2330093309	.2484176177	.2633659967	.2778977917	.2920478482
15+	.3058446500	.3193118223	.3324692198	.3453337311	.3579198822

D-INTEGRAL PROBABILITY M=R+1,B= 8

R	1	2	3	4	5
			A= .75		
20+	.3702402963	.3823060484	.3941269426	.4057117309	.4170682878
25+	.4282037502	.4391246316	.4498369143	.4603461268	.4706574061
30+	.4807755505	.4907050642	.5004501936	.5100149589	.5194031804
35+	.5286185016	.5376644079	.5465442435	.5552612256	.5638184566
40+	.5722189351	.5804655647	.5885611624	.5965084655	.6043101378
45+	.6119687750	.6194869094	.6268670142	.6341115071	.6412227536
50+	.6482030700	.6550547257	.6617799459	.6683809135	.6748597708
55+	.6812186213	.6874595311	.6935845305	.6995956148	.7054947458
60+	.7112838526	.7169648325	.7225395519	.7280098471	.7333775251
65+	.7386443639	.7438121136	.7488824966	.7538572082	.7587379173
70+	.7635262667	.7682238734	.7728323296	.7773532022	.7817880340
75+	.7861383437	.7904056262	.7945913533	.7986969734	.8027239125
80+	.8066735740	.8105473394	.8143465682	.8180725985	.8217267472
85+	.8253103101	.8288245624	.8322707589	.8356501343	.8389639033
90+	.8422132609	.8453993830	.8485234262	.8515865280	.8545898076
95+	.8575343658	.8604212849	.8632516297	.8660264471	.8687467666
100+	.8714136005	.8740279441	.8765907760	.8791030583	.8815657368
105+	.8839797411	.8863459851	.8886653672	.8909387701	.8931670614
110+	.8953510940	.8974917057	.8995897199	.9016459457	.9036611780
115+	.9056361979	.9075717727	.9094686561	.9113275887	.9131492979
120+	.9149344981	.9166838912	.9183981663	.9200780004	.9217240583
125+	.9233369929	.9249174451	.9264660446	.9279834093	.9294701462
130+	.9309268511	.9323541090	.9337524942	.9351225705	.9364648912
135+	.9377799998	.9390684293	.9403307031	.9415673351	.9427788294
140+	.9439656807	.9451283747	.9462673880	.9473831882	.9484762341
145+	.9495469761	.9505958560	.9516233072	.9526297552	.9536156172
150+	.9545813026	.9555272130	.9564537425	.9573612776	.9582501975
155+	.9591208740	.9599736721	.9608089495	.9616270573	.9624283397
160+	.9632131343	.9639817722	.9647345783	.9654718709	.9661939625
165+	.9669011593	.9675937617	.9682720641	.9689363554	.9695869187
170+	.9702240316	.9708479665	.9714589903	.9720573646	.9726433460
175+	.9732171861	.9737791315	.9743294240	.9748683005	.9753959934
180+	.9759127304	.9764187350	.9769142258	.9773994174	.9778745201
185+	.9783397401	.9787952792	.9792413357	.9796781034	.9801057728
190+	.9805245300	.9809345580	.9813360357	.9817291386	.9821140388
195+	.9824909046	.9828599014	.9832211910	.9835749319	.9839212796
200+	.9842603864	.9845924015	.9849174712	.9852357387	.9855473444
205+	.9858524259	.9861511180	.9864435527	.9867298595	.9870101651
210+	.9872845936	.9875532669	.9878163041	.9880738220	.9883259350
215+	.9885727553	.9888143927	.9890509547	.9892825467	.9895092720
220+	.9897312318	.9899485251	.9901612490	.9903694985	.9905733668
			A= .80		
0+	.0182615047	.0456007487	.0681311539	.0872624072	.1041495322
5+	.1194712240	.1336408542	.1469249216	.1595041244	.1715062164
10+	.1830246874	.1941299405	.2048762856	.2153064852	.2254548080
15+	.2353491406	.2450124833	.2544640337	.2637199872	.2727941375
20+	.2816983354	.2904428430	.2990366118	.3074875017	.3158024575
25+	.3239876506	.3320485949	.3399902415	.3478170576	.3555330916
30+	.3631420282	.3706472346	.3780517994	.3853585663	.3925701627
35+	.3996890239	.4067174146	.4136574473	.4205110982	.4272802210
40+	.4339665595	.4405717583	.4470973719	.4535448735	.4599156624
45+	.4662110702	.4724323671	.4785807671	.4846574324	.4906634780
50+	.4965999753	.5024679553	.5082684120	.5140023052	.5196705627
55+	.5252740826	.5308137359	.5362903676	.5417047991	.5470578291
60+	.5523502357	.5575827773	.5627561938	.5678712077	.5729285254
65+	.5779288376	.5828728207	.5877611370	.5925944360	.5973733547
70+	.6020985183	.6067705406	.6113900250	.6159575644	.6204737420

D-INTEGRAL PROBABILITY M=R+1,B= 8

A= .80

R	1	2	3	4	5
75+	.6249391317	.6293542984	.6337197982	.6380361792	.6423039814
80+	.6465237371	.6506959711	.6548212011	.6588999381	.6629326861
85+	.6669199427	.6708621992	.6747599407	.6786136466	.6824237903
90+	.6861908395	.6899152566	.6935974985	.6972380169	.7008372583
95+	.7043956644	.7079136718	.7113917124	.7148302133	.7182295970
100+	.7215902817	.7249126809	.7281972037	.7314442551	.7346542357
105+	.7378275422	.7409645668	.7440656980	.7471313203	.7501618141
110+	.7531575561	.7561189190	.7590462721	.7619399807	.7648004064
115+	.7676279074	.7704228381	.7731855496	.7759163893	.7786157013
120+	.7812838261	.7839211011	.7865278600	.7891044336	.7916511491
125+	.7941683306	.7966562991	.7991153721	.8015458642	.8039480871
130+	.8063223489	.8086689550	.8109882077	.8132804064	.8155458473
135+	.8177848239	.8199976266	.8221845431	.8243458580	.8264818534
140+	.8285928082	.8306789989	.8327406989	.8347781791	.8367917076
145+	.8387815498	.8407479685	.8426912237	.8446115731	.8465092714
150+	.8483845711	.8502377219	.8520689711	.8538785634	.8556667410
155+	.8574337439	.8591798093	.8609051723	.8626100652	.8642947184
160+	.8659593596	.8676042144	.8692295057	.8708354546	.8724222795
165+	.8739901969	.8755394208	.8770701630	.8785826333	.8800770391
170+	.8815535859	.8830124767	.8844539126	.8858780927	.8872852138
175+	.8886754707	.8900490562	.8914061612	.8927469742	.8940716820
180+	.8953804695	.8966735195	.8979510128	.8992131283	.9004600432
185+	.9016919326	.9029089699	.9041113263	.9052991717	.9064726737
190+	.9076319985	.9087773101	.9099087710	.9110265420	.9121307820
195+	.9132216483	.9142992963	.9153638800	.9164155515	.9174544614
200+	.9184807586	.9194945903	.9204961022	.9214854384	.9224627414
205+	.9234281521	.9243818100	.9253238528	.9262544169	.9271736371
210+	.9280816469	.9289785781	.9298645611	.9307397250	.9316041971
215+	.9324581038	.9333015697	.9341347182	.9349576713	.9357705495
220+	.9365734722	.9373665572	.9381499213	.9389236796	.9396879463
225+	.9404428340	.9411884542	.9419249172	.9426523320	.9433708062
230+	.9440804463	.9447813578	.9454736448	.9461574102	.9468327557
235+	.9474997821	.9481585889	.9488092743	.9494519356	.9500866689
240+	.9507135693	.9513327308	.9519442461	.9525482072	.9531447047
245+	.9537338285	.9543156671	.9548903083	.9554578387	.9560183440
250+	.9565719088	.9571186169	.9576585511	.9581917930	.9587184236
255+	.9592385227	.9597521694	.9602594416	.9607604165	.9612551705
260+	.9617437788	.9622263160	.9627028558	.9631734708	.9636382330
265+	.9640972135	.9645504825	.9649981096	.9654401632	.9658767113
270+	.9663078209	.9667335582	.9671539886	.9675691769	.9679791870
275+	.9683840821	.9687839246	.9691787763	.9695686980	.9699537501
280+	.9703339922	.9707094830	.9710802807	.9714464427	.9718080260
285+	.9721650865	.9725176797	.9728658604	.9732096827	.9735492002
290+	.9738844656	.9742155313	.9745424488	.9748652691	.9751840426
295+	.9754988191	.9758096478	.9761165773	.9764196555	.9767189301
300+	.9770144478	.9773062550	.9775943974	.9778789203	.9781598683
305+	.9784372857	.9787112159	.9789817020	.9792487868	.9795125121
310+	.9797729196	.9800300503	.9802839449	.9805346432	.9807821850
315+	.9810266093	.9812679548	.9815062597	.9817415617	.9819738979
320+	.9822033053	.9824298202	.9826534784	.9828743156	.9830923667
325+	.9833076663	.9835202488	.9837301478	.9839373967	.9841420286
330+	.9843440760	.9845435711	.9847405456	.9849350310	.9851270584
335+	.9853166583	.9855038609	.9856886963	.9858711940	.9860513830
340+	.9862292923	.9864049503	.9865783852	.9867496246	.9869186960
345+	.9870856266	.9872504431	.9874131719	.9875738392	.9877324707
350+	.9878890920	.9880437281	.9881964040	.9883471442	.9884959730
355+	.9886429143	.9887879917	.9889312287	.9890726483	.9892122733

D-INTEGRAL PROBABILITY M=R+1,B= 8

R	1	2	3	4	5
			A= .80		
360+	.9893501262	.9894862293	.9896206046	.9897532736	.9898842579
365+	.9900135787	.9901412567	.9902673127	.9903917670	.9905146399

D-INTEGRAL PROBABILITY M=R+1,B= 9

R	1	2	3	4	5
			A=1/10		
0+	.2770083102	.7104757721	.9038486907	.9698845131	.9907159834
			A=1/ 9		
0+	.2500000000	.6699635924	.8786504286	.9580645308	.9857996533
5+	.9952105590	.9983808859	.9994502095	.9998124081	.9999357002
			A=1/ 8		
0+	.2214532872	.6215495643	.8446000418	.9401522358	.9775123513
5+	.9916157190	.9968752561	.9988323372	.9995620372	.9998350905
			A=1/ 7		
0+	.1914062500	.5635533917	.7979703588	.9122892227	.9629832791
5+	.9845596288	.9935827928	.9973321083	.9988885059	.9995356206
			A=1/ 6		
0+	.1600000000	.4941604063	.7334543522	.8678650822	.9364644528
5+	.9699124949	.9858525440	.9933648585	.9968884716	.9985391144
			A=1/ 5		
0+	.1275510204	.4117814338	.6439111904	.7957173384	.8862752084
5+	.9377838615	.9663049035	.9818533752	.9902544354	.9947716813
			A=1/ 4		
0+	.0946745562	.3159905538	.5211878477	.6783660516	.7893920158
5+	.8643826279	.9136508035	.9454395724	.9657029828	.9785139969
10+	.9865686957	.9916144273	.9947678804	.9967359914	.9979634583
			A=1/ 3		
0+	.0625000000	.2097105489	.3608188266	.4950847912	.6074755370
5+	.6983870730	.7702807613	.8262300419	.8692562859	.9020459562
10+	.9268594106	.9455335192	.9595260441	.9699743646	.9777547359
15+	.9835357672	.9878238355	.9910002146	.9933506614	.9950885629
			A= .40		
0+	.0472589792	.1555137761	.2695073037	.3767825665	.4734620001
5+	.5584584714	.6319417274	.6946940364	.7477764427	.7923416004
10+	.8295275614	.8603997090	.8859221374	.9069470442	.9242148966
15+	.9383607313	.9499236240	.9593574583	.9670418489	.9732925531
20+	.9783710104	.9824928538	.9858353563	.9885438515	.9907372065
			A=1/ 2		
0+	.0330578512	.1037226200	.1772012840	.2483217219	.3157327919
5+	.3789174081	.4376825028	.4920085610	.5419833708	.5877642498
10+	.6295532587	.6675797936	.7020879947	.7333275610	.7615470766
15+	.7869892310	.8098874818	.8304638201	.8489273775	.8654736693
20+	.8802843174	.8935271272	.9053564209	.9159135530	.9253275469
25+	.9337158082	.9411848802	.9478312148	.9537419391	.9589956038
30+	.9636629033	.9678073608	.9714859735	.9747498164	.9776446031
35+	.9802112041	.9824861235	.9845019360	.9862876854	.9878692476
40+	.9892696598	.9905094186	.9916067494	.9925778492	.9934371047
			A= .60		
0+	.0244140625	.0723208278	.1197709249	.1650488107	.2082141720
5+	.2494270816	.2888105259	.3264532951	.3624230762	.3967758566
10+	.4295616204	.4608276064	.4906200890	.5189852872	.5459697606
15+	.5716205114	.5959849199	.6191105928	.6410451705	.6618361223
20+	.6815305454	.7001749786	.7178152334	.7344962469	.7502619560
25+	.7651551914	.7792175922	.7924895381	.8050100977	.8168169917

D-INTEGRAL PROBABILITY M=R+1, B= 9

R	1	2	3	4	5
			A= .60		
30+	.8279465691	.8384337947	.8483122463	.8576141210	.8663702480
35+	.8746101086	.8823618602	.8896523657	.8965072257	.9029508136
40+	.9090063128	.9146957559	.9200400644	.9250590896	.9297716537
45+	.9341955909	.9383477886	.9422442274	.9459000219	.9493294590
50+	.9525460366	.9555625012	.9583908835	.9610425343	.9635281577
55+	.9658578444	.9680411032	.9700868909	.9720036419	.9737992959
60+	.9754813244	.9770567563	.9785322023	.9799138780	.9812076260
65+	.9824189369	.9835529692	.9846145683	.9856082848	.9865383911
70+	.9874088977	.9882235687	.9889859362	.9896993141	.9903668107
			A= .65		
0+	.0213117374	.0612502565	.0995166163	.1353835352	.1692870482
5+	.2015789278	.2324933824	.2621866257	.2907664492	.3183104070
10+	.3448768077	.3705114635	.3952519270	.4191302085	.4421745509
15+	.4644106046	.4858622153	.5065519531	.5265014704	.5457317416
20+	.5642632226	.5821159549	.5993096308	.6158636322	.6317970508
25+	.6471286964	.6618770963	.6760604897	.6896968188	.7028037181
30+	.7153985033	.7274981601	.7391193334	.7502783179	.7609910494
35+	.7712730975	.7811396598	.7906055568	.7996852291	.8083927349
40+	.8167417488	.8247455624	.8324170852	.8397688468	.8468129995
45+	.8535613224	.8600252255	.8662157551	.8721435987	.8778190916
50+	.8832522232	.8884526438	.8934296713	.8981922991	.9027492036
55+	.9071087512	.9112790070	.9152677422	.9190824419	.9227303137
60+	.9262182949	.9295530609	.9327410330	.9357883860	.9387010562
65+	.9414847489	.9441449458	.9466869126	.9491157060	.9514361810
70+	.9536529978	.9557706286	.9577933643	.9597253208	.9615704456
75+	.9633325237	.9650151839	.9666219043	.9681560180	.9696207189
80+	.9710190666	.9723539919	.9736283016	.9748446832	.9760057102
85+	.9771138458	.9781714479	.9791807728	.9801439800	.9810631352
90+	.9819402150	.9827771098	.9835756280	.9843374988	.9850643757
95+	.9857578399	.9864194030	.9870505099	.9876525419	.9882268191
100+	.9887746030	.9892970993	.9897954600	.9902707856	.9907241278
			A= .70		
0+	.0187652468	.0523067641	.0832529804	.1115814750	.1379712276
5+	.1628916335	.1866452453	.2094320084	.2313889950	.2526135619
10+	.2731771132	.2931335590	.3125246923	.3313837164	.3497376221
15+	.3676088283	.3850163389	.4019765726	.4185039676	.4346114289
20+	.4503106623	.4656124264	.4805267250	.4950629545	.5092300179
25+	.5230364127	.5364903000	.5495995585	.5623718261	.5748145333
30+	.5869349292	.5987401015	.6102369925	.6214324117	.6323330450
35+	.6429454629	.6532761253	.6633313865	.6731174985	.6826406131
40+	.6919067839	.7009219676	.7096920251	.7182227215	.7265197275
45+	.7345886188	.7424348768	.7500638885	.7574809469	.7646912507
50+	.7716999048	.7785119203	.7851322148	.7915656123	.7978168444
55+	.8038905497	.8097912748	.8155234748	.8210915137	.8264996651
60+	.8317521131	.8368529527	.8418061908	.8466157474	.8512854559
65+	.8558190645	.8602202371	.8644925541	.8686395142	.8726645345
70+	.8765709525	.8803620269	.8840409388	.8876107929	.8910746189
75+	.8944353726	.8976959372	.9008591245	.9039276764	.9069042658
80+	.9097914983	.9125919132	.9153079850	.9179421242	.9204966795
85+	.9229739379	.9253761270	.9277054158	.9299639157	.9321536822
90+	.9342767161	.9363349643	.9383303213	.9402646304	.9421396849
95+	.9439572290	.9457189593	.9474265256	.9490815321	.9506855388
100+	.9522400621	.9537465761	.9552065137	.9566212675	.9579921908
105+	.9593205987	.9606077689	.9618549429	.9630633265	.9642340913
110+	.9653683751	.9664672829	.9675318881	.9685632326	.9695623285
115+	.9705301582	.9714676757	.9723758068	.9732554506	.9741074795
120+	.9749327403	.9757320550	.9765062211	.9772560126	.9779821806

D-INTEGRAL PROBABILITY M=R+1, B= 9

R	1	2	3	4	5
			A= .70		
125+	.9786854537	.9793665391	.9800261227	.9806648699	.9812834261
130+	.9818824177	.9824624519	.9830241177	.9835679866	.9840946126
135+	.9846045330	.9850982689	.9855763254	.9860391926	.9864873453
140+	.9869212442	.9873413357	.9877480526	.9881418147	.9885230285
145+	.9888920884	.9892493766	.9895952636	.9899301082	.9902542587
			A= .75		
0+	.0166493236	.0450083802	.0701036289	.0924244997	.1128042569
5+	.1317791234	.1496878361	.1667528998	.1831267407	.1989177221
10+	.2142053530	.2290495591	.2434965601	.2575827255	.2713371832
15+	.2847836316	.2979416280	.3108275228	.3234551508	.3358363492
20+	.3479813527	.3598990981	.3715974622	.3830834487	.3943633378
25+	.4054428058	.4163270218	.4270207270	.4375282987	.4478538043
30+	.4580010452	.4679735936	.4777748236	.4874079366	.4968759840
35+	.5061818848	.5153284419	.5243183553	.5331542335	.5418386032
40+	.5503739181	.5587625655	.5670068730	.5751091138	.5830715108
45+	.5908962415	.5985854407	.6061412039	.6135655897	.6208606224
50+	.6280282937	.6350705642	.6419893656	.6487866012	.6554641477
55+	.6620238559	.6684675514	.6747970357	.6810140868	.6871204597
60+	.6931178867	.6990080784	.7047927237	.7104734902	.7160520247
65+	.7215299533	.7269088818	.7321903957	.7373760608	.7424674230
70+	.7474660087	.7523733249	.7571908596	.7619200813	.7665624400
75+	.7711193666	.7755922734	.7799825542	.7842915845	.7885207212
80+	.7926713034	.7967446518	.8007420694	.8046648415	.8085142353
85+	.8122915010	.8159978709	.8196345603	.8232027671	.8267036725
90+	.8301384404	.8335082182	.8368141365	.8400573096	.8432388353
95+	.8463597953	.8494212549	.8524242639	.8553698562	.8582590499
100+	.8610928477	.8638722371	.8665981903	.8692716645	.8718936021
105+	.8744649306	.8769865632	.8794593984	.8818843208	.8842622007
110+	.8865938946	.8888802450	.8911220810	.8933202184	.8954754594
115+	.8975885932	.8996603962	.9016916318	.9036830508	.9056353916
120+	.9075493802	.9094257305	.9112651444	.9130683120	.9148359115
125+	.9165686098	.9182670625	.9199319136	.9215637965	.9231633334
130+	.9247311359	.9262678049	.9277739310	.9292500942	.9306968648
135+	.9321148027	.9335044581	.9348663717	.9362010742	.9375090873
140+	.9387909232	.9400470851	.9412780669	.9424843542	.9436664233
145+	.9448247423	.9459597706	.9470719596	.9481617522	.9492295833
150+	.9502758801	.9513010618	.9523055399	.9532897185	.9542539942
155+	.9551987562	.9561243866	.9570312606	.9579197462	.9587902047
160+	.9596429905	.9604784517	.9612969296	.9620987593	.9628842696
165+	.9636537831	.9644076163	.9651460797	.9658694782	.9665781105
170+	.9672722701	.9679522446	.9686183164	.9692707622	.9699098538
175+	.9705358575	.9711490348	.9717496419	.9723379302	.9729141465
180+	.9734785325	.9740313255	.9745727581	.9751030584	.9756224502
185+	.9761311530	.9766293818	.9771173477	.9775952575	.9780633141
190+	.9785217165	.9789706596	.9794103348	.9798409294	.9802626273
195+	.9806756088	.9810800505	.9814761257	.9818640042	.9822438524
200+	.9826158336	.9829801078	.9833368317	.9836861592	.9840282408
205+	.9843632243	.9846912545	.9850124734	.9853270199	.9856350304
210+	.9859366385	.9862319753	.9865211690	.9868043454	.9870816279
215+	.9873531371	.9876189916	.9878793074	.9881341981	.9883837753
220+	.9886281480	.9888674234	.9891017062	.9893310992	.9895557030
225+	.9897756164	.9899909360	.9902017565	.9904081707	.9906102695
			A= .80		
0+	.0148720999	.0389969797	.0593992165	.0769454782	.0925645945
5+	.1068264289	.1200844412	.1325687591	.1444363417	.1557985648
10+	.1667371133	.1773135754	.1875754829	.1975602577	.2072978712
15+	.2168126897	.2261247842	.2352508811	.2442050651	.2529993083

D-INTEGRAL PROBABILITY M=R+1, B= 9

R	1	2	3	4	5
			A= .80		
20+	.2616438737	.2701476275	.2785182840	.2867625991	.2948865262
25+	.3028953408	.3107937430	.3185859412	.3262757215	.3338665054
30+	.3413613981	.3487632293	.3560745876	.3632978502	.3704352077
35+	.3774886857	.3844601639	.3913513916	.3981640020	.4048995248
40+	.4115593961	.4181449686	.4246575194	.4310982575	.4374683301
45+	.4437688284	.4500007930	.4561652177	.4622630545	.4682952163
50+	.4742625809	.4801659936	.4860062698	.4917841972	.4975005385
55+	.5031560329	.5087513976	.5142873303	.5197645095	.5251835967
60+	.5305452371	.5358500610	.5410986843	.5462917101	.5514297289
65+	.5565133196	.5615430502	.5665194786	.5714431528	.5763146115
70+	.5811343851	.5859029953	.5906209564	.5952887750	.5999069505
75+	.6044759758	.6089963371	.6134685145	.6178929821	.6222702083
80+	.6266006559	.6308847825	.6351230406	.6393158776	.6434637362
85+	.6475670544	.6516262659	.6556417996	.6596140806	.6635435296
90+	.6674305632	.6712755942	.6750790313	.6788412798	.6825627408
95+	.6862438123	.6898848882	.6934863592	.6970486126	.7005720321
100+	.7040569982	.7075038880	.7109130756	.7142849316	.7176198235
105+	.7209181159	.7241801701	.7274063443	.7305969939	.7337524712
110+	.7368731254	.7399593031	.7430113478	.7460296000	.7490143975
115+	.7519660754	.7548849658	.7577713979	.7606256986	.7634481915
120+	.7662391979	.7689990362	.7717280221	.7744264688	.7770946867
125+	.7797329836	.7823416647	.7849210327	.7874713876	.7899930268
130+	.7924862452	.7949513352	.7973885868	.7997982872	.8021807214
135+	.8045361717	.8068649181	.8091672381	.8114434068	.8136936969
140+	.8159183785	.8181177197	.8202919858	.8224414400	.8245663432
145+	.8266669538	.8287435280	.8307963196	.8328255802	.8348315591
150+	.8368145034	.8387746577	.8407122648	.8426275648	.8445207961
155+	.8463921944	.8482419936	.8500704254	.8518777190	.8536641019
160+	.8554297993	.8571750342	.8589000277	.8606049987	.8622901640
165+	.8639557385	.8656019348	.8672289638	.8688370342	.8704263527
170+	.8719971240	.8735495510	.8750838345	.8766001733	.8780987645
175+	.8795798030	.8810434820	.8824899929	.8839195248	.8853322655
180+	.8867284006	.8881081139	.8894715874	.8908190015	.8921505346
185+	.8934663633	.8947666625	.8960516055	.8973213636	.8985761065
190+	.8998160022	.9010412171	.9022519157	.9034482611	.9046304144
195+	.9057985354	.9069527821	.9080933109	.9092202767	.9103338326
200+	.9114341303	.9125213200	.9135955500	.9146569676	.9157057181
205+	.9167419455	.9177657923	.9187773995	.9197769067	.9207644518
210+	.9217401716	.9227042011	.9236566742	.9245977233	.9255274792
215+	.9264460716	.9273536287	.9282502773	.9291361429	.9300113497
220+	.9308760206	.9317302770	.9325742393	.9334080263	.9342317557
225+	.9350455440	.9358495063	.9366437565	.9374284073	.9382035702
230+	.9389693555	.9397258721	.9404732281	.9412115301	.9419408836
235+	.9426613932	.9433731619	.9440762921	.9447708847	.9454570395
240+	.9461348555	.9468044304	.9474658608	.9481192423	.9487646694
245+	.9494022357	.9500320335	.9506541544	.9512686887	.9518757259
250+	.9524753544	.9530676616	.9536527340	.9542306572	.9548015156
255+	.9553653928	.9559223715	.9564725335	.9570159596	.9575527297
260+	.9580829228	.9586066171	.9591238897	.9596348172	.9601394749
265+	.9606379376	.9611302790	.9616165722	.9620968893	.9625713016
270+	.9630398796	.9635026932	.9639598111	.9644113017	.9648572321
275+	.9652976690	.9657326783	.9661623250	.9665866734	.9670057873
280+	.9674197293	.9678285617	.9682323460	.9686311428	.9690250121
285+	.9694140134	.9697982052	.9701776455	.9705523917	.9709225003
290+	.9712880274	.9716490283	.9720055577	.9723576697	.9727054176
295+	.9730488544	.9733880321	.9737230024	.9740538163	.9743805242
300+	.9747031759	.9750218206	.9753365069	.9756472830	.9759541963

D-INTEGRAL PROBABILITY M=R+1,B= 9

R	1	2	3	4	5
			A= .80		
305+	.9762572939	.9765566221	.9768522267	.9771441532	.9774324464
310+	.9777171504	.9779983091	.9782759656	.9785501628	.9788209427
315+	.9790883472	.9793524175	.9796131943	.9798707179	.9801250280
320+	.9803761641	.9806241648	.9808690687	.9811109136	.9813497370
325+	.9815855760	.9818184672	.9820484467	.9822755502	.9824998131
330+	.9827212703	.9829399562	.9831559049	.9833691500	.9835797248
335+	.9837876621	.9839929944	.9841957538	.9843959720	.9845936802
340+	.9847889095	.9849816903	.9851720528	.9853600270	.9855456423
345+	.9857289277	.9859099121	.9860886239	.9862650912	.9864393417
350+	.9866114028	.9867813017	.9869490650	.9871147192	.9872782905
355+	.9874398045	.9875992869	.9877567627	.9879122569	.9880657939
360+	.9882173982	.9883670937	.9885149040	.9886608526	.9888049625
365+	.9889472567	.9890877577	.9892264877	.9893634688	.9894987228
370+	.9896322710	.9897641348	.9898943352	.9900228927	.9901498279

D-INTEGRAL PROBABILITY M=R+1,B=10

R	1	2	3	4	5
			A=1/10		
0+	.2500000000	.6884640720	.8947175890	.9667590266	.9897150592
5+	.9968260536	.9990169843	.9996939177	.9999041797	.9999698561
			A=1/ 9		
0+	.2243767313	.6462999586	.8675457843	.9538088207	.9842884585
5+	.9946892956	.9982027673	.9993894147	.9997916130	.9999285639
			A=1/ 8		
0+	.1975308642	.5963379177	.8310656466	.9342703842	.9751666998
5+	.9907142096	.9965339549	.9987037528	.9995136023	.9998168123
			A=1/ 7		
0+	.1695501730	.5370918597	.7815227620	.9040643992	.9592426740
5+	.9829336725	.9928911880	.9970407822	.9987662249	.9994843143
			A=1/ 6		
0+	.1406250000	.4670661354	.7137133245	.8563206017	.9303676037
5+	.9668591571	.9843668610	.9926531578	.9965502891	.9983790210
			A=1/ 5		
0+	.1111111111	.3851667876	.6209257373	.7797337797	.8762736943
5+	.9318845815	.9629461056	.9799822135	.9892260111	.9942110988
			A=1/ 4		
0+	.0816326531	.2916659858	.4961277303	.6573548102	.7735096565
5+	.8530893167	.9059298507	.9402989655	.9623427648	.9763458905
10+	.9851826280	.9907341103	.9942113563	.9963852937	.9977429437
			A=1/ 3		
0+	.0532544379	.1902683289	.3371036791	.4710573093	.5853180744
5+	.6790863316	.7540976274	.8130245429	.8586957044	.8937297840
10+	.9203890435	.9405471267	.9557126031	.9670759718	.9755628927
15+	.9818850381	.9865848061	.9900727602	.9926579901	.9945721867
			A= .40		
0+	.0400000000	.1396211085	.2486937363	.3538092670	.4501918830
5+	.5360820384	.6111648625	.6758828066	.7310659325	.7777160216
10+	.8168779759	.8495646660	.8767155254	.8991765546	.9176937248
15+	.9329145077	.9453940566	.9556037724	.9639408057	.9707375977
20+	.9762709315	.9807702136	.9844248580	.9873907509	.9897958294
25+	.9917448424	.9933233774	.9946012426	.9956352891	.9964717548
			A=1/ 2		
0+	.0277777778	.0920022326	.1610749875	.2292852677	.2948877733
5+	.3570932563	.4155081863	.4699590832	.5204128317	.5669308466

D-INTEGRAL PROBABILITY M=R+1,B=10

R	1	2	3	4	5
			A=1/2		
10+	.6096389085	.6487058668	.6843280740	.7167178412	.7460948510
15+	.7726798062	.7966897924	.8183349655	.8378162629	.8553239066
20+	.8710365121	.8851206614	.8977308225	.9090095271	.9190877339
25+	.9280853214	.9361116673	.9432662798	.9496394550	.9553129412
30+	.9603605946	.9648490170	.9688381664	.9723819374	.9755287061
35+	.9783218404	.9808001726	.9829984371	.9849476721	.9866755882
40+	.9882069045	.9895636546	.9907654650	.9918298061	.9927722204
			A= .60		
0+	.0204081633	.0635672218	.1075936715	.1503137189	.1915317672
5+	.2312607451	.2695283741	.3063559644	.3417602321	.3757568814
10+	.4083630319	.4395985044	.4694863381	.4980528427	.5253273887
15+	.5513420613	.5761312524	.5997312388	.6221797700	.6435156821
20+	.6637785446	.6830083430	.7012451973	.7185291173	.7348997892
25+	.7503963946	.7650574565	.7789207116	.7920230046	.8044002038
30+	.8160871335	.8271175234	.8375239712	.8473379178	.8565896324
35+	.8653082076	.8735215611	.8812564454	.8885384619	.8953920802
40+	.9018406617	.9079064851	.9136107762	.9189737380	.9240145832
45+	.9287515681	.9332020261	.9373824027	.9413082901	.9449944618
50+	.9484549070	.9517028644	.9547508557	.9576107181	.9602936363
55+	.9628101730	.9651702998	.9673834253	.9694584241	.9714036635
60+	.9732270293	.9749359515	.9765374276	.9780380463	.9794440091
65+	.9807611511	.9819949619	.9831506036	.9842329302	.9852465038
70+	.9861956120	.9870842830	.9879163007	.9886952185	.9894243730
75+	.9901068965	.9907457289	.9913436291	.9919031858	.9924268272
			A= .65		
0+	.0177777778	.0536350991	.0889553320	.1225697033	.1546883224
5+	.1855407563	.2152867313	.2440325729	.2718501859	.2987898775
10+	.3248883995	.3501739628	.3746693957	.3983941577	.4213656367
15+	.4435999891	.4651126824	.4859188435	.5060334764	.5254715929
20+	.5442482848	.5623787565	.5798783308	.5967624383	.6130465952
25+	.6287463746	.6438773740	.6584551816	.6724953427	.6860133263
30+	.6990244943	.7115440720	.7235871212	.7351685157	.7463029191
35+	.7570047652	.7672882404	.7771672691	.7866554998	.7957662952
40+	.8045127222	.8129075449	.8209632187	.8286918861	.8361053737
45+	.8432151901	.8500325260	.8565682537	.8628329290	.8688367928
50+	.8745897744	.8801014945	.8853812692	.8904381149	.8952807525
55+	.8999176134	.9043568446	.9086063147	.9126736200	.9165660905
60+	.9202907968	.9238545559	.9272639384	.9305252745	.9336446612
65+	.9366279684	.9394808461	.9422087302	.9448168496	.9473102326
70+	.9496937128	.9519719360	.9541493657	.9562302898	.9582188260
75+	.9601189279	.9619343907	.9636688566	.9653258203	.9669086345
80+	.9684205147	.9698645442	.9712436796	.9725607546	.9738184854
85+	.9750194749	.9761662166	.9772610997	.9783064121	.9793043453
90+	.9802569974	.9811663775	.9820344085	.9828629312	.9836537071
95+	.9844084220	.9851286887	.9858160502	.9864719827	.9870978979
100+	.9876951460	.9882650182	.9888087492	.9893275193	.9898224570
105+	.9902946411	.9907451026	.9911748269	.9915847559	.9919757894
			A= .70		
0+	.0156250000	.0456511882	.0740859610	.1004793639	.1253045620
5+	.1489243182	.1715788927	.1934282684	.2145819803	.2351173244
10+	.2550904326	.2745431686	.2935075464	.3120086413	.3300665554
15+	.3476977751	.3649161262	.3817334568	.3981601339	.4142054068
20+	.4298776774	.4451847020	.4601337435	.4747316860	.4889851222
25+	.5029004194	.5164837698	.5297412286	.5426787426	.5553021713
30+	.5676173033	.5796298669	.5913455387	.6027699486	.6139086834
35+	.6247672884	.6353512681	.6456660859	.6557171631	.6655098774
40+	.6750495612	.6843414995	.6933909279	.7022030302	.7107829360

D-INTEGRAL PROBABILITY M=R+1, B=10

R	1	2	3	4	5
			A= .70		
45+	.7191357192	.7272663951	.7351799189	.7428811835	.7503750178
50+	.7576661853	.7647593822	.7716592361	.7783703052	.7848970766
55+	.7912439659	.7974153162	.8034153974	.8092484056	.8149184630
60+	.8204296173	.8257858416	.8309910345	.8360490198	.8409635472
65+	.8457382918	.8503768547	.8548827639	.8592594737	.8635103661
70+	.8676387511	.8716478671	.8755408819	.8793208934	.8829909302
75+	.8865539524	.8900128529	.8933704576	.8966295270	.8997927567
80+	.9028627783	.9058421608	.9087334112	.9115389759	.9142612413
85+	.9169025350	.9194651271	.9219512309	.9243630039	.9267025493
90+	.9289719168	.9311731032	.9333080544	.9353786656	.9373867828
95+	.9393342036	.9412226783	.9430539110	.9448295605	.9465512413
100+	.9482205247	.9498389397	.9514079738	.9529290743	.9544036488
105+	.9558330667	.9572186594	.9585617220	.9598635134	.9611252577
110+	.9623481449	.9635333316	.9646819421	.9657950689	.9668737739
115+	.9679190887	.9689320157	.9699135287	.9708645737	.9717860698
120+	.9726789093	.9735439593	.9743820615	.9751940334	.9759806687
125+	.9767427382	.9774809901	.9781961508	.9788889253	.9795599982
130+	.9802100337	.9808396765	.9814495525	.9820402689	.9826124148
135+	.9831665619	.9837032651	.9842230625	.9847264763	.9852140130
140+	.9856861639	.9861434057	.9865862007	.9870149972	.9874302301
145+	.9878323212	.9882216794	.9885987015	.9889637718	.9893172635
150+	.9896595380	.9899909459	.9903118270	.9906225110	.9909233171
			A= .75		
0+	.0138408304	.0391646244	.0621333639	.0828224095	.1018747641
5+	.1197309890	.1366754878	.1528968963	.1685249883	.1836521252
10+	.1983460091	.2126575401	.2266258322	.2402815211	.2536490100
15+	.2667480335	.2795947713	.2922026571	.3045829768	.3167453176
20+	.3286979097	.3404478902	.3520015085	.3633642878	.3745411539
25+	.3855365381	.3963544600	.4069985957	.4174723318	.4277788119
30+	.4379209731	.4479015776	.4577232380	.4673884393	.4768995570
35+	.4862588719	.4954685831	.5045308190	.5134476461	.5222210771
40+	.5308530768	.5393455687	.5477004385	.5559195391	.5640046934
45+	.5719576974	.5797803226	.5874743179	.5950414117	.6024833128
50+	.6098017121	.6169982835	.6240746844	.6310325568	.6378735279
55+	.6445992101	.6512112018	.6577110874	.6641004376	.6703808099
60+	.6765537480	.6826207825	.6885834309	.6944431970	.7002015717
65+	.7058600324	.7114200433	.7168830550	.7222505049	.7275238168
70+	.7327044007	.7377936532	.7427929571	.7477036815	.7525271813
75+	.7572647978	.7619178581	.7664876755	.7709755489	.7753827632
80+	.7797105891	.7839602831	.7881330873	.7922302298	.7962529240
85+	.8002023693	.8040797508	.8078862391	.8116229907	.8152911475
90+	.8188918376	.8224261746	.8258952577	.8293001723	.8326419895
95+	.8359217664	.8391405459	.8422993572	.8453992154	.8484411219
100+	.8514260642	.8543550163	.8572289384	.8600487774	.8628154667
105+	.8655299262	.8681930627	.8708057700	.8733689287	.8758834063
110+	.8783500578	.8807697253	.8831432382	.8854714137	.8877550562
115+	.8899949583	.8921919000	.8943466496	.8964599633	.8985325855
120+	.9005652492	.9025586756	.9045135745	.9064306445	.9083105730
125+	.9101540363	.9119617000	.9137342188	.9154722366	.9171763869
130+	.9188472929	.9204855675	.9220918133	.9236666230	.9252105795
135+	.9267242558	.9282082155	.9296630123	.9310891910	.9324872869
140+	.9338578261	.9352013259	.9365182947	.9378092320	.9390746289
145+	.9403149679	.9415307231	.9427223603	.9438903373	.9450351037
150+	.9461571015	.9472567645	.9483345193	.9493907845	.9504259717
155+	.9514404847	.9524347205	.9534090689	.9543639125	.9552996273
160+	.9562165823	.9571151400	.9579956562	.9588584804	.9597039557
165+	.9605324188	.9613442003	.9621396250	.9629190115	.9636826724

D-INTEGRAL PROBABILITY M=B+1,B=10

R	1	2	3	4	5
			A= .75		
170+	.9644309150	.9651640405	.9658823448	.9665861182	.9672756456
175+	.9679512066	.9686130757	.9692615222	.9698968104	.9705191996
180+	.9711289442	.9717262940	.9723114939	.9728847843	.9734464011
185+	.9739965755	.9745355346	.9750635011	.9755806933	.9760873257
190+	.9765836084	.9770697477	.9775459457	.9780124009	.9784693078
195+	.9789168574	.9793552367	.9797846295	.9802052157	.9806171719
200+	.9810206713	.9814158837	.9818029757	.9821821104	.9825534482
205+	.9829171459	.9832733576	.9836222342	.9839639237	.9842985713
210+	.9846263192	.9849473071	.9852616715	.9855695468	.9858710642
215+	.9861663528	.9864555388	.9867387461	.9870160962	.9872877079
220+	.9875536982	.9878141812	.9880692690	.9883190717	.9885636969
225+	.9888032501	.9890378348	.9892675525	.9894925026	.9897127824
230+	.9899284874	.9901397113	.9903465457	.9905490806	.9907474039
			A= .80		
0+	.0123456790	.0338429584	.0524537575	.0686441895	.0831658799
5+	.0965016119	.1089563606	.1207306070	.1319618632	.1427480396
10+	.1531610882	.1632553132	.1730726365	.1826460523	.1920019655
15+	.2011618174	.2101432422	.2189609087	.2276271432	.2361523980
20+	.2445456105	.2528144802	.2609656866	.2690050612	.2769377261
25+	.2847682050	.2925005143	.3001382374	.3076845869	.3151424553
30+	.3225144583	.3298029710	.3370101585	.3441380019	.3511883208
35+	.3581627922	.3650629671	.3718902849	.3786460856	.3853316206
40+	.3919480622	.3984965121	.4049780082	.4113935315	.4177440111
45+	.4240303302	.4302533295	.4364138119	.4425125457	.4485502679
50+	.4545276867	.4604454846	.4663043199	.4721048295	.4778476303
55+	.4835333208	.4891624828	.4947356827	.5002534727	.5057163917
60+	.5111249666	.5164797131	.5217811364	.5270297321	.5322259868
65+	.5373703786	.5424633780	.5475054481	.5524970450	.5574386187
70+	.5623306128	.5671734656	.5719676097	.5767134728	.5814114778
75+	.5860620430	.5906655823	.5952225057	.5997332189	.6041981242
80+	.6086176199	.6129921011	.6173219596	.6216075836	.6258493585
85+	.6300476665	.6342028870	.6383153965	.6423855686	.6464137743
90+	.6504003819	.6543457572	.6582502633	.6621142609	.6659381082
95+	.6697221610	.6734667726	.6771722943	.6808390746	.6844674601
100+	.6880577948	.6916104208	.6951256776	.6986039028	.7020454315
105+	.7054505969	.7088197298	.7121531589	.7154512110	.7187142103
110+	.7219424791	.7251363378	.7282961043	.7314220945	.7345146224
115+	.7375739995	.7406005357	.7435945384	.7465563131	.7494861632
120+	.7523843900	.7552512926	.7580871684	.7608923122	.7636670172
125+	.7664115743	.7691262724	.7718113983	.7744672368	.7770940707
130+	.7796921806	.7822618452	.7848033411	.7873169428	.7898029229
135+	.7922615519	.7946930982	.7970978284	.7994760068	.8018278959
140+	.8041537561	.8064538458	.8087284214	.8109777374	.8132020461
145+	.8154015981	.8175766417	.8197274236	.8218541881	.8239571779
150+	.8260366336	.8280927939	.8301258953	.8321361729	.8341238593
155+	.8360891855	.8380323805	.8399536714	.8418532835	.8437314400
160+	.8455883623	.8474242701	.8492393809	.8510339105	.8528080730
165+	.8545620804	.8562961429	.8580104691	.8597052656	.8613807370
170+	.8630370866	.8646745154	.8662932228	.8678934066	.8694752626
175+	.8710389849	.8725847659	.8741127963	.8756232648	.8771163588
180+	.8785922637	.8800511633	.8814932397	.8829186732	.8843276426
185+	.8857203250	.8870968959	.8884575290	.8898023964	.8911316688
190+	.8924455150	.8937441024	.8950275968	.8962961622	.8975499614
195+	.8987891554	.9000139037	.9012243642	.9024206934	.9036030462
200+	.9047715761	.9059264350	.9070677734	.9081957402	.9093104829
205+	.9104121477	.9115008792	.9125768205	.9136401135	.9146908986
210+	.9157293147	.9167554994	.9177695890	.9187717183	.9197620209

D-INTEGRAL PROBABILITY M=R+1,B=10

A= .80

R	1	2	3	4	5
215+	.9207406288	.9217076730	.9226632829	.9236075867	.9245407114
220+	.9254627825	.9263739243	.9272742600	.9281639112	.9290429986
225+	.9299116414	.9307699578	.9316180646	.9324560774	.9332841107
230+	.9341022778	.9349106908	.9357094605	.9364986968	.9372785082
235+	.9380490023	.9388102854	.9395624627	.9403056384	.9410399154
240+	.9417653958	.9424821804	.9431903690	.9438900603	.9445813520
245+	.9452643408	.9459391222	.9466057909	.9472644404	.9479151634
250+	.9485580513	.9491931949	.9498206837	.9504406063	.9510530505
255+	.9516581031	.9522558498	.9528463755	.9534297642	.9540060990
260+	.9545754619	.9551379343	.9556935964	.9562425278	.9567848071
265+	.9573205120	.9578497195	.9583725055	.9588889453	.9593991132
270+	.9599030829	.9604009271	.9608927178	.9613785260	.9618584222
275+	.9623324759	.9628007559	.9632633304	.9637202664	.9641716307
280+	.9646174890	.9650579063	.9654929470	.9659226747	.9663471522
285+	.9667664418	.9671806050	.9675897025	.9679937944	.9683929402
290+	.9687871987	.9691766280	.9695612854	.9699412279	.9703165114
295+	.9706871917	.9710533234	.9714149610	.9717721579	.9721249674
300+	.9724734417	.9728176327	.9731575917	.9734933693	.9738250155
305+	.9741525799	.9744761114	.9747956584	.9751112686	.9754229894
310+	.9757308674	.9760349489	.9763352794	.9766319042	.9769248678
315+	.9772142143	.9774999872	.9777822298	.9780609844	.9783362933
320+	.9786081980	.9788767395	.9791419587	.9794038955	.9796625898
325+	.9799180807	.9801704070	.9804196070	.9806657186	.9809087793
330+	.9811488260	.9813858953	.9816200233	.9818512458	.9820795980
335+	.9823051148	.9825278308	.9827477798	.9829649957	.9831795117
340+	.9833913606	.9836005750	.9838071869	.9840112281	.9842127300
345+	.9844117235	.9846082392	.9848023074	.9849939580	.9851832206
350+	.9853701243	.9855546979	.9857369701	.9859169689	.9860947222
355+	.9862702575	.9864436019	.9866147824	.9867838254	.9869507572
360+	.9871156036	.9872783903	.9874391426	.9875978854	.9877546434
365+	.9879094411	.9880623025	.9882132514	.9883623115	.9885095058
370+	.9886548574	.9887983890	.9889401230	.9890800816	.9892182866
375+	.9893547597	.9894895222	.9896225953	.9897539997	.9898837562
380+	.9900118850	.9901384062	.9902633398	.9903867053	.9905085221

TABLE J (A = 1)

This table gives the value of the incomplete Dirichlet integral of type 2

$$D_A^{(B)}(R, M) = \frac{\Gamma(M+BR)}{\Gamma^B(R)\Gamma(M)} \int_A^\infty \cdots \int_A^\infty \frac{\prod_{i=1}^{B} x_i^{R-1} \, dx_i}{\left(1 + \sum_{i=1}^{B} x_i\right)^{M+BR}}$$

for

$M = R+1$, $R = 1(1)200$
$A = 1$
$B = 1(1)10$

Note: The symbols M, B, R and A which are all capital here in the table correspond to the same lower case symbols in the text.

D-INTEGRAL PROBABILITY M=R+1, B= 1

R	1	2	3	4	5
			A= 1		
0+	.2500000000	.3125000000	.3437500000	.3632812500	.3769531250
5+	.3872070312	.3952636719	.4018096924	.4072647095	.4119014740
10+	.4159059525	.4194098711	.4225094914	.4252770096	.4277677760
15+	.4300250330	.4320831202	.4339697002	.4357073397	.4373146562
20+	.4388071644	.4401979106	.4414979561	.4427167486	.4438624137
25+	.4449419826	.4459615756	.4469265474	.4478416069	.4487109135
30+	.4495381568	.4503266231	.4510792500	.4517986728	.4524872632
35+	.4531471624	.4537803088	.4543884626	.4549732259	.4555360606
40+	.4560783038	.4566011811	.4571058185	.4575932524	.4580644385
45+	.4585202598	.4589615337	.4593890177	.4598034155	.4602053813
50+	.4605955246	.4609744138	.4613425797	.4617005188	.4620486959
55+	.4623875468	.4627174806	.4630388817	.4633521115	.4636575105
60+	.4639553998	.4642460821	.4645298433	.4648069539	.4650776697
65+	.4653422328	.4656008728	.4658538076	.4661012438	.4663433777
70+	.4665803962	.4668124768	.4670397886	.4672624927	.4674807428
75+	.4676946852	.4679044600	.4681102007	.4683120348	.4685100846
80+	.4687044668	.4688952932	.4690826710	.4692667027	.4694474868
85+	.4696251177	.4697996860	.4699712787	.4701399794	.4703058684
90+	.4704690229	.4706295174	.4707874232	.4709428092	.4710957418
95+	.4712462848	.4713944999	.4715404463	.4716841814	.4718257605
100+	.4719652369	.4721026622	.4722380862	.4723715569	.4725031210
105+	.4726328232	.4727607072	.4728868151	.4730111875	.4731338639
110+	.4732548825	.4733742804	.4734920933	.4736083560	.4737231023
115+	.4738363648	.4739481752	.4740585643	.4741675619	.4742751971
120+	.4743814979	.4744864918	.4745902052	.4746926641	.4747938934
125+	.4748939177	.4749927605	.4750904450	.4751869937	.4752824283
130+	.4753767702	.4754700400	.4755622579	.4756534435	.4757436160
135+	.4758327938	.4759209953	.4760082381	.4760945394	.4761799160
140+	.4762643844	.4763479605	.4764306600	.4765124980	.4765934893
145+	.4766736486	.4767529900	.4768315271	.4769092737	.4769862428
150+	.4770624473	.4771378998	.4772126125	.4772865975	.4773598666
155+	.4774324311	.4775043023	.4775754913	.4776460086	.4777158648
160+	.4777850702	.4778536348	.4779215684	.4779888807	.4780555811
165+	.4781216787	.4781871827	.4782521018	.4783164447	.4783802198
170+	.4784434356	.4785061000	.4785682211	.4786298067	.4786908644
175+	.4787514017	.4788114260	.4788709444	.4789299641	.4789884920
180+	.4790465349	.4791040993	.4791611920	.4792178191	.4792739872
185+	.4793297023	.4793849705	.4794397977	.4794941897	.4795481524
190+	.4796016913	.4796548119	.4797075196	.4797598198	.4798117177
195+	.4798632184	.4799143270	.4799650484	.4800153875	.4800653490

D-INTEGRAL PROBABILITY M=R+1, B= 2

A= 1

R	1	2	3	4	5
0+	.1111111111	.1604938272	.1870141747	.2041186134	.2163156399
5+	.2255787845	.2329241472	.2389347838	.2439722674	.2482743057
10+	.2520044593	.2552794617	.2581851633	.2607863058	.2631327611
15+	.2652636560	.2672101778	.2689975340	.2706463510	.2721736923
20+	.2735938135	.2749187301	.2761586507	.2773223127	.2784172439
25+	.2794499699	.2804261799	.2813508590	.2822283971	.2830626774
30+	.2838571485	.2846148860	.2853386421	.2860308885	.2866938521
35+	.2873295450	.2879397911	.2885262474	.2890904237	.2896336987
40+	.2901573346	.2906624891	.2911502265	.2916215273	.2920772961
45+	.2925183693	.2929455216	.2933594716	.2937608869	.2941503887
50+	.2945285561	.2948959293	.2952530132	.2956002801	.2959381725
55+	.2962671056	.2965874692	.2968996295	.2972039314	.2975006999
60+	.2977902411	.2980728443	.2983487827	.2986183146	.2988816845
65+	.2991391242	.2993908532	.2996370799	.2998780023	.3001138084
70+	.3003446771	.3005707785	.3007922748	.3010093204	.3012220627
75+	.3014306422	.3016351932	.3018358439	.3020327168	.3022259292
80+	.3024155934	.3026018168	.3027847024	.3029643487	.3031408506
85+	.3033142987	.3034847803	.3036523791	.3038171755	.3039792470
90+	.3041386679	.3042955097	.3044498415	.3046017294	.3047512374
95+	.3048984272	.3050433581	.3051860873	.3053266702	.3054651600
100+	.3056016083	.3057360649	.3058685776	.3059991931	.3061279563
105+	.3062549104	.3063800977	.3065035587	.3066253328	.3067454581
110+	.3068639715	.3069809089	.3070963049	.3072101931	.3073226061
115+	.3074335754	.3075431319	.3076513053	.3077581244	.3078636173
120+	.3079678113	.3080707328	.3081724076	.3082728606	.3083721162
125+	.3084701980	.3085671290	.3086629316	.3087576274	.3088512377
130+	.3089437831	.3090352836	.3091257588	.3092152277	.3093037088
135+	.3093912203	.3094777796	.3095634040	.3096481103	.3097319146
140+	.3098148330	.3098968810	.3099780736	.3100584257	.3101379516
145+	.3102166656	.3102945812	.3103717120	.3104480709	.3105236709
150+	.3105985245	.3106726438	.3107460408	.3108187271	.3108907142
155+	.3109620132	.3110326350	.3111025904	.3111718896	.3112405430
160+	.3113085604	.3113759518	.3114427265	.3115088941	.3115744635
165+	.3116394439	.3117038440	.3117676723	.3118309374	.3118936474
170+	.3119558104	.3120174343	.3120785269	.3121390959	.3121991485
175+	.3122586922	.3123177341	.3123762812	.3124343404	.3124919184
180+	.3125490219	.3126056574	.3126618311	.3127175494	.3127728184
185+	.3128276440	.3128820323	.3129359890	.3129895197	.3130426301
190+	.3130953256	.3131476116	.3131994934	.3132509762	.3133020650
195+	.3133527650	.3134030809	.3134530176	.3135025799	.3135517725

D-INTEGRAL PROBABILITY M=R+1, B= 3

A= 1

R	1	2	3	4	5
0+	.0625000000	.1005859375	.1221694946	.1364147067	.1467095254
5+	.1545981715	.1608943286	.1660721339	.1704289035	.1741617700
10+	.1774072968	.1802634909	.1828027668	.1850799507	.1871374219
15+	.1890085290	.1907199250	.1922932035	.1937460722	.1950932085
20+	.1963468962	.1975175057	.1986138613	.1996435260	.2006130241
25+	.2015280183	.2023934506	.2032136565	.2039924572	.2047332360
30+	.2054390006	.2061124350	.2067559434	.2073716863	.2079616115
35+	.2085274806	.2090708909	.2095932947	.2100960161	.2105802649
40+	.2110471491	.2114976857	.2119328100	.2123533840	.2127602036
45+	.2131540049	.2135354698	.2139052313	.2142638777	.2146119566
50+	.2149499787	.2152784204	.2155977272	.2159083163	.2162105783
55+	.2165048800	.2167915656	.2170709591	.2173433654	.2176090716
60+	.2178683490	.2181214532	.2183686261	.2186100965	.2188460809
65+	.2190767845	.2193024019	.2195231178	.2197391076	.2199505381
70+	.2201575678	.2203603477	.2205590217	.2207537268	.2209445938
75+	.2211317474	.2213153069	.2214953861	.2216720939	.2218455344
80+	.2220158074	.2221830084	.2223472289	.2225085565	.2226670755
85+	.2228228667	.2229760074	.2231265724	.2232746330	.2234202582
90+	.2235635141	.2237044645	.2238431707	.2239796918	.2241140846
95+	.2242464040	.2243767030	.2245050325	.2246314417	.2247559783
100+	.2248786880	.2249996153	.2251188029	.2252362923	.2253521235
105+	.2254663352	.2255789649	.2256900488	.2257996222	.2259077189
110+	.2260143720	.2261196133	.2262234739	.2263259835	.2264271715
115+	.2265270660	.2266256944	.2267230832	.2268192583	.2269142447
120+	.2270080668	.2271007481	.2271923118	.2272827800	.2273721746
125+	.2274605166	.2275478266	.2276341246	.2277194299	.2278037616
130+	.2278871380	.2279695771	.2280510963	.2281317127	.2282114428
135+	.2282903028	.2283683084	.2284454750	.2285218175	.2285973506
140+	.2286720884	.2287460448	.2288192335	.2288916676	.2289633599
145+	.2290343232	.2291045697	.2291741114	.2292429601	.2293111272
150+	.2293786238	.2294454610	.2295116495	.2295771996	.2296421216
155+	.2297064254	.2297701209	.2298332176	.2298957248	.2299576517
160+	.2300190072	.2300798001	.2301400388	.2301997319	.2302588875
165+	.2303175136	.2303756181	.2304332088	.2304902930	.2305468784
170+	.2306029720	.2306585809	.2307137122	.2307683727	.2308225689
175+	.2308763075	.2309295949	.2309824374	.2310348410	.2310868120
180+	.2311383562	.2311894794	.2312401873	.2312904856	.2313403797
185+	.2313898751	.2314389770	.2314876906	.2315360210	.2315839732
190+	.2316315522	.2316787628	.2317256096	.2317720974	.2318182308
195+	.2318640141	.2319094519	.2319545485	.2319993080	.2320437348

D-INTEGRAL PROBABILITY M=R+1,B= 4

A= 1

R	1	2	3	4	5
0+	.0400000000	.0702080000	.0880841728	.1001000863	.1088752256
5+	.1156462199	.1210775028	.1255611910	.1293454067	.1325957986
10+	.1354277429	.1379244078	.1401474703	.1421437731	.1439496226
15+	.1455936558	.1470988078	.1484836946	.1497636076	.1509512405
20+	.1520572316	.1530905733	.1540589262	.1549688628	.1558260589
25+	.1566354449	.1574013278	.1581274885	.1588172626	.1594736049
30+	.1600991443	.1606962281	.1612669604	.1618132327	.1623367516
35+	.1628390610	.1633215618	.1637855281	.1642321222	.1646624063
40+	.1650773539	.1654778588	.1658647432	.1662387653	.1666006253
45+	.1669509709	.1672904026	.1676194778	.1679387148	.1682485960
50+	.1685495716	.1688420617	.1691264592	.1694031319	.1696724247
55+	.1699346612	.1701901454	.1704391631	.1706819838	.1709188612
60+	.1711500347	.1713757304	.1715961621	.1718115319	.1720220315
65+	.1722278422	.1724291363	.1726260770	.1728188196	.1730075118
70+	.1731922938	.1733732993	.1735506557	.1737244844	.1738949012
75+	.1740620166	.1742259362	.1743867610	.1745445873	.1746995076
80+	.1748516102	.1750009797	.1751476974	.1752918411	.1754334852
85+	.1755727016	.1757095590	.1758441234	.1759764584	.1761066250
90+	.1762346821	.1763606861	.1764846915	.1766067508	.1767269144
95+	.1768452311	.1769617479	.1770765101	.1771895613	.1773009439
100+	.1774106985	.1775188647	.1776254803	.1777305823	.1778342062
105+	.1779363865	.1780371564	.1781365482	.1782345931	.1783313213
110+	.1784267622	.1785209440	.1786138944	.1787056400	.1787962067
115+	.1788856196	.1789739030	.1790610806	.1791471753	.1792322095
120+	.1793162047	.1793991821	.1794811621	.1795621645	.1796422088
125+	.1797213135	.1797994972	.1798767775	.1799531717	.1800286968
130+	.1801033690	.1801772044	.1802502186	.1803224266	.1803938433
135+	.1804644829	.1805343597	.1806034871	.1806718786	.1807395472
140+	.1808065054	.1808727658	.1809383404	.1810032408	.1810674788
145+	.1811310654	.1811940116	.1812563282	.1813180255	.1813791139
150+	.1814396032	.1814995033	.1815588236	.1816175734	.1816757620
155+	.1817333981	.1817904905	.1818470477	.1819030781	.1819585897
160+	.1820135906	.1820680885	.1821220911	.1821756059	.1822286402
165+	.1822812011	.1823332957	.1823849308	.1824361132	.1824868494
170+	.1825371460	.1825870092	.1826364453	.1826854602	.1827340601
175+	.1827822507	.1828300377	.1828774268	.1829244235	.1829710331
180+	.1830172610	.1830631123	.1831085922	.1831537057	.1831984576
185+	.1832428527	.1832868959	.1833305917	.1833739446	.1834169592
190+	.1834596399	.1835019909	.1835440165	.1835857208	.1836271080
195+	.1836681819	.1837089466	.1837494060	.1837895638	.1838294237

D-INTEGRAL PROBABILITY M=R+1,B= 5

A= 1

R	1	2	3	4	5
0+	.0277777778	.0524262689	.0675513365	.0778733129	.0854765985
5+	.0913767776	.0961288991	.1000641241	.1033936258	.1062592043
10+	.1087600752	.1109680210	.1129364382	.1147059857	.1163082472
15+	.1177681831	.1191058179	.1203374303	.1214764097	.1225338845
20+	.1235191921	.1244402328	.1253037436	.1261155094	.1268805297
25+	.1276031510	.1282871725	.1289359320	.1295523753	.1301391139
30+	.1306984725	.1312325276	.1317431416	.1322319895	.1327005831
35+	.1331502908	.1335823546	.1339979047	.1343979722	.1347835001
40+	.1351553526	.1355143235	.1358611434	.1361964859	.1365209736
45+	.1368351824	.1371396463	.1374348612	.1377212883	.1379993571
50+	.1382694682	.1385319957	.1387872897	.1390356777	.1392774670
55+	.1395129460	.1397423856	.1399660409	.1401841517	.1403969446
60+	.1406046331	.1408074188	.1410054926	.1411990348	.1413882163
65+	.1415731993	.1417541372	.1419311761	.1421044546	.1422741044
70+	.1424402509	.1426030136	.1427625063	.1429188374	.1430721104
75+	.1432224241	.1433698730	.1435145471	.1436565329	.1437959127
80+	.1439327658	.1440671678	.1441991913	.1443289059	.1444563783
85+	.1445816727	.1447048506	.1448259712	.1449450912	.1450622654
90+	.1451775463	.1452909846	.1454026291	.1455125266	.1456207224
95+	.1457272602	.1458321820	.1459355283	.1460373384	.1461376501
100+	.1462364999	.1463339231	.1464299538	.1465246250	.1466179685
105+	.1467100151	.1468007949	.1468903366	.1469786682	.1470658168
110+	.1471518088	.1472366694	.1473204235	.1474030948	.1474847067
115+	.1475652817	.1476448415	.1477234074	.1478009999	.1478776392
120+	.1479533445	.1480281348	.1481020284	.1481750431	.1482471963
125+	.1483185049	.1483889852	.1484586532	.1485275244	.1485956140
130+	.1486629367	.1487295067	.1487953382	.1488604446	.1489248392
135+	.1489885349	.1490515444	.1491138799	.1491755533	.1492365763
140+	.1492969604	.1493567166	.1494158557	.1494743884	.1495323249
145+	.1495896754	.1496464496	.1497026573	.1497583077	.1498134101
150+	.1498679734	.1499220064	.1499755177	.1500285155	.1500810081
155+	.1501330035	.1501845095	.1502355337	.1502860837	.1503361668
160+	.1503857901	.1504349606	.1504836852	.1505319706	.1505798234
165+	.1506272501	.1506742569	.1507208500	.1507670354	.1508128192
170+	.1508582070	.1509032046	.1509478176	.1509920514	.1510359113
175+	.1510794027	.1511225307	.1511653002	.1512077164	.1512497839
180+	.1512915077	.1513328923	.1513739423	.1514146622	.1514550565
185+	.1514951295	.1515348853	.1515743283	.1516134624	.1516522917
190+	.1516908202	.1517290517	.1517669901	.1518046390	.1518420021
195+	.1518790831	.1519158854	.1519524126	.1519886681	.1520246553

D-INTEGRAL PROBABILITY M=R+1,B= 6

R	1	2	3	4	5
			A= 1		
0+	.0204081633	.0410014153	.0540393197	.0630530374	.0697414336
5+	.0749566404	.0791715466	.0826710306	.0856379823	.0881958296
10+	.0904312733	.0924072406	.0941706635	.0957573579	.0971951981
15+	.0985062496	.0997082402	.1008155999	.1018402112	.1027919609
20+	.1036791517	.1045088152	.1052869518	.1060187179	.1067085729
25+	.1073603961	.1079775802	.1085631077	.1091196120	.1096494290
30+	.1101546386	.1106370999	.1110984806	.1115402814	.1119638572
35+	.1123704349	.1127611282	.1131369511	.1134988288	.1138476079
40+	.1141840642	.1145089107	.1148228037	.1151263486	.1154201048
45+	.1157045900	.1159802845	.1162476343	.1165070541	.1167589304
50+	.1170036237	.1172414708	.1174727866	.1176978663	.1179169864
55+	.1181304067	.1183383712	.1185411097	.1187388385	.1189317615
60+	.1191200713	.1193039496	.1194835685	.1196590905	.1198306700
65+	.1199984528	.1201625777	.1203231763	.1204803737	.1206342887
70+	.1207850345	.1209327189	.1210774446	.1212193095	.1213584071
75+	.1214948265	.1216286530	.1217599680	.1218888494	.1220153719
80+	.1221396068	.1222616224	.1223814845	.1224992558	.1226149967
85+	.1227287652	.1228406167	.1229506049	.1230587809	.1231651944
90+	.1232698928	.1233729219	.1234743257	.1235741468	.1236724259
95+	.1237692026	.1238645150	.1239583996	.1240508920	.1241420263
100+	.1242318356	.1243203518	.1244076058	.1244936273	.1245784453
105+	.1246620876	.1247445813	.1248259525	.1249062265	.1249854279
110+	.1250635804	.1251407071	.1252168302	.1252919715	.1253661519
115+	.1254393918	.1255117110	.1255831288	.1256536636	.1257233337
120+	.1257921566	.1258601494	.1259273288	.1259937109	.1260593113
125+	.1261241454	.1261882280	.1262515735	.1263141961	.1263761094
130+	.1264373268	.1264978613	.1265577254	.1266169315	.1266754917
135+	.1267334176	.1267907206	.1268474119	.1269035023	.1269590023
140+	.1270139223	.1270682724	.1271220624	.1271753018	.1272280000
145+	.1272801662	.1273318092	.1273829379	.1274335606	.1274836858
150+	.1275333215	.1275824757	.1276311561	.1276793704	.1277271259
155+	.1277744298	.1278212894	.1278677115	.1279137029	.1279592702
160+	.1280044200	.1280491586	.1280934922	.1281374270	.1281809688
165+	.1282241235	.1282668969	.1283092945	.1283513218	.1283929843
170+	.1284342871	.1284752354	.1285158343	.1285560888	.1285960036
175+	.1286355836	.1286748334	.1287137577	.1287523608	.1287906473
180+	.1288286213	.1288662872	.1289036492	.1289407112	.1289774773
185+	.1290139514	.1290501375	.1290860391	.1291216602	.1291570043
190+	.1291920750	.1292268758	.1292614102	.1292956815	.1293296932
195+	.1293634484	.1293969504	.1294302024	.1294632074	.1294959685

D-INTEGRAL PROBABILITY M=R+1, B= 7

R	1	2	3	4	5
			A= 1		
0+	.0156250000	.0331642032	.0445779275	.0525587216	.0585184962
5+	.0631849468	.0669675780	.0701152377	.0727886499	.0750967691
10+	.0771163948	.0789034240	.0804996412	.0819369959	.0832403944
15+	.0844295814	.0855204397	.0865259125	.0874566685	.0883215924
20+	.0891281515	.0898826748	.0905905668	.0912564745	.0918844194
25+	.0924779015	.0930399836	.0935733590	.0940804074	.0945632399
30+	.0950237374	.0954635817	.0958842818	.0962871961	.0966735516
35+	.0970444593	.0974009284	.0977438777	.0980741461	.0983925009
40+	.0986996459	.0989962279	.0992828423	.0995600384	.0998283240
45+	.1000881693	.1003400102	.1005842518	.1008212708	.1010514184
50+	.1012750221	.1014923879	.1017038022	.1019095332	.1021098324
55+	.1023049359	.1024950658	.1026804309	.1028612281	.1030376428
60+	.1032098503	.1033780160	.1035422964	.1037028395	.1038597858
65+	.1040132681	.1041634127	.1043103394	.1044541621	.1045949891
70+	.1047329236	.1048680635	.1050005026	.1051303300	.1052576309
75+	.1053824865	.1055049745	.1056251691	.1057431413	.1058589592
80+	.1059726877	.1060843893	.1061941238	.1063019484	.1064079181
85+	.1065120859	.1066145023	.1067152162	.1068142744	.1069117220
90+	.1070076023	.1071019572	.1071948267	.1072862497	.1073762634
95+	.1074649039	.1075522058	.1076382027	.1077229269	.1078064095
100+	.1078886809	.1079697700	.1080497050	.1081285133	.1082062211
105+	.1082828538	.1083584362	.1084329922	.1085065447	.1085791164
110+	.1086507287	.1087214029	.1087911592	.1088600174	.1089279967
115+	.1089951158	.1090613927	.1091268449	.1091914894	.1092553429
120+	.1093184213	.1093807403	.1094423150	.1095031602	.1095632903
125+	.1096227191	.1096814604	.1097395272	.1097969324	.1098536886
130+	.1099098080	.1099653023	.1100201832	.1100744619	.1101281494
135+	.1101812564	.1102337933	.1102857702	.1103371972	.1103880838
140+	.1104384395	.1104882734	.1105375947	.1105864119	.1106347338
145+	.1106825685	.1107299244	.1107768093	.1108232311	.1108691973
150+	.1109147153	.1109597925	.1110044359	.1110486525	.1110924490
155+	.1111358321	.1111788082	.1112213837	.1112635649	.1113053577
160+	.1113467682	.1113878021	.1114284651	.1114687628	.1115087007
165+	.1115482840	.1115875181	.1116264081	.1116649589	.1117031755
170+	.1117410627	.1117786253	.1118158677	.1118527947	.1118894106
175+	.1119257198	.1119617265	.1119974350	.1120328493	.1120679735
180+	.1121028116	.1121373674	.1121716446	.1122056472	.1122393786
185+	.1122728425	.1123060425	.1123389820	.1123716643	.1124040929
190+	.1124362710	.1124682019	.1124998886	.1125313343	.1125625421
195+	.1125935149	.1126242557	.1126547673	.1126850527	.1127151145

D-INTEGRAL PROBABILITY M=R+1,B= 8

A= 1

R	1	2	3	4	5
0+	.0123456790	.0275204676	.0376407428	.0447887696	.0501568343
5+	.0543754421	.0578040097	.0606626886	.0630944468	.0651965931
10+	.0670379321	.0686686667	.0701264021	.0714399418	.0726317727
15+	.0737197441	.0747182305	.0756389573	.0764915980	.0772842151
20+	.0780235896	.0787154726	.0793647780	.0799757336	.0805519997
25+	.0810967635	.0816128154	.0821026102	.0825683175	.0830118630
30+	.0834349628	.0838391517	.0842258075	.0845961709	.0849513626
35+	.0852923978	.0856201988	.0859356057	.0862393854	.0865322397
40+	.0868148125	.0870876954	.0873514333	.0876065291	.0878534476
45+	.0880926192	.0883244432	.0885492905	.0887675064	.0889794124
50+	.0891853086	.0893854756	.0895801757	.0897696549	.0899541436
55+	.0901338584	.0903090029	.0904797686	.0906463361	.0908088755
60+	.0909675475	.0911225039	.0912738884	.0914218366	.0915664776
65+	.0917079332	.0918463194	.0919817462	.0921143180	.0922441345
70+	.0923712901	.0924958750	.0926179752	.0927376725	.0928550452
75+	.0929701679	.0930831120	.0931939456	.0933027341	.0934095397
80+	.0935144224	.0936174395	.0937186458	.0938180941	.0939158350
85+	.0940119168	.0941063864	.0941992884	.0942906660	.0943805607
90+	.0944690122	.0945560589	.0946417379	.0947260847	.0948091337
95+	.0948909179	.0949714693	.0950508187	.0951289958	.0952060293
100+	.0952819470	.0953567756	.0954305411	.0955032686	.0955749821
105+	.0956457053	.0957154607	.0957842704	.0958521556	.0959191369
110+	.0959852343	.0960504671	.0961148541	.0961784135	.0962411630
115+	.0963031196	.0963643001	.0964247205	.0964843965	.0965433433
120+	.0966015758	.0966591083	.0967159548	.0967721288	.0968276435
125+	.0968825119	.0969367464	.0969903591	.0970433620	.0970957665
130+	.0971475838	.0971988248	.0972495003	.0972996206	.0973491958
135+	.0973982357	.0974467499	.0974947478	.0975422386	.0975892310
140+	.0976357339	.0976817557	.0977273046	.0977723888	.0978170161
145+	.0978611942	.0979049306	.0979482326	.0979911075	.0980335622
150+	.0980756036	.0981172383	.0981584730	.0981993139	.0982397674
155+	.0982798395	.0983195362	.0983588634	.0983978268	.0984364320
160+	.0984746844	.0985125894	.0985501524	.0985873783	.0986242722
165+	.0986608391	.0986970837	.0987330108	.0987686250	.0988039309
170+	.0988389329	.0988736353	.0989080424	.0989421583	.0989759873
175+	.0990095332	.0990428001	.0990757917	.0991085119	.0991409644
180+	.0991731529	.0992050808	.0992367518	.0992681692	.0992993364
185+	.0993302568	.0993609336	.0993913700	.0994215691	.0994515339
190+	.0994812677	.0995107732	.0995400534	.0995691111	.0995979493
195+	.0996265705	.0996549777	.0996831733	.0997111601	.0997389406

D-INTEGRAL PROBABILITY M=R+1,B= 9

R	1	2	3	4	5
			A= 1		
0+	.0100000000	.0233010784	.0323704653	.0388346048	.0437136890
5+	.0475606379	.0506944611	.0533119976	.0555417177	.0574713838
10+	.0591632266	.0606627548	.0620041208	.0632135239	.0643114454
15+	.0653141609	.0662347909	.0670840475	.0678707751	.0686023503
20+	.0692849816	.0699239373	.0705237215	.0710882110	.0716207640
25+	.0721243058	.0726013984	.0730542967	.0734849942	.0738952609
30+	.0742866745	.0746606467	.0750184449	.0753612107	.0756899756
35+	.0760056741	.0763091556	.0766011938	.0768824950	.0771537060
40+	.0774154199	.0776681823	.0779124953	.0781488226	.0783775928
45+	.0785992028	.0788140210	.0790223895	.0792246268	.0794210297
50+	.0796118752	.0797974222	.0799779128	.0801535741	.0803246189
55+	.0804912472	.0806536469	.0808119951	.0809664584	.0811171939
60+	.0812643503	.0814080677	.0815484788	.0816857091	.0818198777
65+	.0819510974	.0820794752	.0822051128	.0823281070	.0824485496
70+	.0825665280	.0826821258	.0827954222	.0829064932	.0830154109
75+	.0831222444	.0832270598	.0833299200	.0834308855	.0835300141
80+	.0836273609	.0837229792	.0838169196	.0839092309	.0839999599
85+	.0840891516	.0841768491	.0842630938	.0843479257	.0844313830
90+	.0845135027	.0845943202	.0846738700	.0847521848	.0848292964
95+	.0849052356	.0849800319	.0850537137	.0851263086	.0851978433
100+	.0852683433	.0853378336	.0854063380	.0854738799	.0855404815
105+	.0856061648	.0856709505	.0857348592	.0857979105	.0858601234
110+	.0859215166	.0859821079	.0860419147	.0861009538	.0861592417
115+	.0862167941	.0862736265	.0863297539	.0863851907	.0864399511
120+	.0864940488	.0865474971	.0866003089	.0866524968	.0867040731
125+	.0867550497	.0868054381	.0868552497	.0869044953	.0869531857
130+	.0870013313	.0870489422	.0870960282	.0871425991	.0871886640
135+	.0872342323	.0872793127	.0873239139	.0873680445	.0874117126
140+	.0874549264	.0874976936	.0875400220	.0875819190	.0876233920
145+	.0876644480	.0877050940	.0877453369	.0877851833	.0878246397
150+	.0878637124	.0879024076	.0879407314	.0879786897	.0880162884
155+	.0880535330	.0880904292	.0881269823	.0881631976	.0881990804
160+	.0882346357	.0882698685	.0883047836	.0883393859	.0883736799
165+	.0884076702	.0884413614	.0884747578	.0885078637	.0885406832
170+	.0885732205	.0886054797	.0886374647	.0886691794	.0887006275
175+	.0887318128	.0887627390	.0887934096	.0888238282	.0888539982
180+	.0888839229	.0889136057	.0889430499	.0889722586	.0890012350
185+	.0890299821	.0890585030	.0890868006	.0891148778	.0891427375
190+	.0891703825	.0891978156	.0892250394	.0892520566	.0892788698
195+	.0893054816	.0893318945	.0893581110	.0893841334	.0894099643

D-INTEGRAL PROBABILITY M=R+1,B=10

R	1	2	3	4	5
			A= 1		
0+	.0082644628	.0200511242	.0282519248	.0341454081	.0386142113
5+	.0421481594	.0450330843	.0474465616	.0495050272	.0512882917
10+	.0528530943	.0542410138	.0554833063	.0566039830	.0576218376
15+	.0585518201	.0594059934	.0601942138	.0609246254	.0616040249
20+	.0622381359	.0628318182	.0633892282	.0639139454	.0644090716
25+	.0648773098	.0653210282	.0657423117	.0661430041	.0665247430
30+	.0668889886	.0672370475	.0675700934	.0678891834	.0681952730
35+	.0684892282	.0687718361	.0690438135	.0693058156	.0695584419
40+	.0698022425	.0700377232	.0702653500	.0704855531	.0706987300
45+	.0709052493	.0711054528	.0712996580	.0714881607	.0716712363
50+	.0718491421	.0720221186	.0721903906	.0723541690	.0725136516
55+	.0726690240	.0728204611	.0729681271	.0731121771	.0732527570
60+	.0733900050	.0735240513	.0736550193	.0737830257	.0739081812
65+	.0740305906	.0741503535	.0742675645	.0743823135	.0744946859
70+	.0746047632	.0747126228	.0748183387	.0749219814	.0750236181
75+	.0751233130	.0752211276	.0753171205	.0754113480	.0755038637
80+	.0755947192	.0756839637	.0757716447	.0758578074	.0759424954
85+	.0760257506	.0761076131	.0761881214	.0762673128	.0763452229
90+	.0764218861	.0764973354	.0765716027	.0766447188	.0767167132
95+	.0767876144	.0768574500	.0769262465	.0769940296	.0770608241
100+	.0771266537	.0771915418	.0772555105	.0773185816	.0773807760
105+	.0774421137	.0775026145	.0775622973	.0776211804	.0776792816
110+	.0777366182	.0777932069	.0778490638	.0779042046	.0779586447
115+	.0780123987	.0780654811	.0781179058	.0781696863	.0782208358
120+	.0782713670	.0783212923	.0783706238	.0784193733	.0784675522
125+	.0785151714	.0785622419	.0786087742	.0786547784	.0787002646
130+	.0787452423	.0787897212	.0788337103	.0788772186	.0789202548
135+	.0789628276	.0790049450	.0790466154	.0790878465	.0791286460
140+	.0791690214	.0792089801	.0792485293	.0792876758	.0793264266
145+	.0793647882	.0794027672	.0794403699	.0794776024	.0795144710
150+	.0795509814	.0795871395	.0796229509	.0796584212	.0796935557
155+	.0797283597	.0797628384	.0797969969	.0798308401	.0798643729
160+	.0798975999	.0799305259	.0799631553	.0799954925	.0800275421
165+	.0800593081	.0800907949	.0801220063	.0801529466	.0801836195
170+	.0802140290	.0802441788	.0802740725	.0803037138	.0803331063
175+	.0803622534	.0803911586	.0804198251	.0804482562	.0804764552
180+	.0805044253	.0805321694	.0805596907	.0805869921	.0806140766
185+	.0806409470	.0806676061	.0806940567	.0807203016	.0807463433
190+	.0807721845	.0807978277	.0808232756	.0808485305	.0808735949
195+	.0808984712	.0809231617	.0809476688	.0809719947	.0809961416

TABLE K (A = 1)

This table gives the value of the incomplete Dirichlet integral of type 2

$$D_A^{(B)}(R, M) = \frac{\Gamma(M+BR)}{\Gamma^B(R)\Gamma(M)} \int_A^\infty \cdots \int_A^\infty \frac{\prod_{i=1}^{B} x_i^{R-1}\, dx_i}{\left(1 + \sum_{i=1}^{B} x_i\right)^{M+BR}}$$

for

$M = R+2$, $R = 1(1)200$
$A = 1$
$B = 1(1)10$

Note: The symbols M, B, R and A which are all capital here in the table correspond to the same lower case symbols in the text.

D-INTEGRAL PROBABILITY M=R+2,B= 1

R	1	2	3	4	5
			A= 1		
0+	.1250000000	.1875000000	.2265625000	.2539062500	.2744140625
5+	.2905273437	.3036193848	.3145294189	.3238029480	.3318119049
10+	.3388197422	.3450189829	.3505540192	.3555355519	.3600500659
15+	.3641662404	.3679394004	.3714146794	.3746293124	.3776143288
20+	.3803958213	.3829959121	.3854334973	.3877248273	.3898839653
25+	.3919231511	.3938530948	.3956832139	.3974218270	.3990763137
30+	.4006532463	.4021585001	.4035973457	.4049745265	.4062943247
35+	.4075606176	.4087769253	.4099464519	.4110721212	.4121566076
40+	.4132023622	.4142116371	.4151865048	.4161288770	.4170405197
45+	.4179230673	.4187780354	.4196068309	.4204107626	.4211910493
50+	.4219488276	.4226851594	.4234010376	.4240973918	.4247750937
55+	.4254349613	.4260777633	.4267042229	.4273150211	.4279107996
60+	.4284921641	.4290596866	.4296139078	.4301553393	.4306844655
65+	.4312017456	.4317076152	.4322024875	.4326867555	.4331607924
70+	.4336249535	.4340795772	.4345249854	.4349614855	.4353893705
75+	.4358089200	.4362204013	.4366240696	.4370201692	.4374089336
80+	.4377905864	.4381653419	.4385334054	.4388949736	.4392502354
85+	.4395993720	.4399425573	.4402799587	.4406117367	.4409380458
90+	.4412590347	.4415748464	.4418856185	.4421914836	.4424925697
95+	.4427889997	.4430808926	.4433683628	.4436515210	.4439304739
100+	.4442053245	.4444761724	.4447431139	.4450062419	.4452656464
105+	.4455214144	.4457736301	.4460223749	.4462677278	.4465097650
110+	.4467485607	.4469841866	.4472167121	.4474462046	.4476727296
115+	.4478963504	.4481171286	.4483351239	.4485503942	.4487629958
120+	.4489729836	.4491804105	.4493853282	.4495877869	.4497878353
125+	.4499855210	.4501808901	.4503739874	.4505648567	.4507535404
130+	.4509400800	.4511245158	.4513068870	.4514872319	.4516655877
135+	.4518419906	.4520164762	.4521890788	.4523598321	.4525287688
140+	.4526959211	.4528613199	.4530249959	.4531869787	.4533472972
145+	.4535059799	.4536630543	.4538185474	.4539724856	.4541248946
150+	.4542757995	.4544252250	.4545731951	.4547197331	.4548648622
155+	.4550086047	.4551509825	.4552920172	.4554317296	.4555701404
160+	.4557072696	.4558431369	.4559777614	.4561111622	.4562433575
165+	.4563743654	.4565042036	.4566328893	.4567604397	.4568868711
170+	.4570122000	.4571364422	.4572596133	.4573817287	.4575028034
175+	.4576228519	.4577418889	.4578599283	.4579769840	.4580930697
180+	.4582081987	.4583223839	.4584356383	.4585479744	.4586594046
185+	.4587699409	.4588795953	.4589883795	.4590963048	.4592033825
190+	.4593096237	.4594150392	.4595196396	.4596234354	.4597264369
195+	.4598286540	.4599300968	.4600307750	.4601306980	.4602298754

D-INTEGRAL PROBABILITY M=R+2,B= 2

R	1	2	3	4	5
			A= 1		
0+	.0370370370	.0727023320	.0986130163	.1181335275	.1334507918
5+	.1458678119	.1561934850	.1649552126	.1725121177	.1791178669
10+	.1849572418	.1901683959	.1948569411	.1991051729	.2029782933
15+	.2065287203	.2097991402	.2128247152	.2156347079	.2182536967
20+	.2207024989	.2229988804	.2251581082	.2271933855	.2291161976
25+	.2309365893	.2326633894	.2343043940	.2358665171	.2373559148
30+	.2387780892	.2401379755	.2414400145	.2426882150	.2438862063
35+	.2450372835	.2461444457	.2472104296	.2482377385	.2492286670
40+	.2501853231	.2511096473	.2520034292	.2528683224	.2537058576
45+	.2545174540	.2553044294	.2560680097	.2568093368	.2575294758
50+	.2582294216	.2589101047	.2595723968	.2602171149	.2608450260
55+	.2614568511	.2620532684	.2626349166	.2632023981	.2637562810
60+	.2642971022	.2648253693	.2653415625	.2658461368	.2663395234
65+	.2668221313	.2672943490	.2677565455	.2682090717	.2686522614
70+	.2690864328	.2695118886	.2699289178	.2703377960	.2707387864
75+	.2711321402	.2715180978	.2718968889	.2722687334	.2726338420
80+	.2729924162	.2733446495	.2736907272	.2740308272	.2743651203
85+	.2746937704	.2750169351	.2753347659	.2756474084	.2759550028
90+	.2762576840	.2765555817	.2768488213	.2771375231	.2774218035
95+	.2777017746	.2779775446	.2782492177	.2785168949	.2787806734
100+	.2790406473	.2792969075	.2795495419	.2797986354	.2800442703
105+	.2802865262	.2805254801	.2807612065	.2809937777	.2812232637
110+	.2814497324	.2816732494	.2818938787	.2821116819	.2823267192
115+	.2825390487	.2827487271	.2829558091	.2831603481	.2833623959
120+	.2835620029	.2837592178	.2839540884	.2841466607	.2843369797
125+	.2845250893	.2847110319	.2848948489	.2850765807	.2852562666
130+	.2854339446	.2856096522	.2857834254	.2859552998	.2861253096
135+	.2862934885	.2864598692	.2866244836	.2867873629	.2869485373
140+	.2871080366	.2872658895	.2874221244	.2875767688	.2877298495
145+	.2878813928	.2880314243	.2881799691	.2883270517	.2884726959
150+	.2886169250	.2887597620	.2889012292	.2890413483	.2891801407
155+	.2893176273	.2894538284	.2895887640	.2897224537	.2898549166
160+	.2899861714	.2901162363	.2902451293	.2903728679	.2904994693
165+	.2906249504	.2907493275	.2908726169	.2909948343	.2911159953
170+	.2912361149	.2913552081	.2914732895	.2915903733	.2917064735
175+	.2918216039	.2919357779	.2920490087	.2921613094	.2922726925
180+	.2923831705	.2924927557	.2926014600	.2927092953	.2928162730
185+	.2929224046	.2930277010	.2931321734	.2932358323	.2933386883
190+	.2934407517	.2935420327	.2936425413	.2937422873	.2938412802
195+	.2939395296	.2940370446	.2941338346	.2942299083	.2943252746

D-INTEGRAL PROBABILITY M=R+2,B= 3

R	1	2	3	4	5
			A= 1		
0+	.0156250000	.0375976562	.0554361343	.0696072280	.0810893827
5+	.0906036900	.0986442033	.1055525995	.1115710179	.1168755645
10+	.1215974417	.1258365228	.1296702923	.1331598809	.1363542373
15+	.1392930755	.1420089981	.1445290525	.1468758919	.1490686529
20+	.1511236279	.1530547871	.1548741881	.1565923002	.1582182640
25+	.1597601014	.1612248855	.1626188807	.1639476582	.1652161913
30+	.1664289364	.1675898994	.1687026934	.1697705864	.1707965433
35+	.1717832607	.1727331974	.1736486012	.1745315311	.1753838778
40+	.1762073806	.1770036429	.1777741456	.1785202587	.1792432521
45+	.1799443045	.1806245118	.1812848947	.1819264047	.1825499306
50+	.1831563032	.1837463006	.1843206519	.1848800416	.1854251128
55+	.1859564703	.1864746836	.1869802896	.1874737950	.1879556779
60+	.1884263908	.1888863614	.1893359950	.1897756756	.1902057675
65+	.1906266166	.1910385513	.1914418840	.1918369120	.1922239181
70+	.1926031721	.1929749310	.1933394400	.1936969331	.1940476341
75+	.1943917566	.1947295050	.1950610750	.1953866537	.1957064207
80+	.1960205477	.1963291997	.1966325348	.1969307049	.1972238557
85+	.1975121274	.1977956546	.1980745669	.1983489887	.1986190401
90+	.1988848364	.1991464889	.1994041048	.1996577874	.1999076363
95+	.2001537478	.2003962145	.2006351261	.2008705690	.2011026270
100+	.2013313807	.2015569083	.2017792853	.2019985848	.2022148776
105+	.2024282320	.2026387144	.2028463890	.2030513180	.2032535616
110+	.2034531784	.2036502250	.2038447564	.2040368259	.2042264853
115+	.2044137849	.2045987734	.2047814981	.2049620053	.2051403395
120+	.2053165443	.2054906619	.2056627333	.2058327987	.2060008968
125+	.2061670654	.2063313415	.2064937607	.2066543581	.2068131676
130+	.2069702222	.2071255543	.2072791952	.2074311756	.2075815252
135+	.2077302733	.2078774481	.2080230774	.2081671881	.2083098066
140+	.2084509585	.2085906690	.2087289625	.2088658630	.2090013937
145+	.2091355776	.2092684367	.2093999930	.2095302676	.2096592814
150+	.2097870546	.2099136070	.2100389582	.2101631270	.2102861320
155+	.2104079914	.2105287229	.2106483439	.2107668714	.2108843220
160+	.2110007120	.2111160573	.2112303736	.2113436761	.2114559798
165+	.2115672993	.2116776491	.2117870431	.2118954951	.2120030187
170+	.2121096270	.2122153331	.2123201496	.2124240890	.2125271636
175+	.2126293852	.2127307657	.2128313166	.2129310492	.2130299745
180+	.2131281035	.2132254468	.2133220148	.2134178180	.2135128663
185+	.2136071696	.2137007378	.2137935802	.2138857064	.2139771254
190+	.2140678464	.2141578782	.2142472295	.2143359089	.2144239249
195+	.2145112855	.2145979991	.2146840736	.2147695168	.2148543364

D-INTEGRAL PROBABILITY M=R+2, B= 4

R	1	2	3	4	5
			A= 1		
0+	.0080000000	.0226944000	.0357220557	.0465099960	.0554707339
5+	.0630218217	.0694823382	.0750861540	.0800052558	.0843680262
10+	.0882719964	.0917925515	.0949889042	.0979082512	.1005887128
15+	.1030614488	.1053522056	.1074824653	.1094703113	.1113310895
20+	.1130779188	.1147220903	.1162733824	.1177403121	.1191303370
25+	.1204500194	.1217051596	.1229009070	.1240418508	.1251320961
30+	.1261753282	.1271748660	.1281337078	.1290545701	.1299399211
35+	.1307920087	.1316128858	.1324044311	.1331683681	.1339062809
40+	.1346196286	.1353097579	.1359779137	.1366252489	.1372528331
45+	.1378616602	.1384526551	.1390266797	.1395845384	.1401269831
50+	.1406547172	.1411683998	.1416686494	.1421560466	.1426311374
55+	.1430944357	.1435464257	.1439875642	.1444182823	.1448389873
60+	.1452500648	.1456518796	.1460447773	.1464290857	.1468051160
65+	.1471731635	.1475335093	.1478864203	.1482321510	.1485709435
70+	.1489030287	.1492286269	.1495479482	.1498611934	.1501685543
75+	.1504702143	.1507663489	.1510571261	.1513427067	.1516232449
80+	.1518988886	.1521697794	.1524360535	.1526978416	.1529552691
85+	.1532084568	.1534575206	.1537025720	.1539437186	.1541810636
90+	.1544147067	.1546447438	.1548712673	.1550943666	.1553141276
95+	.1555306334	.1557439641	.1559541971	.1561614072	.1563656668
100+	.1565670455	.1567656111	.1569614288	.1571545620	.1573450717
105+	.1575330174	.1577184563	.1579014442	.1580820348	.1582602805
110+	.1584362319	.1586099381	.1587814468	.1589508043	.1591180554
115+	.1592832438	.1594464118	.1596076005	.1597668499	.1599241988
120+	.1600796850	.1602333451	.1603852149	.1605353290	.1606837214
125+	.1608304248	.1609754712	.1611188919	.1612607171	.1614009764
130+	.1615396985	.1616769117	.1618126431	.1619469194	.1620797666
135+	.1622112100	.1623412743	.1624699837	.1625973616	.1627234309
140+	.1628482142	.1629717331	.1630940091	.1632150629	.1633349150
145+	.1634535852	.1635710929	.1636874571	.1638026964	.1639168287
150+	.1640298720	.1641418434	.1642527599	.1643626380	.1644714940
155+	.1645793437	.1646862026	.1647920859	.1648970084	.1650009846
160+	.1651040288	.1652061548	.1653073764	.1654077068	.1655071591
165+	.1656057462	.1657034806	.1658003745	.1658964401	.1659916891
170+	.1660861330	.1661797833	.1662726511	.1663647472	.1664560823
175+	.1665466670	.1666365114	.1667256257	.1668140198	.1669017033
180+	.1669886858	.1670749767	.1671605850	.1672455199	.1673297900
185+	.1674134042	.1674963709	.1675786984	.1676603951	.1677414688
190+	.1678219277	.1679017793	.1679810315	.1680596916	.1681377670
195+	.1682152651	.1682921928	.1683685573	.1684443654	.1685196238

D-INTEGRAL PROBABILITY M=R+2,B= 5

R	1	2	3	4	5
			A= 1		
0+	.0046296296	.0150844002	.0250541959	.0335963453	.0408367856
5+	.0470220514	.0523669062	.0570385548	.0611644478	.0648420544
10+	.0681467402	.0711375538	.0738613802	.0763559234	.0786518637
15+	.0807744392	.0827446181	.0845799799	.0862953863	.0879034970
20+	.0894151717	.0908397858	.0921854819	.0934593703	.0946676918
25+	.0958159498	.0969090191	.0979512350	.0989464683	.0998981874
30+	.1008095104	.1016832499	.1025219502	.1033279192	.1041032568
35+	.1048498776	.1055695318	.1062638232	.1069342242	.1075820898
40+	.1082086690	.1088151155	.1094024968	.1099718024	.1105239510
45+	.1110597967	.1115801351	.1120857080	.1125772084	.1130552842
50+	.1135205420	.1139735506	.1144148439	.1148449236	.1152642615
55+	.1156733022	.1160724645	.1164621439	.1168427138	.1172145272
60+	.1175779182	.1179332033	.1182806824	.1186206399	.1189533462
65+	.1192790578	.1195980188	.1199104615	.1202166071	.1205166663
70+	.1208108400	.1210993200	.1213822893	.1216599228	.1219323878
75+	.1221998440	.1224624445	.1227203358	.1229736583	.1232225465
80+	.1234671295	.1237075311	.1239438702	.1241762609	.1244048129
85+	.1246296317	.1248508187	.1250684716	.1252826842	.1254935471
90+	.1257011473	.1259055690	.1261068930	.1263051975	.1265005578
95+	.1266930468	.1268827346	.1270696892	.1272539760	.1274356585
100+	.1276147979	.1277914535	.1279656827	.1281375408	.1283070818
105+	.1284743575	.1286394185	.1288023135	.1289630898	.1291217935
110+	.1292784689	.1294331593	.1295859066	.1297367513	.1298857331
115+	.1300328902	.1301782599	.1303218783	.1304637806	.1306040010
120+	.1307425727	.1308795279	.1310148982	.1311487140	.1312810052
125+	.1314118006	.1315411285	.1316690163	.1317954908	.1319205779
130+	.1320443031	.1321666910	.1322877658	.1324075509	.1325260693
135+	.1326433432	.1327593945	.1328742444	.1329879137	.1331004225
140+	.1332117907	.1333220375	.1334311817	.1335392418	.1336462356
145+	.1337521808	.1338570944	.1339609933	.1340638937	.1341658118
150+	.1342667631	.1343667628	.1344658261	.1345639675	.1346612013
155+	.1347575415	.1348530018	.1349475956	.1350413361	.1351342361
160+	.1352263080	.1353175644	.1354080171	.1354976781	.1355865588
165+	.1356746707	.1357620247	.1358486318	.1359345026	.1360196476
170+	.1361040769	.1361878008	.1362708288	.1363531708	.1364348362
175+	.1365158342	.1365961740	.1366758644	.1367549143	.1368333322
180+	.1369111266	.1369883056	.1370648775	.1371408503	.1372162316
185+	.1372910293	.1373652508	.1374389036	.1375119950	.1375845320
190+	.1376565217	.1377279710	.1377988867	.1378692753	.1379391435
195+	.1380084976	.1380773441	.1381456890	.1382135385	.1382808985

D-INTEGRAL PROBABILITY M=R+2,B= 6

A= 1

R	1	2	3	4	5
0+	.0029154519	.0107076178	.0186148235	.0255886151	.0316015226
5+	.0367973791	.0413248070	.0453073136	.0488424765	.0520066690
10+	.0548599243	.0574498651	.0598146750	.0619853042	.0639871040
15+	.0658410450	.0675646340	.0691726100	.0706774793	.0720899301
20+	.0734191581	.0746731233	.0758587555	.0769821194	.0780485490
25+	.0790627567	.0800289237	.0809507749	.0818316406	.0826745092
30+	.0834820712	.0842567558	.0850007635	.0857160926	.0864045628
35+	.0870678351	.0877074293	.0883247393	.0889210459	.0894975285
40+	.0900552754	.0905952924	.0911185109	.0916257948	.0921179466
45+	.0925957131	.0930597901	.0935108270	.0939494304	.0943761680
50+	.0947915718	.0951961405	.0955903426	.0959746185	.0963493828
55+	.0967150258	.0970719160	.0974204010	.0977608094	.0980934518
60+	.0984186225	.0987366001	.0990476488	.0993520196	.0996499505
65+	.0999416680	.1002273874	.1005073136	.1007816420	.1010505586
70+	.1013142410	.1015728585	.1018265729	.1020755389	.1023199043
75+	.1025598104	.1027953928	.1030267810	.1032540993	.1034774668
80+	.1036969979	.1039128021	.1041249849	.1043336474	.1045388868
85+	.1047407966	.1049394668	.1051349839	.1053274313	.1055168893
90+	.1057034351	.1058871434	.1060680861	.1062463324	.1064219494
95+	.1065950017	.1067655517	.1069336597	.1070993839	.1072627806
100+	.1074239045	.1075828080	.1077395423	.1078941567	.1080466991
105+	.1081972155	.1083457511	.1084923491	.1086370518	.1087798999
110+	.1089209332	.1090601900	.1091977077	.1093335225	.1094676696
115+	.1096001831	.1097310961	.1098604410	.1099882491	.1101145507
120+	.1102393756	.1103627525	.1104847094	.1106052736	.1107244717
125+	.1108423294	.1109588719	.1110741237	.1111881086	.1113008499
130+	.1114123702	.1115226916	.1116318355	.1117398230	.1118466744
135+	.1119524098	.1120570485	.1121606096	.1122631116	.1123645725
140+	.1124650101	.1125644414	.1126628834	.1127603525	.1128568647
145+	.1129524357	.1130470808	.1131408150	.1132336529	.1133256088
150+	.1134166968	.1135069304	.1135963231	.1136848880	.1137726377
155+	.1138595850	.1139457419	.1140311206	.1141157327	.1141995897
160+	.1142827030	.1143650834	.1144467418	.1145276887	.1146079345
165+	.1146874894	.1147663632	.1148445657	.1149221064	.1149989948
170+	.1150752399	.1151508507	.1152258361	.1153002046	.1153739648
175+	.1154471250	.1155196933	.1155916777	.1156630860	.1157339260
180+	.1158042052	.1158739310	.1159431108	.1160117515	.1160798603
185+	.1161474441	.1162145095	.1162810632	.1163471118	.1164126617
190+	.1164777191	.1165422902	.1166063811	.1166699978	.1167331460
195+	.1167958317	.1168580604	.1169198376	.1169811690	.1170420598

D-INTEGRAL PROBABILITY M=R+2, B= 7

R	1	2	3	4	5
			A= 1		
0+	.0019531250	.0079720169	.0144195756	.0202507019	.0253531723
5+	.0298060698	.0337139817	.0371703935	.0402519211	.0430199049
10+	.0455233211	.0478014731	.0498861587	.0518033381	.0535744024
15+	.0552171351	.0567464462	.0581749351	.0595133277	.0607708173
20+	.0619553338	.0630737578	.0641320918	.0651355994	.0660889179
25+	.0669961512	.0678609459	.0686865550	.0694758912	.0702315708
30+	.0709559522	.0716511672	.0723191484	.0729616531	.0735802823
35+	.0741764991	.0747516432	.0753069441	.0758435325	.0763624503
40+	.0768646595	.0773510498	.0778224456	.0782796118	.0787232598
45+	.0791540514	.0795726039	.0799794936	.0803752591	.0807604048
50+	.0811354033	.0815006981	.0818567058	.0822038184	.0825424048
55+	.0828728127	.0831953703	.0835103874	.0838181568	.0841189556
60+	.0844130462	.0847006773	.0849820848	.0852574926	.0855271133
65+	.0857911493	.0860497930	.0863032275	.0865516275	.0867951594
70+	.0870339819	.0872682467	.0874980985	.0877236757	.0879451107
75+	.0881625301	.0883760552	.0885858022	.0887918825	.0889944027
80+	.0891934654	.0893891688	.0895816076	.0897708723	.0899570502
85+	.0901402253	.0903204782	.0904978867	.0906725255	.0908444668
90+	.0910137800	.0911805321	.0913447876	.0915066090	.0916660563
95+	.0918231876	.0919780590	.0921307248	.0922812373	.0924296472
100+	.0925760034	.0927203535	.0928627431	.0930032168	.0931418176
105+	.0932785872	.0934135658	.0935467926	.0936783056	.0938081416
110+	.0939363362	.0940629240	.0941879387	.0943114129	.0944333783
115+	.0945538656	.0946729048	.0947905248	.0949067540	.0950216198
120+	.0951351488	.0952473671	.0953582999	.0954679718	.0955764067
125+	.0956836279	.0957896581	.0958945193	.0959982332	.0961008206
130+	.0962023020	.0963026972	.0964020258	.0965003066	.0965975581
135+	.0966937982	.0967890446	.0968833144	.0969766242	.0970689905
140+	.0971604292	.0972509557	.0973405854	.0974293330	.0975172131
145+	.0976042398	.0976904269	.0977757880	.0978603363	.0979440848
150+	.0980270460	.0981092323	.0981906559	.0982713285	.0983512617
155+	.0984304668	.0985089550	.0985867370	.0986638234	.0987402248
160+	.0988159511	.0988910124	.0989654185	.0990391789	.0991123029
165+	.0991847998	.0992566785	.0993279478	.0993986163	.0994686926
170+	.0995381849	.0996071013	.0996754499	.0997432384	.0998104744
175+	.0998771656	.0999433193	.1000089427	.1000740430	.1001386271
180+	.1002027018	.1002662739	.1003293499	.1003919363	.1004540395
185+	.1005156657	.1005768211	.1006375115	.1006977430	.1007575213
190+	.1008168522	.1008757411	.1009341938	.1009922154	.1010498114
195+	.1011069869	.1011637472	.1012200972	.1012760418	.1013315860

D-INTEGRAL PROBABILITY M=R+2,B= 8

R	1	2	3	4	5
			A= 1		
0+	.0013717421	.0061537795	.0115279411	.0164974042	.0209027834
5+	.0247808154	.0282056197	.0312492649	.0339731139	.0364274119
10+	.0386528857	.0406825785	.0425434484	.0442576517	.0458435452
15+	.0473164598	.0486892992	.0499730042	.0511769163	.0523090631
20+	.0533763850	.0543849168	.0553399329	.0562460665	.0571074060
25+	.0579275746	.0587097963	.0594569505	.0601716182	.0608561208
30+	.0615125524	.0621428079	.0627486068	.0633315137	.0638929553
35+	.0644342363	.0649565524	.0654610014	.0659485937	.0664202610
40+	.0668768638	.0673191986	.0677480038	.0681639648	.0685677193
45+	.0689598610	.0693409437	.0697114849	.0700719682	.0704228469
50+	.0707645458	.0710974635	.0714219750	.0717384328	.0720471690
55+	.0723484968	.0726427116	.0729300927	.0732109039	.0734853950
60+	.0737538027	.0740163512	.0742732534	.0745247114	.0747709171
65+	.0750120533	.0752482936	.0754798036	.0757067410	.0759292563
70+	.0761474929	.0763615880	.0765716724	.0767778713	.0769803045
75+	.0771790866	.0773743274	.0775661319	.0777546011	.0779398315
80+	.0781219160	.0783009437	.0784770001	.0786501675	.0788205250
85+	.0789881485	.0791531112	.0793154837	.0794753337	.0796327267
90+	.0797877256	.0799403912	.0800907821	.0802389548	.0803849640
95+	.0805288624	.0806707008	.0808105286	.0809483932	.0810843407
100+	.0812184156	.0813506609	.0814811183	.0816098282	.0817368297
105+	.0818621605	.0819858575	.0821079561	.0822284908	.0823474951
110+	.0824650015	.0825810414	.0826956454	.0828088431	.0829206634
115+	.0830311343	.0831402829	.0832481358	.0833547185	.0834600561
120+	.0835641728	.0836670924	.0837688378	.0838694313	.0839688948
125+	.0840672495	.0841645160	.0842607144	.0843558644	.0844499849
130+	.0845430947	.0846352118	.0847263540	.0848165384	.0849057820
135+	.0849941010	.0850815116	.0851680292	.0852536692	.0853384465
140+	.0854223755	.0855054704	.0855877451	.0856692132	.0857498878
145+	.0858297818	.0859089079	.0859872784	.0860649054	.0861418006
150+	.0862179757	.0862934417	.0863682099	.0864422909	.0865156954
155+	.0865884336	.0866605157	.0867319516	.0868027508	.0868729230
160+	.0869424774	.0870114231	.0870797690	.0871475238	.0872146961
165+	.0872812942	.0873473264	.0874128007	.0874777249	.0875421068
170+	.0876059539	.0876692738	.0877320736	.0877943605	.0878561415
175+	.0879174234	.0879782131	.0880385172	.0880983420	.0881576940
180+	.0882165795	.0882750044	.0883329750	.0883904971	.0884475764
185+	.0885042187	.0885604296	.0886162146	.0886715790	.0887265282
190+	.0887810674	.0888352017	.0888889360	.0889422755	.0889952248
195+	.0890477888	.0890999722	.0891517795	.0892032154	.0892542842

D-INTEGRAL PROBABILITY M=R+2,B= 9

R	1	2	3	4	5
		A= 1			
0+	.0010000000	.0048867386	.0094466178	.0137478052	.0176052887
5+	.0210273448	.0240663553	.0267786390	.0292141272	.0314146380
10+	.0334145698	.0352421255	.0369204973	.0384688658	.0399032034
15+	.0412369058	.0424812886	.0436459756	.0447392061	.0457680771
20+	.0467387377	.0476565443	.0485261872	.0493517921	.0501370052
25+	.0508850618	.0515988445	.0522809303	.0529336320	.0535590312
30+	.0541590077	.0547352641	.0552893463	.0558226619	.0563364959
35+	.0568320237	.0573103233	.0577723851	.0582191212	.0586513729
40+	.0590699180	.0594754765	.0598687162	.0602502574	.0606206773
45+	.0609805137	.0613302684	.0616704104	.0620013782	.0623235830
50+	.0626374101	.0629432216	.0632413577	.0635321387	.0638158663
55+	.0640928251	.0643632837	.0646274960	.0648857019	.0651381288
60+	.0653849919	.0656264955	.0658628333	.0660941893	.0663207384
65+	.0665426470	.0667600734	.0669731686	.0671820762	.0673869334
70+	.0675878712	.0677850146	.0679784829	.0681683904	.0683548464
75+	.0685379556	.0687178180	.0688945297	.0690681829	.0692388657
80+	.0694066630	.0695716561	.0697339232	.0698935396	.0700505774
85+	.0702051062	.0703571929	.0705069018	.0706542952	.0707994326
90+	.0709423719	.0710831684	.0712218760	.0713585464	.0714932294
95+	.0716259733	.0717568249	.0718858291	.0720130294	.0721384681
100+	.0722621858	.0723842220	.0725046149	.0726234014	.0727406173
105+	.0728562974	.0729704752	.0730831834	.0731944535	.0733043162
110+	.0734128012	.0735199375	.0736257529	.0737302748	.0738335296
115+	.0739355429	.0740363397	.0741359442	.0742343799	.0743316699
120+	.0744278364	.0745229010	.0746168850	.0747098087	.0748016923
125+	.0748925551	.0749824161	.0750712938	.0751592062	.0752461708
130+	.0753322046	.0754173244	.0755015463	.0755848862	.0756673596
135+	.0757489815	.0758297665	.0759097291	.0759888833	.0760672426
140+	.0761448205	.0762216300	.0762976837	.0763729942	.0764475736
145+	.0765214338	.0765945862	.0766670424	.0767388133	.0768099099
150+	.0768803426	.0769501219	.0770192579	.0770877605	.0771556394
155+	.0772229042	.0772895640	.0773556281	.0774211053	.0774860043
160+	.0775503336	.0776141017	.0776773168	.0777399867	.0778021195
165+	.0778637227	.0779248040	.0779853708	.0780454303	.0781049895
170+	.0781640555	.0782226351	.0782807351	.0783383618	.0783955219
175+	.0784522217	.0785084672	.0785642647	.0786196201	.0786745392
180+	.0787290279	.0787830916	.0788367361	.0788899667	.0789427887
185+	.0789952075	.0790472281	.0790988557	.0791500951	.0792009513
190+	.0792514290	.0793015331	.0793512680	.0794006384	.0794496487
195+	.0794983033	.0795466065	.0795945626	.0796421758	.0796894501

D-INTEGRAL PROBABILITY M=R+2,B=10

A= 1

R	1	2	3	4	5
0+	.0007513148	.0039700406	.0078962116	.0116669354	.0150843250
5+	.0181371477	.0208618876	.0233029892	.0255016048	.0274929985
10+	.0293066009	.0309667872	.0324937541	.0339043040	.0352124951
15+	.0364301647	.0375673450	.0386325932	.0396332530	.0405756642
20+	.0414653303	.0423070540	.0431050468	.0438630189	.0445842535
25+	.0452716671	.0459278612	.0465551640	.0471556666	.0477312531
30+	.0482836262	.0488143291	.0493247643	.0498162095	.0502898317
35+	.0507466994	.0511877929	.0516140132	.0520261908	.0524250920
40+	.0528114253	.0531858473	.0535489669	.0539013501	.0542435239
45+	.0545759792	.0548991744	.0552135380	.0555194710	.0558173491
50+	.0561075250	.0563903298	.0566660748	.0569350534	.0571975418
55+	.0574538006	.0577040759	.0579486003	.0581875939	.0584212650
60+	.0586498110	.0588734193	.0590922676	.0593065247	.0595163511
65+	.0597218997	.0599233157	.0601207378	.0603142979	.0605041221
70+	.0606903306	.0608730383	.0610523550	.0612283856	.0614012308
75+	.0615709866	.0617377452	.0619015949	.0620626205	.0622209031
80+	.0623765207	.0625295483	.0626800579	.0628281185	.0629737969
85+	.0631171568	.0632582601	.0633971659	.0635339315	.0636686119
90+	.0638012603	.0639319278	.0640606638	.0641875161	.0643125308
95+	.0644357522	.0645572234	.0646769859	.0647950798	.0649115441
100+	.0650264163	.0651397327	.0652515286	.0653618381	.0654706942
105+	.0655781290	.0656841733	.0657888574	.0658922104	.0659942605
110+	.0660950353	.0661945613	.0662928645	.0663899700	.0664859021
115+	.0665806846	.0666743405	.0667668922	.0668583614	.0669487694
120+	.0670381367	.0671264833	.0672138288	.0673001921	.0673855916
125+	.0674700454	.0675535711	.0676361855	.0677179054	.0677987470
130+	.0678787260	.0679578579	.0680361576	.0681136398	.0681903188
135+	.0682662085	.0683413225	.0684156740	.0684892759	.0685621411
140+	.0686342817	.0687057098	.0687764372	.0688464753	.0689158356
145+	.0689845288	.0690525657	.0691199569	.0691867125	.0692528426
150+	.0693183571	.0693832654	.0694475770	.0695113011	.0695744466
155+	.0696370223	.0696990369	.0697604987	.0698214160	.0698817968
160+	.0699416491	.0700009806	.0700597989	.0701181114	.0701759254
165+	.0702332480	.0702900862	.0703464468	.0704023367	.0704577623
170+	.0705127300	.0705672463	.0706213174	.0706749493	.0707281479
175+	.0707809192	.0708332689	.0708852026	.0709367259	.0709878441
180+	.0710385626	.0710888867	.0711388215	.0711883720	.0712375431
185+	.0712863398	.0713347667	.0713828286	.0714305302	.0714778758
190+	.0715248700	.0715715170	.0716178213	.0716637870	.0717094183
195+	.0717547192	.0717996938	.0718443459	.0718886795	.0719326983

TABLE L (A = 1)

This table gives the A value of the incomplete Dirichlet integral of type 2

$$P = D_A^{(B)}(R, R) = \frac{\Gamma[(B+1)R]}{\Gamma^{B+1}(R)} \int_A^\infty \cdots \int_A^\infty \frac{\prod_{i=1}^{B} x_i^{R-1}\, dx_i}{\left(1 + \sum_{i=1}^{B} x_i\right)^{M+BR}}$$

for

$P^* = .750, .950, .975, .990, .995, .999$
$R = 1(1)50$
$B = 1(1)10$

Note: The symbols M, B, R and A which are all capital here in the table correspond to the same lower case symbols in the text.

A-VALUES FOR INVERSE PROBLEM D(R,R)=P* B= 1

R	.750	.900	.950	.975	.990	.995	.999
	P* - VALUES						
1	.333333	.111111	.052632	.025641	.010101	.005025	.001001
2	.484454	.243472	.156538	.104118	.062590	.043188	.018714
3	.561125	.327380	.233434	.171828	.118118	.090309	.049926
4	.609913	.386197	.290858	.225568	.165869	.133406	.083018
5	.644639	.430551	.335769	.269049	.206222	.171037	.114235
6	.671066	.465671	.372213	.305131	.240659	.203822	.142764
7	.692095	.494454	.402621	.335730	.270450	.232597	.168642
8	.709371	.518651	.428544	.362141	.296556	.258088	.192130
9	.723908	.539397	.451020	.385269	.319693	.280873	.213526
10	.736372	.557462	.470775	.405764	.340398	.301406	.233102
11	.747220	.573395	.488336	.424110	.359085	.320044	.251096
12	.756778	.587594	.504093	.440669	.376071	.337070	.267710
13	.765286	.600362	.518346	.455725	.391610	.352712	.283114
14	.772927	.611930	.531327	.469500	.405904	.367155	.297450
15	.779840	.622479	.543221	.482173	.419117	.380549	.310840
16	.786136	.632155	.554176	.493888	.431384	.393022	.323387
17	.791902	.641075	.564313	.504765	.442818	.404678	.335178
18	.797210	.649334	.573732	.514902	.453511	.415606	.346290
19	.802119	.657012	.582517	.524383	.463544	.425882	.356788
20	.806675	.664176	.590738	.533277	.472985	.435571	.366728
21	.810921	.670882	.598455	.541645	.481891	.444728	.376161
22	.814890	.677178	.605718	.549539	.490314	.453404	.385129
23	.818611	.683106	.612571	.557003	.498297	.461640	.393672
24	.822110	.688699	.619053	.564076	.505879	.469474	.401823
25	.825409	.693990	.625197	.570791	.513093	.476938	.409613
26	.828524	.699006	.631032	.577180	.519969	.484062	.417068
27	.831474	.703769	.636584	.583269	.526534	.490873	.424214
28	.834273	.708301	.641876	.589081	.532811	.497392	.431071
29	.836933	.712620	.646927	.594637	.538822	.503642	.437659
30	.839465	.716743	.651757	.599955	.544585	.509641	.443997
31	.841880	.720684	.656381	.605054	.550118	.515406	.450101
32	.844185	.724457	.660813	.609948	.555436	.520952	.455984
33	.846391	.728074	.665067	.614650	.560553	.526293	.461661
34	.848503	.731544	.669156	.619174	.565482	.531443	.467144
35	.850527	.734879	.673088	.623530	.570234	.536412	.472445
36	.852471	.738087	.676876	.627730	.574821	.541211	.477572
37	.854339	.741175	.680526	.631782	.579251	.545851	.482537
38	.856136	.744152	.684049	.635695	.583534	.550340	.487347
39	.857866	.747023	.687450	.639477	.587678	.554686	.492012
40	.859534	.749796	.690737	.643136	.591691	.558897	.496538
41	.861144	.752475	.693917	.646678	.595579	.562981	.500932
42	.862697	.755066	.696996	.650109	.599350	.566943	.505201
43	.864199	.757574	.699978	.653435	.603009	.570790	.509352
44	.865651	.760003	.702868	.656663	.606561	.574527	.513389
45	.867057	.762358	.705672	.659795	.610012	.578161	.517318
46	.868418	.764641	.708394	.662838	.613367	.581694	.521144
47	.869738	.766857	.711038	.665796	.616631	.585133	.524871
48	.871017	.769009	.713607	.668672	.619807	.588482	.528504
49	.872259	.771100	.716105	.671470	.622899	.591744	.532047
50	.873465	.773133	.718536	.674195	.625912	.594923	.535504

A-VALUES FOR INVERSE PROBLEM D(R,R)=P* B= 2

R	.750	.900	.950	.975	.990	.995	.999
	P* - VALUES						
1	.166667	.055556	.026316	.012821	.005051	.002513	.000501
2	.316749	.162952	.106199	.071390	.043362	.030088	.013142
3	.404046	.241695	.174889	.130275	.090655	.069806	.039029
4	.462790	.300165	.229290	.179896	.133908	.108490	.068341
5	.505966	.345704	.273232	.221365	.171673	.143401	.096955
6	.539535	.382534	.309644	.256498	.204571	.174448	.123658
7	.566666	.413181	.340483	.286731	.233442	.202088	.148226
8	.589224	.439247	.367073	.313116	.259014	.226834	.170758
9	.608388	.461803	.390334	.336423	.281869	.249135	.191447
10	.624948	.481594	.410930	.357225	.302462	.269365	.210496
11	.639457	.499159	.429350	.375954	.321151	.287829	.228097
12	.652314	.514899	.445965	.392944	.338222	.304774	.244419
13	.663816	.529118	.461062	.408460	.353903	.320404	.259609
14	.674190	.542054	.474866	.422709	.368379	.334886	.273792
15	.683612	.553895	.487558	.435862	.381804	.348358	.287076
16	.692222	.564790	.499285	.448058	.394302	.360937	.299556
17	.700132	.574863	.510168	.459412	.405981	.372721	.311310
18	.707434	.584216	.520305	.470019	.416929	.383793	.322410
19	.714204	.592932	.529782	.479962	.427223	.394227	.332916
20	.720504	.601083	.538670	.489308	.436928	.404082	.342881
21	.726386	.608729	.547028	.498119	.446100	.413413	.352351
22	.731897	.615920	.554910	.506444	.454788	.422266	.361369
23	.737073	.622703	.562360	.514329	.463035	.430683	.369969
24	.741949	.629114	.569417	.521812	.470879	.438700	.378186
25	.746552	.635188	.576116	.528927	.478352	.446348	.386048
26	.750908	.640953	.582488	.535705	.485483	.453657	.393580
27	.755038	.646437	.588558	.542172	.492300	.460651	.400806
28	.758961	.651660	.594350	.548353	.498826	.467354	.407748
29	.762694	.656645	.599886	.554268	.505081	.473786	.414423
30	.766253	.661409	.605185	.559936	.511084	.479965	.420850
31	.769650	.665967	.610263	.565375	.516852	.485909	.427044
32	.772898	.670336	.615135	.570601	.522402	.491632	.433020
33	.776007	.674528	.619817	.575626	.527746	.497148	.438790
34	.778988	.678554	.624319	.580465	.532898	.502470	.444367
35	.781849	.682426	.628654	.585129	.537869	.507609	.449761
36	.784597	.686154	.632832	.589628	.542670	.512577	.454983
37	.787241	.689746	.636863	.593973	.547311	.517382	.460042
38	.789786	.693211	.640755	.598171	.551801	.522034	.464947
39	.792239	.696556	.644516	.602232	.556148	.526541	.469706
40	.794606	.699788	.648153	.606163	.560360	.530911	.474326
41	.796891	.702913	.651674	.609971	.564443	.535150	.478814
42	.799098	.705938	.655084	.613662	.568405	.539266	.483176
43	.801233	.708867	.658390	.617243	.572252	.543264	.487419
44	.803299	.711706	.661596	.620718	.575989	.547150	.491548
45	.805300	.714460	.664709	.624094	.579622	.550930	.495568
46	.807240	.717132	.667731	.627375	.583155	.554608	.499484
47	.809120	.719726	.670669	.630565	.586593	.558189	.503301
48	.810946	.722247	.673525	.633669	.589940	.561677	.507023
49	.812718	.724698	.676303	.636690	.593201	.565077	.510653
50	.814439	.727082	.679008	.639633	.596379	.568392	.514197

A-VALUES FOR INVERSE PROBLEM D(R,R)=P* B= 3

P* - VALUES

R	.750	.900	.950	.975	.990	.995	.999
1	.111111	.037037	.017544	.008547	.003367	.001675	.000334
2	.249356	.129716	.085080	.057478	.035079	.024403	.010697
3	.336887	.203922	.148597	.111323	.077923	.060208	.033846
4	.397815	.260963	.200699	.158349	.118572	.096410	.061098
5	.443454	.306240	.243594	.198409	.154756	.129727	.088248
6	.479383	.343313	.279574	.232761	.186659	.159718	.113900
7	.508684	.374435	.310312	.262574	.214893	.186640	.137699
8	.533212	.401081	.336987	.288759	.240056	.210891	.159659
9	.554163	.424261	.360443	.312005	.262655	.232851	.179915
10	.572349	.444686	.381298	.332835	.283096	.252848	.198635
11	.588341	.462879	.400014	.351653	.301708	.271156	.215985
12	.602556	.479231	.416946	.368774	.318755	.288003	.232114
13	.615309	.494043	.432370	.384446	.334450	.303578	.247157
14	.626838	.507548	.446504	.398870	.348969	.318037	.261228
15	.637331	.519935	.459525	.412210	.362458	.331511	.274430
16	.646939	.531354	.471578	.424600	.375036	.344112	.286850
17	.655781	.541929	.482779	.436151	.386807	.355933	.298564
18	.663956	.551761	.493229	.446959	.397855	.367054	.309638
19	.671545	.560937	.503011	.457101	.408256	.377545	.320130
20	.678617	.569528	.512195	.466647	.418072	.387465	.330092
21	.685228	.577596	.520842	.475654	.427358	.396866	.339568
22	.691428	.585193	.529004	.484173	.436162	.405795	.348598
23	.697258	.592364	.536726	.492249	.444527	.414290	.357217
24	.702754	.599150	.544047	.499920	.452488	.422387	.365457
25	.707948	.605583	.551003	.507220	.460080	.430118	.373346
26	.712867	.611695	.557623	.514179	.467329	.437510	.380910
27	.717534	.617512	.563935	.520823	.474263	.444589	.388170
28	.721971	.623057	.569962	.527177	.480905	.451377	.395148
29	.726196	.628353	.575726	.533261	.487275	.457894	.401862
30	.730227	.633416	.581246	.539096	.493392	.464158	.408329
31	.734077	.638265	.586539	.544697	.499273	.470187	.414565
32	.737759	.642914	.591621	.550081	.504934	.475994	.420584
33	.741287	.647377	.596506	.555262	.510387	.481594	.426397
34	.744671	.651666	.601207	.560253	.515647	.487000	.432019
35	.747920	.655793	.605735	.565065	.520724	.492222	.437458
36	.751043	.659768	.610101	.569710	.525630	.497271	.442725
37	.754048	.663600	.614315	.574196	.530374	.502157	.447830
38	.756943	.667298	.618385	.578534	.534965	.506889	.452781
39	.759734	.670870	.622320	.582730	.539412	.511475	.457586
40	.762428	.674322	.626127	.586794	.543722	.515923	.462252
41	.765030	.677661	.629813	.590733	.547902	.520240	.466786
42	.767545	.680895	.633385	.594551	.551959	.524432	.471194
43	.769978	.684027	.636849	.598257	.555899	.528505	.475483
44	.772333	.687064	.640210	.601855	.559727	.532466	.479658
45	.774615	.690010	.643473	.605351	.563450	.536319	.483724
46	.776827	.692870	.646643	.608749	.567072	.540070	.487685
47	.778973	.695648	.649724	.612055	.570597	.543722	.491547
48	.781057	.698347	.652721	.615272	.574031	.547281	.495314
49	.783080	.700973	.655637	.618404	.577376	.550750	.498990
50	.785047	.703528	.658477	.621455	.580638	.554134	.502578

A-VALUES FOR INVERSE PROBLEM D(R,R)=P* B= 4

R	P* - VALUES .750	.900	.950	.975	.990	.995	.999
1	.083333	.027778	.013158	.006410	.002525	.001256	.000250
2	.211181	.110615	.072838	.049357	.030210	.021049	.009247
3	.297297	.181283	.132679	.099753	.070082	.054266	.030610
4	.358723	.236973	.183022	.144912	.108920	.088763	.056469
5	.405362	.281782	.225036	.183911	.143973	.120961	.082609
6	.442406	.318794	.260584	.217642	.175145	.150194	.107527
7	.472806	.350057	.291138	.247093	.202896	.176592	.130782
8	.498376	.376949	.317775	.273075	.227739	.200474	.152332
9	.520300	.400428	.341281	.296222	.250125	.222173	.172277
10	.539389	.421180	.362242	.317023	.270430	.241984	.190757
11	.556218	.439709	.381098	.335859	.288959	.260163	.207921
12	.571210	.456399	.398192	.353029	.305963	.276923	.223905
13	.584685	.471543	.413790	.368773	.321644	.292442	.238836
14	.596887	.485375	.428107	.383286	.336172	.306868	.252821
15	.608010	.498078	.441314	.396725	.349684	.320329	.265956
16	.618206	.509803	.453553	.409222	.362300	.332931	.278326
17	.627601	.520674	.464940	.420885	.374116	.344764	.290004
18	.636296	.530792	.475574	.431807	.385218	.355906	.301052
19	.644376	.540243	.485537	.442066	.395678	.366426	.311529
20	.651912	.549099	.494899	.451729	.405557	.376380	.321482
21	.658963	.557422	.503719	.460854	.414909	.385820	.330955
22	.665580	.565265	.512051	.469490	.423782	.394791	.339988
23	.671807	.572673	.519939	.477682	.432217	.403331	.348615
24	.677681	.579687	.527422	.485467	.440249	.411475	.356866
25	.683235	.586341	.534535	.492880	.447912	.419255	.364769
26	.688498	.592666	.541309	.499950	.455234	.426698	.372349
27	.693494	.598689	.547770	.506704	.462240	.433828	.379629
28	.698246	.604434	.553943	.513165	.468953	.440668	.386628
29	.702774	.609922	.559848	.519355	.475395	.447237	.393364
30	.707095	.615171	.565506	.525293	.481583	.453555	.399856
31	.711224	.620201	.570934	.530996	.487534	.459636	.406116
32	.715175	.625025	.576148	.536480	.493264	.465496	.412161
33	.718962	.629657	.581160	.541759	.498786	.471149	.418002
34	.722595	.634111	.585986	.546845	.504114	.476607	.423650
35	.726084	.638398	.590635	.551751	.509259	.481881	.429118
36	.729440	.642528	.595120	.556487	.514231	.486982	.434413
37	.732670	.646511	.599449	.561064	.519040	.491920	.439547
38	.735782	.650355	.603632	.565490	.523696	.496703	.444527
39	.738784	.654069	.607677	.569773	.528206	.501340	.449361
40	.741682	.657660	.611592	.573922	.532579	.505838	.454057
41	.744481	.661135	.615384	.577943	.536820	.510204	.458620
42	.747188	.664500	.619059	.581844	.540938	.514445	.463058
43	.749807	.667761	.622623	.585629	.544938	.518567	.467376
44	.752344	.670922	.626082	.589306	.548826	.522575	.471581
45	.754802	.673991	.629441	.592878	.552607	.526476	.475676
46	.757185	.676970	.632705	.596352	.556286	.530273	.479667
47	.759497	.679864	.635878	.599732	.559868	.533971	.483558
48	.761743	.682677	.638965	.603022	.563357	.537576	.487354
49	.763924	.685414	.641970	.606225	.566757	.541090	.491058
50	.766044	.688077	.644896	.609347	.570072	.544518	.494675

A-VALUES FOR INVERSE PROBLEM D(R,R)=P* B= 5

R	P* - VALUES .750	.900	.950	.975	.990	.995	.999
1	.066667	.022222	.010526	.005128	.002020	.001005	.000200
2	.185982	.097885	.064632	.043889	.026918	.018775	.008260
3	.270372	.165714	.121658	.091698	.064591	.050090	.028323
4	.331732	.220220	.170593	.135410	.102051	.083301	.053140
5	.378817	.264543	.211866	.173562	.136227	.114639	.078515
6	.416470	.301402	.247022	.206783	.166823	.143283	.102869
7	.447522	.332685	.277382	.235923	.194187	.169269	.125705
8	.473736	.359691	.303943	.261721	.218766	.192857	.146938
9	.496278	.383336	.327448	.284765	.240973	.214344	.166638
10	.515950	.404282	.348453	.305519	.261159	.234003	.184929
11	.533327	.423020	.367385	.324346	.279611	.252073	.201945
12	.548834	.439925	.384574	.341534	.296569	.268756	.217814
13	.562791	.455286	.400280	.357316	.312228	.284223	.232653
14	.575445	.469333	.414713	.371879	.326750	.298616	.246567
15	.586993	.482247	.428041	.385379	.340272	.312059	.259648
16	.597590	.494178	.440403	.397943	.352906	.324654	.271975
17	.607362	.505249	.451915	.409679	.364749	.336490	.283621
18	.616413	.515562	.462672	.420677	.375884	.347642	.294647
19	.624831	.525201	.472758	.431014	.386380	.358177	.305107
20	.632686	.534239	.482241	.440757	.396300	.368152	.315050
21	.640041	.542739	.491182	.449961	.405696	.377617	.324518
22	.646947	.550752	.499631	.458678	.414615	.386614	.333550
23	.653449	.558326	.507633	.466950	.423097	.395184	.342179
24	.659586	.565499	.515229	.474815	.431178	.403361	.350436
25	.665391	.572308	.522452	.482306	.438890	.411174	.358348
26	.670894	.578782	.529334	.489454	.446262	.418651	.365938
27	.676120	.584950	.535900	.496284	.453318	.425816	.373230
28	.681093	.590834	.542175	.502821	.460082	.432692	.380243
29	.685833	.596457	.548181	.509086	.466573	.439298	.386994
30	.690357	.601839	.553937	.515097	.472811	.445652	.393502
31	.694683	.606995	.559460	.520872	.478812	.451771	.399780
32	.698823	.611943	.564766	.526427	.484592	.457668	.405842
33	.702792	.616696	.569870	.531775	.490163	.463358	.411701
34	.706600	.621267	.574784	.536929	.495539	.468854	.417369
35	.710260	.625667	.579520	.541902	.500732	.474165	.422856
36	.713780	.629907	.584089	.546704	.505752	.479303	.428172
37	.717169	.633997	.588501	.551345	.510608	.484278	.433326
38	.720435	.637946	.592764	.555834	.515310	.489097	.438327
39	.723586	.641762	.596888	.560180	.519866	.493770	.443182
40	.726628	.645452	.600880	.564390	.524283	.498304	.447898
41	.729567	.649023	.604747	.568471	.528570	.502706	.452483
42	.732410	.652482	.608496	.572430	.532731	.506983	.456942
43	.735162	.655834	.612132	.576273	.536775	.511140	.461282
44	.737827	.659086	.615662	.580006	.540705	.515183	.465507
45	.740409	.662241	.619090	.583634	.544528	.519117	.469624
46	.742914	.665306	.622422	.587163	.548248	.522948	.473636
47	.745345	.668283	.625661	.590596	.551871	.526680	.477549
48	.747706	.671178	.628813	.593938	.555400	.530318	.481366
49	.749999	.673995	.631882	.597193	.558840	.533865	.485091
50	.752229	.676736	.634870	.600365	.562195	.537325	.488729

A-VALUES FOR INVERSE PROBLEM D(R,R)=P* B= 6

R	P* - VALUES .750	.900	.950	.975	.990	.995	.999
1	.055556	.018519	.008772	.004274	.001684	.000838	.000167
2	.167823	.088646	.058652	.039891	.024503	.017104	.007534
3	.250497	.154129	.113417	.085649	.060450	.046933	.026586
4	.311567	.207600	.161182	.128184	.096803	.079117	.050578
5	.358838	.251460	.201820	.165634	.130265	.109759	.075338
6	.396852	.288137	.236626	.198423	.160387	.137923	.099238
7	.428325	.319387	.266800	.227294	.187428	.163568	.121733
8	.454974	.346443	.293274	.252925	.211784	.186912	.142706
9	.477944	.370186	.316753	.275871	.233837	.208221	.162205
10	.498027	.391258	.337774	.296573	.253917	.227751	.180340
11	.515796	.410138	.356749	.315380	.272299	.245727	.197233
12	.531673	.427192	.373998	.332571	.289212	.262342	.213006
13	.545980	.442707	.389777	.348373	.304846	.277761	.227769
14	.558965	.456907	.404289	.362968	.319358	.292123	.241622
15	.570824	.469974	.417703	.376508	.332880	.305547	.254655
16	.581716	.482055	.430153	.389118	.345523	.318132	.266946
17	.591767	.493273	.441755	.400905	.357383	.329966	.278564
18	.601083	.503729	.452603	.411957	.368539	.341123	.289568
19	.609751	.513508	.462779	.422351	.379061	.351667	.300013
20	.617845	.522682	.472352	.432151	.389010	.361655	.309946
21	.625427	.531313	.481381	.441415	.398438	.371136	.319408
22	.632549	.539455	.489918	.450191	.407390	.380153	.328438
23	.639257	.547152	.498007	.458522	.415906	.388744	.337068
24	.645591	.554445	.505687	.466446	.424023	.396943	.345327
25	.651585	.561370	.512993	.473997	.431772	.404780	.353244
26	.657269	.567957	.519956	.481203	.439181	.412283	.360841
27	.662669	.574233	.526601	.488092	.446275	.419475	.368141
28	.667809	.580224	.532954	.494686	.453076	.426377	.375163
29	.672709	.585950	.539036	.501007	.459606	.433011	.381926
30	.677387	.591431	.544866	.507074	.465881	.439393	.388445
31	.681861	.596685	.550461	.512904	.471920	.445539	.394736
32	.686145	.601727	.555839	.518512	.477737	.451465	.400811
33	.690251	.606571	.561011	.523913	.483346	.457184	.406685
34	.694194	.611231	.565993	.529120	.488759	.462707	.412367
35	.697982	.615717	.570796	.534144	.493989	.468047	.417869
36	.701627	.620042	.575429	.538997	.499044	.473214	.423200
37	.705137	.624214	.579904	.543687	.503937	.478217	.428370
38	.708520	.628243	.584230	.548225	.508674	.483064	.433386
39	.711784	.632136	.588415	.552618	.513265	.487765	.438257
40	.714936	.635902	.592466	.556875	.517718	.492327	.442990
41	.717983	.639547	.596391	.561002	.522038	.496756	.447590
42	.720929	.643078	.600196	.565006	.526234	.501060	.452066
43	.723782	.646501	.603888	.568894	.530310	.505244	.456422
44	.726545	.649821	.607472	.572670	.534274	.509314	.460664
45	.729223	.653044	.610953	.576341	.538129	.513275	.464797
46	.731821	.656174	.614337	.579912	.541882	.517132	.468826
47	.734342	.659216	.617628	.583386	.545536	.520890	.472754
48	.736791	.662174	.620830	.586769	.549097	.524554	.476588
49	.739170	.665051	.623947	.590064	.552567	.528126	.480330
50	.741484	.667852	.626984	.593276	.555952	.531612	.483984

A-VALUES FOR INVERSE PROBLEM D(R,R)=P* B= 7

R	P* - VALUES .750	.900	.950	.975	.990	.995	.999
1	.047619	.015873	.007519	.003663	.001443	.000718	.000143
2	.153970	.081560	.054051	.036807	.022635	.015811	.006970
3	.235025	.145052	.106935	.080877	.057172	.044428	.025204
4	.295712	.197612	.153703	.122422	.092604	.075761	.048515
5	.343032	.241043	.193788	.159275	.125465	.105821	.072765
6	.381266	.277531	.228281	.191689	.155183	.133579	.096283
7	.413027	.308722	.258280	.220323	.181947	.158936	.118491
8	.439988	.335794	.284664	.245804	.206111	.182070	.139244
9	.463271	.359596	.308108	.268658	.228029	.203226	.158574
10	.483661	.380753	.329129	.289307	.248014	.222643	.176576
11	.501725	.399734	.348127	.308089	.266332	.240536	.193364
12	.517885	.416898	.365416	.325276	.283203	.257091	.209053
13	.532459	.432527	.381246	.341087	.298811	.272468	.223750
14	.545699	.446844	.395818	.355702	.313310	.286800	.237552
15	.557799	.460027	.409294	.369270	.326829	.300204	.250543
16	.568920	.472225	.421811	.381914	.339476	.312778	.262802
17	.579188	.483557	.433482	.393739	.351346	.324608	.274394
18	.588711	.494124	.444400	.404832	.362517	.335766	.285379
19	.597575	.504012	.454646	.415269	.373058	.346315	.295810
20	.605856	.513292	.464289	.425114	.383028	.356312	.305733
21	.613616	.522026	.473387	.434423	.392479	.365805	.315189
22	.620909	.530268	.481993	.443245	.401457	.374836	.324215
23	.627780	.538062	.490149	.451623	.410000	.383443	.332844
24	.634270	.545450	.497896	.459594	.418145	.391659	.341105
25	.640413	.552467	.505268	.467191	.425922	.399515	.349024
26	.646240	.559143	.512294	.474443	.433360	.407037	.356626
27	.651777	.565507	.519003	.481378	.440483	.414250	.363933
28	.657049	.571581	.525417	.488017	.447315	.421174	.370962
29	.662076	.577389	.531559	.494383	.453874	.427829	.377733
30	.666877	.582950	.537448	.500494	.460180	.434233	.384261
31	.671469	.588281	.543102	.506368	.466249	.440401	.390561
32	.675866	.593398	.548536	.512019	.472096	.446350	.396647
33	.680083	.598315	.553764	.517463	.477734	.452091	.402532
34	.684132	.603046	.558800	.522711	.483177	.457637	.408225
35	.688023	.607602	.563655	.527777	.488436	.463000	.413739
36	.691767	.611994	.568341	.532669	.493521	.468189	.419082
37	.695374	.616231	.572867	.537400	.498442	.473214	.424264
38	.698850	.620324	.577242	.541976	.503208	.478085	.429293
39	.702205	.624280	.581475	.546408	.507828	.482808	.434177
40	.705445	.628107	.585574	.550702	.512308	.487392	.438922
41	.708577	.631811	.589546	.554867	.516656	.491844	.443536
42	.711607	.635400	.593397	.558907	.520879	.496169	.448025
43	.714540	.638880	.597133	.562831	.524983	.500375	.452394
44	.717382	.642255	.600761	.566643	.528973	.504467	.456649
45	.720136	.645532	.604285	.570348	.532855	.508449	.460795
46	.722809	.648715	.607711	.573953	.536633	.512328	.464837
47	.725402	.651808	.611043	.577460	.540313	.516107	.468780
48	.727922	.654816	.614286	.580876	.543899	.519791	.472626
49	.730370	.657743	.617443	.584204	.547395	.523384	.476381
50	.732751	.660593	.620518	.587447	.550804	.526890	.480049

A-VALUES FOR INVERSE PROBLEM D(R,R)=P* B= 8

R	.750	.900	.950	.975	.990	.995	.999
	P* - VALUES						
1	.041667	.013889	.006579	.003205	.001263	.000628	.000125
2	.142968	.075907	.050370	.034336	.021135	.014770	.006516
3	.222521	.137679	.101653	.076978	.054486	.042373	.024067
4	.282787	.189427	.147555	.117673	.089133	.072981	.046802
5	.330081	.232461	.187151	.154005	.121475	.102542	.070614
6	.368450	.268763	.221360	.186090	.150843	.129949	.093806
7	.400414	.299884	.251196	.214511	.177365	.155055	.115766
8	.427606	.326951	.277492	.239856	.201358	.178007	.136330
9	.451129	.350789	.300895	.262624	.223155	.199028	.155512
10	.471758	.372006	.321907	.283222	.243056	.218345	.173398
11	.490054	.391062	.340918	.301977	.261315	.236165	.190094
12	.506436	.408310	.358235	.319155	.278147	.252665	.205711
13	.521225	.424028	.374102	.334969	.293729	.268002	.220349
14	.534667	.438436	.388718	.349598	.308214	.282306	.234104
15	.546962	.451712	.402243	.363185	.321727	.295692	.247059
16	.558267	.464001	.414813	.375855	.334376	.308255	.259288
17	.568711	.475424	.426538	.387709	.346252	.320079	.270858
18	.578400	.486081	.437511	.398834	.357434	.331236	.281825
19	.587424	.496057	.447814	.409304	.367988	.341788	.292243
20	.595857	.505423	.457513	.419185	.377975	.351791	.302156
21	.603763	.514241	.466667	.428531	.387444	.361292	.311606
22	.611194	.522565	.475328	.437390	.396442	.370334	.320628
23	.618198	.530439	.483540	.445806	.405007	.378953	.329256
24	.624815	.537904	.491341	.453814	.413174	.387184	.337517
25	.631081	.544996	.498766	.461449	.420974	.395054	.345438
26	.637025	.551746	.505845	.468739	.428436	.402592	.353044
27	.642675	.558180	.512606	.475711	.435583	.409821	.360354
28	.648056	.564324	.519071	.482388	.442439	.416762	.367389
29	.653187	.570199	.525263	.488791	.449023	.423435	.374166
30	.658089	.575825	.531200	.494939	.455353	.429857	.380702
31	.662778	.581220	.536902	.500848	.461447	.436044	.387010
32	.667269	.586399	.542382	.506535	.467319	.442011	.393104
33	.671576	.591377	.547656	.512014	.472982	.447770	.398997
34	.675713	.596166	.552737	.517297	.478449	.453335	.404700
35	.679689	.600779	.557636	.522396	.483732	.458717	.410223
36	.683515	.605227	.562365	.527322	.488841	.463925	.415577
37	.687201	.609519	.566933	.532086	.493786	.468969	.420769
38	.690755	.613665	.571349	.536695	.498576	.473858	.425808
39	.694185	.617672	.575623	.541158	.503219	.478600	.430702
40	.697497	.621549	.579761	.545484	.507723	.483202	.435458
41	.700700	.625303	.583771	.549679	.512094	.487672	.440083
42	.703798	.628940	.587660	.553750	.516340	.492016	.444583
43	.706798	.632467	.591434	.557703	.520466	.496240	.448963
44	.709705	.635888	.595098	.561545	.524478	.500350	.453229
45	.712523	.639210	.598658	.565279	.528382	.504351	.457386
46	.715257	.642436	.602119	.568912	.532182	.508247	.461439
47	.717911	.645573	.605486	.572448	.535884	.512044	.465393
48	.720489	.648623	.608762	.575891	.539490	.515745	.469250
49	.722994	.651591	.611952	.579245	.543007	.519355	.473017
50	.725431	.654480	.615060	.582515	.546437	.522878	.476695

A-VALUES FOR INVERSE PROBLEM D(R,R)=P* B= 9

R	.750	.900	.950	.975	.990	.995	.999
			P* - VALUES				
1	.037037	.012346	.005848	.002849	.001122	.000558	.000111
2	.133965	.071264	.047342	.032298	.019897	.013911	.006140
3	.212132	.131526	.097234	.073709	.052229	.040644	.023109
4	.271967	.182545	.142372	.113660	.086191	.070623	.045344
5	.319189	.225213	.181529	.149531	.118079	.099746	.068776
6	.357637	.261335	.215480	.181322	.147138	.126845	.091681
7	.389750	.292378	.245165	.209552	.173444	.151730	.113424
8	.417119	.319429	.271375	.234772	.197285	.174519	.133821
9	.440831	.343287	.294736	.257460	.218974	.195420	.152873
10	.461651	.364547	.315734	.278008	.238798	.214648	.170656
11	.480134	.383660	.334750	.296736	.257003	.232400	.187271
12	.496698	.400974	.352085	.313902	.273797	.248851	.202822
13	.511661	.416763	.367980	.329716	.289355	.264152	.217409
14	.525271	.431244	.382630	.344352	.303825	.278430	.231122
15	.537726	.444596	.396195	.357955	.317331	.291798	.244044
16	.549183	.456961	.408807	.370644	.329979	.304349	.256247
17	.559773	.468459	.420576	.382520	.341859	.316167	.267795
18	.569602	.479190	.431595	.393671	.353048	.327322	.278747
19	.578759	.489239	.441944	.404169	.363614	.337876	.289152
20	.587319	.498677	.451689	.414079	.373613	.347883	.299056
21	.595345	.507565	.460891	.423455	.383098	.357391	.308500
22	.602893	.515957	.469598	.432345	.392111	.366441	.317518
23	.610009	.523898	.477856	.440792	.400693	.375070	.326143
24	.616733	.531428	.485702	.448832	.408879	.383311	.334404
25	.623101	.538583	.493172	.456499	.416699	.391194	.342327
26	.629144	.545395	.500296	.463821	.424180	.398745	.349934
27	.634889	.551889	.507099	.470824	.431347	.405988	.357248
28	.640361	.558091	.513607	.477533	.438224	.412943	.364288
29	.645580	.564023	.519841	.483967	.444828	.419630	.371070
30	.650567	.569704	.525820	.490145	.451179	.426067	.377611
31	.655338	.575153	.531562	.496085	.457294	.432269	.383926
32	.659908	.580384	.537082	.501802	.463186	.438252	.390027
33	.664292	.585413	.542394	.507310	.468870	.444027	.395927
34	.668502	.590252	.547513	.512623	.474358	.449608	.401638
35	.672550	.594914	.552450	.517751	.479662	.455005	.407169
36	.676446	.599408	.557215	.522705	.484792	.460229	.412531
37	.680199	.603747	.561818	.527497	.489757	.465289	.417731
38	.683818	.607937	.566270	.532133	.494567	.470194	.422779
39	.687312	.611989	.570578	.536624	.499230	.474952	.427683
40	.690686	.615909	.574750	.540976	.503753	.479570	.432448
41	.693949	.619704	.578793	.545198	.508144	.484056	.437082
42	.697105	.623382	.582714	.549295	.512409	.488416	.441590
43	.700162	.626949	.586519	.553273	.516555	.492655	.445980
44	.703123	.630409	.590214	.557139	.520586	.496780	.450255
45	.705995	.633769	.593805	.560898	.524508	.500796	.454422
46	.708781	.637033	.597296	.564555	.528327	.504707	.458485
47	.711486	.640206	.600692	.568115	.532047	.508519	.462447
48	.714114	.643292	.603997	.571582	.535672	.512236	.466315
49	.716668	.646295	.607215	.574959	.539206	.515861	.470090
50	.719152	.649218	.610351	.578252	.542654	.519398	.473778

A-VALUES FOR INVERSE PROBLEM D(R,R)=P* B=10

R	P* - VALUES .750	.900	.950	.975	.990	.995	.999
1	.033333	.011111	.005263	.002564	.001010	.000503	.000100
2	.126429	.067365	.044793	.030582	.018852	.013185	.005823
3	.203314	.126285	.093462	.070913	.050295	.039160	.022285
4	.262721	.176641	.137915	.110202	.083652	.068584	.044081
5	.309843	.218970	.176676	.145661	.115136	.097320	.067176
6	.348334	.254919	.210391	.177187	.143918	.124144	.089828
7	.380555	.285884	.239934	.205243	.170030	.148830	.111377
8	.408064	.312911	.266062	.230348	.193733	.171473	.131624
9	.431929	.336779	.289380	.252961	.215324	.192266	.150559
10	.452904	.358069	.310360	.273462	.235077	.211413	.168250
11	.471542	.377226	.329377	.292162	.253231	.229104	.184792
12	.488257	.394593	.346724	.309314	.269990	.245510	.200285
13	.503366	.410440	.362641	.325126	.285525	.260776	.214825
14	.517118	.424983	.377318	.339767	.299980	.275030	.228500
15	.529707	.438397	.390915	.353380	.313479	.288381	.241391
16	.541294	.450825	.403561	.366085	.326125	.300921	.253570
17	.552007	.462387	.415367	.377980	.338007	.312732	.265099
18	.561954	.473181	.426424	.389151	.349202	.323884	.276036
19	.571224	.483292	.436812	.399673	.359775	.334439	.286430
20	.579893	.492790	.446597	.409607	.369785	.344449	.296325
21	.588023	.501738	.455838	.419008	.379281	.353961	.305763
22	.595670	.510188	.464585	.427924	.388309	.363018	.314777
23	.602881	.518186	.472882	.436398	.396905	.371655	.323400
24	.609697	.525772	.480768	.444465	.405106	.379905	.331660
25	.616153	.532981	.488276	.452159	.412942	.387798	.339583
26	.622280	.539845	.495437	.459508	.420440	.395360	.347193
27	.628107	.546391	.502278	.466538	.427625	.402614	.354509
28	.633657	.552643	.508823	.473273	.434518	.409582	.361552
29	.638952	.558624	.515092	.479734	.441141	.416281	.368338
30	.644012	.564353	.521107	.485939	.447510	.422731	.374884
31	.648853	.569848	.526883	.491905	.453642	.428947	.381204
32	.653492	.575124	.532437	.497648	.459552	.434942	.387311
33	.657942	.580197	.537784	.503182	.465254	.440731	.393218
34	.662216	.585079	.542935	.508519	.470760	.446326	.398935
35	.666326	.589783	.547904	.513672	.476082	.451737	.404473
36	.670281	.594318	.552700	.518652	.481229	.456974	.409842
37	.674093	.598696	.557335	.523467	.486212	.462048	.415050
38	.677768	.602926	.561817	.528127	.491040	.466967	.420105
39	.681316	.607015	.566154	.532642	.495720	.471738	.425016
40	.684744	.610972	.570355	.537017	.500260	.476370	.429789
41	.688058	.614804	.574427	.541261	.504668	.480870	.434431
42	.691265	.618517	.578376	.545380	.508950	.485243	.438948
43	.694371	.622118	.582209	.549381	.513112	.489496	.443345
44	.697380	.625613	.585931	.553269	.517160	.493634	.447629
45	.700298	.629005	.589548	.557049	.521099	.497663	.451804
46	.703129	.632302	.593064	.560728	.524933	.501588	.455875
47	.705878	.635506	.596486	.564308	.528669	.505413	.459846
48	.708549	.638623	.599816	.567795	.532310	.509142	.463721
49	.711145	.641657	.603059	.571192	.535859	.512780	.467505
50	.713670	.644610	.606218	.574505	.539322	.516330	.471201

TABLE M (A < 1)

This table gives the value of the incomplete Dirichlet integral of type 2

$$C_A^{(B)}(R, M) = \frac{\Gamma(M+BR)}{\Gamma^B(R)\Gamma(M)} \int_0^A \cdots \int_0^A \frac{\prod_{i=1}^{B} x_i^{R-1} \, dx_i}{\left(1 + \sum_{i=1}^{B} x_i\right)^{M+BR}}$$

for

$$M = R$$
$$A^{-1} = 3(1)10 \quad \text{and} \quad A = .40(.10).60(.05).80,$$
$$B = 1(1)10$$

The common value of M and R increases from 1 by ones until we reach the line containing .01; the last line is completed.

<u>Note</u> 1: In the table entries less than 10^{-4} are given in exponential notation to avoid too many zeros. Thus .85999\04 represents .000085999

<u>Note</u> 2: The symbols M, B, R and A which are all capital here in the table correspond to the same lower case symbols in the text.

C-INTEGRAL PROBABILITY M=R, B= 1

R	1	2	3	4	5
			A=1/10		
0+	.90909091\01	.23290759\01	.65258831\02	.19074600\02	.57155254\03
			A=1/ 9		
0+	.10000000\00	.28000000\01	.85600000\02	.27280000\02	.89092000\03
			A=1/ 8		
0+	.11111111\00	.34293553\01	.11532795\01	.40395411\02	.14492806\02
			A=1/ 7		
0+	.12500000\00	.42968750\01	.16052246\01	.62389374\02	.24822801\02
			A=1/ 6		
0+	.14285714\00	.55393586\01	.23264116\01	.10150047\01	.45297314\02
			A=1/ 5		
0+	.16666667\00	.74074074\01	.35493827\01	.17632602\01	.89500616\02
			A=1/ 4		
0+	.20000000\00	.10400000\00	.57920000\01	.33344000\01	.19581440\01
5+	.11654205\01	.70035612\02	.42397497\02	.25814628\02	.15791205\02
			A=1/ 3		
0+	.25000000\00	.15625000\00	.10351562\00	.70556641\01	.48927307\01
5+	.34327507\01	.24290144\01	.17299838\01	.12384779\01	.89032793\02
			A= .40		
0+	.28571429\00	.19825073\00	.14470161\00	.10827364\00	.82253663\01
5+	.63136944\01	.48831916\01	.37988455\01	.29689887\01	.23291899\01
10+	.18330194\01	.14463930\01	.11439301\01	.90651819\02	.71963415\02
			A=1/ 2		
0+	.33333333\00	.25925926\00	.20987654\00	.17329675\00	.14484581\00
5+	.12208505\00	.10353925\00	.88231598\01	.75475226\01	.64766173\01
10+	.55722972\01	.48049954\01	.41513678\01	.35927118\01	.31138638\01
15+	.27024093\01	.23481012\01	.20424236\01	.17782578\01	.15496230\01
20+	.13514729\01	.11795332\01	.10301713\01	.90029148\02	.78724791\02
			A= .60		
0+	.37500000\00	.31640625\00	.27520752\00	.24302101\00	.21661802\00
5+	.19434049\00	.17519574\00	.15852955\00	.14388153\00	.13091194\00
10+	.11936089\00	.10902401\00	.99736974\01	.91365246\01	.83797054\01
15+	.76938380\01	.70709311\01	.65041317\01	.59875176\01	.55159373\01
20+	.50848834\01	.46903921\01	.43289619\01	.39974872\01	.36932038\01
25+	.34136434\01	.31565957\01	.29200761\01	.27022985\01	.25016522\01
30+	.23166813\01	.21460681\01	.19886174\01	.18432439\01	.17089605\01
35+	.15848683\01	.14701475\01	.13640502\01	.12658928\01	.11750500\01
40+	.10909494\01	.10130666\01	.94092074\02	.87407046\02	.81211050\02
			A= .65		
0+	.39393939\00	.34329521\00	.30702114\00	.27815289\00	.25402974\00
5+	.23329579\00	.21514486\00	.19904880\00	.18463772\00	.17163967\00
10+	.15984713\00	.14909711\00	.13925855\00	.13022406\00	.12190422\00
15+	.11422359\00	.10711777\00	.10053127\00	.94415862\01	.88729307\01
20+	.83434388\01	.78498113\01	.73891088\01	.69587005\01	.65562219\01
25+	.61795403\01	.58267257\01	.54960257\01	.51858453\01	.48947290\01
30+	.46213451\01	.43644733\01	.41229926\01	.38958715\01	.36821596\01
35+	.34809794\01	.32915198\01	.31130301\01	.29448145\01	.27862274\01
40+	.26366692\01	.24955822\01	.23624475\01	.22367816\01	.21181339\01
45+	.20060838\01	.19002386\01	.18002313\01	.17057187\01	.16163797\01
50+	.15319138\01	.14520394\01	.13764923\01	.13050252\01	.12374057\01
55+	.11734159\01	.11128509\01	.10555184\01	.10012377\01	.94983864\02
			A= .70		
0+	.41176471\00	.36902096\00	.33796150\00	.31288465\00	.29162573\00
5+	.27308855\00	.25662530\00	.24181407\00	.22836097\00	.21605094\00
10+	.20472060\00	.19424209\00	.18451291\00	.17544925\00	.16698148\00
15+	.15905088\00	.15160736\00	.14460776\00	.13801452\00	.13179471\00
20+	.12591924\00	.12036229\00	.11510075\00	.11011388\00	.10538297\00
25+	.10089107\00	.96622739\01	.92563916\01	.88701715\01	.85024305\01

C-INTEGRAL PROBABILITY M=R,B= 1

R	1	2	3	4	5
			A= .70		
30+	.81520798\01	.78181146\01	.74996053\01	.71956907\01	.69055706\01
35+	.66285009\01	.63637881\01	.61107847\01	.58688856\01	.56375244\01
40+	.54161702\01	.52043247\01	.50015200\01	.48073157\01	.46212974\01
45+	.44430746\01	.42722789\01	.41085624\01	.39515967\01	.38010709\01
50+	.36566912\01	.35181792\01	.33852711\01	.32577167\01	.31352790\01
55+	.30177326\01	.29048636\01	.27964688\01	.26923550\01	.25923383\01
60+	.24962439\01	.24039051\01	.23151634\01	.22298677\01	.21478738\01
65+	.20690445\01	.19932487\01	.19203613\01	.18502630\01	.17828398\01
70+	.17179830\01	.16555884\01	.15955566\01	.15377928\01	.14822060\01
75+	.14287093\01	.13772195\01	.13276572\01	.12799462\01	.12340135\01
80+	.11897894\01	.11472070\01	.11062023\01	.10667139\01	.10286829\01
85+	.99205305\02	.95677024\02	.92278266\02	.89004062\02	.85849644\02
			A= .75		
0+	.42857143\00	.39358601\00	.36788243\00	.34689992\00	.32891491\00
5+	.31305874\00	.29882054\00	.28586918\00	.27397508\00	.26297100\00
10+	.25273048\00	.24315492\00	.23416563\00	.22569847\00	.21770034\00
15+	.21012661\00	.20293928\00	.19610572\00	.18959756\00	.18338999\00
20+	.17746113\00	.17179156\00	.16636391\00	.16116261\00	.15617361\00
25+	.15138417\00	.14678270\00	.14235861\00	.13810220\00	.13400454\00
30+	.13005741\00	.12625320\00	.12258485\00	.11904581\00	.11562998\00
35+	.11233166\00	.10914553\00	.10606660\00	.10309019\00	.10021190\00
40+	.97427603\01	.94733386\01	.92125573\01	.89600685\01	.87155432\01
45+	.84786697\01	.82491525\01	.80267112\01	.78110793\01	.76020034\01
50+	.73992426\01	.72025669\01	.70117576\01	.68266057\01	.66469118\01
55+	.64724854\01	.63031442\01	.61387142\01	.59790284\01	.58239272\01
60+	.56732575\01	.55268724\01	.53846312\01	.52463988\01	.51120453\01
65+	.49814461\01	.48544815\01	.47310361\01	.46109991\01	.44942640\01
70+	.43807280\01	.42702923\01	.41628617\01	.40583443\01	.39566518\01
75+	.38576987\01	.37614027\01	.36676846\01	.35764675\01	.34876776\01
80+	.34012433\01	.33170956\01	.32351679\01	.31553956\01	.30777165\01
85+	.30020703\01	.29283987\01	.28566453\01	.27867557\01	.27186771\01
90+	.26523582\01	.25877498\01	.25248039\01	.24634741\01	.24037155\01
95+	.23454846\01	.22887392\01	.22334383\01	.21795424\01	.21270131\01
100+	.20758131\01	.20259063\01	.19772577\01	.19298332\01	.18835999\01
105+	.18385258\01	.17945799\01	.17517320\01	.17099529\01	.16692141\01
110+	.16294881\01	.15907482\01	.15529683\01	.15161232\01	.14801883\01
115+	.14451398\01	.14109546\01	.13776101\01	.13450846\01	.13133567\01
120+	.12824058\01	.12522119\01	.12227554\01	.11940173\01	.11659793\01
125+	.11386233\01	.11119319\01	.10858882\01	.10604757\01	.10356782\01
130+	.10114803\01	.98786669\02	.96482260\02	.94233366\02	.92038588\02
			A= .80		
0+	.44444444\00	.41700960\00	.39668750\00	.37996148\00	.36550691\00
5+	.35265839\00	.34102599\00	.33035783\00	.32047990\00	.31126592\00
10+	.30262071\00	.29447033\00	.28675599\00	.27942993\00	.27245273\00
15+	.26579137\00	.25941784\00	.25330814\00	.24744149\00	.24179975\00
20+	.23636696\00	.23112900\00	.22607328\00	.22118852\00	.21646458\00
25+	.21189227\00	.20746326\00	.20316993\00	.19900532\00	.19496305\00
30+	.19103722\00	.18722239\00	.18351354\00	.17990597\00	.17639533\00
35+	.17297758\00	.16964889\00	.16640573\00	.16324476\00	.16016284\00
40+	.15715701\00	.15422449\00	.15136266\00	.14856903\00	.14584124\00
45+	.14317706\00	.14057437\00	.13803116\00	.13554551\00	.13311560\00
50+	.13073969\00	.12841612\00	.12614330\00	.12391971\00	.12174391\00
55+	.11961451\00	.11753018\00	.11548963\00	.11349166\00	.11153507\00
60+	.10961874\00	.10774158\00	.10590255\00	.10410064\00	.10233487\00
65+	.10060433\00	.98908091\01	.97245299\01	.95615112\01	.94016717\01
70+	.92449332\01	.90912198\01	.89404585\01	.87925782\01	.86475105\01
75+	.85051890\01	.83655492\01	.82285290\01	.80940678\01	.79621072\01

C-INTEGRAL PROBABILITY M=R,B= 1

R	1	2	3	4	5
			A= .80		
80+	.78325903\01	.77054620\01	.75806688\01	.74581587\01	.73378813\01
85+	.72197876\01	.71038300\01	.69899621\01	.68781390\01	.67683169\01
90+	.66604533\01	.65545066\01	.64504365\01	.63482039\01	.62477706\01
95+	.61490992\01	.60521535\01	.59568982\01	.58632990\01	.57713221\01
100+	.56809350\01	.55921057\01	.55048032\01	.54189970\01	.53346576\01
105+	.52517560\01	.51702642\01	.50901545\01	.50114002\01	.49339749\01
110+	.48578531\01	.47830097\01	.47094203\01	.46370610\01	.45659085\01
115+	.44959399\01	.44271330\01	.43594660\01	.42929176\01	.42274669\01
120+	.41630937\01	.40997778\01	.40374999\01	.39762410\01	.39159822\01
125+	.38567055\01	.37983929\01	.37410269\01	.36845905\01	.36290669\01
130+	.35744397\01	.35206928\01	.34678105\01	.34157775\01	.33645786\01
135+	.33141990\01	.32646244\01	.32158405\01	.31678334\01	.31205895\01
140+	.30740956\01	.30283385\01	.29833054\01	.29389838\01	.28953614\01
145+	.28524261\01	.28101661\01	.27685698\01	.27276258\01	.26873230\01
150+	.26476504\01	.26085974\01	.25701534\01	.25323081\01	.24950513\01
155+	.24583733\01	.24222641\01	.23867143\01	.23517145\01	.23172555\01
160+	.22833283\01	.22499240\01	.22170340\01	.21846496\01	.21527625\01
165+	.21213646\01	.20904477\01	.20600039\01	.20300254\01	.20005046\01
170+	.19714341\01	.19428064\01	.19146143\01	.18868507\01	.18595087\01
175+	.18325814\01	.18060621\01	.17799442\01	.17542212\01	.17288867\01
180+	.17039345\01	.16793584\01	.16551524\01	.16313106\01	.16078271\01
185+	.15846962\01	.15619123\01	.15394698\01	.15173634\01	.14955876\01
190+	.14741373\01	.14530072\01	.14321924\01	.14116878\01	.13914885\01
195+	.13715898\01	.13519868\01	.13326750\01	.13136498\01	.12949067\01
200+	.12764413\01	.12582492\01	.12403261\01	.12226680\01	.12052705\01
205+	.11881298\01	.11712418\01	.11546025\01	.11382082\01	.11220550\01
210+	.11061393\01	.10904572\01	.10750053\01	.10597800\01	.10447778\01
215+	.10299953\01	.10154291\01	.10010758\01	.98693224\02	.97299520\02

C-INTEGRAL PROBABILITY M=R, B= 2

R	1	2	3	4	5
			A=1/10		
0+	.15151515\01	.14426288\02	.16355248\03	.20116687\04	.25954756\05
			A=1/ 9		
0+	.18181818\01	.20419370\02	.27189003\03	.39205381\04	.59242635\05
			A=1/ 8		
0+	.22222222\01	.29871056\02	.47358960\03	.81130260\04	.14547126\04
			A=1/ 7		
0+	.27777778\01	.45474680\02	.87242692\03	.18034412\03	.38961003\04
			A=1/ 6		
0+	.35714286\01	.72715470\02	.17207982\02	.43720338\03	.11586114\03
			A=1/ 5		
0+	.47619048\01	.12371388\01	.36954789\02	.11794773\02	.39161243\03
			A=1/ 4		
0+	.66666667\01	.22814815\01	.88441152\02	.36386681\02	.15514612\02
			A=1/ 3		
0+	.10000000\00	.46900000\01	.24375250\01	.13310913\01	.74898501\02
			A= .40		
0+	.12698413\00	.70941330\01	.43249842\01	.27506695\01	.17950609\01
5+	.11922624\01	.80220008\02	.54515652\02	.37343145\02	.25746972\02
			A=1/ 2		
0+	.16666667\00	.11226852\00	.80885899\01	.60216554\01	.45738861\01
5+	.35226073\01	.27408387\01	.21495292\01	.16965324\01	.13460199\01
10+	.10726245\01	.85796973\02	.68850169\02	.55407812\02	.44701949\02
			A= .60		
0+	.20454545\00	.15702533\00	.12672368\00	.10480129\00	.87988022\01
5+	.74647161\01	.63821794\01	.54896563\01	.47449537\01	.41177804\01
10+	.35856290\01	.31313251\01	.27414850\01	.24054990\01	.21148376\01
15+	.18625629\01	.16429747\01	.14513503\01	.12837475\01	.11368549\01
20+	.10078753\01	.89443369\02	.79450523\02	.70635637\02	.62849784\02
			A= .65		
0+	.22266140\00	.17990627\00	.15163170\00	.13047328\00	.11371186\00
5+	.99986562\01	.88500043\01	.78737358\01	.70343208\01	.63060452\01
10+	.56696156\01	.51101444\01	.46158836\01	.41773918\01	.37869657\01
15+	.34382400\01	.31258975\01	.28454557\01	.25931050\01	.23655847\01
20+	.21600871\01	.19741808\01	.18057503\01	.16529469\01	.15141483\01
25+	.13879266\01	.12730201\01	.11683113\01	.10728073\01	.98562422\02
			A= .70		
0+	.24019608\00	.20281334\00	.17734769\00	.15776502\00	.14184370\00
5+	.12847413\00	.11700681\00	.10702220\00	.98230383\01	.90420999\01
10+	.83435622\01	.77151417\01	.71470890\01	.66315166\01	.61619398\01
15+	.57329556\01	.53400095\01	.49792240\01	.46472694\01	.43412643\01
20+	.40586987\01	.37973728\01	.35553484\01	.33309098\01	.31225312\01
25+	.29288510\01	.27486492\01	.25808293\01	.24244030\01	.22784768\01
30+	.21422413\01	.20149611\01	.18959668\01	.17846478\01	.16804463\01
35+	.15828515\01	.14913951\01	.14056473\01	.13252124\01	.12497265\01
40+	.11788538\01	.11122843\01	.10497318\01	.99093126\02	.93563738\02
			A= .75		
0+	.25714286\00	.22557201\00	.20354086\00	.18622403\00	.17184973\00
5+	.15953535\00	.14876542\00	.13920749\00	.13063208\00	.12287275\00
10+	.11580414\00	.10932897\00	.10336992\00	.97864299\01	.92760453\01
15+	.88015199\01	.83592008\01	.79459651\01	.75591184\01	.71963177\01
20+	.68555103\01	.65348867\01	.62328427\01	.59479483\01	.56789231\01
25+	.54246159\01	.51839873\01	.49560956\01	.47400849\01	.45351749\01
30+	.43406522\01	.41558628\01	.39802060\01	.38131285\01	.36541197\01
35+	.35027077\01	.33584551\01	.32209564\01	.30898345\01	.29647385\01
40+	.28453414\01	.27313379\01	.26224429\01	.25183893\01	.24189273\01
45+	.23238224\01	.22328546\01	.21458172\01	.20625157\01	.19827673\01
50+	.19063995\01	.18332497\01	.17631647\01	.16959996\01	.16316177\01

C-INTEGRAL PROBABILITY M=R,B= 2

R	1	2	3	4	5
			A= .75		
55+	.15698896\01	.15106929\01	.14539117\01	.13994363\01	.13471627\01
60+	.12969922\01	.12488313\01	.12025911\01	.11581872\01	.11155395\01
65+	.10745717\01	.10352113\01	.99738947\02	.96104046\02	.92610180\02
			A= .80		
0+	.27350427\00	.24804532\00	.22992914\00	.21543800\00	.20320979\00
5+	.19256756\00	.18311685\00	.17460382\00	.16685356\00	.15973944\00
10+	.15316621\00	.14706002\00	.14136226\00	.13602539\00	.13101030\00
15+	.12628424\00	.12181954\00	.11759251\00	.11358270\00	.10977230\00
20+	.10614568\00	.10268906\00	.99390182\01	.96238095\01	.93222977\01
25+	.90335968\01	.87569049\01	.84914929\01	.82366961\01	.79919063\01
30+	.77565652\01	.75301591\01	.73122141\01	.71022919\01	.68999861\01
35+	.67049193\01	.65167402\01	.63351210\01	.61597557\01	.59903579\01
40+	.58266593\01	.56684077\01	.55153665\01	.53673127\01	.52240363\01
45+	.50853391\01	.49510339\01	.48209436\01	.46949007\01	.45727464\01
50+	.44543301\01	.43395088\01	.42281466\01	.41201145\01	.40152893\01
55+	.39135541\01	.38147971\01	.37189121\01	.36257973\01	.35353558\01
60+	.34474947\01	.33621255\01	.32791633\01	.31985268\01	.31201383\01
65+	.30439233\01	.29698101\01	.28977304\01	.28276181\01	.27594102\01
70+	.26930460\01	.26284671\01	.25656174\01	.25044430\01	.24448920\01
75+	.23869145\01	.23304623\01	.22754891\01	.22219504\01	.21698030\01
80+	.21190055\01	.20695178\01	.20213015\01	.19743192\01	.19285350\01
85+	.18839143\01	.18404233\01	.17980298\01	.17567025\01	.17164111\01
90+	.16771264\01	.16388200\01	.16014647\01	.15650338\01	.15295019\01
95+	.14948441\01	.14610365\01	.14280557\01	.13958793\01	.13644855\01
100+	.13338531\01	.13039616\01	.12747913\01	.12463228\01	.12185376\01
105+	.11914175\01	.11649450\01	.11391031\01	.11138753\01	.10892455\01
110+	.10651982\01	.10417183\01	.10187912\01	.99640255\02	.97453864\02

C-INTEGRAL PROBABILITY M=R, B= 3

R	1	2	3	4	5
			A=1/10		
0+	.34965035\02	.15305092\03	.86170494\05	.54538473\06	.36972971\07
			A=1/ 9		
0+	.45454545\02	.25081108\03	.17668790\04	.13947432\05	.11772389\06
			A=1/ 8		
0+	.60606061\02	.42965348\03	.38542367\04	.38592538\05	.41232084\06
			A=1/ 7		
0+	.83333333\02	.77615398\03	.90482466\04	.11718059\04	.16150207\05
			A=1/ 6		
0+	.11904762\01	.14957943\02	.23205971\03	.39755086\04	.72238955\05
			A=1/ 5		
0+	.17857143\01	.31243650\02	.66326438\03	.15427099\03	.37893708\04
			A=1/ 4		
0+	.28571429\01	.72328064\02	.21702173\02	.70596020\03	.24106411\03
			A=1/ 3		
0+	.50000000\01	.19172222\01	.84314985\02	.39595247\02	.19348639\02
			A= .40		
0+	.69264069\01	.33319147\01	.17980429\01	.10254704\01	.60488458\02
			A=1/ 2		
0+	.10000000\00	.61107778\01	.40734947\01	.28320945\01	.20204279\01
5+	.14672160\01	.10795750\01	.80249310\02	.60143362\02	.45380259\02
			A= .60		
0+	.13149351\00	.94765840\01	.72793481\01	.57673558\01	.46573192\01
5+	.38108645\01	.31489604\01	.26219670\01	.21966511\01	.18497246\01
10+	.15643016\01	.13278104\01	.11306930\01	.96555729\02	.82660576\02
			A= .65		
0+	.14718296\00	.11309019\00	.91675831\01	.76276845\01	.64491617\01
5+	.55137756\01	.47532840\01	.41242712\01	.35972353\01	.31511555\01
10+	.27705093\01	.24435117\01	.21610143\01	.19157864\01	.17020262\01
15+	.15150199\01	.13508959\01	.12064453\01	.10789866\01	.96626345\02
			A= .70		
0+	.16271347\00	.13207872\00	.11205094\00	.97124369\01	.85310471\01
5+	.75627155\01	.67504959\01	.60579348\01	.54600731\01	.49389676\01
10+	.44812343\01	.40765974\01	.37169830\01	.33959252\01	.31081639\01
15+	.28493626\01	.26159046\01	.24047446\01	.22132955\01	.20393435\01
20+	.18809811\01	.17365555\01	.16046263\01	.14839329\01	.13733668\01
25+	.12719497\01	.11788149\01	.10931924\01	.10143957\01	.94181125\02
			A= .75		
0+	.17802198\00	.15151845\00	.13361162\00	.11987159\00	.10869786\00
5+	.99299495\01	.91217290\01	.84156652\01	.77915287\01	.72347218\01
10+	.67343066\01	.62818414\01	.58706535\01	.54953641\01	.51515658\01
15+	.48355967\01	.45443774\01	.42752918\01	.40260971\01	.37948552\01
20+	.35798791\01	.33796917\01	.31929917\01	.30186273\01	.28555739\01
25+	.27029167\01	.25598352\01	.24255914\01	.22995189\01	.21810145\01
30+	.20695307\01	.19645691\01	.18656748\01	.17724321\01	.16844600\01
35+	.16014087\01	.15229565\01	.14488072\01	.13786873\01	.13123440\01
40+	.12495435\01	.11900691\01	.11337197\01	.10803085\01	.10296619\01
45+	.98161791\02	.93602594\02	.89274532\02	.85164472\02	.81260142\02
			A= .80		
0+	.19306184\00	.17122616\00	.15606803\00	.14415893\00	.13426050\00
5+	.12576103\00	.11830548\00	.11166608\00	.10568632\00	.10025321\00
10+	.95281949\01	.90707001\01	.86476454\01	.82548364\01	.78888265\01
15+	.75467433\01	.72261631\01	.69250189\01	.66415313\01	.63741559\01
20+	.61215423\01	.58825018\01	.56559820\01	.54410461\01	.52368559\01
25+	.50426585\01	.48577745\01	.46815888\01	.45135422\01	.43531251\01
30+	.41998716\01	.40533544\01	.39131810\01	.37789894\01	.36504456\01
35+	.35272405\01	.34090873\01	.32957198\01	.31868903\01	.30823677\01
40+	.29819365\01	.28853952\01	.27925551\01	.27032393\01	.26172818\01

C-INTEGRAL PROBABILITY M=R, B= 3

R	1	2	3	4	5
			A= .80		
45+	.25345266\01	.24548270\01	.23780447\01	.23040494\01	.22327181\01
50+	.21639348\01	.20975896\01	.20335785\01	.19718032\01	.19121704\01
55+	.18545917\01	.17989831\01	.17452648\01	.16933610\01	.16431996\01
60+	.15947121\01	.15478331\01	.15025002\01	.14586543\01	.14162386\01
65+	.13751992\01	.13354844\01	.12970450\01	.12598338\01	.12238059\01
70+	.11889182\01	.11551294\01	.11224000\01	.10906923\01	.10599701\01
75+	.10301986\01	.10013445\01	.97337587\02	.94626214\02	.91997385\02

C-INTEGRAL PROBABILITY M=R, B= 4

R	1	2	3	4	5
			A=1/10		
0+	.99900100\03	.23120486\04	.73805030\06	.27351443\07	.11065812\08
			A=1/ 9		
0+	.13986014\02	.43265974\04	.18261147\05	.89055136\07	.47290670\08
			A=1/ 8		
0+	.20202020\02	.85417741\04	.48643291\05	.31826890\06	.22604863\07
			A=1/ 7		
0+	.30303030\02	.17970687\03	.14132074\04	.12681353\05	.12305830\06
			A=1/ 6		
0+	.47619048\02	.40823920\03	.45530203\04	.57452635\05	.78026067\06
			A=1/ 5		
0+	.79365079\02	.10192748\02	.16623494\03	.30344146\04	.59249666\05
			A=1/ 4		
0+	.14285714\01	.28665556\02	.70783108\03	.19284839\03	.55744685\04
			A=1/ 3		
0+	.28571429\01	.94056685\02	.36504062\02	.15325832\02	.67484219\03
			A= .40		
0+	.42624043\01	.18328619\01	.90353380\02	.47556437\02	.26050001\02
			A=1/ 2		
0+	.66666667\01	.37898189\01	.23860678\01	.15780308\01	.10755716\01
5+	.74849937\02	.52896877\02	.37832366\02	.27319599\02	.19885324\02
			A= .60		
0+	.92818946\01	.63853855\01	.47309290\01	.36330813\01	.28522069\01
5+	.22736212\01	.18330948\01	.14910611\01	.12215387\01	.10066691\01
10+	.83374918\02	.69350020\02	.57899969\02	.48499349\02	.40743648\02
			A= .65		
0+	.10629880\00	.78716736\01	.62023087\01	.50359578\01	.41652997\01
5+	.34896580\01	.29516740\01	.25153217\01	.21564340\01	.18580132\01
10+	.16076703\01	.13961212\01	.12162529\01	.10625158\01	.93051537\02
			A= .70		
0+	.11939414\00	.94551673\01	.78464193\01	.66741449\01	.57641354\01
5+	.50311330\01	.44261038\01	.39179095\01	.34854010\01	.31134972\01
10+	.27910413\01	.25095387\01	.22623706\01	.20442817\01	.18510328\01
15+	.16791588\01	.15257947\01	.13885485\01	.12654059\01	.11546577\01
20+	.10548440\01	.96471052\02	.88317359\02	.80929271\02	.74224799\02
			A= .75		
0+	.13351648\00	.11115782\00	.96393587\01	.85258821\01	.76336309\01
5+	.68929927\01	.62637514\01	.57202293\01	.52448752\01	.48250831\01
10+	.44514525\01	.41167629\01	.38153350\01	.35426136\01	.32948850\01
15+	.30690793\01	.28626283\01	.26733616\01	.24994281\01	.23392367\01
20+	.21914101\01	.20547481\01	.19281998\01	.18108394\01	.17018481\01
25+	.16004986\01	.15061422\01	.14181985\01	.13361465\01	.12595167\01
30+	.11876855\01	.11208694\01	.10581202\01	.99932135\02	.94418439\02

C-INTEGRAL PROBABILITY M=R, B= 4

R	1	2	3	4	5
			A= .80		
5+	.12835127\00	.11556340\00	.10564428\00	.97488753\01	.90553197\01
10+	.84523018\01	.79196922\01	.74436961\01	.70143752\01	.66242922\01
15+	.62677117\01	.59401028\01	.56378139\01	.53578524\01	.50977307\01
20+	.48553555\01	.46289464\01	.44169752\01	.42181190\01	.40312246\01
25+	.38552795\01	.36893903\01	.35327636\01	.33846920\01	.32445418\01
30+	.31117429\01	.29857807\01	.28661888\01	.27525435\01	.26444585\01
35+	.25415807\01	.24435867\01	.23501794\01	.22610850\01	.21760513\01
40+	.20948451\01	.20172503\01	.19430667\01	.18721079\01	.18042009\01
45+	.17391842\01	.16769070\01	.16172287\01	.15600175\01	.15051499\01
50+	.14525104\01	.14019901\01	.13534871\01	.13069053\01	.12621542\01
55+	.12191485\01	.11778080\01	.11380567\01	.10998228\01	.10630386\01

C-INTEGRAL PROBABILITY M=R, B= 5

R	1	2	3	4	5
			A=1/10		
0+	.33300033\03	.45003243\05	.89385311\07	.21224245\08	.55986023\10
			A=1/ 9		
0+	.49950050\03	.95068657\05	.26210406\06	.85842788\08	.31124324\09
			A=1/ 8		
0+	.77700078\03	.21348255\04	.83531991\06	.38537858\07	.19601343\08
			A=1/ 7		
0+	.12626263\02	.51521448\04	.29339191\05	.19523277\06	.14250848\07
			A=1/ 6		
0+	.21645022\02	.13555208\03	.11558985\04	.11392201\05	.12240197\06
			A=1/ 5		
0+	.39682540\02	.39623760\03	.52254378\04	.78557795\05	.12774471\05
			A=1/ 4		
0+	.79365079\02	.13207243\02	.27917349\03	.66087270\04	.16749177\04
			A=1/ 3		
0+	.17857143\01	.52066938\02	.18309927\02	.70403310\03	.28578275\03
			A= .40		
0+	.28416028\01	.11182475\01	.51347975\02	.25384728\02	.13127059\02
			A=1/ 2		
0+	.47619048\01	.25577172\01	.15401163\01	.97973415\02	.64458201\02
			A= .60		
0+	.69614209\01	.46181327\01	.33267533\01	.24936148\01	.19153558\01
5+	.14963007\01	.11837349\01	.94570196\02	.76154689\02	.61729227\02
			A= .65		
0+	.81287319\01	.58486822\01	.45076032\01	.35912380\01	.29202367\01
5+	.24085060\01	.20075307\01	.16871668\01	.14274098\01	.12143399\01
10+	.10379196\01	.89070639\02	.76705817\02	.66261918\02	.57397563\02
			A= .70		
0+	.93250996\01	.71904836\01	.58656374\01	.49168583\01	.41912722\01
5+	.36146264\01	.31445142\01	.27541901\01	.24256133\01	.21460076\01
10+	.19059870\01	.16984547\01	.15179198\01	.13600526\01	.12213853\01
15+	.10991026\01	.99089342\02	.89484121\02	.80934329\02	.73304909\02
			A= .75		
0+	.10540775\00	.86261857\01	.73837870\01	.64591538\01	.57265948\01
5+	.51246704\01	.46180361\01	.41842164\01	.38079108\01	.34781750\01
10+	.31868810\01	.29278120\01	.26960985\01	.24878518\01	.22999146\01
15+	.21296886\01	.19750093\01	.18340555\01	.17052808\01	.15873616\01
20+	.14791571\01	.13796778\01	.12880608\01	.12035495\01	.11254773\01
25+	.10532550\01	.98635926\02	.92432371\02	.86673156\02	.81320909\02

C-INTEGRAL PROBABILITY M=R,B= 5

R	1	2	3	4	5
			A= .80		
5+	.10138983\00	.90423255\01	.81999771\01	.75131431\01	.69333793\01
10+	.64327191\01	.59933080\01	.56029368\01	.52528284\01	.49364290\01
15+	.46486958\01	.43856537\01	.41441066\01	.39214409\01	.37154895\01
20+	.35244328\01	.33467272\01	.31810504\01	.30262611\01	.28813664\01
25+	.27454971\01	.26178875\01	.24978600\01	.23848116\01	.22782033\01
30+	.21775518\01	.20824217\01	.19924195\01	.19071884\01	.18264043\01
35+	.17497715\01	.16770196\01	.16079012\01	.15421888\01	.14796733\01
40+	.14201619\01	.13634762\01	.13094516\01	.12579353\01	.12087854\01
45+	.11618702\01	.11170669\01	.10742612\01	.10333463\01	.99422252\02

C-INTEGRAL PROBABILITY M=R, B= 6

R	1	2	3	4	5
			A=1/10		
0+	.12487512\03	.10614137\05	.14050512\07	.22840706\09	.41915402\11
			A=1/ 9		
0+	.19980020\03	.25066529\05	.48097774\07	.11243672\08	.29540590\10
			A=1/ 8		
0+	.33300033\03	.63324030\05	.18032784\06	.61981009\08	.23818759\09
			A=1/ 7		
0+	.58275058\03	.17313072\04	.75131218\06	.38918388\07	.22399800\08
			A=1/ 6		
0+	.10822511\02	.52004171\04	.35426421\05	.28424014\06	.25147462\07
			A=1/ 5		
0+	.21645022\02	.17505321\03	.19349752\04	.24777807\05	.34656041\06
			A=1/ 4		
0+	.47619048\02	.67831814\03	.12612947\03	.26604577\04	.60552077\05
			A=1/ 3		
0+	.11904762\01	.31409481\02	.10188981\02	.36468835\03	.13856780\03
			A= .40		
0+	.20058373\01	.73401980\02	.31809052\02	.14944678\02	.73760980\03
			A=1/ 2		
0+	.35714286\01	.18315097\01	.10635773\01	.65560473\02	.41917520\02
			A= .60		
0+	.54480686\01	.35088995\01	.24706940\01	.18159606\01	.13704289\01
5+	.10532899\01	.82063160\02	.64618997\02	.51321209\02	.41050694\02
			A= .65		
0+	.64698071\01	.45474152\01	.34424285\01	.27008281\01	.21661743\01
5+	.17641071\01	.14531141\01	.12076390\01	.10108759\01	.85123586\02
			A= .70		
0+	.75318112\01	.57026049\01	.45876648\01	.38003055\01	.32053848\01
5+	.27376830\01	.23601761\01	.20496516\01	.17905442\01	.15718961\01
10+	.13857044\01	.12259540\01	.10880181\01	.96827005\02	.86382163\02
			A= .75		
0+	.86242705\01	.69601103\01	.58951670\01	.51110148\01	.44954157\01
5+	.39937281\01	.35746321\01	.32182819\01	.29112153\01	.26438370\01
10+	.24090463\01	.22014319\01	.20167706\01	.18517020\01	.17035080\01
15+	.15699596\01	.14492070\01	.13396993\01	.12401247\01	.11493642\01
20+	.10664571\01	.99057317\02	.92099082\02	.85707976\02	.79828684\02
			A= .80		
0+	.97386859\01	.83052112\01	.73499331\01	.66219219\01	.60322840\01
5+	.55375384\01	.51126353\01	.47416132\01	.44135780\01	.41207135\01
10+	.38571941\01	.36185463\01	.34012510\01	.32024839\01	.30199406\01
15+	.28517140\01	.26962060\01	.25520640\01	.24181316\01	.22934128\01
20+	.21770427\01	.20682661\01	.19664189\01	.18709148\01	.17812330\01

C-INTEGRAL PROBABILITY M=R, B= 6

R	1	2	3	4	5
			A= .80		
25+	.16969092\01	.16175276\01	.15427148\01	.14721337\01	.14054796\01
30+	.13424762\01	.12828717\01	.12264368\01	.11729617\01	.11222543\01
35+	.10741380\01	.10284506\01	.98504229\02	.94377496\02	.90452083\02

C-INTEGRAL PROBABILITY M=R, B= 7

R	1	2	3	4	5
			A=1/10		
0+	.51419169\04	.29116000\06	.27080040\08	.31704101\10	.42526989\12
			A=1/ 9		
0+	.87412587\04	.76231533\06	.10684879\07	.18671849\09	.37184468\11
			A=1/ 8		
0+	.15540016\03	.21463906\05	.46465335\07	.12401893\08	.37486970\10
			A=1/ 7		
0+	.29137529\03	.65787161\05	.22605812\06	.94521579\08	.44426666\09
			A=1/ 6		
0+	.58275058\03	.22294640\04	.12535271\05	.84424696\07	.63344315\08
			A=1/ 5		
0+	.12626263\02	.85261170\04	.81110596\05	.90671813\06	.11167735\06
			A=1/ 4		
0+	.30303030\02	.37818620\03	.63097308\04	.12076850\04	.25113949\05
			A=1/ 3		
0+	.83333333\02	.20206052\02	.61245441\03	.20643723\03	.74223119\04
			A= .40		
0+	.14779854\01	.50874979\02	.21001833\02	.94557776\03	.44888641\03
			A=1/ 2		
0+	.27777778\01	.13702195\01	.77192062\02	.46346499\02	.28934625\02
			A= .60		
0+	.44003631\01	.27647825\01	.19099565\01	.13810011\01	.10269440\01
5+	.77864915\02	.59899173\02	.46602162\02	.36589408\02	.28946181\02
			A= .65		
0+	.53040761\01	.36557499\01	.27262430\01	.21116788\01	.16743129\01
5+	.13492271\01	.11004735\01	.90609689\02	.75177226\02	.62769764\02
			A= .70		
0+	.62552330\01	.46642534\01	.37089487\01	.30421328\01	.25433279\01
5+	.21547140\01	.18436395\01	.15897393\01	.13794275\01	.12031888\01
10+	.10541094\01	.92701895\02	.81796067\02	.72384849\02	.64223713\02
			A= .75		
0+	.72443872\01	.57787485\01	.48515823\01	.41749063\01	.36477130\01
5+	.32209978\01	.28667610\01	.25673161\01	.23107028\01	.20884229\01
10+	.18942076\01	.17232944\01	.15719785\01	.14373209\01	.13169522\01
15+	.12089356\01	.11116686\01	.10238120\01	.94423663\02	.87198233\02
			A= .80		
0+	.82631275\01	.69866458\01	.61432345\01	.55046502\01	.49903175\01
5+	.45609052\01	.41937942\01	.38745961\01	.35935072\01	.33435054\01
10+	.31193663\01	.29170847\01	.27335155\01	.25661392\01	.24129036\01
15+	.22721132\01	.21423503\01	.20224165\01	.19112897\01	.18080907\01
20+	.17120579\01	.16225270\01	.15389153\01	.14607092\01	.13874532\01
25+	.13187420\01	.12542135\01	.11935427\01	.11364373\01	.10826332\01
30+	.10318912\01	.98399411\02	.93874431\02	.89596133\02	.85548014\02

C-INTEGRAL PROBABILITY M=R, B= 8

R	1	2	3	4	5
			A=1/10		
0+	.22852964\04	.90251617\07	.61489310\09	.53954057\11	.54992596\13
			A=1/ 9		
0+	.41135335\04	.26008477\06	.27660742\08	.37465001\10	.58574748\12
			A=1/ 8		
0+	.77700078\04	.80967890\06	.13785919\07	.29508681\09	.72378511\11
			A=1/ 7		
0+	.15540016\03	.27574119\05	.77289816\07	.26826346\08	.10578056\09
			A=1/ 6		
0+	.33300033\03	.10438202\04	.49671320\06	.28748019\07	.18708651\08
			A=1/ 5		
0+	.77700078\03	.44846313\04	.37465719\05	.37251805\06	.41128382\07
			A=1/ 4		
0+	.20202020\02	.22484056\03	.34166578\04	.60157948\05	.11578593\05
			A=1/ 3		
0+	.60606061\02	.13663894\02	.39071592\03	.12509810\03	.42905276\04
			A= .40		
0+	.11260841\01	.36775643\02	.14562617\02	.63221589\03	.29032353\03
			A=1/ 2		
0+	.22222222\01	.10602538\01	.58198812\02	.34165922\02	.20900729\02
			A= .60		
0+	.36416798\01	.22400793\01	.15224308\01	.10854907\01	.79710849\02
			A= .65		
0+	.44485799\01	.30151356\01	.22198509\01	.17006970\01	.13352905\01
5+	.10663823\01	.86249668\02	.70454339\02	.58015174\02	.48091382\02
			A= .70		
0+	.53074705\01	.39061482\01	.30753685\01	.25011979\01	.20753576\01
5+	.17461347\01	.14844618\01	.12722921\01	.10976407\01	.95215142\02
			A= .75		
0+	.62094747\01	.49040749\01	.40862904\01	.34939317\01	.30354162\01
5+	.26664453\01	.23617790\01	.21055184\01	.18869416\01	.16984517\01
10+	.15344602\01	.13907333\01	.12639866\01	.11516222\01	.10515512\01
15+	.96207068\02	.88177601\02	.80949635\02	.74424702\02	.68519327\02
			A= .80		
0+	.71464886\01	.59983277\01	.52451580\01	.46780478\01	.42234464\01
5+	.38455136\01	.35236701\01	.32448447\01	.30001447\01	.27832101\01
10+	.25893165\01	.24148475\01	.22569679\01	.21134099\01	.19823295\01
15+	.18622054\01	.17517676\01	.16499446\01	.15558235\01	.14686207\01
20+	.13876583\01	.13123461\01	.12421671\01	.11766664\01	.11154412\01
25+	.10581337\01	.10044246\01	.95402815\02	.90668725\02	.86217029\02

C-INTEGRAL PROBABILITY M=R,B= 9

R	1	2	3	4	5
			A=1/10		
0+	.10825088\04	.30949876\07	.15976270\09	.10849151\11	.86683322\14
			A=1/ 9		
0+	.20567668\04	.97551802\07	.81166903\09	.87714422\11	.11072557\12
			A=1/ 8		
0+	.41135335\04	.33345911\06	.45887162\08	.80818149\10	.16488830\11
			A=1/ 7		
0+	.87412587\04	.12521588\05	.29315010\07	.86353917\09	.29181177\10
			A=1/ 6		
0+	.19980020\03	.52499947\05	.21567617\06	.10926670\07	.62774009\09
			A=1/ 5		
0+	.49950050\03	.25101843\04	.18709778\05	.16789510\06	.16848070\07
			A=1/ 4		
0+	.13986014\02	.14075607\03	.19709604\04	.32264821\05	.58054082\06
			A=1/ 3		
0+	.45454545\02	.96155561\03	.26132395\03	.80007876\04	.26338235\04
			A= .40		
0+	.88128321\02	.27488331\02	.10497248\02	.44151442\03	.19698630\03
			A=1/ 2		
0+	.18181818\01	.84263690\02	.45219254\02	.26032140\02	.15646096\02
			A= .60		
0+	.30726673\01	.18554633\01	.12432394\01	.87569358\02	.63605723\02
			A= .65		
0+	.37991522\01	.25376418\01	.18475501\01	.14020288\01	.10914430\01
5+	.86484631\02	.69440549\02	.56334342\02	.46085397\02	.37963608\02
			A= .70		
0+	.45804197\01	.33329016\01	.26014089\01	.21001923\01	.17312133\01
5+	.14478520\01	.12240124\01	.10435583\01	.89581702\02	.77337693\02
			A= .75		
0+	.54082522\01	.42343750\01	.35051510\01	.29803860\01	.25764746\01
5+	.22530845\01	.19872931\01	.17646971\01	.15756073\01	.14131757\01
10+	.12723772\01	.11494144\01	.10413487\01	.94586132\02	.86109265\02
			A= .80		
0+	.62749656\01	.52332866\01	.45541996\01	.40453168\01	.36390685\01
5+	.33025789\01	.30169965\01	.27703640\01	.25545574\01	.23637747\01
10+	.21937115\01	.20410781\01	.19032987\01	.17783162\01	.16644597\01
15+	.15603535\01	.14648504\01	.13769836\01	.12959312\01	.12209883\01
20+	.11515460\01	.10870751\01	.10271127\01	.97125172\02	.91913268\02

C-INTEGRAL PROBABILITY M=R, B=10

R	1	2	3	4	5
			A=1/10		
0+	.54125441\05	.11555196\07	.46464776\10	.25071173\12	.16108171\14
			A=1/ 9		
0+	.10825088\04	.39614680\07	.26441339\09	.23344599\11	.24343388\13
			A=1/ 8		
0+	.22852964\04	.14778660\06	.16805774\08	.24868164\10	.43059148\12
			A=1/ 7		
0+	.51419169\04	.60783898\06	.12116230\07	.30838791\09	.90855673\11
			A=1/ 6		
0+	.12487512\03	.28021043\05	.10098341\06	.45453650\08	.23381058\09
			A=1/ 5		
0+	.33300033\03	.14790120\04	.99611707\06	.81621569\07	.75271985\08
			A=1/ 4		
0+	.99900100\03	.91933707\04	.11973169\04	.18377502\05	.31158121\06
			A=1/ 3		
0+	.34965035\02	.69905544\03	.18163150\03	.53455454\04	.16972947\04
			A= .40		
0+	.70502657\02	.21114194\02	.78088180\03	.31938940\03	.13892429\03
			A=1/ 2		
0+	.15151515\01	.68439083\02	.36001001\02	.20370894\02	.12054403\02
			A= .60		
0+	.26337148\01	.15646477\01	.10353068\01	.72143506\02	.51898502\02
			A= .65		
0+	.32925986\01	.21711072\01	.15651753\01	.11777878\01	.91000244\02
			A= .70		
0+	.40078672\01	.28871186\01	.22362871\01	.17937083\01	.14700245\01
5+	.12229089\01	.10287570\01	.87302654\02	.74613406\02	.64144834\02
			A= .75		
0+	.47719872\01	.37076847\01	.30513934\01	.25818243\01	.22221879\01
5+	.19355275\01	.17008869\01	.15051279\01	.13394322\01	.11975813\01
10+	.10750224\01	.96832262\02	.87483144\02	.79246209\02	.71954467\02
			A= .80		
0+	.55777472\01	.46256453\01	.40083192\01	.35476636\01	.31812478\01
5+	.28787369\01	.26227578\01	.24023045\01	.22099085\01	.20402426\01
10+	.18893598\01	.17542471\01	.16325487\01	.15223853\01	.14222329\01
15+	.13308377\01	.12471559\01	.11703091\01	.10995511\01	.10342432\01
20+	.97383409\02	.91784534\02	.86585897\02	.81750799\02	.77246857\02

TABLE N (A < 1)

This table gives the value of the expected waiting time in a multinomial problem in which we stop as soon as the frequency in any cell reaches r; here one cell has probability p and the remaining B cells each have probability q (so that $Bq+p = 1$) and $q = pA$ with $A < 1$.

The formula (cf. (5.63) in the text) whose values are presented here is

$$E(T|GLF) = (1 + BA) \sum_{\alpha=1}^{R} D_A^{(B)}(R, \alpha)$$

for

$$A^{-1} = 3(1)10 \quad \text{and} \quad A = .40(.10).60(.05).80,$$
$$B = 1(1)10$$

The range of R is from 1 by ones to selected upper limits depending on the value of A.

Note: The symbols M, B, R and A which are all capital here in the table correspond to the same lower case symbols in the text.

EXPECTATION E(T/GLF)=(1+BA)SUM(D(R,M)),B= 1

R	1	2	3	4	5
		A=1/10			
0+	1.00000	2.16529	3.28891	4.39651	5.49890
		A=1/ 9			
0+	1.00000	2.18000	3.31860	4.43938	5.55382
		A=1/ 8			
0+	1.00000	2.19753	3.35482	4.49237	5.62212
		A=1/ 7			
0+	1.00000	2.21875	3.39990	4.55937	5.70923
		A=1/ 6			
0+	1.00000	2.24490	3.45731	4.64647	5.82382
		A=1/ 5			
0+	1.00000	2.27778	3.53241	4.76346	5.98037
5+	7.18947	8.39435	9.59696	10.79837	11.99912
		A=1/ 4			
0+	1.00000	2.32000	3.63360	4.92672	6.20428
5+	7.47155	8.73232	9.98900	11.24315	12.49574
		A=1/ 3			
0+	1.00000	2.37500	3.77344	5.16309	6.54037
5+	7.90685	9.26481	10.61631	11.96301	13.30616
10+	14.64670	15.98532	17.32254	18.65872	19.99415
		A=0.40			
0+	1.00000	2.40816	3.86214	5.31952	6.77082
5+	8.21420	9.64999	11.07916	12.50279	13.92184
10+	15.33718	16.74951	18.15942	19.56738	20.97377
15+	22.37890	23.78303	25.18634	26.58900	27.99114
		A=1/ 2			
0+	1.00000	2.44444	3.96296	5.50343	7.04999
5+	8.59656	10.14059	11.68109	13.21777	14.75066
10+	16.27999	17.80602	19.32908	20.84945	22.36742
15+	23.88327	25.39724	26.90953	28.42036	29.92988
20+	31.43826	32.94564	34.45213	35.95784	37.46287
25+	38.96729	40.47119	41.97461	43.47763	44.98028
		A=0.60			
0+	1.00000	2.46875	4.03271	5.63445	7.25428
5+	8.88336	10.51707	12.15284	13.78913	15.42501
10+	17.05992	18.69352	20.32561	21.95609	23.58493
15+	25.21213	26.83772	28.46175	30.08429	31.70540
20+	33.32516	34.94363	36.56089	38.17701	39.79206
25+	41.40611	43.01922	44.63144	46.24285	47.85348
30+	49.46339	51.07263	52.68125	54.28928	55.89677
35+	57.50375	59.11025	60.71631	62.32196	63.92723
40+	65.53214	67.13671	68.74098	70.34495	71.94866
45+	73.55211	75.15533	76.75833	78.36113	79.96374
		A=0.65			
0+	1.00000	2.47750	4.05827	5.68321	7.33146
5+	8.99324	10.66325	12.33833	14.01650	15.69644
10+	17.37728	19.05839	20.73935	22.41986	24.09973
15+	25.77879	27.45697	29.13418	30.81041	32.48562
20+	34.15983	35.83303	37.50524	39.17649	40.84679
25+	42.51617	44.18468	45.85233	47.51916	49.18519
30+	50.85048	52.51503	54.17890	55.84210	57.50466
35+	59.16662	60.82800	62.48883	64.14913	65.80894
40+	67.46826	69.12713	70.78557	72.44359	74.10122
45+	75.75848	77.41538	79.07195	80.72819	82.38413
50+	84.03978	85.69515	87.35025	89.00511	90.65973
55+	92.31412	93.96830	95.62227	97.27605	98.92964
60+	100.58306	102.23631	103.88940	105.54234	107.19513

EXPECTATION E(T/GLF)=(1+BA)SUM(D(R,M)),B= 1

R	1	2	3	4	5
		A=0.70			
0+	1.00000	2.48443	4.07865	5.72241	7.39394
5+	9.08282	10.78320	12.49152	14.20543	15.92333
10+	17.64407	19.36683	21.09100	22.81610	24.54179
15+	26.26779	27.99389	29.71992	31.44575	33.17128
20+	34.89643	36.62115	38.34537	40.06908	41.79224
25+	43.51484	45.23686	46.95830	48.67916	50.39943
30+	52.11911	53.83823	55.55677	57.27475	58.99218
35+	60.70907	62.42544	64.14129	65.85663	67.57148
40+	69.28586	70.99976	72.71322	74.42623	76.13882
45+	77.85099	79.56276	81.27414	82.98514	84.69578
50+	86.40606	88.11599	89.82559	91.53487	93.24384
55+	94.95250	96.66087	98.36896	100.07678	101.78433
60+	103.49163	105.19868	106.90549	108.61207	110.31842
65+	112.02456	113.73049	115.43622	117.14176	118.84711
70+	120.55228	122.25727	123.96209	125.66674	127.37124
75+	129.07559	130.77979	132.48384	134.18776	135.89155
80+	137.59520	139.29874	141.00215	142.70544	144.40863
85+	146.11171	147.81468	149.51755	151.22032	152.92300
90+	154.62559	156.32809	158.03051	159.73284	161.43510
		A=0.75			
0+	1.00000	2.48980	4.09454	5.75315	7.44322
5+	9.15386	10.87883	12.61427	14.35757	16.10691
10+	17.86097	19.61876	21.37950	23.14262	24.90764
15+	26.67418	28.44194	30.21066	31.98015	33.75022
20+	35.52073	37.29157	39.06263	40.83382	42.60508
25+	44.37633	46.14754	47.91864	49.68961	51.46041
30+	53.23102	55.00141	56.77156	58.54146	60.31109
35+	62.08044	63.84950	65.61827	67.38673	69.15489
40+	70.92275	72.69029	74.45752	76.22444	77.99104
45+	79.75734	81.52333	83.28901	85.05439	86.81946
50+	88.58424	90.34872	92.11290	93.87681	95.64042
55+	97.40376	99.16682	100.92961	102.69213	104.45439
60+	106.21639	107.97813	109.73963	111.50088	113.26189
65+	115.02267	116.78321	118.54353	120.30362	122.06349
70+	123.82315	125.58260	127.34184	129.10088	130.85973
75+	132.61838	134.37684	136.13511	137.89320	139.65112
80+	141.40886	143.16643	144.92383	146.68107	148.43815
85+	150.19507	151.95184	153.70845	155.46493	157.22125
90+	158.97744	160.73349	162.48941	164.24519	166.00085
95+	167.75638	169.51178	171.26707	173.02224	174.77729
100+	176.53223	178.28706	180.04179	181.79640	183.55092
105+	185.30533	187.05965	188.81387	190.56799	192.32203
110+	194.07597	195.82983	197.58360	199.33728	201.09089
115+	202.84441	204.59785	206.35122	208.10452	209.85773
120+	211.61088	213.36396	215.11697	216.86991	218.62279
125+	220.37560	222.12835	223.88104	225.63366	227.38623
130+	229.13875	230.89120	232.64361	234.39595	236.14825
135+	237.90050	239.65269	241.40484	243.15694	244.90899
		A=0.80			
0+	1.00000	2.49383	4.10654	5.77645	7.48075
5+	9.20817	10.95225	12.70888	14.47530	16.24952
10+	18.03008	19.81589	21.60607	23.39994	25.19696
15+	26.99666	28.79867	30.60269	32.40844	34.21569
20+	36.02427	37.83399	39.64472	41.45633	43.26871
25+	45.08176	46.89540	48.70954	50.52413	52.33911
30+	54.15441	55.96999	57.78581	59.60183	61.41802
35+	63.23434	65.05077	66.86727	68.68383	70.50043

EXPECTATION E(T/GLF)=(1+BA)SUM(D(R,M)),B= 1

R	1	2	3	4	5
			A=0.80		
40+	72.31705	74.13367	75.95027	77.76684	79.58336
45+	81.39984	83.21625	85.03259	86.84884	88.66500
50+	90.48107	92.29703	94.11289	95.92862	97.74425
55+	99.55974	101.37511	103.19035	105.00546	106.82043
60+	108.63527	110.44996	112.26452	114.07893	115.89319
65+	117.70732	119.52129	121.33512	123.14881	124.96234
70+	126.77573	128.58897	130.40207	132.21502	134.02782
75+	135.84047	137.65298	139.46535	141.27757	143.08965
80+	144.90158	146.71337	148.52502	150.33653	152.14790
85+	153.95913	155.77023	157.58119	159.39201	161.20270
90+	163.01326	164.82368	166.63398	168.44414	170.25418
95+	172.06409	173.87387	175.68353	177.49307	179.30248
100+	181.11178	182.92095	184.73001	186.53895	188.34778
105+	190.15649	191.96509	193.77357	195.58195	197.39021
110+	199.19837	201.00642	202.81437	204.62221	206.42995
115+	208.23759	210.04512	211.85256	213.65990	215.46714
120+	217.27429	219.08134	220.88830	222.69517	224.50194
125+	226.30863	228.11523	229.92173	231.72816	233.53449
130+	235.34075	237.14691	238.95300	240.75901	242.56493
135+	244.37078	246.17654	247.98223	249.78785	251.59339
140+	253.39885	255.20424	257.00956	258.81481	260.61999
145+	262.42509	264.23013	266.03510	267.84001	269.64485
150+	271.44962	273.25433	273.25897	275.06356	276.86808
155+	278.67254	280.47694	280.48128	282.28556	284.08979
160+	285.89396	287.69807	287.70213	289.50613	291.31007
165+	293.11397	293.11781	294.92160	296.72534	296.72903
170+	298.53267	300.33626	302.13980	302.14329	303.94674
175+	305.75014	305.75349	307.55680	309.36006	311.16328
180+	311.16646	312.96959	314.77268	314.77573	316.57874
185+	318.38171	318.38464	320.18753	321.99037	321.99318
190+	323.79596	325.59869	325.60139	327.40405	329.20668
195+	329.20927	331.01182	332.81434	332.81683	334.61929
200+	336.42171	336.42409	338.22645	340.02877	340.03106
205+	341.83333	341.83556	343.63776	345.43993	345.44207
210+	347.24418	349.04627	349.04832	350.85035	350.85235
215+	352.65433	354.45627	354.45819	356.26009	356.26196
220+	358.06380	359.86562	359.86742	361.66919	361.67094
225+	363.47266	365.27436	365.27604	367.07769	367.07933

EXPECTATION E(T/GLF)=(1+BA)SUM(D(R,M)),B= 2

R	1	2	3	4	5
		A=1/10			
0+	1.00000	2.32639	3.57608	4.79242	5.99761
		A=1/ 9			
0+	1.00000	2.35462	3.63470	4.87781	6.10730
		A=1/ 8			
0+	1.00000	2.38800	3.70596	4.98318	6.24363
		A=1/ 7			
0+	1.00000	2.42798	3.79415	5.11606	6.41728
		A=1/ 6			
0+	1.00000	2.47656	3.90555	5.28800	6.64515
5+	7.98982	9.32852	10.66439	11.99892	13.33282
		A=1/ 5			
0+	1.00000	2.53644	4.04941	5.51715	6.95503
5+	8.37572	9.78692	11.19295	12.59620	13.99795
		A=1/ 4			
0+	1.00000	2.61111	4.23920	5.83226	7.39388
5+	8.93336	10.45830	11.97395	13.48373	14.98984
		A=1/ 3			
0+	1.00000	2.70400	4.49184	6.27521	8.03780
5+	9.77916	11.50257	13.21180	14.91017	16.60034
10+	18.28440	19.96390	21.64004	23.31370	24.98551
		A=0.40			
0+	1.00000	2.75720	4.64494	6.55695	8.46461
5+	10.35957	12.24022	14.10740	15.96271	17.80794
10+	19.64475	21.47461	23.29879	25.11834	26.93413
15+	28.74688	30.55716	32.36545	34.17213	35.97752
		A=1/ 2			
0+	1.00000	2.81250	4.81055	6.87305	8.95961
5+	11.05324	13.14602	15.23418	17.31600	19.39084
10+	21.45863	23.51965	25.57431	27.62311	29.66655
15+	31.70516	33.73942	35.76979	37.79667	39.82045
20+	41.84148	43.86007	45.87649	47.89099	49.90379
25+	51.91509	53.92507	55.93388	57.94164	59.94850
30+	61.95455	63.95989	65.96460	67.96875	69.97242
		A=0.60			
0+	1.00000	2.84748	4.91862	7.08551	9.30174
5+	11.54549	13.80496	16.07320	18.34593	20.62038
10+	22.89475	25.16786	27.43892	29.70744	31.97311
15+	34.23576	36.49531	38.75175	41.00513	43.25550
20+	45.50298	47.74766	49.98966	52.22912	54.46614
25+	56.70087	58.93342	61.16392	63.39247	65.61920
30+	67.84422	70.06762	72.28951	74.50998	76.72911
35+	78.94699	81.16371	83.37933	85.59392	87.80755
40+	90.02029	92.23219	94.44330	96.65367	98.86337
45+	101.07242	103.28087	105.48876	107.69613	109.90301
50+	112.10943	114.31543	116.52103	118.72626	120.93115
		A=0.65			
0+	1.00000	2.85958	4.95655	7.16116	9.42530
5+	11.72574	14.04950	16.38864	18.73803	21.09419
10+	23.45468	25.81777	28.18219	30.54702	32.91158
15+	35.27536	37.63799	39.99920	42.35879	44.71663
20+	47.07262	49.42671	51.77887	54.12910	56.47740
25+	58.82379	61.16831	63.51100	65.85190	68.19107
30+	70.52855	72.86440	75.19868	77.53144	79.86274
35+	82.19263	84.52117	86.84841	89.17441	91.49922
40+	93.82288	96.14545	98.46698	100.78750	103.10707
45+	105.42572	107.74350	110.06044	112.37659	114.69198
50+	117.00664	119.32061	121.63392	123.94660	126.25868

EXPECTATION E(T/GLF)=(1+BA)SUM(D(R,M)),B= 2

R	1	2	3	4	5
			A=0.65		
55+	128.57018	130.88114	133.19158	135.50152	137.81099
60+	140.12001	142.42860	144.73678	147.04457	149.35199
65+	151.65905	153.96578	156.27218	158.57828	160.88409
70+	163.18962	165.49489	167.79991	170.10468	172.40923
75+	174.71356	177.01768	179.32161	181.62535	183.92891
			A=0.70		
0+	1.00000	2.86892	4.98603	7.22031	9.52251
5+	11.86841	14.24422	16.64132	19.05398	21.47819
10+	23.91107	26.35048	28.79478	31.24274	33.69338
15+	36.14592	38.59975	41.05437	43.50940	45.96450
20+	48.41943	50.87396	53.32793	55.78120	58.23366
25+	60.68523	63.13582	65.58539	68.03390	70.48131
30+	72.92761	75.37278	77.81682	80.25973	82.70150
35+	85.14215	87.58168	90.02011	92.45746	94.89373
40+	97.32896	99.76316	102.19634	104.62854	107.05977
45+	109.49006	111.91942	114.34789	116.77548	119.20221
50+	121.62812	124.05321	126.47752	128.90106	131.32386
55+	133.74594	136.16732	138.58801	141.00805	143.42744
60+	145.84621	148.26438	150.68196	153.09897	155.51544
65+	157.93137	160.34678	162.76170	165.17613	167.59008
70+	170.00359	172.41665	174.82929	177.24151	179.65333
75+	182.06477	184.47583	186.88653	189.29688	191.70688
80+	194.11656	196.52592	198.93497	201.34373	203.75219
85+	206.16038	208.56830	210.97595	213.38336	215.79052
90+	218.19744	220.60413	223.01061	225.41687	227.82292
95+	230.22877	232.63443	235.03990	237.44519	239.85031
100+	242.25526	244.66004	247.06466	249.46914	251.87346
105+	254.27764	256.68168	259.08559	261.48937	263.89303
			A=0.75		
0+	1.00000	2.87600	5.00844	7.26548	9.59706
5+	11.97829	14.39485	16.83763	19.30052	21.77914
10+	24.27031	26.77159	29.28112	31.79742	34.31930
15+	36.84580	39.37612	41.90962	44.44574	46.98402
20+	49.52408	52.06558	54.60823	57.15180	59.69607
25+	62.24085	64.78600	67.33137	69.87684	72.42232
30+	74.96771	77.51293	80.05792	82.60261	85.14695
35+	87.69090	90.23442	92.77748	95.32005	97.86209
40+	100.40360	102.94456	105.48494	108.02475	110.56396
45+	113.10257	115.64058	118.17798	120.71476	123.25094
50+	125.78650	128.32145	130.85580	133.38953	135.92266
55+	138.45520	140.98714	143.51849	146.04925	148.57944
60+	151.10906	153.63812	156.16661	158.69456	161.22197
65+	163.74884	166.27518	168.80100	171.32631	173.85112
70+	176.37543	178.89926	181.42260	183.94548	186.46788
75+	188.98984	191.51134	194.03241	196.55304	199.07325
80+	201.59304	204.11243	206.63141	209.15000	211.66820
85+	214.18602	216.70347	219.22056	221.73728	224.25366
90+	226.76969	229.28539	231.80075	234.31579	236.83052
95+	239.34493	241.85904	244.37284	246.88636	249.39958
100+	251.91253	254.42520	256.93760	259.44974	261.96162
105+	264.47324	266.98461	269.49575	272.00664	274.51730
110+	277.02773	279.53794	282.04793	284.55771	287.06727
115+	289.57663	292.08579	294.59475	297.10352	299.61210
120+	302.12050	304.62871	304.63675	307.14461	309.65231
125+	312.15983	312.16720	314.67441	317.18146	319.68836
130+	319.69511	322.20171	324.70817	324.71449	327.22068
135+	329.72673	329.73265	332.23844	334.74411	334.74965

EXPECTATION E(T/GLF)=(1+BA)SUM(D(R,M)),B= 2

R	1	2	3	4	5
		A=0.75			
140+	337.25508	339.76038	339.76557	342.27065	344.77562
145+	344.78048	347.28524	349.78989	349.79445	352.29890
150+	352.30326	354.80752	357.31169	357.31577	359.81976
155+	362.32366	362.32748	364.83122	364.83488	367.33845
160+	367.34195	369.84537	372.34872	372.35200	374.85520
165+	374.85834	377.36141	377.36441	379.86734	382.37022
		A=0.80			
0+	1.00000	2.88120	5.02496	7.29885	9.65228
5+	12.05993	14.50709	16.98436	19.48534	22.00544
10+	24.54126	27.09020	29.65021	32.21966	34.79724
15+	37.38185	39.97259	42.56869	45.16951	47.77448
20+	50.38313	52.99504	55.60985	58.22724	60.84693
25+	63.46867	66.09223	68.71742	71.34407	73.97201
30+	76.60110	79.23122	81.86224	84.49406	87.12659
35+	89.75975	92.39344	95.02760	97.66216	100.29706
40+	102.93225	105.56768	108.20330	110.83907	113.47495
45+	116.11090	118.74689	121.38289	124.01888	126.65482
50+	129.29070	131.92649	134.56218	137.19774	139.83315
55+	142.46841	145.10350	147.73841	150.37312	153.00762
60+	155.64190	158.27596	160.90979	163.54337	166.17671
65+	168.80979	171.44261	174.07517	176.70746	179.33947
70+	181.97121	184.60267	187.23385	189.86474	192.49534
75+	195.12566	197.75568	200.38542	203.01486	205.64401
80+	208.27287	210.90144	213.52971	216.15769	218.78538
85+	221.41278	224.03988	226.66670	229.29323	231.91947
90+	234.54542	237.17109	239.79647	242.42157	245.04639
95+	247.67093	250.29519	252.91918	255.54289	258.16633
100+	260.78950	263.41241	266.03504	268.65742	271.27953
105+	273.90138	276.52297	279.14431	281.76540	284.38623
110+	287.00682	289.62716	292.24725	294.86711	297.48672
115+	300.10610	302.72524	305.34415	307.96283	310.58128
120+	313.19950	315.81750	318.43528	321.05285	323.67019
125+	326.28732	328.90424	331.52095	334.13745	334.15375
130+	336.76984	339.38574	342.00143	342.01693	344.63224
135+	347.24735	349.86227	349.87701	352.49156	355.10592
140+	355.12011	357.73412	360.34795	362.96160	362.97508
145+	365.58839	368.20153	368.21451	370.82732	373.43996
150+	373.45245	376.06478	378.67694	378.68896	381.30082
155+	383.91253	383.92409	386.53550	386.54676	389.15788
160+	391.76886	391.77970	394.39039	397.00095	397.01138
165+	399.62167	399.63182	402.24185	404.85175	404.86152
170+	407.47116	410.08068	410.09007	412.69935	412.70850
175+	415.31754	415.32646	417.93526	420.54395	420.55253
180+	423.16100	423.16935	425.77760	428.38574	428.39378
185+	431.00171	431.00953	433.61726	433.62489	436.23241
190+	436.23984	438.84717	441.45441	441.46155	444.06860
195+	444.07556	446.68243	446.68920	449.29589	449.30250
200+	451.90901	451.91544	454.52179	454.52805	457.13424
205+	457.14034	459.74636	462.35230	462.35817	464.96396
210+	464.96967	467.57531	467.58088	470.18637	470.19179
215+	472.79714	472.80242	475.40763	475.41277	478.01785
220+	478.02286	480.62780	480.63268	483.23750	483.24225
225+	485.84694	485.85156	485.85613	488.46064	488.46509
230+	491.06948	491.07381	493.67809	493.68231	496.28647
235+	496.29058	498.89464	498.89864	501.50259	501.50649
240+	504.11034	504.11414	506.71788	506.72158	506.72523
245+	509.32883	509.33239	511.93590	511.93936	514.54278

EXPECTATION E(T/GLF)=(1+BA)SUM(D(R,M)),B= 2

R	1	2	3	4	5
			A=0.80		
250+	514.54615	517.14947	517.15276	517.15600	519.75920
255+	519.76235	522.36547	522.36854	524.97158	524.97457
260+	527.57752	527.58044	527.58332	530.18616	530.18896
265+	532.79173	532.79445	535.39715	535.39981	535.40243
270+	538.00502	538.00757	540.61009	540.61258	540.61504

EXPECTATION E(T/GLF)=(1+BA)SUM(D(R,M)),B= 3

R	1	2	3	4	5
			A=1/10		
0+	1.00000	2.48360	3.86153	5.18773	6.49612
			A=1/ 9		
0+	1.00000	2.52431	3.94839	5.31532	6.66044
			A=1/ 8		
0+	1.00000	2.57209	4.05358	5.47248	6.86452
			A=1/ 7		
0+	1.00000	2.62880	4.18310	5.67015	7.12415
5+	8.56354	9.99667	11.42717	12.85655	14.28546
			A=1/ 6		
0+	1.00000	2.69684	4.34544	5.92483	7.46406
5+	8.98293	10.49191	11.99617	13.49818	14.99914
			A=1/ 5		
0+	1.00000	2.77930	4.55269	6.26177	7.92426
5+	9.55886	11.17774	12.78798	14.39351	15.99649
			A=1/ 4		
0+	1.00000	2.87963	4.82109	6.71897	8.57001
5+	10.38601	12.17824	13.95497	15.72179	17.48234
			A=1/ 3		
0+	1.00000	3.00000	5.16744	7.34552	9.49862
5+	11.62115	13.71605	15.78825	17.84261	19.88327
10+	21.91355	23.93604	25.95271	27.96504	29.97417
			A=0.40		
0+	1.00000	3.06632	5.36987	7.73130	10.09681
5+	12.44815	14.77989	17.09162	19.38493	21.66211
10+	23.92551	26.17736	28.41963	30.65400	32.88190
15+	35.10451	37.32282	39.53763	41.74961	43.95930
			A=1/ 2		
0+	1.00000	3.13280	5.58094	8.14883	10.76713
5+	13.40510	16.04758	18.68673	21.31852	23.94101
10+	26.55346	29.15580	31.74836	34.33168	36.90643
15+	39.47329	42.03297	44.58614	47.13345	49.67550
20+	52.21282	54.74594	57.27529	59.80129	62.32432
25+	64.84471	67.36274	69.87870	72.39281	74.90528
30+	77.41631	79.92606	82.43467	84.94228	87.44901
			A=0.60		
0+	1.00000	3.17326	5.71313	8.41786	11.21116
5+	14.05618	16.93246	19.82771	22.73408	25.64640
10+	28.56119	31.47604	34.38930	37.29985	40.20692
15+	43.11000	46.00880	48.90313	51.79293	54.67821
20+	57.55903	60.43549	63.30771	66.17586	69.04009
25+	71.90057	74.75747	77.61096	80.46122	83.30842
30+	86.15272	88.99428	91.83326	94.66980	97.50405
35+	100.33613	103.16618	105.99433	108.82067	111.64534
40+	114.46842	117.29002	120.11023	122.92914	125.74682
45+	128.56336	131.37883	134.19330	137.00682	139.81947

EXPECTATION E(T/GLF)= (1+BA)SUM(D(R,M)),B= 3

R	1	2	3	4	5
		A=0.60			
50+	142.63129	145.44234	148.25267	151.06233	153.87135
55+	156.67979	159.48767	162.29504	165.10193	167.90836
		A=0.65			
0+	1.00000	3.18689	5.75820	8.51075	11.36649
5+	14.28691	17.25014	20.24259	23.25531	26.28214
10+	29.31869	32.36176	35.40900	38.45863	41.50931
15+	44.56001	47.60995	50.65854	53.70532	56.74996
20+	59.79221	62.83188	65.86885	68.90303	71.93439
25+	74.96291	77.98858	81.01144	84.03153	87.04888
30+	90.06356	93.07564	96.08518	99.09225	102.09693
35+	105.09931	108.09944	111.09743	114.09333	117.08723
40+	120.07920	123.06933	126.05767	129.04430	132.02930
45+	135.01272	137.99464	140.97511	143.95420	146.93196
50+	149.90846	152.88375	155.85788	158.83090	161.80286
55+	164.77380	167.74378	170.71284	173.68101	176.64834
60+	179.61486	182.58062	185.54564	188.50996	191.47361
65+	194.43662	197.39903	200.36085	203.32213	206.28287
70+	209.24311	212.20287	215.16217	218.12103	221.07948
75+	224.03753	226.99520	229.95251	232.90947	235.86611
80+	238.82243	241.77846	244.73420	247.68967	250.64488
		A=0.70			
0+	1.00000	3.19726	5.79264	8.58206	11.48632
5+	14.46584	17.49780	20.56780	23.66612	26.78594
10+	29.92227	33.07137	36.23038	39.39707	42.56968
15+	45.74678	48.92726	52.11018	55.29478	58.48044
20+	61.66665	64.85298	68.03908	71.22466	74.40948
25+	77.59333	80.77606	83.95753	87.13763	90.31627
30+	93.49339	96.66893	99.84285	103.01513	106.18574
35+	109.35467	112.52192	115.68750	118.85140	122.01365
40+	125.17425	128.33324	131.49062	134.64642	137.80068
45+	140.95341	144.10465	147.25442	150.40276	153.54969
50+	156.69526	159.83948	162.98239	166.12403	169.26442
55+	172.40359	175.54158	178.67842	181.81414	184.94877
60+	188.08233	191.21487	194.34639	197.47695	200.60655
65+	203.73523	206.86302	209.98994	213.11601	216.24127
70+	219.36573	222.48942	225.61236	228.73458	231.85609
75+	234.97692	238.09709	241.21661	244.33551	247.45381
80+	250.57153	253.68868	256.80527	259.92134	263.03689
85+	266.15194	269.26651	272.38061	275.49426	278.60746
90+	281.72024	284.83261	287.94457	291.05616	294.16736
95+	297.27820	300.38870	303.49885	306.60867	309.71818
100+	312.82737	315.93627	319.04488	322.15321	325.26127
105+	328.36906	331.47661	334.58390	334.59096	337.69779
110+	340.80440	340.81079	343.91698	347.02296	350.12875
115+	350.13435	353.23976	356.34500	356.35007	359.45497
		A=0.75			
0+	1.00000	3.20500	5.81841	8.63557	11.57651
5+	14.60096	17.68549	20.81514	23.97973	27.17200
10+	30.38659	33.61939	36.86723	40.12756	43.39835
15+	46.67794	49.96494	53.25820	56.55675	59.85978
20+	63.16657	66.47654	69.78915	73.10396	76.42058
25+	79.73866	83.05792	86.37809	89.69893	93.02025
30+	96.34187	99.66363	102.98540	106.30704	109.62845
35+	112.94954	116.27023	119.59043	122.91008	126.22913
40+	129.54752	132.86521	136.18217	139.49835	142.81374
45+	146.12830	149.44202	152.75487	156.06685	159.37794
50+	162.68813	165.99742	169.30581	172.61328	175.91983

EXPECTATION E(T/GLF)=(1+BA)SUM(D(R,M)),B= 3

R	1	2	3	4	5
			A=0.75		
55+	179.22548	182.53021	185.83403	189.13695	192.43897
60+	195.74010	199.04033	202.33968	205.63816	208.93577
65+	212.23253	215.52844	218.82351	222.11775	225.41117
70+	228.70378	231.99559	235.28662	238.57687	241.86636
75+	245.15509	248.44308	251.73033	255.01687	258.30269
80+	261.58782	264.87226	268.15602	271.43912	274.72157
85+	278.00338	281.28455	284.56510	287.84504	291.12439
90+	294.40314	297.68132	300.95893	304.23598	307.51248
95+	310.78845	314.06389	317.33881	320.61323	323.88714
100+	327.16057	330.43352	333.70600	336.97801	340.24958
105+	343.52070	346.79138	350.06164	353.33147	356.60090
110+	359.86993	363.13856	366.40680	369.67467	369.69216
115+	372.95929	376.22606	379.49248	379.50856	382.77431
120+	386.03972	389.30481	389.31959	392.58405	395.84822
125+	395.86208	399.12566	402.38894	402.40195	405.66469
130+	408.92716	408.93936	412.20131	412.21300	415.47445
135+	418.73566	418.74663	422.00736	425.26788	425.27816
140+	428.53824	428.54809	431.80774	435.06719	435.07643
145+	438.33548	438.34433	441.60300	441.61149	444.86979
150+	448.12792	448.13587	451.39366	451.40128	454.65874
155+	454.66603	457.92318	461.18017	461.18701	464.44371
160+	464.45026	467.70668	467.71296	470.96910	470.97511
165+	474.23100	474.23676	477.49239	477.49791	480.75331
170+	480.75859	484.01376	484.01882	487.27377	487.27862
175+	490.53336	490.53800	493.79254	493.79699	497.05134
180+	497.05559	500.30976	500.31384	503.56782	503.57173
			A=0.80		
0+	1.00000	3.21062	5.83712	8.67446	11.64216
5+	14.69948	17.82259	20.99620	24.20983	27.45593
10+	30.72890	34.02442	37.33912	40.67026	44.01565
15+	47.37345	50.74214	54.12041	57.50719	60.90150
20+	64.30255	67.70961	71.12206	74.53934	77.96097
25+	81.38652	84.81560	88.24786	91.68299	95.12071
30+	98.56077	102.00293	105.44700	108.89277	112.34008
35+	115.78877	119.23870	122.68972	126.14173	129.59461
40+	133.04825	136.50257	139.95747	143.41288	146.86872
45+	150.32492	153.78142	157.23816	160.69509	164.15215
50+	167.60930	171.06650	174.52370	177.98088	181.43799
55+	184.89500	188.35188	191.80861	195.26516	198.72151
60+	202.17763	205.63351	209.08912	212.54446	215.99949
65+	219.45422	222.90862	226.36268	229.81640	233.26975
70+	236.72274	240.17535	243.62757	247.07940	250.53083
75+	253.98186	257.43247	260.88267	264.33245	267.78181
80+	271.23074	274.67923	278.12730	281.57493	285.02213
85+	288.46889	291.91521	295.36109	298.80654	302.25154
90+	305.69610	309.14023	312.58392	316.02716	319.46998
95+	322.91235	326.35429	329.79580	333.23688	336.67753
100+	340.11774	343.55754	346.99691	350.43586	353.87439
105+	357.31250	360.75020	364.18749	367.62436	371.06084
110+	374.49690	377.93257	381.36784	384.80272	388.23720
115+	391.67129	395.10500	398.53833	401.97127	405.40384
120+	408.83604	408.86787	412.29933	415.73042	419.16116
125+	419.19154	422.62156	426.05124	426.08056	429.50955
130+	432.93819	436.36649	436.39446	439.82210	443.24942
135+	443.27640	446.70307	450.12942	450.15545	453.58117
140+	457.00659	457.03169	460.45650	463.88101	463.90522
145+	467.32914	470.75277	470.77611	474.19917	474.22195

EXPECTATION E(T/GLF)=(1+BA)SUM(D(R,M)),B= 3

A=0.80

R	1	2	3	4	5
150+	477.64445	481.06668	481.08864	484.51033	484.53175
155+	487.95291	491.37382	491.39446	494.81485	494.83499
160+	498.25489	501.67454	501.69394	505.11311	505.13204
165+	508.55073	511.96919	511.98743	515.40544	515.42322
170+	518.84078	518.85813	522.27526	522.29217	525.70888
175+	529.12537	529.14166	532.55775	532.57364	535.98932
180+	536.00481	539.42011	539.43521	542.85013	542.86486
185+	546.27940	546.29376	549.70794	549.72194	553.13576
190+	553.14941	556.56289	556.57620	559.98934	560.00231
195+	563.41512	563.42777	566.84026	566.85259	570.26477
200+	570.27679	573.68865	573.70037	577.11194	577.12336
205+	580.53464	580.54577	583.95677	583.96762	587.37833
210+	587.38891	590.79936	590.80967	594.21985	594.22990
215+	597.63982	597.64962	601.05929	601.06883	601.07826
220+	604.48756	604.49675	607.90582	607.91477	611.32361
225+	611.33234	614.74095	614.74946	618.15785	618.16614
230+	618.17432	621.58240	621.59037	624.99824	625.00602
235+	628.41369	628.42126	628.42874	631.83612	631.84340
240+	635.25059	635.25769	638.66470	638.67162	638.67845
245+	642.08519	642.09184	645.49841	645.50490	645.51130
250+	648.91762	648.92386	652.33002	652.33610	655.74210
255+	655.74802	655.75387	659.15964	659.16534	662.57096
260+	662.57651	662.58199	665.98740	665.99274	665.99801
265+	669.40322	669.40835	672.81342	672.81843	672.82337
270+	676.22824	676.23306	679.63781	679.64250	679.64713
275+	683.05170	683.05621	686.46066	686.46505	686.46939
280+	689.87367	689.87790	689.88207	693.28619	693.29026
285+	693.29427	696.69823	696.70214	700.10600	700.10981
290+	700.11357	703.51728	703.52094	703.52456	706.92813
295+	706.93165	706.93513	710.33856	710.34195	713.74529

EXPECTATION E(T/GLF)=(1+BA)SUM(D(R,M)),B= 4

R	1	2	3	4	5
		A=1/10			
0+	1.00000	2.63717	4.14534	5.58246	6.99444
		A=1/ 9			
0+	1.00000	2.68944	4.25975	5.75191	7.21326
		A=1/ 8			
0+	1.00000	2.75039	4.39784	5.96029	7.48482
5+	8.99423	10.49781	11.99917	13.49968	14.99988
		A=1/ 7			
0+	1.00000	2.82210	4.56703	6.22173	7.82988
5+	9.41704	10.99513	12.56937	14.14199	15.71392
		A=1/ 6			
0+	1.00000	2.90720	4.77762	6.55718	8.28063
5+	9.97487	11.65473	13.32767	14.99731	16.66539
		A=1/ 5			
0+	1.00000	3.00879	5.04362	6.99791	8.88828
5+	10.73898	12.56687	14.38206	16.19030	17.99475
		A=1/ 4			
0+	1.00000	3.12988	5.38250	7.58874	9.73369
5+	11.83003	13.89240	15.93220	17.95740	19.97327
10+	21.98324	23.98950	25.99342	27.99587	29.99741
		A=1/ 3			
0+	1.00000	3.27090	5.80838	8.38055	10.92756
5+	13.43611	15.90745	18.34712	20.76130	23.15557
10+	25.53457	27.90199	30.26069	32.61287	34.96018
15+	37.30387	39.64486	41.98387	44.32140	46.65784
		A=0.40			
0+	1.00000	3.34626	6.05001	8.85487	11.67791
5+	14.48849	17.27572	20.03703	22.77342	25.48734
10+	28.18172	30.85945	33.52318	36.17528	38.81775
15+	41.45231	44.08039	46.70318	49.32165	51.93662
20+	54.54874	57.15855	59.76648	62.37290	64.97809
		A=1/ 2			
0+	1.00000	3.41975	6.29480	9.35357	12.49532
5+	15.67376	18.86514	22.05665	25.24126	28.41521
10+	31.57671	34.72508	37.86038	40.98304	44.09378
15+	47.19340	50.28277	53.36277	56.43425	59.49802
20+	62.55483	65.60539	68.65035	71.69029	74.72575
25+	77.75721	80.78511	83.80984	86.83176	89.85117
30+	92.86836	95.88358	98.89704	101.90896	104.91950
35+	107.92883	110.93708	113.94438	116.95083	119.95653
		A=0.60			
0+	1.00000	3.46323	6.44340	9.66454	13.01893
5+	16.45341	19.93804	23.45445	26.99076	30.53898
10+	34.09359	37.65072	41.20765	44.76241	48.31363
15+	51.86037	55.40197	58.93801	62.46824	65.99255
20+	69.51091	73.02338	76.53005	80.03109	83.52665
25+	87.01693	90.50214	93.98249	97.45819	100.92948
30+	104.39655	107.85963	111.31893	114.77464	118.22696
35+	121.67608	125.12218	128.56542	132.00599	135.44402
40+	138.87967	142.31309	145.74440	149.17374	152.60121
45+	156.02695	159.45104	162.87361	166.29473	169.71450
50+	173.13300	176.55031	179.96652	183.38168	186.79586
55+	190.20913	193.62154	197.03315	200.44401	203.85417
60+	207.26366	210.67255	214.08085	217.48862	220.89588
		A=0.65			
0+	1.00000	3.47761	6.49300	9.76947	13.19767
5+	16.72279	20.31332	23.94947	27.61803	31.30982
10+	35.01823	38.73840	42.46666	46.20024	49.93699

EXPECTATION E(T/GLF)=(1+BA) SUM(D(R,M)),B= 4

R	1	2	3	4	5
		A=0.65			
15+	53.67525	57.41374	61.15145	64.88759	68.62156
20+	72.35288	76.08121	79.80626	83.52785	87.24584
25+	90.96015	94.67073	98.37755	102.08063	105.78001
30+	109.47572	113.16782	116.85640	120.54152	124.22328
35+	127.90177	131.57707	135.24929	138.91852	142.58486
40+	146.24841	149.90927	153.56754	157.22330	160.87666
45+	164.52770	168.17652	171.82320	175.46783	179.11049
50+	182.75125	186.39021	190.02742	193.66297	197.29693
55+	200.92936	204.56033	208.18989	211.81812	215.44507
60+	219.07079	222.69534	226.31876	229.94112	233.56245
65+	237.18281	240.80223	244.42076	248.03843	251.65529
70+	255.27137	258.88671	262.50134	266.11530	269.72860
75+	273.34129	276.95339	280.56493	284.17594	287.78643
80+	291.39643	295.00597	298.61506	302.22373	305.83199
85+	309.43987	313.04738	316.65454	320.26137	323.86787
		A=0.70			
0+	1.00000	3.48843	6.53043	9.84893	13.33358
5+	16.92851	20.60125	24.33109	28.10403	31.91011
10+	35.74203	39.59426	43.46256	47.34357	51.23463
15+	55.13360	59.03873	62.94858	66.86197	70.77791
20+	74.69559	78.61432	82.53351	86.45269	90.37145
25+	94.28945	98.20639	102.12205	106.03621	109.94871
30+	113.85941	117.76821	121.67501	125.57974	129.48234
35+	133.38278	137.28102	141.17705	145.07085	148.96243
40+	152.85179	156.73894	160.62390	164.50669	168.38733
45+	172.26585	176.14229	180.01666	183.88902	187.75939
50+	191.62781	195.49432	199.35896	203.22177	207.08278
55+	210.94205	214.79960	218.65548	222.50973	226.36239
60+	230.21349	234.06308	237.91120	241.75788	245.60317
65+	249.44709	253.28969	257.13100	260.97106	264.80989
70+	268.64754	272.48405	276.31943	280.15372	283.98696
75+	287.81917	291.65038	295.48063	299.30993	303.13833
80+	306.96584	310.79249	314.61831	318.44332	322.26754
85+	326.09101	329.91374	333.73575	337.55706	341.37771
90+	345.19770	349.01706	352.83580	356.65396	360.47153
95+	364.28855	368.10503	371.92098	375.73642	379.55138
100+	383.36585	387.17986	390.99343	394.80656	394.81927
105+	398.63158	402.44349	406.25502	406.26617	410.07698
110+	413.88743	413.89755	417.70734	421.51682	421.52599
115+	425.33487	429.14346	429.15178	432.95982	432.96761
120+	436.77515	440.58244	440.58950	444.39633	444.40294
125+	448.20933	452.01552	452.02151	455.82731	455.83292
		A=0.75			
0+	1.00000	3.49643	6.55813	9.90782	13.43449
5+	17.08162	20.81611	24.61672	28.46892	32.36229
10+	36.28910	40.24344	44.22067	48.21710	52.22975
15+	56.25616	60.29429	64.34245	68.39917	72.46323
20+	76.53358	80.60930	84.68961	88.77381	92.86132
25+	96.95159	101.04416	105.13863	109.23464	113.33186
30+	117.43001	121.52884	125.62811	129.72764	133.82724
35+	137.92674	142.02602	146.12495	150.22340	154.32128
40+	158.41851	162.51499	166.61066	170.70547	174.79934
45+	178.89224	182.98413	187.07495	191.16470	195.25333
50+	199.34082	203.42716	207.51232	211.59630	215.67908
55+	219.76066	223.84102	227.92018	231.99812	236.07484
60+	240.15036	244.22466	248.29776	252.36966	256.44037
65+	260.50989	264.57823	268.64541	272.71143	276.77631

EXPECTATION E(T/GLF)=(1+BA)SUM(D(R,M)),B= 4

R	1	2	3	4	5
		A=0.75			
70+	280.84005	284.90267	288.96417	293.02458	297.08390
75+	301.14215	305.19934	309.25548	313.31060	317.36469
80+	321.41778	325.46988	329.52101	333.57118	337.62040
85+	341.66868	345.71605	349.76252	353.80809	357.85279
90+	361.89663	365.93963	369.98179	374.02313	378.06367
95+	382.10341	386.14238	390.18058	394.21803	398.25474
100+	402.29073	406.32601	410.36058	414.39447	418.42768
105+	422.46023	426.49213	430.52340	434.55403	438.58405
110+	442.61347	442.64229	446.67053	450.69821	454.72532
115+	454.75188	458.77790	462.80340	462.82838	466.85285
120+	470.87682	470.90030	474.92330	478.94583	478.96790
125+	482.98952	487.01069	487.03143	491.05174	491.07164
130+	495.09112	499.11021	499.12890	503.14720	507.16513
135+	507.18268	511.19988	511.21671	515.23320	515.24935
140+	519.26516	523.28064	523.29581	527.31065	527.32519
145+	531.33943	531.35336	535.36701	539.38038	539.39346
150+	543.40627	543.41882	547.43110	547.44313	551.45490
155+	551.46643	555.47771	555.48876	559.49958	559.51017
160+	563.52054	563.53070	567.54063	567.55036	571.55989
165+	571.56922	575.57835	575.58728	579.59603	579.60460
170+	583.61299	583.62119	587.62923	587.63710	591.64480
175+	591.65234	595.65972	595.66694	599.67401	599.68093
180+	603.68771	603.69434	603.70083	607.70719	607.71341
185+	611.71950	611.72546	615.73130	615.73701	619.74260
190+	619.74807	619.75343	623.75867	623.76380	627.76883
		A=0.80			
0+	1.00000	3.50218	6.57804	9.95012	13.50700
5+	17.19173	20.97080	24.82264	28.73239	32.68936
10+	36.68556	40.71483	44.77232	48.85414	52.95712
15+	57.07861	61.21641	65.36866	69.53375	73.71031
20+	77.89714	82.09321	86.29759	90.50948	94.72815
25+	98.95297	103.18336	107.41882	111.65887	115.90310
30+	120.15114	124.40263	128.65726	132.91474	137.17483
35+	141.43726	145.70184	149.96834	154.23660	158.50643
40+	162.77767	167.05019	171.32385	175.59851	179.87407
45+	184.15042	188.42745	192.70508	196.98322	201.26178
50+	205.54069	209.81989	214.09931	218.37888	222.65856
55+	226.93828	231.21801	235.49769	239.77728	244.05674
60+	248.33604	252.61514	256.89401	261.17261	265.45093
65+	269.72893	274.00659	278.28389	282.56081	286.83733
70+	291.11343	295.38909	299.66429	303.93904	308.21330
75+	312.48707	316.76034	321.03309	325.30532	329.57703
80+	333.84819	338.11881	342.38888	346.65838	350.92733
85+	355.19571	359.46351	363.73075	367.99740	372.26348
90+	376.52897	380.79388	385.05820	389.32193	393.58508
95+	397.84764	402.10962	406.37101	410.63181	414.89202
100+	419.15165	423.41070	427.66917	431.92705	436.18436
105+	440.44109	444.69725	448.95283	453.20785	457.46230
110+	461.71619	465.96951	470.22228	474.47450	478.72616
115+	482.97728	487.22785	487.27788	491.52737	495.77633
120+	500.02476	500.07267	504.32005	508.56691	512.81326
125+	512.85910	517.10443	521.34926	521.39358	525.63742
130+	529.88077	529.92362	534.16600	538.40790	538.44932
135+	542.69028	546.93077	546.97080	551.21037	555.44949
140+	555.48816	559.72638	559.76417	564.00152	568.23843
145+	568.27492	572.51099	576.74663	576.78187	581.01668
150+	581.05110	585.28510	589.51871	589.55193	593.78475

EXPECTATION E(T/GLF)=(1+BA)SUM(D(R,M)),B= 4

A=0.80

R	1	2	3	4	5
155+	593.81719	598.04924	598.08091	602.31221	606.54314
160+	606.57370	610.80390	610.83373	615.06321	615.09234
165+	619.32112	623.54956	623.57766	627.80541	627.83284
170+	632.05993	632.08670	636.31315	636.33927	640.56508
175+	640.59058	644.81577	644.84066	649.06524	649.08952
180+	653.31351	653.33721	657.56062	657.58374	661.80658
185+	661.82914	666.05143	666.07344	670.29519	670.31667
190+	674.53788	674.55884	678.77954	678.79998	683.02017
195+	683.04011	687.25981	687.27927	691.49848	691.51746
200+	695.73621	695.75472	699.97300	699.99106	704.20889
205+	704.22651	708.44390	708.46108	712.67804	712.69480
210+	712.71135	716.92769	716.94383	721.15976	721.17550
215+	725.39105	725.40640	729.62155	729.63652	729.65130
220+	733.86590	733.88032	738.09455	738.10861	742.32249
225+	742.33620	742.34973	746.56310	746.57630	750.78933
230+	750.80220	755.01491	755.02746	755.03985	759.25209
235+	759.26417	763.47610	763.48788	767.69951	767.71100
240+	767.72234	771.93354	771.94459	776.15551	776.16629
245+	776.17694	780.38745	780.39782	784.60807	784.61819
250+	784.62817	788.83804	788.84778	793.05739	793.06688
255+	793.07626	797.28551	797.29465	797.30367	801.51258
260+	801.52138	805.73006	805.73863	805.74710	809.95546
265+	809.96371	809.97185	814.17990	814.18784	818.39568
270+	818.40342	818.41106	822.61861	822.62606	822.63341
275+	826.84067	826.84784	831.05492	831.06191	831.06881
280+	835.27562	835.28234	835.28898	839.49554	839.50201
285+	839.50840	843.71470	843.72093	843.72708	847.93315
290+	847.93914	847.94505	852.15089	852.15666	856.36235
295+	856.36797	856.37352	860.57899	860.58440	860.58974
300+	864.79501	864.80021	864.80534	869.01041	869.01542
305+	869.02036	873.22524	873.23006	873.23481	877.43950
310+	877.44414	877.44871	881.65323	881.65768	881.66208
315+	885.86643	885.87072	885.87495	885.87913	890.08326

EXPECTATION E(T/GLF)=(1+BA)SUM(D(R,M)),B= 5

R	1	2	3	4	5
		A=1/10			
0+	1.00000	2.78736	4.42753	5.97661	7.49257
		A=1/ 9			
0+	1.00000	2.85036	4.56886	6.18760	7.76574
		A=1/ 8			
0+	1.00000	2.92338	4.73889	6.44666	8.10453
5+	9.74220	11.37203	12.99887	14.62457	16.24984
		A=1/ 7			
0+	1.00000	3.00863	4.94624	6.77087	8.53448
5+	10.27004	11.99337	13.71149	15.42739	17.14236
		A=1/ 6			
0+	1.00000	3.10880	5.20262	7.18523	9.09490
5+	10.96565	12.81698	14.65889	16.49631	18.33158
		A=1/ 5			
0+	1.00000	3.22680	5.52333	7.72611	9.84733
5+	11.91616	13.95432	15.97520	17.98657	19.99273
		A=1/ 4			
0+	1.00000	3.36497	5.92596	8.44314	10.88577
5+	13.26587	15.60101	17.90576	20.19062	22.46266
10+	24.72656	26.98529	29.24077	31.49421	33.74637
		A=1/ 3			
0+	1.00000	3.52191	6.42045	9.38515	12.32830
5+	15.22667	18.07861	20.88965	23.66706	26.41779
10+	29.14781	31.86199	34.56415	37.25727	39.94361
15+	42.62488	45.30238	47.97708	50.64970	53.32077
		A=0.40			
0+	1.00000	3.60371	6.69407	9.93621	13.21550
5+	16.48692	19.73286	22.94770	26.13133	29.28609
10+	32.41524	35.52227	38.61051	41.68297	44.74228
15+	47.79071	50.83019	53.86231	56.88842	59.90962
20+	62.92682	65.94076	68.95205	71.96120	74.96861
		A=1/ 2			
0+	1.00000	3.68178	6.96482	10.50183	14.15919
5+	17.87382	21.61247	25.35661	29.09567	32.82368
10+	36.53745	40.23548	43.91734	47.58326	51.23386
15+	54.87003	58.49273	62.10302	65.70192	69.29046
20+	72.86959	76.44021	80.00318	83.55926	87.10916
25+	90.65353	94.19296	97.72798	101.25906	104.78664
30+	108.31110	111.83278	115.35199	118.86902	122.38410
35+	125.89745	129.40928	132.91975	136.42901	139.93721
		A=0.60			
0+	1.00000	3.72697	7.12518	10.84530	14.74727
5+	18.76089	22.84608	26.97799	31.14029	35.32190
10+	39.51502	43.71411	47.91516	52.11528	56.31236
15+	60.50489	64.69178	68.87230	73.04595	77.21243
20+	81.37160	85.52341	89.66791	93.80521	97.93547
25+	102.05889	106.17570	110.28613	114.39043	118.48887
30+	122.58170	126.66919	130.75160	134.82917	138.90216
35+	142.97080	147.03532	151.09596	155.15292	159.20641
40+	163.25662	167.30375	171.34797	175.38945	179.42835
45+	183.46483	187.49902	191.53107	195.56111	199.58925
50+	203.61562	207.64032	211.66346	215.68512	219.70541
55+	223.72440	227.74218	231.75883	235.77441	239.78900
60+	243.80264	247.81542	251.82737	255.83855	259.84901
65+	263.85880	267.86796	271.87653	275.88454	279.89203
		A=0.65			
0+	1.00000	3.74169	7.17782	10.95909	14.94413
5+	19.06115	23.26851	27.53983	31.85737	36.20868

EXPECTATION E(T/GLF)=(1+BA)SUM(D(R,M)),B= 5

R	1	2	3	4	5
			A=0.65		
10+	40.58474	44.97883	49.38586	53.80193	58.22399
15+	62.64966	67.07707	71.50472	75.93144	80.35629
20+	84.77854	89.19761	93.61306	98.02454	102.43180
25+	106.83465	111.23297	115.62668	120.01574	124.40015
30+	128.77992	133.15511	137.52578	141.89199	146.25384
35+	150.61142	154.96484	159.31421	163.65964	168.00124
40+	172.33914	176.67345	181.00428	185.33176	189.65601
45+	193.97713	198.29525	202.61047	206.92290	211.23264
50+	215.53981	219.84450	224.14681	228.44684	232.74467
55+	237.04040	241.33411	245.62588	249.91580	254.20395
60+	258.49039	262.77521	267.05846	271.34022	275.62055
65+	279.89952	284.17717	288.45358	292.72879	297.00285
70+	301.27582	305.54774	309.81867	314.08863	318.35769
75+	322.62587	326.89323	331.15978	335.42558	339.69065
80+	343.95502	348.21874	352.48182	356.74430	361.00620
85+	365.26755	369.52838	373.78871	378.04856	382.30795
90+	386.56691	390.82545	395.08360	399.34137	403.59878
			A=0.70		
0+	1.00000	3.75269	7.21718	11.04438	15.09211
5+	19.28768	23.58847	27.96722	32.40531	36.88951
10+	41.41002	45.95940	50.53189	55.12294	59.72891
15+	64.34684	68.97431	73.60931	78.25020	82.89557
20+	87.54425	92.19527	96.84778	101.50108	106.15456
25+	110.80772	115.46011	120.11138	124.76121	129.40933
30+	134.05553	138.69961	143.34143	147.98084	152.61776
35+	157.25210	161.88379	166.51278	171.13904	175.76255
40+	180.38329	185.00126	189.61646	194.22891	198.83862
45+	203.44562	208.04992	212.65158	217.25061	221.84706
50+	226.44097	231.03238	235.62133	240.20786	244.79204
55+	249.37389	253.95347	258.53083	263.10601	267.67907
60+	272.25004	276.81899	281.38596	285.95099	290.51414
65+	295.07545	299.63496	304.19273	308.74879	313.30319
70+	317.85598	322.40720	326.95689	331.50509	336.05184
75+	340.59719	345.14116	349.68380	354.22515	358.76524
80+	363.30411	367.84180	372.37833	376.91373	381.44806
85+	385.98132	390.51356	395.04480	399.57508	404.10442
90+	408.63285	413.16040	417.68708	422.21294	426.73799
95+	431.26226	435.78577	440.30854	444.83060	449.35197
100+	453.87267	458.39271	458.41213	462.93093	467.44914
105+	471.96677	471.98385	476.50039	481.01640	481.03191
110+	485.54693	490.06147	490.07555	494.58918	494.60238
115+	499.11516	503.62753	503.63950	508.15110	512.66233
120+	512.67320	517.18372	517.19390	521.70376	521.71331
125+	526.22255	530.73149	530.74015	535.24853	535.25664
130+	539.76449	539.77209	544.27945	544.28657	548.79346
			A=0.75		
0+	1.00000	3.76077	7.24606	11.10697	15.20083
5+	19.45436	23.82437	28.28304	32.81128	37.39535
10+	42.02500	46.69240	51.39138	56.11701	60.86528
15+	65.63291	70.41715	75.21568	80.02654	84.84806
20+	89.67880	94.51751	99.36310	104.21462	109.07124
25+	113.93223	118.79692	123.66477	128.53524	133.40788
30+	138.28230	143.15812	148.03502	152.91272	157.79095
35+	162.66948	167.54810	172.42662	177.30487	182.18270
40+	187.05997	191.93657	196.81237	201.68729	206.56124
45+	211.43413	216.30590	221.17649	226.04584	230.91391
50+	235.78064	240.64602	245.50999	250.37255	255.23366

EXPECTATION E(T/GLF)=(1+BA)SUM(D(R,M)),B= 5

R	1	2	3	4	5
			A=0.75		
55+	260.09330	264.95146	269.80813	274.66330	279.51696
60+	284.36910	289.21972	294.06883	298.91642	303.76250
65+	308.60707	313.45014	318.29171	323.13180	327.97042
70+	332.80756	337.64326	342.47751	347.31033	352.14174
75+	356.97175	361.80037	366.62762	371.45351	376.27807
80+	381.10130	385.92322	390.74385	395.56321	400.38132
85+	405.19818	410.01382	414.82826	419.64150	424.45358
90+	429.26451	434.07429	438.88297	443.69053	448.49702
95+	453.30244	458.10681	462.91014	467.71246	472.51378
100+	477.31411	482.11348	486.91190	491.70938	496.50595
105+	501.30161	506.09638	510.89029	515.68333	515.72554
110+	520.51691	525.30748	530.09724	530.13623	534.92444
115+	539.71190	539.74862	544.53461	549.31988	549.35446
120+	554.13834	558.92155	558.95410	563.73599	568.51725
125+	568.54788	573.32789	573.35730	578.13612	582.91436
130+	582.94202	587.71913	587.74569	592.52172	597.29721
135+	597.32219	602.09666	602.12064	606.89413	611.66713
140+	611.68968	616.46176	616.48339	621.25457	621.27533
145+	626.04566	626.06557	630.83508	630.85418	635.62290
150+	640.39123	640.40918	645.17676	645.19398	649.96085
155+	649.97737	654.74355	654.75939	659.52491	659.54011
160+	664.30499	664.31956	669.08384	669.09781	673.86150
165+	673.87491	678.63803	678.65089	683.41348	683.42580
170+	688.18787	688.19969	688.21126	692.97260	692.98369
175+	697.74456	697.75520	702.51561	702.52582	707.28580
180+	707.29558	712.05516	712.06453	712.07371	716.83270
185+	716.84150	721.60011	721.60855	726.36681	726.37489
190+	731.13281	731.14056	731.14815	735.90558	735.91285
195+	740.66997	740.67694	740.68377	745.44045	745.44699
200+	750.20340	750.20967	750.21580	754.97181	754.97770
			A=0.80		
0+	1.00000	3.76654	7.26665	11.15153	15.27819
5+	19.57296	23.99229	28.50802	33.10075	37.75645
10+	42.46464	47.21721	52.00779	56.83123	61.68333
15+	66.56060	71.46011	76.37938	81.31629	86.26899
20+	91.23590	96.21562	101.20692	106.20871	111.22002
25+	116.24001	121.26788	126.30295	131.34460	136.39224
30+	141.44537	146.50351	151.56624	156.63317	161.70394
35+	166.77821	171.85569	176.93610	182.01917	187.10468
40+	192.19239	197.28211	202.37365	207.46683	212.56149
45+	217.65748	222.75465	227.85288	232.95203	238.05201
50+	243.15269	248.25399	253.35580	258.45804	263.56063
55+	268.66349	273.76655	278.86975	283.97301	289.07629
60+	294.17952	299.28267	304.38567	309.48848	314.59106
65+	319.69338	324.79539	329.89706	334.99836	340.09925
70+	345.19972	350.29972	355.39925	360.49827	365.59677
75+	370.69472	375.79210	380.88891	385.98511	391.08071
80+	396.17568	401.27001	406.36369	411.45672	416.54907
85+	421.64074	426.73173	431.82202	436.91161	442.00050
90+	447.08867	452.17613	457.26286	462.34888	467.43416
95+	472.51872	477.60254	482.68563	487.76798	492.84960
100+	497.93049	503.01063	508.09005	513.16872	518.24667
105+	523.32387	528.40035	533.47610	538.55112	543.62541
110+	548.69898	553.77183	558.84396	563.91537	568.98607
115+	574.05607	574.12535	579.19394	584.26182	589.32901
120+	589.39551	594.46133	599.52646	599.59091	604.65468
125+	609.71779	609.78023	614.84201	619.90314	619.96361

EXPECTATION E(T/GLF)= (1+BA)SUM(D(R,M)),B= 5

R	1	2	3	4	5
		A=0.80			
130+	625.02343	630.08262	630.14116	635.19908	640.25636
135+	640.31302	645.36907	650.42450	650.47932	655.53354
140+	655.58716	660.64019	665.69263	665.74449	670.79577
145+	675.84648	675.89662	680.94620	680.99522	686.04369
150+	691.09161	691.13899	696.18583	696.23214	701.27792
155+	701.32318	706.36792	711.41215	711.45588	716.49910
160+	716.54182	721.58405	721.62580	726.66706	726.70784
165+	731.74815	731.78799	736.82737	741.86629	741.90475
170+	746.94277	746.98034	752.01747	752.05416	757.09043
175+	757.12627	762.16168	762.19668	767.23126	767.26544
180+	772.29921	772.33258	777.36555	777.39814	782.43033
185+	782.46214	787.49358	787.52464	792.55532	792.58564
190+	797.61560	797.64520	802.67445	802.70334	807.73189
195+	807.76009	812.78796	812.81549	812.84268	817.86955
200+	817.89610	822.92232	822.94823	827.97382	827.99911
205+	833.02409	833.04876	838.07314	838.09722	843.12100
210+	843.14450	843.16771	848.19064	848.21329	853.23566
215+	853.25776	858.27959	858.30115	858.32245	863.34349
220+	863.36427	868.38480	868.40507	873.42510	873.44488
225+	873.46442	878.48372	878.50278	883.52160	883.54020
230+	888.55856	888.57670	888.59461	893.61231	893.62978
235+	898.64704	898.66409	898.68093	903.69755	903.71398
240+	908.73020	908.74621	908.76203	913.77766	913.79309
245+	918.80833	918.82338	918.83824	923.85292	923.86741
250+	928.88173	928.89587	928.90983	933.92361	933.93723
255+	933.95068	938.96396	938.97707	943.99002	944.00281
260+	944.01544	949.02791	949.04022	949.05239	954.06440
265+	954.07625	959.08797	959.09953	959.11095	964.12223
270+	964.13336	964.14436	969.15522	969.16594	969.17653
275+	974.18699	974.19731	979.20750	979.21757	979.22751
280+	984.23733	984.24702	984.25659	989.26604	989.27537
285+	989.28458	994.29368	994.30267	994.31154	999.32030
290+	999.32894	999.33748	1004.34592	1004.35424	1004.36246
295+	1009.37058	1009.37859	1009.38651	1014.39432	1014.40204
300+	1014.40965	1019.41718	1019.42460	1019.43194	1024.43918
305+	1024.44633	1024.45338	1029.46035	1029.46723	1029.47403
310+	1034.48074	1034.48736	1034.49390	1039.50036	1039.50673
315+	1039.51302	1044.51924	1044.52538	1044.53143	1044.53741
320+	1049.54332	1049.54915	1049.55491	1054.56059	1054.56620
325+	1054.57174	1059.57721	1059.58261	1059.58795	1064.59321
330+	1064.59841	1064.60354	1064.60861	1069.61361	1069.61855

EXPECTATION E(T/GLF)=(1+BA)SUM(D(R,M)),B= 6

R	1	2	3	4	5
			A=1/10		
0+	1.00000	2.93435	4.70816	6.37019	7.99050
			A=1/ 9		
0+	1.00000	3.00736	4.87581	6.62240	8.31790
			A=1/ 8		
0+	1.00000	3.09149	5.07685	6.93160	8.72365
5+	10.48994	12.24617	13.99854	15.74944	17.49979
			A=1/ 7		
0+	1.00000	3.18903	5.32095	7.31764	9.23797
5+	11.12255	12.99140	14.85351	16.71275	18.57078
			A=1/ 6		
0+	1.00000	3.30259	5.62091	7.80916	9.90695
5+	11.95529	13.97866	15.98985	17.99517	19.99771
			A=1/ 5		
0+	1.00000	3.43481	5.99279	8.44683	10.80160
5+	13.09049	15.34013	17.56743	19.78233	21.99042
			A=1/ 4		
0+	1.00000	3.58722	6.45348	9.28345	12.02700
5+	14.69394	17.30428	19.87576	22.42149	24.95054
10+	27.46889	29.98046	32.48773	34.99230	37.49516
			A=1/ 3		
0+	1.00000	3.75669	7.00793	10.36308	13.70379
5+	16.99498	20.23105	23.41690	26.56060	29.67039
10+	32.75358	35.81623	38.86321	41.89833	44.92451
15+	47.94399	50.95847	53.96922	56.97720	59.98311
			A=0.40		
0+	1.00000	3.84319	7.30818	10.98163	14.71532
5+	18.44838	22.15536	25.82690	29.46128	33.06038
10+	36.62764	40.16705	43.68253	47.17776	50.65601
15+	54.12012	57.57252	61.01527	64.45009	67.87842
20+	71.30144	74.72014	78.13531	81.54761	84.95758
			A=1/ 2		
0+	1.00000	3.92430	7.59951	11.60363	15.76930
5+	20.01579	24.29963	28.59602	32.89040	37.17425
10+	41.44271	45.69324	49.92476	54.13717	58.33092
15+	62.50687	66.66606	70.80965	74.93886	79.05488
20+	83.15886	87.25191	91.33507	95.40931	99.47550
25+	103.53448	107.58698	111.63369	115.67521	119.71211
30+	123.74488	127.77397	131.79978	135.82268	139.84298
35+	143.86098	147.87693	151.89107	155.90359	159.91468
			A=0.60		
0+	1.00000	3.97043	7.76858	11.97308	16.41103
5+	20.99466	25.67335	30.41539	35.19978	40.01207
10+	44.84201	49.68223	54.52732	59.37331	64.21726
15+	69.05701	73.89099	78.71805	83.53741	88.34853
20+	93.15110	97.94494	102.73002	107.50638	112.27416
25+	117.03353	121.78473	126.52799	131.26360	135.99185
30+	140.71303	145.42746	150.13542	154.83723	159.53317
35+	164.22354	168.90863	173.58870	178.26402	182.93485
40+	187.60143	192.26399	196.92276	201.57795	206.22977
45+	210.87841	215.52406	220.16689	224.80707	229.44475
50+	234.08008	238.71321	243.34426	247.97337	252.60064
55+	257.22620	261.85015	266.47258	271.09359	275.71327
60+	280.33169	284.94895	289.56511	294.18023	298.79440
65+	303.40765	308.02006	312.63167	317.24254	321.85272
			A=0.65		
0+	1.00000	3.98529	7.82334	12.09366	16.62239
5+	21.32036	26.13542	31.03434	35.99462	41.00038

EXPECTATION E(T/GLF)=(1+BA)SUM(D(R,M)),B= 6

R	1	2	3	4	5
		A=0.65			
10+	46.04000	51.10482	56.18821	61.28504	66.39128
15+	71.50376	76.61995	81.73782	86.85574	91.97243
20+	97.08683	102.19811	107.30562	112.40883	117.50735
25+	122.60087	127.68916	132.77207	137.84950	142.92138
30+	147.98770	153.04848	158.10375	163.15358	168.19805
35+	173.23726	178.27130	183.30030	188.32437	193.34366
40+	198.35828	203.36838	208.37409	213.37554	218.37288
45+	223.36625	228.35577	233.34159	238.32383	243.30262
50+	248.27809	253.25036	258.21956	263.18580	268.14920
55+	273.10987	278.06791	283.02343	287.97653	292.92732
60+	297.87588	302.82230	307.76668	312.70910	317.64963
65+	322.58837	327.52539	332.46075	337.39454	342.32681
70+	347.25764	352.18708	357.11519	362.04204	366.96768
75+	371.89215	376.81552	381.73783	386.65913	391.57946
80+	396.49886	401.41738	406.33505	411.25192	416.16802
85+	421.08338	425.99803	430.91202	435.82536	440.73809
90+	445.65024	450.56182	455.47288	460.38342	465.29348
95+	465.30308	470.21223	475.12096	475.12929	480.03723
		A=0.70			
0+	1.00000	3.99632	7.86399	12.18327	16.77981
5+	21.56363	26.48172	31.49996	36.59501	41.75015
10+	46.95299	52.19409	57.46608	62.76315	68.08059
15+	73.41460	78.76202	84.12024	89.48710	94.86074
20+	100.23963	105.62244	111.00806	116.39554	121.78404
25+	127.17288	132.56144	137.94922	143.33578	148.72073
30+	154.10376	159.48460	164.86300	170.23878	175.61178
35+	180.98185	186.34890	191.71283	197.07357	202.43107
40+	207.78529	213.13621	218.48382	223.82811	229.16909
45+	234.50676	239.84117	245.17232	250.50025	255.82500
50+	261.14660	266.46510	271.78055	277.09299	282.40247
55+	287.70905	293.01278	298.31371	303.61189	308.90740
60+	314.20027	319.49057	324.77835	330.06367	335.34659
65+	340.62716	345.90544	351.18148	356.45533	361.72706
70+	366.99671	372.26434	377.52999	382.79373	388.05559
75+	393.31563	398.57390	403.83044	409.08530	414.33853
80+	419.59017	424.84026	430.08885	435.33597	440.58168
85+	445.82600	451.06899	456.31066	461.55108	466.79026
90+	472.02824	477.26507	482.50076	487.73536	492.96890
95+	498.20141	503.43292	508.66345	513.89305	519.12172
100+	524.34951	529.57644	529.60253	534.82781	540.05230
105+	540.07603	545.29902	550.52129	550.54286	555.76377
110+	560.98401	561.00362	566.22262	566.24102	571.45884
115+	576.67610	576.69282	581.90901	581.92469	587.13988
120+	592.35458	592.36882	597.58262	597.59597	602.80890
125+	608.02142	608.03355	613.24529	613.25666	618.46767
130+	618.47832	623.68864	623.69863	628.90831	628.91767
135+	634.12674	634.13552	639.34401	639.35224	644.56020
		A=0.75			
0+	1.00000	4.00438	7.89361	12.24853	16.89447
5+	21.74097	26.73451	31.84045	37.03497	42.30088
10+	47.62534	52.99849	58.41255	63.86131	69.33969
15+	74.84351	80.36929	85.91406	91.47533	97.05095
20+	102.63906	108.23806	113.84653	119.46326	125.08714
25+	130.71724	136.35270	141.99276	147.63675	153.28408
30+	158.93420	164.58663	170.24094	175.89673	181.55366
35+	187.21141	192.86970	198.52827	204.18688	209.84532
40+	215.50341	221.16097	226.81786	232.47392	238.12904

EXPECTATION E(T/GLF)=(1+BA)SUM(D(R,M)),B= 6

R	1	2	3	4	5
			A=0.75		
45+	243.78310	249.43601	255.08766	260.73798	266.38690
50+	272.03435	277.68028	283.32462	288.96735	294.60842
55+	300.24780	305.88545	311.52136	317.15550	322.78786
60+	328.41842	334.04717	339.67411	345.29924	350.92254
65+	356.54402	362.16368	367.78153	373.39756	379.01180
70+	384.62424	390.23490	395.84378	401.45090	407.05628
75+	412.65992	418.26185	423.86207	429.46061	435.05748
80+	440.65269	446.24627	451.83824	457.42861	463.01740
85+	468.60463	474.19032	479.77450	485.35717	490.93837
90+	496.51810	502.09640	507.67327	513.24875	518.82285
95+	524.39559	529.96699	535.53708	541.10586	546.67337
100+	552.23962	557.80463	563.36843	568.93102	574.49244
105+	580.05269	585.61180	591.16979	591.22667	596.78247
110+	602.33720	607.89089	607.94354	613.49517	619.04581
115+	619.09547	624.64417	630.19193	630.23876	635.78467
120+	641.32969	641.37383	646.91710	652.45953	652.50112
125+	658.04190	658.08187	663.62105	669.15946	669.19711
130+	674.73402	674.77019	680.30564	685.84039	685.87445
135+	691.40783	691.44054	696.97260	697.00402	702.53480
140+	708.06497	708.09454	713.62351	713.65190	719.17972
145+	719.20698	724.73368	724.75985	730.28549	730.31061
150+	735.83523	741.35934	741.38296	746.90611	746.92879
155+	752.45100	752.47276	757.99408	758.01497	763.53543
160+	763.55547	769.07511	769.09434	774.61318	774.63163
165+	780.14971	780.16741	780.18476	785.70174	785.71838
170+	791.23468	791.25064	796.76628	796.78159	802.29659
175+	802.31127	807.82566	807.83975	813.35355	813.36706
180+	813.38030	818.89326	818.90596	824.41839	824.43056
185+	829.94248	829.95416	829.96559	835.47679	835.48776
190+	840.99849	841.00901	846.51931	846.52939	846.53926
195+	852.04893	852.05840	857.56767	857.57675	857.58564
200+	863.09434	863.10286	868.61121	868.61938	868.62738
205+	874.13521	874.14288	879.65039	879.65775	879.66495
210+	885.17200	885.17890	890.68566	890.69228	890.69876
			A=0.80		
0+	1.00000	4.01011	7.91460	12.29466	16.97542
5+	21.86609	26.91280	32.08063	37.34545	42.68981
10+	48.10061	53.56774	59.08322	64.64061	70.23464
15+	75.86092	81.51578	87.19609	92.89918	98.62272
20+	104.36470	110.12335	115.89712	121.68464	127.48466
25+	133.29610	139.11797	144.94938	150.78954	156.63771
30+	162.49323	168.35551	174.22399	180.09816	185.97757
35+	191.86178	197.75041	203.64310	209.53951	215.43932
40+	221.34227	227.24806	233.15647	239.06726	244.98022
45+	250.89514	256.81185	262.73016	268.64991	274.57095
50+	280.49315	286.41636	292.34046	298.26534	304.19088
55+	310.11699	316.04356	321.97051	327.89775	333.82521
60+	339.75279	345.68045	351.60810	357.53569	363.46317
65+	369.39046	375.31753	381.24433	387.17080	393.09691
70+	399.02262	404.94789	410.87268	416.79696	422.72070
75+	428.64387	434.56645	440.48841	446.40972	452.33036
80+	458.25032	464.16957	470.08809	476.00588	481.92291
85+	487.83916	493.75464	499.66932	505.58319	511.49624
90+	517.40847	523.31986	529.23041	535.14012	541.04896
95+	546.95695	552.86407	558.77031	564.67569	570.58019
100+	576.48381	582.38654	588.28840	594.18937	600.08945
105+	605.98865	611.88697	617.78440	623.68094	629.57661

EXPECTATION E(T/GLF)=(1+BA)SUM(D(R,M)),B= 6

R	1	2	3	4	5
		A=0.80			
110+	635.47139	641.36529	647.25831	653.15046	659.04174
115+	659.13214	665.02168	670.91035	676.79816	676.88512
120+	682.77122	688.65647	688.74088	694.62445	700.50718
125+	700.58907	706.47014	712.35039	712.42982	718.30844
130+	724.18624	724.26325	730.13946	736.01487	736.08950
135+	741.96335	747.83642	747.90872	753.78026	759.65104
140+	759.72106	765.59034	765.65887	771.52667	777.39374
145+	777.46009	783.32571	783.39063	789.25484	795.11834
150+	795.18116	801.04328	801.10472	806.96548	807.02558
155+	812.88501	818.74377	818.80189	824.65936	824.71619
160+	830.57238	830.62794	836.48288	836.53720	842.39091
165+	842.44401	848.29651	854.14842	854.19974	860.05048
170+	860.10063	865.95022	865.99924	871.84770	871.89560
175+	877.74295	877.78976	883.63604	883.68177	889.52698
180+	889.57167	895.41584	895.45950	901.30265	901.34531
185+	907.18746	907.22912	913.07030	913.11100	918.95122
190+	918.99097	924.83026	924.86909	930.70745	930.74537
195+	930.78285	936.61988	936.65647	942.49263	942.52837
200+	948.36368	948.39858	954.23306	954.26713	960.10080
205+	960.13406	960.16694	965.99942	966.03151	971.86322
210+	971.89455	977.72551	977.75610	983.58632	983.61618
215+	983.64568	989.47483	989.50363	995.33208	995.36019
220+	1001.18797	1001.21541	1001.24251	1007.06929	1007.09575
225+	1012.92189	1012.94771	1018.77322	1018.79843	1018.82332
230+	1024.64792	1024.67221	1030.49622	1030.51993	1030.54335
235+	1036.36649	1036.38935	1042.21193	1042.23423	1042.25626
240+	1048.07803	1048.09953	1053.92076	1053.94174	1053.96246
245+	1059.78293	1059.80315	1065.62312	1065.64284	1065.66233
250+	1071.48157	1071.50058	1071.51936	1077.33791	1077.35622
255+	1083.17432	1083.19219	1083.20984	1089.02727	1089.04449
260+	1089.06150	1094.87830	1094.89489	1100.71128	1100.72746
265+	1100.74345	1106.55924	1106.57483	1106.59023	1112.40544
270+	1112.42046	1112.43530	1118.24995	1118.26442	1118.27872
275+	1124.09283	1124.10677	1129.92054	1129.93414	1129.94757
280+	1135.76083	1135.77392	1135.78686	1141.59963	1141.61225
285+	1141.62471	1147.43701	1147.44916	1147.46116	1153.27301
290+	1153.28471	1153.29627	1159.10768	1159.11895	1159.13008
295+	1164.94107	1164.95193	1164.96264	1170.77323	1170.78368
300+	1170.79400	1176.60420	1176.61426	1176.62420	1182.43402
305+	1182.44371	1182.45328	1188.26274	1188.27207	1188.28129
310+	1194.09039	1194.09938	1194.10825	1194.11701	1199.92567
315+	1199.93421	1199.94265	1205.75098	1205.75921	1205.76734
320+	1211.57536	1211.58328	1211.59110	1217.39883	1217.40645
325+	1217.41399	1223.22142	1223.22877	1223.23602	1223.24318
330+	1229.05024	1229.05723	1229.06412	1234.87092	1234.87764
335+	1234.88428	1240.69083	1240.69730	1240.70369	1240.71000
340+	1246.51622	1246.52237	1246.52845	1252.33444	1252.34036
345+	1252.34621	1252.35198	1258.15768	1258.16330	1258.16886

EXPECTATION E(T/GLF)=(1+BA)SUM(D(R,M)),B= 7

R	1	2	3	4	5
		A=1/10			
0+	1.00000	3.07835	4.98727	6.76320	8.48825
		A=1/ 9			
0+	1.00000	3.16070	5.18065	7.05634	8.86974
5+	10.66003	12.44215	14.22143	15.99972	17.77768
		A=1/ 8			
0+	1.00000	3.25510	5.41184	7.41515	9.34219
5+	11.23745	13.12022	14.99818	16.87431	18.74973
		A=1/ 7			
0+	1.00000	3.36383	5.69140	7.86211	9.94037
5+	11.97457	13.98922	15.99544	17.99807	19.99918
		A=1/ 6			
0+	1.00000	3.48938	6.03292	8.42914	10.71682
5+	12.94381	15.13980	17.32053	19.49391	21.66377
		A=1/ 5			
0+	1.00000	3.63400	6.45281	9.16048	11.75128
5+	14.26205	16.72433	19.15874	21.57758	23.98784
		A=1/ 4			
0+	1.00000	3.79842	6.96671	10.11079	13.15805
5+	16.11459	19.00241	21.84229	24.65008	27.43693
10+	30.21027	32.97501	35.73430	38.49014	41.24380
		A=1/ 3			
0+	1.00000	3.97790	7.57413	11.31734	15.05643
5+	18.74287	22.36607	25.92978	29.44256	32.91382
10+	36.35217	39.76492	43.15801	46.53614	49.90295
15+	53.26124	56.61316	59.96031	63.30390	66.64485
		A=0.40			
0+	1.00000	4.06789	7.89691	11.99591	16.18186
5+	20.37679	24.54655	28.67736	32.76544	36.81193
10+	40.82028	44.79481	48.74005	52.66029	56.55942
15+	60.44088	64.30764	68.16224	72.00680	75.84313
20+	79.67271	83.49676	87.31630	91.13216	94.94504
		A=1/ 2			
0+	1.00000	4.15103	8.20488	12.66624	17.33348
5+	22.10756	26.93429	31.78214	36.63221	41.47311
10+	46.29809	51.10338	55.88713	60.64874	65.38844
15+	70.10698	74.80542	79.48501	84.14709	88.79303
20+	93.42416	98.04179	102.64715	107.24140	111.82560
25+	116.40075	120.96777	125.52748	130.08065	134.62796
30+	139.17002	143.70741	148.24063	152.77013	157.29631
35+	161.81955	166.34016	170.85843	175.37464	179.88900
		A=0.60			
0+	1.00000	4.19763	8.38059	13.05696	18.02075
5+	23.16627	28.43202	33.77919	39.18187	44.62209
10+	50.08699	55.56724	61.05591	66.54787	72.03926
15+	77.52721	83.00955	88.48472	93.95157	99.40930
20+	104.85737	110.29547	115.72342	121.14120	126.54889
25+	131.94662	137.33460	142.71309	148.08237	153.44275
30+	158.79457	164.13816	169.47386	174.80202	180.12298
35+	185.43708	190.74465	196.04601	201.34147	206.63135
40+	211.91594	217.19551	222.47035	227.74070	233.00682
45+	238.26895	243.52730	248.78211	254.03356	259.28186
50+	264.52719	269.76972	275.00962	280.24705	285.48214
55+	290.71505	295.94591	301.17483	306.40194	311.62735
60+	316.85116	322.07346	327.29436	332.51393	337.73227
65+	342.94944	348.16553	353.38059	358.59469	363.80789
70+	369.02025	374.23182	379.44265	384.65279	389.86228

EXPECTATION E(T/GLF)=(1+BA)SUM(D(R,M)),B= 7

R	1	2	3	4	5
		A=0.65			
0+	1.00000	4.21250	8.43687	13.18289	18.24402
5+	23.51340	28.92811	34.44785	40.04519	45.70067
10+	51.40000	57.13245	62.88975	68.66548	74.45458
15+	80.25301	86.05753	91.86554	97.67493	103.48400
20+	109.29135	115.09587	120.89666	126.69301	132.48433
25+	138.27019	144.05023	149.82422	155.59195	161.35332
30+	167.10825	172.85671	178.59873	184.33432	190.06357
35+	195.78655	201.50337	207.21414	212.91900	218.61807
40+	224.31151	229.99946	235.68206	241.35949	247.03189
45+	252.69941	258.36223	264.02048	269.67433	275.32392
50+	280.96940	286.61093	292.24865	297.88269	303.51319
55+	309.14028	314.76411	320.38478	326.00243	331.61717
60+	337.22912	342.83839	348.44509	354.04932	359.65119
65+	365.25078	370.84820	376.44354	382.03687	387.62829
70+	393.21788	398.80570	404.39185	409.97639	415.55938
75+	421.14090	426.72101	432.29977	437.87723	443.45346
80+	449.02851	454.60243	460.17526	465.74707	471.31788
85+	476.88776	482.45673	488.02483	493.59212	499.15862
90+	504.72436	510.28939	515.85373	521.41741	521.43047
95+	526.99293	532.55482	532.56617	538.12699	543.68732
		A=0.70			
0+	1.00000	4.22349	8.47839	13.27582	18.40902
5+	23.77050	29.29657	34.94611	40.69088	46.51058
10+	52.39014	58.31802	64.28520	70.28451	76.31018
15+	82.35747	88.42247	94.50194	100.59314	106.69377
20+	112.80187	118.91577	125.03405	131.15547	137.27899
25+	143.40370	149.52881	155.65364	161.77762	167.90024
30+	174.02107	180.13973	186.25591	192.36934	198.47978
35+	204.58703	210.69095	216.79138	222.88823	228.98140
40+	235.07083	241.15647	247.23827	253.31622	259.39030
45+	265.46052	271.52688	277.58941	283.64813	289.70306
50+	295.75425	301.80174	307.84556	313.88578	319.92245
55+	325.95561	331.98532	338.01166	344.03466	350.05441
60+	356.07095	362.08436	368.09469	374.10202	380.10640
65+	386.10790	392.10659	398.10252	404.09577	410.08639
70+	416.07444	422.06000	428.04311	434.02385	440.00226
75+	445.97842	451.95237	457.92417	463.89387	469.86155
80+	475.82723	481.79099	487.75287	493.71291	499.67118
85+	505.62772	511.58257	517.53579	523.48741	529.43749
90+	535.38606	541.33317	547.27886	553.22316	559.16613
95+	565.10780	571.04819	576.98736	582.92534	588.86215
100+	594.79784	594.83244	600.76598	606.69849	606.72999
105+	612.66053	618.59013	618.61881	624.54661	630.47355
110+	630.49965	636.42494	642.34945	642.37320	648.29621
115+	654.21850	654.24010	660.16102	660.18129	666.10093
120+	666.11995	672.03838	677.95623	677.97352	683.89027
125+	683.90649	689.82220	689.83742	695.75216	695.76643
130+	701.68026	701.69365	707.60662	713.51917	713.53133
135+	719.44311	719.45451	725.36556	725.37625	731.28660
140+	731.29663	737.20634	737.21574	743.12484	743.13365
		A=0.75			
0+	1.00000	4.23148	8.50848	13.34306	18.52835
5+	23.95647	29.56331	35.30726	41.15965	47.09973
10+	53.11199	59.18446	65.30773	71.47422	77.67777
15+	83.91326	90.17644	96.46368	102.77190	109.09848
20+	115.44112	121.79783	128.16687	134.54673	140.93604
25+	147.33362	153.73840	160.14945	166.56591	172.98703

EXPECTATION E(T/GLF)=(1+BA)SUM(D(R,M)),B= 7

R	1	2	3	4	5
			A=0.75		
30+	179.41212	185.84058	192.27186	198.70546	205.14093
35+	211.57786	218.01588	224.45467	230.89392	237.33336
40+	243.77273	250.21182	256.65042	263.08834	269.52542
45+	275.96150	282.39645	288.83014	295.26247	301.69332
50+	308.12262	314.55027	320.97621	327.40037	333.82270
55+	340.24314	346.66164	353.07818	359.49272	365.90522
60+	372.31567	378.72404	385.13032	391.53450	397.93656
65+	404.33650	410.73431	417.13000	423.52357	429.91501
70+	436.30434	442.69156	449.07668	455.45971	461.84066
75+	468.21955	474.59638	480.97118	487.34396	493.71474
80+	500.08353	506.45036	512.81524	519.17819	525.53923
85+	531.89840	538.25569	544.61115	550.96478	557.31661
90+	563.66667	570.01498	576.36155	582.70642	589.04960
95+	595.39112	601.73100	608.06927	614.40594	620.74104
100+	627.07460	633.40664	639.73717	646.06623	652.39383
105+	658.71999	665.04475	671.36812	671.44013	677.76079
110+	684.08012	684.14816	690.46492	696.78041	703.09467
115+	703.15772	709.46956	715.78023	715.83974	722.14812
120+	728.45538	728.51154	734.81662	734.87064	741.17362
125+	747.47558	747.52653	753.82649	753.87548	760.17352
130+	766.47063	766.51681	772.81210	772.85649	779.15002
135+	785.44270	785.48453	791.77554	791.81575	798.10516
140+	798.14379	804.43166	810.71878	810.75517	817.04083
145+	817.07578	823.36004	823.39362	829.67653	829.70878
150+	835.99038	836.02136	842.30171	842.33146	848.61061
155+	848.63918	854.91717	854.94460	861.22148	861.24782
160+	867.52362	867.54891	873.82368	873.84796	880.12174
165+	880.14504	886.41787	886.44024	892.71215	892.73362
170+	899.00466	899.02526	905.29545	905.31522	905.33460
175+	911.60358	911.62217	917.89038	917.90822	924.17570
180+	924.19282	930.45958	930.47601	930.49210	936.75786
185+	936.77330	943.03842	943.05323	949.31773	949.33194
190+	949.34586	955.60949	955.62284	961.88591	961.89872
195+	961.91126	968.17354	968.18557	974.44735	974.45889
200+	980.72019	980.73126	980.74210	987.00271	987.01311
205+	987.02329	993.28326	993.29302	999.55258	999.56194
210+	999.57111	1005.83009	1005.83888	1012.09749	1012.10593
215+	1012.11418	1018.37227	1018.38018	1018.38794	1024.64553
			A=0.80		
0+	1.00000	4.23714	8.52969	13.39032	18.61205
5+	24.08674	29.74999	35.55993	41.48760	47.51201
10+	53.61741	59.79159	66.02493	72.30964	78.63935
15+	85.00876	91.41342	97.84952	104.31382	110.80351
20+	117.31612	123.84952	130.40179	136.97127	143.55645
25+	150.15599	156.76868	163.39344	170.02928	176.67532
30+	183.33072	189.99477	196.66676	203.34610	210.03219
35+	216.72452	223.42259	230.12597	236.83423	243.54699
40+	250.26389	256.98460	263.70881	270.43623	277.16659
45+	283.89965	290.63517	297.37292	304.11271	310.85435
50+	317.59764	324.34243	331.08855	337.83586	344.58422
55+	351.33349	358.08355	364.83429	371.58559	378.33736
60+	385.08949	391.84190	398.59450	405.34721	412.09995
65+	418.85265	425.60525	432.35767	439.10987	445.86177
70+	452.61334	459.36452	466.11526	472.86552	479.61525
75+	486.36443	493.11300	499.86094	506.60822	513.35479
80+	520.10065	526.84575	533.59007	540.33359	547.07629
85+	553.81815	560.55915	567.29926	574.03848	580.77679

EXPECTATION E(T/GLF)=(1+BA)SUM(D(R,M)),B= 7

R	1	2	3	4	5
		A=0.80			
90+	587.51418	594.25062	600.98612	607.72065	614.45422
95+	621.18680	627.91840	634.64900	641.37860	648.10718
100+	654.83476	661.56132	668.28685	675.01136	681.73484
105+	688.45729	695.17870	701.89908	708.61843	715.33674
110+	722.05402	728.77026	735.48547	742.19964	748.91279
115+	749.02491	755.73599	762.44606	762.55510	769.26313
120+	775.97013	782.67613	782.78112	789.48511	796.18809
125+	796.29008	802.99108	809.69110	809.79013	816.48819
130+	823.18527	823.28139	829.97655	836.67076	836.76401
135+	843.45633	850.14770	850.23815	856.92767	857.01627
140+	863.70396	870.39074	870.47663	877.16162	877.24572
145+	883.92894	890.61129	890.69277	897.37338	897.45315
150+	904.13207	910.81014	910.88739	917.56380	917.63940
155+	924.31418	924.38815	931.06133	931.13371	937.80530
160+	944.47612	944.54616	951.21543	951.28395	957.95171
165+	958.01873	964.68501	964.75055	971.41537	971.47947
170+	978.14286	978.20554	984.86752	984.92881	991.58942
175+	991.64934	998.30859	998.36718	1005.02510	1005.08237
180+	1011.73900	1011.79498	1018.45033	1018.50505	1025.15914
185+	1025.21262	1031.86549	1031.91776	1038.56943	1038.62051
190+	1045.27100	1045.32091	1051.97025	1052.01902	1058.66723
195+	1058.71488	1065.36198	1065.40853	1072.05455	1072.10004
200+	1078.74499	1078.78942	1078.83334	1085.47675	1085.51964
205+	1092.16204	1092.20395	1098.84536	1098.88629	1105.52674
210+	1105.56671	1105.60622	1112.24526	1112.28384	1118.92196
215+	1118.95964	1125.59688	1125.63367	1125.67003	1132.30596
220+	1132.34146	1138.97654	1139.01121	1145.64547	1145.67932
225+	1145.71276	1152.34581	1152.37847	1159.01074	1159.04262
230+	1159.07412	1165.70524	1165.73600	1172.36638	1172.39641
235+	1179.02607	1179.05537	1179.08433	1185.71294	1185.74120
240+	1185.76913	1192.39672	1192.42398	1199.05091	1199.07751
245+	1199.10380	1205.72976	1205.75542	1212.38076	1212.40580
250+	1212.43053	1219.05497	1219.07911	1225.70296	1225.72652
255+	1225.74979	1232.37278	1232.39549	1232.41793	1239.04009
260+	1239.06199	1239.08362	1245.70498	1245.72609	1252.34693
265+	1252.36753	1252.38787	1259.00796	1259.02781	1259.04742
270+	1265.66678	1265.68592	1265.70481	1272.32348	1272.34191
275+	1278.96012	1278.97811	1278.99588	1285.61343	1285.63076
280+	1285.64788	1292.26480	1292.28150	1292.29800	1298.91429
285+	1298.93039	1298.94629	1305.56199	1305.57750	1305.59281
290+	1312.20794	1312.22289	1312.23764	1318.85222	1318.86662
295+	1318.88083	1325.49488	1325.50875	1325.52244	1332.13597
300+	1332.14934	1332.16253	1338.77556	1338.78844	1338.80115
305+	1345.41370	1345.42610	1345.43835	1352.05044	1352.06239
310+	1352.07418	1352.08583	1358.69734	1358.70870	1358.71992
315+	1365.33100	1365.34194	1365.35275	1371.96343	1371.97397
320+	1371.98438	1378.59465	1378.60481	1378.61483	1385.22473
325+	1385.23451	1385.24417	1385.25370	1391.86312	1391.87242
330+	1391.88160	1398.49067	1398.49962	1398.50847	1405.11720
335+	1405.12582	1405.13434	1405.14275	1411.75105	1411.75925
340+	1411.76735	1418.37535	1418.38324	1418.39104	1418.39874
345+	1425.00635	1425.01386	1425.02127	1431.62859	1431.63582
350+	1431.64296	1431.65001	1438.25697	1438.26385	1438.27063
355+	1444.87734	1444.88396	1444.89049	1444.89695	1451.50332

EXPECTATION E(T/GLF)=(1+BA)SUM(D(R,M)),B= 8

R	1	2	3	4	5
		A=1/10			
0+	1.00000	3.21952	5.26489	7.15565	8.98581
		A=1/ 9			
0+	1.00000	3.31062	5.48345	7.48941	9.42125
5+	11.32528	13.21943	15.11015	16.99966	18.88877
		A=1/ 8			
0+	1.00000	3.41454	5.74399	7.89734	9.96016
5+	11.98474	13.99418	15.99778	17.99915	19.99968
		A=1/ 7			
0+	1.00000	3.53350	6.05777	8.40434	10.64169
5+	12.82610	14.98682	17.13727	19.28335	21.42757
		A=1/ 6			
0+	1.00000	3.66982	6.43901	9.04530	11.52457
5+	13.93124	16.30038	18.65094	20.99251	23.32977
		A=1/ 5			
0+	1.00000	3.82536	6.90411	9.86743	12.69654
5+	15.43091	18.10695	20.74916	23.37233	25.98497
		A=1/ 4			
0+	1.00000	4.00000	7.46701	10.92610	14.27948
5+	17.52816	20.69557	23.80544	26.87643	29.92185
10+	32.95070	35.96896	38.98048	41.98773	44.99229
		A=1/ 3			
0+	1.00000	4.18756	8.12163	12.25037	16.38822
5+	20.47187	24.48481	28.42910	32.31350	36.14846
10+	39.94385	43.70823	47.44866	51.17077	54.87896
15+	58.57665	62.26646	65.95035	69.62982	73.30600
		A=0.40			
0+	1.00000	4.28016	8.46372	12.98283	17.61870
5+	22.27537	26.90917	31.50135	36.04566	40.54223
10+	44.99430	49.40648	53.78378	58.13108	62.45291
15+	66.75331	71.03580	75.30340	79.55870	83.80386
20+	88.04068	92.27066	96.49506	100.71489	104.93099
		A=1/ 2			
0+	1.00000	4.36464	8.78539	13.69514	18.85770
5+	24.15528	29.52249	34.92075	40.32651	45.72526
10+	51.10817	56.47007	61.80817	67.12131	72.40938
15+	77.67296	82.91310	88.13108	93.32835	98.50642
20+	103.66681	108.81100	113.94041	119.05638	124.16018
25+	129.25298	134.33586	139.40982	144.47576	149.53451
30+	154.58682	159.63336	164.67476	169.71155	174.74425
35+	179.77328	184.79906	189.82194	194.84224	199.86025
40+	204.87622	209.89038	214.90292	219.91405	224.92390
		A=0.60			
0+	1.00000	4.41141	8.96624	14.10361	19.58428
5+	25.28437	31.13129	37.07892	43.09627	49.16169
10+	55.25963	61.37864	67.51022	73.64798	79.78706
15+	85.92381	92.05546	98.17993	104.29568	110.40161
20+	116.49693	122.58114	128.65391	134.71513	140.76477
25+	146.80294	152.82981	158.84562	164.85065	170.84524
30+	176.82973	182.80449	188.76988	194.72630	200.67414
35+	206.61376	212.54556	218.46990	224.38715	230.29766
40+	236.20177	242.09981	247.99211	253.87896	259.76066
45+	265.63750	271.50974	277.37763	283.24143	289.10136
50+	294.95766	300.81051	306.66013	312.50671	318.35042
55+	324.19143	330.02991	335.86600	341.69984	347.53158
60+	353.36134	359.18923	365.01538	370.83989	376.66285
65+	382.48437	388.30454	394.12343	399.94112	405.75770
70+	411.57322	417.38776	423.20138	429.01413	434.82606

EXPECTATION E(T/GLF)=(1+BA)SUM(D(R,M)), B= 8

R	1	2	3	4	5
		A=0.65			
0+	1.00000	4.42623	9.02363	14.23384	19.81753
5+	25.64989	31.65705	37.79147	44.02068	50.32152
10+	56.67692	63.07400	69.50280	75.95555	82.42608
15+	88.90945	95.40167	101.89953	108.40039	114.90210
20+	121.40290	127.90138	134.39636	140.88690	147.37224
25+	153.85177	160.32502	166.79162	173.25128	179.70382
30+	186.14910	192.58704	199.01761	205.44082	211.85670
35+	218.26534	224.66682	231.06125	237.44876	243.82948
40+	250.20358	256.57121	262.93252	269.28770	275.63691
45+	281.98032	288.31812	294.65047	300.97756	307.29954
50+	313.61660	319.92891	326.23662	332.53990	338.83891
55+	345.13381	351.42475	357.71187	363.99532	370.27524
60+	376.55177	382.82504	389.09517	395.36229	401.62653
65+	407.88798	414.14678	420.40303	426.65682	432.90827
70+	439.15746	445.40450	451.64946	457.89245	464.13354
75+	470.37281	476.61034	482.84620	489.08047	495.31320
80+	501.54448	507.77436	514.00290	520.23017	526.45620
85+	532.68107	538.90482	545.12749	551.34915	557.56982
90+	563.78956	570.00840	576.22639	582.44357	582.45996
95+	588.67561	594.89054	594.90480	601.11840	607.33139
100+	607.34378	613.55560	619.76689	619.77765	625.98793
		A=0.70			
0+	1.00000	4.43713	9.06575	14.32937	19.98875
5+	25.91864	32.04452	38.31810	44.70613	51.18468
10+	57.73586	64.34601	71.00438	77.70244	84.43326
15+	91.19117	97.97146	104.77021	111.58412	118.41038
20+	125.24660	132.09075	138.94107	145.79605	152.65438
25+	159.51493	166.37673	173.23891	180.10076	186.96161
30+	193.82092	200.67822	207.53307	214.38513	221.23408
35+	228.07968	234.92168	241.75991	248.59420	255.42444
40+	262.25050	269.07232	275.88982	282.70297	289.51172
45+	296.31606	303.11599	309.91151	316.70264	323.48940
50+	330.27182	337.04993	343.82379	350.59343	357.35891
55+	364.12029	370.87762	377.63097	384.38041	391.12599
60+	397.86778	404.60586	411.34030	418.07117	424.79853
65+	431.52246	438.24303	444.96032	451.67440	458.38533
70+	465.09319	471.79805	478.49999	485.19906	491.89533
75+	498.58889	505.27978	511.96809	518.65386	525.33718
80+	532.01809	538.69666	545.37296	552.04703	558.71895
85+	565.38876	572.05653	578.72230	585.38614	592.04809
90+	598.70820	605.36653	612.02313	618.67803	625.33130
95+	631.98298	638.63310	645.28172	651.92887	658.57461
100+	665.21896	665.26197	671.90367	678.54411	678.58332
105+	685.22134	691.85820	691.89393	698.52857	705.16215
110+	705.19471	711.82626	718.45684	718.48648	725.11521
115+	725.14305	731.77004	738.39619	738.42153	745.04608
120+	745.06988	751.69293	758.31527	758.33692	764.95789
125+	764.97821	771.59789	771.61696	778.23544	778.25334
130+	784.87068	784.88747	791.50374	791.51950	798.13477
135+	798.14956	804.76388	804.77775	811.39119	811.40420
140+	818.01680	818.02901	824.64083	824.65228	831.26337
145+	831.27411	837.88450	837.89457	844.50432	844.51377
		A=0.75			
0+	1.00000	4.44502	9.09613	14.39812	20.11183
5+	26.11175	32.32300	38.69690	45.19978	51.80728
10+	58.50112	65.26721	72.09443	78.97382	85.89807
15+	92.86113	99.85793	106.88415	113.93615	121.01077

EXPECTATION E(T/GLF)=(1+BA)SUM(D(R,M)),B= 8

R	1	2	3	4	5
		A=0.75			
20+	128.10526	135.21725	142.34466	149.48564	156.63858
25+	163.80202	170.97470	178.15546	185.34327	192.53723
30+	199.73648	206.94030	214.14799	221.35896	228.57265
35+	235.78855	243.00621	250.22522	257.44520	264.66581
40+	271.88674	279.10769	286.32842	293.54869	300.76828
45+	307.98700	315.20468	322.42115	329.63627	336.84990
50+	344.06193	351.27224	358.48075	365.68735	372.89199
55+	380.09457	387.29505	394.49336	401.68947	408.88331
60+	416.07487	423.26410	430.45099	437.63550	444.81761
65+	451.99733	459.17463	466.34950	473.52194	480.69196
70+	487.85954	495.02470	502.18743	509.34775	516.50566
75+	523.66118	530.81432	537.96509	545.11351	552.25959
80+	559.40335	566.54482	573.68400	580.82092	587.95560
85+	595.08807	602.21834	609.34644	616.47239	623.59621
90+	630.71794	637.83759	644.95519	652.07076	659.18434
95+	666.29594	673.40559	680.51332	687.61915	694.72311
100+	701.82523	708.92553	716.02403	723.12077	730.21576
105+	737.30904	744.40063	751.49055	751.57884	758.66550
110+	765.75058	765.83410	772.91607	779.99653	787.07549
115+	787.15299	794.22905	801.30368	801.37691	808.44878
120+	815.51929	815.58847	822.65634	822.72293	829.78826
125+	836.85235	836.91521	843.97688	844.03737	851.09670
130+	858.15489	858.21197	865.26795	865.32285	872.37669
135+	879.42948	879.48126	886.53203	886.58182	893.63063
140+	893.67850	900.72544	907.77145	907.81657	914.86080
145+	914.90417	921.94669	921.98837	929.02923	929.06928
150+	936.10854	936.14703	943.18476	943.22175	950.25800
155+	950.29353	957.32835	957.36248	964.39594	964.42873
160+	971.46086	971.49236	978.52322	978.55347	985.58311
165+	985.61216	992.64063	992.66853	999.69587	999.72265
170+	1006.74890	1006.77463	1013.79983	1013.82452	1013.84872
175+	1020.87243	1020.89566	1027.91842	1027.94072	1034.96257
180+	1034.98398	1042.00495	1042.02550	1042.04563	1049.06536
185+	1049.08468	1056.10361	1056.12215	1063.14032	1063.15811
190+	1063.17555	1070.19263	1070.20936	1077.22575	1077.24180
195+	1077.25752	1084.27293	1084.28802	1091.30280	1091.31728
200+	1091.33146	1098.34535	1098.35895	1105.37228	1105.38533
205+	1105.39812	1112.41064	1112.42291	1119.43492	1119.44669
210+	1119.45821	1126.46950	1126.48055	1126.49138	1133.50198
215+	1133.51236	1140.52253	1140.53249	1140.54225	1147.55180
220+	1147.56116	1147.57032	1154.57929	1154.58808	1161.59669
		A=0.80			
0+	1.00000	4.45059	9.11745	14.44620	20.19768
5+	26.24619	32.51662	38.96004	45.54255	52.23955
10+	59.03254	65.90721	72.85222	79.85843	86.91833
15+	94.02567	101.17520	108.36244	115.58354	122.83518
20+	130.11445	137.41881	144.74601	152.09406	159.46117
25+	166.84575	174.24636	181.66171	189.09064	196.53206
30+	203.98502	211.44862	218.92205	226.40458	233.89550
35+	241.39420	248.90008	256.41260	263.93128	271.45563
40+	278.98523	286.51969	294.05861	301.60166	309.14852
45+	316.69887	324.25242	331.80893	339.36813	346.92978
50+	354.49368	362.05961	369.62738	377.19680	384.76771
55+	392.33994	399.91334	407.48777	415.06308	422.63917
60+	430.21589	437.79315	445.37083	452.94883	460.52705
65+	468.10542	475.68383	483.26222	490.84050	498.41861
70+	505.99647	513.57402	521.15121	528.72798	536.30427

EXPECTATION E(T/GLF)=(1+BA)SUM(D(R,M)),B= 8

R	1	2	3	4	5
		A=0.80			
75+	543.88002	551.45521	559.02977	566.60367	574.17687
80+	581.74933	589.32102	596.89189	604.46193	612.03110
85+	619.59937	627.16672	634.73313	642.29857	649.86301
90+	657.42646	664.98887	672.55024	680.11056	687.66979
95+	695.22795	702.78500	710.34094	717.89577	725.44946
100+	733.00201	740.55342	748.10368	755.65277	763.20070
105+	770.74747	778.29305	785.83747	793.38070	800.92275
110+	808.46362	816.00330	823.54180	831.07911	838.61523
115+	838.75018	846.28394	853.81651	861.34791	861.47813
120+	869.00717	876.53504	876.66175	884.18728	891.71166
125+	891.83487	899.35694	906.87785	906.99762	914.51625
130+	922.03374	922.15011	929.66535	937.17947	937.29248
135+	944.80439	952.31519	952.42490	959.93352	960.04107
140+	967.54753	975.05293	975.15727	982.66055	982.76279
145+	990.26398	997.76414	997.86328	1005.36140	1005.45850
150+	1012.95460	1020.44971	1020.54382	1028.03695	1028.12911
155+	1035.62030	1035.71054	1043.19982	1043.28815	1050.77555
160+	1058.26202	1058.34757	1065.83220	1065.91593	1073.39876
165+	1073.48070	1080.96176	1081.04194	1088.52125	1088.59970
170+	1096.07730	1096.15405	1103.62997	1103.70505	1111.17931
175+	1118.65276	1118.72540	1126.19724	1126.26828	1133.73854
180+	1133.80802	1141.27673	1141.34468	1148.81187	1148.87831
185+	1148.94401	1156.40897	1156.47320	1163.93672	1163.99952
190+	1171.46161	1171.52301	1178.98371	1179.04372	1186.50306
195+	1186.56172	1194.01972	1194.07706	1201.53374	1201.58978
200+	1209.04518	1209.09995	1209.15409	1216.60761	1216.66052
205+	1224.11282	1224.16452	1231.61563	1231.66614	1239.11607
210+	1239.16543	1239.21422	1246.66244	1246.71011	1254.15722
215+	1254.20378	1261.64981	1261.69530	1269.14026	1269.18469
220+	1269.22861	1276.67201	1276.71491	1284.15731	1284.19921
225+	1284.24062	1291.68154	1291.72199	1299.16195	1299.20145
230+	1306.64049	1306.67906	1306.71718	1314.15485	1314.19208
235+	1321.62886	1321.66521	1321.70113	1329.13663	1329.17170
240+	1336.60636	1336.64061	1336.67445	1344.10789	1344.14093
245+	1344.17358	1351.60584	1351.63771	1359.06921	1359.10033
250+	1359.13107	1366.56145	1366.59147	1374.02113	1374.05043
255+	1374.07938	1381.50798	1381.53624	1381.56417	1388.99175
260+	1389.01901	1396.44593	1396.47254	1396.49882	1403.92479
265+	1403.95044	1403.97578	1411.40082	1411.42556	1411.45000
270+	1418.87414	1418.89799	1426.32155	1426.34483	1426.36782
275+	1433.79054	1433.81298	1433.83515	1441.25705	1441.27868
280+	1441.30005	1448.72116	1448.74202	1448.76262	1456.18297
285+	1456.20307	1456.22293	1463.64255	1463.66192	1463.68106
290+	1471.09997	1471.11865	1471.13709	1478.55532	1478.57332
295+	1478.59110	1486.00866	1486.02601	1486.04314	1493.46007
300+	1493.47679	1493.49330	1500.90961	1500.92572	1500.94163
305+	1508.35735	1508.37288	1508.38821	1515.80336	1515.81831
310+	1515.83309	1523.24768	1523.26210	1523.27633	1530.69040
315+	1530.70428	1530.71800	1530.73155	1538.14493	1538.15815
320+	1538.17120	1545.58409	1545.59682	1545.60940	1553.02182
325+	1553.03408	1553.04620	1560.45816	1560.46998	1560.48165
330+	1560.49318	1567.90456	1567.91580	1567.92691	1575.33787
335+	1575.34870	1575.35940	1575.36996	1582.78039	1582.79069
340+	1582.80087	1590.21092	1590.22084	1590.23064	1597.64032
345+	1597.64988	1597.65932	1597.66864	1605.07784	1605.08694
350+	1605.09591	1612.50478	1612.51354	1612.52218	1612.53072
355+	1619.93916	1619.94748	1619.95571	1619.96383	1627.37185

EXPECTATION E(T/GLF)=(1+BA)SUM(D(R,M)),B= 8

R	1	2	3	4	5
			A=0.80		
360+	1627.37977	1627.38759	1634.79531	1634.80294	1634.81047
365+	1634.81791	1642.22526	1642.23251	1642.23967	1649.64675

EXPECTATION E(T/GLF)=(1+BA)SUM(D(R,M)),B= 9

R	1	2	3	4	5
			A=1/10		
0+	1.00000	3.35802	5.54106	7.54755	9.48318
			A=1/ 9		
0+	1.00000	3.45732	5.78428	7.92164	9.97244
5+	11.99042	13.99668	15.99885	17.99960	19.99986
			A=1/ 8		
0+	1.00000	3.57008	6.07339	8.37820	10.57756
5+	12.73179	14.86805	16.99735	19.12399	21.24961
			A=1/ 7		
0+	1.00000	3.69844	6.42027	8.94439	11.34196
5+	13.67716	15.98421	18.27902	20.56859	22.85594
			A=1/ 6		
0+	1.00000	3.84450	6.83952	9.65779	12.33024
5+	14.91757	17.46041	19.98109	22.49099	24.99571
			A=1/ 5		
0+	1.00000	4.00969	7.34729	10.56801	13.63752
5+	16.59712	19.48802	22.33869	25.16659	27.98184
			A=1/ 4		
0+	1.00000	4.19310	7.95554	11.73022	15.39181
5+	18.93494	22.38392	25.76531	29.10059	32.40533
10+	35.69020	38.96231	42.22627	45.48508	48.74062
			A=1/ 3		
0+	1.00000	4.38725	8.65253	13.16418	17.70087
5+	22.18330	26.58826	30.91556	35.17393	39.37465
10+	43.52885	47.64632	51.73527	55.80229	59.85260
15+	63.89027	67.91839	71.93936	75.95497	79.96658
			A=0.40		
0+	1.00000	4.48181	9.01136	13.94540	19.02879
5+	24.14676	29.24552	34.30081	39.30354	44.25257
10+	49.15076	54.00288	58.81435	63.59063	68.33687
15+	73.05770	77.75721	82.43894	87.10592	91.76070
20+	96.40543	101.04191	105.67163	110.29581	114.91548
			A=1/ 2		
0+	1.00000	4.56713	9.34446	14.69462	20.34670
5+	26.16382	32.06910	38.01655	43.97775	49.93486
10+	55.87678	61.79676	67.69101	73.55767	79.39624
15+	85.20706	90.99108	96.74960	102.48417	108.19640
20+	113.88797	119.56053	125.21570	130.85501	136.47990
25+	142.09173	147.69176	153.28113	158.86092	164.43209
30+	169.99553	175.55204	181.10235	186.64711	192.18692
35+	197.72231	203.25375	208.78168	214.30648	219.82849
40+	225.34802	230.86535	236.38071	241.89434	247.40642
			A=0.60		
0+	1.00000	4.61389	9.52936	15.11810	21.10762
5+	27.35567	33.77835	40.32207	46.95065	53.63863
10+	60.36764	67.12407	73.89773	80.68092	87.46773
15+	94.25367	101.03529	107.80998	114.57577	121.33120
20+	128.07522	134.80710	141.52636	148.23274	154.92613
25+	161.60656	168.27417	174.92916	181.57181	188.20245

EXPECTATION E(T/GLF)=(1+BA)SUM(D(R,M)),B= 9

R	1	2	3	4	5
			A=0.60		
30+	194.82144	201.42917	208.02603	214.61245	221.18885
35+	227.75566	234.31329	240.86217	247.40271	253.93530
40+	260.46034	266.97820	273.48926	279.99386	286.49234
45+	292.98503	299.47225	305.95429	312.43144	318.90396
50+	325.37213	331.83618	338.29635	344.75287	351.20594
55+	357.65576	364.10254	370.54643	376.98762	383.42627
60+	389.86252	396.29652	402.72841	409.15831	415.58635
65+	422.01263	428.43726	434.86035	441.28199	447.70227
70+	454.12127	460.53907	466.95574	473.37137	479.78600
			A=0.65		
0+	1.00000	4.62860	9.58755	15.25185	21.34936
5+	27.73718	34.33031	41.07385	47.93016	54.87229
10+	61.88034	68.93919	76.03715	83.16503	90.31552
15+	97.48274	104.66192	111.84917	119.04131	126.23574
20+	133.43029	140.62321	147.81304	154.99860	162.17891
25+	169.35320	176.52084	183.68133	190.83429	197.97943
30+	205.11655	212.24550	219.36620	226.47863	233.58279
35+	240.67873	247.76653	254.84628	261.91810	268.98215
40+	276.03856	283.08750	290.12915	297.16369	304.19129
45+	311.21216	318.22648	325.23445	332.23626	339.23210
50+	346.22216	353.20663	360.18570	367.15956	374.12838
55+	381.09235	388.05163	395.00640	401.95682	408.90306
60+	415.84527	422.78360	429.71820	436.64923	443.57680
65+	450.50107	457.42216	464.34020	471.25531	478.16761
70+	485.07721	491.98423	498.88877	505.79092	512.69080
75+	519.58849	526.48409	533.37768	540.26935	547.15918
80+	554.04724	560.93362	567.81839	574.70161	581.58334
85+	588.46366	595.34263	602.22030	609.09673	615.97198
90+	622.84609	629.71911	636.59110	643.46210	643.48215
95+	650.35129	657.21957	657.23702	664.10368	670.96958
100+	670.98477	677.84926	684.71309	684.72629	691.58889
			A=0.70		
0+	1.00000	4.63939	9.63007	15.34945	21.52579
5+	28.01591	34.73431	41.62544	48.65095	55.78313
10+	63.00131	70.28958	77.63547	85.02900	92.46209
15+	99.92807	107.42144	114.93756	122.47254	130.02305
20+	137.58627	145.15975	152.74141	160.32942	167.92223
25+	175.51846	183.11692	190.71658	198.31654	205.91601
30+	213.51429	221.11080	228.70502	236.29649	243.88482
35+	251.46967	259.05076	266.62784	274.20070	281.76915
40+	289.33307	296.89232	304.44681	311.99646	319.54123
45+	327.08107	334.61595	342.14587	349.67084	357.19085
50+	364.70594	372.21613	379.72147	387.22199	394.71775
55+	402.20880	409.69520	417.17701	424.65431	432.12715
60+	439.59562	447.05979	454.51973	461.97551	469.42723
65+	476.87495	484.31876	491.75873	499.19494	506.62748
70+	514.05642	521.48185	528.90384	536.32246	543.73781
75+	551.14994	558.55895	565.96490	573.36787	580.76793
80+	588.16516	595.55962	602.95138	610.34051	617.72708
85+	625.11116	632.49281	639.87209	647.24906	654.62380
90+	661.99634	669.36677	676.73512	684.10147	691.46585
95+	698.82834	706.18897	713.54780	720.90488	728.26026
100+	735.61399	735.66611	743.01666	750.36570	750.41327
105+	757.75940	765.10414	765.14752	772.48960	779.83039
110+	779.86995	787.20831	794.54550	794.58155	801.91651
115+	809.25040	809.28325	816.61509	816.64596	823.97588
120+	824.00488	831.33298	838.66022	838.68662	846.01221

EXPECTATION E(T/GLF)=(1+BA)SUM(D(R,M)),B= 9

R	1	2	3	4	5
			A=0.70		
125+	846.03700	853.36103	853.38432	860.70688	860.72874
130+	868.04993	868.07045	875.39034	882.70961	882.72828
135+	890.04636	890.06388	897.38086	897.39730	904.71323
140+	904.72866	912.04361	912.05809	919.37212	919.38571
145+	919.39887	926.71161	926.72396	934.03591	934.04750
			A=0.75		
0+	1.00000	4.64717	9.66062	15.41937	21.65194
5+	28.21503	35.02287	42.01955	49.16637	56.43526
10+	63.80514	71.25972	78.78615	86.37405	94.01495
15+	101.70183	109.42877	117.19079	124.98361	132.80353
20+	140.64738	148.51236	156.39603	164.29623	172.21104
25+	180.13878	188.07793	196.02714	203.98518	211.95098
30+	219.92353	227.90196	235.88544	243.87327	251.86476
35+	259.85933	267.85641	275.85552	283.85619	291.85801
40+	299.86060	307.86361	315.86673	323.86968	331.87217
45+	339.87399	347.87490	355.87470	363.87322	371.87029
50+	379.86575	387.85947	395.85133	403.84121	411.82901
55+	419.81463	427.79801	435.77905	443.75770	451.73390
60+	459.70760	467.67875	475.64731	483.61324	491.57653
65+	499.53715	507.49507	515.45028	523.40277	531.35253
70+	539.29956	547.24385	555.18540	563.12423	571.06032
75+	578.99370	586.92437	594.85234	602.77762	610.70024
80+	618.62021	626.53754	634.45225	642.36437	650.27392
85+	658.18092	666.08538	673.98735	681.88683	689.78386
90+	697.67846	705.57065	713.46047	721.34794	729.23309
95+	737.11595	744.99654	752.87489	760.75103	768.62499
100+	776.49680	784.36649	792.23408	800.09960	807.96309
105+	815.82457	823.68406	831.54161	831.64723	839.50096
110+	847.35281	855.20283	855.30104	863.14747	870.99214
115+	871.08508	878.92631	886.76588	886.85379	894.69008
120+	902.52477	902.60790	910.43948	918.26954	918.34810
125+	926.17520	926.25085	934.07508	941.89791	941.96936
130+	949.78947	949.85825	957.67572	965.49192	965.55685
135+	973.37054	973.43302	981.24430	981.30441	989.11337
140+	996.92119	996.97789	1004.78351	1004.83805	1012.64153
145+	1012.69398	1020.49542	1020.54585	1028.34530	1028.39379
150+	1036.19134	1036.23796	1044.03366	1044.07847	1051.87241
155+	1051.91548	1059.70771	1059.74910	1067.53969	1067.57947
160+	1075.36847	1075.40669	1083.19417	1083.23090	1091.01690
165+	1091.05219	1098.83679	1098.87069	1106.65392	1106.68649
170+	1114.46841	1114.49969	1122.28035	1122.31040	1130.08985
175+	1130.11871	1130.14699	1137.92470	1137.95186	1145.72848
180+	1145.75456	1153.53011	1153.55516	1161.32969	1161.35374
185+	1161.37730	1169.15038	1169.17300	1176.94517	1176.96688
190+	1184.73816	1184.75900	1184.77942	1192.54943	1192.56904
195+	1200.33825	1200.35706	1200.37550	1208.14356	1208.16125
200+	1215.92859	1215.94557	1215.96220	1223.72850	1223.74446
205+	1231.51010	1231.52542	1231.54043	1239.30513	1239.31953
210+	1247.08363	1247.09745	1247.11098	1254.87424	1254.88723
215+	1254.89995	1262.66241	1262.67461	1270.43656	1270.44827
220+	1270.45974	1278.22097	1278.23197	1278.24274	1286.00330
225+	1286.01363	1286.02376	1293.78367	1293.79338	1301.55289
			A=0.80		
0+	1.00000	4.65265	9.68199	15.46807	21.73952
5+	28.35293	35.22235	42.29166	49.52195	56.88493
10+	64.35933	71.92866	79.57986	87.30238	95.08756
15+	102.92819	110.81818	118.75235	126.72627	134.73608

EXPECTATION E(T/GLF)=(1+BA)SUM(D(R,M)),B= 9

R	1	2	3	4	5
		A=0.80			
20+	142.77842	150.85032	158.94919	167.07271	175.21880
25+	183.38561	191.57148	199.77489	207.99448	216.22901
30+	224.47734	232.73844	241.01136	249.29522	257.58922
35+	265.89263	274.20475	282.52495	290.85264	299.18729
40+	307.52837	315.87542	324.22799	332.58567	340.94808
45+	349.31486	357.68567	366.06019	374.43813	382.81920
50+	391.20315	399.58972	407.97870	416.36985	424.76298
55+	433.15789	441.55439	449.95232	458.35151	466.75180
60+	475.15306	483.55515	491.95793	500.36129	508.76510
65+	517.16926	525.57367	533.97823	542.38284	550.78741
70+	559.19188	567.59614	576.00014	584.40380	592.80706
75+	601.20985	609.61211	618.01380	626.41484	634.81520
80+	643.21483	651.61369	660.01172	668.40890	676.80518
85+	685.20053	693.59492	701.98831	710.38069	718.77201
90+	727.16225	735.55140	743.93942	752.32630	760.71201
95+	769.09655	777.47989	785.86201	794.24291	802.62257
100+	811.00097	819.37811	827.75398	836.12856	844.50184
105+	852.87383	861.24452	869.61389	877.98194	886.34867
110+	894.71407	903.07815	911.44089	919.80230	928.16238
115+	928.32112	936.67852	945.03459	953.38932	953.54272
120+	961.89479	970.24552	970.39493	978.74301	987.08977
125+	995.43521	995.57933	1003.92215	1012.26365	1012.40385
130+	1020.74276	1020.88037	1029.21669	1037.55173	1037.68550
135+	1046.01799	1054.34922	1054.47919	1062.80791	1071.13539
140+	1071.26162	1079.58663	1079.71041	1088.03297	1096.35433
145+	1096.47448	1104.79343	1104.91120	1113.22779	1121.54321
150+	1121.65746	1129.97056	1130.08251	1138.39332	1138.50300
155+	1146.81155	1155.11899	1155.22532	1163.53055	1163.63469
160+	1171.93775	1172.03974	1180.34066	1180.44052	1188.73934
165+	1188.83711	1197.13385	1205.42957	1205.52427	1213.81797
170+	1213.91067	1222.20238	1222.29311	1230.58286	1230.67166
175+	1238.95949	1239.04638	1247.33233	1247.41735	1255.70144
180+	1255.78462	1264.06690	1264.14828	1272.42877	1272.50837
185+	1280.78711	1280.86497	1289.14198	1289.21815	1297.49347
190+	1297.56795	1305.84162	1305.91446	1314.18650	1314.25773
195+	1314.32817	1322.59782	1322.66670	1330.93480	1331.00214
200+	1339.26872	1339.33456	1347.59965	1347.66401	1355.92765
205+	1355.99056	1364.25276	1364.31426	1364.37506	1372.63517
210+	1372.69460	1380.95335	1381.01143	1389.26884	1389.32561
215+	1389.38172	1397.63719	1397.69202	1405.94622	1405.99981
220+	1414.25277	1414.30513	1414.35688	1422.60804	1422.65860
225+	1430.90858	1430.95798	1439.20681	1439.25508	1439.30278
230+	1447.54993	1447.59653	1455.84259	1455.88812	1455.93311
235+	1464.17758	1464.22153	1472.46496	1472.50789	1472.55032
240+	1480.79224	1480.83368	1489.07463	1489.11510	1489.15509
245+	1497.39462	1497.43368	1505.67227	1505.71042	1505.74811
250+	1513.98536	1514.02217	1514.05854	1522.29448	1522.33000
255+	1530.56510	1530.59978	1530.63405	1538.86791	1538.90137
260+	1538.93443	1547.16710	1547.19938	1555.43127	1555.46279
265+	1555.49393	1563.72470	1563.75510	1563.78513	1572.01481
270+	1572.04414	1572.07311	1580.30173	1580.33002	1588.55796
275+	1588.58557	1588.61284	1596.83979	1596.86642	1596.89272
280+	1605.11871	1605.14439	1605.16975	1613.39482	1613.41957
285+	1613.44404	1621.66820	1621.69207	1621.71566	1629.93896
290+	1629.96198	1629.98472	1638.20718	1638.22938	1638.25130
295+	1646.47296	1646.49436	1646.51549	1654.73637	1654.75700
300+	1654.77737	1662.99750	1663.01739	1663.03703	1671.25643

EXPECTATION E(T/GLF)=(1+BA)SUM(D(R,M)),B= 9

R	1	2	3	4	5
			A=0.80		
305+	1671.27560	1671.29453	1679.51323	1679.53171	1679.54996
310+	1687.76798	1687.78579	1687.80338	1696.02076	1696.03792
315+	1696.05487	1704.27162	1704.28816	1704.30450	1704.32063
320+	1712.53657	1712.55232	1712.56787	1720.78323	1720.79841
325+	1720.81340	1729.02820	1729.04282	1729.05726	1737.27153
330+	1737.28562	1737.29954	1737.31328	1745.52686	1745.54027
335+	1745.55351	1753.76659	1753.77951	1753.79227	1753.80488
340+	1762.01733	1762.02962	1762.04176	1770.25376	1770.26560
345+	1770.27730	1778.48886	1778.50027	1778.51154	1778.52267
350+	1786.73367	1786.74453	1786.75525	1794.96584	1794.97630
355+	1794.98663	1794.99683	1803.20691	1803.21686	1803.22669
360+	1803.23640	1811.44598	1811.45545	1811.46480	1819.67403
365+	1819.68315	1819.69216	1819.70105	1827.90984	1827.91851
370+	1827.92708	1836.13554	1836.14389	1836.15214	1836.16029

EXPECTATION E(T/GLF)=(1+BA)SUM(D(R,M)),B=10

R	1	2	3	4	5
			A=1/10		
0+	1.00000	3.49398	5.81581	7.93890	9.98036
5+	11.99375	13.99802	15.99937	17.99980	19.99994
			A=1/ 9		
0+	1.00000	3.60100	6.08318	8.35303	10.52331
5+	12.65544	14.77389	16.88754	18.99953	21.11095
			A=1/ 8		
0+	1.00000	3.72200	6.40014	8.85774	11.19440
5+	13.47862	15.74183	17.99688	20.24881	22.49955
			A=1/ 7		
0+	1.00000	3.85901	6.77904	9.48232	12.04118
5+	14.52774	16.98140	19.42068	21.85380	24.28430
			A=1/ 6		
0+	1.00000	4.01390	7.23474	10.26674	13.13389
5+	15.90284	18.61991	21.31097	23.98933	26.66158
			A=1/ 5		
0+	1.00000	4.18767	7.78290	11.26252	14.57438
5+	17.76076	20.86756	23.92734	26.96035	29.97842
			A=1/ 4		
0+	1.00000	4.37865	8.43329	12.52387	16.49551
5+	20.33521	24.06762	27.72197	31.32259	34.88739
10+	38.42877	41.95506	45.47168	48.98218	52.48879
			A=1/ 3		
0+	1.00000	4.57821	9.16856	14.06046	18.99582
5+	23.87832	28.67729	33.38983	38.02429	42.59272
10+	47.10739	51.57934	56.01793	60.43077	64.82392
15+	69.20210	73.56898	77.92736	82.27936	86.62658
			A=0.40		
0+	1.00000	4.67424	9.54202	14.88611	20.41457
5+	25.99324	31.55758	37.07741	42.54048	47.94408
10+	53.29056	58.58471	63.83232	69.03940	74.21165
15+	79.35432	84.47209	89.56900	94.64856	99.71374
20+	104.76704	109.81056	114.84606	119.87498	124.89851
25+	129.91765	134.93320	139.94583	144.95608	149.96440
			A=1/ 2		
0+	1.00000	4.76007	9.88476	15.66806	21.80428
5+	28.13719	34.57811	41.07345	47.58966	54.10540

EXPECTATION E(T/GLF)=(1+BA)SUM(D(R,M)),B=10

R	1	2	3	4	5
			A=1/ 2		
10+	60.60713	67.08643	73.53836	79.96028	86.35120
15+	92.71120	99.04107	105.34211	111.61588	117.86413
20+	124.08866	130.29130	136.47383	142.63797	148.78535
25+	154.91751	161.03588	167.14178	173.23643	179.32097
30+	185.39639	191.46365	197.52359	203.57696	209.62447
35+	215.66674	221.70432	227.73773	233.76741	239.79377
40+	245.81718	251.83796	257.85640	263.87276	269.88727
			A=0.60		
0+	1.00000	4.80667	10.07287	16.10441	22.59555
5+	29.38550	36.37895	43.51469	50.75122	58.05920
10+	65.41733	72.80979	80.22462	87.65274	95.08715
15+	102.52246	109.95452	117.38014	124.79687	132.20288
20+	139.59682	146.97772	154.34490	161.69796	169.03667
25+	176.36099	183.67098	190.96681	198.24875	205.51711
30+	212.77226	220.01459	227.24453	234.46254	241.66907
35+	248.86457	256.04952	263.22438	270.38960	277.54563
40+	284.69290	291.83185	298.96288	306.08639	313.20278
45+	320.31241	327.41564	334.51281	341.60426	348.69028
50+	355.77119	362.84726	369.91877	376.98597	384.04912
55+	391.10844	398.16415	405.21646	412.26558	419.31169
60+	426.35496	433.39557	440.43367	447.46941	454.50293
65+	461.53438	468.56386	475.59151	482.61743	489.64173
70+	496.66450	503.68585	510.70585	517.72460	524.74216
75+	531.75861	538.77403	545.78847	552.80200	559.81467
			A=0.65		
0+	1.00000	4.82125	10.13165	16.24108	22.84461
5+	29.78109	36.95430	44.30185	51.78084	59.36049
10+	67.01796	74.73587	82.50072	90.30189	98.13087
15+	105.98079	113.84609	121.72218	129.60531	137.49237
20+	145.38081	153.26848	161.15365	169.03486	176.91091
25+	184.78084	192.64385	200.49932	208.34673	216.18570
30+	224.01594	231.83724	239.64946	247.45252	255.24639
35+	263.03110	270.80669	278.57324	286.33088	294.07973
40+	301.81995	309.55169	317.27515	324.99051	332.69796
45+	340.39771	348.08996	355.77493	363.45283	371.12386
50+	378.78823	386.44616	394.09786	401.74353	409.38337
55+	417.01758	424.64636	432.26989	439.88837	447.50199
60+	455.11091	462.71530	470.31535	477.91122	485.50306
65+	493.09103	500.67527	508.25594	515.83318	523.40712
70+	530.97788	538.54561	546.11043	553.67244	561.23176
75+	568.78851	576.34280	583.89471	591.44436	598.99183
80+	606.53721	614.08061	621.62209	629.16174	636.69963
85+	644.23585	651.77046	659.30354	666.83514	674.36533
90+	681.89418	689.42174	696.94806	704.47321	704.49723
95+	712.02016	719.54207	727.06299	727.08297	734.60204
100+	734.62026	742.13765	749.65425	749.67011	757.18524
105+	757.19969	764.71348	772.22664	772.23920	779.75120
			A=0.70		
0+	1.00000	4.83190	10.17444	16.34036	23.02545
5+	30.06848	37.37286	44.87567	52.53337	60.31446
10+	68.19537	76.15795	84.18792	92.27387	100.40650
15+	108.57816	116.78248	125.01413	133.26857	141.54197
20+	149.83102	158.13289	166.44512	174.76558	183.09243
25+	191.42406	199.75906	208.09619	216.43438	224.77269
30+	233.11029	241.44646	249.78056	258.11204	266.44043
35+	274.76531	283.08630	291.40311	299.71546	308.02313
40+	316.32592	324.62368	332.91626	341.20357	349.48552

EXPECTATION E(T/GLF)=(1+BA)SUM(D(R,M)),B=10

R	1	2	3	4	5
			A=0.70		
45+	357.76205	366.03310	374.29867	382.55872	390.81326
50+	399.06230	407.30586	415.54397	423.77667	432.00400
55+	440.22603	448.44280	456.65438	464.86083	473.06224
60+	481.25867	489.45020	497.63691	505.81888	513.99620
65+	522.16895	530.33721	538.50108	546.66063	554.81597
70+	562.96716	571.11431	579.25749	587.39680	595.53232
75+	603.66414	611.79234	619.91700	628.03820	636.15604
80+	644.27058	652.38192	660.49012	668.59526	676.69743
85+	684.79670	692.89313	700.98680	709.07779	717.16616
90+	725.25198	733.33531	741.41623	749.49480	757.57107
95+	765.64512	773.71699	781.78676	789.85447	797.92019
100+	805.98397	814.04585	814.10590	822.16417	830.22070
105+	830.27554	838.32875	846.38036	846.43043	854.47899
110+	862.52609	862.57177	870.61607	870.65904	878.70070
115+	886.74111	886.78028	894.81827	894.85510	902.89080
120+	910.92542	910.95898	918.99152	919.02306	927.05363
125+	927.08327	935.11199	943.13983	943.16682	951.19297
130+	951.21832	959.24289	959.26669	967.28976	967.31212
135+	975.33378	975.35477	983.37511	983.39482	991.41392
140+	991.43242	999.45035	999.46772	1007.48454	1007.50084
145+	1015.51664	1015.53194	1023.54676	1023.56112	1031.57502
150+	1031.58850	1031.60155	1039.61419	1039.62643	1047.63829
			A=0.75		
0+	1.00000	4.83958	10.20508	16.41120	23.15416
5+	30.27274	37.67016	45.28323	53.06811	60.99295
10+	69.03384	77.17226	85.39358	93.68599	102.03983
15+	110.44707	118.90097	127.39581	135.92669	144.48939
20+	153.08025	161.69605	170.33399	178.99157	187.66660
25+	196.35711	205.06135	213.77775	222.50491	231.24154
30+	239.98651	248.73877	257.49738	266.26149	275.03032
35+	283.80316	292.57937	301.35836	310.13958	318.92256
40+	327.70682	336.49198	345.27764	354.06346	362.84913
45+	371.63435	380.41887	389.20244	397.98484	406.76586
50+	415.54532	424.32305	433.09890	441.87272	450.64440
55+	459.41381	468.18086	476.94544	485.70747	494.46688
60+	503.22359	511.97756	520.72872	529.47703	538.22245
65+	546.96494	555.70448	564.44104	573.17460	581.90514
70+	590.63266	599.35713	608.07857	616.79697	625.51232
75+	634.22464	642.93392	651.64019	660.34344	669.04369
80+	677.74096	686.43527	695.12662	703.81504	712.50056
85+	721.18319	729.86295	738.53988	747.21399	755.88532
90+	764.55388	773.21971	781.88283	790.54327	799.20107
95+	807.85624	816.50883	825.15886	833.80636	842.45137
100+	851.09391	859.73401	868.37171	877.00703	885.64002
105+	894.27069	902.89909	911.52524	920.14917	920.27092
110+	928.89051	937.50799	937.62337	946.23669	954.84798
115+	954.95728	963.56460	972.16998	972.27346	980.87505
120+	989.47480	989.57272	998.16885	1006.76321	1006.85584
125+	1015.44676	1024.03600	1024.12359	1032.70955	1032.79391
130+	1041.37670	1049.95794	1050.03767	1058.61589	1058.69265
135+	1067.26796	1067.34185	1075.91435	1084.48547	1084.55525
140+	1093.12370	1093.19084	1101.75671	1101.82132	1110.38470
145+	1110.44686	1119.00783	1127.56763	1127.62629	1136.18381
150+	1136.24022	1144.79554	1144.84980	1153.40300	1153.45517
155+	1162.00633	1162.05649	1170.60568	1170.65391	1179.20120
160+	1179.24757	1187.79303	1187.83760	1196.38130	1196.42414
165+	1204.96614	1205.00731	1213.54768	1213.58724	1222.12603

EXPECTATION E(T/GLF)=(1+BA)SUM(D(R,M)),B=10

R	1	2	3	4	5
			A=0.75		
170+	1222.16406	1222.20133	1230.73786	1230.77367	1239.30878
175+	1239.34318	1247.87691	1247.90996	1256.44236	1256.47411
180+	1265.00523	1265.03573	1265.06562	1273.59492	1273.62363
185+	1282.15176	1282.17934	1290.70636	1290.73284	1290.75879
190+	1299.28421	1299.30913	1307.83355	1307.85747	1316.38092
195+	1316.40389	1316.42640	1324.94846	1324.97007	1333.49124
200+	1333.51199	1333.53232	1342.05223	1342.07175	1350.59086
205+	1350.60959	1350.62795	1359.14592	1359.16354	1367.68080
210+	1367.69770	1367.71426	1376.23049	1376.24638	1376.26196
215+	1384.77721	1384.79215	1393.30679	1393.32113	1393.33518
220+	1401.84894	1401.86242	1401.87563	1410.38856	1410.40123
225+	1410.41364	1418.92580	1418.93771	1427.44937	1427.46080
230+	1427.47199	1435.98295	1435.99368	1436.00420	1444.51450
			A=0.80		
0+	1.00000	4.84496	10.22644	16.46035	23.24314
5+	30.41354	37.87465	45.56310	53.43487	61.45794
10+	69.60821	77.86698	86.21944	94.65361	103.15964
15+	111.72935	120.35580	129.03312	137.75625	146.52078
20+	155.32290	164.15922	173.02677	181.92291	190.84528
25+	199.79174	208.76039	217.74952	226.75754	235.78304
30+	244.82472	253.88138	262.95194	272.03540	281.13082
35+	290.23736	299.35423	308.48070	317.61609	326.75976
40+	335.91113	345.06966	354.23482	363.40616	372.58321
45+	381.76555	390.95281	400.14461	409.34060	418.54046
50+	427.74388	436.95058	446.16028	455.37274	464.58770
55+	473.80494	483.02426	492.24543	501.46829	510.69263
60+	519.91830	529.14512	538.37295	547.60165	556.83107
65+	566.06108	575.29156	584.52240	593.75349	602.98471
70+	612.21598	621.44719	630.67826	639.90910	649.13964
75+	658.36980	667.59951	676.82869	686.05729	695.28524
80+	704.51249	713.73898	722.96467	732.18949	741.41342
85+	750.63639	759.85838	769.07934	778.29924	787.51805
90+	796.73572	805.95224	815.16757	824.38168	833.59455
95+	842.80617	852.01650	861.22552	870.43322	879.63958
100+	888.84458	898.04821	907.25045	916.45129	925.65072
105+	934.84873	944.04530	953.24044	962.43413	971.62636
110+	980.81713	990.00643	999.19426	1008.38061	1017.56549
115+	1026.74888	1026.93078	1036.11121	1045.29014	1045.46759
120+	1054.64355	1063.81803	1072.99102	1073.16253	1082.33256
125+	1091.50110	1091.66818	1100.83378	1109.99791	1110.16058
130+	1119.32178	1128.48153	1128.63983	1137.79669	1146.95210
135+	1147.10608	1156.25863	1165.40976	1165.55947	1174.70777
140+	1174.85467	1184.00017	1193.14429	1193.28702	1202.42838
145+	1202.56837	1211.70701	1220.84429	1220.98024	1230.11485
150+	1230.24813	1239.38010	1248.51075	1248.64011	1257.76817
155+	1257.89496	1267.02046	1267.14471	1276.26769	1276.38943
160+	1285.50993	1294.62920	1294.74725	1303.86408	1303.97972
165+	1313.09416	1313.20742	1322.31950	1322.43042	1331.54019
170+	1331.64880	1340.75628	1340.86263	1349.96786	1350.07198
175+	1359.17500	1368.27693	1368.37778	1377.47755	1377.57627
180+	1386.67392	1386.77053	1395.86611	1395.96066	1405.05419
185+	1405.14671	1414.23823	1414.32876	1414.41831	1423.50689
190+	1423.59450	1432.68116	1432.76687	1441.85164	1441.93548
195+	1451.01841	1451.10042	1460.18153	1460.26174	1469.34107
200+	1469.41953	1478.49711	1478.57383	1487.64970	1487.72473
205+	1487.79892	1496.87228	1496.94482	1506.01656	1506.08749
210+	1515.15762	1515.22696	1524.29553	1524.36333	1524.43036

EXPECTATION E(T/GLF)=(1+BA)SUM(D(R,M)),B=10

R	1	2	3	4	5
			A=0.80		
215+	1533.49663	1533.56216	1542.62694	1542.69099	1551.75431
220+	1551.81692	1551.87881	1560.94000	1561.00050	1570.06030
225+	1570.11942	1579.17787	1579.23564	1579.29276	1588.34922
230+	1588.40503	1597.46021	1597.51474	1597.56865	1606.62194
235+	1606.67462	1615.72669	1615.77815	1615.82902	1624.87931
240+	1624.92901	1633.97813	1634.02668	1634.07467	1643.12211
245+	1643.16899	1652.21532	1652.26112	1652.30638	1661.35111
250+	1661.39532	1670.43902	1670.48220	1670.52488	1679.56706
255+	1679.60874	1679.64993	1688.69064	1688.73087	1697.77063
260+	1697.80992	1697.84875	1706.88712	1706.92504	1706.96251
265+	1715.99954	1716.03613	1725.07229	1725.10802	1725.14333
270+	1734.17821	1734.21269	1734.24676	1743.28042	1743.31368
275+	1743.34655	1752.37902	1752.41111	1752.44282	1761.47415
280+	1761.50510	1770.53569	1770.56591	1770.59577	1779.62528
285+	1779.65443	1779.68323	1788.71169	1788.73981	1788.76759
290+	1797.79504	1797.82216	1797.84895	1806.87542	1806.90157
295+	1806.92741	1815.95294	1815.97816	1816.00308	1825.02770
300+	1825.05202	1825.07605	1834.09978	1834.12324	1834.14640
305+	1843.16929	1843.19191	1843.21424	1852.23631	1852.25811
310+	1852.27965	1861.30093	1861.32195	1861.34271	1870.36323
315+	1870.38349	1870.40351	1870.42328	1879.44282	1879.46212
320+	1879.48118	1888.50001	1888.51861	1888.53699	1897.55514
325+	1897.57307	1897.59079	1906.60828	1906.62557	1906.64264
330+	1906.65950	1915.67616	1915.69262	1915.70887	1924.72493
335+	1924.74079	1924.75645	1933.77192	1933.78721	1933.80231
340+	1933.81722	1942.83195	1942.84650	1942.86087	1951.87506
345+	1951.88908	1951.90293	1951.91661	1960.93012	1960.94347
350+	1960.95665	1969.96966	1969.98252	1969.99522	1979.00777
355+	1979.02016	1979.03240	1979.04448	1988.05642	1988.06821
360+	1988.07986	1988.09136	1997.10272	1997.11394	1997.12502
365+	2006.13597	2006.14678	2006.15745	2006.16800	2015.17841
370+	2015.18870	2015.19886	2024.20889	2024.21880	2024.22859
375+	2024.23825	2033.24780	2033.25722	2033.26653	2033.27573
380+	2042.28481	2042.29378	2042.30264	2042.31139	2051.32003

ABCDEFGHIJ—AMS/AP—898765